FLUID
MECHANICS
MEASUREMENTS

Second Edition

FLUID MECHANICS MEASUREMENTS

Second Edition

Edited by

Richard J. Goldstein

University of Minnesota

Publishers since 1798

USA	Publishing Office:	Taylor & Francis 1101 Vermont Avenue, NW, Suite 200 Washington, DC 20005-3521 Tel: (202) 289-2174 Fax· (202) 289-3665
	Distribution Center:	Taylor & Francis 1900 Frost Road, Suite 101 Bristol, PA 19007-1598 Tel: (215) 785-5800 Fax: (215) 785-5515
UK		Taylor & Francis Ltd. 1 Gunpowder Square London EC4A 3DE Tel: 0171 583 0490 Fax: 0171 583 0581

FLUID MECHANICS MEASUREMENTS, Second Edition

1 2 3 4 5 6 7 8 9 0   BRBR   9 8 7 6

This book was set in Times Roman by Graphic Composition, Inc. The editors were Christine Williams and Carol Edwards. Cover design by Michelle Fleitz. Printing and binding by Braun-Brumfield, Inc.

A CIP catalog record for this book is available from the British Library.
⊗ The paper in this publication meets the requirements of the ANSI Standard
Z39.48-1984 (Permanence of Paper)

Library of Congress Cataloging-in-Publication Data
Fluid mechanics measurements / edited by Richard J. Goldstein.—2nd
   ed.
      p.   cm.
      Includes bibliographical references and index.

   1. Fluid dynamic measurements.   I. Goldstein, Richard J.
TA357.5.M43F58   1996
620.1′064′0287—dc20   95-46165
                                                                          CIP

ISBN 1-56032-306-X

CONTENTS

Contributors   xv

Preface   xix

Preface to the First Edition   xxiii

1   What Do We Measure, and Why?   *Roger E. A. Arndt*   1

1.1   Introduction   1

1.2   The Need for Flow Measurements   2

1.3   What Do We Need to Know?   4

1.4   Examples of Fluid Mechanics Measurements   4

    1.4.1   Measurement of Sediment Load in a Stream   4

    1.4.2   Wind-Tunnel Studies   6

    1.4.3   Propeller Vibration   7

    1.4.4   Aeroacoustics   8

    1.4.5   Turbulent Mixing Layer   11

    1.4.6   Summary   13

1.5   Outline of the Theory of Fluid Mechanics   13

    1.5.1   Inviscid Flow   13

    1.5.2   Viscous Flow and Turbulence   18

    1.5.3   Turbulence   20

1.6   Spatial and Temporal Resolution in Measurements   25

1.7   Correlation of Data and Signal Analysis   29

    1.7.1   Classification of Deterministic Data   29

    1.7.2   Random Data and Signal Analysis   30

1.8   Planning and Carrying Out an Experimental Program   47
        1.8.1   Coherent Structures and Acoustic Radiation   47
        1.8.2   Vortex Cavitation Research   48
        1.8.3   Hydrodynamic Design of a Very Large Water Tunnel   55
        Nomenclature   60
        References   63

2   **Physical Laws of Fluid Mechanics and Their Application to
        Measurement Techniques**   *E. R. G. Eckert*   65

2.1   Introduction   65
2.2   Basic Laws of Mechanics   66
2.3   Inviscid Incompressible Fluids   68
2.4   Viscous Flow Navier-Stokes Equations   72
2.5   Flow Visualization   73
2.6   Similarity Analysis   81
        2.6.1   One-Component Fluids   81
        2.6.2   Two-Component Fluids   85
        2.6.3   Heat Transfer   87
        2.6.4   Natural Convection   89
        2.6.5   Two-Phase Flow   90
        2.6.6   Gravity Boiling on a Horizontal Surface   91
2.7   Viscous Fluids   92
2.8   Natural or Free Convection   98
2.9   Inviscid Compressible Fluids   102
2.10   Viscous and Aerodynamic Heating   106
        Nomenclature   112
        References   113

3   **Thermal Anemometers**   *Leroy M. Fingerson and Peter
        Freymuth*   115

3.1   Introduction   115
3.2   Strengths, Limitations, and Comparisons with Laser Velocimeters   117
3.3   Hot-Wire Sensors   121
3.4   Probe Supports and Mounting   122
3.5   Control Circuit   123
3.6   Calibration of a Hot-Wire Anemometer   125
3.7   Heat Transfer from Fine Wires   126
        3.7.1   High-Speed Flows   130
        3.7.2   Conduction to Walls   130
3.8   Conduction to the Supports   131
3.9   Angle Sensitivity and Support Interference   134

3.10	Measuring Mean Velocity, Velocity Components, and Temperature	136
	3.10.1 One Component Using a Single Hot Wire	136
	3.10.2 Two Components Using an X Probe	137
	3.10.3 Three Components	138
	3.10.4 Multiposition Measurements	140
	3.10.5 Nonisothermal Flows	141
3.11	Dynamics of the Constant-Temperature Hot-Wire Anemometer	143
	3.11.1 Frequency Response	143
	3.11.2 Optimization and Electronic Testing	144
	3.11.3 Large Velocity Fluctuations	146
	3.11.4 Dynamic Effects of Conduction Losses to the Supports	147
	3.11.5 Attenuation of Heat Waves	150
	3.11.6 Finite Resolution Due to Finite Sensor Size	150
3.12	Noise in Constant-Temperature Thermal Anemometry	151
3.13	Film Sensors	152
	3.13.1 Cylindrical Film Sensors	154
	3.13.2 Noncylindrical Film Sensors	159
3.14	Constant-Current Operation	163
3.15	Other Measurement Techniques and Applications Using the Constant-Temperature Anemometer	164
	3.15.1 Aspirating Probe	164
	3.15.2 Pressure Measurement	165
	3.15.3 Split-Film Sensors	165
	3.15.4 Nonresearch Applications	166
3.16	Conclusion	166
	Nomenclature	167
	References	170
4	Laser Velocimetry *Ronald J. Adrian*	175
4.1	Introduction	175
4.2	Basic Principles	177
	4.2.1 Doppler Shift of Light Scattered by Small Particles	177
	4.2.2 Optical Heterodyne Detection	179
	4.2.3 Basic Optical Systems	180
4.3	The Dual-Beam LDV	184
	4.3.1 Practical Dual-Beam Optics	184
	4.3.2 Characteristics of the Dual-Beam Signal	185
4.4	The Reference-Beam LDV	198
4.5	Multivelocity-Component Systems	200
4.6	Photodetectors	203
	4.6.1 Detector Characteristics	203
	4.6.2 Photoemission Statistics	204
	4.6.3 Shot Noise	205

4.7	Signal-to-Noise Ratio Effects	208
4.8	Scattering Particles	210
4.9	Properties of the Random Light Flux	213
	4.9.1 Signal Representation	213
	4.9.2 Random Doppler Light Flux	214
	4.9.3 Statistical Properties of $g(\mathbf{x}, t, D)$	215
	4.9.4 Correlation and Power Spectrum	220
	4.9.5 Burst Density	228
	4.9.6 High-Burst-Density Signals	231
4.10	Signal Processors	237
	4.10.1 Amplitude Correlators	238
	4.10.2 Photon Correlators	240
	4.10.3 Spectrum Analysis	242
	4.10.4 Frequency Trackers	243
	4.10.5 Frequency Counters	247
	4.10.6 Burst-LDV Signal Processors	249
	4.10.7 Selection of Signal Processors	251
4.11	Data Processing	252
	4.11.1 Time-Averaging Processors	252
	4.11.2 Time-Resolving Signal Processors	253
	4.11.3 Fringe Biasing	262
4.12	Fiber-Optic and Laser Diode Systems	262
4.13	Phase Doppler Velocimetry	265
4.14	Laser Measurement of Velocity Fields	266
4.15	Double-Pulsed Planar Laser Velocimetry	268
	4.15.1 System Parameters	268
	4.15.2 Accuracy and Spatial Resolution	270
	4.15.3 Images and Speckle	271
	4.15.4 Image Density in the PIV Mode	272
	4.15.5 Interrogation Analysis	273
	4.15.6 Automated Analysis	280
	4.15.7 Results and Applications	282
4.16	Applications of LDV and PIV	287
	Nomenclature	288
	References	293
5	**Volume Flow Measurements** *G. E. Mattingly*	**301**
5.1	Introduction	301
5.2	Classification of Metering Devices	302
5.3	Selected Meter Performance Characteristics	304
	5.3.1 Orifice Meters	304
	5.3.2 Venturi Tubes and Flow Nozzles	309
	5.3.3 Elbow Meters	309

5.3.4	Pitot Tubes	310
5.3.5	Laminar Flowmeters	312
5.3.6	Turbine Meters	312
5.3.7	Rotameters	314
5.3.8	Target Meters	315
5.3.9	Thermal Flowmeters	316
5.3.10	Weirs and Flumes	317
5.3.11	Magnetic Flowmeters	318
5.3.12	Acoustic Flowmeters	319
5.3.13	Vortex-Shedding Meters	320
5.3.14	Laser Flowmeters	320
5.3.15	Coriolis-Acceleration Flowmeters	321
5.3.16	Flow-Conditioning Devices	321
5.4	Proving—Primary and Secondary Standards	323
5.4.1	Liquid Flow: Static Weighing Procedure	323
5.4.2	Liquid Flow: Dynamic Weighing Procedure	325
5.4.3	Gas Flow: Static Procedure	327
5.4.4	Gas Flow: Dynamic Procedure	329
5.4.5	Ballistic Calibrators	330
5.4.6	NBS Facilities and Secondary Standards	332
5.5	Traceability to National Flow Standards—Measurement Assurance Programs for Flow	334
5.5.1	Static Traceability	337
5.5.2	Dynamic Traceability	337
5.5.3	Measurement Assurance Programs	338
5.5.4	The Role of Flow Conditioning in the Artifact Package	345
5.5.5	Test Program	346
5.5.6	Data Analysis	347
Appendix 5.A	Ideal Performance Characteristics for Differential-Pressure-Type Meters: Incompressible Fluids	351
Appendix 5.B	Ideal Performance Characteristics for Differential-Pressure-Type Meters: Compressible Fluids	352
Appendix 5.C	Real, Compressible Orifice-Flow Calculation	353
Appendix 5.D	Diverter Evaluation and Correction	353
Appendix 5.E	Empirical Formulas for Orifice Discharge Coefficients	357
Appendix 5.F	Pressure Measurements	357
5.F.1	Sensing Static Pressure	358
5.F.2	Sensing Total Pressure	359
Appendix 5.G	Temperature Measurement and Recovery Factor	360
Appendix 5.H	Analysis of Variance with Two Flowmeters in Series	362
Nomenclature		363
References		365

6   Flow Visualization by Direct Injection   *Thomas J. Mueller*   367

6.1   Introduction   367
6.2   Aerodynamic Flow Visualization   369
   6.2.1   Smoke Tube Method   369
   6.2.2   Smoke-Wire Method   397
   6.2.3   Helium Bubble Method   408
   6.2.4   Concluding Remarks   418
6.3   Hydrodynamic Flow Visualization   418
   6.3.1   Dye Method   419
   6.3.2   Hydrogen Bubble Method   432
   6.3.3   Concluding Remarks   445
6.4   Conclusions   445
   References   445

7   Optical Systems for Flow Measurement: Shadowgraph, Schlieren, and Interferometric Techniques   *Richard J. Goldstein and T. H. Kuehn*   451

7.1   Introduction   451
7.2   Schlieren System   455
   7.2.1   Analysis by Geometric or Ray Optics   455
   7.2.2   Applications and Special Systems   463
7.3   Shadowgraph System   471
7.4   Interferometers   474
   7.4.1   Basic Principles   474
   7.4.2   Fringe Pattern with Mach-Zehnder Interferometer   478
   7.4.3   Examples of Interferograms   483
   7.4.4   Design and Adjustment   485
   7.4.5   Errors in a Two-Dimensional Field   490
   7.4.6   Other Inteferometers   491
   7.4.7   Holography   493
   7.4.8   Three-Dimensional Measurements   495
7.5   Conclusion   500
   Nomenclature   500
   References   502

8   Fluid Mechanics Measurements in Non-Newtonian Fluids   *Christopher W. Macosko and Paulo R. Souza Mendes*   509

8.1   Introduction   509
8.2   Material Functions   510

	8.2.1	Some Mechanics Concepts	511
	8.2.2	Steady Shear Flows	512
	8.2.3	Transient Shear	515
	8.2.4	Material Functions in Extension	520
8.3	Constitutive Relations		523
	8.3.1	General Viscous (Reiner-Rivlin) Fluid	524
	8.3.2	Plastic Behavior	529
	8.3.3	Linear Viscoelasticity	531
	8.3.4	Nonlinear Viscoelasticity	535
	8.3.5	Discussion	539
8.4	Rheometry		541
	8.4.1	Shear Rheometers	542
	8.4.2	Extensional Rheometry	551
8.5	Measurements in Complex Flows		559
	8.5.1	Pressure Measurements	560
	8.5.2	Velocity Measurements	561
	8.5.3	Flow Birefringence	564
	Nomenclature		568
	References		570

9	Measurement of Wall Shear Stress *Thomas J. Hanratty and Jay A. Campbell*		575
9.1	Introduction		575
9.2	Direct Measurements		579
9.3	Preston Tube		582
9.4	Stanton Gage		586
9.5	Sublayer Fence		589
9.6	Analysis of Heat or Mass Transfer Probes		590
	9.6.1	Design Equation for a Two-Dimensional Mass Transfer Probe	590
	9.6.2	Limitations of the Design Equation	592
	9.6.3	Nonhomogeneous Two-Dimensional Laminar Flows	594
	9.6.4	Frequency Response	595
	9.6.5	Turbulence Measurements	600
9.7	Effect of Configuration of Mass Transfer Probes		604
	9.7.1	Circular Probes	604
	9.7.2	Slanted Transfer Surface	606
	9.7.3	Other Methods to Measure Direction	608
	9.7.4	Sandwich Elements	609
	9.7.5	Time-Varying and Reversing Flows	610
9.8	Heat Transfer Probes		610
	9.8.1	Analysis	610
	9.8.2	Use in Turbulent Flows	612
	9.8.3	Design Considerations for Turbulent Flows	614

	9.8.4	Analysis of Unsteady Probe Performance	614
	9.8.5	Compressible Flows	615
9.9		Experimental Procedures for Mass Transfer Probes	616
	9.9.1	Electrochemical Cell	616
	9.9.2	Electrolyte	618
	9.9.3	Counterelectrode	620
	9.9.4	Test Electrode	620
	9.9.5	Flow Loop	621
	9.9.6	Measurement of Fluid Properties	622
	9.9.7	Instrumentation	623
9.10		Experimental Procedures for Heated-Element Probes	624
	9.10.1	Principles of Operation	624
	9.10.2	Instrumentation	625
	9.10.3	Calibration	625
	9.10.4	Insertion of the Probe in the Wall	627
	9.10.5	Basic Probe Design	628
9.11		Application of Mass Transfer Probes	630
	9.11.1	Turbulent Flow in a Pipe	630
	9.11.2	Flow Around a Cylinder	632
9.12		Other Methods	634
	9.12.1	Determination from Velocity Measurements	634
	9.12.2	Oil-Film Gage	635
	9.12.3	Modified Laser-Doppler Method	636
	9.12.4	Pulsed-Wall Probe	636
	Nomenclature		637
	References		640

10	Acquiring and Processing Turbulent Flow Data *T. W. Simon*		649
10.1		Introduction	649
10.2		Nature of the Hot-Wire Signal	650
	10.2.1	Random Versus Deterministic	651
	10.2.2	Stationary Versus Nonstationary Processes	652
	10.2.3	Ergodic Versus Nonergodic Processes	656
	10.2.4	Analysis of the Random Ergodic Process	658
10.3		Techniques of Signal Processing	672
	10.3.1	Filtering	673
	10.3.2	Linearization	674
	10.3.3	Corrections	677
	10.3.4	Spectral Analysis	680
	10.3.5	Digital Signal Processing	681
	10.3.6	Analog Signal Processing	685
	10.3.7	Uncertainty	688

10.4	Sampling	690
	10.4.1 Analog-to-Digital Conversion	690
	10.4.2 Sampling Rate	692
10.5	Digital Computer Systems	695
10.6	Conclusions	696
	Nomenclature	697
	References	698
	Index	701

CONTRIBUTORS

Ronald J. Adrian
Department of Theoretical and Applied Mechanics
University of Illinois at Urbana
Urbana, IL 61801

Roger E. A. Arndt
Department of Civil Engineering
University of Minnesota
Minneapolis, MN 55455

Jay A. Campbell
Upjohn Company
Mail Stop 1500-91-2
Kalamazoo, MI 49001

E. R. G. Eckert
Department of Mechanical Engineering
University of Minnesota
111 Church Street SE
Minneapolis, MN 55455

Leroy M. Fingerson
TSI, Inc.
500 Cardigan Road
P.O. Box 64204
St. Paul, MN 55164

Peter Freymuth
Department of Aerospace Engineering Sciences
University of Colorado
Boulder, CO 80302

Richard J. Goldstein
Department of Mechanical Engineering
University of Minnesota
111 Church Street SE
Minneapolis, MN 55455

Thomas J. Hanratty
Department of Chemical Engineering
205 Roger Adams Laboratory
1209 West California
University of Illinois at Urbana
Urbana, IL 61801

T. H. Kuehn
Department of Mechanical Engineering
University of Minnesota
111 Church Street SE
Minneapolis, MN 55455

Christopher W. Macosko
Department of Chemical Engineering & Materials Science
University of Minnesota
304 Admundson Hall
421 Washington Avenue SE
Minneapolis, MN 55455

G. E. Mattingly
National Institute of Standards and Technology
Fluid Flow Group
Fluid Mechanics Building, Room 105
Gaithersburg, MD 20899

Paulo R. Souza Mendes
Department of Mechanical Engineering
Pontificia Universidad Catolica—RJ
Rio de Janeiro, RJ 22453-900, Brazil

Thomas J. Mueller
Aerospace and Mechanical Engineering Department
365 Fitzpatrick Hall of Engineering
University of Notre Dame
Notre Dame, IN 46556

T. W. Simon
Department of Mechanical Engineering
University of Minnesota
111 Church St. SE
Minneapolis, MN 55455

PREFACE

Fluid mechanics continues to be of great interest to scientists and engineers in a number of research and applied areas. Fluid flow is important in almost all disciplines of engineering, in many industries, and in fields related to astronomy, biology, chemistry, geology, meteorology, oceanography, and physics.

The advent of sophisticated numerical methods to study and predict flows has not diminished the requirement for measurements; rather, in many ways it has enhanced this need. The development of turbulence models requires much experimental input and increases the need for detailed experiments to provide verification in different flows. Measurements are required to increase our understanding of the physical processes in flowing systems, particularly turbulent and separated flows, as well as to determine flow quantities needed in a variety of industrial applications. Sophisticated new techniques increase our measurement capabilities not only in fundamental studies but in a number of manufacturing and operational systems, including aircraft, power plants, chemical processing plants, etc.

Many different flow parameters are often required. These include the instantaneous magnitude of velocity, the vector velocity itself, spatial and temporal correlations, turbulent shear stress, and overall volume or mass flow. There is often a need now for more complete and more accurate information than had been required previously. New variations on classical instrumentation as well as relatively new instrumentation and techniques are being used. This new edition has many changes from the original volume with chapters deleted and added and with major additions in almost every remaining chapter.

Chapter 1 has been expanded and updated with material on turbulent flow analysis including the Proper Orthogonal Decomposition (POD) that provides an unbiased method to deduce large-scale coherent structures in complex turbulent flows. Chapter 2 reviews the physical laws or principles on which many flow instruments are based. It demonstrates how results obtained by measurements can be generalized from dimensional analysis, similarity, and analogy considerations.

Chapters 3 and 4 describe techniques used in measuring local velocity. Chapter 3 deals with hot-wire and hot-film systems. Though these have been used for more than 60 years in local measurement of velocity and for a variety of spatial and temporal correlations, introduction of significant innovations has improved our ability to use thermal sensors. Thus, probes with three orthogonal sensors to determine the total velocity vector have become commonplace. Dynamic corrections have been improved and the increased capability of computers for data reduction has led to significant advances. In the mid-1960s laser velocimetry, described in Chapter 4, was first used to make non-obtrusive measurements of fluid velocity. In the present edition the description of laser Doppler velocimeters has been extended to include such advances as burst-signal analysis, phase-Doppler techniques, and the latest optical systems including fiberoptics and laser diodes. Material has also been added that describes particle image velocimetry, a technique that permits simultaneous measurement of velocity vectors at thousands of locations.

Accurate determination of the total volume or mass-flow rate of fluid is often required in applications such as custody transfer of valuable fluids or control systems for industrial processes. Consider, for example, the consequences of a 1% error in measurement of the oil flow from the Alaska fields (about 580,000,000 barrels in 1994) at a price of, say, $18 per barrel. Different types of instruments for volume and mass-flow measurements, including the general conditions under which they are best used, are examined in Chapter 5.

The next two chapters cover flow visualization for both qualitative and quantitative studies. Our understanding of a particular flow field can be enhanced by a low visualization even if quantitative velocity information is not obtained. Chapter 6 considers discrete particle seeding systems while Chapter 7 reviews instruments that make use of a variation of the index of refraction of light in a fluid to visualize the flow and often give quantitative information. Although the direct injection of particles into a flow field (Chapter 6) is an old method of visualization, it remains one of the most valuable and widely used methods today. It is often the easiest and quickest method to obtain a qualitative understanding of a complex flow field. New material in Chapter 6 includes the use of titanium tetrachloride and laser light sheets, and a comparison of water tunnel studies with dye injection to wind tunnel studies with titanium tetrachloride injection, used for the same generic model flow. Optical diagnostic systems (Chapter 7) are inertialess and can be used to measure extremely rapid phenomena such as the passage of a shock wave. Recent advances include tomographic reconstruction methods for analysis of three-dimensional flows and electronic image processing.

Many flows of interest today are not those of a classical Newtonian fluid. Chapter 8 examines characteristics of the flow of non-Newtonian fluids and special experimental techniques for measuring unique properties, including time-dependent viscosity and elasticity, normal stresses in shear, and elongational viscosity. A section is also devoted to stress birefringence methods.

Chapter 9 covers techniques used in the measurement of wall shear stress. This quantity is often needed to determine drag on solid boundaries, to measure turbulence close to a wall, and to test theoretical analyses. Recent uses of inverse methods to analyze the behavior of flush-mounted mass or heat-transfer probes are discussed. These open the possibility of making measurements in regions with haphazard flows, such as separated zones. Newly developed methods for measuring wall shear stresses are presented. Sixty-nine new references are cited.

Care and forethought must be given to the acquisition and processing of a complex signal, such as that taken from instrumentation within a turbulent flow. In recent years, signal processing equipment has vastly improved and techniques to extract the maximum amount of useful information from the signal have similarly progressed. The brand new Chapter 10 presents current techniques and, for perspective, some classical techniques for acquiring and processing time-varying signals, such as the analog voltage from a hot-wire bridge.

The book is not intended to provide details of current commercial systems. Rather the authors strive to provide the general properties and uses, as well as the limitations, of a number of key measurement techniques. We expect this to be useful to those who make fluid mechanics measurements themselves and to those who use results of measurements in their work.

R.J. Goldstein

PREFACE TO THE FIRST EDITION

Interest in fluid mechanics by scientists and engineers continues to increase. Fluid flow is important in almost all fields of engineering, in many industries, and in many studies related to oceanography, meteorology, astronomy, chemistry, geology, and physics.

The advent of sophisticated numerical methods to study and predict fluid flows has not diminished the requirement for measurements but rather has enhanced this need. The development of turbulence models requires much experimental input and eventually verification in many different flows. Flow measurements are required to increase our understanding of the physical processes in flowing systems, particularly turbulent flow and three-dimensional flows, as well as to determine flow quantities needed in a variety of industrial applications.

Many different flow parameters are often required. These include the instantaneous magnitude of velocity, the vector velocity itself, spatial and temporal correlations, turbulent shear stress, and overall volume or mass flow. There is often a need now for more complete and more accurate information than had been required previously. New variations on classical instrumentation as well as relatively new instrumentation and techniques are being used.

Chapter 1 starts this volume with a discussion of the need for different types of measurements—in particular, how the data acquired are used in turbulent flow studies. The second chapter reviews the physical laws or principles that are the bases of many flow instruments.

Chapters 3 through 5 describe techniques used in measuring local velocity. Differential pressure measurements are used in many ways. Perhaps most familiar is using the difference between total and static pressure to determine velocity or volume flow. Chapter 4 deals with hot-wire and hot-film systems. These have been used for more than 50 years in local measurements of velocity and also for a variety of spatial and temporal correlations. In the mid-1960s, laser velocimetry, described in Chapter 4, was first used to make nonobtrusive measurements of velocity. Since that time, a number of specialized instruments and data acquisition and processing devices have been developed for this technique.

The total volume or mass flow rate of a fluid is often required in industrial applications. Different types of instruments for flow measurement, including the general conditions under which they are best used, are examined in Chapter 6.

The next two chapters cover flow visualization for both qualitative and quantitative studies. Our understanding of a particular flow field can be greatly enhanced by a visualization system even if no quantitative information is obtained. Chapter 7 considers discrete particle and seeding systems while Chapter 8 reviews systems that make use of the variation of the index of refraction of a fluid to visualize flow.

Many flows of interest today are not those of a classical single-phase Newtonian fluid. Chapter 8 examines characteristics of the flow of non-Newtonian fluids, including determination of physical properties of such fluids. Chapter 9 considers two-phase flow systems with emphasis on the different types of instrumentation to yield information on liquid-gas flows.

Chapter 11 describes techniques used in the measurement of wall shear stress. This quantity is often needed for determination of pressure drop as well as for examining the transport mechanisms in diverse flow regimes and geometries.

Much of the material in the present volume was first used at the University of Minnesota as part of a week-long course for practicing engineers and scientists reviewing the latest techniques for measurements in flowing fluids. Considerable assistance in the formation of the course was provided by R. Arndt, E. Eckert, and L. Fingerson. The final volume owes much to the efforts of K. Sikora, whose patience and persistence encouraged the authors to produce a completed manuscript in a reasonable period of time.

R. J. Goldstein

WHAT DO WE MEASURE, AND WHY?

Roger E. A. Arndt

1.1 INTRODUCTION

The purpose of this chapter is to acquaint the reader with the need for flow measurements, to provide some insight concerning various applications, and to introduce the methodology and philosophy of flow measurements. A distinction is drawn between measurements having direct application (e.g., flow rate in a pipe) and measurements that are used indirectly for the correlation of primary data, for the verification of theory, and for the tuning of mathematical models. Some emphasis is placed on the need to understand fluid mechanics for the design of experiments, for the interpretation of results, and for estimating deterministic errors due to flow modification by instrumentation placed in the flow. Examples of diagnostic techniques in research problems are drawn from various fields. The principles of fluid mechanics are reviewed to demonstrate the need for measurements, as well as to provide a basis for describing various systematic errors inherent in flow measurement devices.

Various limitations on flow measurement techniques are described in general terms. The need to understand the trade-off between temporal response and spatial resolution in a flowing system is underscored. Finally, an overview of data analysis is given to provide the reader with some insight into the interrelationship between the flow field, the measurement device, and the results.

The details of actual measurement techniques and data reduction are given in following chapters.

1.2 THE NEED FOR FLOW MEASUREMENTS

It would be impossible in this chapter to describe all the various types of flow measurements and their applications. However, an attempt is made to describe the overall methodology and philosophy of fluid measurement techniques and their application. One can think of several general needs for flow measurements. In some cases, the data are useful of themselves, as for example, the rate of flow in a pipe or the mass flux of contaminants in a river. In these cases, the flow quantity desired could be measured either directly or indirectly and it would be the final result. In other cases, flow measurements are necessary for correlation of dependent variables. For example, it is well known that the lift and drag of various vehicles and structures are dependent on the density and velocity of the flow. In this case, a measurement of the velocity would be utilized to correlate force measurements. Erosion rate, heat transfer, ablation, etc., are also known to be functions of velocity, and again, velocity would be an independent variable. The measurement of velocity and pressure in a flowing system can also be useful as a diagnostic for determining various quantities. For example, velocity measurements are often used in problems related to noise and vibration and as a diagnostic in heat and mass transfer research. Measurements of velocity and pressure are also used for the verification of theory and, in addition, a great deal of experimental data is necessary for calibrating mathematical models of various types. For example, there are numerous mathematical models describing the diffusion of pollutants from a point source into a large body of fluid, e.g., sewer and smokestack fallout. In these cases, a basic understanding of turbulent mixing is necessary, and appropriate mathematical models are dependent on careful measurements of turbulence.

Physical modeling is still very useful in many branches of engineering, ranging from wind-tunnel tests of airplanes and other aerospace vehicles, buildings, diffusion of pollutants, windmills, and even snow fences to hydraulic models of entire dams, reaches of rivers, and cooling-water intakes and outlets. Many of these applications require specialized instrumentation. An example of the detail inherent in many hydraulic models is shown in Fig. 1.1.

The many applications of fluid flow measurements cover a broad spectrum of activities. For example, one could think of various wind-tunnel studies related to lift and drag, vibration, and noise radiation. In engine and compressor research involving both reciprocating and rotating machinery and in the broad area of hydraulic engineering research, many different types of flow measurements are necessary. These measurements would be made in water tunnels, tow tanks, and flumes, for example. Agricultural research is now becoming more and more involved with basic fluid mechanics, and many types of fluid mechanics measurements are related to erosion, wind effects on crops, transpiration rates for irrigation systems, and meteorological effects on buildings (including the effects of wind loading, both steady and unsteady, on grain storage buildings, grain elevators, and other structures).

Figure 1.1 Model of Guri Dam in Venezuela. Upon completion of the second stage of development, this will be one of the world's largest hydroelectric sites (10,000 MW). (Courtesy St. Anthony Falls Hydraulic Laboratory, University of Minnesota)

Ocean engineering requires an extensive amount of experimentation. Measurements are required, for example, in determining the drag and seakeeping characteristics of a ship, and very detailed wake surveys are necessary for the design of the propeller. There are many other types of ocean engineering measurements, requiring a broad spectrum of fluid mechanics measurements. These include wave forces on offshore structures and air-sea interaction studies for understanding the development of waves. Flow measurements are required for fundamental research in many other fields, such as ballistics, combustion, magnetohydrodynamics (MHD), explosion studies, and heat transfer. In addition, many basic types of testing require fluid mechanics measurements, including cooling studies, the design of hydraulic systems, engine tests, flow calibration facilities, and various types of hydraulic machinery for handling pump flow and suspensions.

Fluid mechanics measurements are also extremely important in biomedical research related to pulmonary function, blood flow, urine flow, heart-valve models, biological cell movements, and other studies. Many flow measurements are involved in the broad area of process control measurement, such as in engines, refinery systems, semiconductor doping processes, and pilot-plant studies.

1.3 WHAT DO WE NEED TO KNOW?

Although it would seem to be obvious, we should have a clear picture of the requirements before undertaking any physical measurement. In many cases, the application of a particular fluid flow measurement device is inappropriate, simply because the user of the instrument has not taken the time to clearly define the need. For example, we must first know the purpose of the measurement. Is the flow quantity that is measured to be used as is? Second, we must understand the fluid mechanics of the problem. Almost all types of measurement techniques depend on the nature of the flow, and this in turn, governs instrument selection. We must also understand the physical principles involved in flow measurement. Almost all fluid flow measurements are indirect, in that the techniques rely on the physical interpretation of the quantity measured; e.g., a Pitot tube senses pressure, and the velocity is inferred from the measured pressure using the Bernoulli equation. When we measure pressure with a pressure transducer, we sense deflection of a membrane or diaphragm, usually electronically, and we interpret the deflection of this membrane in terms of pressure. Hot-wire instrumentation is dependent on the physical laws relating heat transfer with flow velocity.

Surprising as it sounds, we also need to know the answer before we begin. That is to say, we must have a reliable estimate of what the results should be. This is often overlooked, and sometimes measurements that are grossly incorrect are taken as gospel simply because the physical situation has not been carefully evaluated. It can also be said that one cannot wisely purchase flow measurement equipment without considering specific experiments. *Almost all sophisticated equipment purchased for general or unspecified use is a waste of money.*

1.4 EXAMPLES OF FLUID MECHANICS MEASUREMENTS

As already mentioned, there is an infinite variety of fluid mechanics measurements and applications. However, to provide some insight into the various types of measurements, a few examples are given below.

1.4.1 Measurement of Sediment Load in a Stream

To measure the total quantity of suspended material in a stream, both the velocity and the concentration at several depths must be measured and the results integrated:

$$G = b \int_0^d cu \, dy \tag{1}$$

The measurement equipment necessary to evaluate Eq. (1) is shown in Fig 1.2. The sediment sampler shown in Fig. 1.2*a* is the result of extensive development by the

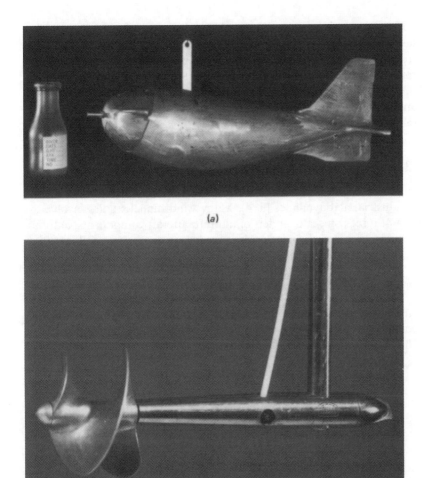

(a)

(b)

Figure 1.2 (a) Sediment sampler developed by U.S. Geological Survey. (b) Typical propeller meter used in hydraulic engineering studies

U.S. Interagency Group on Sedimentation [1, 2]. By correctly positioning exhaust ports at a point where the pressure is the same as that in the free stream, the velocity in the intake nozzle is made equal to the local stream velocity. Proper sizing of fins is necessary for stability, the design being an outgrowth of standard procedures for tail-plane design in aerodynamics. The device is designed to automatically carry out the integration procedure in Eq. (1) during lowering and raising at a uniform rate. It is an interesting example of the application of fluid mechanical principles to a specialized measurement problem. It is typical to use a propeller meter of the type shown in Fig. 1.2b for measurements of flow velocity in large channels, streams, etc. These propeller meters are based on the simple principle that the rotational speed of the meter is proportional to the local flow velocity. The rotational speed is sensed by a counter, and the number of counts per second is displayed on a recorder. Such devices require periodic calibration, and this is usually accomplished in a laboratory flume.

1.4.2 Wind-Tunnel Studies

Wind tunnels are used for a myriad of investigations, ranging from fundamental research to industrial aerodynamics [3, 4]. Many wind-tunnel studies involve the determination of forces on scale models of aircraft, aircraft components, automobiles, or buildings. Forces such as lift or drag are known to obey the following law of similitude:

$$F = \tfrac{1}{2}\rho C_N V^2 S \tag{2}$$

where S is a surface area or cross-sectional area, depending on the application. The force coefficient C_N is known to be a function of several nondimensional parameters. The major ones in aerodynamics are

$$\text{Reynolds number} = \frac{\text{inertial force}}{\text{viscous force}} = \frac{\rho V I}{\mu} \tag{3}$$

$$\text{Mach number} = \frac{\text{inertial force}}{\text{elastic force}} = \frac{V}{a_0} \tag{4}$$

To correlate data, V is typically measured with a Pitot or Prandtl tube, and the temperature and pressure are determined with appropriate instrumentation (see Chap. 2). The forces and moments on the model are usually determined with a specially designed balance. The density is usually computed from the measured temperature and pressure. Typical wind-tunnel data are shown in Fig. 1.3. These data, adapted from National Advisory Committee for Aeronautics (NACA) data for an airfoil section, display the variation of section lift coefficient and moment coefficient, defined as

$$C_l = \frac{L}{\tfrac{1}{2}\rho U^2 c} \tag{5}$$

$$C_m = \frac{M}{\tfrac{1}{2}\rho U^2 c^2} \tag{6}$$

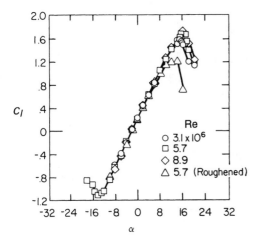

Figure 1.3 Typical airfoil data obtained in a wind tunnel

Note that the primary parameter is the angle of attack, with the influence of Reynolds number only being noted at high angles of attack, where stall occurs. The maximum lift coefficient is strongly dependent on Reynolds number. This is due to the separation of the boundary layer at the leading edge of the airfoil. Separation phenomena are known to be dependent on such characteristics of the boundary layer at separation as shape factor and whether it is laminar or turbulent.

1.4.3 Propeller Vibration

Marine propellers must operate in the confused flow in the wake of a ship. Because a propeller rotates through a nonuniform velocity field, unsteady thrust and side forces are produced. The unsteadiness of the propulsor can in some cases be extreme, producing unwanted vibration within the ship. In some cases the propeller vibration produced is so extreme that piping is shaken loose and the crew find working on the ship extremely uncomfortable. There have been cases of newly built ships that have been boycotted by the seamen's union, creating the need for an extremely expensive retrofit program. The overall solution to this problem is complex. Figure 1.4 illustrates the problem and its analysis.

As shown, a propeller produces unsteady thrust and side forces that are periodic in time with a frequency equal to the blade passing frequency (number of blades multiplied by the rotational speed). The reason for the periodic fluctuations is that each section of a propeller blade must rotate through the wake of the ship. The axial velocity at each position is then a variable, resulting in a fluctuation in angle of attack and unsteady lift, as shown in the simplified velocity diagram for a blade section (Fig. 1.4). Velocity measurements are necessary in the analysis of the problem. Sample velocity data are shown in the lower portion of Fig. 1.4 as a

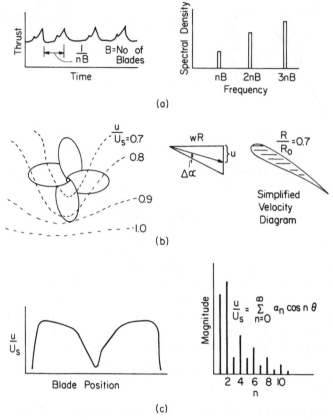

Figure 1.4 Use of wake velocity measurements to isolate a ship-propeller vibration problem

function of angular position. These data can be Fourier analyzed in terms of harmonic components. The harmonics of the unsteady thrust can then be determined from information on the harmonic content of the velocity field. Specialized total-pressure probes have been developed for the collection of these data. The problem is of such significance that highly specialized and expensive laser-Doppler instrumentation is being developed for measurements of this type.

1.4.4 Aeroacoustics

Increasing concern with aircraft noise, as well as other sources of noise of aerodynamic origin (automobiles, trucks, and air-conditioning systems, for example), has placed emphasis on the interrelationship between aerodynamics and acoustics.

The pioneering work of Lighthill [5] forms the basis for much of the current research on turbulence as a source of sound. One practical question relates to the distribution and intensity of sound sources within a turbulent flow. Information of this type can be obtained, in principle, by velocity measurements alone. In research, a more common method is to cross correlate (with appropriate time delay) a velocity signal (usually determined with a hot-wire or laser-Doppler velocimeter) with a sound signal determined by a microphone placed in the radiated sound field.

The first method depends on a simplified theory and is extremely useful in certain applications, such as the determination of sound sources in the region of a jet. The principle is based on Lighthill's theory. In simplified form, the intensity per unit volume of sound source is given by

$$\frac{I}{V_e} \sim \frac{V_e \omega^4 T_{ij}^2}{r^2 \rho_0^2 a_0^5} \tag{7}$$

In terms of near-field pressure p, an alternative form of Eq. (7) is

$$\frac{I}{V_e} \sim \frac{V_e \omega^4}{r^2} \frac{\overline{p^2}}{\rho_0^2 a_0^5} \tag{8}$$

Work by Kraichnan [6], Davies et al. [7], and Lilley [8] indicate that the following simplifications can be used:

$$p \cong \rho u l \frac{\partial U}{\partial y} \tag{9}$$

$$\omega \cong \frac{\partial U}{\partial y} \cong \frac{u}{l} \tag{10}$$

$$V_e \cong l^3 \tag{11}$$

where l is a characteristic scale of turbulence usually defined by the integral scale. (More is said about integral scale in later chapters.) Hence a simplified method for determining acoustic noise sources in a turbulent flow is based on the following equation [which follows from substitution of Eqs. (9)–(11) into Eq. (8)]:

$$\frac{I}{V_e} \sim \frac{l^5}{a_0^5 r^2} \overline{\rho u^2} \left(\frac{\partial U}{\partial y}\right)^6 \tag{12}$$

Hence measurement of integral scale l, turbulence intensity $\overline{u^2}$, and mean velocity gradient $\partial U/\partial y$ can give an estimate of the sound-source distribution in a turbulent flow. Direct verification of Eq. (12) is difficult; however, its basis, Eq. (9), has been verified directly. An example is shown in Fig. 1.5. The upper portion of the figure illustrates the theoretical distribution of pressure intensity within a turbulent jet. The lower portion of the figure contains pressure data obtained with a special pressure probe developed by Arndt and Nilsen [10].

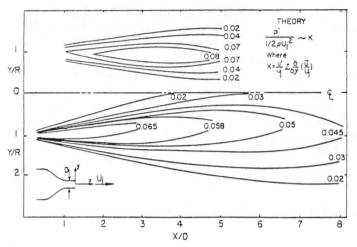

Figure 1.5   Measured pressure fluctuations in a turbulent jet. (After Barefoot [9])

Cross-correlation techniques (discussed in detail in later chapters) are useful in aeroacoustic research. The pressure p_a in the far field is given by the volume integral

$$p_a = \frac{1}{4\pi a_0^2 r} \int_{\text{vol}} \left[\frac{\partial^2}{\partial t^2} (\rho u_r^2) \right]_{t'} dy \tag{13}$$

where $t' = t - r/a_0 =$ delay time
    $u_r =$ component of velocity fluctuation in direction of observer
Since the acoustic intensity I is related to the mean square acoustic pressure, Eq. (13) can be modified to yield

$$I = \frac{\overline{p_a^2}}{\rho_0 a_0} \simeq \frac{V_e}{4\pi \rho_0 a_0^3 r} \frac{\partial}{\partial t} \overline{(\rho u_r^2 p_a)}_{t = t'} \tag{14}$$

or, equivalently,

$$I \simeq \frac{V_e}{4\pi \rho a_0^3 r} \frac{\partial}{\partial t} \overline{(pp_a)}_{t = t'} \tag{15}$$

where p is the pressure in the flow. Hence sound sources can be determined by cross correlation of a near-field pressure or velocity signal and a far-field velocity signal. The principle is illustrated in Fig. 1.6 for the case of airfoil noise due to pressure fluctuations on the surface of the foil and velocity fluctuations in the wake. A typical instrumentation package is also illustrated.

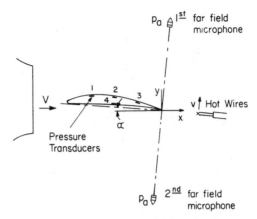

Experimental set-up of the airfoil into the flow

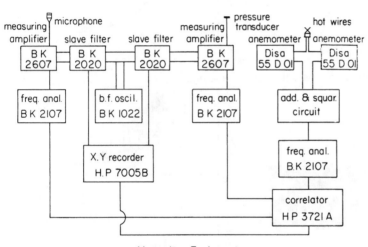

Measuring Equipment

Figure 1.6 Use of cross-correlation techniques to isolate noise sources on and in the vicinity of an airfoil. (Courtesy Professor G. Comte-Bellot, Ecole Central de Lyon, Ecully, France)

1.4.5 Turbulent Mixing Layer

As a final example, consider fundamental research into the properties of a turbulent mixing layer formed at the interface between two fluids of different densities and different velocities, as illustrated in Fig. 1.7. This is an interesting example, since it illustrates how single-point measurements can be misleading. As shown in the upper portion of the figure, mean velocity and density vary uniformly across the mixing layer, with turbulence intensity peaking somewhere in the central por-

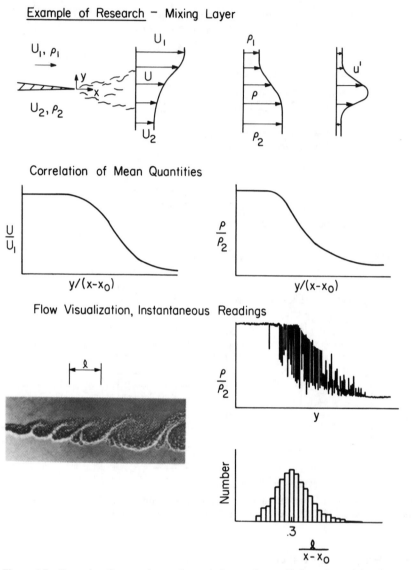

Figure 1.7 Example of research on the turbulence characteristics of a mixing layer. (Adapted from [11])

tion of the layer. Correlation of mean velocity and density in a manner implied by classical theory results in similar profiles, as shown in the figure. Instantaneous readings of density (shown in the lower portion of the figure) indicate that sharp gradients in density exist. Further, a shadowgraph of the mixing layer indicates the existence of a quasi-orderly structure. The wavelength of the structure varies with distance x and has a statistical variation. It is the statistical variation (or "jitter") that has led to the misinterpretation of hot-wire data over the years. In fact, hindsight tells us that observations at one or two spatially fixed stations will include realizations from a large number of such structures at various stages in their life history. Time and space averages in the classical sense will tend to "smear out" the essential features of the quasi-orderly structure in the flow [12].

1.4.6 Summary

Examples of the use of flow instrumentation have been drawn from a variety of fields to illustrate the interrelated roles of flow instrumentation, the theory of fluid mechanics, and research and development. The need is evident for a clear understanding of the problem at hand, the principles of fluid mechanics, the principles of operation of flow instrumentation, and the elements of statistical analysis.

1.5 OUTLINE OF THE THEORY OF FLUID MECHANICS

1.5.1 Inviscid Flow

The reader is assumed to have a basic understanding of fluid mechanics. However, the mathematical principles are briefly reviewed here. It is sufficient for this overview to limit the discussion to incompressible fluid mechanics, to illustrate the interrelationship between theory and measurement principles.

In a Eulerian frame of reference, conservation of mass and momentum yield (in standard tensor notation)

$$\frac{\partial U_i}{\partial x_i} = 0 \tag{16}$$

$$\rho \frac{DU_i}{Dt} = \rho X_i + \frac{\partial \tau_{ji}}{\partial x_j} \tag{17}$$

$$\frac{D}{Dt} \equiv \frac{\partial}{\partial t} + U_i \frac{\partial}{\partial x_i} \tag{18}$$

where x_i are the components of a body force field. In "frictionless" flow the only term in the stress tensor τ_{ji} is the pressure p. Hence

$$\rho \frac{DU_i}{Dt} = X_i - \frac{\partial p}{\partial x_i} \tag{19}$$

where p is positive for compression. Equation (19) can be written in vector notation as

$$\rho \left[\frac{\partial \mathbf{V}}{\partial t} + (\mathbf{V} \cdot \nabla)\mathbf{V} \right] = \mathbf{X} - \nabla p \tag{20}$$

Using the vector identity

$$\nabla(\mathbf{V} \cdot \mathbf{V}) = 2(\mathbf{V} \cdot \nabla)\mathbf{V} + 2\mathbf{V} \times (\nabla \times \mathbf{V}) \tag{21}$$

Eq. (20) can be written in the form

$$\rho \left[\frac{\partial \mathbf{V}}{\partial t} + \nabla \frac{V^2}{2} - \mathbf{V} \times (\nabla \times \mathbf{V}) \right] = \mathbf{X} - \nabla p \tag{22}$$

When the flow is irrotational ($\nabla \times \mathbf{V} = 0$), Eq. (22) can be integrated. A velocity potential ϕ exists such that

$$\mathbf{V} = -\nabla \phi \tag{23}$$

Limiting ourselves to conservative force fields, we can also write

$$\mathbf{X} = -\nabla \Omega \tag{24}$$

Thus

$$\nabla \left(-\frac{\partial \phi}{\partial t} + \frac{V^2}{2} + \frac{p}{\rho} + \Omega \right) = 0 \tag{25}$$

Equation (25) leads to

$$-\frac{\partial \phi}{\partial t} + \frac{V^2}{2} + \frac{p}{\rho} + \Omega = F(t) \tag{26}$$

where $F(t)$ is an arbitrary function of time. For steady flow,

$$\frac{V^2}{2} + \frac{p}{\rho} + \Omega = \text{const} \tag{27}$$

If the only conservative force field is that due to gravity, then

$$\Omega = gz \tag{28}$$

Equation (22) can also be integrated for inviscid rotational flow. The result is the same as Eq. (27), except that the *constant is different for each streamline.*

The velocity potential ϕ satisfies Laplace's equation,

$$\nabla^2 \phi = 0 \tag{29}$$

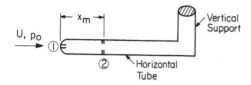

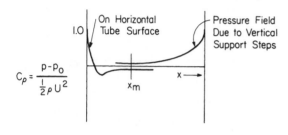

Figure 1.8 Fluid-dynamic considerations in the design of a Prandtl tube

This follows directly from the definition of ϕ and continuity,

$$\mathbf{V} = -\nabla\phi \qquad \nabla \cdot \mathbf{V} = 0$$

The many techniques for the solution of Eq. (29) are beyond the scope of this review. In each case, the principle is the same: ϕ is determined from Eq. (29) for the given boundary conditions, the velocity field is determined from Eq. (23), and finally, the pressure is deduced from the Bernoulli equation. An illustration of the technique is the design of a probe for measuring velocity in a uniform flow. The pressure sensed at the nose is the total pressure:

$$p_t = p_0 + \tfrac{1}{2}\rho V^2 \tag{30}$$

The problem is to find a position on the probe where the pressure in p_0. The difference in the pressures at these two points can be used to determine the flow velocity:

$$V = \sqrt{\frac{2(p_t - p_0)}{\rho}} \tag{31}$$

A simple probe, consisting of a blunt-nosed cylinder aligned with the flow and a vertical stem, is illustrated in Fig. 1.8. The pressure varies from p_t at the nose to values less than p_0 a short distance downstream and then increases asymptotically to p_0 further downstream. Since the stem of the probe causes an increase in pressure, it is comparatively easy to find a position where the vacuum due to the nose

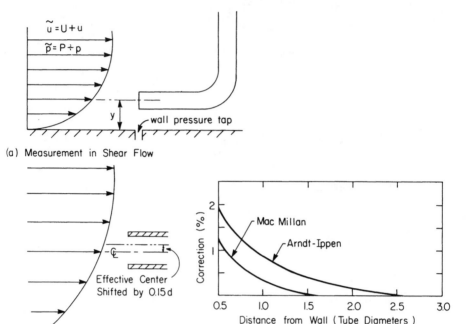

(a) Measurement in Shear Flow

(b) Shear Flow Displaces Effective
Center of the Probe

(c) Presence of wall creates Flow Distortion
around Tube

Figure 1.9 Effects of turbulence velocity gradients and wall proximity on measurements of total pressure

of the cylinder is balanced by the increased pressure due to the stem. Details are given in [13].

Measurements of flow velocity are not usually made in the ideal uniform flow for which a Prandtl tube is designed. A typical situation is illustrated in Fig. 1.9. The measurement is being made in turbulent shear flow. Often, under these circumstances, it is convenient to measure total pressure with a total-head tube, and static pressure with a wall tap. There are three different flow effects on the total-pressure measurement: The time mean total pressure does not correspond to the time mean velocity because of the turbulence; the effective center of the probe is shifted toward the high-velocity region; and the wall creates an asymmetrical flow around the tube. The first of these effects can be estimated from the Bernoulli equation. The instantaneous total pressure is

$$\tilde{p}_t = \tfrac{1}{2}\rho(U + u)^2 + P + p \tag{32}$$

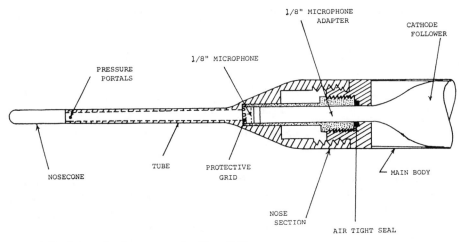

Figure 1.10 Schematic of pressure probe. (From [14])

where U and P are the time mean velocity and pressure, respectively, and u and p are the instantaneous fluctuations about the mean. Expanding Eq. (32) and taking the average yield (where overbars denote time averages)

$$P_t = \frac{\rho}{2} U^2 + \frac{\rho}{2} \overline{u^2} + P \qquad (33)$$

Thus

$$\frac{U}{[2(P_t - P/\rho]^{1/2}} = \left(1 + \frac{\overline{u^2}}{U^2}\right)^{-1/2} = 1 - \frac{1}{2}\frac{\overline{u^2}}{U^2} + \text{h.o. terms} \qquad (34)$$

The other two problems are more complex. However, measured corrections are given in Fig. 1.9.

George et al. [14] have extended these ideas to the problem of measuring fluctuating pressures in a jet. In this case the error in measuring fluctuating pressure with a "static" pressure probe as shown in Fig. 1.10 is given by

$$\frac{1}{\rho} (p_m - p) = A[2Uu + (u^2 - \overline{u^2})] + B[(v^2 - \overline{v^2}) + (w^2 - \overline{w^2})] \qquad (35)$$

The coefficient A measures the sensitivity to streamwise velocity fluctuations, while B measures the sensitivity to crosswise fluctuations. The numerical values for A and B are a function of the shape of the probe. Inviscid analysis, similar to that outlined for the Prandtl probe, can be used to calculate these constants. George et al. [14] found values of $A = -0.0075$ and $B = -0.15$ for the probe shown in Fig. 1.10.

The error in the mean square static-pressure fluctuation is readily obtained from Eq. (35) as

$$\frac{1}{\rho^2}\,(\overline{p_m^2} - \overline{p^2}) = \frac{1}{\rho}\,[A(2\,U\overline{pu} + \overline{pu^2}) + B(\overline{pv^2} + \overline{pw^2})]$$
$$+ \{A^2[4U^2\overline{u^2} + 4U\overline{u^3} + \overline{(u^2 - \overline{u^2})}]$$
$$+ AB\,[2U(\overline{uv^2} + \overline{uw^2}) + \overline{(u^2 - \overline{u^2})\,(v^2 - \overline{v^2})} \tag{36}$$
$$+ \overline{(u^2 - \overline{u^2})\,(w^2 - \overline{w^2})}] + B^2\,[\overline{[(v^2 - \overline{v^2})^2}$$
$$+ \overline{(v^2 - \overline{v^2})\,(w^2 - \overline{w^2})} + \overline{(w^2 - \overline{w^2})^2}]\}$$

Making further assumptions about pressure-velocity and velocity-velocity correlations (discussed below), an estimate of the error in measuring pressure spectra can also be determined as a function of turbulence intensity and normalized wave number, $k1$. Further details can be found in [14] and Chap. 3.

1.5.2 Viscous Flow and Turbulence

For incompressible flow the stress tensor can be split into isotropic and anisotropic components:

$$\tau_{ji} = -p\delta_{ji} + \tau_{ji}^* \tag{37}$$

$$\tau_{ii}^* \equiv 0 \tag{38}$$

Newtonian fluids display a linear relationship between τ_{ji}^* and the rate of deformation:

$$\tau_{ij}^* = 2\mu e_{ij} \tag{39}$$

$$e_{ij} = \frac{\partial u_i}{\partial x_j} + \frac{\partial u_j}{\partial x_i} \tag{40}$$

where μ is the viscosity. Thus

$$\tau_{ji} = -p\delta_{ji} + 2\mu \left(\frac{\partial u_i}{\partial x_j} + \frac{\partial u_j}{\partial x_i} \right) \tag{41}$$

Substitution into the momentum equation yields

$$\rho \frac{Du_i}{D_t} = \rho X_i - \frac{\partial p}{\partial x_i} + \mu \frac{\partial^2 u_i}{\partial x_j\,\partial x_j} \tag{42}$$

or

$$\frac{DV}{Dt} = X - \frac{1}{\rho}\,\nabla p + \nu \nabla^2 V \tag{43}$$

Equations (42) and (43) are the Navier-Stokes equations, which form the basis for theoretical fluid mechanics. They are nonlinear, and no general solution is known.

In certain cases the equations reduce to a linear form, such as for flow through a straight channel, where the convective acceleration terms are identically zero. Another example is Stokes' first problem, which considers a viscous fluid in the half-plane $y \geqslant 0$, bounded by a wall at $y = 0$, which is suddenly accelerated to a velocity U. In this case the problem reduces to

$$\frac{\partial u}{\partial t} = \nu \frac{\partial^2 u}{\partial y^2} \tag{44}$$

$$\frac{\partial u}{\partial x} = 0 \tag{45}$$

with

$$u(y, 0) = 0$$

$$u(0, t) = U_0 \tag{46}$$

$$u(\infty, t) = 0$$

The solution is

$$\frac{u}{U_0} = \operatorname{erfc} \frac{y}{\sqrt{4\nu t}} \tag{47}$$

For $y > \sqrt{4\nu t}$, u is negligible. This leads to the observation that viscous effects are limited to a thin layer of fluid of thickness $\sqrt{\nu t}$. To an observer moving with velocity U, the thickness of this layer is

$$\delta = 4\sqrt{\frac{\nu x}{U}} \tag{48}$$

Boundary layers, jets, and wakes all display this behavior. In such cases the Navier-Stokes equations and the continuity equation take the form

$$\frac{\partial u}{\partial t} + u\frac{\partial u}{\partial x} + v\frac{\partial u}{\partial y} = -\frac{1}{\rho}\frac{\partial p}{\partial x} + \nu\frac{\partial^2 u}{\partial y^2} \tag{49}$$

$$\frac{\partial u}{\partial x} + \frac{\partial v}{\partial y} = 0 \tag{50}$$

There are two equations and three unknowns, namely, u, v, and p. The problem is solved by, first, neglecting viscous effects and solving for the pressure p. Gradients in pressure normal to the boundary layer flow are negligible, and hence p is considered known in solving the system of equations (49) and (50). (Refer to [15] for details.)

1.5.3 Turbulence

Turbulent flows are highly complex. Since many measurements are made in turbulent flow, some features are reviewed here. The uninitiated should, however, refer to the several textbooks on the topic. (See, for example, [16–19].) Turbulence can be regarded as a highly disordered motion resulting from the growth of instabilities in an initially laminar flow. Recent information, however, indicates that turbulence may be more orderly than has been previously thought possible [12].

It is convenient to decompose the velocity into a mean and fluctuating part. For this discussion the mean motion will be a function of position only:

$$U = \lim_{T \to \infty} \frac{1}{T} \int_0^T \tilde{u}(t)dt \tag{51}$$

and

$$\tilde{u} = U + u \tag{52}$$

Similarly, for pressure,

$$\tilde{p} = P + p \tag{53}$$

and so on. The turbulent velocity field u is made up of a broad range of scales. The timescale for the large eddies is proportional with l/u, where l is a characteristic length in the flow. Most of the turbulent kinetic energy is associated with large eddies. At the other end of the spectrum, viscous dissipation occurs. The timescale for viscous dissipation can be estimated from the diffusion equation

$$\frac{\partial u}{\partial t} = \nu \frac{\partial^2 u}{\partial y^2} \tag{54}$$

which can be written as

$$\frac{u}{T_m} \sim \nu \frac{u}{l^2} \tag{55}$$

or

$$T_m \sim \frac{l^2}{\nu} \tag{56}$$

Hence, timescales for turbulent and molecular diffusion have the ratio

$$\frac{T_l}{T_m} = \frac{l/U}{l^2/\nu} = \text{Re} \tag{57}$$

where Re is the Reynolds number, Vl/ν. An equation for turbulent diffusion can be written in analogy with Eq. (54):

$$\frac{\partial u}{\partial t} = v_t \frac{\partial^2 u}{\partial y^2} \tag{58}$$

where

$$v_t = \frac{l^2}{T_t} = ul \tag{59}$$

The ratio of turbulent to molecular diffusivity is then

$$\frac{v_t}{v} = \frac{ul}{v} = \mathrm{Re} \tag{60}$$

For turbulent flows the characteristic length l is proportional to δ or the width of a jet or wake, i.e., a transverse length scale. The smallest scales in a turbulent flow are controlled by viscosity. The small-scale motion has much smaller characteristic timescales than the large-scale motion and may be considered statistically independent. We may think of the small-scale motion as being controlled by the rate of supply of energy and the rate of dissipation. It is probably reasonable to assume that the two rates are equal. Thus we consider the dissipation rate per unit mass ε and the kinematic viscosity v. From purely dimensional reasoning, we can form length, time, and velocity scales:

$$\eta = \left(\frac{v^3}{\varepsilon}\right)^{1/4} \qquad \tau = \left(\frac{v}{\varepsilon}\right)^{1/2} \qquad u_v = (v\varepsilon)^{1/4} \tag{61}$$

A Reynolds number formed by these parameters is unity:

$$\frac{u_v \eta}{v} = 1 \tag{62}$$

which indicates that viscous effects dominate the dissipation mechanism and length scales adjust themselves to the energy supply. For equilibrium flow, we can assume that the rate of supply of energy from the large eddies is inversely proportional to their timescale. Thus

$$\varepsilon \sim \frac{u^3}{l} \tag{63}$$

This yields

$$\frac{\eta}{l} = \left(\frac{v^3 l}{u^3 l^4}\right)^{1/4} = \frac{1}{\mathrm{Re}^{3/4}} \tag{64}$$

We can also say that

$$\frac{\tau u}{l} = \left(\frac{ul}{v}\right)^{-1/2} = \frac{1}{\mathrm{Re}^{1/2}} \tag{65}$$

Hence the timescale of the small-eddy structure is much smaller than the large-scale eddies. Since vorticity is proportional to $1/t$, we find that the vorticity is concentrated in the small scales.

It is generally accepted that turbulent flows can be described by the Navier-Stokes equations. Using the decomposition described by Eqs. (52) and (53) in Eq. (42) and taking the time mean results in

$$U_j \frac{\partial U_i}{\partial x_j} = \frac{1}{\rho} \frac{\partial}{\partial x_j} (-P\delta_{ij} + 2\mu e_{ij} - \rho \overline{u_i u_j}) \tag{66}$$

$$\frac{\partial U_i}{\partial x_i} = 0 \tag{67}$$

The important feature of Eq. (66) is the additional apparent stress tensor, $-\rho \overline{u_i u_j}$, the components of which are the basis for the so-called closure problem in turbulent flow computations. Subtraction of Eq. (66) from the fully expanded Navier-Stokes equations yields a system of equations for the turbulent fluctuations:

$$\frac{\partial u_i}{\partial t} + U_j \frac{\partial u_i}{\partial x_j} + u_j \frac{\partial U_i}{\partial x_j} + u_j \frac{\partial u_i}{\partial x_j} - \overline{u_j \frac{\partial u_i}{\partial x_j}} = -\frac{1}{\rho} \frac{\partial p}{\partial x_i} + v \frac{\partial u_i}{\partial x_j \partial x_j} \tag{68}$$

$$\frac{\partial u_j}{\partial x_j} = 0 \tag{69}$$

$$U_j \frac{\partial}{\partial x_j} \left(\frac{1}{2} U_i U_i \right) = \frac{\partial}{\partial x_j} \left(\underbrace{-\frac{P}{\rho} U_j}_{(1)} + \underbrace{2vU_i e_{ij} - \overline{u_i u_j} U_i}_{(2)} \right) - \underbrace{2v e_{ij} e_{ij}}_{(3)} + \underbrace{\overline{u_i u_j} e_{ij}}_{(4)} \tag{70}$$

$$U_j \frac{\partial}{\partial x_j} \left(\frac{1}{2} \overline{u_i u_i} \right) = -\frac{\partial}{\partial x_j} \left(\underbrace{\frac{\overline{p u_j}}{\rho}}_{(1)} - \underbrace{2v \overline{u_i e'_{ij}} + \frac{1}{2} \overline{u_i u_i u_j}}_{(2)} - \underbrace{2v \overline{e'_{ij} e'_{ij}}}_{(3)} - \underbrace{\overline{u_i u_j} e_{ij}}_{(4)} \right) \tag{71}$$

In each equation, (1) denotes pressure work, (2) denotes transport terms, (3) denotes dissipation terms, and (4) denotes production. Also, e_{ij} is the mean strain tensor, and e'_{ij} the fluctuation strain tensor. The important point is that the production term in Eq. (70) is negative, whereas in Eq. (71) it is positive. In other words, the turbulence is being produced at the expense of the mean flow.

In a steady, homogeneous, pure-shear flow in which all averaged quantities except U_i are independent of position and in which e_{ij} is constant, Eq. (71) reduces to

$$- \overline{u_i u_j} e_{ij} = 2v \overline{e'_{ij} e'_{ij}} \tag{72}$$

This says simply that the rate of production of turbulent energy by Reynolds stresses equals the rate of viscous dissipation. This is an idealized situation, and in most shear flows, production and dissipation do not balance, although they are usually of the same order of magnitude.

The case of isotropic turbulence is of interest for two reasons: First, it is the simplest case of turbulence and is amenable to analytical treatment; second, the small-scale structure is approximately isotropic. This type of turbulence is homogeneous (the same at every point) and statistically independent of orientation and location of axis.

As a result of the definitions, we have

$$\overline{u^2} = \overline{v^2} = \overline{w^2} \tag{73}$$

$$\overline{uv} = \overline{vw} = \overline{wu} = 0 \tag{74}$$

For truly isotropic conditions the mean velocity is zero. It can be shown that this type of flow can be completely described by two functions:

$$R_u(r) = \frac{\overline{u_1 u_2}}{\overline{u^2}} \tag{75}$$

and

$$R_v(r) = \frac{\overline{v_1 v_2}}{\overline{v^2}} \tag{76}$$

where $u, v = x, y$ components of velocity
   r = separation distance between points 1 and 2
The longitudinal and lateral velocity coordinates are not independent but are related in the form

$$R_v = R_u + \frac{r}{2} \frac{\partial R_u}{\partial r} \tag{77}$$

Based on the correlation function R_u, we can define a length scale as the radius of curvature of $R_u(0)$:

$$-\frac{1}{\lambda_T^2} = \frac{\partial^2 R_u}{\partial r^2} \bigg|_{r \to 0}$$

where λ_T is related to the radius of the smaller eddies and is referred to as the Taylor microscale. Successive differentiation of Eq. (77) yields the relation (see page 290 in [20])

$$2 \overline{\left(\frac{\partial u_i}{\partial x_i}\right)^2} = \overline{\left(\frac{\partial u_j}{\partial x_i}\right)^2} \text{ (no sum)} \tag{78}$$

It can be shown that

$$\frac{\overline{u^2}}{\lambda_T^2} = \overline{\left(\frac{\partial u}{\partial x}\right)^2}$$ (79)

In isotropic turbulence the dissipation term can be written as

$$\varepsilon = 2\nu\overline{e_{ij}e_{ij}} = 15\nu\overline{\left(\frac{\partial u}{\partial x}\right)^2}$$ (80)

From Eq. (79),

$$\varepsilon = 15\nu\frac{\overline{u^2}}{\lambda_T^2}$$ (81)

If we can approximate the fine-scale turbulence in a shear flow with this model of isotropic turbulence, it is possible to obtain an approximate energy balance in terms of the Taylor microscale. Approximating the production term by

$$-\overline{u_iu_j}e_{ij} \simeq A\frac{u^3}{l}$$ (82)

we obtain

$$A\frac{u^3}{l} = 15\nu\frac{u^2}{\lambda_T^2}$$ (83)

or

$$\frac{x}{l} = \left(\frac{15}{A}\right)^{1/2}\left(\frac{ul}{\nu}\right)^{-1/2}$$ (84)

Since $ul/\nu \gg 1$, we conclude that $\lambda_T/l \ll 1$.

We can reason that the timescale of the large-eddy structure is proportional to l/u and the timescale of the strain-rate fluctuations is λ_T/u. The ratio of the two timescales is then

$$\frac{l}{\lambda_T} \sim \frac{ul}{\nu}$$ (85)

An important feature of turbulent flows that can be deduced from this discussion is that *production is greatest at large scales, and dissipation is greatest at small scales.*

Isotropic flow is obviously an idealization. Most practical flows, such as boundary layers, jets, and wakes, are shear flows. Hence information concerning the Reynolds stress tensor is necessary. The many different techniques that have evolved for calculating these flows are beyond the scope of this review (cf. [18, 19]).

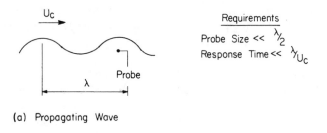

(a) Propagating Wave

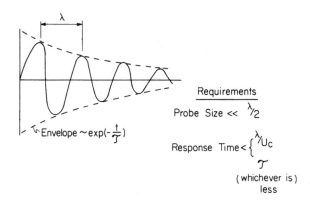

(b) Decaying, traveling wave

Figure 1.11 Spatial versus temporal resolution

1.6 SPATIAL AND TEMPORAL RESOLUTION IN MEASUREMENTS

In general, the following information is determined in a measurement program:

1. Physical properties such as ρ_0 and μ_0; these are not functions of the flow field
2. Scalar fields such as temperature and pressure
3. Vector fields such as velocity

To select instrumentation, we need to estimate the scale of the flow and the scale of any disturbances in the flow (e.g., wavelength λ and boundary layer thickness δ). We also need to know the frequencies of any disturbances.

One of the problems we face is the distinction between spatial resolution and the temporal response of our instrumentation. This is illustrated in Fig. 1.11. Consider first a probe placed in a disturbance of wavelength λ, propagating with speed

U_c. We have two criteria. First the probe must be small enough to resolve the spatial extent of the disturbance; i.e., the probe size must be less than $\lambda/2$. Second, since the probe is fixed in space, it must have adequate frequency response; i.e., the response time must be less than λ/U_c.

Consider the more complex case of resolving a wavelike disturbance whose amplitude is decaying. The example cited is one of exponential decay, $a \sim \exp(-t/\Im)$. We still have the spatial requirement (probe size $< \lambda/2$), but our response-time requirement is twofold: The response time must be less than λ/U_c or $\Im$, whichever is less. This is a classical problem in the measurement of turbulence, where time-varying disturbances are convected by a fixed probe. The Taylor hypothesis assumes the turbulence to be frozen:

$$\left(\frac{\partial}{\partial t}\right)_{\text{measured}} \cong \left(U_c \frac{\partial}{\partial x}\right)_{\text{field}} \tag{86}$$

For this to be true,

$$\frac{l}{U_c} \ll \frac{l}{u} \sim \Im$$

or

$$\frac{u}{U_c} \ll 1$$

Hence Taylor's hypothesis is limited to low-intensity turbulent flows. A somewhat sophisticated way around this problem is to move the probe in a direction opposite to the flow, thereby artificially creating a high U_c. Such is the case when meteorological data are collected with flow instrumentation mounted in the wing of an airplane.

An example of the trade-off between temporal and spatial resolution is illustrated in Fig. 1.12. The problem is to measure unsteady wall pressure with a disturbance wavelength λ. Placing a pressure transducer flush with the wall ensures maximum frequency response. However, the diameter of the probe must be less than half a wavelength. If this is not the case, the transducer can be placed in a chamber with a pinhole leading to the surface. Here spatial resolution is achieved at the expense of temporal response. There are two methods of analysis for determining temporal response under these conditions. If the test fluid is compressible, the system can be analyzed as a Helmholtz resonator. The natural frequency f_c of the pinhole cavity system is given by

$$\frac{2\pi f_c l_T}{a_0} \tan \frac{2\pi f_c l_T}{a_0} = \frac{\pi R^2 l_T}{V_c} \quad \text{where } V_c = \text{cavity volume}$$
$$R = \text{pinhole radius}$$
$$a_0 = \text{speed of sound in the medium} \tag{87}$$

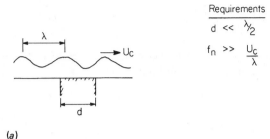

Requirements

$$d \ll \frac{\lambda}{2}$$

$$f_n \gg \frac{U_c}{\lambda}$$

(a)

Requirements

$$d' \ll \frac{\lambda}{2}$$

$$f_n \gg \frac{U_c}{\lambda}$$

Additional   Problems

(b)

Attenuation of system natural frequency by increased added mass, change in stiffness of system.

Assumptions

Can replace transducer diaphragm by a spring loaded piston.

Flow is one dimensional.

Viscous effects are negligible.

(c)

Figure 1.12   Trade-off between spatial and temporal resolution. (*a*) Measurement of wall pressure with a flush-mounted transducer. (*b*) Same measurement using a cavity-mounted transducer. A pinhole leads from the surface to the cavity. (*c*) Idealized system for analysis

Equation (87) fails to predict an attenuation in incompressible fluids ($a_0 \rightarrow \infty$). A lesser known mechanism, that of inertial damping, can result in a significant attenuation of frequency response [21]. The analysis is illustrated in the lower portion of Fig 1.12. If the flow velocity in the tube is V, then $\phi = Vs$, and the unsteady Bernoulli equation (26) can be written in the form

$$\frac{1}{g}\frac{\partial V}{\partial t} = -\frac{\partial H}{\partial s} \tag{88}$$

$$H = \frac{p}{\gamma} + \frac{V^2}{2g} + z \tag{89}$$

As shown, the transducer is replaced by a spring-loaded piston. If the flow velocity in the chamber is $\dot{x}$, then V is determined from continuity:

$$V = \frac{A_2}{A_1} \dot{x} \tag{90}$$

The equation of motion for flow in the tube is then

$$\frac{1}{g} \frac{A_2}{A_2} \ddot{x} = -\frac{1}{\gamma} \frac{\partial p}{\partial s} \tag{91}$$

The force balance on the piston is

$$p_c A_2 = -Kx \tag{92}$$

If the pressure is zero at the end of the tube, then

$$-\frac{\partial p}{\partial s} = \frac{p_c}{l} = -\frac{Kx}{lA_2} \tag{93}$$

Hence

$$\frac{1}{g} \frac{A_2}{A_1} \ddot{x} + \frac{K}{\gamma l A_2} x = 0 \tag{94}$$

$$\ddot{x} + \frac{K}{\rho l A_2} \frac{A_1}{A_2} x = 0 \tag{95}$$

The solution to Eq. (95) is

$$2\pi f_c = \sqrt{\frac{K}{\rho l A_2} \frac{A_1}{A_2}} \tag{96}$$

which should be compared with Eq. (88). Further details are given in [21]. In making unsteady pressure measurements in liquids, the problem can be further complicated by the presence of air bubbles in the cavity, which would significantly reduce the stiffness of the system.

The important point here is that we are generally limited to a Eulerian frame of reference. This makes it difficult to distinguish between a truly temporal change in a flow quantity and a pseudo-temporal change due to a spatially varying disturbance convected past the fixed probe. It is essential that there be a proper balance between the spatial resolution and temporal response of the instrumentation.

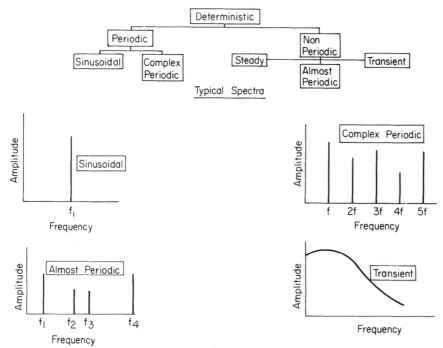

Figure 1.13 Classification of deterministic data

1.7 CORRELATION OF DATA AND SIGNAL ANALYSIS

The elements of data correlation and signal analysis are presented here as an introduction to the subject. Further discussion of this important and complex topic is given in subsequent chapters and in other texts (e.g., [22]).

In a broad sense, data fall into two categories, namely, deterministic and random. The distinction between the two is somewhat nebulous, since any physical process can be contaminated by unknown factors and therefore become not strictly deterministic. A physical process is deterministic if everything is known about it. We extend this definition to imply that data are deterministic if the influence of random perturbations on a physical process being measured is minimal.

1.7.1 Classification of Deterministic Data

The classification of deterministic data is illustrated in Fig. 1.13. Steady data are classified by amplitude. Sinusoidal data are classified by amplitude and frequency. Complex periodic data can usually be decomposed into harmonics of the fundamental frequency, which are characterized by their amplitude and phase relationship. This is done through a Fourier series:

$$\tilde{u}(t) = \sum_{n=-\infty}^{\infty} C_n e^{in(t/T)}$$

$$C_n = \frac{1}{T}\int_0^T \tilde{u}(t)e^{-in(t/T)}dt$$

where T is the fundamental period.

Two or more sine waves will be periodic only when the ratios of all possible pairs of frequencies form rational numbers like 2/7 and 3/7, as opposed to $3/\sqrt{50}$. In the case of almost-periodic data, a series of the form

$$\tilde{u}(t) = \Sigma A_n \sin(2\pi f_n t + \Phi_n)$$

can usually be used to describe the function. The ratio $\tilde{u}_n/\tilde{u}_m$ is not a rational number in all cases.

1.7.2 Random Data and Signal Analysis

Many practical flow problems involve turbulent flow. The signal is not deterministic and is therefore classified as random. Thus we are forced into the realm of statistics to quantify measurements. This discussion is limited to stationary random functions. Stationarity is loosely defined as the case in which the various statistical functions describing the signal are time invariant.

1.7.2.1 Basic description. The simplest method of describing a function that varies randomly with time is to determine its mean value.

$$U = \lim_{T\to\infty} \frac{1}{T}\int_0^T \tilde{u}(t)dt \qquad (97)$$

The mean value does not indicate how much $\tilde{u}(t)$ is fluctuating about its mean value. A measure of the extent of deviations from the mean is the variance, or mean square value,

$$\overline{u^2} = \lim_{T\to\infty} \frac{1}{T}\int_0^T (\tilde{u} - U)^2\, dt \qquad (98)$$

An estimate of the percentage of time during which $\tilde{u}(t)$ can lie within a well-defined range is given by the probability density. With reference to Fig. 1.14, the probability density is determined as follows.

The total time during which $\tilde{u}(t)$ lies in the range $u_1 \leqslant \tilde{u}(t) \leqslant u_2$ is given by

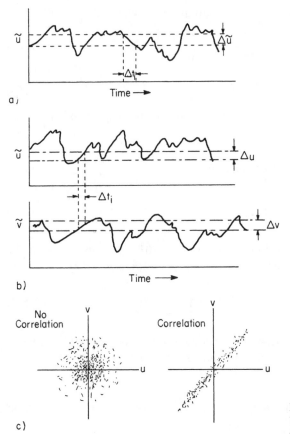

Figure 1.14 Determination of probability and joint probability density. Correlation of two variables

$$T_u = \sum_{i=1}^{\infty} \Delta t_i \tag{99}$$

where Δt_i is the length of time the function takes on values between u_1 and u_2 as it becomes larger than u_1 or less than u_2. The probability of this occurring is then

$$\text{Prob } \{u \leqslant \tilde{u}(t) \leqslant u + \Delta u\} = \lim_{T \to \infty} \frac{T_u}{T} \tag{100}$$

A probability density function is given by

$$p(u) = \lim_{\Delta u \to 0} \frac{\text{Prob } \{u \leqslant \tilde{u}(t) \leqslant u + \Delta u\}}{\Delta u} \tag{101}$$

$$p(u) = \lim_{\Delta u \to 0} \lim_{T \to \infty} \frac{1}{T} \frac{T_u}{\Delta u} \tag{102}$$

A good deal of random data follow a Gaussian distribution,

$$p(u) = \frac{1}{\sqrt{2\pi}\sqrt{\sigma}} \exp\left(-\frac{u^2}{2\sigma^2}\right) \tag{103}$$

$$\sigma \equiv \sqrt{\overline{u^2}} \tag{104}$$

For this situation the time mean and variance are related to the probability density in the following way:

$$U = \int_{-\infty}^{\infty} \tilde{u} p(u) d\tilde{u} \tag{105}$$

$$\overline{u^2} = \int_{-\infty}^{\infty} (\tilde{u}^2 - U^2) p(u) du \tag{106}$$

From the definition of the probability density, Eq. (101), it is obvious that the probability that $\tilde{u}$ will fall between two values, say A and B, is given by

$$\text{Prob} \{A \leq \tilde{u} \leq B\} = \int_A^B p(u) d\tilde{u} \tag{107}$$

The cumulative probability function describes the probability that $\tilde{u}(t)$ is less than or equal to some value u. It is equal to the integral of the probability density function, taken from $-\infty$ to u:

$$P(u) = \text{Prob} \{\tilde{u}(t) \leq u\} = \int_{-\infty}^{u} p(\xi) d\xi \tag{108}$$

Thus for a Gaussian distribution,

$$P(u) = \frac{1}{2\pi\sigma} \int_{-\infty}^{u} e^{-\xi^2/2\sigma^2} d\xi \tag{109}$$

There is no closed-form solution for the integral in Eq: (109). Hence tables of $P(u)$ have been prepared through numerical integration. These are found in many references, such as *Standard Mathematical Tables,* published by the Chemical Rubber Publishing Company.

1.7.2.2 Joint probability density. We quite often wish to investigate the interdependence of several random variables. The joint probability density function describes the probability that two functions each will asssume values within certain limits at any instant of time.

In a manner similar to the computation for a single variable, we consider two continuous functions of time, say u and v. The total time during which both functions are within prescribed limits is given by

$$T_{uv} = \Sigma \, \Delta t_i \tag{110}$$

where Δt_i is the time both functions lie within a given band during each fluctuation, as shown in Fig. 1.14. The probability of this occurring is

$$\text{Prob } \{u < \tilde{u}(t) < u + \Delta u, \; v < \tilde{v}(t) < v + \Delta v\} = \lim_{T \to \infty} \frac{T_{uv}}{T} \tag{111}$$

$$p(u,v) = \lim_{\substack{\Delta u \to 0 \\ \Delta v \to 0}} \frac{\text{Pr } \{ \}}{\Delta u \, \Delta v} \tag{112}$$

$$p(u,v) = \lim_{\substack{\Delta u \to 0 \\ \Delta u \to 0}} \lim_{T \to \infty} \frac{1}{T} \frac{T_{uv}}{\Delta u \, \Delta v} \tag{113}$$

If the two functions $\tilde{u}$ and $\tilde{v}$ are statistically independent, then

$$P(u,v) = P(u)P(v) \tag{114}$$

In this case, a three-dimensional plot is required to display the data. (See, for example, page 95 in [23].)

In a manner similar to that for a single variable, we define a joint probability function $P(u,v)$ as follows:

$$P(u,v) = \text{Pr } [u(t) \leq u, \; v(t) \leq v] = \int_{-\infty}^{u} \int_{-\infty}^{v} p(\xi,\eta)d\xi \, d\eta \tag{115}$$

where ξ and η are dummy variables. If the variances in u and v are equal and u and v are statistically independent, a simplification occurs for Gaussian distributed functions such that the probability density is

$$p(u,v) = \frac{1}{2\pi\sigma} \exp\left(-\frac{u^2 + v^2}{2\sigma^2}\right) = \frac{1}{2\pi(r')^2} \exp\left[-\frac{r^2}{2(r')^2}\right] \tag{116}$$

The joint probability function is given by

$$P(r) = \int_{0}^{2\pi} \left\{ \int_{0}^{r} \frac{1}{2\pi} \exp\left[-\frac{r^2}{2(r')^2}\right] \frac{r \, dr}{(r')^2} \right\} d\theta \tag{117}$$

or

$$P(r) = \int_{0}^{r^2/(r')^2} e^{-\xi} \, d\xi = 1 - \exp\left[-\frac{r^2}{(r')^2}\right] \tag{118}$$

$P(r)$ is, for example, the probability of being less than or equal to a distance r from a given target, assuming deviations in all directions are equally likely. Note that there is a closed-form solution for $P(r)$, which was not true for a single variable.

We now want to investigate the relationship between two variables. Figure 1.14c can be generated by inputting two random signals $\tilde{u}$ and $\tilde{v}$ to the x and y axes of an oscilloscope. The shaded area represents the shape of the spot we would see on the oscilloscope. In the case where we consider $\tilde{v}$ and $\tilde{u}$ to be linearly related, we seek an equation of the form

$$v_c = Au + B \tag{119}$$

where A and B are constants. Their values are determined from the requirement

$$\Sigma(\tilde{v} - \tilde{v}_c)^2 = \text{minimum} \tag{120}$$

where $\tilde{v}$ represents the raw data and v_c is computed from Eq. (119). A best fit curve results in minimum error. This error is found to be equal to

$$E = \overline{u^2}(1 - R_{uv}^2) \tag{121}$$

with

$$R_{uv} = \frac{\overline{(\tilde{u} - U)(\tilde{v} - V)}}{\sqrt{\overline{u^2}}\sqrt{\overline{v^2}}} \tag{122}$$

where R_{uv} is a correlation function that lies in the range $-1 \leqslant R \leqslant 1$. The numerator in Eq. (122) represents the time mean of the product of the two functions $(\tilde{u} - U)$ and $(v - V)$.

1.7.2.3 Autocorrelation. One type of correlation is that which determines the length of past history that is related to a given event. This is formed by looking at the correlation between a function of time and the same function at a later time τ:

$$R_\tau = \frac{\overline{\tilde{u}(t)\tilde{u}(t + \tau)}}{\overline{u^2}} \tag{123}$$

An important distinction between a periodic function and a random function is that as $\tau \to \infty$, $R_\tau \to 0$ for a random variable but R_τ is periodic for a periodic function. This is easily illustrated. Let

$$\tilde{u}(t) = \sin \omega t$$

Then

$$R_\tau = \lim_{T\to\infty} \frac{(1/T)\displaystyle\int_0^T \sin \omega t \sin \omega(t + \tau)\,dt}{(1/T)\displaystyle\int_0^T \sin^2 \omega t\,dt} \tag{124}$$

$$R_\tau = \cos \omega\tau \tag{125}$$

1.7.2.4 Spectral analysis. An important part of the description of random variables is the determination of the distribution of energy content with frequency. This is defined by the power spectral density

$$S(\omega) = \lim_{\substack{\Delta\omega \to 0 \\ T \to \infty}} \frac{1}{T\Delta\omega} \int_0^T u_\omega^2(t)dt \tag{126}$$

where u_ω is that part of the signal contained within a bandwidth of $\Delta\omega$ centered at ω. Equation (126) is not a very precise definition mathematically, but it is very clear from a physical point of view. A precise mathematical definition will follow.

The natural inclination of an engineer is to reduce a time-dependent signal into its harmonic components. If the signal is periodic, a Fourier series may be used:

$$\tilde{u}(t) = \sum_{-\infty}^{\infty} C_n e^{in\omega_0 t} \tag{127}$$

where

$$C_n = \frac{1}{T_0} \int_{-T_0/2}^{+T_0/2} \tilde{u}(t)e^{-in\omega_0 t}dt \tag{128}$$

$$T_0 = \frac{2\pi}{\omega_0} \tag{129}$$

If the function is *not* periodic, we have $T_0 \to \infty$, $\omega_0 \to 0$. Thus the spectrum coefficients C_n approach a continuous function of ω rather than discrete values of $n\omega_0$. Thus we let ω_0 and $\Delta\omega \to 0$, and $n\omega_0 \to \omega$ and use the Fourier-integral representation

$$F(\omega) = \frac{1}{2\pi} \int_{-\infty}^{\infty} \tilde{u}(t)e^{-i\omega t}dt \tag{130}$$

$$\tilde{u}(t) = \int_{-\infty}^{\infty} F(\omega)e^{i\omega t}d\omega \tag{131}$$

For $F(\omega)$ to exist, it is necessary that

$$\int_{-\infty}^{\infty} |(\tilde{u}(t)|^2 \, dt = 2\pi \int_{-\infty}^{\infty} |F(\omega)|^2 \, d\omega \tag{132}$$

The integrals in Eq. (132) must be finite. The spectrum density is normally defined as

$$S(\omega) = |F(\omega)|^2 \, d\omega \tag{133}$$

In the case of a stationary random function, the integral on the left-hand side of (132) is infinite. Thus it would appear impossible to define a spectral density function using Eq. (133). This is overcome through an artifice. The signal is considered to be analyzed for an amount of time T and is assumed equal to zero for the rest of the time. Then

$$\int_{-\infty}^{\infty} |\tilde{u}(t)|^2 \, dt = T\overline{u^2} \tag{134}$$

Using our definitions, the following may be written:

$$\overline{u^2} = \lim_{T\to\infty} \frac{1}{T} \int_{-T/2}^{T/2} \tilde{u}^2 \, dt = \frac{1}{T} \int_{-\infty}^{\infty} \tilde{u}(t) \left[\int_{-\infty}^{\infty} F(\omega)e^{i\omega t}d\omega \right] dt \tag{135}$$

$$\overline{u^2} = \frac{1}{T} \int_{-\infty}^{\infty} F(\omega) \left[\int_{-\infty}^{\infty} \tilde{u}(t)e^{i\omega t}dt \right] d\omega \tag{136}$$

$$\overline{u^2} = \frac{1}{T} \int_{-\infty}^{\infty} F(\omega)2\pi F^*(\omega)d\omega \tag{137}$$

Equation (137) can be rewritten in the form

$$\overline{u^2} = 4\pi \int_{0}^{\infty} \frac{|F(\omega)|^2}{T} \, d\omega \tag{138}$$

By definition of the spectrum,

$$\overline{u^2} = \int_{0}^{\infty} S(\omega)d\omega \tag{139}$$

we arrive at a modified relationship between the spectrum and the Fourier transform of $\tilde{u}(t)$:

$$S(\omega) = \frac{4\pi|F(\omega)|^2}{T} \tag{140}$$

To determine the spectrum of a pure tone, the Dirac delta function must be introduced. This is defined by

$$\int_{-\infty}^{\infty} \tilde{u}(t) \, \delta(t - t_0)dt = \tilde{u}(t_0) \tag{141}$$

The Fourier transform of the Dirac delta function is given by

$$F(\omega) = \frac{1}{2\pi} \int_{-\infty}^{\infty} \delta(t - t_0)e^{-i\omega t}dt = \frac{1}{2\pi}e^{-i\omega t_0} \tag{142}$$

Similarly, the inverse transform of the delta function is given by

$$\int_{-\infty}^{\infty} \delta(\omega - \omega_0)e^{i\omega t}d\omega = e^{i\omega_0 t} \tag{143}$$

Noting Eqs. (142) and (143), one obtains

$$\delta(\omega - \omega_0) = \frac{1}{2\pi} \int_{-\infty}^{\infty} e^{i\omega_0 t}e^{-i\omega t}dt \tag{144}$$

Thus, if a function can be described by a Fourier series,

$$\tilde{u}(t) = \sum_{0}^{\infty} A_n e^{i(n\omega_0 t - \phi_n)} \tag{145}$$

$$F(\omega) = \sum_{0}^{\infty} A_n e^{-i\phi_n} \delta(\omega - n\omega_0) \tag{146}$$

It can be inferred that the power spectral density of a periodic function is a collection of equally spaced spikes, with the area under these spikes proportional to $|A_n|^2$.

1.7.2.5 Relationship between the autocorrelation and the spectral density function. Intuitively, one could expect a relationship between the autocorrelation and the spectral density function. By definition, the autocorrelation is given by

$$R_\tau = \lim_{T \to \infty} \frac{1}{T} \int_{-T/2}^{T/2} \tilde{u}(t)\tilde{u}(t + \tau)dt \tag{147}$$

We can write Eq. (147) in terms of a Fourier integral

$$R_\tau = \lim_{T \to \infty} \frac{1}{T} \int_{-T/2}^{+T/2} \tilde{u}(t) \int_{-\infty}^{\infty} F(\omega)e^{i\omega(t + \tau)}d\omega \, d\tau \tag{148}$$

where we replace t with $t + \tau$ in the second integration. Rearranging Eq. (148) gives

$$R_\tau = \lim_{T \to \infty} \frac{1}{T} \int_{-\infty}^{\infty} F(\omega)e^{i\omega\tau} \, dt \int_{-T/2}^{+T/2} \tilde{u}(t)e^{i\omega\tau}d\omega \tag{149}$$

or

$$R_\tau = \lim_{T \to \infty} \frac{2}{T} \int_{-\infty}^{\infty} F(\omega)F^*(\omega)e^{i\omega t}d\omega \tag{150}$$

where

$$F^*(\omega) = \frac{1}{2\pi} \int_{-\infty}^{\infty} \tilde{u}(t)e^{i\omega t}dt \tag{151}$$

$$F^*(\omega) = \frac{1}{2\pi} \int_{-T/2}^{+T/2} \tilde{u}(t)e^{i\omega t}dt \tag{152}$$

for $\tilde{u}(t) = 0$ except when $-T/2 \leq x \leq +T/2$. It follows that

$$R_\tau = \int_{-\infty}^{\infty} \frac{2\pi|F(\omega)|^2}{T} e^{i\omega t}d\omega \tag{153}$$

or

$$R_\tau = \int_{-\infty}^{\infty} \frac{1}{2}S(\omega)e^{i\omega t}d\omega \tag{154}$$

$$S(\omega) = \frac{1}{\pi} \int_{-\infty}^{\infty} R(\tau)e^{-i\omega\tau}d\tau \tag{155}$$

Since $S(\omega)$ is an even function,

$$S(\omega) = \frac{2}{\pi} \int_{0}^{\infty} R(\tau) \cos \omega\tau \, d\tau \tag{156}$$

In terms of frequency f, we obtain

$$S(f) = 2\pi S(\omega) \tag{157}$$

such that

$$\overline{u^2} = \int_{0}^{\infty} S(f)df \tag{158}$$

and

$$S(f) = 4 \int_{0}^{\infty} R(\tau) \cos 2\pi f\tau \, d\tau \tag{159}$$

$$R(\tau) = \int_{0}^{\infty} S(f) \cos 2\pi f\tau \, df \tag{160}$$

Thus the power spectrum and the autocorrelation are Fourier-transform pairs. Functions that are transform pairs have an inverse spreading relationship. For example, if the power spectrum is wide, the energy content in the signal is spread out

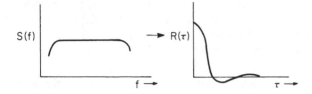

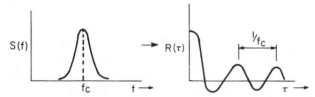

Figure 1.15   Relationship between spectrum and autocorrelation

over a wide range of frequencies with little coherence. The autocorrelation will be
nonzero only over a limited range of delay time. Conversely, if the power spectrum
is very narrow, the autocorrelation function has nonzero values for very large values of τ. (One can visualize this inverse spreading by noting $\omega \sim 1/\tau$.) To be precise,
the autocorrelation will oscillate with decreasing amplitude and with a frequency
equal to the frequency at which the spectrum is centered. This is illustrated in Fig.
1.15. As an example, consider band-passed white noise:

$$S'(\omega) = \begin{cases} 0 & 0 < \omega < \omega_1 \\ 1 & \omega_1 < \omega < \omega_2 \\ 0 & \omega_2 < \omega < \infty \end{cases} \tag{161}$$

Then

$$R(\tau) = \int_{\omega_1}^{\omega_2} \cos \omega\tau \, d\omega = \left(\frac{\sin \omega\tau}{\tau}\right)_{\omega_1}^{\omega_2} \tag{162}$$

$$= \frac{2}{\tau} \sin \frac{\omega_2 - \omega_1}{2} \tau \cos \frac{\omega_2 + \omega_1}{2} \tau$$

$$R(\tau) = \frac{2}{\tau} \sin \frac{\Delta\omega\tau}{2} \cos \omega_0\tau \tag{163}$$

where

$$\omega_0 = \frac{\omega_1 + \omega_2}{2} \qquad \Delta\omega = \omega_2 - \omega_1$$

The autocorrelation depends on the bandwidth $\Delta\omega$ of the filter and the center frequency ω_0. As the bandwidth approaches that for a pure tone of frequency ω_0, $R(\tau) = \cos \omega_0\tau$. For finite bandwidth, the autocorrelation consists of $\cos \omega_0\tau$ with an amplitude modulation equal to $(2/\tau) \sin (\Delta\omega\tau/2)$.

1.7.2.6 Cross-correlation and cross-power spectrum. Consider two functions of time $\tilde{u}_1(t)$ and $\tilde{u}_2(t)$. The cross correlation is

$$R_{12}(\tau) = \lim_{T \to \infty} \frac{1}{T} \int_{-T/2}^{+T/2} \tilde{u}_1(t)\tilde{u}_2(t + \tau) \, dt \qquad (164)$$

As an example, $\tilde{u}_1$ and $\tilde{u}_2$ could be a velocity component at two points in a flow. The cross-power spectrum is the Fourier transform of the cross-correlation function. R_{12} is generally an odd function. For example, in the case of velocity, the correlation will peak at a value of delay time corresponding to some characteristic eddy size divided by some convective velocity. Hence the Fourier transform will have a real and an imaginary part. The real part is called the cospectrum:

$$C_{12}(f) = 2 \int_{-\infty}^{\infty} R_{12}(\tau) \cos 2\pi f\tau \, d\tau \qquad (165)$$

The imaginary part is denoted as the quadrature spectrum:

$$Q_{12}(f) = 2 \int_{-\infty}^{\infty} R_{12}(\tau) \sin 2\pi f\tau \, d\tau \qquad (166)$$

The cospectrum represents a narrow-band correlation between $\tilde{u}_1$ and $\tilde{u}_2$. Similarly, the quadrature spectrum is a narrow-band correlation of $\tilde{u}_1$ and $\tilde{u}_2$ with a $90°$ phase shift.

1.7.2.7 Signal analysis in turbulence research. Turbulence is usually characterized by its intensity $\overline{u^2}$, its spectrum $S(f)$, and various correlation parameters as deemed appropriate. As an example, consider the measurement of turbulence in the development region of a round jet (zero to six diameters from the nozzle). The data are illustrated in Fig. 1.16. As shown, measurements are made in the mixing region of a jet of diameter $2R$. The width of the mixing region varies linearly with distance x from the nozzle. This is illustrated by the fact that the mean velocity profiles are similar when the distance from the center of the mixing region $(r - r_{1/2})$ is normalized with respect to x. The turbulence is a maximum in the center of the mixing region and also follows a similarity law. The measured Reynolds stress is expressed in terms of a correlation coefficient:

$$R_{uv} = -\frac{\overline{uv}}{\sqrt{\overline{u^2}} \sqrt{\overline{v^2}}} \qquad (167)$$

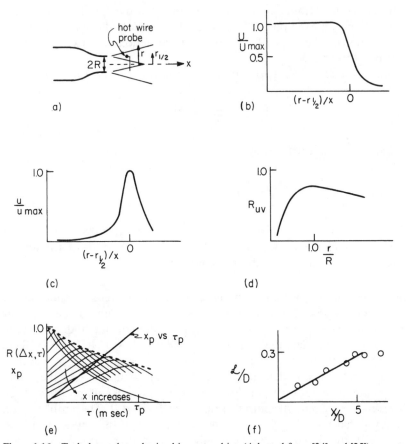

Figure 1.16 Turbulence data obtained in a round jet. (Adapted from [24] and [25])

Sample data are shown in Fig. 1.16*d*.

Finally, a quantitative measure of the spatial structure is obtained from the cross correlation

$$R_{12} = \frac{\overline{u_1(x,t)u_2(x + \Delta x,t + \tau)}}{\sqrt{\overline{u_1^2}}\,\sqrt{\overline{u_2^2}}} \tag{168}$$

where u_1 and u_2 are longitudinal velocity components in the center of the mixing region at x and Δx, respectively. As Δx is increased, the correlation peaks at a different delay time τ_p. The peak value of the correlation also decays with increasing spacing between the points x_1 and x_2. This corresponds to the true evolution of the turbulence. In fact, the envelope of the peaks is the autocorrelation that would be measured in a frame of reference moving with the speed of convection of the turbulence past a fixed point of reference. If x_p is plotted against τ_p, a linear plot results, as shown. The slope of the line is the convection velocity U_c, which in this case, is found to be $0.54U_0$.

The value of the two-point correlation at zero time delay is defined as

$$R_{\Delta x} = \frac{\overline{u_1(x,t)u_2(x + \Delta x,t)}}{\sqrt{\overline{u_1^2}}\,\sqrt{\overline{u_2^2}}} \tag{169}$$

An integral length scale is defined by

$$l = \int_0^\infty R_{\Delta x}\, dx \tag{170}$$

The results are plotted in Fig. 1.16f. This shows that the integral scale is a linear function of x, in agreement with the observation that the width of the mixing region varies linearly with distance from the nozzle.

1.7.2.8 Orthogonal decomposition of turbulence data. The material outlined above summarizes standard procedures in signal analysis. With the rapid increase in our understanding of turbulence in the past 15 years, many new methods have been developed to further elucidate the structure of turbulent flows [19]. Many of these approaches are variations of *conditional sampling* or averaging, that is, the data sampling is preconditioned to depend on a certain event occurring, e.g., the amplitude of a signal exceeding a certain threshold value. Such techniques have been used to investigate coherent structures in turbulent flows [12]. The problem with these routines is that they rely on a preconceived notion of the physics involved. Hence the results are biased by sampling only events that are believed to be important by an individual experimenter.

An unbiased technique is the orthogonal decomposition scheme first proposed by Lumley [26]. In this procedure, the largest eddies in a turbulent flow are identified as the eigenvalue and associated eigenfunction of the proper orthogonal decomposition of the random fluctuations in velocity. This reduces to a simple harmonic decomposition when the independent variable is homogeneous, e.g., the azimuthal variation of turbulent velocity fluctuations in a round jet. Further details are found in [27].

As an example of the application of this powerful method, Long and Arndt [28] were able to show that this technique could be applied to the pressure field in the close vicinity of a round turbulent jet. The analysis of a scaled quantity considerably reduces the quantity of data to be analyzed. It can further be shown that the pressure spectrum is dominated by the larger scales in the velocity field [14]. Thus orthogonal decomposition of the pressure field should result in a clear definition of the orderly structure in the flow.

By taking advantage of axisymmetry, the pressure signal can be expressed as

$$p(x,\theta,t) = \sum_{(n=1)}^{\infty} \lambda_n (\omega, m)\xi_n (\omega, m)\psi_n(x, \omega, m) \tag{171}$$

where λ_n are the eigenvalues and ψ_n are orthogonal eigenfunctions, ω is frequency, and m represents the decomposition in the θ direction into modal components, i.e., the eigenfunctions reduce to $\cos m\theta$ in the azimuthal direction. The weighting functions, $\xi_n(\omega, m)$ are not known explicitly, but their integral properties are

$$E(\xi_n, \xi_m) = \delta_{nm} \tag{172}$$

where $E(\ )$ is the expected value. It is shown in [27] that eigenvalues and eigenfunctions can be obtained from the cross-spectral density of the pressure field:

$$\int \Phi(x, x', m, \omega)\psi_n(x', m, \omega) \, dx' = |\lambda_n(\omega, m)|^2\psi_n(x, m, \omega) \tag{173}$$

where Φ is the model decomposition of the cross-spectral density matrix of the pressure fluctuations ϕ

$$\Phi(x, x', m, \omega) = \frac{2}{\pi} \int_0^\pi \phi(x, x', \theta, \omega) \cos m\theta \, d\theta \tag{174}$$

$$\Phi = (x, x', 0, \omega) = \frac{1}{\pi} \int_0^\pi \phi(x, x', \theta, \omega) \cos m\theta \, d\theta \tag{175}$$

For the discrete locations illustrated in Fig. 1.17a, Eq. (173) reduces to the following matrix equation:

$$C_{ij}D_{ij}E_{j(n)} = |\lambda_n|^2E_{i(n)} \tag{176}$$

C_{ij} is the matrix of complex cospectrum coefficients corresponding to the ith and j^{th} axial locations and E_i is the complex eigenvector at the ith location. D_{ij} is a diagonal matrix of transducer separations. The details of solving Eq. (176) are found in [28]. Some manipulation is required, since, depending on the choice of D_{ij}, there is no guarantee that the eigenvalues will be real. In this example, the decomposition is applied on the surface of an imaginary cone whose axis is coincident with the jet axis (Fig. 1.17a). The surface of the cone passes through the nozzle lip. The semi-angle of the cone is chosen as $10°$ to take into account the spreading of the jet. The required input for the orthogonal decomposition is the cross-spectral density matrix of pressure fluctuations between $x = 0.5$ diameter and $x = 3.0$ diameters. This range includes at least 90% of the energy in the frequency range of interest. Eight axial probe locations and four azimuthal intervals are chosen to provide good information up to the $m = 2$ mode.

The spacing of the four azimuthal intervals is also shown in Fig. 1.17a. With the assumption that the jet structure is periodic in azimuthal angle, it is not necessary to make measurements on the upper half of the jet cross section. This simplification reduces the number of measurements by a factor of 2, the only drawback

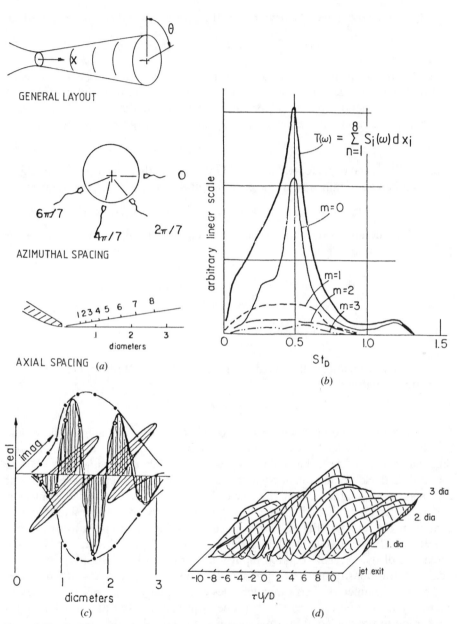

GENERAL LAYOUT

$6\pi/7$

$4\pi/7$

$2\pi/7$

AZIMUTHAL SPACING

AXIAL SPACING

1 2 3 4 5 6 7 8

diameters

(a)

arbitrary linear scale

$$T(\omega) = \sum_{n=1}^{8} S_i(\omega)\, d\, x_i$$

$m=0$

$m=1$

$m=2$

$m=3$

St_D

(b)

real

imag

dicmeters

(c)

3 dia

2 dia

1. dia

jet exit

$\tau U/D$

(d)

Figure 1.17 (*a*) Schematic of microphone positions for measurement of the near-field pressure surrounding a turbulent jet. (*b*) Comparison of each mode of the first eigenvalue of pressure with the spatially averaged pressure power spectrum. (*c*) The first eigenvector of the $m = 0$ mode. (*d*) Shot-effect reconstruction of the $m = 0$ mode of the pressure fluctuations

being that the positive and negative spinning modes ($m = \pm 1$) cannot be distinguished from one another; they are lumped into the same value. This is justified
by the assumption that in a well-designed jet experiment there is no swirl. The
eight axial locations are spaced to take advantage of the observation that small
scales occur near the jet axis and large scales occur further downstream. The spacing between the probes is logarithmic and has been designed to roughly satisfy the
folding frequency criteria, i.e., there should be at least two probe locations per
wavelength. Only two microphones are needed to measure the cross-spectral density matrix. Details can be found in Long [29].

Using 8 axial locations requires that 144 cross spectra be measured. Of these
144 cross spectra, 8 are the power spectra that result when $i = j$ and $\theta = 0$. The
summation of these spectra results in a total power spectrum $T(\omega)$:

$$T(\omega) = \sum_{i=1}^{\infty} S_i(\omega) \, dx_i \qquad (177)$$

Long and Arndt [28] refer to this spatially averaged spectrum as the "baseline"
situation. Each eigenvalue λ_n is a function of mode m and frequency ω. The spectrum $T(\omega)$ is equivalent to the double summation of all the eigenvalues, i.e.,

$$T(\omega) = \sum_{m=1}^{\infty} \sum_{n=1}^{\infty} |\lambda_n(\omega, m)|^2 \qquad (178)$$

Thus the relative importance of each eigenvalue can be found by comparing it to
$T(\omega)$. This is shown in Figure 1.17b, where the coherent energy in the largest eigen-
value, $|\lambda_1(\omega, m)|^2$, is compared to $T(\omega)$. In this example, the peak value of the spec-
trum is centered at the column mode frequency $St_d = 0.5$. The sum of the coherent
energy at this frequency, represented by the largest eigenvalue for each mode, is
98% of the total at $St_d = 0.5$. Summing up across the frequency band of interest
shows that 50% of the energy associated with the pressure fluctuations is contained
in the $m = 0$ mode, 23% in the $m = 1$ mode, 14% in the $m = 2$ mode, and 5% in
the $m = 3$ mode. It is safe to say that most of the energy associated with the
pressure fluctuations can be represented by the first four modes of the largest eigenvalue. In fact, Long and Arndt [28] found that the second eigenvalue of the axisymmetric mode was only 5% of the first eigenvalue of the axisymmetric mode. Hence
the series representation in Eq. (171) converges very rapidly. It is beyond the scope
of this example to draw further conclusions from the data in Fig. 1.17b.

The eigenvectors represent the spatial signal form associated with any particular frequency mode–number combination. In view of the eigenvalue spectrum of
Fig. 1.17b, the most important eigenvector is that which corresponds to the column
mode frequency of $St_d = 0.5$ and the axisymmetric or $m = 0$ mode because this is
the most energetic mode. This eigenvector is shown in Fig. 1.17c. It turns out that
this form is representative of the axial wave form associated with any particular
frequency mode–number combination. They all show the characteristics of amplification, saturation, and subsequent decay of an instability wave within about three
wavelengths. The amplitude of the eigenvector is shown as the heavy line through

the solid circles. Because each eigenvector is a complex quantity, this amplitude is found by the square root of the sum of the squares of the real and imaginary parts, i.e.,

$$A = (RE^2 + IM^2)^{1/2} \tag{179}$$

The real part is shown in the vertical plane, and the imaginary part is shown in the horizontal plane. This procedure can be repeated for the measured eigenvectors. By noting the frequency and the measured wavelength near the peak, the phase velocity can be determined:

$$U_p = \omega/k = f\lambda \tag{180}$$

For the data in this example, $U_p/U_j = 0.58$, which agrees very closely with the measured convection velocity in the example of section 1.7.2.7.

The utility of this procedure is that a turbulent flow can be reconstructed from the measured cospectra. In most experiments where correlation techniques are used, a fixed probe is located at a specific point, and another probe is traversed. This method can produce correlation lengths and amplitudes for individual frequency components, but it cannot determine the phase relationship between various frequency components; all components are phase referenced to the location of the fixed probe. This is not the case for this technique. If, on average, there is a phase relationship between any two frequency components, the orthogonal decomposition will preserve it. It is shown in [26] that this information can be used as an input to the shot-effect decomposition to reconstruct a typical signal or characteristic event.

The shot effect was considered by Rice [30] for studying the statistics of vacuum tube noise where a pulse is emitted every time an electron reaches the anode. He calculated the statistical variation of a sequence of pulses that have constant amplitude and a well-defined shape but whose arrival times at the anode occur randomly. In particular, he showed that independent of how well the pulses are defined, the probability distribution of the signal will be normal because of the random arrival times. The randomness of arrival implies that the individual pulses are independent of one another and the central limit theorem guarantees the normality of the distribution. In other words, the moments of the distribution associated with the pulse are indeterminant from standard statistical measurements.

The mechanics of reconstructing a typical pressure signal from the measured cospectra are outlined in [29]. This method can reconstruct both the temporal and spatial variation of a typical signal, as shown in Fig. 1.17d, which is the reconstruction of the $m = 0$ mode. The spatial growth and decay of disturbances are clearly evident. Note also that there is a coalescing of various pressure ridges at approximately $x/d = 1$. Flow visualization indicates that this is where the first set of pairing events occurs [12]. Similar reconstructions of higher modes indicate that there is a large initial growth of a pulse in the $m = 1$ mode just downstream of a pairing event in the $m = 0$ mode. As shown in [28], many other features of the turbulent jet mixing process can be deduced from these data.

This brief summary illustrates the usefulness of the orthogonal decomposition scheme. It is a relatively simple tool that can be used to deduce the large-scale structure in an unbiased manner. The method did not gain immediate attention at the time of its introduction to the turbulence community (1970) because of the problems associated with collecting, storing, and manipulating large amounts of experimental data. These impediments have largely been eliminated by the significant advances in digital computation and data acquisition systems over the last decade. The method is quite general and can be used to analyze a variety of turbulent flows. The technique has also been applied recently to the analysis of numerically generated turbulence data.

1.8 PLANNING AND CARRYING OUT AN EXPERIMENTAL PROGRAM

The need for flow measurements, principles of fluid mechanics as they relate to measurements, and the correlation and analyses of data have all been outlined. The actual application of these principles to a given situation will vary considerably, depending on the nature of the study involved. The purpose of this book is to provide a balanced overview of the instrumentation and analysis techniques that are available. In this section, a few examples are provided in order to give the reader some experience in selecting the proper combination of instrumentation and data acquisition equipment for a given job. Specific step-by-step instructions cannot be given, since each application will require a different approach.

To illustrate the philosophy behind planning a successful measurement program, three examples are given. Two are from fundamental research in aeroacoustics and cavitation. The third example relates to hydrodynamic design of two very large and complicated water tunnels, specifically designed for advanced research in marine engineering. Examples from both fundamental and applied research are given, since the focus of the measurement program is entirely different, as are the constraints on time and cost and the scope of the experimental data to be generated.

1.8.1 Coherent Structures and Acoustic Radiation

The realization that turbulent flows can have an underlying orderly structure was discussed briefly in section 4.5. An example is given in Fig. 1.7. It is well known that the fine-grained turbulence shown in the photo becomes more dominate with increasing Re [11]. In fact, the orderly structure is much more difficult to detect in high Re flows. This is one of the reasons why many new signal analysis techniques have been developed in recent years, such as the orthogonal decomposition scheme discussed in section 1.7.2.8. It has also become increasingly evident that noise radiation from turbulent flows may be influenced by a complex interaction between the orderly structure and the fine-scale turbulence [25].

In reviewing the literature on jet noise, two points become very meaningful in terms of planning a significant program of experimental research on coherent structures and noise.

1. Subsonic jet noise scales with the fifth power of Mach number (Ma); therefore, most previous experiments were made at high Ma and also at high Re.
2. Coherent structures are difficult to visualize at high Re. Thus, almost all experimental research on coherent structures was carried out with low-Re flows, typically at Re that were 1–2 orders of magnitude less than for noise studies.

It becomes obvious that if the interaction between coherent structures and small-scale turbulence is an important noise source, then flow noise should be considered to be a function of both Ma and Re, i.e., for a turbulent jet of diameter d, the acoustic pressure in the far field, p_a, is given by

$$\frac{\overline{p_a^2}}{\rho^2 U^2} \frac{\dfrac{r^2}{d}}{\dfrac{\Delta f d}{U}} = f(\text{Ma}, \text{Re}) \tag{181}$$

where

$$\text{Ma} = U/a_0$$
$$\text{Re} = Ud/v$$

The measurement problem is to carry out experiments in Ma-Re space such that Ma can be varied at constant Re and variations with Re can be studied at constant Ma. Obviously, this can be accomplished with a series of different-sized nozzles that are geometrically similar. The problem for sound measurements is that frequency scales with U/d and sound level scales with Ma^5. Thus there is a lower limit on Ma that is determined by microphone sensitivity. There is a lower limit on frequency that is determined by the sound absorption properties of the anechoic chamber used for the studies, and there is an upper limit on frequency that is determined by both the frequency response of the microphones and limitations on the data rate capabilities and storage capacity of the data acquisition system. This means that there is a well-defined area in the Re-Ma plane in which experiments can be carried out. A reasonable number of different-sized nozzles must then be selected to cover this area. This is illustrated in Fig. 1.18. Further details can be found in [31].

1.8.2 Vortex Cavitation Research

A second example for planning and carrying out a measurement program is selected from the field of cavitation. Loosely defined, cavitation is the formation of

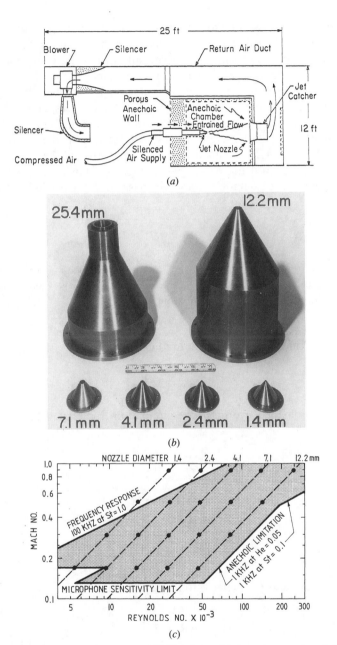

(a)

(b)

(c)

Figure 1.18 (a) Jet noise research apparatus. (b) Nozzles used in jet noise experiment. (c) Useful range of Mach number and Reynolds number (shaded area). Dotted lines are simultaneous values of Mach number and Reynolds number for each nozzle

the vapor phase in a liquid, due to a hydrodynamic reduction pressure. Cavitation is a problem in many applications involving the flow of liquids, such as pumps, turbines, valves, marine propellers, and hydraulic structures. It can increase drag of marine vehicles or decrease the efficiency of turbomachinery. Noise and vibration are also significant problems. Cavitating flows can be highly erosive, since the vapor bubbles that are formed can implode on solid surfaces, creating very high pressures and temperatures. Further details can be found in [32].

One example that is of current interest is cavitation in the tip vortices of a marine propeller (Fig. 1.19a). Similar vortices occur in axial flow pumps and can be a significant problem if noise is of concern (see section 1.8.3). The important scaling parameter is the cavitation index, which is simply a modified Euler number:

$$\sigma = \frac{(p_0 - p_v)}{\frac{1}{2}\rho U^2} \tag{182}$$

where p_0 is a reference value of pressure. Cavitation will occur if this parameter is lower than the incipient value, σ_i. The incipient value of the cavitation index is related to the minimum pressure in the flow,

$$\sigma_i = -C_{pm} \tag{183}$$

where, for a Rankine vortex with a core radius r_c,

$$C_{pm} = -2\left(\frac{\Gamma}{2\pi r_c U}\right)^2 \tag{184}$$

For an elliptically loaded hydrofoil, the circulation Γ is related to the lift coefficient by

$$\frac{\Gamma}{UC_0} = -\frac{C_l}{2} \tag{185}$$

Thus

$$\sigma_i = KC_l^2 \tag{186}$$

$$K = 2\left[\frac{c_0}{4\pi r_c}\right]^2 \tag{187}$$

where the lift coefficient C_l is related to the aspect ratio AR by

$$C_l = 2\pi\frac{\text{AR}}{\text{AR} + 2}(\alpha - \alpha_0) \tag{188}$$

where α is the angle of attack and α_0 is the angle of attack at zero lift. Thus, selecting a hydrofoil of elliptic planform simplifies the problem. A hydrofoil having an aspect ratio of 3 and a uniform section, NACA 66_2–415, $a = 0.8$ was selected for study in a water tunnel. This particular section was chosen to ensure that vortex cavitation would occur at higher values of σ than surface cavitation. This was the

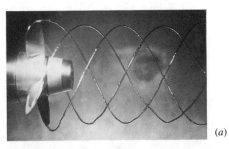

(a)

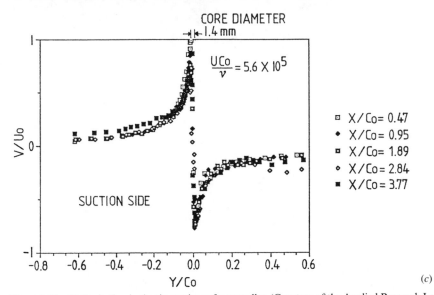

(b)

(c)

Figure 1.19 (a) Cavitation in the tip vortices of a propeller. (Courtesy of the Applied Research Laboratory.) (b) Tip vortex on a hydrofoil of aspect ratio 3 with a NACA 66_2–415 cross section. (Courtesy of the Saint Anthony Falls Hydraulic Laboratory.) (c) Measured tangential velocity in the tip vortex shown in Figure 1.19b

51

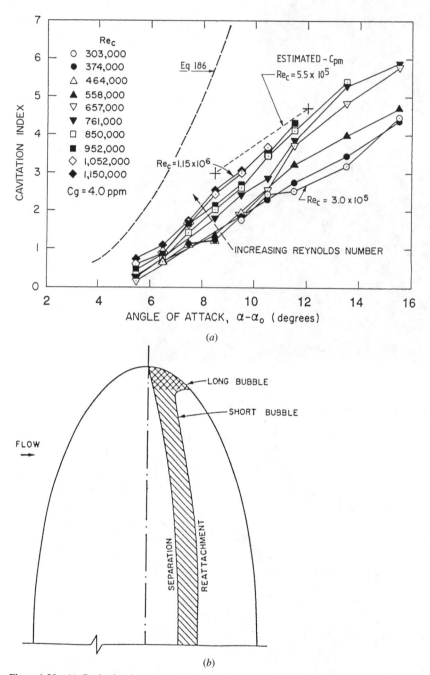

Figure 1.20 (*a*) Cavitation inception data for a tip vortex. (*b*) Calculated boundary layer separation and reattachment for a hydrofoil at a Reynolds number of 5.3×10^5 and $\alpha - \alpha_0 = 9.5$. (*c*) Oil drop

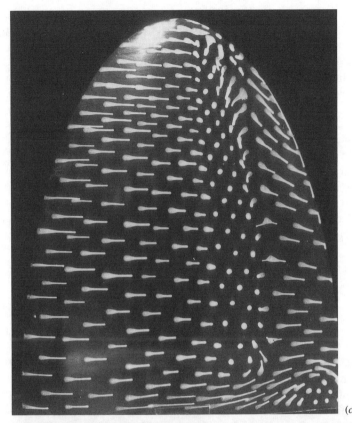

(c)

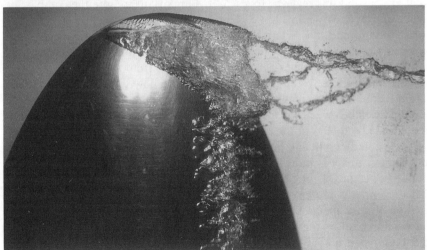

(d)

flow visualization for the condition calculated in Fig. 1.20*b*. (*d*) The same conditions as Fig. 1.20*b* with σ well below the incipient point. Note cavitation in the separation bubble

53

case, as shown in Fig. 1.19b. The unknown parameter in Eq. (187) is the core radius. The dependence of this parameter on Re is a complex issue and is beyond the scope of this example. One can make estimates of its magnitude from information available in the literature [33].

The question of actually measuring r_c is difficult, since vortex flows are known to be very sensitive to any intrusive instrumentation; hence a laser-Doppler system is selected. Based on estimates of the magnitude of r_c, a beam expander is required to achieve the necessary spatial resolution. Sample data are shown in Fig. 1.19c. Using the measured value of r_c, the value of K in Eq. (186) is 31.6, which does not agree with the measured data shown in Fig. 1.20a. In fact, the dependence on angle of attack is clearly to a lesser power than $(\alpha - \alpha_0)^2$.

If the vortex can be assumed to be axisymmetric, the pressure in the vortex can be found by directly integrating the Navier-Stokes equations using the measured profile. For this case the momentum equation is simply

$$\frac{dp}{dr} = \frac{\rho v_\theta^2}{r} \tag{189}$$

Thus

$$\sigma_i = 2 \int_0^\infty \frac{v_\theta^2}{rU^2} \, dr \tag{190}$$

Using this technique, the velocity measurements agree more closely with measured cavitation data, as shown in Fig. 1.20a. Careful inspection of the data indicates that there are still numerous discrepancies between the measured cavitation data and the measured value of C_{pm}. Most of these discrepancies are related to complexities that are beyond this review. However, it is important to note the complex dependence on Re. Because an elliptic planform was selected, the flow is essentially two-dimensional, and it is relatively easy to make boundary layer calculations. These computations (Fig. 1.20b) indicate that in the Re range of this experiment, a laminar separation bubble exists, which transitions from a "short" to a "long" separation bubble close to the tip [34]. An extensive flow visualization study in a wind tunnel confirms this analysis (Fig. 1.20c). Further confirmation is given when σ is lowered well below σ_i. As shown in Fig. 1.20d, the laminar separation bubble is apparent at about 60% of the chord. Also apparent are secondary vortices, created by the complex viscous flow near the tip, which roll up with the primary vortex. Further analysis indicates that this secondary vorticity is responsible for the braided structure of the primary tip vortex shown in Fig. 1.19b.

This example is presented in some detail to give the reader a grasp of the application of fundamentals to the design of an experiment, as well as to illustrate how a combination of further analysis and experimentation can help to clarify apparent discrepancies between theory and experiment. It is also important to note that in this particular study a well thought out program of flow visualization was of immeasurable benefit. In fact, the value of flow visualization in any program should not be overlooked.

Figure 1.21   Display model of the U.S. Navy Cavitation Channel, which is the world's largest water tunnel. (Photo courtesy of the David Taylor Research Center)

1.8.3 Hydrodynamic Design of a Very Large Water Tunnel

The cavitation problem in section 1.8.2 illustrates the complexity of cavitation phenomena. For this reason, several very large cavitation research facilities have recently been constructed around the world. Two examples are the large cavitation channel (LCC) completed in the United States in 1992, and the hydrodynamic and cavitation research tunnel (HYKAT), recently completed in Hamburg, Germany. Both facilities are described here, since each design was based on a combination of physical and mathematical modeling. The HYKAT project was based on previous investigations of the LCC [35, 36].

Figure 1.21 is a display model of the LCC, illustrating the general features of a water tunnel. Elbow 4 is in the upper left corner. Just downstream is the turbulence management system followed by the contraction and the test section. The substantial kinetic energy contained in the test section flow must be recovered in the diffuser before being turned in elbow 1 (upper right). The flow is turned again in elbow 2 before entering the pump (lower leg just below the entrance to the diffuser). After leaving the pump, the flow is further diffused before being turned in elbow 3 and once again in elbow 4 to complete the circuit. Each elbow is mitred and contains a series of turning vanes to ensure separation-free turning of the flow.

The two facilities are compared in Table 1.1. Both facilities have similar design problems, in that the nozzle and diffuser for each are nonsymmetrical and require additional attention to design, since the literature does not contain information on either nonsymmetrical diffusers or contractions. Very low noise levels were desired in each facility. This means that careful attention must be paid to cavitation in all parts of the tunnel, with special consideration being given to the pump. Critical components in the design are listed in Table 1.2.

Table 1.1 Comparison of LCC and HYKAT research facilities

	LCC	HYKAT
Nozzle contraction ratio	6	4
Test section type	Closed	Closed
Test section dimensions (m)		
Length	12	11
Width	3	2.8
Height	3	1.6
Maximum velocity (m/s)	15	12
Pressure (bar)		
Maximum	6	2.5
Minimum	0.15	0.15
Power (kW)	10,500	1,625

The implementation of such a project is a classic example of the need to identify, at a very early stage in the process, those features of the design that require further investigation. It is then necessary to determine the best tools for investigating these problems. In the preliminary design of the HYKAT, inviscid flow calculations coupled with boundary layer computations were used to optimize the contraction shape and number of turning vanes in each elbow. Computer analysis provided information on cavitation, boundary layer separation, uniformity of velocity, pressure distribution, and turning angle for these components. The turbulence management system was analyzed using existing semi-empirical theory. Preliminary analysis yielded the configuration shown in Fig. 1.22a.

After the preliminary design phase, a decision must be made on whether computer models alone will suffice for the final design or whether a physical model is necessary. In the case of HYKAT, the decision was made to base the final design on a combination of physical and mathematical modeling. Experience with the previous design of the LCC indicated that the HYKAT was sufficiently different to warrant further investigation. Component interaction effects were deemed important. In addition, the pump inflow velocity distribution must be sufficiently smooth to ensure a cavitation-free design.

Commitment to a measurement program implies a commitment in both time and money. Therefore it is extremely important to focus an applied experimental research program on those issues that cannot be studied with existing computer routines. In the case of HYKAT, the major issues were diffuser flow, component interaction, and the turbulence management system. Based on the requirements set forth in Table 1.2, various positions were established where measurements of pressure, mean velocity, and turbulence can be made. This is shown schematically in Fig. 1.22b. These measurements are supplemented by flow visualization using boundary tufts and smoke as a tracer, since air was selected as the test medium. Modeling in air substantially reduced costs and the time necessary for the measure-

Table 1.2 Critical components in water-tunnel design

Component	Requirement
Elbow 4	Velocity uniformity
	Turbulence levels
Honeycomb	Flow straightening
	Turbulence management
	Noise
Contraction	Velocity uniformity
	Turbulence management
	Separation
Test section	Velocity uniformity
	Turbulence level
Diffuser	Separation and stability
	Cavitation
	Exit velocity and profile
Elbow 1	Cavitation
	Noise
	Dynamic loads
Elbow 2	Velocity distribution into pump intake

ment program. Only three-fourths of the tunnel circuit is modeled, to simplify model construction. The physical model was used both to verify performance and to obtain design data for the contraction, diffuser, turning vanes, pump inflow, and turbulence management system. Further mathematical modeling was used to optimize the turning vane design and to evaluate its performance with measured inflow velocity profiles. A large eddy simulation was used to evaluate the flow in the diffuser. This numerical model provided guidance on the selection of the diffuser, taking into account the sensitivity to flow distortions and boundary layer flow at the test section exit. The physical model was also used to "tune" the mathematical model of the diffuser. The calibrated mathematical model was then used to extrapolate model test data to predict prototype performance. Thus the design process involves a fairly complex interaction between physical and mathematical modeling.

Verification of the accuracy of an inviscid Euler code for contraction design is shown in Fig. 1.22c. The results of large eddy simulation of the diffuser flow are shown in Fig. 1.22d. This figure displays the calculated effect of inlet flow distortion (expressed as a percentage of the average inlet flow velocity) on the velocity profile at the outlet. This is an important consideration for the turning vane design for elbows 1 and 2 and is an important factor in the final shape of the inlet velocity profile for the pump intake. This is illustrated in Fig. 1.22e, where the results from the velocity measurement program are shown schematically. This information was extremely useful as input to the mathematical program for determining changes in turning vane configuration, etc. These configuration changes were then tested in

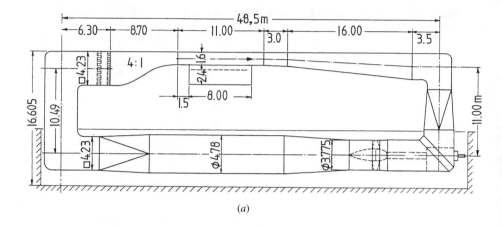

(a)

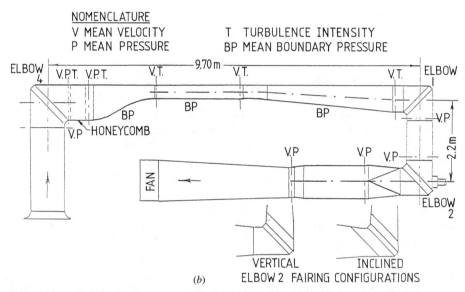

NOMENCLATURE
V MEAN VELOCITY   T TURBULENCE INTENSITY
P MEAN PRESSURE   BP MEAN BOUNDARY PRESSURE

(b)   ELBOW 2 FAIRING CONFIGURATIONS

Figure 1.22 (*a*) Schematic of the hydrodynamic and cavitation research tunnel (HYKAT) of the Hamburg Ship Model Tank, Hamburg, Germany. (*b*) Physical model (1:5) of HYKAT. (*c*) Experimental verification of the Euler computer code. (*d*) Comparison of calculated diffuser outlet velocity profile with experimental data obtained with a normalized inlet flow distortion of −5%. A large eddy computer simulation was used to make the calculations. (*e*) Measured velocity profiles at critical cross sections in the tunnel

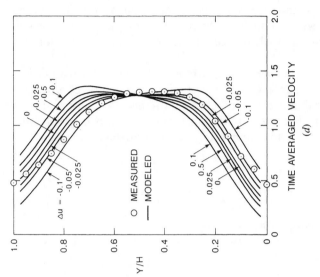

$\Delta u = -0.1$ -0.05   -0.025

0.025   0.5   0.1   0

-0.025   -0.05   -0.1

0.1   0.5   0.025   0

○ MEASURED

── MODELED

Y/H

TIME AVERAGED VELOCITY

(*d*)

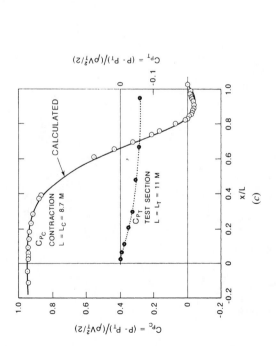

$C_{P_C} = (P - P_1)/(\rho V_1^2/2)$

$C_{P_T} = (P - P_1)/(\rho V_1^2/2)$

CALCULATED

C_{P_C}
CONTRACTION
$L = L_C = 8.7\ M$

C_{P_T}
TEST SECTION
$L = L_T = 11\ M$

x/L

(*c*)

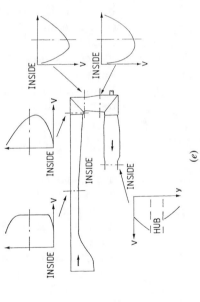

INSIDE   V

INSIDE   V

INSIDE   V

INSIDE

INSIDE

INSIDE   V

HUB   y

V

(*e*)

59

the physical model. This method is a substantial improvement over the "cut and try" process of strictly physical modeling. Operating the physical and mathematical models interactively substantially reduced the time and cost necessary for arriving at a final design.

This example illustrates the need to clearly identify what information is required, and to determine where a focused measurement program will provide the most benefit. Time constraints are usually very tight (in this case, construction had begun before completion of the research), and it is essential to have an overall understanding of the project before embarking on a measurement program.

NOMENCLATURE

A_n	amplitude of nth harmonic
A_0	speed of sound in undisturbed medium
A_1, A_2	cross-sectional areas
AR	aspect ratio
b	breadth of river or flow passage
c	concentration, chord length
$\mathbf{C}_{ij}$	cospectrum
C_l	lift coefficient
C_m	moment coefficient
C_n	Fourier coefficient
C_N	normal-force coefficient
C_p	pressure coefficient, $= \dfrac{(p - p_0)}{\frac{1}{2} \rho U_2}$
d	depth, diameter
$\mathbf{D}_{ij}$	diagonal matrix
e_{ij}	strain rate
e'_{ij}	fluctuating strain rate
E	error
$E( )$	expected value
$\mathbf{E}_i$	complex eigenvector
f	frequency (Hz)
F	force
$F(t)$	unspecified function of time
$F(\omega)$	Fourier transform
g	acceleration due to gravity
G	mass flux of suspended material
H	total head, defined in Eq. (89)
I	acoustic intensity

k	wave number, constant
K	spring constant
l	characteristic or integral length
l_T	tube length
L	lift
M	moment
Ma	Mach number
p	pressure
p_a	acoustic pressure
p_t	total pressure
p_v	vapor pressure
$p(u)$	probability density of $\tilde{u}$
$p(u, v)$	joint probability density
P	time mean pressure
P_t	time mean total pressure
$P(u)$	cumulative probability of $\tilde{u}$
Q_{12}	quadrature spectrum
r	radius, separation distance, distance to observer
r'	standard deviation in radial direction
r_c	core radius
$r_{1/2}$	radial position in jet where velocity is one-half the centerline velocity
R	pipe or jet radius
R_{12}	cross-correlation function
R_u, R_v	spatial correlation functions in isotropic turbulence
$R_{\Delta x}$	spatial correlation
R_τ	autocorrelation
Re	Reynolds number, $= Ul/2$
s	distance along streamline
S	surface area
$S(\omega)$, $S(f)$	power spectral density
St_d	Strouhal number, $= fd/U_j$
t	time
t'	retarded time, $= t - r/a_0$
T	characteristic time
T_{ij}	turbulent stress tensor
$T(\omega)$	spatially averaged power spectrum
$\mathfrak{I}$	characteristic decay time
$\tilde{u}_1, \tilde{u}_2, \tilde{u}_3$	velocity in tensor notation
u_1, u_2, u_3	velocity fluctuations in tensor notation
u, v, w	velocity fluctuations in Cartesian coordinates
U_1, U_2, U_3	time mean velocity in tensor notation
U, V, W	time mean velocity in Cartesian coordinates

v, v_θ	tangential velocity component
V	velocity along streamline
V_c	cavity volume
V_e	effective source volume
x_1, x_2, x_3	coordinates in tensor notation
x, y, z	Cartesian coordinates
α	angle of attack
γ	specific weight
Γ	circulation
δ	Kronecker delta function
$\delta(t - t_0)$	Dirac delta function
Δ	incremental change
ε	dissipation rate
η	Kolmogorov length scale
λ	wavelength
λ_n	eigenvalue
λ_T	Taylor microscale
μ	viscosity
ν	kinematic viscosity
ξ	dummy variable
ξ_n	weighting function
ρ	density
σ	standard deviation, cavitation number
τ	delay time Kolmogorov timescale
τ_{ij}	stress tensor
τ_{ij}^*	nonisotropic stress tensor
ϕ	cross-spectral density matrix
Φ	model decomposition of cross-spectral density matrix
ψ_n	orthogonal eigenfunctions
ω	radian frequency
Ω	force potential

Subscripts

a	acoustic
c	convection
i	incipient
m	molecular, maximum
p	phase
t	total, turbulent
v	viscous
0	undisturbed medium

REFERENCES

1. V. A. Vanoni (ed.), *Sedimentation Engineering,* American Society of Civil Engineering, Manual 54, chap. 2, 1975.
2. M. E. Nelson and P. C. Benedict, Measurement and Analysis of Suspended Sediment Loads in Streams, *Trans. ASCE,* vol. 116, pp. 891–918, 1951.
3. A. Pope and J. J. Harper, *Low Speed Wind Tunnel Testing,* chap. 1, Wiley, New York, 1966.
4. J. E. Cermak, Aerodynamics of Buildings, in M. van Dyke, W. G. Vincenti, and J. V. Weyhausen (eds.), *Annual Review of Fluid Mechanics,* vol. 8, p. 75, Annual Reviews, Palo Alto, Calif., 1976.
5. M. J. Lighthill, On Sound Generated Aerodynamically, I, General Theory, *Proc. R. Soc. London Ser. A,* vol. 211, no. 1107, pp. 564–587, 1952.
6. R. H. Kraichnan, Pressure Field Within Homogeneous, Anisotropic Turbulence, *J. Acoust. Soc. Am.,* vol. 28, no. 1, p. 64, 1956.
7. P. O. A. L. Davies, M. J. Fisher, and M. J. Barratt, Characteristics of the Turbulence in the Mixing Region of a Round Jet, *J. Fluid Mech.,* vol. 15, pt. 3, pp. 337–367, 1963.
8. G. M. Lilley, On the Noise from Air Jets, British Aerospace Research Council Rep. ARC-20, 376, N 40; FM-2724, September 1958.
9. G. Barefoot, Fluctuating Pressure Characteristics in the Mixing Region of Perturbed and Unperturbed Round Free Jet, M.S. thesis, The Pennsylvania State University, University Park, 1972.
10. R. E. A. Arndt and A. W. Nilsen, On the Measurement of Fluctuating Pressure in the Mixing Zone of a Round Free Jet, ASME Paper 71-FE-31, May 1971.
11. G. L. Brown and A. Roshko, On Density Effects and Large Structure in Turbulent Mixing Layers, *J. Fluid Mech.,* vol. 64, pt. 4, pp. 775–816, 1974.
12. J. Laufer, New Trends in Experimental Turbulence Research, in M. van Dyke, W. G. Vincenti, and J. V. Weyhausen (eds.), *Annual Review of Fluid Mechanics,* vol. 7, p. 307, Annual Reviews, Palo Alto, Calif., 1975.
13. L. Prandtl and O. G. Tietjens, *Applied Hydro- and Aeromechanics,* p. 230, 1934 (reproduction available from Dover, New York, 1957).
14. W. K. George, P. D. Buether, and R. E. A. Arndt, Pressure Spectra in Turbulent Free Shear Flows, *J. Fluid Mech.,* vol. 148, pp. 155–191, 1984.
15. H. Schlichting, *Boundary Layer Theory,* 7th ed., McGraw-Hill, New York, 1979.
16. H. Tennekes and J. L. Lumley, *A First Course in Turbulence,* MIT Press, Cambridge, Mass., 1972.
17. J. O. Hinze, *Turbulence,* 2nd ed., McGraw-Hill, New York, 1975.
18. P. Bradshaw (ed.), Turbulence, in *Topics in Applied Physics,* vol. 12, Springer-Verlag, Berlin, 1978.
19. W. K. George and R. E. A. Arndt (eds.), *Recent Advances in Turbulence,* Hemisphere, Washington, D.C., 1988.
20. H. Rouse (ed.), *Advanced Mechanics of Fluids,* chap. 6, Wiley, New York, 1959.
21. R. E. A. Arndt and A. T. Ippen, Turbulence Measurements in Liquids Using an Improved Total Pressure Probe, *J. Hydraul. Res.,* vol. 8, no. 2, pp. 131–158, 1970.
22. J. S. Bendat and A. G. Piersol, *Measurement and Analysis of Random Data,* Wiley, New York, 1968.
23. E. J. Richards and D. J. Mead (eds.), *Noise and Acoustic Fatigue in Aeronautics,* chap. 4, Wiley, New York, 1968.
24. E. D. von Frank, Turbulence in the Mixing Region of a Perturbed and Unperturbed Free Jet, M.S. thesis, The Pennsylvania State University, University Park, 1970.
25. N. C. Tran, Turbulence Characteristics in the Mixing Region of a Screen Perturbed Jet, M.S. thesis, The Pennsylvania State University, University Park, 1973.
26. J. L. Lumley, *Stochastic Tools in Turbulence,* Academic, San Diego, Calif., 1970.
27. R. E. A. Arndt and W. K. George, Investigation of Large Scale Coherent Structures in a Jet and Its Relevance to Jet Noise, *Proc. 2nd Symp. on Trans. Noise,* Raleigh, N.C., 1974.
28. D. F. Long and R. E. A. Arndt, The Orthogonal Decomposition of Pressure Fluctuations Surrounding a Turbulent Jet, *Proc. Turbulent Shear Flow Conf.,* Ithaca, N.Y., 1985.

29. D. F. Long, Noise Radiation and Coherent Structure in Turbulent Jets: Viscous and Compressibility Effects, Ph.D. dissertation, University of Minnesota, Minneapolis-St. Paul, 1985.
30. S. O. Rice, Mathematical Analysis of Random Noise, *Bell Syst. Tech.,* vol. 23, p. 24, 1944.
31. D. F. Long and R. E. A. Arndt, Jet Noise at Low Reynolds Number, *AIAA J.,* vol. 22, pp. 187–193, 1984.
32. R. E. A. Arndt, Cavitation in Fluid Machinery and Hydraulic Structures, *Annu. Rev. Fluid Mech.,* vol. 13, pp. 273–328, 1981.
33. B. W. McCormick, On Cavitation Produced by a Vortex Trailing from a Lifting Surface, *J. Basic Eng.,* vol. 84, pp. 369–379, 1962.
34. R. E. A. Arndt, V. H. Arakeri, and H. Higuchi, Some Observations of Tip Vortex Cavitation, *J. Fluid Mech.,* vol. 229, pp. 269–289, 1991.
35. J. M. Wetzel and R. E. A. Arndt, Hydrodynamic Design Considerations for Hydroacoustic Facilities: I. Flow Quality, *J. Fluid Eng.,* vol. 116, no. 2, 1994.
36. J. M. Wetzel and R. E. A. Arndt, Hydrodynamic Design of Hydroacoustic Design Facilities: II. Pump Design Factors, *J. Fluid Eng.,* vol. 116, no. 2, 1994.

PHYSICAL LAWS OF FLUID MECHANICS AND THEIR APPLICATION TO MEASUREMENT TECHNIQUES

E. R. G. Eckert

2.1 INTRODUCTION

Measurement techniques and special instrumentation are developed to obtain quantitative information on parameters that are deemed important for the understanding and prediction of fluid flow processes. One tries to measure, as directly as possible, the quantities that one is interested in. Frequently, however, this cannot be done, and one has to make use of some basic physical laws to obtain the required parameter. In determining the velocity of a fluid, for instance, with a Pitot tube, one has to use Bernoulli's equation or, if a hot-wire instrument is the sensor, the velocity is obtained from the electric signal by a heat transfer relation. One also has to have a good understanding of the flow situation to ensure, by an order of magnitude estimate of the possible error, that no extraneous effect influences the desired result of the measurement.

For these reasons, the basic physical laws of fluid mechanics will be reviewed in this chapter. It is useful to discuss them approximately in the order in which they were discovered during the last two centuries.

Much of the behavior of fluids in flow was learned by observation, using special tricks to make the detailed local movements of the fluid elements visible, a process now known as flow visualization. This method is still a valuable tool when one is careful in the interpretation of the observations.

Figure 2.1 Portrait of Sir Isaac Newton

Much progress was made in systematizing fluid mechanics when it was realized that very general relations can be obtained by using dimensionless parameters like Reynolds or Prandtl numbers as a base. Dimensional or similarity analysis that leads to these parameters is the most powerful tool in fluid mechanics and will be treated in considerable detail.

In most of this chapter, fluids will be considered incompressible and as having constant transport properties. Flow of gases as an example of a compressible fluid is discussed in sections 2.9 and 2.10. In most engineering situations, the flow encountered is unsteady and it will be considered as such unless it is specifically mentioned to be steady.

2.2 BASIC LAWS OF MECHANICS

The basic laws of mechanics as derived by Isaac Newton (Fig. 2.1) are described in his book *Philosophiae Naturalis Principia Mathematica* published in 1686. The most important law states that a force F applied to an isolated body causes the body to change its velocity V in such a way that the velocity change in time t is proportional to the force. The proportionality factor is called mass (m). If several forces simultaneously act on the body, then all of them have to be considered. Forces and velocities are vectors, and Newton's law therefore is described by a vector equation:

$$\sum \mathbf{F} = m\frac{d\mathbf{V}}{dt} \tag{1}$$

A difficulty arises in using this law for a fluid because in almost all cases a fluid can be considered a continuum in which there are no well-defined masses to which Newton's law can be applied. To get around this difficulty, we imagine an arbitrary part of the fluid surrounded by what is called a control surface and isolated from the rest of the fluid. Two different approaches are used in applying Newton's law to the movement of this fluid portion. In the Lagrangian approach, the control volume is considered to move with the fluid, and its history is studied as it moves along. In the second or Eulerian approach, the control surface is considered at rest relative to a selected coordinate system, and fluid moves through it. The second approach has advantages, especially when we can find a view of a coordinate system from which the flow appears steady.

The mass contained in the control volume varies in the Eulerian approach, and this makes it necessary to introduce the mass conservation law. This law states that no mass is created or destroyed, at least as long as nuclear processes that transform mass into energy are not involved. Such processes will be excluded here.

Gravitational, electric, magnetic, centrifugal, and Coriolis forces can act on the mass within the control volume. Such forces are called body forces. The mass also interacts with its surroundings through the control surface. The interaction can be described by introducing forces, called surface forces, applied by the surroundings and acting on the control surfaces. A simple example makes this clear. Consider a fluid at rest in a gravitational field and select a control volume in the shape of a small cylinder with its axis parallel to the gravity vector, as shown in Fig. 2.2. Gravity acts on the mass in the control surface and would cause this mass to be accelerated in the direction of the gravity vector if no other forces were present. Actually, the mass does not move, and surface forces must therefore counter the gravitational force as indicated in the figure. A fluid is defined as a medium in which surface forces act only normal to the surface, as long as it is at rest. Therefore no forces parallel to the gravity vector act on the cylindrical surface of the cylinder. The force per unit area is called pressure. The gravity force must be in equilibrium with the pressure forces acting on the upper and lower plane surfaces of the control volume.

Additional forces appear when the fluid is in motion. They are called viscous forces, and they may have components normal to and parallel to the surfaces of the control volume. In most situations, the components parallel to the surface dominate. They oppose a relative shear motion of the fluid layers on both sides of the control surface and depend on the transverse velocity gradient. They start with

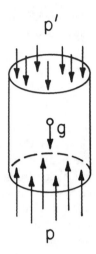

Figure 2.2   Forces on a fluid at rest in a gravitational field

the value zero for zero velocity gradient according to the definition of a fluid, and they increase with increasing velocity gradient. Newton's assumption that the viscous stresses (force per unit area) are proportional to the velocity gradient describes the actual dependence extremely well for many fluids. These are called Newtonian. Some fluids, however, do not follow this law, especially those treated in rheology.

All of the forces mentioned above have to be inserted into Eq. (1), which also has to be changed somewhat because there may be an influx to or outflux from the control volume. This causes the mass within the control volume to change in time. Newton's law therefore now has the form

$$\sum \mathbf{F} = \frac{d}{dt}(m\mathbf{V}) \tag{2}$$

The term $m V$ is called momentum, and the equation is called the momentum equation. Centrifugal and Coriolis forces have to be considered only when the flow is studied relative to a rotating coordinate system. Normal fluids do not respond to electric and magnetic fields. Gravity causes movement in the fluid only when the fluid density varies locally. Only pressure and viscous forces have to be inserted into the left side of Eq. (2) when the forces mentioned above are excluded.

2.3 INVISCID INCOMPRESSIBLE FLUIDS

Viscous forces in many fluids are quite small. This suggests that relations that are useful approximations to reality can be obtained in an approach that neglects the effect of viscosity altogether and considers the fluid involved as inviscid. This was

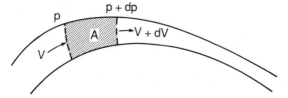

Figure 2.3 Path tube of an inviscid fluid

the approach during the early period of the development of fluid mechanics, and it will be taken up in this section. The only surface forces to be considered are then caused by fluid pressure.

The mass of the fluid per unit volume, its density ρ, will be considered constant. This is well fulfilled for liquids but is also a good approximation for gases as long as the velocities occurring are small in comparison with the sound velocity.

A good way to describe a flow field is to trace lines along which the fluid flows, the so-called path lines. A group of path lines starting at a closed curve then forms a path tube. Such a path tube with a small cross-sectional area is sketched in Fig. 2.3. Mass conservation requires that the same mass (or, at constant density, the same volume) flows through any cross section of the path tube. This is expressed by the statement

$$\frac{\partial}{\partial s}(vA) = 0 \tag{3}$$

where s denotes the direction of the path tube and v the velocity in this direction. The derivation of the relations obtained from Newton's law will not be presented in detail but will only be sketched. In the Lagrangian approach, one follows the control volume (dashed lines in Fig. 2.3) as it moves along with the fluid. Equation (1) written in the flow direction leads to

$$\frac{\partial p}{\partial s} + \rho \frac{Dv}{Dt} = 0 \tag{4}$$

The equation states that the force per unit volume $\partial p/\partial s$ in the flow direction is equal to the mass per unit volume ρ times the change of velocity per unit of time. Dv/Dt describes the velocity change in time of the fluid mass and is called the substantive derivative. It can be composed of two parts

$$\frac{Dv}{Dt} = \frac{\partial v}{\partial t} + v \frac{\partial v}{\partial s} \tag{5}$$

the first one describing the partial derivation of the velocity in time (at a fixed location) and the second one describing the change of velocity because the fluid mass has changed its location. With this relation, Eq. (4) can be written

$$\frac{\partial p}{\partial s} + \rho v \frac{\partial v}{\partial s} + \rho \frac{\partial v}{\partial t} = 0 \tag{6}$$

This equation was derived by Leonhard Euler (1707–1783) and is called the Euler equation.

The momentum equation in the direction n normal to the tube is

$$\frac{\partial p}{\partial n} - \rho \frac{v^2}{r} + \rho \frac{\partial v_n}{\partial t} = 0 \tag{7}$$

where r denotes the radius of curvature of the path tube. The second term on the left-hand side expresses a centrifugal force caused by the fact that the velocity vector turns through an angle when the path lines are curved; v_n is the velocity component in the n direction. The velocity component v_n has the value zero where the flow is steady.

In steady flow the last term in Eq. (6) drops out, and the equation can be integrated along a path line, which now coincides with a streamline. In this way, one obtains the Bernoulli equation (Daniel Bernoulli, 1700–1782). Equation (6) simplifies for steady flow to

$$p + \rho \frac{v^2}{2} = \text{const} \tag{8}$$

and Eq. (7) to

$$\frac{\partial p}{\partial n} = \rho \frac{v^2}{r} \tag{9}$$

The pressure p in the above equations can, in principle, be measured by an instrument that moves along with the fluid. This, however, is usually impossible, and the pressure is measured via a small hole in a wall arranged in such a way that the wall does not disturb the flow. The upper part of Fig. 2.4 shows two ways in which this pressure, also called the static pressure, can be measured. On the left-hand side of Fig. 2.4, the flow passes through a straight tube. Correspondingly, no centrifugal force acts on the fluid particles moving along streamlines, and the static pressure p is constant across the duct cross section according to Eq. (9).

One may also define a total pressure p_t, which is obtained when the flow velocity at a selected location is reduced to zero. According to Eq. (8), the total pressure is

$$p_t = p + \rho \frac{v^2}{2} \tag{10}$$

A way to measure the total pressure is indicated in the lower left-hand panel of Fig. 2.4. From a measurement of the static and the total pressure, the velocity of the fluid can be obtained according to the equation

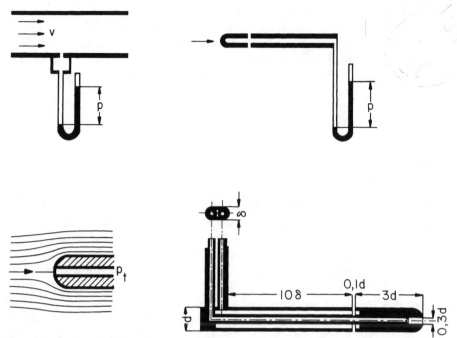

Figure 2.4 Static and total pressure sensors

$$v = \sqrt{\frac{2(p_t - p)}{\rho}} \tag{11}$$

The two pressure measurements can be combined in one sensor, as shown in the lower right panel of Fig. 2.4. The dimensions indicated in the figure have been determined by Ludwig Prandtl in such a way that the accuracy of the pressure measurements is optimized.

In unsteady flow the total pressure defined by Eq. (10) can be introduced into Eq. (4) to result in the simple relation

$$\frac{Dp_t}{Dt} = \frac{\partial p}{\partial t} \tag{12}$$

This equation states that the variation of the total pressure of a fluid particle as it moves along a path tube is equal to the variation in time of the static pressure at a fixed location. For steady flow this pressure derivative in time of the static pressure is zero and the total pressure of a fluid particle is constant as it moves along a stream tube.

For an inviscid fluid with constant density, there is no possibility to convert internal energy into mechanical (kinetic or pressure) energy. Equation (8) can also be interpreted as a conservation equation of mechanical energy.

Many fluids have a thermal conductivity of the same order of magnitude as the viscosity. For those fluids, heat conduction can be neglected when the effect of viscosity is neglected. The enthalpy or, with constant specific heat, the temperature T cannot change:

$$\frac{DT}{Dt} = 0 \qquad (13)$$

Conservation of internal energy requires that the enthalpy of the fluid or, with constant specific heat, the temperature T of the inviscid fluid remains constant as the fluid particle moves along.

The Pitot or Prandtl tube is actually used to measure the velocity in real viscous fluids, and the question arises whether Eq. (11), by which the velocity is calculated from the pressure measurements, can really be applied. It will be discussed in section 2.7 that, at high Reynolds numbers (Re), the effect of viscosity is restricted to a very thin boundary layer along the surface of the tube, whereas the viscosity influence can be neglected outside this layer. The pressure is transmitted through this thin layer without any measurable change. The use of Eq. (11) is therefore justified at high Re. However, a correction must be made when the Re based on the tube diameter is small. This was established by Barker, who obtained the following relation from his experiments [1]:

$$\frac{p_t - p}{\rho v^2/2} = 1 + \frac{C}{Re} \qquad (14)$$

The value of the constant C depends on the shape of the head of the tube. Values between 3 and 5.6 are reported in the literature. A correction must also be applied to Eq. (11), where large lateral velocity gradients occur over a distance of the order of the tube diameter.

It will also have been noted that the reading of the Pitot tube evaluated with Eq. (11) holds for steady flow. The equation is, however, also a good approximation to reality for unsteady flow as long as the term $\partial v/\partial t$ in Eq. (6) is small in comparison with the convective term $v\, \partial v/\partial s$. For such flows, called quasi-steady, the reading of the Pitot tube can be evaluated with Eq. (11).

2.4 VISCOUS FLOW NAVIER-STOKES EQUATIONS

The preceding paragraph mentioned a flow situation where a correction has to be applied to a relation obtained for an inviscid fluid to account for the effect of the viscosity of the real fluid. Another situation where viscosity has a large effect is flow through a tube. An inviscid fluid encounters no pressure drop in steady flow, and no energy is expended in moving the fluid.

The equations describing the flow of a real fluid had to be expanded to include the effect of fluid viscosity. This was done by N. Navier in 1827 [2] and by G. G. Stokes in 1845 [3]. These equations are referred to as the Navier-Stokes equations. Written in a Cartesian coordinate system, for a fluid with constant properties and in the absence of body forces, they assume the following form.

Mass continuity

$$\frac{\partial u}{\partial x} + \frac{\partial v}{\partial y} + \frac{\partial w}{\partial z} = 0 \tag{15}$$

Momentum equation

$$\rho\left(\frac{\partial u}{\partial t} + u\frac{\partial u}{\partial x} + v\frac{\partial u}{\partial y} + w\frac{\partial u}{\partial z}\right) = -\frac{\partial p}{\partial x} + \mu\left(\frac{\partial^2 u}{\partial x^2} + \frac{\partial^2 u}{\partial y^2} + \frac{\partial^2 u}{\partial z^2}\right) \tag{16}$$

$$\rho\left(\frac{\partial v}{\partial t} + u\frac{\partial v}{\partial x} + v\frac{\partial v}{\partial y} + w\frac{\partial v}{\partial z}\right) = -\frac{\partial p}{\partial y} + \mu\left(\frac{\partial^2 v}{\partial x^2} + \frac{\partial^2 v}{\partial y^2} + \frac{\partial^2 v}{\partial z^2}\right) \tag{17}$$

$$\rho\left(\frac{\partial w}{\partial t} + u\frac{\partial w}{\partial x} + v\frac{\partial w}{\partial y} + w\frac{\partial w}{\partial z}\right) = -\frac{\partial p}{\partial z} + \mu\left(\frac{\partial^2 w}{\partial x^2} + \frac{\partial^2 w}{\partial y^2} + \frac{\partial^2 w}{\partial z^2}\right) \tag{18}$$

The symbols in these equations are explained in the Nomenclature.

These equations are mathematically so involved that a deciding advance in their solutions was achieved only when high-speed electronic computers became available. But even with supercomputers, solutions cannot be obtained in situations that involve very high Re. In this case, research has to rely primarily on experimentation, but the Navier-Stokes equations are still useful because they make it possible to generalize experimental results by the application of dimensional or similarity analysis. This will be discussed in section 2.6.

2.5 FLOW VISUALIZATION

Osborne Reynolds was the first to clarify the difference between laminar and turbulent flow and the nature of transition from one to the other by observing the details of the motions within the moving fluid. He devised the setup (reproduced from his original publication in Fig. 2.5) and used it for the study and demonstration of the characteristics of the flow through a tube with circular cross section. He discussed the results in a lecture to the Royal Society in 1884 [4].

A trough with vertical glass walls can be recognized in Fig. 2.5. Located inside the trough is a horizontal tube, also manufactured of glass, with a bell-mouth entrance. Water fills the trough and is discharged through the tube by opening of a valve on the vertical pipe downstream of the tube. This is done after the water is left standing in the trough for some time to make sure that any disturbances in the

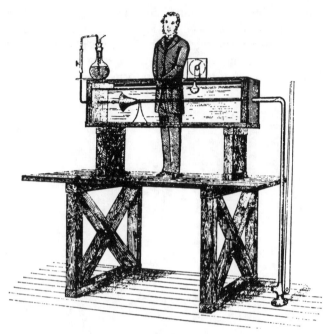

Figure 2.5 Osborne Reynolds demonstrates transition to turbulence [4]

water have died down. A streak of colored fluid taken from the bottle above the trough can be inserted into the water flow at the entrance of the tube. The water flow rate is measured by the descent of the water surface.

Figure 2.6 presents sketches made by Reynolds as he observed the flow when the flow rate was gradually increased. Up to a certain flow rate, the streak of colored water moved straight through the tube, as shown in Fig. 2.6a. A further small increase of the flow rate, however, changed the picture to that shown in Fig. 2.6b. The streak started to fluctuate irregularly at some distance from the entrance, to form vortices, and to mix with the noncolored water. Reynolds concluded that two characteristic forms of flow exist in the tube: a steady flow as shown in Fig. 2.6a, today called laminar, and an unsteady irregular flow, which he called sinuous and today we call turbulent. He observed the character of that flow also by short-time illumination with an electric spark and documented what he observed in Fig. 2.6c. He also concluded from a series of such measurements and from a comparison of the expressions for inertia to viscous forces that a dimensionless parameter,

$$\mathrm{Re} = \frac{\rho V r}{\mu} \tag{19}$$

where ρ is density, V is velocity, r is tube radius, and μ is viscosity, determines the character of the flow. Today we call this the Reynolds number. It describes the

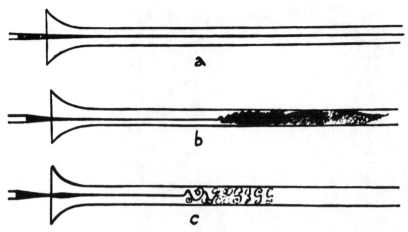

Figure 2.6 Reynolds' sketches of laminar and turbulent flow

transition from laminar to turbulent flow. Reynolds also evaluated all of the experimental results on pressure drop in the flow through tubes that were available at that time and demonstrated that the resistance R to the flow, made dimensionless with its kinetic energy ρV^2, can be expressed as a function of Re:

$$\frac{R}{\rho V^2} = \mathrm{Re}^{n-2} \tag{20}$$

where the parameter n has a value of 1 for laminar flow, 1.723 for turbulent flow, and 2 for turbulent flow through tubes with rough walls. Osborne Reynolds, in this way, brought order to all the measurements describing fluid flow. He also contributed to progress in fluid mechanics through other investigations: developing the hydrodynamic theory of lubrication, developing the time-averaged Reynolds equations describing turbulent flow, developing the analogy between fluid friction and heat transfer, and making model studies of the flow in estuaries of rivers. In his later years (Fig. 2.7) he attempted to develop a theory describing the nature of ether, which physicists believed was filling the universe, so that light in the form of electromagnetic waves could be transmitted through space.

This method to make fluid flow visible by the injection of a colored fluid has been developed and used often since Reynolds' time. Figure 2.8 shows the flow around a cylinder with circular cross section made visible by the injection of a number of colored fluid streaks upstream of the cylinder. The separation of the flow and the formation of a vortex street downstream of the cylinder can be recognized. If one would actually observe the flow in time, one would find that the streaks of colored fluid maintain nearly fixed positions in time upstream and sideways of the cylinder, showing that this flow is nearly steady. However, the streaks in the vortex street fluctuate continually as the vortices move downstream, indicat-

Figure 2.7 Portrait of Osborne Reynolds in his later years. The original hangs in the University of Manchester

ing that the flow in this region is unsteady. Such a photo not only provides a good impression of the character of the flow but also leads to more detailed information in the region in which the flow is steady when the fluid is incompressible. From mass continuity, one concludes that the velocity is slow where the distance between neighboring streaks is large and that it is large where the streaks move closer together. In steady flow the streak lines also describe the path along which fluid particles move in time.

However, one must be more careful in interpreting the photo when the flow is unsteady. For this situation, the streaks of colored fluid obtained by instantaneous exposure of the flow, called streak lines, are different from the path lines, which describe the path traveled in time by individual fluid particles. One could determine path lines if one would emit the colored fluid in puffs, make a movie of the flow, and identify the location of specific fluid puffs in the sequence of pictures.

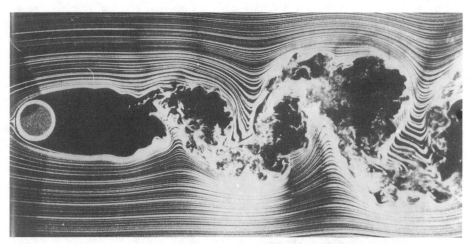

Figure 2.8 Streak lines revealing flow around cylinder at Re = 10,000. (Courtesy of Th. Corke and H. Nagil)

For a flow situation that can be described analytically, path lines and streak lines can be calculated. This was done by M. Kurosaka [5]. Figure 2.9, taken from his publication, presents streak lines and path lines for a specific situation. A vortex is carried by a uniform mainstream from left to right. The three dashed lines are path lines, all starting at the same point on the vertical line through point 0, but starting at increasing time steps from line a to line c. They are of the nature of epiycloids. The solid line represents a streak line starting at the same point and

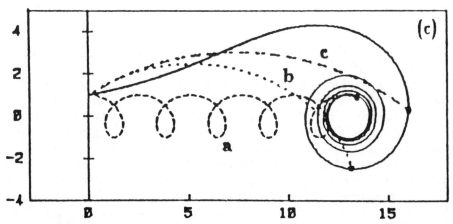

Figure 2.9 Streak lines (dashed curves) and path line (solid curves) of flow around a vortex moving to right. (From [5])

Figure 2.10 Portrait of Ludwig Prandtl

photographed at the moment at which the vortex has reached the location in the right half of the picture.

Flow visualization in gas streams is often achieved by injection of streaks of smoke. In such an experiment, one has to be aware of the fact that the smoke diffuses much more slowly than velocity differences diffuse in a fluid flow that is to be represented by the flow visualization experiment. Both diffusion processes would have the same speed where the Schmidt number $Sc = v/D = 1$ (where v is kinematic viscosity of the fluid, and D is the mass diffusion coefficient of the smoke particles into the fluid). Actually, smoke particles diffusing into air have Schmidt numbers of the order of 10^5 and 10^6. The mass diffusion is correspondingly slow [6]. This also explains the long persistence time of the condensation trails behind an airplane in a quiet sky.

Ludwig Prandtl, whose portrait is shown in Fig. 2.10, used another method of

Figure 2.11   Ludwig Prandtl with his water tunnel

flow visualization in his early studies, which may well have led him to the concept of the boundary layer. In 1903, he designed and constructed the water tunnel shown in Fig. 2.11. The horizontal trough in Fig. 2.11 is open on top and filled with water. A horizontal wall separates the water into an upper and lower part. The water can be made to flow from right to left in the upper part and return in the lower part driven by a paddlewheel, which is seen operated by Ludwig Prandtl in Fig. 2.11. A powder of aluminum flakes is strewn onto the surface of the water, and the model to be studied is inserted into the flow. In this way, photos of the kind shown in Fig. 2.12 [7] can be obtained with a short but finite illumination time, so that the flakes are depicted as short lines indicating the direction of the velocity vector. Streamlines can be obtained by drawing curves in such a way that the velocity vectors are everywhere tangent to those lines. In the steady part of the photo, streamlines coincide with path lines and streak lines, but all differ in the unsteady portion. The two photos in Fig. 2.12 show the flow directed normal to the axis of a cylinder at two different times after the flow has been started impulsively from rest. The upper panel shows the flow at a short time after its start, and the lower panel after a longer time, when the flow has become quasi-steady. The

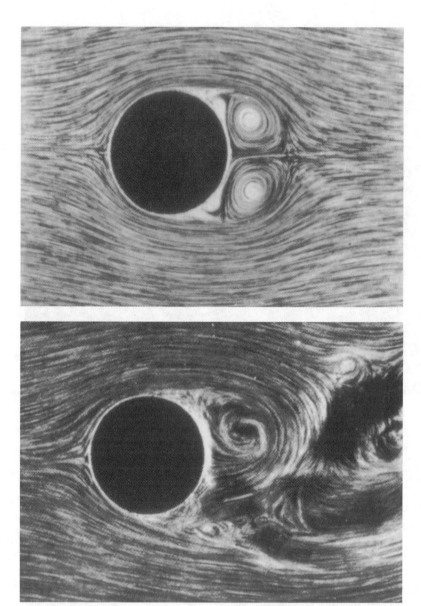

Figure 2.12   Streamlines around a cylinder at two different times after the flow has been started impulsively. (From [7])

photos therefore indicate that a certain time is required before the vortices start separating from the cylinder and forming the vortex street.

Figure 2.13 presents a series of photos [8] depicting turbulent flow through a two-dimensional channel with the walls located at the horizontal rims of the individual photos. The camera was moved with different speeds relative to the flow. In the top panel the camera was stationary relative to the channel, and in the bottom panel its velocity was equal to the maximum fluid velocity at the centerline of the channel. Figure 2.13 demonstrates that the appearance of the flow is quite different, depending on the relative movement of the camera to the flow. It also shows that large-scale vortices and fluctuations occur in turbulent channel flow.

Flow visualization is a very useful tool when we correctly interpret what is observed. It has advanced our understanding of flow processes tremendously. A large number of flow visualization photos for viscous flow processes have been collected in books, for instance, *An Album of Fluid Motion* by Milton Van Dyke [9]. This type of book is a basis for anyone interested in fluid flow. The photos would have only limited utility if the flow is visualized just for the specific condition for which the photo was taken. However, the photos describe a range of similar flow conditions, as has already been demonstrated by Osborne Reynolds. Figure 2.7, for instance, describes the flow of any incompressible Newtonian fluid around a cylinder of any size as long as the Re formed with the approach velocity V, the cylinder diameter d (or radius r), and the kinematic viscosity v has the value 10,000. In some cases, other restrictions apply, and the dimensionless parameters describing those cases are obtained by similarity analysis. Similarity analysis is possibly the most important tool in fluid mechanics and indispensable for the interpretation of the results of measurements. It will therefore be discussed in detail in the following section for constant-property Newtonian fluids. The treatment of fluids with variable properties and of non-Newtonian fluids can be found in the literature.

2.6 SIMILARITY ANALYSIS

2.6.1 One-Component Fluids

The following equations describe the flow of a Newtonian constant-property fluid (constant density) in a Cartesian coordinate system. The flow will be considered unsteady. The mass conservation equation reads

$$\frac{\partial u}{\partial x} + \frac{\partial v}{\partial y} + \frac{\partial w}{\partial z} = 0 \tag{21}$$

A list of nomenclature is found at the end of the chapter. There are three momentum equations for the three space coordinates. Only one will be presented here because the other two have exactly the same form as this one. They have been listed in section 2.4.

Figure 2.13 Streamlines revealing turbulent flow in a water channel photographed with different camera speeds. (From [8])

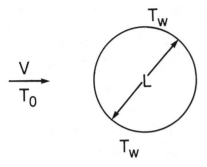

Figure 2.14 Flow normal to a sphere

$$\rho\left(\frac{\partial u}{\partial t} + u\frac{\partial u}{\partial x} + v\frac{\partial u}{\partial z} + w\frac{\partial u}{\partial z}\right) = -\frac{\partial p}{\partial x} + \mu\left(\frac{\partial^2 u}{\partial x^2} + \frac{\partial^2 u}{\partial y^2} + \frac{\partial^2 u}{\partial z^2}\right) \tag{22}$$

The system of Eqs. (21) and (22) was developed by N. Navier in 1827 and by G. G. Stokes (1819–1903). In writing down the boundary conditions, we will assume that we are dealing with flow over a sphere with the diameter L and with a uniform approach velocity V, which starts impulsively at time $t = 0$ (Fig. 2.14). Upstream at $t > 0$.

$$u = V \qquad v = 0 \qquad w = 0 \tag{23}$$

on the surface,

$$u = v = w = 0 \tag{24}$$

and the initial condition at $t < 0$ is

$$u = v = w = 0 \tag{25}$$

In a similarity analysis, we restrict consideration to a group of geometrically similar objects that differ only by their scale. Any length dimension L can be used to specify the object. In the present example, all spheres are similar, but the discussion applies as well to groups of other forms, for instance airfoils, of similar shape.

We will now proceed to make the equations dimensionless using parameters that are prescribed to the problem (V, L). The new independent variables are

$$x' = x/L \qquad y' = y/L \qquad z' = z/L \qquad t' = t/(L/V) \tag{26}$$

No characteristic time is prescribed because the flow is started impulsively; therefore, such a time has been formed with the prescribed parameters V and L. This term L/V has been used in Eq. (26) to define the dimensionless time t'. The dependent variables are also made dimensionless:

$$u' = u/V \qquad v' = v/V \qquad w' = w/V \qquad p' = \Delta p/\rho V^2 \tag{27}$$

Equation (22) shows that we are only concerned with pressure differences Δp and such a pressure difference is not prescribed. However, a characteristic term with the dimension of a pressure difference is ρV^2, which has been used to define a dimensionless pressure in Eq. (27). Equations (21) and (22) now take on the form

$$\frac{\partial u'}{\partial x'} + \frac{\partial v'}{\partial y'} + \frac{\partial w'}{\partial z'} = 0 \tag{28}$$

$$\frac{\partial u'}{\partial t'} + u'\frac{\partial u'}{\partial x'} + v'\frac{\partial u'}{\partial y'} + w'\frac{\partial u'}{\partial z'} = -\frac{\partial p'}{\partial x'} + \frac{1}{\mathrm{Re}}\left(\frac{\partial^2 u'}{\partial x'^2} + \frac{\partial^2 v'}{\partial y'^2} + \frac{\partial^2 w'}{\partial z'^2}\right) \tag{29}$$

The Reynolds number,

$$\mathrm{Re} = \rho VL/\mu = VL/\nu \tag{30}$$

has been introduced into Eq. (29). The only boundary condition with a value different from zero is upstream for $t > 0$:

$$u' = 1 \qquad v' = w' = 0 \tag{31}$$

The dimensionless dependent variables would be obtained by a solution of the flow equations with the appropriate boundary conditions. It is not always possible to obtain solutions, especially when the flow changes to turbulence. We can, however, predict the parameters on which the solution will depend. It is described by the equations

$$\left.\begin{array}{c} u/V \\ v/V \\ w/V \\ \Delta p/\partial V^2 \end{array}\right\} = f\left(\frac{x}{L}, \frac{y}{L}, \frac{z}{L}, \frac{t}{L,V}, \mathrm{Re}\right) \tag{32}$$

It is convenient to express any parameter that depends on fluid flow in a dimensionless way. The drag coefficient c_D, for instance, is a dimensionless expression for the drag D of the sphere defined by equation

$$c_\mathrm{D} = \frac{D}{(\rho V^2/2)(\pi L^2/4)} \tag{33}$$

The pressure drop experienced by a fluid flowing through a tube with a diameter d, length L, and mean velocity V is determined by the friction factor

$$c_f = \frac{\Delta p D}{L}\frac{2}{\rho V^2} \tag{34}$$

Any of these dimensionless parameters have to be functions of the dimensionless parameters listed on the right side of Eq. (32). The following functional relation therefore must exist and must hold for laminar as well as for turbulent flow:

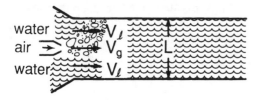

Figure 2.15   Two-phase flow through a tube

$$\left.\begin{array}{c} c_D \\ c_f \end{array}\right\} = f\left(\frac{t}{L/V}, \mathrm{Re}\right) \tag{35}$$

The form of the function f is different for the sphere and for the tube, and it is also different for laminar and for turbulent flow.

The derivation of this equation has been based on logic and is therefore exact if the flow equations are valid and if the boundary conditions are stated properly. This, however, may not always be simple. Osborne Reynolds has already demonstrated that the relation for the pressure drop of a fluid flowing through a pipe follows different relations when the walls of the pipe are rough than when they are smooth. In terms of the similarity analysis, this means that a rough sphere is not geometrically similar to a sphere with a smooth surface and additional parameters have to be used in the equations above to describe the roughness in dimensionless terms. This is impossible in a strict sense because, in general, one does not know the roughness characteristic in all its details. At this point one has to rely on experience. For instance, it was found that, at smaller Re, roughness does not influence the flow (such a surface is called hydraulically smooth). Otherwise, certain types of roughness have been defined, and the ratio of the mean roughness height to the characteristic length L is used as a dimensionless parameter describing the scale of the roughness for a specific type. A similar difficulty can occur at the upstream boundary, where a uniform approach velocity V has been prescribed. Sometimes the approaching flow is turbulent, and in a strict sense, one would have to know the turbulence characteristics in all its details, which is not possible. Here again, empirical relations for the dimensionless parameters describing the upstream flow have to be introduced.

2.6.2 Two-Component Fluids

We consider a process illustrated in Fig. 2.15 in which a liquid and a gas are injected into a tube with velocity V_1 and V_g, respectively. Depending on flow conditions, the gas jet will, after it leaves the tube delivering the gas, be separated from the liquid by a straight interface, which, however, soon becomes unstable and breaks up into bubbles or other irregular forms varying along the tube length. The flow of each of the fluids is described again by the Navier-Stokes equations, so that now two sets of equations, Eqs. (21) and (22), describe the flow process. The appearance of the flow in the tube, of course, is irregular and varying but probably not more so than the appearance of turbulent flow, and this does not concern us at this point. The following boundary conditions apply:

Liquid at entrance at $t > 0$ $\qquad\qquad \overline{u_1} = V_1$ (36a)

Liquid at wall $\qquad\qquad u_1 = v_1 = w_1 = 0$ (36b)

Gas at entrance at $t > 0$ $\qquad\qquad \overline{u_g} = V_g$ (36c)

Gas at wall $\qquad\qquad u_g = v_g = w_g = 0$ (36d)

A boundary condition has to be satisfied at the interface between the gas and liquid. If the interface is planar, then the pressure on the liquid side has to be equal to the pressure on the gas side of the interface. When the surface is curved, the surface tension σ will create a pressure difference on both sides of the interface, given by equation

$$\Delta p = \frac{\sigma}{r_1 + r_2} \tag{37}$$

($r_1, r_2, \ldots$, principal radii of curvature). By introducing a dimensionless pressure difference, $\Delta p' = \Delta p / \rho V_2$, the equation transforms to

$$\Delta p' = \frac{\sigma}{p_\ell L V_\ell^2} \frac{1}{r_1' + r_2'} \tag{38}$$

The dimensionless radii of curvature r_1 and r_2 are not prescribed to the problem but internally determined in the flow process. This leaves the parameter called the Weber number:

$$\text{We} = \frac{\rho L V^2}{\sigma} \tag{39}$$

Similarity of the entrance velocity profile requires V_g / V_ℓ as an additional parameter.* The equation

$$c_f = f\left(\frac{\rho_\ell V_\ell L}{\mu_\ell}, \frac{\rho_g V_g L}{\mu_g}, \frac{\sigma}{\rho_\ell l_\ell V^2}, \frac{V_g}{V_\ell}\right) = f\left(\text{Re}_\ell, \text{Re}_g, \text{We}, \frac{V_g}{V_\ell}\right) \tag{40}$$

then describes the relation for the pressure drop of two-phase flow. The functional dependence does not change when the product or quotient of two of the parameters is replacing one of them. The relation for the pressure drop can therefore also be written in the form

---

*Actually, the shear stress also has to be required to have the same value on both sides of the interface. This can be neglected for fluids with small viscosity.

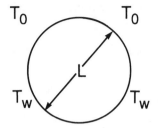

Figure 2.16   Natural convection around a sphere

$$c_f = f\left(\frac{\rho V_\ell L}{\mu}, \frac{V_g}{V_\ell}, \frac{\sigma}{\rho L V_\ell^2}, \frac{\rho_\ell \mu_g}{\rho_g \mu_\ell}\right) \tag{41}$$

In most situations, the density ratio ρ_g/ρ_ℓ exhibits a larger deviation from the value 1 than the viscosity ratio. This is often empirically taken care of in Eq. (41) by replacing $\rho_\ell \mu_g/\rho_g \mu_\ell$ by the density ratio ρ_ℓ/ρ_g, resulting in the relation

$$c_f = f\left(\mathrm{Re}_\ell, \frac{\rho_\ell V_\ell}{\rho_g V_g}, \mathrm{We}, \frac{\rho_\ell}{\rho_g}\right) \tag{42}$$

2.6.3 Heat Transfer

Flow of a fluid with constant properties is not influenced by heat transfer because the Navier-Stokes equations can, in principle, be solved without any assumption as to temperature differences in the fluid. Heat transfer in the fluid caused by some prescribed temperature difference is superimposed to fluid flow but does not influence it. In this section, heat transfer under the condition mentioned above will be analyzed. It is described by an energy equation developed by J. Fourier in 1822 in the book *Theorie Analytique de la Chaleur.* In a Cartesian coordinate system the equation has the form

$$\rho c_p\left(\frac{\partial T}{\partial t} + u\frac{\partial T}{\partial x} + v\frac{\partial T}{\partial y} + w\frac{\partial T}{\partial z}\right) = k\left(\frac{\partial^2 T}{\partial x^2} + \frac{\partial^2 T}{\partial y^2} + \frac{\partial^2 T}{\partial z^2}\right) \tag{43}$$

To determine boundary conditions, we will consider again the example of a sphere with a diameter L exposed to flow with a velocity V and will prescribe that at time $t = 0$ the surface temperature of the cylinder is raised from T_0 to T_w (Fig. 2.16). The approaching fluid temperature is T_0. The boundary conditions therefore are

upstream and on the surface at $t < 0$ $\qquad T = T_0$ $\tag{44}$

on the surface at $t > 0$ $\qquad T = T_w$ $\tag{45}$

Equation (43) retains its forms when it is written in a temperature difference $T - T_0$ with the wall difference

$$\theta_w = T_w - T_0 \qquad (46)$$

and a dimensionless temperature difference

$$\theta' = \theta/\theta_w \qquad (47)$$

The energy equation takes on the form

$$\frac{\partial\theta'}{\partial t'} + u'\frac{\partial\theta'}{\partial x'} + v'\frac{\partial\theta'}{\partial y'} + w'\frac{\partial\theta'}{\partial z'} = \frac{1}{Pe}\left(\frac{\partial^2\theta}{\partial x'^2} + \frac{\partial\theta'}{\partial y'^2} + \frac{\partial^2\theta'}{\partial z'^2}\right) \qquad (48)$$

with the parameter

$$Pe = \frac{VL}{\alpha} \qquad (49)$$

which is called the Péclet number. The dimensionless boundary conditions are

$$\theta' = 0 \qquad (50a)$$

upstream and at $t < 0$

$$\theta' = 1 \qquad (50b)$$

on the surface at $t > 0$. In writing down the functional form of the solution to Eq. (48) with the boundary conditions Eqs. (50a) and (50b), it must be considered that the dimensionless velocities u', v', w' appear in Eq. (48). The dimensionless solution therefore has the following form, since Eq. (32) determines the dimensionless velocities,

$$\frac{T - T_0}{T_w - T_0} = f\left(\frac{x}{L}, \frac{y}{L}, \frac{z}{L}, \frac{t}{L/V'}, Re, Pe\right) \qquad (51)$$

which can be changed to the more convenient form

$$\frac{T - T_0}{T_w - T_0} = f\left(\frac{x}{L}, \frac{y}{L}, \frac{z}{L}, \frac{t}{L/V'}, Re, Pr\right) \qquad (52)$$

with the new parameter

$$Pr = \frac{Pe}{Re} = \frac{\upsilon}{\alpha} \qquad (53)$$

which is called the Prandtl number, and contains only properties. It is therefore a dimensionless property. The Nusselt number as a dimensionless expression of the average heat transfer coefficient $h = q_w/(T_w - T_0)$ on the surface of the cylinder is then described by the equation

$$\overline{Nu} = \frac{hL}{k} = \frac{q_w L}{k(T_w - T_0)} = f(\text{Re}, \text{Pr}) \tag{54}$$

2.6.4 Natural Convection

Natural convection flow is generated by buoyancy forces. Therefore a term $g(\rho - \rho_0)$ has to be added to Eq. (22). In other terms of this equation, the density ρ will still be considered constant, which requires that $\rho - \rho_0 \ll \rho$. This restriction was introduced by Boussinesq and is called the Boussinesq assumption.

As an example, natural convection flow around a sphere of diameter L will be considered, which is located in a quiescent fluid of density ρ_0. The density of the fluid adjacent to the sphere surface is raised to a value ρ_w at time $t = 0$. Equations (21), enlarged (22), and (43) and boundary conditions can again be made dimensionless to obtain the dimensionless parameters for this process. Another, shorter way will be selected here. No prescribed velocity is available in this natural convection process. However, one can form a term with the dimensional velocity. Using the prescribed buoyancy condition, one obtains

$$V_n = \sqrt{\frac{g(\rho_w - \rho_0)L}{\rho_0}} \tag{55}$$

A natural convection Reynolds number can be formed

$$\text{Re}_n = \frac{\rho V_n L}{\mu} = \sqrt{\frac{g\rho(\rho_n - \rho_0)L^3}{\mu^2}} \tag{56}$$

The temperature T appears as the unknown in Eq. (43). The density in a process that is described by this equation is assumed to vary as a function of temperature. With the volumetric expansion coefficient β and the assumption of small density differences, the equation $\rho - \rho_0 = \beta\rho(T - T_0)$ expresses the density difference, and with this expression, Eq. (56) changes to (Fig. 2.16)

$$\text{Re}_n = \sqrt{\frac{g\beta(T_0 - T_w)L^3}{\nu^2}} \tag{57}$$

The functional relation for the average Nusselt number over the surface of the sphere is obtained by replacing in Eq. (54) the forced convection Re by the natural convection Re_n:

$$Nu = f(\text{Re}_n, \text{Pr}) \tag{58}$$

Conventionally, natural convection heat transfer and flow are presented as functions of a term called the Grashof number, which is simply

$$\text{Gr} = \text{Re}_n^2 \tag{59}$$

and Eq. (58) can also be written

$$\text{Nu} = f(\text{Gr, Pr}) \tag{60}$$

2.6.5 Two-Phase Flow

Two-phase flow is different from two-component flow, in that a phase change occurs at the interface. A liquid and its vapor will be considered in this discussion. The equation describing the phase change at the interface is

$$\rho h_{lv} v_\ell = k\frac{\partial}{\partial n}(T_\ell - T_v) \tag{61}$$

in which ρv_ℓ is the mass flux at the interface in the direction n normal to the interface. The two temperatures in this equation are the liquid and the vapor temperature at the interface, and the thermal conductivity k is assumed to have the same value for the liquid and the vapor.

The equation is nondimensionalized, introducing

$$v_i' = \frac{v_i}{V_1} \qquad n' = \frac{n}{L} \qquad \theta_\ell' = \frac{T_\ell - T_0}{T_w - T_0} \qquad \theta_v' = \frac{T_v - T_0}{T_w - T_0}$$

It assumes the form

$$\rho h_{\ell v} v_i' = k\frac{T_w - T_0}{L}\frac{\partial}{\partial n'}(\theta_\ell' - \theta_v') \qquad v_i' = \frac{k(T_w - T_0)}{\rho h_{\ell v} V_\ell L}\frac{\partial}{\partial n'}(\theta_\ell' - \theta_v') \tag{62}$$

which produces a dimensionless parameter

$$\frac{k(T_w - T_0)}{\rho h_{\ell v} V_\ell L} = \frac{c_p(T_w - T_0)}{h_{\ell v}}\frac{1}{\text{Pe}}$$

The new parameter

$$\frac{c_p(T_w - T_0)}{h_{\ell v}} = \text{Ja} \tag{63}$$

called the Jacob number, must be added to Eq. (42), resulting in the expression

$$c_f = f\left(\text{Re}_\ell, \frac{\rho_\ell V_\ell}{\rho_v V_v}, \text{We}, \frac{\rho_\ell}{\rho_v}, \text{Ja}, \text{Pr}\right) \tag{64}$$

The Prandtl number Pr is added because flow and heat transfer are now interdependent. The functional relation for heat transfer is also

$$\text{Nu} = f\left(\text{Re}_\ell, \frac{\rho_\ell V_\ell}{\rho_v V_v}, \text{We}, \frac{\rho_\ell}{\rho_v}, \text{Ja}, \text{Pr}\right) \tag{65}$$

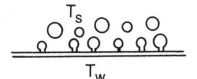

T_s

T_w

Figure 2.17   Pool boiling on a horizontal surface

Equations (64) and (65) are inconvenient because of the large number of dimensionless parameters. However, in special processes this number usually is reduced. This will be demonstrated for a specific two-phase flow situation.

2.6.6 Gravity Boiling on a Horizontal Surface

A liquid at saturation temperature T_s is brought to boiling by a heat flux q_w from the surface of the horizontal plate sketched in Fig. 2.17. Bubbles form intermittently on the surface and are moved by buoyancy through the bulk of the liquid. This bubble formation can be observed to be a very local process, so that the length dimension L of the surface does not enter as a boundary condition influencing the process as long as L is not very small. Therefore, only two parameters appear in this process, namely, the heat flux per unit area q_w at the surface and the difference between the temperature of the wall surface T_w and the saturation temperature T_s in the bulk of the fluid. The number of parameters in Eq. (65) can be reduced because no velocities are prescribed to the process, resulting in the relation

$$\text{Nu} = f\left(\text{Re}_{\text{n},\ell}^2, \text{We}, \text{Ja}, \text{Pr}, \frac{\rho_\ell}{\rho_v}\right) \tag{66}$$

A dimensionless expression for L has to be formed, since no length dimension has influence on the process. This is done in the following equation:

$$L = \frac{\mu c_p (T_w - T_s)}{q_w} \tag{67}$$

The liquid velocity in the Reynolds number of Eq. (66) as well as the dimensionless parameters in this equation assume the form

$$V_n^2 = \frac{g(\rho_\ell - \rho_v)L}{\rho_\ell} \tag{68}$$

$$\text{Nu} = \frac{q_w L}{k(T_w - T_s)} = \frac{\mu c_p}{k} = \text{Pr} \tag{69}$$

$$\text{Re}_{\text{n}'\ell}^2 = \frac{g\rho_\ell(\rho_\ell - \rho_v)L^3}{\mu^2} \tag{70}$$

$$\text{We} = \frac{\rho_\ell V_n^2 L}{\sigma} \tag{71}$$

$$\text{Ja} = \frac{c_p(T_w - T_s)}{h_{\ell v}} \tag{72}$$

After eliminating L and V_n^2 and some transformations, one obtains

$$\frac{c_p(T_w - T_s)}{h_{\ell v}} = f\left(\frac{\sigma^3}{g(\rho_\ell - \rho_v)\rho_\ell^2 v^4}, \frac{qv}{\sigma h_{\ell v}}, \text{Pr}, \frac{\rho_\ell}{\rho_v}\right) \tag{73}$$

This is the functional relation between the dimensionless parameters with which all of the various phases of boiling can be expressed, namely, natural convection, nucleate, transition, and film boiling. Equation (73) holding for any of these processes can be simplified if it is established by experiment that one or the other of the dimensionless parameters has a negligible influence. In film boiling, for instance, the surface tension can be neglected and can be eliminated from Eq. (73) by dividing the first term by the second term within the parentheses because the interface is essentially planar. On the other hand, new parameters like geometry, surface roughness, or surface chemistry may influence the process.

Similarity analysis is especially useful for those areas in fluid mechanics where no solutions of the conservation equations (mass conservation, Navier-Stokes, energy) can yet be obtained. That is the case for almost all two-phase flow problems in engineering where boiling is of importance. Turbulent flows also cannot be analyzed starting with the basic Navier-Stokes equations, and analyses using the time-integrated Reynolds equations have to use turbulence models derived from experiments. In all these situations, similarity analysis provides the tool by which results obtained by experiments or computer modeling can be generalized. This will become clear in the chapter on viscous flow.

2.7 VISCOUS FLUIDS

A large amount of research has been devoted to flow processes of viscous fluids. Only the most basic accomplishments can be discussed in this section.

Water at 22°C has a kinematic viscosity of 1×10^{-6} m²/s. At a moderate velocity of 1 m/s and a characteristic length of 1 cm of an object exposed to the flow, the Reynolds number is of the order of 10^4. Air at 20°C and 1 bar has a kinematic viscosity of 15×10^{-6} m²/s. At a velocity of 15 m/s and the same characteristic length of 1 cm of an object exposed to the flow, the Reynolds number has the same value of 10^4. The Reynolds number expresses the ratio of inertia to viscous forces. Viscous forces are therefore small in the flow of normal fluids in engineering devices that have, in general, Reynolds numbers larger than those indicated above, and they can be neglected except in regions where very large velocity gradients

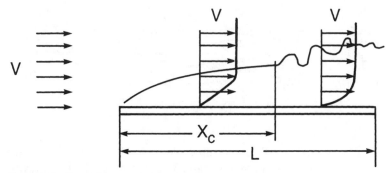

Figure 2.18 Transition to turbulence in a boundary layer

occur. This is the case in boundary layers adjacent to the surface of solid bodies exposed to flow and in free shear layers formed where fluid streams with different velocities meet, whereas in the bulk the fluid can be considered inviscid.

A flow over a flat plate exposed to a longitudinal fluid stream of constant velocity is sketched in Fig. 2.18. An inviscid fluid would not be disturbed at all by the thin plate. It has been found, however, that fluid particles adhere to the plate surface and therefore have a velocity of zero when the plate is at rest. This would create an infinitely large velocity gradient normal to the surface of the plate, which nature abhors. The fluid therefore changes this situation by creating a boundary layer in which the velocity increases continually from zero to the value V in the mainstream. The boundary layer grows in thickness as the fluid moves in the downstream direction. The thickness of the boundary layer in Fig. 2.18 is very exaggerated. It is, for a laminar boundary layer on a flat plate, given by the equation $\delta/x = 4/\sqrt{Re}$. At $Re = 10^4$, this results in $\delta/x = 0.04$. When the boundary layer has reached a certain thickness, it becomes unstable, wavelike motions and vortices arise, and the flow becomes three-dimensional, with irregular motion, which is called "turbulence." The turbulence creates a transverse transfer of momentum and makes the velocity profile fuller, with a steep increase near the wall and a gradual transition to the mean stream velocity in the outer portion of the boundary layer. The velocity profile in the laminar part of the boundary layer has the shape indicated in the figure, with a velocity increase that is close to linear over a considerable part of the boundary layer and a more rapid transition into the constant velocity in the mainstream. The critical length x_c, which denotes the extent of a laminar boundary layer, is characterized by a critical value of the Reynolds number Re_{x_c} formed with the mainstream velocity V and this length x_c. The value of Re_{x_c} varies, depending on disturbances fed into the boundary layer from the mainstream. For average conditions, it has a value

$$Re_{x_c} = \frac{V_{x_c}}{\nu} = 5 \times 10^5 \qquad (74)$$

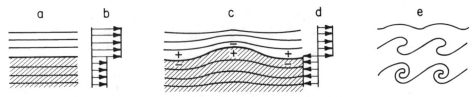

Figure 2.19   Instability of two inviscid fluid streams moving side by side. (From [10])

The reason why a boundary layer becomes unstable is closely connected with the instability of the flow of two inviscid fluid streams with different velocities moving side by side. The instability of such a flow was explained by Ludwig Prandtl [10] with the help of the sketch in Fig. 2.19. Figure 2.19a indicates the two fluid streams and Fig. 2.19b the velocity profile with a sudden change of the velocities in the plane separating the two streams. For an observer moving with the velocity of this separation plane, which will have a velocity approximately half way between the velocities of the two streams, the velocity profile is sketched in Fig. 2.19d. Some outside disturbance may cause the separation surface to assume a slightly wavy shape as indicated in Fig. 2.19c. The streamlines in the neighborhood of the separation surface will assume the shape sketched in this figure and will cause the pressure, according to Bernoulli's equation, to be higher where the distance between streamlines has increased and lower where the distance has decreased. This creates transverse pressure gradients, which will cause the amplitude of the waves to increase in time. The top of the waves moves into a region where the streamwise velocity is positive, and the valley into a region of negative velocity, thus transforming the waves as indicated in Fig. 2.19e and creating a rollup and formation of vortices. In an inviscid fluid this instability occurs at any velocity, whereas in a shear layer or boundary layer the flow becomes unstable only when the Reynolds number exceeds a critical value.

Another typical behavior of a boundary layer is flow separation, which occurs when the mainstream decelerates in flow direction or, equivalently, when the pressure in the mainstream increases in flow direction according to Bernoulli's equation. Such a flow occurs, for instance, in a diffuser. The explanation of the separation was given by Ludwig Prandtl [11] in the famous paper, presented in 1904, in which he also developed the concept of the boundary layer. Figure 2.20, which explains the flow separation, is taken from that paper. The velocity profile at the far left has the shape of a laminar boundary layer profile, as is also sketched in Fig. 2.18. When the fluid moves into a region of increasing pressure, kinetic energy is converted into pressure energy. The fluid in the mainstream has sufficient kinetic energy to penetrate into this region. However, the kinetic energy of the fluid in the boundary layer is weaker and is used up faster, especially in the regions close to the wall. This causes a deformation of the velocity profiles in the boundary layer as shown in Fig. 2.20. At some location along the surface, the velocity close to the

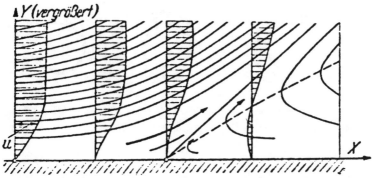

Figure 2.20 Separation of a laminar boundary layer moving into a region of increasing pressure. (From [7])

wall becomes zero, and further downstream becomes negative. The fluid near the wall thus starts moving in an upstream direction, pushing the boundary layer and the main flow away from the wall. This is the process of separation. One has to expect that a turbulent boundary layer with a fuller velocity profile will be able to delay the separation. This is borne out by experiments.

The reader may have wondered why the flow in the boundary layer was described as being independent of the length L of the plate. The reason why a Reynolds number based on this length has not appeared in our discussion is explained by the fact that the boundary layer is very thin; therefore, no information in the flow is transmitted upstream. Such a flow is referred to as being of parabolic form. The situation is different downstream of the point where separation has occurred because in this region the flow moves partially in a direction opposite to the approaching main flow and in this way transmits information upstream. Flow in this region is said to be of an elliptic nature. This flow depends on the Reynolds number VL/ν in addition to x/L or Vx/ν.

Separation of the flow occurs on all blunt bodies at sufficiently high Reynolds numbers. A cylinder in steady cross flow may be discussed as an example. The character of such a flow at different Re is sketched in Fig. 2.21. At a very low Re, smaller than 10, streamlines around the cylinder have the shape shown in the left-hand figure. The effect of viscosity extends over a fairly wide distance from the cylinder, the flow does not have the character of a boundary layer, and no separation occurs. The thickness of the viscous region decreases as Re increases, and in the range of Re between 10 and 60 the flow starts separating on the downstream side of the cylinder, creating a separation bubble in which the fluid rotates as indicated in the figure but does not participate in the downstream movement. The flow is still completely steady, and two counterrotating vortices are formed in the separation bubble, which becomes larger as Re increases. At Re > 60, the appearance of the flow changes. The vortices caused by the flow separation start moving

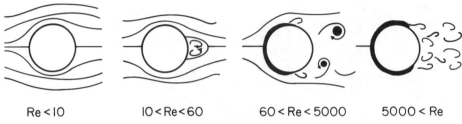

Re < 10   10 < Re < 60   60 < Re < 5000   5000 < Re

Figure 2.21   Flow around a cylinder at various Reynolds numbers

away from the cylinder in a downstream direction. This occurs in the way that vortices separate alternately from the upper and from the lower side of the cylinder. In moving downstream, they form what is called a "vortex street." The flow has now become unsteady. The formation of the vortex street has been explained by von Kármán, who demonstrated that, in an inviscid fluid, only a specific arrangement of the vortices is stable. This formation is shown in Fig. 2.22 [12]. It agrees well with the vortex street in the flow of a real fluid downstream of a cylinder, with the difference that in a real fluid, viscous forces dominate in the core of the vortices

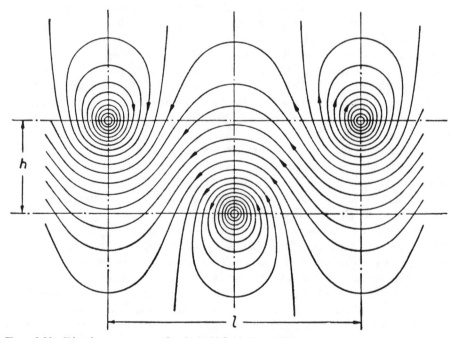

Figure 2.22   Kármán vortex street of an inviscid fluid. (From [12])

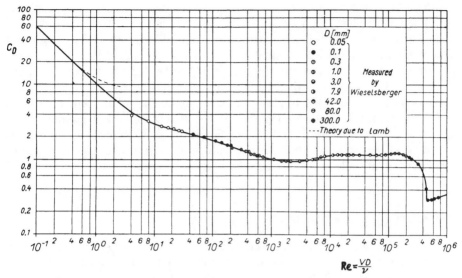

Figure 2.23 Drag coefficient of a circular cylinder in cross flow. (From [7])

and the core increases as the vortices move downstream. At some point the vortices become unstable, break up into smaller vortices, and finally transform to the irregular pattern of turbulence. This process is accelerated with increasing Re, as indicated in the right-hand side of Fig. 2.21.

Figure 2.12 presented two photos of the flow around a cylinder for the condition that the fluid was originally at rest and then was started suddenly. The upstream velocity thus changed from zero stepwise to a constant value V. The two photos were taken at different times after the start of the flow. The development in time of the boundary layer on the cylinder surface and the separation of the flow occur qualitatively in the same sequence, as shown in Fig. 2.21, when the Reynolds number is reinterpreted as Re = $V^2 t / v$, with t denoting time.

The character of the flow around a cylinder is reflected in the drag coefficient c_D. Figure 2.23 presents the coefficient defined by the equation

$$c_D = \frac{D}{dL\rho V^2/2} \tag{75}$$

as a function of Re according to measurements by Wieselsberger. The symbol D in the above equation denotes the drag that a flow normal to the cylinder with a velocity V exerts on the cylinder. The drag in Eq. (75) is divided by the cross-sectional area dL of the cylinder and the kinetic energy in the upstream $p V^2/2$. For Re up to 6×10^{-1}, no separation occurs on the downstream side of the cylinder, and the drag coefficient is described by an analysis by Lamb. At Re > 10^3 the drag coefficient is independent of Re, indicating that the vortex street is fully developed and the drag increases with the square of the upstream velocity V. Remarkable is the fact that the drag coefficient then drops quite suddenly to a value lower by a factor of 4 at a critical Re around 4×10^5. This is due to an interaction between the transition and the separation processes. Up to this critical Re, the boundary layer around the surface of the cylinder is laminar until it separates at an angle of 84°. For values of Re $> 4 \times 10^5$, however, the boundary layer has become turbulent before it reaches the point of separation. The separation is delayed and, consequently, the drag is reduced. This explanation is due to L. Prandtl, who also demonstrated experimentally that the critical Re of 4×10^5 is reduced when the boundary layer is artificially made turbulent by a trip wire.

2.8 NATURAL OR FREE CONVECTION

Natural or free convection flow is generated by buoyancy forces. Additional terms describing these forces have to be added to the Navier-Stokes equations (16)–(18). Similarity analysis, as sketched in section 2.6.4, then reveals that the dimensionless parameters Re_n or Gr and Pr describe such flows. They will be considered for the condition that the density differences creating buoyancy forces are caused by temperature differences and that a fluid with Pr = 0.72 (air) is involved. The temperature and velocity fields in natural convection flow are interrelated. Detailed quantitive information on the temperature field can be obtained with a Mach-Zehnder interferometer, and such interferograms also provide qualitative information about the flow field.

Figure 2.24 provides a sketch of such an interferometer. It consists of two mirrors and two semitransparent plates in a rectangular arrangement. Monochromatic light is emitted by a lamp, made parallel by a lens, and is split into two beams by the first semitransparent plate. The two beams are reflected on the mirrors and united again into a single beam by the second semitransparent plate. The adjustment of the interferometer will be discussed for the condition that all four plates are exactly parallel. In this case the two light beams arrive at the second semitransparent plate in phase and create on the screen a field of uniform brightness.

Now a heated model may be placed into one of the light beams as shown in Fig. 2.24. This increases the temperature of the surrounding air, creating a density field in its neighborhood. The optical refraction index in air is proportional to the air density, and this causes the phase of the light waves to vary locally, creating a number of interference fringes on the screen. With the described adjustment of the instrument, the fringes are lines of constant refraction index and therefore of constant density and constant temperature (pressure differences in natural convection flows are usually negligible).

Figure 2.25 is a photograph of the interferogram that is created on the screen of the interferometer when a vertical plate heated to a uniform temperature is

MACH– ZEHNDER   INTERFEROMETER

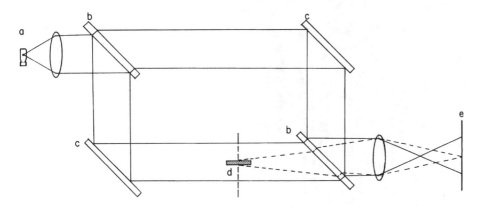

a. Monochromatic Light Source
b. Semitransparent Plates
c. Mirrors
d. Model
e. Screen

Figure 2.24   Mach-Zehdner interferometer

placed into its light beam [13]. The plate is represented by the black shadow. The black interference fringes surrounding the plate describe the temperature field in the air surrounding the plate and thus indicate the extent of the thermal boundary layer. The conditions were thus that $Re_n = 200$ at a distance of 0.01 m from the lower plate end ($Gr_x = 4 \times 10^4$) and $Pr = 0.72$. In air, the velocity boundary layer has almost the same thickness as the thermal boundary layer. The velocity profile can, of course, not be obtained from the interferogram. It has a shape different from that in forced flow because it has a value of zero at the surface as well as outside the boundary layer. It therefore goes through a maximum within the boundary layer.

Natural convection boundary layer flow also becomes unstable when Re_n reaches a critical value. It then forms waves that are quite regular at the beginning but become irregular as the flow moves up along the plate, and finally transforms into fine-grained, three-dimensional turbulent flow. Figure 2.26 presents three phases of this transition process [14]. The left-hand panel (at $Re_n \sim 10^9$) shows the regular sinuous wave, and the middle panel presents a later phase when the waves have already become irregular. In these two panels the flow is still two-dimensional. This, however, is probably not the case in the right-hand panel, where the flow appears to have become three-dimensional as indicated by the washed-out regions in the interferogram. The transition occurs still further downstream. The critical

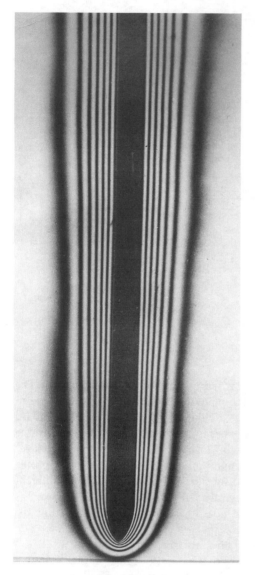

Figure 2.25 Natural convection on a heated vertical plate at $Pr = 0.72$, $Gr_{lcm} = 4 \times 10^4$, and $Re_n = 200$. (From [13])

Reynolds number for transition was found experimentally to have the value $Re_n \sim 3 \times 10^4$ $(Gr \sim 10^9)$.

Figure 2.27 presents a temperature field in forced flow normal to a cylinder with circular cross section at $Re = 218$ [15]. The flow approaches from the left, and the development of the boundary layer and the formation of the separation bubble can readily be recognized. The interference fringes again present lines of

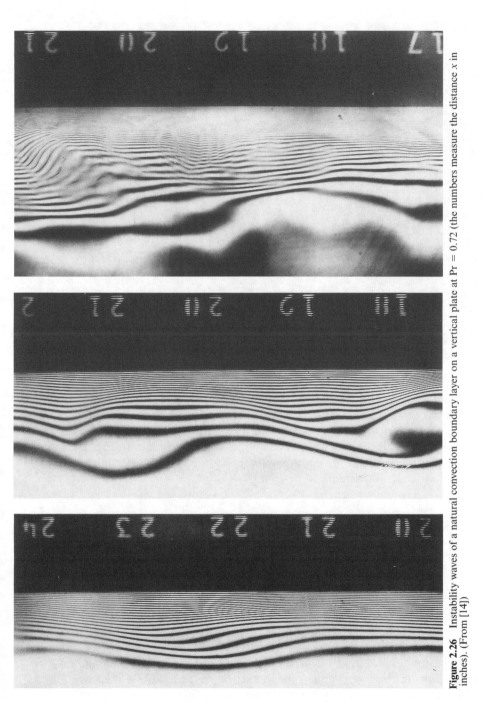

Figure 2.26 Instability waves of a natural convection boundary layer on a vertical plate at Pr = 0.72 (the numbers measure the distance x in inches). (From [14])

101

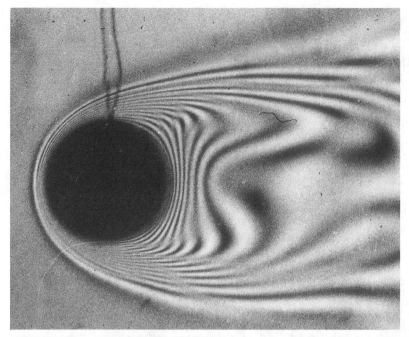

Figure 2.27 Temperature field in flow around a cylinder at Pr = 0.72 and Re = 218. (From [15])

constant temperature created by heating of the cylinder. This has caused some influence of buoyancy, and as a consequence, the fringes in the separation bubble are not symmetrical around a horizontal plane through the axis of the cylinder. Such a flow is called combined forced and natural convection. In Fig. 2.27 it was found that the influence of buoyancy forces on the heat transfer at the cylinder surface was still quite small.

2.9 INVISCID COMPRESSIBLE FLUIDS

Only ideal gases are considered in this section. Pressure variations may cause considerable density variations in a compressible medium. Such a medium can give rise to pressure waves that travel with the velocity of sound:

$$a = \sqrt{\frac{dp}{d\rho}} \tag{76}$$

In an incompressible fluid, $a = \infty$ and a pressure wave travels with infinite velocity. The change in thermodynamic state in such a pressure wave is isentropic. Since $p/\rho^\gamma = $ const for an ideal gas, Eq. (76) takes on the form

$$a = \sqrt{\gamma \frac{p}{\rho}} \qquad (77)$$

Similarity analysis shows that inviscid compressible flow depends on geometry and on the Mach number, $\text{Ma} = V_0/a$, which is the ratio of a characteristic flow velocity V_0 to the sound velocity a. One distinguishes between subsonic and supersonic flow, depending on whether Ma is smaller or larger than 1.

The velocity ratio V_0/a is named after Ernst Mach [16], who demonstrated, by a Schlieren photo of the flow around a bullet in flight, the special character of supersonic flow creating shock waves.

The mass conservation equation for steady flow along a stream tube (Fig. 2.3) is now

$$\frac{\partial}{\partial s}(\rho v A) = 0 \qquad (78)$$

The momentum equation (8) is valid for compressible fluids when the density ρ is considered variable:

$$\frac{\partial}{\partial s}\left(\frac{v^2}{2} + u + \frac{p}{\rho}\right) = 0 \qquad (79)$$

which now includes the internal energy u in addition to the mechanical energy. For an ideal gas with constant specific heat, this becomes

$$i = u + \frac{p}{\rho} = c_p T \qquad (80)$$

$$\frac{V^2}{2} + c_p T = \text{const} = c_p T_t \qquad (81)$$

where T_t is total temperature. Equation (81) indicates that the total temperature T_t in steady, compressible flow is constant. In unsteady, compressible flow the total temperature is expressed by

$$\rho c_p \frac{DT_t}{Dt} = \frac{\partial p}{\partial t} \qquad (82)$$

which is the analog to Eq. (12) in incompressible flow.

The temperature T is again the quantity that, in principle, is measured by an instrument moving along with the fluid. It is shown later that this temperature cannot be measured except by a specially calibrated instrument. One can, however, measure the total temperature T_t that, according to Eq. (81), is obtained when the velocity in the fluid is decreased to zero. An instrument with which this tempera-

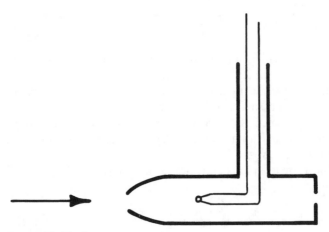

Figure 2.28   Total temperature sensor

ture is measured to a satisfactory approximation is shown in Fig. 2.28. The flow approaching the instrument is slowed at the entrance of the horizontal tube, and the temperature there is measured by a sensor, for instance, a thermocouple. An erroneous measurement would be obtained if the tube were closed at its down-stream end because the fluid at rest in the tube would lose heat to the tube walls. Therefore a small amount of fluid is vented through an orifice at the downstream end of the tube. This flow is adjusted, so that the overall error, which results both from heat loss and because the kinetic energy $v^2/2$ does not completely vanish, is minimized.

The manner in which the flow velocity is determined from pressure measurements depends on whether the flow is subsonic or supersonic. For subsonic flow, the deceleration of the flow ahead of a total pressure p_t tube is isentropic; integration of the momentum equation (4) and use of the relation

$$\frac{p}{\rho^{\gamma}} = \text{const} \tag{83}$$

result, for steady flow and subsonic velocities, in the equation

$$v = \sqrt{\frac{2\gamma}{\gamma - 1} RT_t \left[1 - \left(\frac{p}{p_t} \right)^{(\gamma-1)/\gamma} \right]} \tag{84}$$

For supersonic flow, the deceleration ahead of the tube occurs partially in the form of a shock, as indicated in Fig. 2.29. The stream tube that ends at the mouth of the measuring tube passes through a normal shock; its conservation equations, in terms of the control volume shown in Fig. 2.30, are as follows.

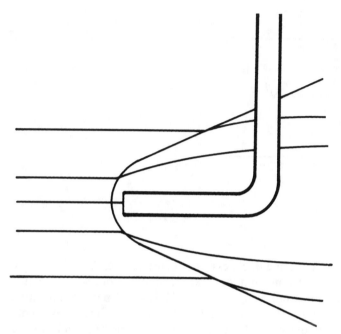

Figure 2.29 Pitot tube in supersonic flow

Mass:

$$\rho v = \rho' v'$$

Momentum:

$$\rho v(v - v') = p' - p$$

Energy:

$$\frac{v^2}{2} + c_p T = \frac{(v')^2}{2} + c_p T' \qquad T_t = T_t' \tag{85}$$

These equations, together with the state equation of an ideal gas,

$$\frac{p}{\rho} = RT = R\left(T_t - \frac{v_0^2}{2c_p}\right) \tag{86}$$

are sufficient to allow calculation of the velocity, pressure, and temperature down-stream of the shock, provided the values upstream of the shock are known.

The equations result in an entropy increase when the velocity decreases, and the pressure increases in passage through the shock. The deceleration in a shock is irreversible because of the effects of heat conduction and dissipation within the

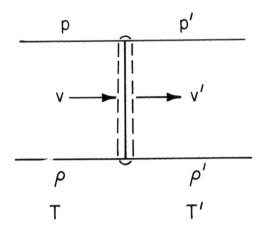

Figure 2.30 Plane shock wave

shock, which has a small but finite thickness. As a consequence, the pressure indicated by the Pitot tube is different from the total pressure that one would obtain if the deceleration ahead of the tube were isentropic. This is shown in a temperature-entropy diagram in Fig. 2.31. The state ahead of the tube is indicated by point 1 in the figure. Deceleration in the shock changes the state to point 2, and subsonic deceleration to zero velocity downstream of the shock causes the state to change to point 3. Point 4 would be reached if the deceleration were isentropic. The figure indicates that the Pitot pressure p_p measured by the Pitot tube is smaller than the total pressure p_t, whereas the total temperature T_t has the same value at points 3 and 4.

Equation (86) describes the connection among pressure, density, and temperature for a gas at thermodynamic equilibrium, i.e., at rest. Its use for a gas in motion implies that this equilibrium persists locally, an assumption called "local thermodynamic equilibrium." This is a valid assumption in most cases, but exceptions do exist (for instance, in high-temperature plasma flow).

2.10 VISCOUS AND AERODYNAMIC HEATING

Viscous and aerodynamic heating have not been considered up to now. These cause considerably large temperature differences in gas streams with velocities of the order of the sound velocity.

Figure 2.32 presents the results of measurements on an unheated cylinder oriented longitudinally in a high-speed airflow [17]. A boundary layer forms on the surface of the cylinder and, in the Re range plotted on the abscissa of the figure, the boundary layer thickness is small in comparison to the cylinder radius. The length dimension in the Reynolds number Re_x is the distance x from the rounded

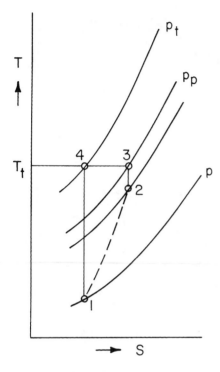

Figure 2.31 Relation between total, static, and pitot pressure

front end of the cylinder. The parameter r on the ordinate is a dimensionless expression for the difference between the temperature T_r, called recovery temperature, of the cylinder surface and the temperature T_0 in the free stream with the velocity V_0. It is called recovery factor and is defined by

$$r = \frac{T_r - T_0}{V_0^2/2c_p} \tag{87}$$

The figure shows that the temperature difference $T_r - T_0$ is constant along the cylinder surface in the laminar flow region. It starts increasing when the boundary layer becomes turbulent at Re $= 5 \times 10^5$ and approaches asymptotically a value of 0.90 at large Re.

Viscous heating, the energy dissipation by viscous forces, adds a number of terms on the right-hand side of Eq. (43), which have the form

$$\mu \frac{\partial^2}{\partial y^2}\left(\frac{u^2}{2}\right)$$

Similarity analysis using the terms describing viscous dissipation and those describing heat conduction results in a new dimensionless parameter,

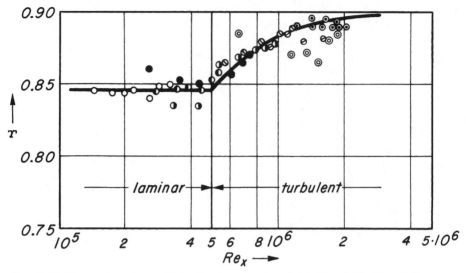

Figure 2.32 Temperature recovery factor for a boundary layer in air (Pr = 0.72). (From [17])

$$Na = \frac{\mu V^2}{k \Delta T} \tag{88}$$

which for highly viscous fluids, has been given the name Nahme number. The recovery factor r in Eq. (87) can also be expressed by the Nahme number and the Prandtl number

$$r = \frac{Pr}{Na} = \frac{1}{Ec} \tag{89}$$

The ratio of Nahme to Prandtl number is referred to in the literature as Eckert number. An analysis of longitudinal flow over a flat plate [18] has established that the recovery factor in Eq. (87) depends on the Prandtl number only for a laminar and a turbulent boundary layer. This is in agreement with the experimental results shown in Fig. 2.32. The results of the analysis for a laminar boundary layer on a flat plate in longitudinal high-speed flow are well approximated by

$$r = \sqrt{Pr} \qquad 0.5 < Pr < 10 \tag{90}$$

For a larger range of Pr up to 1000, the recovery factor is presented in Fig. 2.33 [19]. For turbulent Pr the relation

$$r = \sqrt[3]{Pr} \tag{91}$$

was found to represent the experimental values measured in gases.

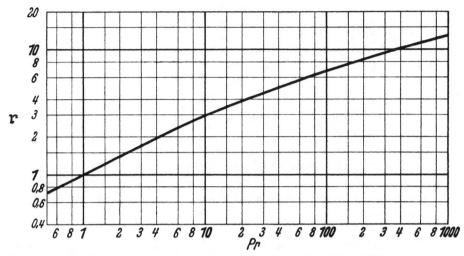

Figure 2.33   Laminar temperature recovery factor for a boundary layer. (From [19])

The recovery temperature can be measured with a sensor similar to the static pressure tube in Fig. 2.4 [20], the difference being that the tube is closed and its wall temperature is measured by a thermocouple. This has the advantage that the probe can be minimized, which is especially important if one plans measurements in thin films. Equation (90) or Fig. 2.33 can also be used to determine the temperature T_p, provided the velocity V_0 is known and one has established that the boundary layer is laminar. Reynolds numbers in oil films of bearings are of the order of 1000, and the flow over the probe will still have the character of a boundary layer.

The situation is different if one wants to make measurements in polymer processing. The Prandtl number then is extremely high, of the order of 10^5–10^9, and the velocity is quite small, which results in Reynolds numbers of the order of 10^{-3}–10^{-1}. This kind of flow does not produce boundary layers but is of a character called "creeping flow." The flow field extends far upstream and to the side of a probe similar to the Pitot tube. An analysis has established that in such a flow the recovery temperature of an unheated cylinder in longitudinal flow is not constant along the cylinder surface but increases in flow direction, provided internal heat conduction in the probe is small. A sensor will have to be designed in such a way that the temperature of the probe tip is measured and that heat conduction in the probe away from the probe tip is minimized. Under this condition, the results of an analysis [21] can be approximated by

$$T_{sp} - T_0 = C \, \frac{\mathrm{Pr}^{0.14}}{\mathrm{Re}^{0.57}} \, \frac{V_0^2}{c_p} \tag{92}$$

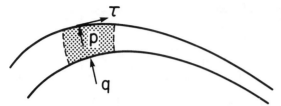

Figure 2.34   Forces and heat flux at the boundary of a fluid particle moving along a path tube

at Pr between 10^6 and 10^8 and Re between 10^{-4} and 10^{-1}. This equation may be used to calculate the recovery temperature T_{sp} with an accuracy that should be sufficient for an exploration as to whether viscous heating has to be considered in such measurements.

In gases, velocities have to be of the order of the sound velocity to generate considerable temperature differences by viscous dissipation. Such velocities are, in general, connected with large pressure differences in the flow field, and temperature differences of an order of magnitude comparable to those by dissipation are caused by expansion or compression of the gas. This effect requires additional terms in Eqs. (16)–(18). The interaction of these effects causes aerodynamic heating. Compression and expansion does not occur in longitudinal flow over a surface because the pressure in the flow field is constant (Figs. 2.32 and 2.33). It has been pointed out in section 2.9 that, in steady flow of an inviscid fluid, the total temperature T_t, which expresses the sum of kinetic and internal energy, stays constant along a streamline. With a uniform approach the total temperature then is constant in the whole flow field. This holds also for regions outside of boundary layers or free sheer layers in a real compressible fluid. Within the boundary layers the situation is different because energy is also transferred by conduction and by sheer forces. This will be explained using Fig. 2.34, which sketches a stream tube in steady flow. In an inviscid fluid, conduction q and energy transfer by sheer stresses τ are absent. Pressure forces acting on the walls of the stream tube do not perform work because the tube walls do not move normal to their surfaces in steady flow. Therefore the total energy, composed of kinetic and internal energy, stays constant along the flow and, with constant specific heat c_p, the total temperature is constant also. The situation is different in a boundary layer. How the temperature in a stream tube within the boundary layer varies depends on whether conduction or viscous energy transport is larger. Figure 2.35 (right) presents a total temperature profile in a boundary layer along an adiabatic plate. Analysis shows that, for a gas with a Pr = 1, the two energy transports cancel each other everywhere within the boundary layer. Consequently, the total temperature is constant throughout the boundary layer and equal to the total temperature in the inviscid free stream. Most gases have Pr < 1, and for those the total temperature close to the wall is smaller than the total temperature outside the boundary layer and is larger in the outer region.

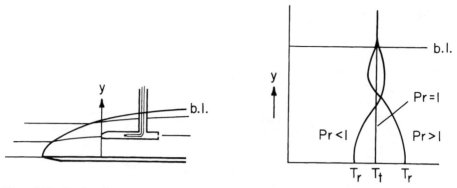

Figure 2.35 Total temperature profiles in a laminar boundary layer and their measurement by a total temperature sensor

For a fluid with Pr > 1, the total temperature in the boundary layer has the opposite character. At the wall surface, the velocity is zero, and there is no difference between the total and the static temperature of the fluid. The temperature of the adiabatic wall surface is equal to the total temperature of the fluid adjacent to the wall, and is therefore the recovery temperature. Figure 2.35 (left) indicates how the local total temperature in the boundary layer could be measured by a total temperature probe as sketched in Fig. 2.28.

The energy recovery process in gases (a compressible medium) is somewhat different from that in an incompressible fluid because of the interaction of the temperature variation by compression or expansion with frictional heating, and the resulting effect is called aerodynamic heating, as distinguished from viscous heating in an incompressible medium. It is advantageous to describe the recovery temperature in a gas by a dimensionless parameter called the energy separation factor e, defined as

$$e = \frac{T_{t_0} - T_r}{V_0^2/2c_p} \tag{93}$$

because the total temperature T_{t_0} is constant in the flow field outside the viscous (boundary or shear) layer. The parameter in Eq. (93) is connected with the recovery factor in Eq. (87) by the relation

$$e = 1 - r \tag{94}$$

The discussion in this section assumed the flow to be steady. The instruments described can also be used in quasi-steady flow. In unsteady flow, energy can be transferred from path line to path line by pressure forces because the walls of the path tubes now move normal to their surface. This is important to consider for an understanding of energy transfer in turbomachinery [22].

NOMENCLATURE

A	surface
c_D	drag coefficient
c_f	friction factor
c_p	specific heat
d	diameter
D	drag, diffusion coefficient
e	energy separation factor
Ec	Eckert number
F	force
g	gravitational acceleration
Gr	Grashof number, $= [g\rho(\rho - \rho_0)L^3/\mu^2$
h_{lv}	heat of evaporation
Ja	Jacob number, $= [c_p(T_w - T_0)]/h_{\ell v}$
k	thermal conductivity
L	reference length
m	mass
Na	Nahme number, $= \mu V^2/k\Delta T$
p	pressure
p_t	total pressure
Pe	Péclet number, $= VL/\alpha$
Pr	Prandtl number, $= \nu/\alpha$
q	heat flux
r	recovery factor
r	radius of curvature
R	resistance
Re	Reynolds number, $= \rho VL/\mu = VL/\nu$
Re_{x_c}	critical Reynolds number
Sc	Schmidt number, $= \nu/D$
t	time
t'	dimensionless time
T	temperature
T_r	recovery temperature
T_t	total temperature
u, v, w	velocity components
V	reference velocity
We	Weber number, $= \rho VL^2/\sigma$
x_c	critical length of a laminar boundary layer
x, y, z	coordinates
α	thermal diffusivity
γ	exponent
Δ_p	pressure difference

θ	temperature difference, $= T - T_0$
μ	viscosity
ν	kinematic viscosity, $= \mu/\rho$
ρ	density
σ	surface tension
τ	shear stresses

Subscripts

g	gas
i	interface
ℓ	liquid
n	natural convection
s	saturation
w	wall
0	reference

REFERENCES

1. M. Barker, On the Use of Very Small Pitot Tubes for Measuring Wind Velocity, *Proc. R. Soc. London Ser. A,* vol. 101, p. 435, 1922.
2. M. Navier, Memoire sur les Lois du Mouvement des Fluides, *Mem. Acad. Sci.,* vol. 6, pp. 389–416, 1987.
3. A. G. Stokes, On the Theories of Internal Friction of Fluids in Motion, *Trans. Cambridge Philos. Soc.,* vol. 8, pp. 287–305, 1845.
4. O. Reynolds, On the Experimental Investigation of the Circumstances Which Determine Whether the Motion of Water Shall be Direct or Sinuous, and the Law of Resistance in Parallel Channels, *Philos. Trans. R. Soc.,* vol. 174, pp. 435–982, 1883.
5. M. Kurosaka and P. Sundaram, Illustrative Examples of Streaklines in Unsteady Vortices: Interpretational Difficulties Revisited, *Phys. Fluids,* vol. 29, pp. 3474–3477, 1986.
6. J. M. Cimbala, H. M. Nagib, and A. Roshko, Large Structure in the Far Wakes of Two-Dimensional Bluff Bodies, *J. Fluid Mech.,* vol. 190, pp. 265–298, 1988.
7. H. Schlichting, *Boundary Layer Theory,* p. 222, McGraw-Hill, New York, 1960.
8. J. Nikuradse, Kinematographische Aufnahme einer Turbulenten Strömung, *Z. Angew. Math. Mech.,* vol. 9, p. 495, 1929.
9. M. van Dyke, *An Album of Fluid Motion,* Parabolic Press, Stanford, Calif., 19
10. L. Prandtl, *Abriss der Stroemungslehre,* p. 43, Friedr. Vieweg und Sohn, Braunschweig, Germany, 1931.
11. L. Prandtl, Über Flüssigkeitsbewegung bei sehr kleiner Reibung, Verhandlungen des III, *Int. Math. Kongr.,* pp. 484–491, Teubner, Leipzig, 1905.
12. T. v. Kármán, *Nachr. Wiss. Ges. Goettingen Math. Phys. Kl.,* p. 509, 1911, p. 547, 1912.
13. E. R. G. Eckert and E. Soehngen, Studies on Heat Transfer in Laminar Free Convection with Zehnder-Mach Interferometer, AF Tech. Rep. 5747.
14. E. R. G. Eckert and E. Soehngen, Interferometric Studies on the Stability and Transition to Turbulence of a Free-Convection Boundary Layer, AF Tech. Rep. 5747.
15. E. R. G. Eckert and E. Soehngen, Distribution of Heat-Transfer Coefficients Around Circular Cylinders in Crossflow at Reynolds Numbers from 20 to 500, *Trans. Am. Soc. Mech. Eng.,* vol. 74, p. 343, 1952.

16. E. Mach, *Photography of Projectile Phenomena in Air,* 1884.
17. E. R. G. Eckert and Weise, Messungen der Temperatur-verteilung auf der Oberfläche schnell auge-strömter unbeheizter Körper, *Jahrb. Dtsch. Luftfahrtforsch.,* vol. II, pp. 25–31, 1940.
18. E. Pohlhausen, Der Wärmeaustausch zwischen festen Körpern und Flüssigkeiten mit kleiner Reibung und kleiner Wärmeleitung, *Z. Angew. Math. Mech.,* vol. 1, p. 115, 1921.
19. E. R. G. Eckert and O. Drewitz, Der Wärmeübergang an eine mit grosser Geschwindigkeit längs angeströmte Platte, *Forsch. Ing. Wes.,* vol. 11, pp. 118–124, 1940.
20. E. R. G. Eckert, Temperaturemessung in schnell strömenden Gasen, *Z. Ver. Dtsch. Ing.,* vol. 84, pp. 813–817, 1940.
21. E. R. G. Eckert and J. N. Shadid, Viscous Heating of a Cylinder with Finite Length by a High Viscosity Fluid in Steady Longitudinal Flow, 1: Newtonian Fluids, *Int. J. Heat Mass Transfer,* vol. 32, pp. 321–334, 1989.
22. E. R. G. Eckert, Cross Transport in Fluid Streams, *Wärme Stoffübertragung,* vol. 21, pp. 73–81, 1987.

THERMAL ANEMOMETERS

Leroy M. Fingerson and Peter Freymuth

3.1 INTRODUCTION

Thermal anemometers measure fluid velocity by sensing the changes in heat transfer from a small, electrically heated sensor exposed to the fluid motion. Their generally small size and good frequency response makes them especially suitable for studying flow details, particularly in turbulent flow.

In many applications, fluid temperature, composition, and pressure are constant, so the only variable affecting heat transfer is fluid velocity. When other parameters vary, accurate velocity measurements with a thermal sensor become more difficult. At the same time, sensitivity of thermal sensors to other fluid parameters presents the possibility of measuring more than just velocity by using more than one sensor.

A simple thermal anemometer is shown in Fig. 3.1, along with some typical sensor and operating parameters. Assuming a linear relation between temperature and resistance (usually adequate in thermal anemometry), the resistance R of the sensor can be represented as

$$R = R_r[1 + \alpha(T_m - T_r)] \tag{1}$$

where R_r is resistance at reference temperature T_r, T_m is average sensor temperature, and α is temperature coefficient of resistance. The value of α is critical, since if the sensor did not vary in resistance with temperature, there would be no signal from a thermal anemometer. For convenience, the fluid temperature T_a is often

115

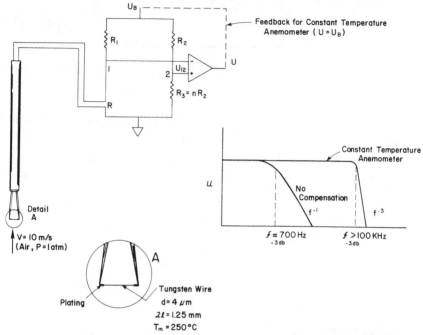

Figure 3.1 Basic elements of a hot-wire anemometer

used for the reference temperature T_r. The value of α depends on the reference temperature used.

In Fig. 3.1, if the resistance of resistor R_1 is large in comparison with that of the sensor (R), then the current I is nearly constant, and any increase in heat transfer rate from the sensor to its surroundings will cause the sensor to cool. Because of the temperature coefficient of resistance α, this cooling will cause a decrease in R, a decrease in U_{12}, and a decrease in amplifier output U. A decrease in heat transfer between sensor and fluid would have the opposite effect. When the changes in heat transfer are caused by changes in fluid velocity, the result is a thermal anemometer.

The system of Fig. 3.1 (without the feedback) is an uncompensated constant-current hot-wire anemometer. According to Freymuth [1, 2], systems of this nature were first considered during the last two decades of the nineteenth century. The earliest work frequently referred to in the current literature is that by King [3]. Subsequently, there have been over 2500 publications up to 1992 relating to thermal anemometry techniques.

Advances have taken place in sensors, electronic control circuits, and data analysis. Small-diameter wires are still frequently used as sensors and, in fact, are still optimal for much research work in fluid mechanics. Film sensors, introduced by Lowell and Patton [4] and Ling [5], have been a major addition to sensor technology, especially for liquids or gases with particles.

Since transistors became available, this constant-current operation of thermal sensors has been largely replaced by constant-temperature operation. Stability criteria and techniques for checking response are now well understood for constant-temperature systems. In addition, constant-temperature systems work equally well for both wire and film sensors, and frequency compensation is not as sensitive to mean velocity.

Digital techniques—either in specialized instruments or with general-purpose computers—have significantly expanded capabilities for analyzing data from thermal anemometers. The nonlinear output is no longer a limitation. Correlation, spectrum, and amplitude probability density distributions are all readily obtainable. Special techniques for conditional sampling have also been developed.

Many excellent researchers have made significant contributions to thermal anemometry and should be given credit in any complete review of the subject. However, in this chapter the focus is on the problems of and procedures for getting good measurements with current thermal anemometer and data analysis techniques. The reader is referred to Freymuth [1] and a review paper by Comte-Bellot [6] for a more thorough review of the literature.

In the next section, a brief comparison is made between the thermal anemometer and the laser velocimeter, since both instruments measure the time history of the flow at a point in space. Then a rather complex discussion is given of the characteristics of hot-wire sensors operated in a constant-temperature circuit. Finally, film sensors are compared with hot-wire sensors.

3.2 STRENGTHS, LIMITATIONS, AND COMPARISONS WITH LASER VELOCIMETERS

To measure velocity details in a flowing fluid, the ideal instrument should

1. Have high-frequency response to accurately follow transients
2. Be small in size for an essentially point measurement
3. Measure a wide velocity range
4. Measure only velocity, and work in a wide range of temperature, density, and composition
5. Measure velocity components and detect flow reversal
6. Have high accuracy
7. Have high resolution (low noise)
8. Create minimal flow disturbance

9. Be low in cost
10. Be easy to use

For many years, only the hot-wire anemometer satisfied enough criteria to be used extensively in turbulence studies. Pitot probes, flow visualization, and other techniques complemented hot-wire data but generally could not provide details as well as the hot wire.

For most applications, items 4, 5, 6, and 10 are perhaps the weakest areas of thermal anemometers. To be even more specific, the primary practical limitations are fragility and sensitivity to contamination. From a theoretical point of view, the limitations include the facts that (1) velocity is not measured directly but is deduced from a measurement of convection heat transfer from the sensor, (2) normal configurations limit the turbulence intensity that can be measured accurately, and (3) heat losses from the sensor other than by convection can cause errors. The importance of these limitations depends on the application. Relative to accuracy, thermal anemometer measurements are very repeatable when the conditions are reproduced exactly. It is the effects of contamination and of variables other than flow on the heat transfer that cause inaccuracies.

Another limitation of thermal anemometers that is common to all devices that aim for point measurements is their inability to fully map velocity and vorticity fields that depend strongly on space coordinates and simultaneously on time. It would require spatial arrays of thousands of probes that have to work without mutual interference. This limitation is strongly felt in emerging areas of research like unsteady separated flows and field mapping of turbulence (in contrast to averaged turbulent fields). Particle image velocimetry [7] has been effective for quantitative evaluation of two-dimensional spatial velocity fields with promise for three-dimensional measurements and even time history. Flow visualization reinforced by computer simulations is also promising, with point measuring devices providing a few quantitative reference points.

Since its introduction in 1964 [8], the use of the laser velocimeter to measure velocity details of flowing fluids has expanded rapidly. Since it, too, satisfies many of the "ideal instrument" criteria, a comparison between it and the thermal anemometer follows. It should be emphasized from the start that, rather than replacing thermal anemometers, the laser velocimeter complements their use. As is often the case in making measurements, it is not a question of the best instrument but rather which instrument will perform best for the specific application.

The laser velocimeter uses Doppler-shifted light scattered from particles in the flow to deduce the velocity of those particles. If the particles are small enough to follow the flow, then the flow velocity is measured. A window to the flow must be provided for both the incident light and the scattered light, but no probe needs to enter the flow field. Of course, there must be enough particles of the appropriate size and concentration that the desired statistical data can be determined.

In the following, the thermal anemometer and laser velocimeter are compared on the basis of 10 criteria for the ideal instrument.

1. *Frequency response.* In current practice, the thermal anemometer is definitely superior. Measurements to several hundred kilohertz (kHz) are quite easily obtained, with 1 MHz being feasible.

   Theoretically, a laser velocimeter could approach the response of a thermal anemometer. Practically, spectra up to only about 30 kHz have been measured. In many applications the problem is inadequate size and concentration of the scattering centers (particles). At sufficiently high frequencies, electronic limitations also start to enter.

2. *Size.* A hot-wire-type thermal sensor is typically 5 μm in diameter by about 2 mm long, although wires as small as 1 μm by 0.2 mm are feasible.

   For a laser velocimeter, measuring volumes of 50 μm by 0.25 mm are common, while a 5- by 5-μm measuring volume is achievable in very small test sections. If the distance from the focusing lens to the measuring point is long (e.g., over 400 mm), then small measuring volumes are difficult to achieve. They may also be impractical in some flows, owing to movement of the incident beams caused by refractive index variations along the path of the incident beams. *laze better for low velocity*

3. *Velocity range.* Both techniques have a very wide velocity range. The laser has the advantage at very low velocities because the "free convection" effects that affect hot-wire readings are usually not a problem.

   Although both instruments measure high-speed (compressible) flows, laser data are easier to interpret because they are sensitive only to particle velocity and no calibration is required. At the same time, it can be difficult to provide particles that follow the flow, scatter enough light, and have high enough concentration for spectral measurements.

   The thermal anemometer is sensitive to recovery temperature, Mach number (Ma), and Reynolds number in compressible flows. Measurements in transonic flows require considerable calibration, while for $Ma > 1.5$, Mach-number independence makes measurements more feasible.

4. *Measure only velocity over wide temperature, density, and composition ranges.* The laser velocimeter measures only the velocity of the scattering center (particle), and it measures it in a known direction (pure cosine response). The hot-wire sensor measures heat transfer to the environment. This can be a plus, since for example, by using two sensors, both temperature and velocity fluctuations can be measured. Generally, though, it is preferable to be sensitive to velocity only.

   Both instruments will operate over wide ranges of temperature, density, and composition. The maximum temperature for thermal anemometers is limited by the maximum operating temperature of the sensor. There is no similar limit for the laser velocimeter. At low density, measurements become more difficult for both. Conduction losses become excessive, and slip flow effects complicate hot-wire anemometry, while in laser velocimetry there are problems in finding particles that both follow the flow and scatter enough light.

The laser must "see" into the flow. In liquid metals, for example, thermal anemometers can be used, but generally not laser velocimeters.

5. *Component resolution.* The hot wire can be used to resolve one, two, or all three components of a flow field by using one, two, or three sensors, respectively. However, it appears limited to low turbulence intensities even with rather sophisticated data reduction procedures. Film sensors give the potential for measuring at any turbulence intensity [9], but the complexity of present techniques and probe size limit the applications.

The laser velocimeter can resolve components and, with frequency shifting, can detect flow reversals. While it is more difficult to obtain the third component, there are systems available.

6. *Accuracy.* Hot-wire results can be very repeatable, so accuracy is really a function of how closely the calibration conditions are reproduced in the flow to be measured. In practice, contamination, temperature changes, and other factors generally limit accuracy to a few percent.

The laser velocimeter can give very high accuracy (0.1%) in carefully controlled experiments. In many practical measurements, refractive index variations, limited accuracy on beam-crossing angle, and signal processor limitations make a value of 1% more realistic.

7. *Resolution.* The hot wire is clearly superior, since it can have a very low noise level. Resolution of 1 part in 10,000 is easily accomplished, while with a laser velocimeter, 1 part in 1000 is difficult with present technology.

8. *Flow disturbance.* Since only light needs to enter the flow, the laser is clearly better. The size and concentration of particles normally required do not measurably alter the flow field.

9. *Cost.* The hot wire is still lower in cost by a significant factor, but new technology could change that in the future.

10. *Ease of use.* At present, the laser velocimeter is probably more difficult to set up and start getting valid data with. Once it is set up, the laser velocimeter may be easier to use, since there are no fragile sensors to get dirty, break, or shift calibration. A difficulty in data interpretation with the laser velocimeter is the fact that discrete measurements are made (on each measurable particle), which gives a discontinuous output.

As a general rule, if other instruments (such as a Pitot tube or pressure transducer) cannot give the detailed measurements required, one should consider a hot-wire (or hot-film) anemometer. If high temperatures, moving objects in the flow, proximity to walls, dirt in the flow, high turbulence intensities, or some other prob-

Table 3.1 Some properties of common hot-wire materials

	Tungsten	Platinum	80% Platinum, 20% iridium
Temperature coefficient of resistance α, $°C^{-1}$	0.0045	0.0039	0.0008
Resistivity, $\Omega \cdot cm$	5.5×10^{-6}	10×10^{-6}	31×10^{-6}
Ultimate tensile strength, Kg/mm^2 (lb/in^2)	420 (60×10^4)	24.6 (3.5×10^4)	100 (14.22×10^4)
Thermal conductivity, cal/(cm $\cdot °C$)	0.47	0.1664	0.042

lem makes hot wires difficult or impossible to use, then a laser velocimeter should be considered.

In summary, thermal anemometers can theoretically be used in almost any fluid flow situation. However, sensor fragility, calibration shifts due to contamination, or difficulties in separating out variables make many potential applications difficult. The most common and easiest measurements with thermal anemometers are in constant-temperature gases near atmospheric pressure, at relatively low turbulence intensities, and at flow velocities low enough that the assumption of incompressibility is adequate. But when the need is sufficient, good measurements can be made over a much wider range of conditions.

3.3 HOT-WIRE SENSORS

Common hot-wire materials are tungsten, platinum, and platinum-iridium (80% Pt, 20% Ir). These materials are used partly because of their properties but also because of their availability in the small diameters of interest in hot-wire anemometry. Table 3.1 gives some of their properties. Ideally, values of α, resistivity, and tensile strength should be high, and conductivity low; low thermal conductivity reduces conduction losses to the supports.

Tungsten is desirable because of its high temperature coefficient of resistance and high strength. The major disadvantage is the rate at which it oxidizes, especially at temperatures above about 300°C [10]. Tungsten wires are available commercially in diameters as small as 2.5 μm. They are made by first drawing and then etching to the final diameter.

Platinum is available in very small sizes (0.5 μm), has a good temperature coefficient, and does not oxidize. It would be the ideal wire if it were not so weak, especially at high temperatures. At high wire temperatures and high air velocities, aerodynamic drag alone can cause the stress on a platinum wire to exceed its limit.

Platinum-iridium is a compromise wire that does not oxidize, has better strength than platinum, but has a low temperature coefficient of resistance. It finds application where the wire temperature will be too high for tungsten and where platinum is too weak.

These wires have all been used since 1950 [10] and probably long before that. Newer materials might make superior hot wires, but no extensive study has been reported recently. In any case, the three materials in Table 3.1 seem to satisfy most requirements for hot-wire sensors. Sandborn [11] provides considerable details on wire properties and mounting procedures.

In the selection of wire diameter and length, many conflicting criteria come into play. For example, as regards sensor length, a short sensor is desired to maximize spatial resolution and minimize aerodynamic stress. A long sensor is desired to minimize conduction losses to supports, provide a more uniform temperature distribution, and minimize support interference. With regard to sensor diameter, a small diameter is desired to eliminate output noise due to separated flow around the sensor (sensor-generated flow fluctuations), maximize the time response of the wire due to lower thermal inertia and higher heat transfer coefficient, maximize spatial resolution, and improve the signal-to-noise ratio at high frequencies. A large diameter is desired to increase strength and reduce contamination effects due to particles in the fluid.

For research work, 2.5- to 5-μm wires are the most common, with the length-to-diameter ratios $2\ell/d = 100$–600. Commercially mounted wires typically have $2\ell/d \simeq 300$. The hot-wire sensor in Fig. 3.1 has a $2\ell/d$ ratio of 312. Possible errors due to the effects listed above are discussed below.

3.4 PROBE SUPPORTS AND MOUNTING

Figure 3.2a shows a typical probe support for a hot-wire probe. The ends of tungsten wires are generally electroplated to isolate the sensing element from the supports. With platinum wire, the Wollaston process used to make the wire leaves a silver coating on the ends, since only the silver on the sensor portion is etched away.

It is preferable to have both wire supports electrically insulated from the body and therefore from the test section. In locations with strong electromagnetic fields or in water flows, an outside ground shield connecting the electronics and probe, independent of the sensor leads, can often reduce background noise.

The plug-in tip shown in Fig. 3.2a is a convenience. Wires are fragile, so it is useful to have replacement wires already mounted when running tests. At the same time, in clean gas a hot wire will last indefinitely if it is not physically damaged or burned out by a malfunction in the control circuit. This should not happen with modern transistorized constant-temperature anemometers.

A wide variety of probe supports have been designed, including miniature versions with much smaller bodies, as shown in Fig. 3.2b. Sandborn [11] discusses

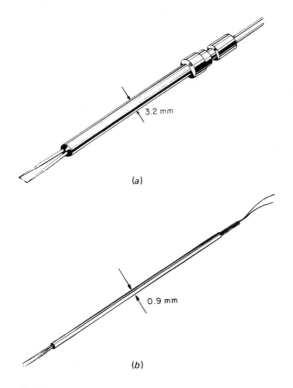

(a)

0.9 mm

(b)

Figure 3.2 Examples of probes for thermal anemometers. (*a*) Typical plug-in probe. (*b*) Subminiature probe

some special considerations in the design of supports for measurements in supersonic flow.

3.5 CONTROL CIRCUIT

Figure 3.1 shows a very simple electric circuit for heating a hot-wire probe. The next step in the evolution of hot-wire circuitry was to provide open-loop frequency compensation for the hot wire. This was easy to do electrically, since hot-wire response is similar to a resistance-capacitance circuit (first order). Until the introduction of transistorized circuitry, the compensated constant-current circuit was the most common for high-frequency hot-wire anemometry.

Since about 1960, the application of constant-temperature systems has increased, owing to both improved circuitry and improved understanding of the system's characteristics. These, combined with its applicability to both films and wires as well as to a wide range of flows, have caused it to essentially displace the constant-current approach. Therefore, in the following, we concentrate on the constant-temperature-type control circuit.

Adding a feedback line converts the bridge circuit and amplifier of Fig. 3.1 to

THERMAL ANEMOMETERS

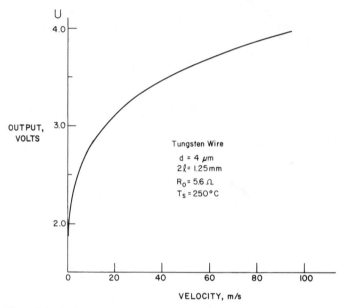

Figure 3.3 Typical calibration curve for a hot-wire anemometer

a constant-temperature system. The feedback from the output of the high-gain amplifier to the top of the bridge acts to maintain the resistance of the sensor (and hence the average temperature) essentially constant. The step-by-step operation is as follows.

1. A velocity increase past the hot wire cools it, lowering the temperature, the resistance R, and the voltage at point 1.
2. The lowered voltage at the negative input to the amplifier causes the input voltage U_{12} to increase.
3. The increased amplifier input voltage U_{12} increases the output voltage of the amplifier U.
4. The increased voltage on the bridge U increases the current through the sensor.
5. The increased current heats the sensor, resulting in a decrease in U_{12} until the entire system is again in equilibrium.

All this takes place almost simultaneously, so an increase in velocity is seen as an increase in the voltage U. This voltage is generally used as the anemometer output. In some cases the voltage across the fixed resistors in series with the sensor is used as a direct measure of sensor current I, independent of the sensor operating resistance R.

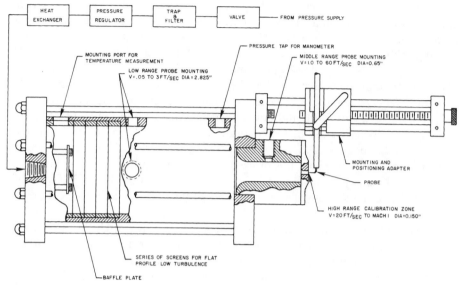

Figure 3.4 Calibration system using air

Figure 3.1 also shows a sketch of the amplitude response of the anemometer. By adding the feedback, the frequency response (-3 dB point) can be increased from about 700 Hz for the wire and environment shown to over 100 kHz. The latter depends on the characteristics of the bridge and amplifier as well as the hot wire and its environment.

3.6 CALIBRATION OF A HOT-WIRE ANEMOMETER

A typical calibration curve for a hot-wire anemometer is given in Fig. 3.3. The two outstanding characteristics are that the output is very nonlinear with velocity and the sensitivity decreases as velocity increases. It turns out that the sensitivity as a percent of reading stays nearly constant. This helps make the hot-wire anemometer useful over a very wide range of flow velocities.

Figure 3.4 shows a typical calibration system. By measuring the total pressure upstream, the velocity in the nozzle can be precisely calculated with the simple relation

$$V = \sqrt{2gh_t}$$

where g is the gravitational constant and h_t is total pressure in height of flowing fluid. The relation is exact for the centerline velocity of a properly designed nozzle and incompressible flow.

Once a calibration curve similar to Fig. 3.3 is obtained, good measurements in the unknown environment can be made directly, providing (1) the fluid temperature, composition, and density in the unknown environment are the same as those during calibration or, if not, are corrected for; (2) the turbulence intensity is below about 20%; and (3) the flow is incompressible. With these restrictions, the single hot-wire anemometer can be used as the basic transducer to measure mean velocity, turbulence intensity, turbulence spectrum, waveform, and flow transients. Mean flow, true waveform, and flow transients must be interpreted by converting voltage to velocity by use of the calibration curve. Spectrum and turbulence are interpreted by measuring the slope of the curve at the operating point. Electronic linearizing greatly simplifies this process and is common. It can also reduce errors due to nonlinearity effects, but this is certainly not assured [12, 13]. Typically, a fast analog-to-digital converter is used, and all the above functions, including linearizing, are performed with a digital computer.

The above list of limiting factors applies only to this straightforward application of hot-wire anemometry. As the complexity of the instrument, calibration, and/or data reduction technique increases, the potential range of applications of the basic thermal anemometer can become very broad.

3.7 HEAT TRANSFER FROM FINE WIRES

If an actual velocity calibration can be made that covers the test conditions, then an analytical relation between flow velocity and anemometer output may not be required. Situations where an equation is useful or necessary include complex measurement situations such as compressible flow, where a complete set of calibration curves is impractical, and investigating or correcting for various error sources that are difficult to determine experimentally. In both cases, some calibration should still be done on each sensor. The reproducibility of thermal sensors is typically not sufficient to give identical calibration curves for two different sensors of the same type.

It is useful to think of heat transfer from the sensor to its environment H as the fundamental variable. From Fig. 3.1, for a constant-temperature anemometer, H is related to the anemometer output voltage U by the equation

$$H = P = \frac{U^2 R}{(R + R_1)^2} = \phi + K \tag{2}$$

where P is electrical power input to the sensor. Equation (2) is valid as long as R represents the resistance of the sensor only.

The more difficult relation is that between H and flow velocity V. H consists of two parts, as follows.

1. Convective heat transfer ϕ between the heated portion of the sensor and the flowing fluid.

2. Conduction heat transfer K between the heated portion of the sensor and its supports (K is generally transferred by convection from the supports to the flowing fluid).

As is shown later, K influences the dynamic response as well as the steady state heat balance on the sensor. Radiation is usually negligible, except for measurements in rarefied gases. The quantity of primary interest here is the convective heat transfer ϕ; conduction is treated below as a deviation from the ideal. For the hot wire,

$$\phi = \text{Nu } \pi 2\ell k_f (T_m - T_a) \tag{3}$$

where

$\text{Nu} = \text{Nusselt number} = h_c d/k_f$
$d = \text{sensor diameter}$
$h_c = \text{heat transfer coefficient for convection}$
$k_f = \text{thermal conductivity of the fluid at the "film" temperature}$
$\quad T_f = (T_m + T_a)/2$
$T_m = \text{mean cylinder temperature}$
$T_a = \text{ambient temperature of the fluid}$
$2\ell = \text{length of sensitive area of the hot wire}$

The problem is to find a representative expression for the Nusselt number in terms of the fluid and sensor parameters. A general expression for Nu would be

$$\text{Nu} = f(\text{Re}, \text{Pr}, \alpha_1, \text{Gr}, \text{Ma}, \gamma_h, a_T, \frac{2\ell}{d}, \frac{k_f}{k_w}) \tag{4}$$

where

$\text{Re} = \text{Reynolds number} = \rho V d/\mu$
$V = \text{free stream velocity}$
$\rho = \text{fluid density}$
$\mu = \text{fluid viscosity}$
$\text{Pr} = \text{Prandtl number} = \mu C_p/k$
$C_p = \text{specific heat of the fluid at constant pressure}$
$\alpha_1 = \text{angle between free stream flow direction and flow normal to the cylinder}$
$\text{Gr} = \text{Grashof number} = \rho^2 g \beta_v (T_m - T_a) d^3/\mu^2$
$g = \text{gravitational constant}$
$\beta_v = \text{volume coefficient of expansion} = 1/T_a$
$\text{Ma} = \text{Mach number} = V/\gamma_h R_0 T_a^{1/2}$
$R_0 = \text{gas constant}$
$\gamma_h = C_p/C_v$
$C_v = \text{specific heat at constant volume}$
$a_T = \text{overheat ratio or temperature loading} = (T_m - T_a)/T_a$
$k_w = \text{thermal conductivity of sensor material}$
$k_f = \text{thermal conductivity of fluid}$

Fortunately, most applications permit a significant reduction in the number of parameters that must be included. The simplest expression is often referred to as "King's law" [3]:

$$\text{Nu} = A' + B' \, \text{Re}^{0.5} \tag{5}$$

where A and B are empirical constants, usually determined by calibration.

Although Eq. (5) does not accurately represent Nu over a wide range of velocities, it is still frequently used in calculations because of its simplicity. A more accurate expression for air is that of Collis and Williams [14]:

$$\text{Nu} = (A + B \, \text{Re}^n)(1 + \frac{a_T}{2})^{0.17} \tag{6}$$

where

$A = 0.24$	$B = 0.56$	$n = 0.45$	$0.02 < \text{Re} < 44$
$A = 0$	$B = 0.48$	$n = 0.51$	$44 < \text{Re} < 140$

At this point, it is useful to discuss why the remaining parameters of Eq. (4) are ignored in Eqs. (5) and (6). Equations (5) and (6) are derived for forced convection normal to the wire ($\alpha = 0$) and for incompressible flow. Since Pr depends only on fluid properties, limiting the discussion to air eliminates Pr as a variable in Eq. (4). It was found [14] that buoyancy effects (free convection) are important in air only if $\text{Gr}^{1/3} > \text{Re}$. For the hot wire of Figure 3.1, this occurs at a forced convection velocity of 5.2 cm/s. This value can be reduced further by lowering the sensor temperature (overheat). For higher velocities, Gr can be ignored.

For high-velocity or low-density flows, Ma and C_p must be considered as variables. For low density, the most relevant parameter is the Knudsen number, which is

$$\text{Kn} = \frac{\lambda}{d} = \left(\frac{\pi \gamma_h}{2}\right)^{0.5} (\text{Ma/Re}) \tag{7}$$

where λ is the molecular mean free path. Three ranges of Kn are usually defined, as follows.

Continuum flow

$$\text{Kn} < 0.01$$

Slip flow

$$0.01 < \text{Kn} < 1$$

Free molecular flow

$$\text{Kn} > 1$$

Again for the conditions of Fig. 3.1, Kn ≃ 0.02, which is in the slip-flow region. While most measurements with fine wires are in slip flow, continuum-flow assumptions are usually adequate as long as density changes are small.

At high velocity, two fluid temperatures are normally defined: static temperature T_{st} and total temperature T_0. They are related by (assuming C_p constant)

$$T_0 - T_{st} = \frac{V^2}{2C_p} \simeq \frac{V^2}{2000} \qquad (8)$$

for air, with T in degrees Kelvin and V in meters per second. The total temperature is the temperature the fluid attains when brought to rest; it is both the most convenient reference temperature in thermal anemometry and the easiest temperature to measure in high-speed flows. However, when exposed to a high-velocity airstream, a hot wire does not always attain the total temperature but instead equilibrates at a recovery temperature T_{re} between T_0 and T_{st}. The recovery factor η is defined as

$$T_{re} = \eta T_0$$

From [15], for Kn < 0.1 and Ma < 1, the recovery factor η is greater than 0.98. It should be noted that, experimentally, when the "cold-wire" temperature is measured in the flow stream, it is the recovery temperature that is measured, and T_{re} should be substituted for T_a in heat transfer equations.

The value of a_T appears in Eq. (4) to account for the change in fluid properties with temperature. For incompressible flow, this effect is minimized by using the film temperature T_f when selecting fluid properties. However, a weak effect is still present, as shown in Eq. (6).

The aspect ratio $2\ell/d$ enters Eq. (4) because three-dimensional aerodynamic effects near the prongs may affect flow. Champagne [16] discusses this as it relates to Eq. (6) for Re < 44 and a platinum wire. While n was unaffected, B decreased by 5% and A doubled when $2\ell/d$ was decreased from 10^3 to 10^2.

Finally, k_f/k_w may enter Eq. (4) because the temperature distribution along and around the sensor depends on it. In actual practice, this effect is not generally taken into account, and in any case, it is considered as a "perturbation" or correction to relations such as Eq. (5). Of course, both $2\ell/d$ and k_f/k_w are very important parameters when heat conduction to the supports K is considered. However, here we are concerned only with convective heat transfer ϕ.

Equation (6) applies for air only. A more general equation often used is that proposed by Kramers [17]:

$$Nu = 0.42Pr^{0.26} + 0.57Pr^{0.53} Re^{0.50} \qquad (9)$$

which covers the range $0.71 \leq Pr \leq 525$ and $2 \leq Nu \leq 20$. Again, fluid properties are to be selected at the film temperature T_f. Equation (9) covers a wide range of fluids and is very useful for liquids and gases other than air [where the more precise Eq. (6) should be used].

3.7.1. High-Speed Flows

Measurements in high-speed flows present a number of special problems, and the reader is referred to publications on the subject [11, 15, 18–22]. Some comments may, however, be helpful.

In high-speed flows, Eq. (3) should be reformulated as

$$\phi = \text{Nu} \; \pi 2 \ell k_0 (T_m - T_{re}) \tag{10}$$

where k_0 is the thermal conductivity of the gas at the stagnation temperature T_0. According to Kovasznay et al. [18], it is advantageous to formulate dependence in terms of the Reynolds number:

$$\text{Re}_0 = \frac{\rho_\infty V_\infty d}{\mu_0}$$

where the subscript ∞ refers to free-stream conditions and the subscript 0 refers to stagnation conditions. In supersonic continuous flow in air (Ma > 1.5, Kn < 0.01), Kovasznay et al. [18] found experimentally that Nu is independent of Ma and for vertical flow, incidence can be represented by a relation

$$\text{Nu} = \left(A \, \text{Re}_0^{1/2} - B \right) \left(1 - C \frac{T_m - T_{re}}{T_0} \right) \tag{11}$$

where $A = 0.58$, $B = 0.8$, and $C = 0.18$. However, according to [11] and [15], temperature loading effects are much smaller, that is, C may be negligible, and Nu is about 20% too high [23]. Because of Ma independence (Fig. 3.5), the anemometer in supersonic continuum flow is sensitive to $\rho_\infty V_\infty$ and to T_0, just as the incompressible-flow anemometer is sensitive to ρV and T_a.

In the low-density flow regime, as shown in Fig. 3.5, the Ma dependence of Nu sets in strongly as a parameter. In this regime, Nu becomes proportional to Re rather than $\text{Re}^{1/2}$, and heat transfer becomes mainly sensitive to density and rather insensitive to velocity; as a consequence, hot wires have marginal value as anemometers in the molecular flow regime. In the low-density regime the recovery factor η is a function of Re and Ma [15].

The influence of aspect ratio on Nu seems to be sufficiently small to have received no particular attention, and the influence of Ma on directionality also remains uninvestigated.

3.7.2 Conduction to Walls

As a hot wire approaches a wall, the temperature and velocity field around the wire is modified, owing to the presence of the solid surface. Any modification of this boundary layer will change the relation between velocity and the output voltage of the anemometer.

According to Wills [24], the primary parameter is b/a, where b is the distance

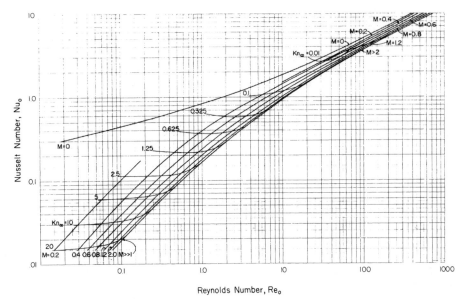

Figure 3.5 Empirical correlation of cylinder heat transfer at low Reynolds number. (From [15])

from the wall and a is the sensor radius. For $b/a > 500$, the effect is negligible (2 mm for a 4-μm-diameter wire). For the range tested (Re < 1), if Re > 0.1, the wall effect seemed to be a constant correction on the A term of Eq. (6). As $b/a \to \infty$, Wills [24] found an A of 0.26 rather than 0.24 [14]. This increased to 0.32 at $b/a = 100$, and to 0.47 at $b/a = 20$. Since only the A term was affected, the percent error decreases as velocity increases.

3.8 CONDUCTION TO THE SUPPORTS

Any heat transfer K from the sensor to the supports by conduction is a "loss" and a potential error source. Again, for steady state measurements this conduction loss will be included in the calibration process. However, calibration under actual operating conditions is not always possible. Perhaps even more important, conduction losses cause dynamic effects that are difficult to measure experimentally. While in this section our concern is only with steady state heat transfer, in section 3.11.4 the strong relation between steady state and dynamic effects is shown.

Figure 3.6 shows the temperature profiles of a hot-wire sensor for various values of the parameter $\sqrt{C_0}\ell$. From [25], $\sqrt{C_0}\ell$ can be calculated from

$$\sqrt{C_0}\ell = \frac{2\ell/d\,[(k_a/k_w)\mathrm{Nu}]^{1/2}}{\{1 + \alpha(T_m - T_a) + [\alpha(T_m - T_a)/(\xi - 1)]\}^{1/2}} \tag{12a}$$

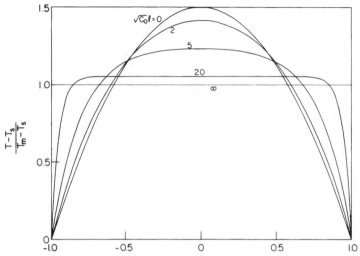

Figure 3.6 Temperature distribution along hot-wires with constant mean temperature for various values of the parameter $\sqrt{C_0}\ell$

where $\xi = \sqrt{C_0}\ell \coth \sqrt{C_0}\ell$. For values of $\sqrt{C_0}\ell > 2$, Eq. (12a) quickly converges by calculating an initial value of $\sqrt{C_0}\ell$ using only the first two terms in the denominator for an estimate of ξ. For the sensor of Fig. 3.1, $\sqrt{C_0}\ell = 2.93$.

In the case of small overheats, $\alpha(T_m - T_a) \ll 1$ and $\alpha(T_m - T_{st}) \ll 1$, Eq. (12a) can be written

$$\sqrt{C_0}\ell = \frac{2\ell}{d}\left(\frac{k_a}{k_w} \text{Nu}\right)^{1/2} \tag{12b}$$

Therefore the sensor temperature profile (and resulting conduction losses) depends on the length-to-diameter ratio $2\ell/d$, the thermal-conductivity ratio between the fluid and the sensor, and the Nusselt number.

The question of primary interest is the amount of heat transferred to the fluid by convection, as compared with the total. Let us introduce the ratio

$$\varepsilon = \frac{\phi}{\phi + K} \tag{13}$$

In [25] the solution was

$$\varepsilon = 1 \tag{14}$$

$$- \frac{(T_m - T_s)/(T_m - T_a)}{(\xi)[1 + \alpha(T_m - T_a)] + [(T_m - T_s)/(T_m - T_a)] - 1 + \alpha(T_a - T_s)}$$

which, for small overheat and $T_s \simeq T_a$ reduces to

$$\varepsilon = 1 - \frac{1}{\xi} \tag{15}$$

For the conditions of Fig. 3.1 (assuming $T_a = T_s$), $\varepsilon = 0.826$. In other words, about 17% of the heat generated in the sensor by the electric current is conducted to the supports, while over 82% is transferred directly to the fluid by convection.

Equation (13) can be analyzed further to determine the effect of changes in the fluid parameters on ε. Two cases may be considered:

$$\varepsilon_1' = \frac{d\phi}{d\phi + dK}\bigg|_{T_a}, \ T_s = \text{const} \tag{16}$$

$$\varepsilon_2' = \frac{d\phi}{d\phi + dK}\bigg|_{N_u} = \text{const} \qquad T_a = T_s \qquad \sqrt{c_0}\ell = \text{const} \tag{17}$$

In the first case, Nu (velocity) varies, while in the second, fluid temperature T_a is the variable. The calculated value of ε_1' is, for the general case,

$$\varepsilon_1' = 1 - \frac{T_m - T_s}{T_m - T_a}\left(\xi - 2 + \frac{C_0\ell^2}{\sinh^2 \sqrt{C_0\ell}}\right)\bigg/ \bigg\{ 2[1 + \alpha(T_m - T_a)](\xi - 1)^2 \tag{18a}$$

$$+ \left[\frac{T_m - T_s}{T_m - T_a} + \alpha(T_m - T_s)\right](\xi - 2 + \frac{C_0\ell^2}{\sinh^2 \sqrt{C_0\ell}})\bigg\}^{-1}$$

For the low-overheat condition $\alpha(T_m - T_a) \ll 1$ and $T_s = T_a$,

$$\varepsilon_1' = 1 - \frac{\xi - 2 + (C_0\ell^2/\sinh^2 \sqrt{C_0\ell})}{2(\xi - 1)^2 + \xi - 2 + (C_0\ell^2/\sinh^2 \sqrt{C_0\ell})} \tag{18b}$$

For both the general case and the low-overheat case,

$$\varepsilon_2' = \varepsilon \tag{19}$$

Equations (18a) and (19) are valid only for a constant-temperature anemometer, since T_m was assumed constant. In Eq. (18a) the support temperature T_s is constant, while in Eq. (19) a changing environment temperature T_a affected the support temperature T_s.

Figure 3.7 is a plot of ε_1' and ε_2' as a function of $\sqrt{C_0}\ell$ for the low-overheat condition. It is evident that low values of $\sqrt{C_0}\ell$ (high conduction losses to the support) do not have as large an influence on fluctuation measurements of velocity as they do on fluctuation measurements of temperature. It turns out that this fact is of particular significance when one considers the effect of conduction losses on the dynamic response of thermal anemometers (section 3.11.4).

The sensitive portion on many hot-wire sensors is isolated from the supports by a plated area, as shown in Fig. 3.1. The support effectively starts at the plating, which provides a very small fragile support. In this case, $T_s > T_a$, where T_s is now

THERMAL ANEMOMETERS

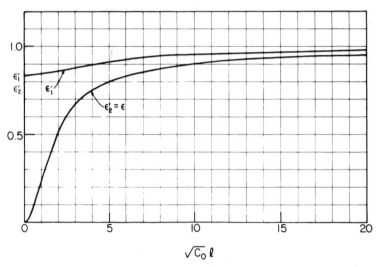

Figure 3.7 Heat-loss fluctuation ratios ε_1' and ε_2' in isothermal and in nonisothermal constant-velocity flow as functions of the Biot number $\sqrt{C_0}\ell$ (small-overheat case) of the hot-wire

the temperature of the junction between the plated area and the sensitive area of the sensor. This, however, can only decrease conduction losses K, resulting in ε and ε_2' being closer to 1. The effect on ε_1' is not so clear, and it is discussed further in section 3.11.4.

In high-speed and low-density flows, the end-loss equations are valid as long as the recovery temperature T_{re} is substituted for ambient temperature T_a in calculating the heat loss from the sensor. Also, the support temperature should be calculated in the same way, although the larger size of the supports would give $\eta = 1$ and $T_s = T_0$ for most flow environments.

3.9 ANGLE SENSITIVITY AND SUPPORT INTERFERENCE

For an infinitely long wire, the angle sensitivity of the hot wire is expressed (Fig. 3.8) as

$$V_{\text{eff}} = V \cos \alpha_1 \tag{20}$$

where V_{eff} is the effective cooling velocity past the sensor. Equation (20) essentially states that the velocity along the sensor ($V_T = V \sin \alpha_1$) has no cooling effect on the sensor and that the sensor is rotationally symmetrical in both construction and

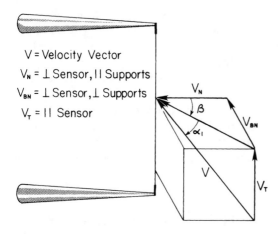

V = Velocity Vector
V_N = ⊥ Sensor, ‖ Supports
V_{BN} = ⊥ Sensor, ⊥ Supports
V_T = ‖ Sensor

Figure 3.8 Velocity components at sensor

response. In many calculations and experiments, Eq. (20) is adequate, maintains simplicity, and depending on probe and sensor design, can be quite accurate [26].

Because the sensor has a finite length, there is heat transfer due to the flow parallel to the sensor (V_T). To account for this, a second term is added [16, 23] to give

$$V_{\text{eff}} = V\sqrt{\cos^2 \alpha_1 + k_T^2 \sin^2 \alpha_1} \tag{21}$$

where k_T is an empirically determined factor. Although k_T is not truly a constant for all velocities and values of α_1, for a limited velocity range and angles from 0° to 60°, a fixed value works quite well. Champagne [16] found that k_T decreases nearly linearly with $2\ell/d$ from a value of $k_T = 0.2$ at $2\ell/d = 200$ to zero at $2\ell/d = 600$–800. Similar results were obtained with both platinum and tungsten wires. Other equations for yaw sensitivity [27–29] have been suggested and may prove more accurate in certain applications. For the sensor of Fig. 3.1, $2\ell/d = 312$ and $k_T \simeq 0.15$.

Equation (21) assumes that the response of the sensor is rotationally symmetric. Comte-Bellot et al. [26, 30] showed that aerodynamic effects from both the support needles and the probe body affected the readings, the minimum reading occurring with the probe parallel to the flow, and the maximum with the probe perpendicular to the flow.

To account for the support interference, an equation of the following form has been suggested [31]:

$$V_{\text{eff}} = \sqrt{V_N^2 + k_T^2 V_T^2 + k_N^2 V_{BN}^2} \tag{22}$$

where V_{BN} is the velocity vector perpendicular to both the sensor and the support prongs. The value of k_N can range from 1.0 to 1.2, depending on the design of the probe support and needles [29]. As is true of k_T, k_N is not constant for all angles and velocities, but careful use can improve accuracy as compared with assuming that $k_N = 1$ (theoretical value). It was found in [31] that plating the wire ends as shown in Figs. 3.1 and 3.8 reduced the values of k_T and k_N.

Equations (21) and (22) are given here to provide concrete examples. There is no intent to imply that they always represent the best functional relationships.

It should be emphasized that, in using a single calibrated sensor in turbulence intensities under 20%, good accuracy can be obtained without the above equations. When multisensor probes are used in large turbulence intensities, or when the sensor orientations during calibration and use are different, the above considerations become important.

3.10 MEASURING MEAN VELOCITY, VELOCITY COMPONENTS, AND TEMPERATURE

The most common measurement is the use of a single hot-wire probe, perpendicular to the flow, to measure mean velocity $\overline{V}$ and fluctuations in the mean flow direction $\overline{v_1^2}$. Two components are often measured with an X probe, while three components can be measured by adding a third sensor or by rotating the X probe. In addition, both temperature and velocity can be obtained by operating two parallel sensors at different temperatures. In all these measurements there are limitations that must be observed.

3.10.1 One Component Using a Single Hot Wire

In Fig. 3.8, assume the sensor is oriented in the flow stream, so that $V_N = V_1$, $V_T = V_2$, and $V_{BN} = V_3$, where V_1, V_2, V_3 are the desired orthogonal velocity components of the velocity vector V. From Eq. (22), the effective cooling velocity past the sensor is

$$V_{\text{eff}} = \sqrt{V_1^2 + k_T^2 V_2^2 + k_N^2 V_3^2} \tag{23}$$

If, further, the mean flow is in the V_1 direction, then $\overline{V_2} = \overline{V_3} = 0$. If the fluctuations are v_1, v_2, and v_3, then

$$V_{\text{eff}} = \sqrt{(\overline{V_1} + v_1)^2 + k_T^2 v_2^2 + k_N^2 v_3^2} \tag{24}$$

Since k_T is small and $k_N \simeq 1$, this can be approximated by

$$V_{\text{eff}} = \sqrt{(\overline{V_1} + v_1)^2 + v_3^2} \tag{25}$$

If we neglect v_3, then

$$\overline{V_1} = \overline{V} = \overline{V}_{\text{eff}} \tag{26}$$

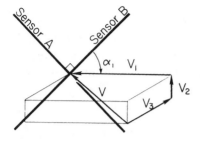

Figure 3.9 Configuration of X probe

$$\sqrt{v_1^2} = \sqrt{v^2} \tag{27}$$

Traditionally, the value of $\overline{V}_1$ is obtained with a mean value (averaging) meter, and $\sqrt{v_1^2}$ is obtained with an ac coupled true root-mean-square (rms) meter. Now digital data processing is typically used.

When $\sqrt{v_1^2}/\overline{V} = 0.2$, the error due to ignoring v_3 is about 2% for isotropic, normally distributed, and normally correlated turbulence [13]. The mean velocity error is also about 2%.

3.10.2 Two Components Using an X Probe

The cross wire or X probe is frequently used to measure two velocity components (V_1 and V_2). In this case, two sensors (A and B, Fig. 3.9) are made sensitive to V_2 by orienting them at an angle to the mean flow. With the two sensors in the $v_1 v_2$ plane and oriented at 90° to each other, Eq. (22) can be written for each sensor:

$$V_{A,eff}^2 = (V_1 \cos \alpha_1 - V_2 \sin \alpha_1)^2 + k_T^2(V_1 \sin \alpha_1 + V_2 \cos \alpha_1)^2 + k_N^2 V_3^2 \tag{28}$$

$$V_{B,eff}^2 = (V_1 \sin \alpha_1 + V_2 \cos \alpha_1)^2 + k_T^2(V_1 \cos \alpha_1 - V_2 \sin \alpha_1)^2 + k_N^2 V_3^2 \tag{29}$$

where α_1 is the angle between V_1 and sensor B.

The coordinates are usually selected such that $\overline{V}_3 = 0$. If the sensors are sufficiently long that $k_T \to 0$ and $k_N \to 1$, then Eqs. (28) and (29) reduce to

$$V_{A,eff}^2 = (V_1 \cos \alpha_1 - V_2 \sin \alpha_1)^2 + v_3^2 \tag{30}$$

$$V_{B,eff}^2 = (V_1 \sin \alpha_1 + V_2 \cos \alpha_1)^2 + v_3^2 \tag{31}$$

In large turbulence intensities, Eqs. (30) and (31), like Eq. (25), cannot be further reduced. However, if v_3 is small, then the last term can be ignored. Further, orienting the sensors so that $\alpha_1 = 45°$ and rearranging Eqs. (30) and (31) give

$$V_1 = 2^{-1/2}(V_{A,eff} + V_{B,eff}) \tag{32}$$

$$V_2 = 2^{-1/2}(V_{A,eff} - V_{B,eff}) \tag{33}$$

With these simplifying assumptions, summing the linearized output voltages of the two constant-temperature anemometers gives V_1, and differencing them gives V_2.

In most applications the orientation is such that the mean flow is in the V_1 direction, so that $\overline{V_2} = 0$. The results are then

$$\overline{V} = 2^{-1/2}\overline{(V_{A,eff} + V_{B,eff})} \tag{34}$$

$$\overline{v_1^2} = \tfrac{1}{2}\,\overline{(v_{A,eff} + v_{B,eff})^2} \tag{35}$$

$$\overline{v_2^2} = \tfrac{1}{2}\,\overline{(v_{A,eff} - v_{B,eff})^2} \tag{36}$$

$$\overline{v_1 v_2} = \tfrac{1}{2}\,\overline{(v_{A,eff} + v_{B,eff})\,(v_{A,eff} - v_{B,eff})} \tag{37}$$

These are the equations used in most measurements with X probes.

Neglecting v_3 gives an error of about 8% when the turbulence intensity is 20%, with the same flow field as discussed for the single wire [13]. It should be emphasized that $k_T \neq 0$ can also significantly influence the accuracy of the results. Finally, the thermal wake from one senor can influence the other [32]. All these considerations suggest the use of sensors with a high $2\ell/d$ ratio, as well as isolating the sensitive area from the supports by plating the wire ends.

3.10.3 Three Components

It is often desirable to measure all three components of the flow. This gives more information about the flow field and provides data on V_3 to improve the accuracy of measurement of V_1 and V_2.

In Fig. 3.9, adding a third sensor C (not shown) whose axis is at an angle α_3 to V_1 and in the $V_1 V_3$ plane gives

$$V^2_{C,eff} = (V_1 \sin\alpha_3 + V_3 \cos\alpha_3)^2 + k_T^2(V_1 \cos\alpha_3 - V_3 \sin\alpha_3)^2 + k_N^2 V_2^2 \tag{38}$$

Equations (28), (29), and (38) now give three equations in three unknowns that, in theory, can be solved for V_1, V_2, and V_3. Once these instantaneous components are available, all the turbulence parameters can be calculated. Again, the equations can be simplified if the sensors are long so $k_T \to 0$ and $k_N \to 1$, if the sensors are oriented so $\overline{V_2} = \overline{V_3} = 0$ and α_1 and α_3 are 45°. Fabris [33] used this approach for making three-component velocity measurements.

The maximum turbulence intensity that can be measured is still limited. For example, with ideal sensors, flow reversals cannot be detected no matter how many sensors are used. In fact, with the X probe of Fig. 3.9, if a velocity vector in the plane of the sensors crosses the axis of sensor B, this will not be detected. In other words, even with two-dimensional flow ($V_3 = 0$), the velocity vector must be limited to a single quadrant for valid measurements with an X probe. Tutu and

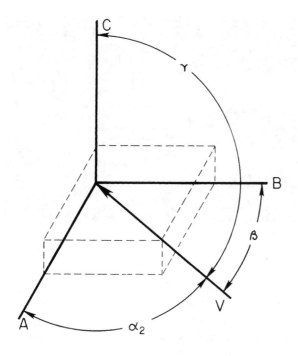

$$V^2_{A,eff} = V^2 \; (\sin^2 \alpha_2 + k^2 \cos^2 \alpha_2)$$

$$V^2_{B,eff} = V^2 \; (\sin^2 \beta + k^2 \cos^2 \beta)$$

$$V^2_{C,eff} = V^2 \; (\sin^2 \gamma + k^2 \cos^2 \gamma)$$

$$V^2 = \frac{V^2_A + V^2_B + V^2_C}{2 + k^2}$$

Figure 3.10 Direction sensitivity using three mutually perpendicular sensors

Chevray [34] report that the combined influence of v_3 and rectification can cause errors in $\overline{v_1 v_2}$ of 28% when an X probe is used in turbulence intensities of 35%. Similar restrictions hold for the measurement of three-dimensional flows with three sensors. Maciejewski and Moffat [35] suggest techniques for reducing the error caused by rectification.

Olin and Kiland [9] provide a way out of this dilemma by using three orthogonal split-film cylindrical sensors. The splits were used to detect the octant, while the effective velocity readings were used to detect where the vector was within the octant. Figure 3.10 shows the configuration and equations [36], where $k_N \rightarrow 1$ was assumed. Atmospheric measurements have been made with this probe by Tielman et al. [37]. The size of the probe (~1 cm) restricts its use where spatial resolution is important.

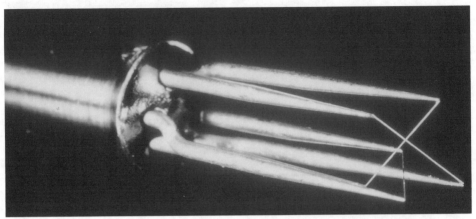

Figure 3.11 Lekakis probe with three orthogonal sensors positioned for good spatial resolution and minimum interference

Recently Lekakis et al. [38] have taken the basic orthogonal configuration of Fig. 3.10 and characterized it with detailed calibrations on a specific probe. The probe design shown in Fig. 3.11 keeps the three sensors within a sphere of diameter 2.6 mm. There is no wake interference between sensors when the velocity vector is in the octant that includes the vector parallel to and toward the support.

The algorithm developed provides a fast solution to Eq. (22) when written for the three sensors. While the orthogonal configuration was found optimum, the algorithm provides for deviations that can occur due to limitations on sensor mounting precision.

The maximum turbulence intensity that can be measured is determined by the fact that the solutions are unique only if the velocity vector is limited to one octant.

3.10.4 Multiposition Measurements

As an alternative to the X probe, two sets of measurements can be taken by orienting a single sensor in each of the two positions. This procedure will not permit the measurement of instantaneous values, but it will measure $\overline{v_1^2}$, $\overline{v_2^2}$, and $\overline{v_1 v_2}$. Using one sensor eliminates the need to match hot-wire sensitivities but does add to the stability requirements of both the flow field and the instrumentation. As with X probes, the usual data reduction procedure includes the assumption of small fluctuations.

Fujita and Kovasznay [39] used a continuously rotating straight-wire probe to improve the accuracy when using a single sensor to measure $\overline{v_1^2}$, $\overline{v_2^2}$, and $\overline{v_1 v_2}$. Bissonnette and Mellor [40] made similar measurements of all six second moments ($\overline{v_1^2}$, $\overline{v_2^2}$, $\overline{v_3^2}$, $\overline{v_1 v_2}$, $\overline{v_1 v_3}$, $\overline{v_2 v_3}$) using a slanted wire. DeGrande [41] used the same basic technique but discrete positions rather than continuous rotation. Kool [42] extended this technique to periodically unsteady turbomachinery flow.

The technique has been extended to measure large turbulence intensities by squaring the signal to eliminate the binomial expansion [43]. This appears to permit the measurement of $\overline{v_2^2}$, $\overline{v_3^2}$, and $\overline{v_1 v_2}$ without neglecting higher-order terms. The primary difficulty is in solving individually for $\overline{V_1^2}$ and $\overline{v_1^2}$, which does require some assumptions. Although attempts have been made to extend these results further [44], lack of experimental data and the fundamental problems of sensing flow reversal suggest caution.

3.10.5 Nonisothermal Flows

A thermal anemometer is sensitive to fluid temperature changes. These may be either slow changes due to mean temperature changes or high-frequency temperature changes resulting from a heat source in the flow or compressibility effects.

To examine temperature effects, it is useful to combine Eqs. (6) and (3) to examine the case for airflow:

$$\phi = 2\ell\pi k_f(A + B\ Re^n)(1 + \frac{a_T}{2})^{0.17}(T_m - T_a) \qquad (39)$$

For velocities for which the A term is small, one can write

$$\frac{\Delta V}{V} \simeq \frac{\Delta T_a/n}{T_m - T_a} \qquad (40)$$

Therefore the velocity error ΔV due to a change in fluid temperature ΔT_a is minimized by maintaining a high overheat $T_m - T_a$. Often this precaution is sufficient to keep errors due to temperature changes within acceptable limits.

In Eq. (39) the following substitutions, based on fluid-property dependence on temperature, can be made [14]:

$$k \sim (T_m + T_a)^{0.8} \qquad (41)$$

$$\rho \sim (T_m + T_a)^{-1} \qquad (42)$$

$$\mu \sim (T_m + T_a)^{0.76} \qquad (43)$$

Substituting $n = 0.45$ into Eq. (39) gives [45]

$$\phi = [A_1(T_m + T_a)^{0.8} + B_1 V^{0.45}]\left(1 + \frac{a_T}{2}\right)^{0.17}(T_m - T_a) \qquad (44)$$

It is interesting to compare this with the simple relation

$$\phi = H_1(V)(T_m - T_a) \qquad (45)$$

With $T_m = 230°C$ and $T_a = 23°C$, for a 50°C increase in T_a, the velocity calculated using Eq. (45) is within $\pm 3\%$ of that using Eq. (44), for the range 6–100 m/s. That is why the common technique of manually setting T_m to maintain $T_m - T_a = \text{const}$ works well for many measurements when the test temperature is different from that during calibration. This gives the same value of $\phi(\simeq P)$ for the same velocity.

A more convenient technique is to replace R_3 (Fig. 3.1) with a temperature-sensitive resistor exposed to the flow. For convenience, this is set to maintain U constant and independent of temperature, rather than ϕ. Again, this technique is used primarily to correct for low-frequency temperature changes and utilizes a rather large sensor that can approach the environmental temperature. Trying to compensate in this manner for fast temperature fluctuations with a small sensor presents several problems. One of the most serious is the effect on the anemometer output U of the thermal capacity of the velocity sensor itself as it is heated and cooled to follow temperature changes.

To compensate for higher frequency changes, an alternative technique is to measure the temperature separately and then correct the output data. For small temperature changes, Eq. (45) is adequate, while for large temperature changes a more complex formula can be used with computer data reduction. The temperature sensor must be small enough that it will follow the temperature changes in the flow. For example, the wire in Fig. 3.1 will follow 700 Hz at 10 m/s. By using 0.625-μm-diameter wires, Fabris [33] was able to follow 4500 Hz at 6.5 m/s. For even better response, frequency compensation could be used as long as velocity changes are small.

As frequency increases, the interest is not only in correcting velocity measurements, but also in measuring statistical parameters such as the rms of temperature fluctuations and the cross correlation of temperature and velocity [46]. Of course, a single temperature sensor can also be used with X probes and other multisensor probes [33].

For higher frequencies the use of two sensors at different overheats can be effective. From Eq. (45),

$$\phi_1 = H_1(V)(T_{m1} - T_a) \tag{46}$$

$$\phi_2 = H_2(V)(T_{m2} - T_a) \tag{47}$$

Then, if $H_1(V) = H_2(V)$,

$$H(V) = \frac{\phi_1 - \phi_2}{T_{m1} - T_{m2}} \tag{48}$$

$$T_a = T_{m1} - \frac{\phi_1}{H(V)} \tag{49}$$

With this technique, the only frequency limitation is that of the constant-temperature anemometers themselves. For maximum sensitivity, $T_{m1} - T_{m2}$ should be large. Although more complex equations can be used with computer data reductions, Eqs. (48) and (49) should give good results for modest temperature fluctuations.

As with velocity components, second moments can be determined by using a single sensor at more than one overheat [47]. In fact, this technique has been used extensively in compressible flows. To improve accuracy, several overheats are generally used [18, 20].

3.11 DYNAMICS OF THE CONSTANT-TEMPERATURE HOT-WIRE ANEMOMETER

As shown in Fig. 3.1, adding the feedback loop to maintain T_{m} constant extends the frequency range from about 700 Hz to over 100 kHz. One of our concerns in this section is to establish the upper frequency limit for a given sensor, environment, amplifier, and bridge. A related problem is adjusting the system properly, so the amplitude response is as flat as possible. Finally, dynamic effects that are not compensated by the feedback system need to be considered. These include heat waves along the wire, and temperature fluctuations of the probe support. In addition, the spatial resolution of the sensor, due to its finite size, and possible boundary layer effects can limit the effective frequency response to velocity fluctuations.

All these effects could be tested experimentally if one could generate velocity and temperature fluctuations of known amplitude over a wide range of frequencies. Since this is not practical for routine measurements, an electrical test is used to optimize the system. The interpretation of the electrical test is based on a theoretical model of the anemometer. Other effects can be examined theoretically to obtain a measure of their importance.

Freymuth [48] has developed a detailed theory of electronic sine wave and square wave testing for the constant-temperature hot-wire anemometer. He uses a third-order linear equation that is consistent with the concept of two adjustable controls for optimizing the response of the anemometer. As it turns out, the optimization suggested by the linear equations also provides optimal response for large fluctuations [49].

3.11.1 Frequency Response

Figure 3.12 is a block diagram of a constant-temperature anemometer. It is similar to Fig. 3.1 except for the addition of U_{T}, the electronic test signal. It is this test signal that is used to optimize the frequency response by adjusting the two controls.

Freymuth [48] solved the third-order equation for three cases, but the one of most interest is the "maximally flat" response. In this case the cutoff frequency is

$$f_{\mathrm{cut}} = \frac{1}{2\pi}\left(\frac{G/M''}{M}\right)^{1/3} \tag{50}$$

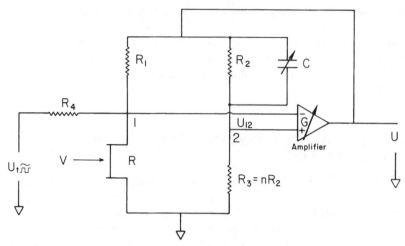

Figure 3.12   Constant-temperature anemometer control circuit

where M'' is a second-order time constant, G is the amplifier gain, and

$$M = \frac{(n_b + 1)^2}{2n_b} \frac{R}{R - R_a} \frac{c}{H(V)} \qquad (51)$$

is the time constant of the wire, where c is the thermal capacity of the wire.

Equation (51) indicates that reducing the sensor thermal capacity c, increasing the overheat $(R - R_a)/R$, or increasing, the convection function $H(V)$ will increase f_{cut}. For example, since $c \simeq d^2 \ell$ and $H(V) \simeq d^{1/2} \ell$, it follows that $f_{cut} \simeq d^{-1/2}$. Therefore, f_{cut} is less sensitive to wire diameter than the time constant M of the wire given above.

3.11.2 Optimization and Electronic Testing

Figure 3.13 is a qualitative representation of the results of Freymuth's theory. For a properly adjusted hot-wire anemometer, a constant-amplitude electrical sine wave input (U_t in Fig. 3.12) will give an output signal U, whose amplitude versus frequency graph is similar to curve 1a. Velocity fluctuations for the same anemometer will give an output whose amplitude and frequency follow curve 1ac. Finally, the uncompensated sensor will give the results represented by curve 1. The important point is that f_{cut} can be established directly from curve 1a, so curve 1ac need not be established experimentally.

Figure 3.14 shows experimental data taken using the sine wave test and a TSI model 1050 anemometer [50]. It can be seen that the experimental points follow the theory closely, even to an $\sim f^{-2}$ attenuation after the cutoff frequency. For curve

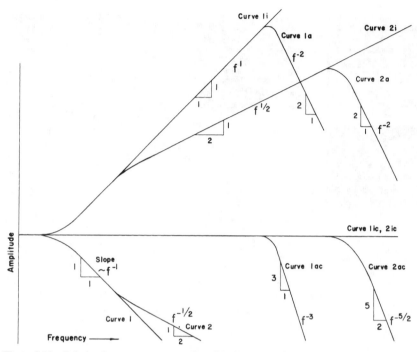

Figure 3.13 Relative frequency response (logarithmic scale) of hot-wire and cylindrical hot-film sensors. *Lower curves:* Response to velocity fluctuations. *Curve 1:* Hot-wire with uncompensated constant-current operation. *Curve 2:* Hot film with uncompensated constant-current operation. *Curves 1ic, 2ic:* Ideal response of both hot wire and hot film with constant-temperature operation. *Curve 1ac:* Actual constant-temperature hot-wire system with optimized third-order response. *Curve 2ac:* Constant-temperature hot-film system with optimized 5/2-order response. *Upper curves:* Response to sine wave test on constant-temperature anemometer. *Curve 1i:* Ideal response with hot wire. *Curve 2i:* Ideal response with hot film. *Curve 1a:* Actual hot-wire system with optimized third-order response. *Curve 2a:* Actual hot-film system with optimized 5/2-order response

1, f_{cut} is ~98 kHz, while for curve 2, it is ~238 kHz. For the condition of Fig. 3.14, M is 3.95×10^{-8} (W·s)/°C. Putting these values into Eq. (5) gives a value of M''/G of 3×10^{-15} s² for curve 1, and 2×10^{-16} s² for curve 2.

The response of the sensor without frequency compensation can also be determined from Fig. 3.14. Since curves 1 and 2 are horizontal at an abscissa 2.38, the 3-dB point for the wire would be at $2.38\sqrt{2} = 3.37$. The frequency corresponding to this amplitude is about 700 Hz. At the higher velocity (curve 3 in Fig. 3.14), this frequency increases to about 1200 Hz.

Although the sine wave test is very helpful in analyzing a constant-temperature anemometer, it is more convenient to observe the response to a step change. For this, Freymuth [48] has also analyzed the expected output for a step input of cur-

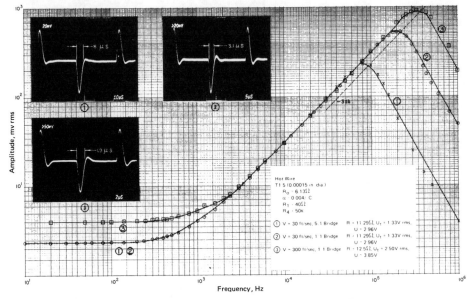

Figure 3.14 Sine wave and square-wave tests on TSI T1.5 hot-wire with airflow

rent. Repetitive step inputs for U are easily provided by a square-wave generator. For the maximally flat case, the output should be a pulse that has an undershoot of 13% relative to the maximum. The inserts in Fig. 3.14 show the appropriate output when the system is properly adjusted. The cutoff frequency is

$$f_{cut} = 1/1.3t \qquad (52)$$

where t is measured from the start of the pulse to where it has decayed to within 3% of its initial value. For curve 1, f_{cut} is calculated to be 96 kHz, and for curve 2 it is 247 kHz. This checks well with the sine wave results.

For the anemometer of Fig. 3.12, the optimization should be done at the mean velocity expected during measurements when the fluctuations are small. When large changes in mean velocity are expected, optimization should be done at the maximum velocity. As the velocity decreases, f_{cut} will also decrease, but much more slowly $(f_{cut}V)$ than the decrease in maximum frequency required $(f_{max}V^{7/4})$ in turbulent flow measurement [51, 52]. The shape of the output pulse should be adjusted carefully, using the controls. Improper adjustment will give a response curve (to velocity) that is not flat over the frequency range.

3.11.3 Large Velocity Fluctuations

An important characteristic of the constant-temperature anemometer is its ability to respond to large fluctuations. Freymuth [49] has analyzed the dynamic effects of large fluctuations. The results indicate that fluctuations of 50% of the mean value at a frequency of $f_{cut}/2$ will give an error in the mean of less than 0.1%, an error in the mean square of about 2.3%, and an error in the skewness of 0.3. It is this last error that is significant, since this compares with a value for isotropic turbulence of about 0.6. Alternatively, if the frequency is $f_{cut}/10$, then even the skewness error is only about 0.02, and the others are truly negligible.

To summarize, errors due to nonlinearities in the dynamic response to large amplitudes are negligible for most measurement conditions. This is especially true if f_{cut} is maintained as large as possible, so that the large-amplitude fluctuations are at frequencies of less than 10% of f_{cut}. This is also the range in which phase shifts are less than 12°. It is then safe to say that the "foolproof" or "safe" dynamic range of the thermal anemometer is the range below 10% of its cutoff frequency. Of course, in most applications a much larger range gives good results.

3.11.4 Dynamic Effects of Conduction Losses to the Supports

Even though T_m is held constant by the feedback electronics, other effects can influence response. Thermal lag of the probe support in nonisothermal flow is one of these, since the conduction losses to the supports depend on the support temperature T_s. At low frequencies, T_s will follow T_a, but for high-frequency temperature fluctuations, T_s will remain at some average temperature, owing to the heat capacity of the support. The frequency at which this happens can be expressed as

$$f_s = H_s(V)/2\pi c_s \qquad (53)$$

where $H_s(V)$ is the heat transfer per unit of temperature difference to the support and c_s is the heat capacity of the support. At frequencies well below f_s, $T_s = T_a$; at frequencies well above f_s, $T_s = \overline{T}_a$. This phenomenon results in an attenuation of the amplitude response above f_s (Fig. 3.15a). In the case of high-speed flows, T_a would be replaced by the recovery temperature of the support.

Since at high frequencies, $dK = 0$, one can write

$$\left.\frac{dH(f \gg f_s)}{dH(f \ll f_s)}\right|_{Nu=const} = \frac{d\phi + dK = 0}{d\phi + dK} = \varepsilon_2' \qquad (54)$$

The value of ε_2' is given by Eq. (19) and, for the low-overheat case, by Fig. 3.7. Thus for short wires with high end losses (small $\sqrt{C_0}\ell$), the dynamic error at high frequencies can be significant when compared with low-frequency or steady state conditions. For the conditions of Fig. 3.1, $\varepsilon_2' = 0.826$, giving an amplitude error of about 17% for temperature fluctuations.

Heat waves along the cylinder in isothermal flow can also cause deviation from the ideal response. As velocity changes, the value of $\sqrt{C_0}\ell$ changes, causing a change in temperature distribution along the wire (Fig. 3.6). According to [25], the frequency f_ℓ at which this occurs can be estimated as

$$f_\ell \simeq 6.4D/\ell^2 \qquad (55)$$

THERMAL ANEMOMETERS

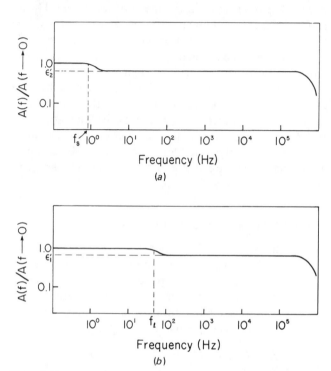

Figure 3.15 Dynamic effect of conduction loss to supports. (*a*) Effect of temperature changes of supports at Nu = const, T_a varying. (*b*) Effect of heat waves along sensor at T_a = const, Nu (velocity) varying

where D is the thermal diffusivity of the wire material. At frequencies well below f_ℓ the wire will have time for the temperature distribution to equilibrate. At frequencies well above f_ℓ, the wire will have a temperature distribution represented by some average velocity $(\sqrt{C_0}\ell)$. Again, this results in an attenuation of the amplitude response above f_ℓ (Fig. 3.15*b*). For the sensor of Fig. 3.1, $f_\ell \simeq 82$ Hz.

The ratio of the fluctuations at high frequencies to those at low frequencies can be represented as

$$\frac{dH(f \gg f_\ell)}{dH(f \ll f_\ell)} = \left.\frac{d\phi + dK = 0}{d\phi + dK}\right|_{T_a,T_M=\text{const}} = \varepsilon_1' \tag{56}$$

where ε_1' is given by Eqs. (18*a*) and (18*b*) and in Fig. 3.7 (for the low overheat case). For the wire of Fig. 3.1, $\varepsilon_1' = 0.953$, giving an amplitude error of less than 5% for velocity fluctuations.

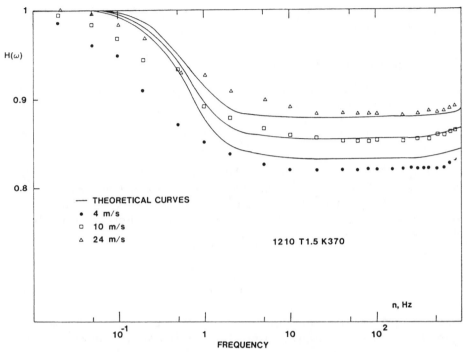

Figure 3.16 Amplitude transfer function of 1210-T1.5 probe: Standard-wire probe, plated sensor ends, 1.25-mm active length. (From [54])

This calculation assumes a relatively massive support whose temperature T_s is not significantly affected by the heat K conducted from the wire. However, the sensitive portion on most hot-wire and cylindrical hot-film sensors is isolated from the supports by a plated area. Effectively, the support starts at the plating, which can be described as a very small, fragile support. In this case, $T_s > T_a$, where T_s is now the temperature of the junction between the plated area and the sensitive area of the sensor. This can only decrease conduction losses K, resulting in ε and ε_2' being closer to 1. For ε_1' and ε_2' it adds another time constant that is between f_ℓ and f_s. Finally, according to Beljaars [53], the plated ends actually decrease the value of ε_1' for short sensors, so for this particular parameter, large supports, where $T_s \rightarrow T_a$, are desirable. On the other hand, plated ends are very helpful in reducing the aerodynamic influences of the support needles and probe body [29].

Experimental data have been collected by Paranthoen et al. [54] on the influence of the supports in a flow with temperature fluctuations. Figure 3.16 shows data for a hot wire similar to that in Fig. 3.1. At 10 m/s, the asymptotic value of ε_1' is about 0.85 (compared with 0.826 calculated previously). The higher number may be due to the influence of the plated ends on the hot wire. From Figure 3.16, it appears that the value of f_s (Fig. 3.15a) is about 0.2 Hz for the particular support configuration tested.

3.11.5 Attenuation of Heat Waves

The thermal boundary layer surrounding the sensors of medium thickness δ_T readjusts to new flow conditions only if

$$f \leq f_\delta \simeq \frac{D_a}{\pi \delta_T^2} \qquad (57)$$

where D_a is the thermal diffusivity of the fluid, and where dynamic flow effects have been neglected. For $f > f_\delta$, heat waves will be more and more attenuated, and thus the sensitivity of the sensor decreases. According to Clark [55], an increase in sensitivity may be observed prior to a decrease for bulky sensors in water or blood flow as a consequence of dynamic flow effects.

For thermal boundary layers,

$$\delta_T \simeq \left(\frac{D_a L}{V}\right)^{1/2} \qquad (58)$$

where L is a characteristic dimension of the sensor (equal to d for cylindrical sensors). Thus, by combining Eqs. (57) and (58), one arrives at an estimate for f_δ of

$$f_\delta \simeq \frac{V}{2\pi d} \qquad (59)$$

For the situation in Fig. 3.1, $f_\delta = 400 \times 10^3$ Hz, which is above the range of concern for $V = 10$ m/s. On the other hand, for an extremely bulky sensor used, for instance, in water or blood flow, $d = 1$ mm, $V = 1$ m/s, and $F_\delta \simeq 160$ Hz. Such a sensor might not be acceptable for some tasks and then would have to be replaced with a smaller one.

A similar attenuation effect can occur for a coated sensor at frequencies

$$f > f_{coat} \simeq \frac{D_{coat}}{\pi \delta_{coat}^2} \qquad (60)$$

where D_{coat} is the thermal diffusivity of coating material and δ_{coat} is the thickness of coating material.

This discussion of boundary layer lag is intended to be introductory. For a detailed study, the reader is referred to the thesis by Lueck [56].

3.11.6 Finite Resolution Due to Finite Sensor Size

To accurately follow flow fluctuations in flows that are not two-dimensional, the sensor length must be small in comparison with the wavelength of the maximum frequencies of interest. This kinematic effect was first considered by Frenkiel [57], and in more detail by Uberoi and Kovasznay [58] and Wyngaard [59, 60]. The results show that for one-dimensional spectra of turbulence, errors of the order of 5% show up for $f_s \simeq 0.08 V/2\ell$; at $f_{20} \simeq 0.5 V/2\ell$, the error is about 20%. Again, in Fig. 3.1 the value of f_s is 640 Hz, and f_{20} is 3200 Hz. This turns out to be one of the most restrictive conditions on the frequency response of thermal sensors.

3.12 NOISE IN CONSTANT-TEMPERATURE THERMAL ANEMOMETRY

The noise in any measurement system limits the minimum measurable change. This is important in hot-wire anemometry because measurements at low turbulence intensities and wide signal bandwidth are often desired. In addition to the usual increase in noise due to bandwidth, at fixed bandwidth the electronic compensation for the thermal lag of the sensor causes the noise amplitude to increase with center frequency the same as the sine wave signal of curve 1a of Fig. 3.13.

Noise can be classified as avoidable and unavoidable. Electronic pickup from power lines, radio or television stations, and magnetic fields can usually be eliminated by proper shielding, grounding, etc. However, Johnson noise from the bridge resistors and electronic noise generated by the bridge amplifier cannot be eliminated.

An analysis of the electronic noise in thermal anemometers [61] yields, as the input noise to the amplifier within a small frequency range,

$$\overline{u_{12,n}^2} = [K_a^2 + 4k_b T_a R \frac{(1 + n_b)(1 + R_2/R_1) + (T_m - T_a)/T_a}{(1 + n_b)^2}]\Delta f \qquad (61)$$

where K_a is the equivalent input noise of the amplifier, in V Hz$^{-1/2}$, and k_b is Boltzmann's constant. It is assumed that all resistors are at temperature T_a, except the sensor, which is at T_m.

With current amplifier technology, values of K_a of the order of 1.5×10^{-9} V Hz$^{-1/2}$ are attainable. The second term of Eq. (61) for the hot wire and bridge of Fig. 3.1 (with $n_b = 0.36$ and $R_2/R_1 = 1$) is $(0.6 \times 10^{-9}$ V Hz$^{-1/2})^2$. Therefore

$$\sqrt{\overline{u_{12,n}^2}} = 1.69 \times 10^{-9} \, \Delta f^{1/2} V$$

and the sensor and bridge resistors contribute only about 12% of the total noise.

Often R_2/R_1 is increased to reduce the current drain through R_2 and R_3 (Fig. 3.1). However, if $R_2/R_1 = 20$, the above calculation would give $\sqrt{\overline{u_{12,n}^2}} = 2.45 \times 10^{-9} \, \Delta f^{1/2}$ V. The sensor and bridge resistors now contribute over 60% of the noise.

In the range where thermal lag dominates and where the influence of noise is most critical, the output noise of the constant-temperature anemometer is

$$u_n = \frac{\pi(n_b + 1)^2}{n_b} \frac{R}{R - R_0} \frac{c}{H_1(V)} f \sqrt{u_{12,n}^2} \tag{62}$$

The velocity signal is

$$u_v = \frac{n_b + 1}{n_b} [H_1(V)(T_m - T_a)R]^{1/2} \frac{dH_1}{2H_1 \, dV} v \tag{63}$$

If we let $H_1(V) = B\sqrt{V}$, then

$$u_v = \frac{n_b + 1}{4n_b} [H_1(T_m - T_a)R]^{1/2} \frac{v}{V} \tag{64}$$

and the signal-to-noise ratio of the wire ($N_w = u_v/u_n$) becomes

$$N_w = \frac{\alpha R_a}{n_b + 1} \frac{B^{3/2} v^{3/4} (T_m - T_a)^{3/2}}{2\pi fc} \frac{v}{2R^{1/2}} \frac{1}{V} \frac{1}{\sqrt{u_{12,n}^2}} \tag{65}$$

Therefore, for a given v/V and equivalent input noise, to maximize the signal-to-noise ratio, one should do the following.

1. Operate at high overheat to maximize $T_m - T_a$.
2. Use a wire material with a high-temperature coefficient of resistance.
3. Keep n_b small in comparison with 1.
4. Use a thin wire to minimize thermal capacity c.

The most difficult measurements are at high velocities, where high frequencies are required. To get total output noise, Eq. (62) must be integrated over the frequency range to be observed after substitution for $\sqrt{u_{12,n}^2}$ from Eq. (61). For the wire of Fig. 3.1, this results in about a 4% noise contribution when measuring 0.1% turbulence with a 0- to 50-kHz bandwidth. At 10 m/s, a bandwidth of 2 kHz is more appropriate. Then less than 0.01% turbulence intensity can be measured with negligible noise contribution. Therefore it is very important to use a low-pass filter set at the maximum frequency of interest in the flow to minimize the noise contribution, even though the anemometer should be tuned for maximum response.

3.13 FILM SENSORS

The concept of film sensors was introduced in 1955 [3, 4] and has become a major addition to sensor technology. Film sensors have been particularly useful for measurements in liquids or gases with particle contamination and in conducting liquids, especially water.

As shown in Fig. 3.17a, the hot wire is a homogeneous, electrically conducting material. On cylindrical film sensors (Fig. 3.17b) the substrate is an electrical insu-

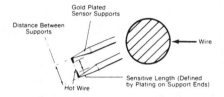

a CROSS SECTION OF HOT-WIRE SENSOR

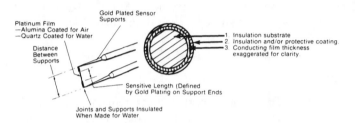

b CROSS SECTION OF HOT-FILM SENSOR

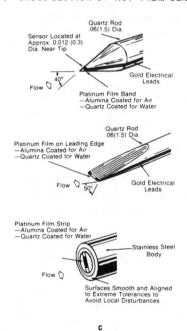

c

Figure 3.17 Thermal sensor configuration. (*a*) Cross section of hot-wire sensor. (*b*) Cross section of hot-film sensor. (*c*) Shapes of film sensors

153

lator, with the conducting film deposited on the surface. This permits selection of the substrate for its strength and low thermal conductivity, while the conducting film can be selected for its electrical properties. In addition, since the sensitive portion is on the surface, the frequency response of a film sensor is superior to that of a hot wire of similar dimensions when operated with a constant-temperature anemometer. A thin overlayer of insulating material is often added for electrical insulation from the fluid or to protect the metal film from erosion by particles in the fluid.

The construction of film sensors permits flexibility in shape, as shown in Fig. 3.17c. While the shapes in Fig. 3.17c have advantages in terms of strength and ability to remain clean in contaminated fluids, they do give rise to some special problems in frequency response [57] not shared by the simple cylindrical configuration.

3.13.1 Cylindrical Film Sensors

Cylindrical film sensors (Fig. 3.17) share most of their measurement characteristics with hot wires. From the performance viewpoint, the primary advantage of hot-wire sensors is that, with present technology, they can be made much smaller in diameter. A typical hot wire is 4 μm in diameter, while typical film sensors are 50 μm in diameter. Using larger-diameter hot wires results in either excessive length or excessive conduction losses to the supports. The low thermal conductivity of film sensor substrates (typically less than 1% of that of hot-wire materials) permits relatively short sensors, while conduction losses to the supports are maintained at permissible levels.

The larger diameter of film sensors has the following advantages for air measurements.

1. In most applications, particles in the fluid will not break and cannot strain the sensor.
2. Smaller particles will not intercept the sensor because of its larger diameter.
3. Since the sensor is rigid, it is always effectively "taut," greatly improving the repeatability on angular sensitivity for X probes and other multisensor measurements.

On the other hand, the primary disadvantages are as follows.

1. At Re > 150 (V > 50 m/s for a 50-μm-diameter sensor in air), self-generated "turbulence intensity" may limit the performance of large-diameter cylindrical sensors for the measurement of low-turbulence intensities.
2. The recommended operating temperature of film sensors is below 370°C, with an absolute maximum of 760°C.

Another factor is the generally greater cost of film sensors. However, if particles in the fluid are breaking wires or contamination is causing signal calibration shifts, the longer life of film sensors more than compensates for their higher initial cost.

In tap water, seawater, and other conducting liquids, film sensors are used almost to the exclusion of hot wires for velocity measurements. The ability to isolate the sensor from the fluid with a thin overlayer of insulating material is the primary reason for this. Of course, the problem of self-generated turbulence at high Re remains, and will occur at much lower velocities in water than in air. In addition, at sufficiently high velocities, cavitation can occur. Even so, cylindrical films are used extensively.

3.13.1.1 Construction and operation. To date, the most common substrate materials used for cylindrical films have been pyrex and quartz because they can be drawn to small diameters. Platinum and nickel are the most common conducting films. Protective layers of pyrex, quartz, and alumina have been used. Alumina is particularly effective in preventing erosion of the sensor, since it is harder than most contaminants in air.

Regarding the length-to-diameter requirements, the conflicting criteria given in section 3.3 are valid for cylindrical films. The discussions of sections 3.4 and 3.5 also hold for film sensors, with one exception. Film sensors cannot be mounted on stiff supports if they are to be exposed to shocks or large transients in the flow. The needle supports must have some flex, since the films themselves are quite rigid.

The basic operation of the constant-temperature anemometer is the same for both films and wires. Constant-current compensation has not been used with films, owing to the difficulty of matching the frequency characteristics of films (curve 2 in Fig. 3.13). Calibration procedures are, of course, identical.

3.13.1.2 Heat transfer. The heat transfer equations cited for hot wires are also valid for cylindrical films. The formula $GR^{1/3} > Re$ for calculating when free convection is important is not affected by sensor diameter. However, the Knudsen number for a 50-μm film, if substituted for the wire in Fig. 3.1, is 0.0016. This is well into the continuum flow region, whereas the wire was in slip flow. Similarly, the recovery factor for the larger film sensors is closer to 1.0.

One difference from hot wires is that the larger diameter of film sensors and their small thermal conductivity guarantee that the surface temperature of the film sensor is not uniform circumferentially. In fact, the operation of split films [9] depends on this. At the same time, the influence of this phenomenon for normal cylindrical film sensors has not been investigated.

Because of the larger diameters of film sensors, conduction to the walls will influence readings at larger distances from the wall than for fine hot wires.

3.13.1.3 Conduction to the supports. The equations of section 3.8 hold for films, with the film substrate properties substituted for the wire properties. The electrically conductive film is so thin that its heat conduction along the sensor is less than 2% of the total for a 50-μm-diameter film sensor. Of course, if very small diameter film sensors were used, this would no longer be true.

Again, substituting a film sensor 50 μm in diameter and 1 mm long in Fig. 3.1 gives the following comparisons with the hot wire, according to the equations of section 3.8:

Parameter	Hot wire	Film
$2\ell/d$	333	20
$\sqrt{C_0}\ell$	2.93	3.48
ε	0.826	0.820
ε_1'	0.953	0.931

Therefore the effects of conduction losses to the supports are similar for a tungsten wire and a film sensor whose $2\ell/d$ ratio is approximately one-sixteenth that of the tungsten wire.

3.13.1.4 Angle sensitivity and support interference. Work by Friehe and Schwarz [27] and Jorgenson [31] indicates that cylindrical film sensors and hot wires give very similar directional response. This is somewhat surprising, since the $2\ell/d$ ratio of film sensors is generally lower by a factor of 10–20. One advantage of the film is its rigidity, giving the potential for better repeatability of measurements. In any case, the discussion in section 3.9 is valid for cylindrical film sensors as well as hot wires. The same is true for section 3.10, with one exception. As is pointed out in the next section, compared with hot wires, the hot films make rather slow sensors for the direct measurements of temperature (no frequency compensation) because of their larger diameters.

3.13.1.5 Dynamic response. Freymuth [62] has extended the dynamic theory of the constant-temperature hot-wire anemometer to constant-temperature cylindrical hot-film anemometers. The result gives the following value for the cutoff frequency:

$$f_{\text{cut}} = \frac{0.9}{2\pi} \left(\frac{G/M''}{M\omega_c^{1/2}} \right)^{2/5} \tag{66}$$

Above ω_c, the heat transfer of the film is dominated by the skin effect. Its definition is

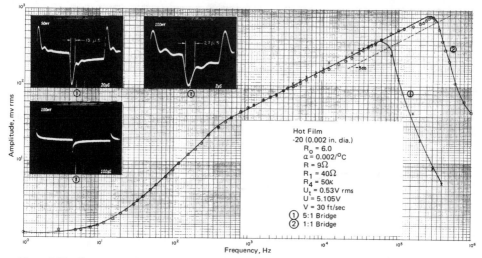

Figure 3.18   Sine wave and square-wave tests on TSI-20 hot-film sensor at 30-ft/s airflow

$$\omega_c = \frac{4D_{su}}{d^2} \tag{67}$$

where D_{su} is the diffusivity of the substrate. As a consequence of the influence of ω_c, $f_{cut} \simeq d^{1/5}$. This implies a cutoff frequency nearly independent of diameter, but one that actually increases as the diameter increases. This is in contrast to the hot wire, for which $f_{cut} \simeq d^{-1/2}$.

3.13.1.6 Optimization and electronic testing. Figure 3.13 gives a qualitative representation of Freymuth's dynamic theory for both hot-wire and hot-film sensors. Curve 2 is the response of an uncompensated film sensor showing the expected $f^{-1/2}$ response at high frequencies. Curve 2a is the response to sine wave testing when operated with a constant-temperature anemometer, and curve 2ac is the response of the same system to velocity changes.

Figure 3.18 shows experimental data taken via the sine wave test on a TSI model 1050 anemometer [50] with a 50-μm-diameter hot-film sensor. Again, the experimental points follow the theory closely. However, both sets of points deviate from the $f^{1/2}$ line by a few percent. This was predicted by [62]. Although this implies some lack of flatness, even for an optimally adjusted system, the error of a few percent is generally small enough to be negligible in high-frequency measurements. If it is important, the sine wave test can be used to determine what the error is, and the appropriate corrections can be made to the data.

Again, the response of the sensor alone can be taken from Fig. 3.18; it is found

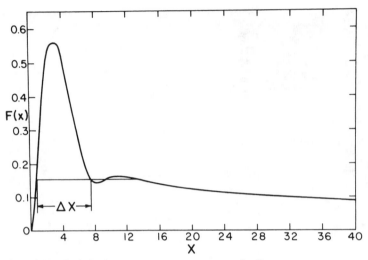

Figure 3.19 Optimized response to square-wave test for film sensors

to be about 13 Hz, compared with about 700 Hz for the hot wire. At the same time, f_{cut} for the film is about 77 kHz, which is comparable to 96 kHz found for the hot wire. This shows the influence of the $f^{1/2}$ region in making the film sensor easier to compensate to high frequencies. The value of f_c from Eq. (67) is 216 Hz. This compares with the 350 Hz from Fig. 3.18.

The theory for square-wave testing of film sensors has been developed [64]. Figure 3.19 shows the optimized output for a step input. With $t (= \Delta x)$ measured as shown in Fig. 3.19,

$$f_{cut} = \frac{1}{1.04t} \simeq \frac{1}{t} \tag{68}$$

The ratio of the level midway between the minimum and maximum of the resonance to the peak height should be 0.28. The $1/\sqrt{x}$ "tail" is an inherent characteristic of the step response of film sensors; hot wires do not exhibit such a tail.

3.13.1.7 Large velocity fluctuations. The nonlinear effects of large velocity fluctuations for film sensors are very similar to those for hot wires [64]. Therefore the earlier comments concerning hot wires are appropriate: for more details, the reader should refer to [64].

3.13.1.8 Dynamic effects of conduction losses to the supports. Again, the data and discussion for hot wires are appropriate for film sensors. In section 3.13.1.3, values of ε_1' and ε_2' for a film were shown to be similar to those for a wire with a much larger $2\ell/d$ ratio. Figure 3.15 represents qualitatively the expected behavior for a film, with the value of f_s being similar for a similar support. However, the value of f_ℓ for the 50-μm-diameter film sensor is about 5.4 Hz.

For film sensors the larger diameter (and therefore larger total heat transfer), along with the somewhat flexible supports, makes the assumption $T_s = T_a$ even more suspect than for wires. In addition, film sensors are isolated from the supports by heavy plating. Therefore the comments at the end of section 3.11.4 are particularly appropriate for cylindrical film sensors.

3.13.1.9 Other dynamic effects. The ability of the thermal boundary layer to respond to new flow conditions is inversely proportional to diameter, Eq. (59). Therefore, while the 4-μm hot-wire of Fig. 3.1 would have a boundary layer response of 400 kHz, for a 50-μm film sensor exposed to the same conditions, the response is 32 kHz. This is still more than adequate for most applications. However, it does point out that even in gases, there may be conditions under which the thermal boundary of a large film sensor would attenuate high frequencies.

Arguments regarding spatial resolution are based on sensor length and are identical for both films and wires.

3.13.1.10 Noise. As pointed out in section 3.12, output noise increases with frequency in the same manner as sinusoidal electronic test signals. From Fig. 3.13 it can be observed that the rate of increase of noise with a film sensor will be less than that with a hot wire at high frequencies, owing to the $f^{1/2}$ response region. At the same time, the large diameter of the film sensor provides significant noise amplification between 10 and 10^2 Hz. Figure 3.20 shows a comparison of the signal-to-noise ratio for a 4-μm-diameter wire and a 50-μm-diameter film exposed to 30 ft/s in air. The larger noise value for the film at about 1 kHz is not a problem for almost all practical applications of film sensors.

3.13.2 Noncylindrical Film Sensors

Figure 3.17c shows three configurations of noncylindrical film sensors. The cone-shaped sensor is the most frequently used, with the flush mount often used to examine boundary layers and shear stress. The shapes of these sensors avoid the self-generated turbulence that causes a problem with the larger cylindrical sensors. In addition, they can be used in high-speed flows with no cavitation problems. Finally, their configuration also minimizes problems due to contaminants in the flow.

The major difficulty with noncylindrical sensors has been proper interpretation of the amplitude response as a function of frequency. At the time they were originally suggested, the dynamic effects of conduction losses were not considered. As pointed out earlier, even with cylindrical sensors, these effects can be of the order of 10% in air. The design of noncylindrical films substantially increases these effects in gases. Fortunately, the greatest need for noncylindrical sensors is in liq-

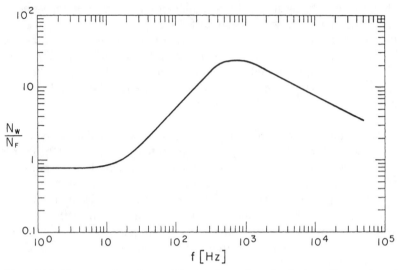

Figure 3.20 Frequency dependence of the signal-to-noise ratio of a 4-μm diameter hot wire compared with that of a 50-μm diameter film sensor for 30-ft/s air velocity

uids, where the high convective heat transfer reduces the dynamic effect of conduction losses to the order of 10%.

3.13.2.1 Conduction losses. For cylindrical hot wires and hot films, end losses are an important part of the heat transfer. For noncylindrical hot-film sensors, heat transfer into areas not covered by the film plays a similar role and is appropriately called "side losses." An estimation of such losses is complicated by the three-dimensional nature of heat transfer through the substrate. Bellhouse and Schultz [63] explained the phenomena with a simple, empirical one-dimensional model. Freymuth [65] combined this model with the anemometer response to predict the results of a test signal. While these theoretical results are recommended for clarifying the phenomena and predicting trends, for purposes of this review, some actual test data seem appropriate.

Figure 3.21 shows amplitude versus frequency for three conical sensors for two different velocities. These experimental results were obtained by comparing the spectrum from a hot wire with that of the noncylindrical sensor in a turbulent flow field [66].

For a given sensor the final asymptote depends primarily on Biot number. According to an extrapolation in [66], the asymptote would be over 0.9 at 3 m/s for water compared with the value of ~ 0.2 for air shown in Fig. 3.21.

The use of noncylindrical films for research measurements in gases has been limited by the difficulties in interpreting transient data. Although wires and cylin-

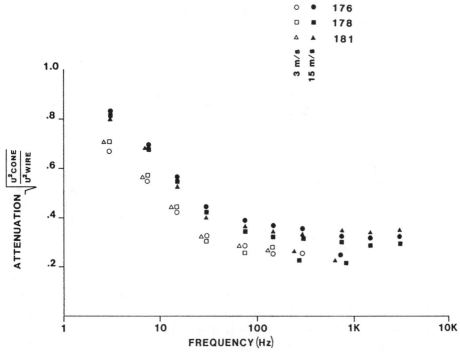

Figure 3.21 Experimental data on amplitude response versus frequency for three conical sensors in air

drical films also have conduction losses, they are small (less than 10%), can be estimated quite accurately, and can be made smaller by increasing the value of $2\ell/d$. For noncylindrical sensors in air, as shown in Fig. 3.21, these effects are often 50% or more.

The ideal solution is a support material whose thermal conductivity approaches zero. Thin-film gages utilizing plastic have been tried [67]. Certainly, the construction of noncylindrical sensors opens the possibility of a wide range of materials. This flexibility is, of course, limited by the need for very high stability of the conductive film for good measurements.

3.13.2.2 Electronic testing. Figure 3.22 shows experimental data from a sine wave test on a conical sensor. As with the cylindrical film sensors, the flatness of the response at high frequencies can be determined by how well the sine wave results follow the $f^{1/2}$ line. As can be observed in Fig. 3.22, they follow quite well from 200 Hz to 100 kHz, or a range of about 500:1. It is below 200 Hz that attenuation due to side losses occurs. This is consistent with Fig. 3.21. In other words, to completely interpret this part of the curve requires a detailed calculation of the transient heat transfer in the substrate of a dynamic calibration such as Fig. 3.21.

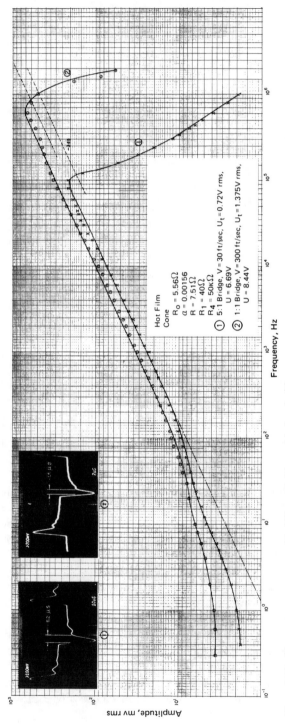

Figure 3.22 Tests on a cone sensor with different bridges and air velocities

162

The value of f_{cut} for the cone is about 122 kHz, compared with 77 kHz for the 50-μm-diameter cylindrical film. This is at least partial confirmation of the rather surprising prediction that f_{cut} increases with diameter ($\sim d^{1/5}$). The 3-dB point of the cone sensor by itself is, from Fig. 3.21, about 2.5 Hz in air at 30 ft/s.

A first theoretical attempt at sine wave testing according to the Bellhouse-Schultz model is that of Freymuth [65]. It remains to be seen whether sine wave testing can become a quantitative tool for assessing the dynamic side-loss as well as end-loss effects of heat conduction for noncylindrical films and short hot wires. Up to now, such effects could only be assessed quantitatively by "direct calibration," in which the probe is exposed to appropriate velocity or temperature fluctuations and its frequency response is directly measured. Relevant methods of direct calibration are probe shaking [68], probe rotation of a slanted wire such that there is an oscillation of angle of attack [69], exposure to a pulsating flow [70], and exposure to a turbulent flow with known spectrum [71], originally proposed by Bellhouse and Schultz [63].

3.13.2.3 General Comments. While the flexibility in shape of noncylindrical sensors can be very promising, applications have been limited because of difficulties in interpreting transient and, in some cases, even steady state signals. Therefore the volume of data available on noncylindrical sensors is very small in comparison with that for cylindrical configurations.

For steady state measurements, calibrations are performed. However, the calibration cannot be extended reliably using heat transfer relations because of inadequate data as well as, perhaps, the lack of repeatability between sensors. Similarly, the angle sensitivities of noncylindrical sensors have not been studied extensively. Ling [5] gathered some data on wedge sensors, and Frey [72] has some data on cones. The cone sensor is relatively insensitive to changes in flow direction.

3.14 CONSTANT-CURRENT OPERATION

In constant-temperature operation there must always be a flow of heat from the sensor to its environment. This means that the sensor temperature must always be above the environment temperature. If the environment temperature is changing, then the maximum environment temperature determines the minimum sensor temperature.

The constant-current anemometer does not have this limitation. Therefore, in high-speed flows where several overheats are used to separate the modes, the constant-current anemometer is still often used. At very low heating currents the compensated constant-current hot-wire anemometer essentially becomes a fast thermometer.

As pointed out by Freymuth [61], the noise is identical in constant-current and constant-temperature systems if a similar amplifier and bridge are used. Since

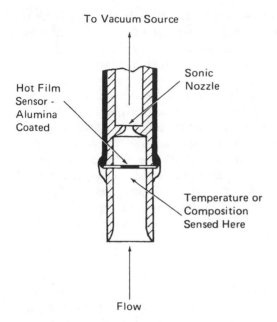

To Vacuum Source

Hot Film
Sensor -
Alumina
Coated

Sonic
Nozzle

Temperature or
Composition
Sensed Here

Flow

Figure 3.23   Aspirating probe

noise increases as overheat decreases, Eq. (65), there may be a minimum overheat at which an adequate signal-to-noise ratio is determined. If this overheat is at the point where $T_m > T_a$ at all times, then the advantage of the constant-current system would seem to vanish. Then the ability to measure large fluctuations, ease of optimizing response, ability to compensate films for frequency, and even the simplification of the equations [22] would favor constant-temperature operation.

3.15 OTHER MEASUREMENT TECHNIQUES AND APPLICATIONS USING THE CONSTANT-TEMPERATURE ANEMOMETER

The usual application of the thermal anemometer is the measurement of the details of flow velocity and direction. However, the unusual sensitivity, small size, and good response have led to the adaptation of the anemometer for other measurements.

3.15.1 Aspirating Probe

If the flow past a thermal sensor is controlled by a sonic nozzle, then velocity past the sensor depends only on gas composition, pressure, and temperature. In other

THERMAL ANEMOMETERS

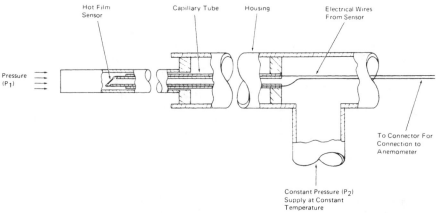

Figure 3.24 Pressure probe

words, the effect of environmental velocity has been removed. Velocity will influence pressure if the probe is pointed directly upstream.

Figure 3.23 is a sketch of an aspirating probe with the sensor in front of the sonic nozzle. Blackshear and Fingerson [73] used a probe of this type for concentration measurements in a helium jet mixing with air. Brown and Rebollo [74] used a smaller probe with the sensor downstream of the sonic nozzle for measuring binary mixtures. A similar probe has been used to detect methane in spill experiments [75]. In constant-composition flows, the same probe configuration can be used for temperature measurements.

3.15.2 Pressure Measurement

Pressure is another flow variable of interest. Figure 3.24 shows a configuration that has been used to measure pressure fluctuations. Spencer and Jones [76], Planchon [77], and Jones et al. [78] used a probe of this type to measure static-pressure fluctuations after some modifications to the tip. Remenyik and Kovasznay [79] and Wills [80] used similar systems to measure wall-pressure fluctuations using hot wires.

3.15.3 Split-Film Sensors

Figure 3.25 shows a split-film sensor and the associated concept. These devices have found applications in sensing angle of attack [81] as well as in some turbulence

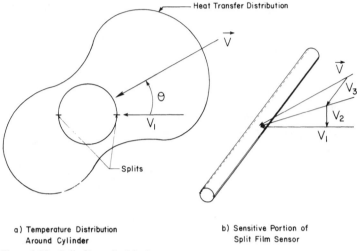

a) Temperature Distribution
   Around Cylinder

b) Sensitive Portion of
   Split Film Sensor

Figure 3.25 Split-film cylindrical sensor

measurements [82, 83]. As pointed out by Spencer and Jones [84] and Young [71], the normal 300-μm-diameter split films suffer from attenuation at higher frequencies. Spencer and Jones [84] did not find this to be true for a 50-μm split film.

3.15.4 Nonresearch Applications

Thermal sensors are finding increasing application in more routine test and control applications. Figure 3.26 shows a basic configuration for measuring the total mass flow of a gas. Variations of this configuration are used to measure very low pressure differences by measuring the flow through a "leak." Examples are room-pressure controls and controls to maintain the face velocity on fume hoods. These are applications where the sensitivity to very low flows is especially useful.

By using relatively large sensors, the sensitivity to contamination and fragility can be reduced. These larger sensors have been used to measure flow in ducts as well as on hand-held anemometers to measure air distribution in rooms. Several companies are presently even using thermal sensors to measure the airflow in automobile engines to control the fuel/air ratios.

3.16 CONCLUSION

The constant-temperature hot-wire anemometer satisfies most of the criteria for an ideal tool for point measurements of flow velocity. Its primary disadvantages are practical ones such as fragility and calibration shifts due to contamination. In addition, to maintain accuracy, all heat transfer effects must be considered.

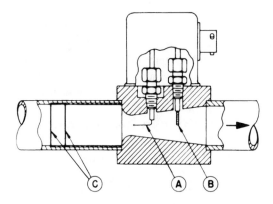

Figure 3.26 Mass flow transducer

The cylindrical hot-film sensor overcomes some of the practical problems of the fine hot-wire sensor and is quite well understood. In fact, its operation is basically identical to that of a hot wire in most respects.

The noncylindrical hot-film sensor can further reduce the fragility and contamination effects, but at present, the interpretation of the dynamic response is difficult, especially in gases. However, noncylindrical film sensors such as surface sensors permit measurements not readily available using hot wires.

NOMENCLATURE

a	radius of sensor
a_T	overheat ratio or temperature loading, $= (T_m - T_a)/T_a$
A	constant in various heat transfer correlation equations
b	distance from sensor to wall
B	constant in various heat transfer correlation equations
c	thermal capacity of sensor
C	constant in heat transfer equation for supersonic flow
C_p	specific heat of fluid at constant pressure
C_v	specific heat of fluid at constant volume
C_0	parameter that, when multiplied by ℓ^2, normalizes equation for conduction losses to the supports
d	sensor diameter
D	thermal diffusivity
D_a	thermal diffusivity of the fluid
D_{coat}	thermal diffusivity of coating material on sensor
D_{su}	thermal diffusivity of substrate of film sensors
f	frequency (Hz)

f_{coat}	frequency where attenuation due to a coating on the sensor becomes significant
f_{cut}	frequency where amplitude has decreased 3 dB (cutoff frequency)
f_ℓ	transition frequency where longitudinal temperature distribution is no longer in equilibrium with the environment
f_s	transition frequency where sensor supports (prongs) no longer follow changes in environment temperature
f_δ	transition frequency where thermal boundary layer around sensor is no longer in equilibrium
f_5	frequency where spatial resolution limits cause turbulence measurement error of 5%
f_{20}	frequency where spatial resolution limits cause turbulence measurement error of 20%
g	gravitational constant
G	amplifier gain
Gr	Grashof number, $= \rho^2 g_c \beta_N (T_m - T_a) d^3/\mu^2$
h_c	convective heat transfer coefficient
h_t	total pressure in height of flowing fluid
H	heat transfer from sensor to its environment
I	electrical current in sensor
k	thermal conductivity
k_a	equivalent input noise of amplifier
k_b	Boltzmann's constant
k_N	factor for deviation of measured velocity with flow normal to supports (prongs)
k_T	factor for cosine law deviation of flow tangential to sensor
K	heat conducted to sensor supports
Kn	Knudsen number, $= \lambda/d$
ℓ	half length of sensor
L	characteristic length, $= d$ for cylindrical sensor
M	time constant of the sensor
M''	second-order time constant of bridge-amplifier system
Ma	Mach number
n	exponent of Reynolds number (velocity) in heat transfer correlations
n_b	ratio of sensor resistance R to resistor in series with sensor R_1 when bridge is in balance
N_w	signal-to-noise ratio of the wire
Nu	Nusselt number, $= h_c d/k$
P	electrical power to sensor
Pr	Prandtl number, $= \mu C_p/k$
R	sensor resistance at temperature T_m
R_a	sensor resistance at temperature T_a
R_r	resistance of sensor at reference temperature T_r

R_0	gas constant
R_1	resistor in series with sensor
R_2	bridge resistor (see Fig. 3.1 or 3.10)
Re	Reynolds number, $= Vpd/\mu$
Re_0	Reynolds number for heat transfer correlation in supersonic flow, $= \rho_\infty V_\infty d_\infty/\mu_0$
t	time
T_a	ambient temperature of the fluid
T_f	film temperature, $= (T_m + T_a)/2$
T_m	mean temperature of sensitive area of sensor
T_r	reference temperature
T_{re}	recovery temperature
T_s	support temperature
T_{st}	static temperature of fluid
T_0	total temperature of fluid
u_n	output voltage due to noise
u_v	output voltage due to velocity changes
$u_{12,n}$	voltage due to noise at amplifier input
U	output voltage of constant-temperature anemometer
U_B	voltage impressed on bridge of anemometer, $= U$ for constant-temperature anemometer
U_T	test signal
U_{12}	bridge off-balance voltage, equal to amplifier input voltage
v	small fluctuations in velocity V (subscripts as for V)
V	velocity vector at sensor location
$V_{A,eff}, V_{B,eff}, V_{C,eff}$	effective velocity as seen by sensors A, B, and C, respectively
V_{BN}	velocity vector normal to sensor and perpendicular to supports
V_{eff}	effective cooling velocity past sensor, equivalent value of V_N
V_N	velocity vector normal to sensor and parallel to supports
V_T	velocity vector tangent to sensor axis
V_1, V_2, V_3	orthogonal components of V relative to flow facility
V_∞	velocity at free stream conditions in high-speed flow
α	temperature coefficient of resistance of sensor
α_1	angle between velocity vector and sensor axis
β_v	volume coefficient of expansion, $= 1/T_a$
γ_h	ratio of specific heats, $= C_p/C_v$
δ_{coat}	thickness of coating material on sensor
δ_T	thermal boundary layer thickness
ε	ratio of convective to total heat transfer from sensor to surroundings
ε_1'	change in ε due to changes in Nu with T_a constant
ε_2'	change in ε due to changes in T_a with Nu constant
η	recovery factor in high-speed flows

λ molecular mean free path

μ dynamic viscosity of fluid

ξ $= \sqrt{C_0 \ell} \coth \sqrt{C_0 \ell}$

π constant, $= 3.1416$

ρ density of fluid

ϕ convective heat transfer between sensor and surrounding fluid

ω frequency, $= 2\pi f$ (radians/s)

ω_c transition frequency where skin effects start to dominate on cylindrical film sensors

Subscripts

a   conditions at ambient temperature T_a

f   conditions at film temperature T_f

s   sensor support

su   substrate

w   wire

0   stagnation conditions in high-speed flow

∞ free stream conditions in high-speed flow

REFERENCES

1. P. Freymuth, A Bibliography of Thermal Anemometry, TSI Incorporated, 1992.
2. P. Freymuth, History of Thermal Anemometry, in M. P. Cheremisinoff (ed.), *Handbook of Fluids in Motion,* pp. 79–91, Ann Arbor Publishers, 1983.
3. L. V. King, On the Convection of Heat from Small Cylinders in a Stream of Fluid: Determination of the Convection Constants of Small Platinum Wires, with Applications to Hot-Wire Anemometry, *Proc. R. Soc. London,* vol. 90, pp. 563–570, 1914.
4. H. H. Lowell and N. Patton, Response of Homogeneous and Two-Material Laminated Cylinders to Sinusoidal Environmental Temperature Change, with Applications to Hot-Wire Anemometry and Thermocouple Pyrometry, NACA TSN 3514, 1955.
5. S. C. Ling, Measurements of Flow Characteristics by the Hot-Wire Technique, Ph.D. thesis, University of Iowa, Iowa City, 1955.
6. G. Comte-Bellot, Hot-Wire Anemometry, *Annu. Rev. Fluid Mech.,* vol. 8, pp. 209–231, 1976.
7. C. L. Landreth, R. J. Adrian, and C. S. Yao, Double-Pulsed Particle Image Velocimeter with Directional Resolution for Complex Flows, *Exp. Fluids,* vol. 6, pp. 119–128, 1988.
8. Y. Yeh and H. Cummins, Localized Fluid Flow Measurement with a HeNe Laser Spectrometer, *Appl. Phys. Lett.,* vol. 4, p. 176, 1964.
9. J. G. Olin and R. B. Kiland, Split-Film Anemometer Sensors for Three-Dimensional Velocity-Vector Measurement, *Proc. Symp. on Aircraft Wake Turbulence,* pp. 57–79, Plenum, New York, 1971.
10. H. H. Lowell, Design and Application of Hot-Wire Anemometers for Steady-State Measurements at Transonic and Supersonic Air Speeds, NACA TN 2117, 1950.
11. V. A. Sandborn, *Resistance Temperature Transducers,* Metrology Press, Fort Collins, Colo., 1972.

12. W. G. Rose, Some Corrections to the Linearized Response of a Constant-Temperature Hot-Wire Anemometer Operated in a Low-Speed Flow, *Trans. ASME J. Appl. Mech.*, vol. 29, pp. 554–558, 1962.

13. S. P. Parthasarathy and D. J. Tritton, Impossibility of Linearizing Hot-Wire Anemometer for Turbulent Flows, *AIAA J.*, vol. 1, pp. 1210–1211, 1963.

14. D. C. Collis and M. J. Williams, Two-Dimensional Convection from Heated Wires at Low Reynolds Numbers, *J. Fluid Mech.*, vol. 6, pp. 357–389, 1959.

15. C. F. Dewey, A Correlation of Convective Heat Transfer and Recovery Temperature Data for Cylinders in Compressible Flow, *Int. J. Heat Mass Transfer*, vol. 8, pp. 245–252, 1965.

16. F. H. Champagne, Turbulence Measurements with Inclined Hot Wires, Rep. 103, Boeing Scientific Research Laboratories, Flight Science Laboratory, 1965.

17. H. Kramers, Heat Transfer from Spheres to Flowing Media, *Physics*, vol. 12, pp. 61–80, 1946.

18. L. S. G. Kovasznay and S. I. A. Toernmark, Heat Loss of Wires in Supersonic Flow, Bumblebee Ser. Rep. 127, 1950.

19. L. V. Baldwin, V. A. Sandborn, and J. C. Lawrence, Heat Transfer from Transverse and Yawed Cylinders in Continuum, Slip and Free Molecular Air Flows, *Trans. ASME Ser. CJ. Heat Transfer*, vol. 82, pp. 77–78, 1960.

20. M. V. Morkovin, Fluctuations and Hot-Wire Anemometry in Compressible Flows, AGARDO Graph 24, 1956.

21. M. V. Morkovin and R. E. Phinney, Extended Application of Hot-Wire Anemometry to High-Speed Turbulent Boundary Layers, Rep. AFOSR TN 58-469, Johns Hopkins Univ., Dep. of Aeronautics, 1958.

22. C. L. Ko, D. K. McLaughlin, and T. R. Troutt, Supersonic Hot-Wire Fluctuation Data Analysis with a Conduction End-Loss Correction, *J. Phys. E*, vol. 11, pp. 488–494, 1978.

23. J. O. Hinze, *Turbulence*, 2nd ed., McGraw-Hill, New York, 1975.

24. J. A. B. Wills, The Correction of Hot-Wire Readings for Proximity to a Solid Boundary, *J. Fluid Mech.*, vol. 12, pp. 388–396, 1962.

25. P. Freymuth, Engineering Estimates of Heat Conduction Loss in Constant-Temperature Thermal Sensors, *TSI Q.* vol. 5, no. 3, pp. 3–8, 1979.

26. G. Comte-Bellot, A. Strohl, and E. Alcaraz, On Aerodynamic Disturbances Caused by Single Hot-Wire Probes, *J. Appl. Mech. Trans. ASME*, vol. 38, pp. 767–774, 1971.

27. C. H. Friehe and W. H. Schwarz, Deviations from the Cosine Law for Yawed Cylindrical Anemometer Sensors, *J. Appl. Mech. Trans. ASME*, vol. 35, pp. 655–662, 1968.

28. J. C. Bennet, Measurement of Periodic Flow in Rotating Machinery, *AIAA 10th Fluid and Plasmadynamic Conf.*, pp. 770–713, 1977.

29. R. E. Drubka, J. Tan-Atichat, and H. M. Nagib, On Temperature and Yaw Dependence of Hot Wires, IIT Fluids and Heat Transfer Rep. R77-1, Illinois Institute of Technology, 1977.

30. A. Strohl and G. Comte-Bellot, Aerodynamic Effects Due to Configuration of X-Wire Anemometers, *J. Appl. Mech. Trans. ASME*, vol. 40, pp. 661–666, 1973.

31. F. E. Jorgensen, Directional Sensitivity of Wire and Fiber Film Probes, DISA Information 11, pp. 31–37, 1971.

32. F. E. Jerome, D. E. Guitton, and R. P. Patel, Experimental Study of the Thermal Wake Interference Between Closely Spaced Wires of an X-Type Hot-Wire Probe, *Aeronaut. Q.*, vol. 22, pp. 119–126, 1971.

33. G. Fabris, Probe and Method for Simultaneous Measurement of "True" Instantaneous Temperature and Three Velocity Components in Turbulent Flow, *Rev. Sci. Instrum.*, vol. 49, pp. 654–664, 1978.

34. N. K. Tutu and R. Chevray, Cross-Wire Anemometry in High Intensity Turbulence, *J. Fluid Mech.*, vol. 71, pp. 785–800, 1975.

35. P. K. Maciejewski and R. J. Moffat, Interpreting Orthogonal Triple-Wire Data from Very High Turbulence Flows, *J. Fluids Eng.*, vol. 116, pp. 463–468, 1994.

36. L. M. Fingerson, Practical Extensions of Anemometer Techniques, *Adv. Hot Wire Anemometry*, pp. 258–275, 1968.
37. H. W. Tielmann, K. P. Fewell, and H. L. Wood, An Evaluation of the Three-Dimensional Split Film Anemometer for Measurements of Atmospheric Turbulence, College of Engineering, Virginia Polytechnia Institute and State University, Blacksburg, 1973.
38. I. C. Lekakis, R. J. Adrian, and B. G. Jones, Measurement of Velocity Vectors with Orthogonal and Non-Orthogonal Triple-Sensor Probes, *Exp. Fluids* (in press).
39. H. Fujita and L. S. G. Kovasznay, Measurement of Reynolds Stress by a Single Rotated Hot-Wire Anemometer, *Rev. Sci. Instrum.*, vol. 39, pp. 1351–1355, 1968.
40. L. R. Bissonnette and S. L. Mellor, Experiments on the Behaviour of an Axisymmetric Turbulent Boundary Layer with a Sudden Circumferential Strain, *J. Fluid Mech.* vol. 63, pt. 2, 1974.
41. DeGrande, Three Dimensional Incompressible Turbulent Boundary Layers, Ph.D. thesis, Vrye Univ., Brussels, 1977.
42. P. Kool, Determination of the Reynolds-Stress Sensor with a Single Slanted Hot Wire in Periodically Unsteady Turbomachinery Flow, ASME Publication 79-GT-130, 1979.
43. W. Rodi, A New Method of Analyzing Hot-Wire Signals in Highly Turbulent Flow and Its Evaluation in a Round Jet, DISA Information 17, pp. 9–18, 1975.
44. M. Acrivelellis, An Improved Method for Determining the Flow Field of Multidimensional Flows of Any Turbulence Intensity, DISA Information 23, pp. 11–16, 1978.
45. P. Freymuth, Hot-Wire Anemometer Thermal Calibration Errors, *Instrum. Contr. Syst.*, vol. 43, no. 10, pp. 82–83, 1970.
46. Townsend, The Diffusion of Heat Spots in Isotropic Turbulence, *Proc. R. Soc. London Ser. A*, vol. 209, pp. 418–430, 1951.
47. S. Corrsin, Extended Application of the Hot-Wire Anemometer, *Rev. Sci. Instrum.*, vol. 18, pp. 469–471, 1947.
48. P. Freymuth, Frequency Response and Electronic Testing for Constant-Temperature Hot-Wire Anemometers, *J. Phys. E*, vol. 10, pp. 705–710, 1977.
49. P. Freymuth, Further Investigation of the Non-Linear Theory for Constant-Temperature Hot-Wire Anemometers, *J. Phys. E*, vol. 10, pp. 710–713, 1977.
50. P. Freymuth and L. M. Fingerson, Electronic Testing of Frequency Response for Thermal Anemometers, *TSI Q.*, vol. 3, pp. 5–12, 1977.
51. M. S. Uberoi and P. Freymuth, Spectra of Turbulence in Wake Behind Circular Cylinder, *Phys. Fluids*, vol. 12, pp. 1359–1363, 1969.
52. M. S. Uberoi and P. Freymuth, Turbulent Energy: Balance and Spectra of Axisymmetric Wake, *Phys. Fluids*, vol. 3, pp. 2205–2210, 1970.
53. A. C. M. Beljaars, Dynamic Behaviour of the Constant-Temperature Anemometer Due to Thermal Inertia of the Wire, *Appl. Sci. Res.*, vol. 32, pp. 509–518, 1976.
54. P. Paranthoen, J. C. Lecordier, and C. Petit, Dynamic Sensitivity of the Constant-Temperature Hot-Wire Anemometer to Temperature Fluctuations, *TSI Q.*, vol. IX, no. 3, 1983.
55. C. Clark, Thin Film Gauges for Fluctuating Velocity Measurements in Blood, *J. Phys. E.*, vol. 7, pp. 548–556, 1974.
56. R. G. Lueck, Heated Anemometry and Thermometry in Water, Ph.D. thesis, Dep. of Physics and Inst. of Oceanography, Univ. of British Columbia, Vancouver, B.C., 1979.
57. F. N. Frenkiel, Etude Statistique De La Turbulence, 1: Measure De La Turbulence Avec Un Fil Chaud Non-Compense, 2: Influence De La Longueur D'Un Fil Chaud Compense Sur La Measure De La Turbulence, ONEEA Rapp. Tech. 37, 1948.
58. M. A. Uberoi and L. S. G. Kovasznay, On Mapping and Measurement of Random Fields, *Q. Appl. Math.*, vol. 10, pp. 375–393, 1953.
59. J. C. Wyngaard, Measurement of Small-Scale Turbulence Structure with Hot Wires, *J. Phys. E*, vol. 1, pp. 1105–1108, 1968.
60. J. C. Wyngaard, Spatial Resolution of the Vorticity Meter and Other Hot-Wire Arrays, *J. Phys. E*, vol. 2, pp. 983–987, 1969.

61. P. Freymuth, Noise in Hot-Wire Anemometers, *Rev. Sci. Instrum.*, vol. 39, pp. 550–557, 1968.
62. P. Freymuth, Calculation of Square Wave Test for Frequency Optimized Hot-Film Anemometers, *J. Phys. E*, vol. 14, pp. 238–240, 1981.
63. B. J. Bellhouse and D. L. Schultz, The Determination of Fluctuating Velocity in Air with Heated Thin Film Gauges, *J. Fluid Mech.*, vol. 29, pp. 289–295, 1967.
64. P. Freymuth, Extension of the Non-Linear Theory to Constant-Temperature Hot-Film Anemometers, *TSI Q.*, pp. 3–6, 1987.
65. P. Freymuth, Sine Wave Testing on Noncylindrical Hot-Film Anemometers According to Bellhouse-Schultz Model, *J. Phys. E. Sci. Instrum.*, 1979.
66. E. W. Nelson and J. A. Borgos, Dynamic Response of Conical and Wedge Type Hot Films: Comparison of Experimental and Theoretical Results, *TSI Q.*, vol. IX, 1983.
67. W. J. McCrosky and E. J. Durbin, Flow Angle and Shear Stress Measurements Using Heated Films and Wires, ASME Paper 71-WA/FE-17, 1971.
68. C. Salter and W. G. Raymer, Direct Calibration of Compensated Hot-Wire Recording Anemometer (by Dynamic Calibration), ARC R&M Rep. 1628, 1934.
69. J. R. Weske, A Hot-Wire Circuit with Very Small Time Lag, NACA Tech. Note TN 881, 1943.
70. H. H. Lowell, Early (1944–1952) Hot-Wire Anemometer Developments at NACA Lewis Research Center, *Adv. Hot Wire Anemometry*, pp. 29–37, 1968.
71. M. F. Young, Calibration of Hot-Wires and Hot-Films for Velocity Fluctuations, Dep. of Mechanical Engineering, Stanford University, Rep. TMC-3, 1976.
72. H. R. Frey, An Investigation of Instrumentation and Techniques for Observing Turbulence in and Above the Oceanic Bottom Boundary Layer, New York University, School of Engineering and Science, 1970.
73. P. L. Blackshear and L. M. Fingerson, Rapid Response Heat Flux Probes for High Temperature Gases, *ARS J.*, pp. 1709–1715, Nov. 1962.
74. G. L. Brown and M. R. Rebollo, A Small Fast-Response Probe to Measure Composition of a Binary Gas Mixture, *AIAA J.*, vol. 10, pp. 649–652, 1972.
75. Ronald Koopman, Data presented at meeting, Gas Research Institute, Chicago, Aug. 27, 1979.
76. B. W. Spencer and B. G. Jones, A Bleed Type Pressure Transducer for In-Stream Measurement of Static-Pressure Fluctuations, *Rev. Sci. Instrum.*, vol. 42, pp. 450–454, 1971.
77. H. P. Planchon, The Fluctuating Static-Pressure Field in a Round Jet Turbulent Mixing Layer, Ph.D. thesis, Univ. of Illinois, Urbana-Champaign, 1974.
78. B. G. Jones, R. J. Adrian, C. K. Nithianandan, and H. P. Planchon Jr., Spectra of Turbulent Static-Pressure Fluctuations in Jet Mixing Layers, *AIAA J.*, vol. 17, no. 5, pp. 449–457, 1979.
79. C. J. Remenyik and L. S. G. Kovasznay, The Orifice Hot-Wire Probe and Measurements of Wall Pressure Fluctuations, *Proc. Heat Transfer Fluid Mech. Inst.*, Stanford University Press, 1962.
80. J. A. B. Wills, A Traversing Orifice-Hot-Wire Probe for Use in Wall Pressure Correlation Measurements, National Physical Laboratory Aero Rep. 1155, 1965.
81. M. D. Mach, H. C. Seetharam, W. G. Kuhn, and J. T. Bright Jr., Aerodynamics of Spoiler Control Devices, AIAA Aircraft Systems and Technology Meeting, New York, Aug. 20–22, 1979.
82. W. H. Wentz Jr. and H. C. Seetharam, Split-Film Anemometer Measurements on an Airfoil with Turbulent Separated Flow, *Proc. Fifth Biennial Symp. on Turbulence*, pp. 31–33, University of Missouri, Rolla, Oct. 1977.
83. C. T. Crowe, D. E. Stock, M. R. Wells, and A. Barriga, Application of Split-Film Anemometry to Low-Speed Flows with High Turbulence Intensity and Recirculation as Found in Electrostatic Precipitators, *Proc. Fifth Biennial Symp. on Turbulence*, pp. 117–123, University of Missouri, Rolla, Oct. 1977.
84. B. W. Spencer and B. G. Jones, Turbulence Measurements with the Split-Film Anemometer Probe, *Proc. Symp. on Turbulence in Liquids*, pp. 7–15, University of Missouri, Rolla, Oct. 4–6, 1971.

FOUR

LASER VELOCIMETRY

Ronald J. Adrian

4.1 INTRODUCTION

Laser-Doppler velocimetry (LDV) is the method of measuring fluid velocities by detecting the Doppler frequency shift of laser light that has been scattered by small particles moving with the fluid. The technique was originally discussed in a pioneering paper by Cummins et al. [1], in which they measured the Brownian motion of an aqueous suspension of micron-sized particles by observing the spectrum of the scattered light. In these measurements the quantity of interest was the broadening of the laser light spectrum due to the random particle motion. However, they also observed a net shift in the frequency of the light, an effect that they attributed to small convection currents that generated mean velocities in their water cell. Hence, almost inadvertently, they performed the first measurements of fluid velocity by laser-Doppler velocimetry. Shortly thereafter, Yeh and Cummins [2] carried out an experiment intended expressly to demonstrate the measurement of fluid velocities.

The LDV concept rapidly attracted the attention of numerous experimental fluid dynamicists, and within a few years, various research groups had communicated the results of successful LDV measurements of laminar water flow in square ducts [3, 4], laminar water flow in round ducts [5], laminar gas flow [6, 7], turbulent water flow in pipes [8], and wind-tunnel turbulence [9]. At this stage of develop-

This work was supported by the National Science Foundation grant ATM 89-00509 and the Office of Naval Research grant N00014–93-1-0552.

ment, measurements were performed using spectrum analysis of the scattered light, a technique that required relatively long averaging times and hence precluded velocity tracking, e.g., measurement of rapidly fluctuating velocity as a function of time. Even so, the results were very encouraging. The technique offered numerous advantages. It was nonintrusive, so it could be used in flows that were hostile to material probes or that would be altered by the presence of a material probe; it did not depend on the thermophysical properties of the fluid, in contrast to thermal probes or chemical probes; it allowed the unambiguous measurement of one or more components of the velocity vector, independent of the fluctuation intensity (e.g., flow reversals could be sensed); it offered reasonably good spatial resolution; and it appeared to be capable of tracking very high frequency fluctuations of the flow velocity, provided that sufficiently fast electronics could be developed. On top of all this, the achievable accuracies were impressive. Goldstein and Kreid [4] reported 0.1% absolute accuracy for measurements of flow development in a square duct.

There were, of course, many problems to be solved. The very weak intensities of light scattered by small particles resulted in noisy signals that were difficult to analyze. The random locations of the scattering particles in the fluid created new types of data analysis problems, since data arrived randomly and often far too infrequently to resolve the time history of the velocity. The optical equipment was sometimes sensitive to vibrations, and electronics that were developed to follow the Doppler frequency often failed to do so, tracking instead noise and extraneous radio frequency signals. Even so, the natural appeal of the laser velocimeter led to an intense period of development over the ensuing decade. This effort has produced, at the time of this writing, an experimental technique that is routinely used in hundreds of research laboratories and industrial applications throughout the world, is available through several commercial manufacturers of scientific equipment (TSI, Inc., St. Paul; Dantec, Denmark, for example), and has reached a high level of maturity.

The power and versatility of LDV are probably best conveyed by a list (admittedly incomplete) of the types of flows in which it has been used successfully. These include supersonic flow; recirculating flow; natural free convection; flow in internal combustion engines, steam turbines, and gas turbines; chemically reacting flows including premixed flames and diffusion flames; jets; drag-reducing polymer flows; rotating flows; helicopter rotor studies; two-phase flows; atmospheric turbulence; arterial flow; capillary blood flow, both simulated and in vivo; ocean-bottom flows; high-temperature plasmas; and magnetohydrodynamic (MHD) channel flows. The range of flow velocities measured extends from less than 10 μm/s to 1 km/s.

In view of the widespread interest in LDV, it is not surprising that an abundant body of scientific literature exists on the topic. In 1974 a literature survey by Durst and Zaré [10] included over 600 papers, and the total number of papers has easily quintupled since then. For the newcomer to the field the task of assimilating all this information is clearly formidable. Fortunately, there are a number of books

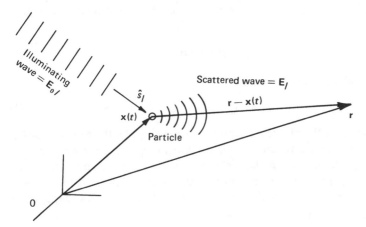

Figure 4.1 Light scattered by a moving particle

that deal in whole or in part with the subject and that are relatively comprehensive as of the time of their writing [11–15]. In addition, the proceedings of numerous workshops and symposia contain a wealth of information on the theory and application of laser velocimetry [16–24].

From this brief description of the extensive scope of LDV, it should be clear that an exhaustive and intensive discussion of the subject here is precluded by space limitations. Instead, we attempt to survey those topics that are of most significance to the understanding and successful application of laser velocimetry as comprehensively as possible, leaving out many of the background details. Often this means that final results are presented with little detailed derivation. In these cases the reader interested in the details is referred to appropriate papers in the literature. However, certain topics are so fundamental to the field that a more or less complete discussion seems advisable, and these topics are presented in detail.

4.2 BASIC PRINCIPLES

4.2.1 Doppler Shift of Light Scattered by Small Particles

The fundamental phenomenon in LDV is the Doppler shift of light that is scattered from small (typically, 0.1–10 μm) particles. The usual situation is shown in Fig. 4.1, where the ith particle located at $\mathbf{x}_i(t)$ scatters the light wave with complex electric vector $\mathbf{E}_{li}$ from an incident illuminating beam $\mathbf{E}_{0l}$. Here the subscript l is used to denote the lth illuminating beam. For reasons that are explained below, the illuminating beam may be assumed to be a plane wave, linearly polarized in the spatial region where it illuminates the particle. Its frequency is ω_{0l}, its direction of propagation is $\hat{\mathbf{s}}_l$ (unit vector), its wave number is $k = 2\pi/\lambda$, its direction of linear polarization is $\hat{\mathbf{p}}_l$, and its intensity is I_{0l} (W/m²). It may be represented by the complex wave

$$E_{0l} = \sqrt{I_{0l}(\mathbf{x})}e^{j\Phi_{0l}(\mathbf{x})}\hat{\mathbf{p}}_l \tag{1}$$

where $\Phi_{0l}(\mathbf{x})$ is the phase evaluated at $\mathbf{x}$,

$$\Phi_{0l}(\mathbf{x}) = \omega_{0l}t - k\hat{\mathbf{s}}_l \cdot \mathbf{x} \tag{2}$$

and it is understood that the physical electric vector is given by the real part of $\mathbf{E}_{0l}$. The particle scatters a light wave from $\mathbf{E}_{0l}$ in all directions, and a point $\mathbf{r}$ is said to be in the far field of the particle if the distance $r = |\mathbf{r}|$ is much greater than both the wavelength of the light and the mean diameter of the particle. In the far field the scattered light wave is a spherical wave, regardless of the particle's shape, given by [25]

$$E_{li} = \sqrt{I_{0l}(\mathbf{x}_i)}\frac{\sigma_{li}}{k|\mathbf{r} - \mathbf{x}_i|} e^{j[\Phi_{0l}(\mathbf{x}_i)-k|\mathbf{r}-\mathbf{x}_i|]} \tag{3}$$

where σ_{li} is a scattering coefficient for the ith particle that specifies the intensity, phase shift, and polarization of the scattered wave relative to the illuminating wave $\mathbf{E}_{0l}$.

Equation (3) is simplified by noting that under normal conditions the region illuminated by the incident wave is very small because the wave has been focused. Hence $|\mathbf{x}_i| \ll |\mathbf{r}|$, the vectors $\mathbf{r} - \mathbf{x}_i$ and $\mathbf{r}$ are nearly parallel, and by simple geometry, $|\mathbf{r} - \mathbf{x}_i| \cong r - \mathbf{x}_i \cdot \hat{\mathbf{r}}$ in the far field. Then

$$E_{li} = \sqrt{I_{0l}}\frac{\sigma_{li}}{kr} e^{j\Phi_{li}} \tag{4}$$

$$\Phi_{li} = \omega_{0l}t - kr + k\mathbf{x}_i \cdot (\hat{\mathbf{r}} - \hat{\mathbf{s}}_l) \tag{5}$$

where the factor $\mathbf{x}_i \cdot \hat{\mathbf{r}}$ must be retained in the phase but can be ignored safely in the denominator. Equation (4) implies that the scattered wave is approximately a spherical wave diverging from the origin (since $|\mathbf{x}_i|$ is small) whose phase depends on the particle position through the term $k\mathbf{x}_i \cdot \hat{\mathbf{s}}_l$. Now, the instantaneous frequency of a nearly sinusoidal signal is defined as the time derivative of its phase. From Eq. (5), this frequency is

$$\dot{\Phi}_{li} = \omega_{0l} + k\mathbf{v}_i(t) \cdot (\hat{\mathbf{r}} - \hat{\mathbf{s}}_l) \tag{6}$$

where the last term is the Doppler shift caused by the particle motion and

$$\mathbf{v}_i(t) = \dot{\mathbf{x}}_i(t) \tag{7}$$

is the velocity of the ith particle. Equation (6) gives the frequency in radians per second. In units of hertz, the frequency is denoted by

$$\nu_{li} = \frac{\dot{\Phi}_{li}}{2\pi} \tag{8a}$$

$$= \nu_{0l} + \frac{\mathbf{v}_i \cdot (\hat{\mathbf{r}} - \hat{\mathbf{s}}_l)}{\lambda} \tag{8b}$$

where ν_{0l} is the frequency of the illuminating beam.

The total Doppler shift is the sum of a shift associated with the particle's component of velocity away from the incident wave, $-\mathbf{v}_i \cdot \hat{\mathbf{s}}_l$, and the particle's component of velocity toward the observer at $\mathbf{r}$, $\mathbf{v}_i \cdot \hat{\mathbf{r}}$. Consequently, the Doppler shift depends linearly on the component of velocity in the direction of $\hat{\mathbf{r}} - \hat{\mathbf{s}}_l$. This is one of the more desirable features of LDV because it gives the experimenter the freedom to study a single component by choosing $\hat{\mathbf{s}}$ and $\hat{\mathbf{r}}$ accordingly. Note further that the Doppler shift has a sign associated with it, so that flow in the positive $\hat{\mathbf{r}} - \hat{\mathbf{s}}_l$ direction can be distinguished from flow in the negative $\hat{\mathbf{r}} - \hat{\mathbf{s}}_l$ direction.

4.2.2 Optical Heterodyne Detection

The maximum Doppler shift occurs where $\hat{\mathbf{r}} = -\hat{\mathbf{s}}$, and a typical value is 4 MHz/(m s^{-1}). Usually, $\hat{\mathbf{r}}$ is more nearly parallel to $\hat{\mathbf{s}}$, and a typical value in practice is about 0.4 MHz/(m s^{-1}). Thus, for example, the Doppler frequency produced by a 500-m/s gas flow would be around 200 MHz, while the shift observed in a 1-mm/s convection flow would be about 400 Hz. While these frequencies can be measured readily and accurately by modern electronics, they are, nonetheless, miniscule compared with the basic frequency of the laser light, which is of the order of 10^{14} Hz. For example, a relatively large Doppler shift of, say, 10^8 Hz, represents a change in the light frequency of only one part in a million. [Since the fractional change in the wavelength would be the same, changes in $k = 2\pi/\lambda$ were ignored, implicitly, in Eq. (3).] Hence, while direct spectroscopic detection of Doppler shift is possible in high-speed flows [26, 27], it is not used in lower-speed flows.

The method used in almost all LDV is to subtract the ω_{0l} term from the total frequency, leaving a signal that oscillates at the Doppler shift frequency. The technique whereby this is accomplished is called optical heterodyne detection, or optical mixing. The essence of this technique is revealed by the simple trigonometric identity $\sin \omega_1 t \sin \omega_2 t = \frac{1}{2}[\cos (\omega_1 + \omega_2)t + \cos (\omega_1 - \omega_2)t]$. Thus, by multiplying two light waves, a signal that oscillates at their difference frequency can be obtained. In actuality, the multiplication is performed by combining two light waves on the surface of a photodetector. Since the photodetector is a square-law device, its output will be of the form $(\sin \omega_1 t + \sin \omega_2 t)^2$, from which the cross product $\sin \omega_1 t \sin \omega_2 t$ is obtained. The output of the photodetector does not contain the sum-frequency component of this cross product because that frequency, of the order of 10^{14} Hz, is much higher than the frequency response of any available detector. Thus the output oscillates only at frequency $\omega_1 - \omega_2$.

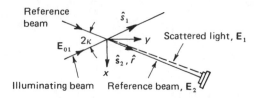

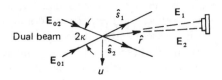

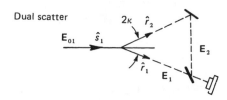

Figure 4.2 Three modes of heterodyne detection in laser-Doppler velocimetry. The geometry of each mode is shown such that the Doppler difference frequency $v_D = 2u \sin \kappa/\lambda$. (From [48])

4.2.3 Basic Optical Systems

There are three distinct types of LDV optical systems, corresponding to three different methods of combining the Doppler shift phenomenon with the optical heterodyne technique to produce a flow-measuring instrument: the reference-beam system, the dual-beam system, and the dual-scatter system. Figure 4.2 depicts the optical geometries corresponding to these three types schematically, and Fig. 4.3 shows typical optical systems used in practice. Many minor variations are, of course, possible. To facilitate comparisons, each of these systems is configured to measure the same component of velocity, and the coordinates are always defined with respect to the optical system so that the origin is imbedded at the center of the measurement volume and the system measures the x component of velocity u. The y axis lies in the plane of the light beams and is referred to as the "axis" of the system.

The basic principle of the dual-beam LDV is to illuminate a scattering particle with two plane light waves, $\mathbf{E}_{01}$ and $\mathbf{E}_{02}$, propagating in two different directions, $\hat{s}_1$ and $\hat{s}_2$, respectively. The ith particle scatters two waves, $\mathbf{E}_{1i}$ from $\mathbf{E}_{01}$ and $\mathbf{E}_{2i}$ from $\mathbf{E}_{02}$, and the frequencies of these waves in the scattering direction $\hat{r}$ are

$$v_{1i} = v_{01} + \frac{\mathbf{v}_i \cdot (\hat{\mathbf{r}} - \hat{\mathbf{s}}_1)}{\lambda} \tag{9a}$$

$$v_{2i} = v_{02} + \frac{\mathbf{v}_i \cdot (\hat{\mathbf{r}} - \hat{\mathbf{s}}_2)}{\lambda} \tag{9b}$$

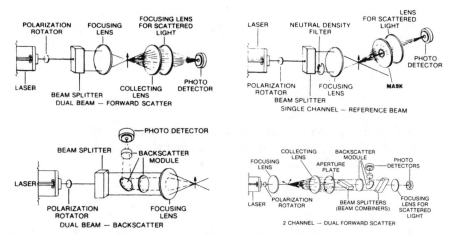

Figure 4.3 Practical LDV systems for the dual-beam, reference-beam, and dual-scatter modes. (Courtesy TSI, Inc.)

The frequency difference is

$$\nu_{1i} - \nu_{2i} = \nu_S + \nu_{Di} \tag{10}$$

where

$$\nu_S = \nu_{01} - \nu_{02} \tag{11}$$

is a constant frequency difference determined by the illuminating-beam frequencies, and

$$\nu_{Di} = \frac{\mathbf{v}_i \cdot (\hat{\mathbf{s}}_2 - \hat{\mathbf{s}}_1)}{\lambda} \tag{12}$$

is the difference between the Doppler shifts. This difference is independent of the scattering direction $\hat{\mathbf{r}}$, so the heterodyne frequency is the same at every point on the photodetector and independent of the detector's location. It is convenient to follow [28] and write Eq. (12) in the form

$$\nu_{Di} = \frac{\mathbf{K} \cdot \mathbf{v}_i}{2\pi} \tag{13}$$

where

$$\mathbf{K} = \frac{2\pi(\hat{\mathbf{s}}_2 - \hat{\mathbf{s}}_1)}{\lambda} \tag{14}$$

is a wave vector in the direction $\hat{\mathbf{s}}_2 - \hat{\mathbf{s}}_1$. Then, if we always agree to let $u_i(t)$ be the component of $\mathbf{v}_i$ in the $\hat{\mathbf{s}}_2 - \hat{\mathbf{s}}_1$ direction, we have

$$v_{Di} = \frac{Ku_i(t)}{2\pi} \qquad (15)$$

where

$$|\mathbf{K}| = K = \frac{4\pi \sin \kappa}{\lambda} \qquad (16)$$

by simple geometry. Thus v_{Di} depends only upon κ, λ, and the single velocity component u_i, which lies in the plane of the illuminating beams and is perpendicular to their bisector. In terms of circular frequencies,

$$\omega_{1i} - \omega_{2i} = \omega_S + Ku_i(t) \qquad (17)$$

where $\omega_S = 2\pi v_S$.

Ordinarily, the illuminating beams are obtained by splitting the original laser output of frequency v_0, so that $v_{01} = v_{02} = v_0$. Then $v_{1i} - v_{2i} = v_{Di}$, and the heterodyne frequency of the photodetector output is directly proportional to u_i. In this case, however, a sign change in u_i, and hence v_{Di}, simply corresponds to a 180° phase change in the sinusoidal heterodyne signal; this is difficult to detect electronically, particularly when the signal is contaminated by noise. Consequently, whenever bipolar velocities are anticipated, e.g., in reversing flows or in high-intensity turbulence, it is common practice to generate a frequency difference v_S between the illuminating beams so that fluctuations about zero velocity correspond to frequency fluctuations about the shift frequency v_S. The signal then looks much like an FM radio signal, with v_S corresponding to the carrier frequency. In addition to resolving the polarity of u_i, frequency shifting offers other advantages that are discussed below.

Frequency shifting is accomplished in practice by splitting the original laser beam and shifting the frequency of one of the beams, say beam 1, by v_S, so that $v_{01} = v_0 + v_S$ and $v_{02} = v_0$. Frequency shifts can be produced by electro-optic Pockels cells and Kerr cells [29], rotating diffraction gratings [30, 31], or acoustic-optic Bragg cells [32], the latter being the most commonly used by far. Typical shifts available from glass Bragg cells are rather large, in the 10- to 80-MHz range, but they can be very accurately controlled, and when the basic shift is too large in comparison with the Doppler frequency, the signal from the photodetector can be electronically shifted downward to a more convenient value. For example, it is common to use a 40-MHz Bragg cell with electronic downmixing to produce effective shift frequencies as low as a few kilohertz.

The reference-beam LDV, shown in Figs. 4.2 and 4.3, uses a single illuminating beam from which a light wave is scattered with frequency

$$v_{1i} = v_{01} + \frac{\mathbf{v}_i \cdot (\hat{\mathbf{r}} - \hat{\mathbf{s}}_1)}{\lambda} \qquad (18)$$

in the $\hat{\mathbf{r}}$ direction. A photodetector signal that oscillates at the Doppler shift frequency is obtained by optically mixing the scattered wave with a reference wave traveling in the $\hat{\mathbf{r}}$ direction whose frequency is ν_{02}. Typically, $\nu_{01} = \nu_0$ and $\nu_{02} = \nu_0 - \nu_S$, so that the heterodyne frequency is

$$\nu_{1i} - \nu_{2i} = \nu_S + \mathbf{v}_i \cdot \frac{\hat{\mathbf{r}} - \hat{\mathbf{s}}_1}{\lambda} \tag{19a}$$

$$= \nu_S + \nu_{Di} \tag{19b}$$

By choosing $\hat{\mathbf{r}}$ to be the same as $\hat{\mathbf{s}}_2$ in the dual-beam system, the reference-beam system is made to measure exactly the same component of velocity.

The dual-scatter LDV uses a single illuminating beam of frequency ν_{01}, as in the reference-beam LDV, but heterodyne detection of the Doppler shift is accomplished by mixing the light wave $\mathbf{E}_{1i}$ scattered in the $\hat{\mathbf{s}}_1$ direction with the light wave $\mathbf{E}_{2i}$ scattered in the $\hat{\mathbf{s}}_2$ direction. In fact, if there is a single particle in the illuminating beam, $\mathbf{E}_{1i}$ and $\mathbf{E}_{2i}$ are but portions of the same scattered wave. Their frequencies are

$$\nu_{1i} = \nu_{01} + \mathbf{v}_i \cdot \frac{\hat{\mathbf{r}}_1 - \hat{\mathbf{s}}_1}{\lambda} \tag{20a}$$

$$\nu_{2i} = \nu_{01} + \mathbf{v}_i \cdot \frac{\hat{\mathbf{r}}_2 - \hat{\mathbf{s}}_1}{\lambda} \tag{20b}$$

with heterodyne difference frequency

$$\nu_{1i} - \nu_{2i} = \mathbf{v}_i \cdot \frac{\hat{\mathbf{r}}_1 - \hat{\mathbf{r}}_2}{\lambda} \tag{21a}$$

$$= \nu_{Di} \tag{21b}$$

Hence the heterodyne frequency is independent of the direction of the illuminating beam.

While the heterodyne frequency equations for the dual-beam, reference-beam, and dual-scatter systems are very similar, there are, in fact, significant differences among other properties of the signals produced by these systems. In particular, the strengths and signal-to-noise ratios of the signals may differ substantially in a given application. These differences arise primarily because of differences in the efficiency with which scattered light is collected and heterodyne mixing is accomplished. It is shown below that the size of the scattered-light collecting aperture is severely restricted for both the reference-beam system and the dual-scatter system, the former because large apertures cause poor mixing efficiency, and the latter because the Doppler frequency shift varies over the aperture [cf. Eq. (21a)]. Thus, only the dual-beam system is capable of effectively utilizing a large light-collecting aperture to produce strong signals, and for this reason, it is the most commonly used type of LDV (although the other types can be superior in certain applications). Consequently, the ensuing discussion concentrates on the dual-beam system.

4.3 THE DUAL-BEAM LDV

4.3.1 Practical Dual-Beam Optics

Before discussing the theory of the dual-beam anemometer, it is best to look at the way these systems are constructed for practical applications. Figure 4.3 shows the most commonly used configurations. The intense, highly collimated light beam from a continuous wave (CW) gas laser, usually helium-neon or argon ion, is divided into two parallel beams of equal power by a beam splitter. The parallelism of the beams is critical, so the beam splitter is either a precisely constructed single-piece unit or a multicomponent unit that provides the very fine adjustments needed to maintain parallelism. Passing the parallel beams through a good-quality focusing lens causes the beams to intersect in the focal plane of the lens and simultaneously focuses the illuminating beams to small spots coincident with the intersection. In the region of the focal spots the light beams are approximately cylinders with diameters determined ideally by the diffraction-limited spot size, and the light waves are approximately plane parallel. The spatial region from which measurements are obtained is essentially the intersection of these beams, since a scattering particle must scatter light from *both* illuminating beams to produce a heterodyne signal.

Scattered light is collected by a set of receiving lenses and focused through a pinhole aperture onto a photodetector that may be either a photomultiplier tube or a photodiode. The process of heterodyne detection takes place on the active surface of the photodetector, where the two scattered light waves are "mixed" together. The illuminating beams are normally blocked from the detector to avoid swamping the weak scattered-light signals. In principle, if only one particle were present in the fluid, one could replace the receiving lens with a photodetector of equal area and achieve the same results as with the lens-pinhole system because the flux of light energy through the light-collecting aperture would be the same as the flux through the pinhole. However, in practice, the photodetector also receives light scattered from particles distributed throughout the illuminating beams, light from extraneous sources such as background radiation (room lights, radiation from flames, etc.), and extraneous laser light from flare and reflections at the surfaces of the transmitting optical elements and test-section windows. The function of the lens-pinhole combination is to reject most of this light by allowing only light from a small region around the beam intersection to enter the photodetector. The size of this region is essentially the diameter of the image of the pinhole in the plane of the beam intersection. It is usually chosen to be slightly smaller than the beam intersection itself, but when the beam intersection is undesirably large, the pinhole may be made even smaller, so that only a portion of the intersection can be "seen" by the detector. In this case the measurement volume is the intersection of the pinhole field of view with the illuminating beam intersection.

If the collecting lens is centered on the axis of the system, e.g., the y axis in Fig. 4.2, the configuration is called coaxial. "Forward scatter" and "back scatter" refer to locations of the collecting aperture that receive light scattered forward from the illuminating beams and scattered backward, respectively. Thus the first dual-beam system in Fig. 4.3 is a coaxial forward-scatter system. While coaxial systems enjoy a certain symmetry, it is often better to use "off-axis" light collection (e.g., not centered on the y axis) to reduce the amount of extraneous light seen by the detector and reduce the size of the measurement volume in the y direction.

The photodetector signal is a series of short, random bursts of Doppler-frequency sine waves caused by particles passing through the measurement volume. For setup purposes it is always advisable to observe these signals on an oscilloscope, but for the purpose of data acquisition, special electronic signal processors are needed to measure the frequencies of the Doppler bursts and convert them into either a voltage proportional to frequency or a digital number proportional to frequency. Likewise, because of the random times at which particles enter the measurement volume, special data processing methods may be needed to properly extract the desired information, such as mean velocity, root-mean-square (rms) velocity, and power spectrum. Signal processing and data processing are discussed in sections 4.10 and 4.11.

4.3.2 Characteristics of the Dual-Beam Signal

The dual-beam LDV can produce two types of signals: "incoherent" and "coherent." The incoherent signal occurs when a *single* particle scatters two light waves, one from each illuminating beam, and these waves subsequently mix on the photodetector. The coherent signal occurs when *at least two* particles reside in the measurement volume simultaneously. Then each particle produces an incoherent signal but, in addition, each *pair* of particles produces a coherent signal. The coherent signal is the result of the light wave scattered from one beam by the first particle mixing with light scattered from the other beam by the second particle, and vice versa. Since the scattered light waves originate from two different points in space, they may not mix efficiently unless the particles are so close together that the waves appear to come from a small, coherently illuminated region. Hence the term "coherent." The incoherent signal is more useful than the coherent signal in most cases, so it is examined first and in the most detail.

4.3.2.1 Single-particle signal. Consider the case of the ith particle crossing the illuminating beams with trajectory $\mathbf{x}_i(t)$, as in Fig. 4.4, and assume that the only light reaching the collecting aperture is that which is scattered by the particle, i.e., background light and laser flare are negligible.

In the region of the beam intersection, each focused illuminating beam is ap-

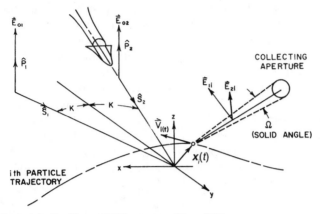

Figure 4.4 Dual-beam LDV geometry. (From [48])

proximately a plane wave with variable intensity I_{0l}, where $l = 1, 2$; thus $\mathbf{E}_{1i}$ and $\mathbf{E}_{2i}$, the light waves scattered from beams 1 and 2, respectively, are represented in the far field by Eq. (4) with $l = 1$ and 2. The scattering coefficients σ_{1i} and σ_{2i} will differ, in general, because each depends upon the scattering angle relative to the direction of the illuminating beam and its polarization, for example, $\sigma_{li} = \sigma_{li}(\hat{\mathbf{p}}_l, \hat{\mathbf{s}}_l, \hat{\mathbf{r}})$ [33, 34]. At the point $\mathbf{r}$ the total electric vector is

$$\mathbf{E}_i = \mathbf{E}_{1i} + \mathbf{E}_{2i} \qquad (22)$$

and the intensity, defined as the flux of light energy per unit area averaged over a time much longer than the light period but much shorter than the period of the heterodyne frequency, is given by [32]

$$I = \mathbf{E}_i \cdot \mathbf{E}_i^* \qquad (23)$$

where the asterisk denotes complex conjugate. Equations (4), (5), (14), (22), and (23) give

$$I_i(\mathbf{r}, \mathbf{x}_i, t) = \mathbf{E}_{1i} \cdot \mathbf{E}_{1i}^* + \mathbf{E}_{2i} \cdot \mathbf{E}_{2i}^* + \mathbf{E}_{2i} \cdot \mathbf{E}_{1i}^* + \mathbf{E}_{1i} \cdot \mathbf{E}_{2i}^* \qquad (24a)$$

$$= \frac{I_{01} \, \boldsymbol{\sigma}_{1j} \cdot \boldsymbol{\sigma}_{1i}^*}{k^2 r^2} + \frac{I_{02} \, \boldsymbol{\sigma}_{2i} \cdot \boldsymbol{\sigma}_{2i}^*}{k^2 r^2} + \frac{2\sqrt{I_{01}I_{02}}}{k^2 r^2} \, \mathrm{Re} \, (\boldsymbol{\sigma}_{1i} \cdot \boldsymbol{\sigma}_{2i}^* \, e^{j\Phi_i(t)}) \qquad (24b)$$

where

$$\Phi_i(t) = \Phi_i[\mathbf{x}_i(t), t] = 2\pi\nu_s t + k\mathbf{x}_i(t) \cdot (\hat{\mathbf{s}}_2 - \hat{\mathbf{s}}_1) \qquad (25a)$$

$$\Phi_i(t) = \omega_s t + Kx_i(t) \qquad (25b)$$

and the intensities I_{01} and I_{02} are evaluated at the location of the particle $\mathbf{x}_i(t)$. The photodetector output is proportional to the light flux hitting the detector, and this in turn, will be equal to the light flux J_i through the light-collecting aperture, which subtends the solid angle Ω, assuming that no pinhole blocks the light:

$$J_i(t) = \int_\Omega I(\mathbf{r}, \mathbf{x}_i, t)\mathbf{r}^2 d\Omega \tag{26a}$$

After some algebraic manipulation, this becomes

$$J_i(t) = \frac{1}{k^2} \{I_{01}P_{1i} + I_{02}P_{2i} + \sqrt{I_{01}I_{02}}\, D_i \cos[\Phi_i(t) - \Psi_i]\} \tag{26b}$$

where P_{1i}, P_{2i}, and D_i represent integrals over Ω of the scattering-coefficient products in the first, second, and third terms of Eq. (24b), respectively. (Here, P_1, P_2, D, and Ψ are the same as $\bar{P}_1$, $\bar{P}_2$, $\bar{D}$, and $\bar{\Psi}$ in [33, 34].) Parameter Ψ_i is an integrated phase shift arising from a phase-angle difference between $\mathbf{E}_{1i}$ and $\mathbf{E}_{2i}$ [33].

The first two terms in Eq. (26b) represent the fluxes of light scattered from beams 1 and 2 individually. That is, if beam 2 were absent ($I_{02} = 0$), the detector would still see a light flux $I_{01}P_{1i}$ as the particle crossed beam 1, and vice versa. These terms will usually be small when $\mathbf{x}_i$ is at the edge of the illuminating beam, where I_{0l} is small, increase to a maximum when $\mathbf{x}_i$ is on the axis of the beam, where I_{0l} is maximum, and then decrease again. Hence a particle will generate a pulse as it crosses each beam. Historically, the sum of these pulses is called the *pedestal*. The pedestal light flux is denoted by

$$J_{P_i}(t) = k^{-2}\{I_{01}[\mathbf{x}_i(t)]P_{1i} + I_{02}[\mathbf{x}_i(t)]P_{2i}\} \tag{27}$$

The third term in Eq. (26b) is the heterodyne mixing term that arises from the interference of $\mathbf{E}_{1i}$ and $\mathbf{E}_{2i}$. Frequently, it is called the Doppler signal or Doppler light flux, since it oscillates at the Doppler difference frequency. It is denoted by

$$J_{D_i}(t) = a[\mathbf{x}_i(t)]D_i \cos[\Phi_i(t) - \Psi_i] \tag{28}$$

where

$$a(\mathbf{x}_i) = k^{-2}\sqrt{I_{01}(\mathbf{x}_i)I_{02}(\mathbf{x}_i)} \tag{29}$$

The amplitude of $J_{D_i}(t)$ must always be less than or equal to $J_{P_i}(t)$; otherwise the total light flux would be negative when the cosine term is equal to -1, an obvious impossibility.

A useful quantity is the visibility of the signal, defined as the ratio of the amplitude of the Doppler signal to the amplitude of the pedestal:

$$V_i(t) = \frac{\sqrt{I_{01}I_{02}}\, D_i}{I_{01}P_{1i} + I_{02}P_{2i}} \tag{30}$$

This quantity is a function of time because I_{01} and I_{02} change as the particle crosses the intensity profiles of the beams. However, when $I_{01}[\mathbf{x}_i(t)] = I_{02}[\mathbf{x}_i(t)]$, the value of the visibility is

$$V_i|_{I_{01} = I_{02}} \equiv \overline{V}_i = \frac{D_i}{P_{1i} + P_{2i}} \tag{31}$$

which depends only upon the geometry of the optical system and the scattering properties of the particle and is independent of time. For this reason, $\overline{V}_i$ is often used to characterize the signal, i.e., the efficiency with which the scattered light waves from a single particle mix to produce the heterodyne signal. $\overline{V}_i = 1$ represents 100% efficiency and, of course, $\overline{V}_i \leqslant 1$ always. It is sometimes convenient to express $J_i(t)$ in the form

$$J_i(t) = J_{P_i}(t) \{1 + V_i(t) \cos [\Phi_i(t) - \Psi_i]\} \tag{32}$$

The spatial resolution of the dual-beam LDV is determined in part by the distribution of light intensity at the intersection of the focused laser beams Almost universally, one uses a laser operating in the TEM_{00} mode, meaning that the laser cavity sustains a purely longitudinal standing-wave oscillation along its axis, with no transverse modes. In this case the output of the laser has an axisymmetric intensity profile that is very nearly a Gaussian function of radial distance from the axis. The diameter of the beam is measured between the points where the intensity is $1/e^2$ of the peak intensity at the centerline and is denoted by D_{e-2}. Typically, D_{e-2} is about 1 mm. The output beam diverges such that in the far field it appears to be a spherical wave originating from a point source that is located at the front mirror of the laser, or perhaps somewhat further back, depending upon the design of the laser cavity. The full divergence angle θ_{e-2} is of the order of a few milliradians, so the beam is essentially parallel by most standards.

The propagation of Gaussian laser beams and their focusing characteristics are discussed in detail in an excellent review paper by Kogelnik and Li [35]. The report by Weichel and Pedrotti [36] is also a concise summary of the relevant equations. The behavior is shown in Fig. 4.5. The effect of an ideal thin lens with focal length f (i.e., a lens with zero aberration) is to convert the spherical wave diverging from the laser into a converging spherical wave whose radius of curvature first decreases as though the wave were converging to a point at s_1, and then increases until it is infinite exactly at the point s_1 (i.e., the wave is planar), where the beam coincidentally has a minimum diameter d_{e-2}. This focal point is located at

$$s_1 = f + \frac{s_0 - f}{(s_0/f - 1)^2 + (\pi D_{e-2}^2/4f\lambda)^2} \tag{33}$$

and the diameter of the focal spot is given by

$$\frac{1}{d_{e-2}^2} = \frac{1}{D_{e-2}^2} \left(1 - \frac{s_0}{f}\right)^2 + \left(\frac{\pi D_{e-2}}{4f\lambda}\right)^2 \tag{34}$$

Equation (33) states that the minimum beam diameter does not occur exactly in the focal plane of the lens unless $s_0 = f$. Since the lens in a dual-beam LDV will

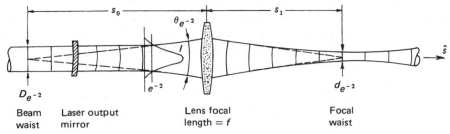

Figure 4.5 Properties of a focused Gaussian beam

cause the beams to cross at f when the beams are initially parallel, errors can occur if $s_1 \neq f$ because the wave fronts will not be planar in the beam intersection. These errors have been analyzed in [37]. For typical LDV parameters the errors are negligible for focal lengths up to several hundred millimeters, and the following approximate equations are satisfactory:

$$s_1 \simeq f \tag{35}$$

$$d_{e-2} \simeq \frac{4f\lambda}{\pi D_{e-2}} \tag{36}$$

This last equation is especially important. It implies that small measurement volumes require either large beam diameters or short focal lengths. Practically, these equations are also valid for LDV systems of long focal length, since to maintain spatial resoultion and high focal-spot intensity, one commonly employs beam expansion before the beam splitter to increase the effective value of D_{e-2}.

In the neighborhood of the focal spot, the exact equations show [35] that the focused Gaussian beam is essentially a plane wave whose diameter is nearly constant and whose intensity distribution is Gaussian. It can be represented by

$$\mathbf{E}_{0l} = \sqrt{I_{0l}}\, e^{j(\omega_0 l t - k\hat{\mathbf{s}}_l \cdot \mathbf{x})}\, \hat{\mathbf{p}}_l \tag{37}$$

wherein

$$I_{0l} = \frac{8P_{0l}}{\pi d_{e-2}^2} \exp\!\left(\frac{-8\zeta^2}{d_{e-2}^2}\right) \tag{38}$$

where P_{0l} is total beam power and ζ is radial distance from the centerline of the beam.

The region of nearly constant diameter and nearly plane-wave behavior, called the beam waist, extends for approximately $(f/D_{e-2})d_{e-2}$ on either side of the focal spot. Hence, it is very long in comparison with the spot diameter whenever the lens F number, f/D_{e-2}, is large, i.e., almost always.

Equation (37) is identical to Eq. (1), so the earlier assumption about illuminating the particle with a plane wave is justified in the region of the beam waist. We note that the intensity distribution in the beam waist of a focused Gaussian beam is Gaussian. Also, the peak intensity at the centerline is inversely proportional to d_{e-2}^2; that is, the illuminating-beam intensity is inversely proportional to the square of the focusing-lens F number.

Conventionally, the LDV measurement volume has been defined as the region in which the amplitude of the Doppler signal is greater than $1/e^2$ of the maximum amplitude that the particle could produce, i.e., all points $\mathbf{x}$ such that

$$a(\mathbf{x}_i)D_i \geq e^{-2}\, a(\mathbf{0})D_i \qquad (39)$$

Using Eq. (38), it can be shown that [38]

$$a(\mathbf{x}_i) = k^{-2}\,\sqrt{I_{01}(0)I_{02}(0)}\,\exp\left[-\frac{8}{d_{e-2}^2}\left(x_i^2\cos^2\kappa + y_i^2\sin^2\kappa + z_i^2\right)\right] \qquad (40)$$

so that the amplitude ratio will exceed e^{-2} whenever the magnitude of the exponential is less than 2. This defines an ellipsoidal measurement volume as shown in Fig. 4.6, with axes in the x, y, and z directions given by

$$d_{\mathrm{m}} = \frac{d_{e-2}}{\cos\kappa} \qquad (41)$$

$$l_{\mathrm{m}} = \frac{d_{e-2}}{\sin\kappa} \qquad (42)$$

$$h_{\mathrm{m}} = d_{e-2} \qquad (43)$$

respectively. The volume enclosed by the ellipsoid is

$$V_D = \frac{\pi d_{e-2}^3}{6\cos\kappa\sin\kappa} \qquad (44)$$

Note that the volume becomes very long in the y direction when $\sin\kappa$ is small. Typical dimensions in a system that has not been designed specifically for high spatial resolution are $d_{\mathrm{m}} \approx 0.1$ mm, $l_{\mathrm{m}} \approx 0.8$ mm, and $h_{\mathrm{m}} \approx 0.1$ mm. With special design, these values can be reduced by more than 1 order of magnitude.

The e^{-2} definition of the measurement volume (mv) provides a convenient basis for comparing different LDV systems: moreover, it is an appropriate definition when certain types of signal processors (e.g., spectrum analyzers or correlators) are used, or if there are many particles contributing to the signal simultaneously. However, if the signal is a series of nonoverlapping Doppler bursts from single particles and if the signal processor is a counter or a frequency tracker, it is inappropriate because these processors will not process the signal at all, unless the amplitude exceeds a fixed threshold voltage level at the input to the processor. The

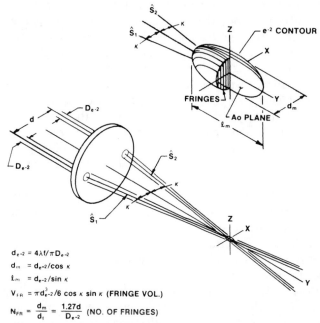

$d_{e\text{-}2} = 4\lambda f / \pi D_{e\text{-}2}$

$d_m = d_{e\text{-}2}/\cos \kappa$

$\ell_m = d_{e\text{-}2}/\sin \kappa$

$V_{FR} = \pi d_{e\text{-}2}^3/6 \cos \kappa \sin \kappa$ (FRINGE VOL.)

$N_{FR} = \dfrac{d_m}{d_f} = \dfrac{1.27d}{D_{e\text{-}2}}$ (NO. OF FRINGES)

Figure 4.6   Geometry of the nominal LDV measurement volume. (Courtesy TSI, Inc.)

threshold level is a characteristic of the particular processor, and it can often be adjusted by the operator, essentially by varying the gain of the signal-processor preamplifier. It can be related to a threshold light-flux level $J_{\min}$ with knowledge of the proportionality between the light flux and the voltage at the processor input.

In this case, the mv is defined by requiring

$$a(\mathbf{x}_i)D_i \geq J_{\min} \qquad (45)$$

This again defines an ellipsoidal volume, but the lengths of the major axes are now

$$
\begin{bmatrix} l_x \\ l_y \\ l_z \end{bmatrix} =
\begin{cases}
\dfrac{d_{e\text{-}2}}{\sqrt{2}} \left(\ln \dfrac{a(0)D_i}{J_{\min}} \right)^{1/2}
\begin{bmatrix} \dfrac{1}{\cos \kappa} \\ \dfrac{1}{\sin \kappa} \\ 1 \end{bmatrix} & a(0)D_i \geq J_{\min} \\[4ex]
\begin{bmatrix} 0 \\ 0 \\ 0 \end{bmatrix} & a(0)D_i < J_{\min}
\end{cases}
\qquad (46)
$$

Thus the size of the mv depends upon the laser power and the illuminating-beam focusing [through I_{01} (0) and I_{02} (0)], the scattering properties of the particle and the light-collecting efficiency (through D_i), and the threshold voltage-detection level, photodetector sensitivity, and postdetector amplification (through J_{min}). When polydisperse suspensions of particles are used, e.g., natural aerosol or hydrosol, the scattering characteristics will vary dramatically, with larger particles usually (but not always) producing bigger signals. A large particle might produce a detectable signal when its trajectory is far from the center of the mv, while a small particle might not produce a detectable signal unless its trajectory passes through a region that is rather smaller than the nominal $1/e^2$ volume. Finally, certain particles may scatter so weakly as to produce no detectable signals, even when they pass through the center of the mv.

To summarize, if the photodetector views the entire beam intersection, and if the particles are polydisperse, then the mv is not a precisely defined region. Rather, it can be thought of as an expandable volume whose size is determined by the value of D_i for each particle. This uncertainty is eliminated if the particles are monodisperse or if the measurements are interpreted statistically.

If the particle follows the fluid motion, its trajectory $x_i(t)$ is a Lagrangian trajectory of the fluid, and the velocity v_i measured by the LDV is a Lagrangian fluid velocity. However, since the LDV measures this velocity only when the particle is in a small neighborhood of a fixed point in the flow, the measurement also represents the Eulerian velocity at that point. This is true so long as the mv is small in comparision with the size scale of the fluid motion. Thus, one can view the individual particle LDV signal as a sequence of Eulerian velocity samples that are obtained at random points in time because the particles are originally located at random points in space.

The form of the signal burst produced by the ith particle can be found as a function of time and velocity by substituting the equation for its trajectory,

$$\mathbf{x}_i(t) = \mathbf{x}_i(0) + \int_0^t \mathbf{v}_i(\xi) \, d\xi \tag{47}$$

into Eq. (26b) and using the Gaussian intensity distribution described by Eq. (40). When the mv is so small that the particle trajectory is essentially a straight line during its crossing of the mv, it is easy to show that $J_i(t)$ always consists of two Gaussian pedestals plus a burst of Doppler sine wave at frequency v_{Di}. The envelope of the sine wave is always a Gaussian function of time, for any particle trajectory.

Some typical signals are shown in Fig. 4.7 for various trajectories and signal visibilities. Note that the individual pedestals coincide for trajectories in the xz plane, so that they appear to be a single pedestal (trajectories a and b). While this is the situation that is most often sketched in the LDV literature, it should be remembered that it is a special case. The pedestal signals vary slowly relative to

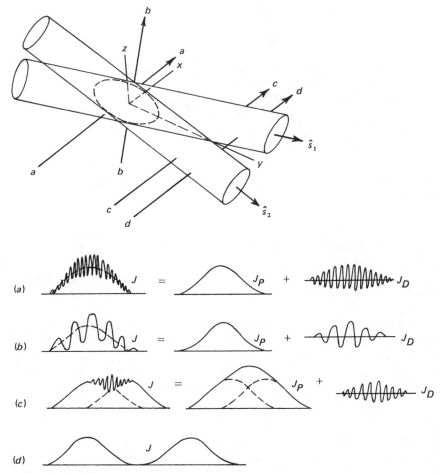

Figure 4.7 Typical LDV signals produced by various partial trajectories decomposed into pedestals plus Doppler bursts

the rate at which the Doppler sine wave oscillates, so they consist of low-frequency components that can be removed from the total signal by high-pass filtering, leaving only the Doppler signal. Decomposing the total signals in this way yields the separate pedestal and Doppler components shown in Fig. 4.7.

4.3.2.2 Fringe model of the dual-beam LDV. The fringe model of the dual-beam LDV is an alternative explanation of the operation of the dual-beam LDV that avoids reference to the Doppler shift effect. This model, first proposed by Rudd [39], is useful because it is usually much easier to visualize than the Doppler shift

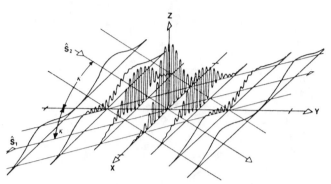

Figure 4.8 Intensity distribution of the fringes in the dual-beam measurement volume. (Adapted from [40])

model. Unfortunately, it is not entirely correct, and it can predict unrealistic results, a fact that was not realized in some of the early work. Even so, it does yield correct results in many regards, and certain concepts based on the fringe model have become so ingrained in the LDV literature that a discussion of the model is warranted.

The fringe model is based on the observation that the light waves of the illuminating beams will interfere to form a set of interference fringes at the intersection of the beams. We can see this mathematically by calculating the intensity of the sum of the two illuminating waves, $I_{00}(\mathbf{x}) = (\mathbf{E}_{01} + \mathbf{E}_{02}) \cdot (\mathbf{E}_{01}^* + \mathbf{E}_{02}^*)$. The algebra is exactly the same as in Eq. (24a) with $\mathbf{E}_{li}$ replaced by E_{0l}. The result is

$$I_{00}(\mathbf{x}) = I_{01}(\mathbf{x}) + I_{02}(\mathbf{x}) + 2\sqrt{I_{01}(\mathbf{x})I_{02}(\mathbf{x})} \cos(\omega_S t + 2kx \sin \kappa) \qquad (48)$$

This intensity distribution is drawn in Fig. 4.8 assuming Gaussian illuminating beams with $P_{01} = P_{02}$ and $\omega_{01} = \omega_{02}$. The interference fringes, i.e., the last term in Eq. (48), lie in planes parallel to the yz plane, like a deck of cards. The spacing between fringes is given by $2k\, d_f \sin \kappa = 2\pi$, so that

$$d_f = \frac{\lambda}{2 \sin \kappa} \qquad (49)$$

Now suppose that a very small particle crosses this fringe pattern and scatters light such that the intensity of the scattered light is proportional to I_{00}. Then the scattered light flux will oscillate sinusoidally as the particle crosses the fringes, and the frequency of the oscillation will be

$$\nu_D = \frac{u}{d_f} = \frac{2u \sin \kappa}{\lambda} \qquad (50)$$

since d_f/u is the time for the particle to cross one fringe. The velocity components v and w are, of course, irrelevant, since they represent motion *parallel* to the fringes.

This frequency is exactly the same as that obtained from Doppler shift considerations, and it will be noted that the wave number defined in Eq. (16) is just $K = 2\pi/d_f$. Moreover, the signal characteristics are also similar. From Fig. 4.8 the intensity consists of two pedestals (the intensity profiles of each beam) and a Gaussian amplitude-modulated sine wave (the interference term).

If $\omega_{02} - \omega_{01} = 2\pi\nu_S$, the cosine term in Eq. (48) implies that the fringe pattern moves with velocity $u_f = \nu_S d_f$, while the envelope of the cosine, that is, the mv, remains fixed. In effect, the interference of two equal-frequency waves creates a standing-wave interference pattern, while the interference of two unequal-frequency waves creates a traveling-wave pattern. The velocity u_f is such that fringes would sweep across a stationary particle at a rate equal to the shift frequency ν_S. If the particle moves with (or against) this fringe motion, the crossing rate is lower (or higher) than the frequency difference by the amount u/d_f. Here too, the results are identical to the Doppler shift theory.

It will be seen later that the number of fringes in the mv, defined as

$$N_{FR} = \frac{d_m}{d_f} \qquad (51)$$

is an important parameter that characterizes many of the LDV signal properties.

Thus far, the fringe model is entirely adequate. The deficiency is manifested when one attempts to calculate the signal strength when the particle is not small in comparison with d_f. In the context of the fringe model, the signal should be proportional to the total light flux striking the particle, i.e., the integral of $I_{00}(\mathbf{x})$ over the particle surface. Farmer [41] has performed this calculation by integrating I_{00} over a disk whose diameter equals the particle diameter d_p. A primary result of this calculation is that the peak visibility $\overline{V}$ of the Doppler signal depends only on the ratio d_p/d_f of the particle diameter to the fringe spacing when both illuminating beams have equal intensities and polarizations in the z direction. This dependence is shown by the dashed line in Fig. 4.9. For small d_p/d_f the visibility equals unity, implying that the amplitude of the Doppler signal is just proportional to the Doppler component of $I_{00}(\mathbf{x})$, as expected. However, when $d_p/d_f = 1.22$, corresponding to the particle being covered equally by one bright fringe and one dark fringe, the visibility vanishes because the integral of I_{00} remains constant as the particle moves, implying zero Doppler signal amplitude. $\overline{V}$ also vanishes at each value of d_p/d_f that corresponds to an integer number of whole fringes.

These implications of the fringe model are not, in general, correct. For example, the data points in Fig. 4.9, obtained by using 9.8-μm-diameter particles in back scatter and varying the fringe spacing, exhibit a maximum at about the same d_p/d_f value that the fringe model predicts as a minimum. The reason for this discrepancy is that the fringe model deals with intensity at the particle, but in reality, detection of the light intensity occurs far away, at the photodetector. The scattering process is an intervening step between illumination and detection that must be accounted for in the correct analysis. Calculation of $\overline{V}$ using Mie scattering theory

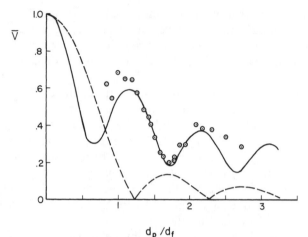

d_p / d_f

Figure 4.9 Peak signal visibility in terms of particle diameter and fringe spacing: - - - Diffraction theory (From [41]); —— Mie scattering theory (From [34]); ⊙ experimental data

for spherical particles to evaluate the scattering coefficients in Eq. (24b) yields the solid curve in Fig. 4.9, and this does agree well with the data [34]. Scattering theory predicts that the signal visibility depends, in general, upon the values of $I_{02}(\mathbf{x})/I_{01}(\mathbf{x})$, $\hat{\mathbf{p}}_1 \cdot \hat{\mathbf{p}}_2$, $\hat{\mathbf{s}}_l \cdot \hat{\mathbf{p}}_m$, d_p/d_f, $2\pi d_p/\lambda$, the location and geometry of the solid angle for light collection, and the ratio of the refractive index of the particle to the refractive index of the fluid.

The relationship between the fringe model and the scattering model is discussed in [34, 42, 43], where it is shown that the fringe model is correct in the scalar diffraction-theory limit $2\pi d_p/\lambda \gg 1$, provided that light is collected in forward scatter, and the light-collecting aperture is very large. Thus, when d_p and d_f are large enough, e.g., a 20-μm particle in 10-μm fringes, the fringe model is accurate, in agreement with physical intuition. But when the particle diameter is too small ($2\pi d_p/\lambda < 60$), it fails, together with our intuitive understanding of scattering on such small scales.

4.3.2.3 Signal coherence. Suppose there are two particles in the mv of the dual-beam system, instead of one. Call these particles i and j. Particle i will scatter two waves: $\mathbf{E}_{1i}$ from beam 1 and $\mathbf{E}_{2i}$ from beam 2. Likewise, particle j will scatter $\mathbf{E}_{1j}$ and $\mathbf{E}_{2j}$. Heterodyne mixing of $\mathbf{E}_{1i}$ with $\mathbf{E}_{2i}$ produces the single-particle signal discussed in section 4.3.2.1, and likewise for $\mathbf{E}_{1j}$ mixing with $\mathbf{E}_{2j}$. However, there will be additional signals due to $\mathbf{E}_{1j}$ mixing with $\mathbf{E}_{2i}$ and $\mathbf{E}_{1i}$ mixing with $\mathbf{E}_{2j}$. These signals will not be as strong as the single-particle signal because the scattered waves originate from two points in space $\mathbf{x}_i$ and $\mathbf{x}_j$, which may not, in general, be coincident. This loss of signal strength is referred to as *spatial incoherence*.

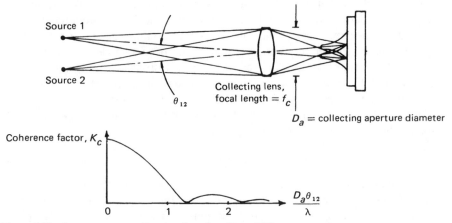

Figure 4.10   Source-aperture coherence function. (From [48])

The easiest way to understand this effect is to consider the mixing that occurs at the detector surface when light waves from two separated sources are collected and focused onto the surface, as shown in Fig. 4.10. If the light waves look like spherical waves coming from point sources (such as a micron-sized particle), diffraction at the lens produces an image of the source whose diameter is of the order of $\lambda f_c / D_a$, where D_a is the diameter of the light-collecting aperture. Hence, if the images of the two sources are separated by more than the diffraction-limited spot size, they will not overlap, and no heterodyne mixing can occur. In this case no Doppler signal will be observed, although there will be pedestals associated with each particle crossing each beam.

At the other extreme, if the angular separation θ_{12} is zero, the images overlap perfectly, and there is perfectly coherent heterodyne mixing so long as the particles are both in focus at the detector. The efficiency of mixing for partial overlap is defined as the amplitude of the mixing signal divided by the amplitude for perfectly coherent mixing. It is denoted by K_c and is commonly called the *coherence factor* [44, 45]. K_c depends on the ratio of the image separation to the image diameter, $f_c \theta_{12} / (f_c \lambda / D_a) = D_a \theta_{12} / \lambda$, as shown qualitatively in Fig. 4.10. The dual-source signal will have poor mixing efficiency when either D_a or θ_{12} is large. Since the two particles must be within the mv, the maximum possible value of θ_{12} is about d_{e-2}/f_c. The maximum diameter of the light-collecting aperture that will still yield efficient mixing is $D_c \approx f_c \lambda / d_{e-2} \approx (f_c/f) D_{e-2}$. That is, if $f_c \approx f$, the coherent aperture diameter D_c is about the same size as the initial laser beam diameter, which is only a few millimeters!

Obviously, one would like to use a large light-collecting aperture to receive as much scattered light energy as possible, but from the foregoing arguments the loss of mixing efficiency associated with increasing D_a may offset the gain in collected light energy to the extent that the Doppler signal amplitude actually decreases as the aperture is increased. This result pertains only to the dual-source signal, which somewhat confusingly, is called the coherent signal because the aperture must satisfy the spatial coherence requirements. The signal from a single particle is called incoherent because the waves scattered from it always appear to originate from the same point and hence always satisfy the coherence requirement; that is, $K_c = 1$ for the coherent signal, regardless of the collecting-aperture diameter.

Suppose that the scattering particle concentration C (number per unit volume), and hence the mean number of particles in the mv,

$$N_D = CV_D \tag{52}$$

can be adjusted. When $N_D \ll 1$, the probability of two or more particles in V_D is negligible, and the only signals will be the incoherent signals from individual particles. When $N_D \gg 1$, there will be about N_D^2 pairs of particles producing N_D^2 coherent signals plus N_D incoherent signals. Now, is it better to use a small coherent aperture and many particles so that the N_D^2 coherent signals will mix efficiently, or is it better to use a large aperture and few particles so that each particle produces a large signal? The answer depends, of course, on the specifics of the application, but in the vast majority of applications the large-aperture, incoherent signal provides the best result. In fact, the coherent signal is usually used only when the particle concentration is naturally or unavoidably large. It should be noted that the coherent signal can be reduced to negligible amplitude simply by making the collecting aperture much larger than D_c. Then, even if $N_D \gg 1$, the total signal will be the sum of many incoherent signals. Quantitative analyses of the coherent and incoherent signals are presented in [44, 45].

4.4 THE REFERENCE-BEAM LDV

As discussed in section 4.2.3, the reference-beam LDV mixes light E_{1p} scattered from an illuminating beam, with a reference beam to detect the frequency difference. The analysis of the reference-beam signal is the same as the analysis of the dual-beam signal, except that $E_2(\mathbf{r})$, the reference wave evaluated on the surface of the photodetector, is a spherical wave that diverges from the focal point of the reference beam. For best performance this focal point *appears* to be located at the center of the mv, and in practice, this is usually accomplished by actually placing it there, as in Figs. 4.2 and 4.3. Thus the reference-beam LDV may look rather similar to the dual-beam LDV, but this is not necessary. For example, the reference beam may be passed around the test section and still be made to originate apparently from the center of the mv by using a system of mirrors, lenses, and beam splitters, as in the system described in [46]. Foreman [47] has shown that when this is done, the path-length difference between the reference-beam and the illuminating-beam scattered-light path must be an integer multiple of twice the laser cavity length to achieve efficient heterodyne mixing with lasers operating in multiple longitudinal modes. The best efficiency occurs for zero path-length difference. If the difference must be large, then the laser should be operated in a single longitudinal mode, i.e., "single frequency." The arrangement shown in Fig. 4.3 allows equal path lengths.

The principal drawback of the reference-beam LDV is that it must satisfy a coherent aperture condition, much like the dual-particle signal of the dual-beam LDV. This is explained by letting source 1 in Fig. 4.10 represent the focus of the reference beam, fixed at the center of the mv, and letting source 2 be a scattering particle. The arguments presented for the coherence aperture of the dual-particle signal apply equally well to this situation, with the result that the coherence aperture diameter is just the diameter attained by the illuminating beam at a distance f_c from the mv. Increasing the aperture diameter beyond this would produce increasingly inefficient mixing, but in compensation, more light would be collected. In general, the mixing efficiency ultimately decreases faster than the scattered-light flux increases, so that the best aperture diameter is not more than a few times D_c. Since this is relatively small, reference-beam systems are not usually used unless the aperture must subtend a small solid angle for other reasons, e.g., in a long-range system wherein D_a is constrained by cost whereas f_c is large, or unless the particle concentration is large. Drain [44] indicates that when N_D is large, the reference-beam signal-to-noise ratio is twice the signal-to-noise ratio obtained from a dual-beam coherent signal.

If one considers the usual system, as shown in Fig. 4.3, the illuminating beam and the reference beam are both Gaussian beams focused to the same spot size. Hence the diameter of the reference beam at the light-collecting aperture matches the coherence aperture diameter and is so small that the scattering coefficient σ_1 is essentially constant over the solid angle Ω. (A simple, approximate criterion for σ_1 to be effectively constant over Ω is that the particle diameter must be rather less than the focal-spot diameter, which is usually the case [48].) Analysis of this system shows that the Doppler signal is given by [38]

$$J_{D_i} = \frac{32\sqrt{P_{01}P_{02}}\,\mathrm{Re}\,(\sigma_{1i}\cdot\hat{\mathbf{p}}_2)}{d_{e-2}^2\,k^2}\exp\left[-8\left(\frac{x_i^2}{d_m^2}+\frac{y_i^2}{l_m^2}+\frac{z_i^2}{h_m^2}\right)\right]\cos\left[\Phi_i(t)-\Psi_i\right] \quad (53)$$

where Φ_i is given by Eq. (25a) (wherein $\hat{\mathbf{s}}_2$ is the direction of the reference beam), and Ψ_i is a phase shift determined by σ_{1i}. Comparison of Eqs. (28) and (53) shows that the single-particle reference-beam Doppler signal is identical in form to the single-particle dual-beam signal, aside from a constant factor in the amplitude. In particular, the reference-beam mv is the same as in Fig. 4.6, provided that the reference- and illuminating-beam focal spots are equal and the reference beam is not blocked at the collecting aperture. Of course, the total reference-beam signal also contains two pedestals, each of Gaussian shape.

The amplitude of the reference-beam Doppler signal is proportional to the square root of the reference-beam power P_{02}. Hence it can be increased almost without limit by the simple expedient of increasing P_{02}. This is a useful feature when the scattered-light flux is small in comparison with a background radiation level (e.g., atmospheric air measurements in daylight). Even in laboratory applications, the scattered-light power is so weak (of the order of 10^{-12}–10^{-6} W) that the reference beam must be attenuated substantially to reach a level that is comparable to the scattered-light power. The system in Fig. 4.3 includes a variable neutral-density filter for this purpose.

The dual-scatter system is used so infrequently (because of small-aperture limitations and alignment difficulties) that it will not be discussed here.

4.5 MULTIVELOCITY-COMPONENT SYSTEMS

It is possible to measure two or even three components of the velocity vector by using several single-component systems. Although this seems a simple task, it is complicated by the necessity of identifying which one-component system the scattered-light signal is coming from. This is easy with the reference-beam and dual-scatter systems because the direction of the velocity component to be measured is determined by the direction in which scattered light is collected. But when using the dual-beam approach, one must supply two basic illuminating beams plus at least one additional illuminating beam for each additional velocity component. Then the scattered-light waves from each beam will heterodyne with the waves from every other beam to produce several signals, and the problem is to identify the signals coming from each pair of beams. (Alternatively, one can imagine one set of fringes being formed by each beam pair, and the subsequent complexity of the signals produced by a particle crossing all the fringe systems simultaneously.)

The identification problem is solved by labeling each pair of beams using some kind of optical property. The three possible methods are different colors, different frequencies, and different polarizations. For the purpose of measuring the u and w components of velocity perpendicular to the optic axis of the LDV system, there are also two possibilities: a four-beam arrangement and a three-beam arrangement that can be oriented in two ways. The various combinations are shown schematically in Fig. 4.11, wherein the beams appear as they would be seen when viewed from behind the focusing lens, looking toward the mv.

The four-beam, two-color, dual-beam LDV is shown in the upper left-hand square of Fig. 4.11. The blue and green lines from an argon ion laser are each split and focused to form two mutually orthogonal dual-beam systems. Scattered light is collected by a pair of receiving optics/photodetector subsystems, one accepting only blue light and the other accepting only green light. The outputs of the blue and green detectors then correspond to the velocity components measured by the blue and green illuminating-beam pairs, u and w, respectively.

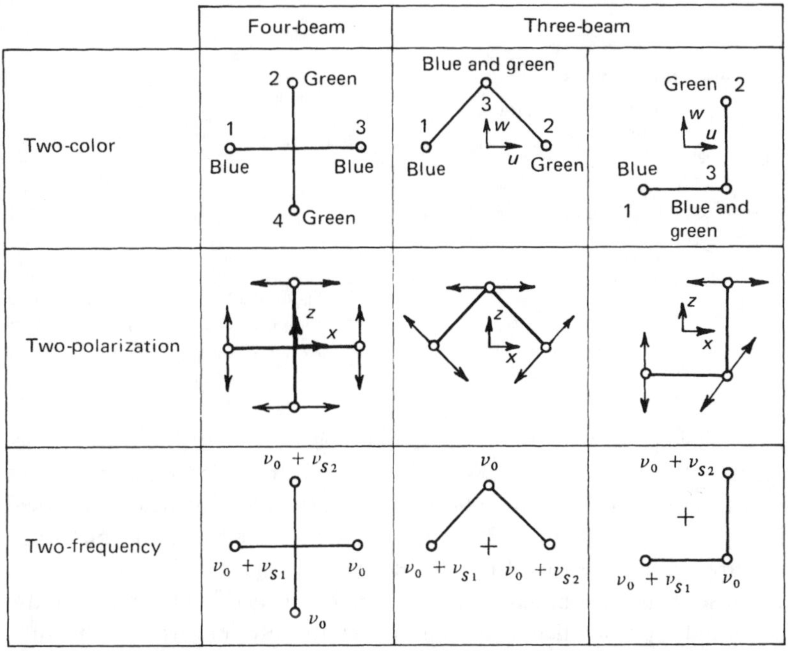

	Four-beam	Three-beam	
Two-color			
Two-polarization			
Two-frequency			

Figure 4.11   Illuminating-beam geometries for two-component LDV systems

While simple in principle, the two-color system is more complicated than two one-color systems, largely because of problems associated with keeping the blue and green channels separate as the light propagates through the optical system. A typical two-color design using a prism to separate the colors at the laser head is shown in Fig. 4.12. Assuming that the primary focusing lens is achromatically corrected, the only condition needed for coincidence of the measurement volumes of the blue and green pairs is that the blue and green lines be parallel ahead of the beam splitters (which, presumably, maintain parallelism). This is accomplished by fine tilting adjustments of the mirrors that deflect the beams toward the beam splitters. Light scattered from both beam pairs is collected by the same lens. Thereafter a dichroic mirror transmits the blue light and reflects the green light to accomplish separation. The various polarization rotators in the system are used to align the laser polarization for optimal operation of the other optical elements (see [49] for applications details).

The two-color, three-beam systems in Fig. 4.11 eliminate one beam by combining blue and green light into a single beam. Light scattered from beam 1 and the blue portion of beam 3 mixes to form a blue heterodyne signal, and light scattered from beam 2 and the green portion of beam 3 forms a green heterodyne signal. In the second of the three-beam arrangements (upper right-hand corner of Fig. 4.11),

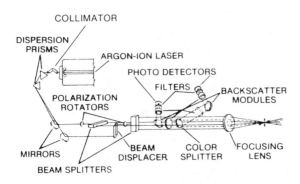

COLLIMATOR

DISPERSION PRISMS

ARGON-ION LASER

PHOTO DETECTORS

FILTERS

BACKSCATTER MODULES

POLARIZATION ROTATORS

MIRRORS

BEAM DISPLACER

COLOR SPLITTER

FOCUSING LENS

BEAM SPLITTERS

2 CHANNEL — 2 COLOR BACKSCATTER

Figure 4.12 Typical two-color, two-component LDV system for back scatter. (Courtesy TSI, Inc.)

the blue signal measures u and the green signal measures w. In the other three-beam arrangement the blue signal measures $u + w$, and the green signal measures $u - w$. Summing and differencing the blue and green signal frequencies then yield u and w. This type of three-beam arrangement is sometimes advantageous when measurements are made close to a wall lying in the xy plane.

Polarization separation is based on the principle that two light waves that are polarized in mutually perpendicular directions cannot interfere with each other. [This is the meaning of the dot product in Eq. (23).] The first two-polarization system shown in Fig. 4.11 uses two pairs of illuminating beams that are cross polarized. If the scattered light retains the same polarizations as the incident illuminating beams, the signals can be separated by passing the scattered light through a polarization cube or a similar device that reflects one polarization and transmits the other. The difficulty with this technique is that the scattering process does in fact depolarize the light, especially when the scattering particles are large (of the order of 10-μm diameter). Its advantage is simplicity. In particular, it does not require a multicolor laser. The three-beam polarization systems employ one beam whose direction of polarization has vector components in the directions of each of the other two beams.

The four-beam frequency separation system in Fig. 4.11 produces signals centered at $\nu_{S2} - \nu_{S1}$, ν_{S2}, 0, and ν_{S1} by heterodyne mixing of the light scattered from beams 1 and 2, 2 and 3, 3 and 4, and 1 and 4, respectively. By choosing ν_{S1} and ν_{S2} judiciously, these four carrier frequencies can be separated sufficiently to permit signal separation by band-pass filtering. For example, $\nu_{S1} = 30$ MHz and $\nu_{S2} = 40$ MHz yield carrier frequencies of 10, 30, 40, and 0 MHz. Discarding the 0- and 10-MHz signals leaves the 30-MHz signal from pair 1-3, which measures u, and the 40-MHz signal from pair 2-4, which measures w. This signal can be separated as long as the Doppler shifts for u and w do not exceed 5 MHz. Thus this technique is limited to moderate flow rates, roughly less than 20 m/s. Crossway et al. [50] describe a system of this type that uses a two-dimensional Bragg cell, and a three-beam system that uses a single Bragg cell driven at 30 and 40 MHz is described in [51].

Three-component LDV systems present special problems because the v component of velocity (the "on-axis" component) is awkward to measure directly with a dual-beam system. One approach is to use a reference-beam system to measure v and a two-component dual-beam system to measure u and w. The v component is measured by illumination along the y axis and light collection along the negative y axis, that is, $2\kappa = 180°$. In this way, all the light-collecting optics can also be kept on the same side of the test section as the transmitting optics.

When simultaneous measurements of u, v, and w are not required, one can use a single-component LDV tilted at three different angles to measure three components that are not necessarily orthogonal. The mean components U, V, and W can be determined from these data, and in certain instances the higher-order moments such as rms or Reynolds stress can also be calculated [52, 53].

4.6 PHOTODETECTORS

The photodetector system converts J_{tot}, the total flux of light energy striking the detector, into a voltage $e(t)$ that is the input to the signal processor. The total light flux is the sum of the background light flux J_B (e.g., room lights, radiation from the fluid, reflections, and flare), the pedestal light flux J_P, the Doppler light flux J_D, and, if appropriate, the reference-beam light flux J_R:

$$J_{tot} = J_B + J_P + J_D + J_R \qquad (54)$$

Ideally, $e = S\tilde{J}_{tot}$, where S is the sensitivity of the detector system in volts per watt, and $\tilde{J}_{tot}$ denotes the value of $J_{tot}(t)$ after it has been filtered by high-pass and/or low-pass filters following the detector. However, real detector systems always contain noise. Part of the noise is generated in the electronics between the photodetector and the signal processor, and part is generated within the detector itself. The latter, called shot noise, is inherent in the photodetection process, and it places a fundamental limit on the signal-to-noise ratio.

4.6.1 Detector Characteristics

The photomultiplier tube (PMT) uses the photoelectric effect wherein photons striking a coating of photoemissive material on the photocathode cause electrons to be emitted from the material. The quantum efficiency of the photocathode is defined as the mean number of electrons emitted per photon. Since the mean number of photons per second is J_{tot}/hv_0, the mean emission rate, i.e., the mean number of photoemissions per second, is given by

$$\dot{\varepsilon} = \frac{J_{tot}\eta_q}{hv_0} \qquad (55)$$

where η_q is the quantum efficiency. In general, η_q depends on the wavelength of the light. The photocurrent is amplified within the PMT by accelerating the electrons from the cathode through a high-voltage field and impacting them on a dynode that emits more than one electron, on the average, for each electron striking it. Repeating the amplification process through several stages of dynodes yields gains between 10^3 and 10^7, depending on the number of stages in the dynode chain and the applied voltage, which is usually 1–3 kV.

The PMT behaves as a light-controlled current source with a very high source impedance. Its output current at the anode is converted to a voltage by passing it through a "load resistor" whose value for high-frequency work is typically $R_L = 50\ \Omega$. The unfiltered anode current is a series of current pulses, each pulse being caused by a single cathode photoemission, but consisting of many electrons after dynode amplification. The width of these pulses is determined, first, by the time spread associated with electrons taking different paths through the dynode chain, and, second, by the anode capacitance C_a. In combination with the load resistance R_L that is connected to the anode, this capacitance forms a low-pass RC filter with time constant $R_L C_a$ and frequency response $1/2\pi R_L C_a$. With care, the pulse width may be made to approach the transit-broadened value, but normally the filtering by the anode capacitance dominates and the pulse width is of the order of $R_L C_a$. Tubes with frequency response up to 200 MHz are used in LDV. In applications requiring less frequency response, the output signal is filtered further to remove noise above the maximum frequency of interest. It may also be high-pass filtered to remove the signal pedestals.

The photodiode is a light-sensitive semiconductor junction whose reverse-biased conductivity changes with the incident light flux. By passing a constant current through the reverse-biased diode, the conductivity variations are converted to a voltage signal with properties much like those of the PMT signal. The chief differences between the PMT and the photodiode are (1) the photodiode works well at relatively high light levels, whereas the PMT will burn out if the light flux is too great; (2) the PMT works well at low light levels because of its high gain, whereas the ordinary photodiode has no internal gain or, in the case of avalanche photodiodes, very little gain (of the order of 10^2); (3) the quantum efficiency for good photodiodes is about 80% in the visible wavelengths, but only about 20% for good PMTs; and (4) photodiodes are much smaller and require smaller power supplies. PMTs are usually used in LDV because of their high sensitivities, but photodiodes should be considered for forward-scatter situations in which the intensity is high.

4.6.2 Photoemission Statistics

The photoemission process is inherently random, and this property greatly influences the approaches taken to extract the Doppler-shift information from the detector signal. For this reason, we discuss the PMT signal statistics in considerable detail. Similar considerations apply to photodiodes, with some changes in the details.

The primary source of randomness in the PMT signal is the fact that the time t_j at which the jth photoelectron is emitted is a random variable. If it is assumed for simplicity that J_{tot} is constant, the probability that the emission time t_j occurs in a small time interval $(t_1, t_1 + dt)$ is the same as the probability that it occurs in any other time interval $(t_2, t_2 + dt)$ and is proportional to dt. Hence the probability of $n = k$ emissions in Δt seconds is given by the Poisson distribution,

$$\text{Prob } \{n = k\} = \frac{e^{-\dot{\varepsilon} \Delta t}(\dot{\varepsilon} \, \Delta t)^k}{k!} \tag{56}$$

with rate parameter $\dot{\varepsilon}$. Here n is a random variable. Some well-known properties of this distribution [54] are that the mean number of emissions in Δt is $\langle n \rangle = \dot{\varepsilon} \, \Delta t$; the rms fluctuation in the number of emissions in Δt is $\langle (n - \langle n \rangle)^2 \rangle^{1/2} = (\dot{\varepsilon} \, \Delta t)^{1/2} = \langle n \rangle^{1/2}$; and the probability density given by Eq. (56) is a maximum at $k = \dot{\varepsilon} \, \Delta t$ for large $\dot{\varepsilon} \, \Delta t$. Thus, while the rms fluctuation in n increases as $\langle n \rangle^{1/2}$, the *relative* rms fluctuation decreases as $\langle (n - \langle n \rangle)^2 \rangle^{1/2}/\langle n \rangle = \langle n \rangle^{-1/2}$, and vanishes in the limit $\langle n \rangle \to \infty$. In this limit, the number of emissions in Δt is a deterministic value proportional to J_{tot}. However, for more modest mean emission rates the number fluctuates randomly about $\langle n \rangle$, and these fluctuations are the source of shot noise in $e(t)$.

4.6.3 Shot Noise

The foregoing results can be extended readily to allow for time-varying light fluxes by defining a time-dependent emission-rate parameter

$$\dot{\varepsilon}(t) = \frac{\eta_q J_{tot}(t)}{h\nu_0} \tag{57}$$

and treating the statistics of the emission process in the sense of conditional averages, i.e., the mean number of emissions at time t given the value of $J_{tot}(t)$, the rms fluctuations given $J_{tot}(t)$, etc. [55].

As sketched in Fig. 4.13, the cathode current can be represented as a sequence of current impulses (with charge q_0) at random times t_j such that the number of impulses per second is large when $J_{tot}(t)$ is large. The dynode chain broadens the jth impulse because of travel-time spread and amplifies it by a random amount g_j so that the total anode current is represented by $\sum_j g_j h_p(t - t_j)$, where $h_p(t)$ is essentially the impulse response function of the dynode chain. Here again the pulse density is proportional to $J_{tot}(t)$. With the exception of photon correlation spectroscopy, which is covered later, one does not usually work directly with the raw anode current. Instead, the photocurrent is converted to a voltage and smoothed by analog filters to give $e(t)$.

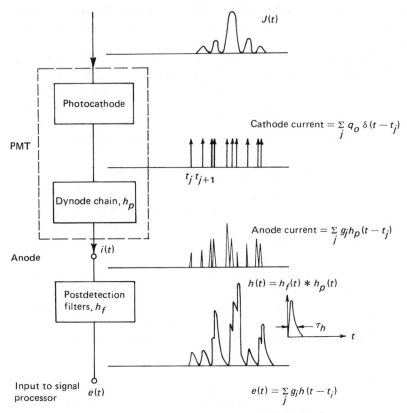

Figure 4.13 Structure of the photomultiplier tube and the PMT signal

Suppose that we let $g_j h(t - t_j)$ denote the filtered output of the system in response to a photoemission at t_j. The function $h(t)$ is the impulse response of the complete system consisting of the PMT, the load resistor, and any additional filters that operate on the signal, e.g., low-pass or high-pass filters applied to the PMT output. [If $h_f(t)$ represents the impulse response of the postdetector filters, $h(t)$ is the convolution of $h_p(t)$ with $h_f(t)$.] Then the filtered voltage signal will be the sum of the responses to the individual photoemissions:

$$e(t) = \sum_{j=-\infty}^{\infty} g_j h(t - t_j) \tag{58}$$

This signal is sketched in Fig. 4.13.

The degree to which $e(t)$ corresponds to $J_{tot}(t)$ depends on the value of $\dot{\varepsilon}\tau_h$, where τ_h is the characteristic time constant of $h(t)$, defined by

$$\tau_h = \frac{1}{2\pi \, \Delta f} \tag{59}$$

where Δf is the bandwidth of the PMT-filter system. Physically, $\dot{\varepsilon}\tau_h$ represents the average number of photoemissions in one time constant. If $\dot{\varepsilon}\tau_h \ll 1$, the average time between photoemissions $\dot{\varepsilon}^{-1}$ is large in comparison with τ_h, and the filtered pulses rarely overlap. In this case, referred to as "low photon density," $e(t)$ would be a sequence of random pulses much like the anode current sketched in Fig. 4.13. Signal processors based on conventional frequency-demodulation techniques are developed essentially for modulated sine waves and would fail, utterly, to measure the Doppler frequency of this type of signal. Instead, one must use techniques such as photon correlation that work directly on the number of emissions per second.

In the other extreme, when $\dot{\varepsilon}\tau_h \gg 1$, the pulses overlap to produce a more continuous signal, and $e(t)$ becomes a fairly good representation of $J_{\text{tot}}(t)$, although randomness in the emission times still results in some differences. These differences are called the shot noise. This case is called "high photon density," and conventional frequency demodulators (counters and trackers) work well on signals that fall into this class. The signal sketched in Fig. 4.13 actually corresponds to the intermediate case in which $\dot{\varepsilon}\tau_h$ is of the order of 1. It is a recognizable, albeit ragged, replica of $J_{\text{tot}}(t)$.

It should be noted that $\dot{\varepsilon}\tau_h$ is determined in equal measure by the optical system (through J_{tot}) and the bandwidth Δf. For example, the signal produced by a particular LDV system may be of low photon density when a 200-MHz bandwidth ($= \Delta f$) is used, but of high photon density when a 1-MHz bandwidth is used, quite simply because band-pass filtering rejects those components of the shot noise whose frequencies lie outside the pass band.

As in turbulent flow, it is convenient to describe the high-photon-density signal in terms of a mean value plus random fluctuations about the mean. However, time averaging must be avoided in defining the mean because a time average of $e(t)$ would not contain any information about the temporal structure (and hence the frequency) of $J_{\text{tot}}(t)$. The appropriate type of average is the conditional average of $e(t)$ given the value of J_{tot}. It can be shown that this average is given [55] by

$$\langle e(t) | J_{\text{tot}}(t) \rangle = K_1 \int_{-\infty}^{\infty} \dot{\varepsilon}(\xi) h(t - \xi) \, d\xi \tag{60a}$$

or

$$\langle e(t) | J_{\text{tot}}(t) \rangle = S \tilde{J}_{\text{tot}} \tag{60b}$$

where K_1 is a proportionality constant, S is the sensitivity of the detector, and $\tilde{J}_{\text{tot}}$ is the filtered value of J_{tot}:

$$\tilde{J}_{\text{tot}}(t) = \int_{-\infty}^{\infty} J_{\text{tot}}(\xi) h(t - \xi) \, d\xi \tag{61}$$

For example, if the signal is high-pass filtered to remove the pedestals, $\tilde{J}_{\text{tot}}$ is just the Doppler signal because the high-pass filter will also remove any dc components. Equation (60a) assumes that "dark currents," i.e., photoemissions that occur spontaneously, independent of J_{tot}, are negligible. The total signal is

$$e(t) = \langle e(t)|J_{tot}(t)\rangle + e_N(t) \tag{62}$$

where the total noise $e_N(t)$ is the sum of the shot noise in the detector e_n, the electronics noise generated in the postdetector filters e_E, any RF noise picked up in the lines e_{RF}, and any spurious heterodyne signals e_H. (Including the latter in the noise term means that the spurious heterodyne light flux should not be included in $\tilde{J}_{tot}$.)

The electronics noise is independent of the light flux, and its power is usually constant in time. It is a combination of Johnson noise, whose power spectrum is "white," i.e., constant power spectral density up to the filter cutoff frequency, and semiconductor surface noise whose power spectrum is inversely proportional to frequency. The RF pickup may include local radio and television broadcasts, but these are usually weak and can be shielded against. The primary source of RF pickup is the oscillator that drives the Bragg cell, and this can be quite troublesome because it appears to the signal processor as a very steady, continuous Doppler signal corresponding to zero velocity. Spurious heterodyne signals occur when laser light enters the photodetector from two or more sources, such as reflections or flare from optical elements, flare from test-section windows, or scattering from particles either inside or outside the mv. When these sources are strong enough, the heterodyne signal may be large even though the coherence factors are small.

The shot noise is unique because it is an intrinsic part of the detection process. The expected value of the square shot noise given $J_{tot}(t)$ is [48, 55]

$$\langle e_n^2(t)|J_{tot}(t)\rangle = K_2 \langle g_i^2 \rangle \int_{-\infty}^{\infty} \dot{\varepsilon}(\xi) h^2(t - \xi)\, d\xi \tag{63a}$$

$$\simeq 2 \frac{h\nu_o}{\eta_q} \Delta f S^2 J_{tot} \tag{63b}$$

where $K_2 = $ const and Eq. (63b) is a satisfactory working formula.

4.7 SIGNAL-TO-NOISE RATIO EFFECTS

At present there are a number of signal-processing techniques that will extract the Doppler frequency from relatively noisy signals ranging from high-photon-density signals that are only slightly contaminated by shot noise to low-photon-density signals that barely resemble $\tilde{J}_{tot}(t)$. One generally obtains the maximum amount of flow information per second when the signal-to-noise ratio (SNR) is high, and therefore it is advisable to strive for the best SNR that can be obtained within the constraints of the experiment. In this section we discuss the most important factors that affect the SNR.

The light flux J_{tot} in Eq. (63b) is the total flux of light energy striking the PMT, only a part of which is useful, i.e., the reference beam and/or the Doppler-shifted scattered-light fluxes. The pedestal light flux is an inevitable part of the heterodyne detection method, but the background light flux J_B is nonessential. Moreover, it is thoroughly detrimental, in that it contributes nothing to the signal power while increasing the shot-noise power, often substantially.

One of the first tasks in the setup of an LDV system is the minimization of extraneous background light. Even with commercial LDV systems, this task falls squarely upon the shoulders of the user because extraneous light is primarily dependent on the flow apparatus. The pinhole aperture in front of the PMT rejects light that is not in the field of view of the pinhole, and narrow-band optical interference filters are often used to block light that is not close to the laser wavelength. Then the primary sources of extraneous light are laser flare (e.g., the nonspecular reflection and diffraction of the laser light where it enters or leaves an optical surface) and reflections. These can be very large. For example, a typical 1-μm scattering particle illuminated by a 10-mW beam may scatter about 10^{-8} W through a back-scatter-collecting aperture. In contrast, 4% of a laser beam will be reflected at an air-glass interface, corresponding to 4×10^{-4} W from a 10-mW beam. If only $\frac{1}{4}$ % of the reflected light were to be collected through the pinhole, it would still be 100 times greater than the light flux from the particle, and the shot-noise power would be 100 times greater than necessary. Hence, very substantial improvements in the SNR can be achieved by minimizing laser flare and reflections.

The methods used to reduce J_B are (1) tilt the optics so that reflections miss the PMT pinhole; (2) use clean optical surfaces, including the test-section windows, or use antireflection-coated surfaces; (3) block stray light with black tape before it enters the collecting lens; and (4) place the pinhole far away from the collecting lens so that stray light that is not in focus is collected over a smaller solid angle. If background light has been reduced to a negligible level, the crossed laser beams will be seen clearly against a dark background when viewed through the receiving optics. If the background is brighter than the light scattered from the beams, extraneous light will contribute appreciably to the shot noise.

In situations such as measurements very close to a wall, the flare from the wall can be very difficult to eliminate by any of these methods because it originates so close to the mv. One proposed technique is to use fluorescent particles that would fluoresce, as they crossed the fringe pattern, at a wavelength *different* from the laser wavelength; then flare could be discriminated against by a narrow-band optical filter at the fluorescence wavelength. This technique would work well, provided that care is taken to avoid coating the wall with fluorescent dye via particle deposition. Alternatively, the wall can be painted with fluorescent paint to provide high absorption and low flare. Stevenson et al. [56] discuss the fluorescent-particle technique.

The ideal dual-beam LDV signal is obtained when all background light, spurious heterodyne signals, and electronics noise are negligible, and a single scattering particle resides at the center of the mv, where the illuminating beams, assumed to be of equal powers $P_{01} = P_{02} = P_0$, have maximum intensities. In this case the light flux on the detector is just the light flux scattered from the particle, given by Eq. (26b). The signal $e(t)$ still contains shot noise due to the scattered light itself, and this noise imposes a fundamental limit upon the maximum attainable SNR. The signal power and the noise power are conveniently defined by time averaging over one Doppler cycle. The period-averaged signal power is $0.5S^2 a^2(0)D_i^2 = 0.5S^2 \overline{V}_i^2 J_{P_i}(\mathbf{x}_i = \mathbf{0})$, and the period-averaged noise power is $2(h\nu_0/\eta_q)\Delta f S^2 J_{P_i}(\mathbf{x}_i = \mathbf{0})$, yielding

$$\text{SNR}_{\text{peak}_i} = \frac{\eta_q}{4h\nu_0\,\Delta f}\, J_P(\mathbf{x}_i = \mathbf{0})\overline{V}_i^2 \tag{64a}$$

$$\text{SNR}_{\text{peak}_i} = \frac{\pi^2}{256}\,\frac{\eta_q P_0}{h\nu_0\,\Delta f}\left(\frac{D_a D_{e-2}}{f_c f}\,\frac{d_{pi}}{\lambda}\right)^2 G_i \overline{V}_i^2 \tag{64b}$$

Equation (64b) pertains to a spherical particle with diameter d_{pi}, and the "scattering gain"

$$G_i = \frac{2(P_{1i} + P_{2i})}{k^2 d_{pi}^2\,\Omega} \tag{65}$$

is the ratio of the actual flux of light scattered through Ω, to the flux that would go through Ω if energy $[I_{01}(0)+I_{02}(0)]\pi\,d_{pi}^2/4$ were scattered isotropically. The primary implication for Eq. (64b) is that the peak SNR for the ith particle is proportional to the scattering power J_{P_i} of the particle and the square of the visibility of the heterodyne signal $\overline{V}_i^2$. Thus, doubling the signal visibility by an appropriate choice of particles is as effective as quadrupling the laser power.

The factor in parentheses in Eq. (64b) contains the first-order effects of the LDV geometry, and it implies that the SNR decreases as the fourth power of the focal distance if $f_c \simeq f$, which is the case in most applications. Thus, long-range measurements usually suffer from very low SNRs. One remedy is to expand the unfocused illuminating-beam diameter so as to keep D_{e-2}/f constant. Beam expansion is generally desirable because it also produces smaller measurement volumes for the same focal distance. However, the limitations of physical size and the alignment accuracy required to cross two small focal spots at large distances usually make beam expansions greater than about 10:1 difficult.

4.8 SCATTERING PARTICLES

The scattering particles are the basic source of the Doppler signal, and their importance in the overall performance of an LDV system should not be underestimated. They can have more influence on the quality of the signal than any other component of the system. For example, the signal strength can be increased by a factor of 10^2–10^4 by increasing the particle diameter from several tenths of a micron to several microns. Similar improvements in scattered-light intensity can be achieved by observing the light scattered in the forward-scatter direction rather than the back-scatter direction. Improvements of these orders of magnitude are difficult, expensive, or perhaps impossible to achieve by increasing the laser power or otherwise improving the optical system.

The most important properties of an individual scattering particle are the SNR that it produces and its aerodynamic size, which is a measure of its ability to follow the flow. From the discussion in section 4.7, good SNR requires that the particle is an effective scatterer (large J_{P_i}) and that it scatters light waves that can mix efficiently (large $\overline{V}_i$). In addition, the concentration and uniformity of the particle population play important roles. Ideally, one would like particles that have the same density as the fluid, large effective area in regard to scattering power, very uniform properties from particle to particle, and easily controlled concentration. They should also be inexpensive. In reality, there are no known particles that satisfy all these requirements.

A wide variety of particles have been used in LDV applications. Naturally occurring aerosols and hydrosols are, of course, the most convenient, and often they can yield satisfactory results. However "natural" does not always imply "best," and this is certainly true in LDV. Natural particles usually have a very wide size distribution (cf. Fig. 4.14), so that many of the particles are too small to produce measurable signals, while others may be too large to follow rapidly accelerating flows. Also, their concentrations may not be controlled. Since these factors influence the design of the LDV optics, signal processing, and data reduction, it is often simplest and most satisfactory to artificially seed the flow with appropriately chosen particles.

In water flows with forward-scattered light, the natural hydrosol is ordinarily quite satisfactory, even with low-power lasers in the 5- to 25-mW range. In back scatter, high laser power and seeding with 5-μm or larger particles is needed to achieve strong signals. Plastic spheres are available in precisely sized batches with specific gravities very close to unity, so that settling is not a problem [57, 58], but they are not suitable for large flow systems because of their expense. Plastic spheres with 0.5-μm diameters are produced by Dow Chemical in bulk lots for paper coating. These are very inexpensive and quite suitable for forward scatter. Silicon carbide particles back scatter effectively because of their high refractive index, but they settle too quickly for use in flows below about 1 m/s. Other particles that have been used in water include milk, latex paint pigment (usually too small), and various fine powders.

Scattering power is not as low in air as it is for back scatter in water, and it is relatively easy to get good signals unless the velocity is very high or the focal length of the transmitting lens is large (more than 0.5 to 1 m without beam expansion). Figure 4.14 shows the number density of particles in a typical urban aerosol versus

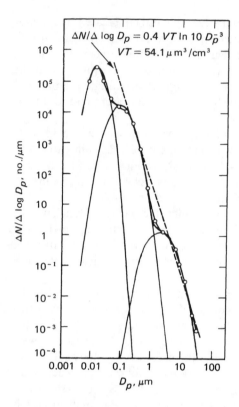

Figure 4.14 Urban aerosol number density distribution versus aerodynamic diameter. (From [59].)

aerodynamic diameter D_p. The distribution is actually trimodal, but the sum of these three modes is often described by a power law. The mode in the range 0.1–1 μm is fairly universal and can be used for LDV estimates. When the natural aerosol is unsatisfactory, one tries to seed the flow with the largest particles that are still capable of following the flow accelerations. Typically, 1- to 2-μm particles are required for most low- to moderate-speed airflows. Particles may be liquid droplets such as water, various oils (vegetable oil, motor oil, etc.), or solid particles such as sieved grinding powder [48, 60, 61].

The sizes given above are guidelines for typical low-speed flows. For a particular flow it is prudent to make some estimate of the particle response time and the flow acceleration to estimate the maximum allowable particle size. For gases (particle density ≫ gas density), a simple first-order model of the particle's response to a step change in velocity yields [60] the time constant

$$t_1 = \frac{d_p^2 \, \rho_p}{18 \mu_f} \qquad (66)$$

The 3-dB frequency at which the particle would follow a sinusoidal velocity variation with an amplitude of 0.707 of the fluid-velocity amplitude is $f_{3\text{-dB}} = 1/2\pi t_1$. This

Table 4.1 Properties of common scattering particles in air

Particles*	n_p	ρ_p	0.5 μm		1 μm		2 μm	
			t_1, μs	$f_{3\text{-dB}}$, kHz	t_1, μs	$f_{3\text{-dB}}$, kHz	t_1, μs	$f_{3\text{-dB}}$, kHz
Silicon carbide	2.6	3.3	2.55	62	10.2	16	41	3.9
Alumina	1.76	3.8	2.93	54	11.7	14	47	3.0
Polystyrene	1.6	1.05	0.81	196	3.24	52	13	13
Peanut oil	1.47	0.91	0.70	227	2.81	57	11.2	14
Microballoons		0.23	0.18	897	0.71	224	2.84	56

*For air, settling velocity = $9.8t_1$ (m/s).

frequency should be larger than the frequency of the velocity fluctuations seen by the particle. In turbulent flows the appropriate fluid frequency is the Lagrangian frequency, since the particle motion is approximately Lagrangian. The time constants and response frequencies of particles that are commonly used in air are given in Table 4.1

In liquid flows the velocities are usually so small that the primary limitation on particle size comes from the settling velocity rather than the ability to follow the flow. Settling velocities for various particles in water are given in Table 4.2. References [62, 63] contain useful information on natural hydrosols.

The scattering properties of small particles are very sensitive to size, refractive index, and scattering angle, so it is difficult to make simple, general statements about their behavior. The computed results for dioctyl phthalate (DOP) particles in air (shown in Fig. 4.15) illustrate this complexity. On average, the following inferences are valid: (1) big particles scatter more than small particles; (2) back scatter is very weak compared to forward scatter for the 0.5- to 20-μm range; (3) high ratios $m = n_p/n$ of the particle refractive index to the fluid refractive index yield better scattering; and (4) particle diameters larger than the fringe spacing do not necessarily imply low signal visibility. They often yield better signals than the smaller particles, especially in back scatter. Computer codes such as the one used to compute Fig. 4.15 are available for the computation of the LDV signal properties in terms of the parameters of the LDV system and the particle [33, 64, 65]; accurate evaluation of particle-size effects requires the use of such codes because of the highly complex nature of the light-scattering phenomenon. For rougher estimates that yield order-of-magnitude accuracies, a procedure is outlined in [66].

4.9 PROPERTIES OF THE RANDOM LIGHT FLUX

4.9.1 Signal Representation

The signal $e(t)$ is random because the photodetector introduces random noise and because the scattering particles are located at random points in the flow. Further, the particles of a polydisperse suspension have random scattering coefficients, and

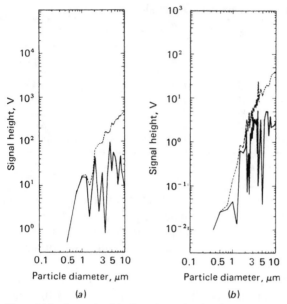

Figure 4.15 Strength of the Doppler and pedestal signals as a function of particle diameter for typical (*a*) forward-scatter and (*b*) back-scatter LDV systems. (——, Doppler height; - - - pedestal height; focal length = 120 mm, $m = 1.48$, e.g., DOP in air.) (From [64])

in turbulent flows their velocities are also random. These random elements pervade all aspects of signal processing and data analysis in LDV, and a thorough description of their statistics is essential. Fortunately, it is possible to isolate certain of these elements. It is shown in section 4.6 that the photodetector noise is either electronics noise that is statistically independent of the light flux, or shot noise that can be described in terms of conditional statistics of $e(t)$, given the value of the light flux. Unconditional statistics of $e(t)$, corresponding to long time averages, can be obtained by averaging over all random values of the light flux. Thus the statistical description of the signal from the photodetector filter system can be completed by analyzing the statistics of the light flux separately from the statistics of the shot noise.

4.9.2 Random Doppler Light Flux

In general, the total Doppler light flux is a sum of the individual bursts caused by each particle:

$$J_D(t) = \sum_i J_{D_i}(t) \tag{67a}$$

$$= \sum_i a[\mathbf{x}_i(t)]D_i \cos (\Phi_i - \Psi_i) \tag{67b}$$

where i extends over all particles in the flow, and $a(\mathbf{x})$ and $\Phi_i = \Phi[\mathbf{x}_i(t), t]$ are given by Eqs. (29) and (25a), respectively. (The amplitude factor a accounts for the presence of a particle in the mv.) If we let

$$g(\mathbf{x}, t, D) = \sum_i \delta[\mathbf{x} - \mathbf{x}_i(t)] \, \delta(D - D_i) \tag{68}$$

Eq. (67b) becomes

$$J_D(t) = \int_{\text{all } \mathbf{x}} \int_0^\infty a(\mathbf{x})D \cos \Phi(\mathbf{x}, t)g(\mathbf{x}, t, D)d^3\mathbf{x} \, dD \tag{69}$$

where, without loss of generality, the random phase shifts Ψ_i can be ignored because the signal phases $\Phi(\mathbf{x},t)$ are also random. The function $g(\mathbf{x}, t, D)$ indicates the presence of a particle at the point $\mathbf{x}$ with scattering amplitude D. When $J_D(t)$ is cast into this form, all the random properties of the particles are contained in the function g.

4.9.3 Statistical Properties of $g(\mathbf{x}, t, D)$

Since all the random characteristics of the light flux are contained in $g(\mathbf{x}, t, D)$, this function's properties are derived in full detail. The analysis generalizes [67, 68] by including the effects of nonuniform concentration of particles, both in space and in time, and the distribution of signal amplitudes as determined by the particle size distribution and the optics of the LDV system.

Consider a volume V that is fixed in space (i.e., a "control volume"), and the material volume $V_m(t)$ that coincides with V at time t (Fig. 4.16). At any instant there may be many different particles in V_m, each with a different value of D_i. We shall refer to all particles that produce Doppler signal amplitudes such that

$$D < D_i < D + \delta D \tag{70}$$

as "D-type" particles, where D is any arbitrarily prescribed numerical value and δD is a small range of amplitudes. At time t the number of D-type particles in V is just equal to the number of D-type particles in $V_m(t)$, and this is given by

$$n(V, t, D) = \int_V \int_D^{D+\delta D} g(\mathbf{x}, t, D)d^3\mathbf{x} \, dD \tag{71}$$

since the integral of $\delta[(\mathbf{x} - \mathbf{x}_i(t)] \, \delta(D - D_i)$ yields a 1 if $\mathbf{x}_i(t) \in V$ and $D_i \in (D, D + \delta D)$. This is also equal to the number of D-type particles in $V_m(t)$, denoted by $n[V_m(t), D]$, and if the particles follow the fluid motion, the number in V_m must be the same for all time; that is, $n[V_m(t), D] = n[V_m(t'), D]$. This assumption implies that both particle lag and particle diffusion by Brownian motion are negligible.

The statistics of g are determined by the statistics of n, and these in turn, can be found under the assumption that the location of any particular particle, for

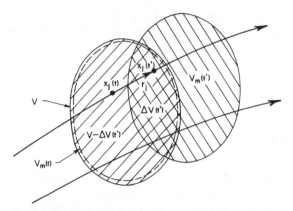

Figure 4.16 Motion of the particle population through the measurement volume.

example, the ith particle, is entirely random. This assumption requires that the relative separation vectors between all pairs of particles are random, but it does not require that the particle concentration be constant. Let the probability that the ith particle is of type D and that it is located in V at time t be p_0, given by

$$p_0 = \int_V \int_D^{D+\delta D} \lambda(\mathbf{x}, t, D) d^3\mathbf{x} \, dD \tag{72}$$

where λ is the probability density of particles in $\mathbf{x}$-D space. Then, if the total number of particles of all types in the entire fluid volume is $\mathfrak{n}$, the probability that any n particles of type D are located in V is the binomial distribution [54]

$$\text{Prob } \{n \text{ } D\text{-type particles in } V \text{ at time } t\} = \frac{\mathfrak{n}!}{n!(\mathfrak{n} - n)!} p_0^n (1 - p_0)^{\mathfrak{n}-n} \tag{73}$$

The average number of D-type particles in V is

$$N(V, t, D) \equiv \langle n(V, t, D) \rangle \tag{74a}$$

$$= \sum_{n=0}^{\infty} n \frac{\mathfrak{n}!}{n!(\mathfrak{n} - n)!} p_0^n (1 - p_0)^{\mathfrak{n}-n} \tag{74b}$$

$$= \mathfrak{n} p_0 \tag{74c}$$

The average concentration of D-type particles, i.e., the number per unit volume per unit amplitude D, is denoted by $C(\mathbf{x}, t, D)$ and is given by

$$C(\mathbf{x}, t, D) = \lim_{\substack{V \to 0 \\ \delta D \to 0}} \frac{\langle n(V, t, D) \rangle}{V \, \delta D} \tag{75a}$$

$$= \lim_{\substack{V \to 0 \\ \delta D \to 0}} \frac{\mathfrak{n}}{V \, \delta D} \int_V \int_D^{D+\delta D} \lambda(\mathbf{x}, t, D) d^3\mathbf{x} \, dD \tag{75b}$$

$$= \mathfrak{n}\lambda(\mathbf{x}, t, D) \tag{75c}$$

where V shrinks to a point at $\mathbf{x}$.

When the total fluid volume is large in comparison with V, as is almost always true, $\mathfrak{n}$ will be much greater than $\langle n \rangle$, and Eq. (73) can be simplified by taking its limit as $\mathfrak{n} \to \infty$, keeping $N = \langle n \rangle = \mathfrak{n}p_0$ fixed. The result is the Poisson distribution [54, 69]

$$\text{Prob } \{n \text{ } D\text{-type particles in } V\} = \frac{N^n e^{-N}}{n!} \tag{76}$$

where, from Eqs. (72), (74c), and (75c),

$$N = \langle n(V, t, D) \rangle = \int_V \int_D^{D+\delta D} C d^3\mathbf{x} \, dD \tag{77}$$

is the mean number of D-type particles in the limit $\delta D \to 0$. The range δD may be replaced in this equation by any finite range ΔD to obtain the mean number in $(D, D + \Delta D)$; inserting this mean number into Eq. (76) gives the probability of finding n particles in V whose amplitudes are anywhere in $(D, D + \Delta D)$.

The expected value of $g(\mathbf{x}, t, D)$ is found by taking the ensemble average of Eq. (71), dividing by V and δD, and taking the limit as $V \to 0$ and $\delta D \to 0$:

$$\lim_{V, \delta D \to 0} \frac{\langle n(V, t, D) \rangle}{V \, \delta D} = \lim_{V, \delta D \to 0} \frac{1}{V \, \delta D} \int_V \int_D^{D+\delta D} \langle g(\mathbf{x}, t, D) \rangle \, d^3\mathbf{x} \, dD \tag{78a}$$

$$= \langle g(\mathbf{x}, t, D) \rangle \tag{78b}$$

Hence from Eq. (77),

$$\langle g(\mathbf{x}, t, D) \rangle = C(\mathbf{x}, t, D) \tag{79}$$

The space-time correlation of g, defined as $\langle g(\mathbf{x}, t, D)g(\mathbf{x}', t', D') \rangle$, is also needed to describe the behavior of $J_D(t)$. It can be calculated by observing that

$$\langle n(V, t, D)n(V, t', D') \rangle = \left\langle \int_V \int_D^{D+\delta D} g(\mathbf{x}, t, D) d^3\mathbf{x} \, dD \right.$$

$$\left. \times \int_V \int_{D'}^{D'+\delta D} g(\mathbf{x}', t', D') d^3\mathbf{x}' \, dD' \right\rangle \tag{80a}$$

$$= \int_V \int_V \int_D^{D+\delta D} \int_{D'}^{D'+\delta D} \langle g(\mathbf{x}, t, D)g(\mathbf{x}', t', D') \rangle$$

$$\times d^3\mathbf{x} \, d^3\mathbf{x}' \, dD \, dD' \qquad (80b)$$

and calculating $\langle n(V, t, D)n(V, t', D') \rangle$. A fundamental assumption implicit in Eq. (72) is that the number of D particles in V is statistically independent of the number of D' particles in V if $D \neq D'$. (This assumption would be violated if, for example, the particles were so large that the presence of one particle precluded the presence of any other particle in V, as can occur in two-phase flows. These situations are excluded from the present analysis.) Then, if $D \neq D'$,

$$\langle n(V, t, D)n(V, t', D') \rangle = \langle n(V, t, D) \rangle \langle n(V, t', D') \rangle \qquad D \neq D' \qquad (81)$$

If, on the other hand, $D' = D$, it is implicit in Eq. (72) that the number of D particles in one material volume is statistically independent of the number of D particles in any other nonoverlapping fluid volume. Thus, with reference to Fig. 4.16, the number $n(V, t, D)$ in V at t will be independent of the number $n(V, t', D)$ in V at t' if $V_m(t')$ does not overlap any part of V, that is, if the material volume in V at t has been completely replaced by a new material volume at t'. However, for sufficiently short time differences $t' - t$, $V_m(t')$ and V must share a common volume $\Delta V_m(t')$, and $n(V, t', D)$ will be correlated with $n(V, t, D)$ because at time t', V will contain D particles that also resided in V at time t. Thus the correlation time is of the order of the time required to sweep a new volume into V. This case $(D = D')$ can be analyzed by writing $n(V, t, D)$ and $n(V, t', D)$ in terms of nonoverlapping material volumes $V_m(t') - \Delta V(t')$, $\Delta V(t')$, and $V - \Delta V(t')$:

$$n(V, t, D) = n[V_m(t), D] \qquad (82a)$$

$$= n[V_m(t'), D] \qquad (82b)$$

$$= n[V_m(t') - \Delta V(t'), D] + n[\Delta V(t'), D] \qquad (82c)$$

$$n(V, t', D) = n[V - \Delta V(t'), D] + n[\Delta V(t'), D] \qquad (83)$$

where Eq. (82b) follows from Eq. (82a) because the particles are assumed to follow the motion of the material volume. Multiplying Eqs. (82c) and (83) and averaging yield an expression for $\langle n(V, t, D)n(V, t', D) \rangle$ in which the average of each cross product is equal to the cross product of the averages (because the numbers in nonoverlapping volumes are statistically independent), with the exception of $\langle n^2(\Delta V(t'), D) \rangle$. For a Poisson distribution this last average is given by

$$\langle n^2[\Delta V(t'), D] \rangle = \langle n[\Delta V(t'), D] \rangle^2 + \langle n[\Delta V(t'), D] \rangle \qquad (84)$$

The final expression,

$$\langle n(V, t, D)n(V, t', D) \rangle = \langle n(V, t, D) \rangle \langle n(V, t', D') \rangle + \langle n(\Delta V(t'), D) \rangle \qquad (85)$$

follows after some algebraic manipulation.

The last term in Eq. (85) is the average number of particles in V at time t' that were also in V at time t, and it depends upon the statistics of the fluid motion as well as those of the particle locations. Consider a fluid volume dx^3, located at $\mathbf{x}$ at time t, that is mapped by the motion into a volume $d^3\mathbf{x}'$ at $\mathbf{x}'$ at time t'. For incompressible flow $d^3\mathbf{x} = d^3\mathbf{x}'$ and the mean number of particles in $d^3\mathbf{x}$ or $d^3\mathbf{x}'$ is just

$$\int_D^{D+\delta D} C(\mathbf{x}, t, D) \, d^3\mathbf{x} \, dD$$

The probability that $\mathbf{x}'$ lies somewhere in V can be expressed as

$$\int_V f(\mathbf{x}', t'; \mathbf{x}, t) d^3\mathbf{x}'$$

where $f(\mathbf{x}', t'; \mathbf{x}, t) \, d^3\mathbf{x}'$ is defined as the probability that the fluid particle at $(\mathbf{x}, t)$ maps into $(\mathbf{x}', \mathbf{x}' + d^3\mathbf{x}')$ at t'. The average number in $d^3\mathbf{x}$ that are in V at t' is just the mean number in $d^3\mathbf{x}$ times the probability that this number is in V at t',

$$\int_D^{D+\delta D} C(\mathbf{x}, t, D) d^3\mathbf{x} \, dD \int_V f(\mathbf{x}', t'; \mathbf{x}, t) \, d^3\mathbf{x}'$$

and the average number in the total overlap volume is obtained by integrating over all $d^3\mathbf{x}$:

$$\langle n[\Delta V(t'), D] \rangle = \int_V \int_V \int_D^{D+\delta D}$$

$$\times C(\mathbf{x}, t, D) \, f(\mathbf{x}', t'; \mathbf{x}, t) d^3\mathbf{x} \, d^3\mathbf{x}' \, dD \qquad D = D' \quad (86a)$$

$$= \int_V \int_V \int_D^{D+\delta D} \int_{D'}^{D'+\delta D} C(\mathbf{x}, t, D)$$

$$\times f(\mathbf{x}', t'; \mathbf{x}, t) \, \delta(D' - D) \, d^3\mathbf{x} \, d^3\mathbf{x}' \, dD \, dD' \qquad (86b)$$

Combining Eqs. (76a), (77), (79), (85), and (86b), dividing the result by $V \, \delta D$, and taking the limit as $V \to 0$ and $\delta D \to 0$ yield

$$\langle g(\mathbf{x}, t, D)g(\mathbf{x}', t', D') \rangle = C(\mathbf{x}, t, D) \, C(\mathbf{x}', t', D') \qquad (87)$$
$$+ \, C(\mathbf{x}, t, D) f(\mathbf{x}', t'; \mathbf{x}, t) \, \delta(D' - D)$$

The time required for particles to cross the mv is clearly an upper bound on the maximum time difference $t' - t$ that is of interest in Eq. (87), and in most applications of laser velocimetry this transit time is small in comparison with the Lagrangian time scales of the flow. Hence the particle velocities are essentially constant, and the probability that a particle moves from $\mathbf{x}$ at time t to $\mathbf{x}'$ at time t'

is just the probability that $\mathbf{u}(\mathbf{x},t) = \mathbf{U}(\mathbf{x},t) + \mathbf{u}'(\mathbf{x},t) = (\mathbf{x}' - \mathbf{x})/(t' - t)$, where $\mathbf{U}$ is the mean velocity and $\mathbf{u}'$ is the fluctuating velocity. Letting Prob $\{\mathbf{c}' < \mathbf{u}'(\mathbf{x}, t) < \mathbf{c}' + d\mathbf{c}'\} = f_{u'}(\mathbf{c}', \mathbf{x}, t) \, d^3\mathbf{c}'$ and Prob $\{\mathbf{c} < \mathbf{u}(\mathbf{x}, t) < \mathbf{c} + d\mathbf{c}\} = f_u(\mathbf{c}, \mathbf{x}, t) \, d^3\mathbf{c}$ be the probability density functions for the velocity fluctuation and the total velocity, respectively, we have

$$f(\mathbf{x}', t'; \mathbf{x}, t) \, d^3\mathbf{x}' = f_u\left(\mathbf{c} = \frac{\mathbf{x}' - \mathbf{x}}{t' - t}, \mathbf{x}, t\right) \frac{d^3\mathbf{x}'}{|t' - t|^{-1}} \tag{88}$$

for small values of $t' - t$, wherein $d^3\mathbf{x}' = d^3\mathbf{c}|t' - t|^{-1}$. Then

$$\langle g(\mathbf{x}, t, D) \, g(\mathbf{x}', t', D')\rangle = C(\mathbf{x}, t, D) \, C(\mathbf{x}', t', D')$$
$$+ \, C(\mathbf{x}, t, D) f_u\left(\mathbf{c} = \frac{\mathbf{x}' - \mathbf{x}}{t' - t}, \mathbf{x}, t\right)$$
$$\times \, |t' - t|^{-1} \, \delta(D' - D) \tag{89}$$

In steady laminar flow, $f_u(\mathbf{c}) = \delta[\mathbf{c} - \mathbf{U}(\mathbf{x}, t)]$. That is, the fluctuations vanish, and the factor $f_u/(t' - t)$ reduces to $\delta[\mathbf{x}' - \mathbf{x} - \mathbf{U}(\mathbf{x}, t)(t' - t)]$. In general, if $t' = t$, Eq. (89) reduces to $\langle g(\mathbf{x}, t, D) g(\mathbf{x}', t, D')\rangle = C(\mathbf{x}, t, D')C(\mathbf{x}', t, D') + C(\mathbf{x}, t, D)\delta(\mathbf{x} - \mathbf{x}') \cdot \delta(D - D')$.

4.9.4 Correlation and Power Spectrum

The single time moments of the light flux $\langle(J_D - \langle J_D\rangle)^n\rangle$ contain useful statistical information, but they do not yield information on the Doppler frequency, which is the variable of primary interest. For example, the mean of J_D, found by averaging Eq. (69), taking the average inside the integral, and using Eq. (79), is

$$\langle J_D(t)\rangle = \iint a(\mathbf{x})D \cos \Phi(\mathbf{x}, t) \, C(\mathbf{x}, t, D) \, d^3\mathbf{x} \, dD \tag{90}$$

It can be shown that this mean is virtually zero for any reasonable distributions for $a(\mathbf{x})$ and $C(\mathbf{x}, t, D)$. Specifically, if C is constant and $a(\mathbf{x})$ is Gaussian, $\langle J_D\rangle \approx \exp(-\pi^2 N_{FR}^2/8)$, which is of the order of 10^{-5} when $N_{FR} = 3$. Thus we may always take J_D to be a zero-mean random variable. Likewise, higher-order moments such as $\langle J_D^2\rangle$ are also independent of the Doppler frequency.

To extract frequency information, it is necessary to use statistical moments involving the signal's values at two or more times. The simplest quantities that accomplish this purpose are the correlation function

$$R_{J_D J_D}(\tau) = \langle J_D(t)J_D(t + \tau)\rangle \tag{91}$$

and the associated power spectrum or power spectral density, defined by

$$S_{J_D}(\omega) = \frac{1}{2\pi} \int_{-\infty}^{\infty} e^{-j\omega\tau} R_{J_D J_D}(\tau) \, d\tau \qquad (92)$$

where $\omega = 2\pi f$. This equation assumes that $\langle J_D \rangle = 0$. The inverse Fourier transform of S_{J_D} gives $R_{J_D J_D}$ in terms of the power spectrum,

$$R_{J_D J_D}(\tau) = \int_{-\infty}^{\infty} e^{j\omega\tau} S_{J_D}(\omega) \, d\omega \qquad (93)$$

from which it follows that the mean square is the area under the power spectral density curve

$$\langle J_D^2 \rangle = R_{J_D J_D}(0) = \int_{-\infty}^{\infty} S_{J_D} \, d\omega \qquad (94)$$

As is well known [53], $S_{J_D} \, d\omega$ represents the contribution that Fourier components of $J_D(t)$ in the range $(\omega, \omega + d\omega)$ make to the mean square value of J_D; that is, $S_{J_D} \, d\omega$ is the signal "power" in a band of frequencies with width $d\omega$. All the quantities above are independent of t if J_D is a stationary random process, i.e., if the statistics of the flow and the particle population are steady in time. Stationarity is assumed throughout this section. In particular, this implies that $C(\mathbf{x}, t, D) = C(\mathbf{x}, D)$ and $f_u(\mathbf{c}, \mathbf{x}, t) = f_u(\mathbf{c}, \mathbf{x})$. Stationarity also implies that $R_{J_D J_D}(-\tau) = R_{J_D J_D}(\tau)$ and $S_{J_D}(-\omega) = S_{J_D}(\omega)$. The only difference between negative and positive frequencies is a 180° phase shift.

The general equation for the autocorrelation is found by forming the product $J_D(t)J_D(t + \tau)$ using the representation in Eq. (69), averaging inside the integrals in this product, and using Eq. (87):

$$R_{J_D J_D}(\tau) = \iint C(\mathbf{x})\langle D^2(\mathbf{x})\rangle \, f(\mathbf{x}', t \qquad (95)$$
$$+ \tau; \mathbf{x}, t)a(\mathbf{x})a(\mathbf{x}') \cdot \cos \Phi(\mathbf{x}, t) \cos \Phi(\mathbf{x}', t + \tau)d^3 \, \mathbf{x} d^3\mathbf{x}'$$

where

$$C(\mathbf{x}) = \int_0^{\infty} C(\mathbf{x}, D) \, dD \qquad (96)$$

is the total concentration of particles, and

$$\langle D^2 \rangle = \frac{\int_0^{\infty} D^2 C(\mathbf{x}, D) \, dD}{C(\mathbf{x})} \qquad (97)$$

is the mean square scattering amplitude averaged over all particles. A simpler working formula, which is still accurate and general enough to encompass all effects of practical interest, is

$$R_{J_D J_D} \simeq \frac{1}{2} \iint C(\mathbf{x}) \langle D^2(\mathbf{x}) \rangle f_u(\mathbf{c}, \mathbf{x}) a(\mathbf{x}) a(\mathbf{x} + \mathbf{c}\tau) \tag{98}$$
$$\cos{(\omega_S \tau + K c_1 \tau)} d^3 \mathbf{x} \, d^3 \mathbf{c}$$

This equation uses the approximation in Eq. (88) and ignores a term of the order of $\exp{(-N_{FR}^2)}$ that arises from the cosine product in Eq. (95). It is valid whenever the particle velocities are essentially constant during the times that the particles reside in the mv.

Before investigating the case of turbulent flow, embodied in Eq. (98), it is instructive to consider the simpler situation in which the flow is steady, laminar, and uniform (that is, $\mathbf{u} = \mathbf{U} = \text{const}$), the particle population is uniform [that is, $C = \text{const}$ and $\langle D^2(\mathbf{x}) \rangle = \text{const}$], and the mv is Gaussian and given by Eq. (40). Then,

$$R_{J_D J_D} = \langle J_D^2 \rangle \exp{\left(-\frac{\Delta\omega_A^2}{2} \tau^2 \right)} \cos{(\omega_S + Ku)\tau} \tag{99}$$

and

$$S_{J_D} = \frac{1}{(8\pi)^{1/2}} \frac{\langle J_D^2 \rangle}{\Delta\omega_A} \left\{ \exp{\left[-\frac{(\omega - \omega_S - Ku)^2}{2 \, \Delta\omega_A^2} \right]} \right.$$
$$\left. + \exp{\left[-\frac{(\omega + \omega_S + Ku)^2}{2 \, \Delta\omega_A^2} \right]} \right\} \tag{100}$$

where

$$\Delta\omega_A = 2\sqrt{2} \left(\frac{u^2}{d_m^2} + \frac{v^2}{l_m^2} + \frac{w^2}{h_m^2} \right)^{1/2} \tag{101}$$

The autocorrelation is almost a replica of the individual Doppler bursts. It oscillates at the frequency $\omega_S + Ku$, and its amplitude envelope is a Gaussian function because of the assumed Gaussian intensity distribution in the mv. The e^{-1} width of the Gaussian envelope is $2\sqrt{2}\Delta\omega_A$. The reader can verify from Eq. (40) that for any straight trajectory the transit time of a particle between the points where the amplitude is e^{-2} of the maximum amplitude for *that trajectory* is

$$\left(\frac{u^2}{d_m^2} + \frac{v^2}{l_m^2} + \frac{w^2}{h_m^2} \right)^{-1/2}$$

and hence that $2\sqrt{2}\Delta\omega_A$ corresponds to the e^{-2} transit time.

The power spectrum of J_D, represented by the dashed line in Fig. 4.17, shows that signal power is concentrated around $\pm(\omega_S + Ku)$ in a band of frequencies that is about $\Delta\omega_A$ wide. More precisely, the first moment of S_{J_D}

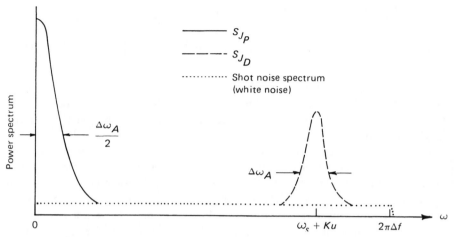

Figure 4.17 Power spectrum of the signal from the photodetector of an LDV.

$$\mu_1 \equiv \frac{\displaystyle\int_0^\infty \omega S_{J_D}\, d\omega}{\displaystyle\int_0^\infty S_{J_D}\, d\omega} \qquad (102a)$$

$$= \omega_S + Ku \qquad (102b)$$

and the second central moment is

$$\mu_2 \equiv \frac{\displaystyle\int_0^\infty (\omega - \mu_1)^2 S_{J_D}\, d\omega}{\displaystyle\int_0^\infty S_{J_D}\, d\omega} \qquad (103a)$$

$$= \Delta\omega_A^2 \qquad (103b)$$

where $\Delta\omega_A$ is given by Eq. (101). As noted above, $\Delta\omega_A$ is inversely proportional to the transit time. Since the Doppler frequency is constant for all particles in this case, it must be concluded that $\Delta\omega_A$ does not represent any variation of the Doppler frequency in time or in space. Rather, $\Delta\omega_A$ is solely a consequence of the finite duration of the individual Doppler bursts, as measured by their transit times, and for this reason, the phenomenon is referred to as transit-time broadening. If the bursts were very long, corresponding to d_m, l_m, $h_m \to \infty$, then $\Delta\omega_A$ would approach zero, and the power spectrum would consist of Dirac delta functions at $\omega_S + Ku$ and $-\omega_S - Ku$.

The bandwidth as a fraction of the Doppler shift is

$$\frac{\Delta\omega_A}{\omega_D} = \frac{2^{1/2}}{\pi N_{FR}} \left(1 + \frac{v^2}{u^2}\tan^2\kappa + \frac{w^2}{u^2}\sec^2\kappa\right)^{1/2} \tag{104}$$

which is independent of the magnitude of u if the direction of u is fixed. The fractional bandwidth is minimized by aligning the mv so that $v = w = 0$. Then, typical values of $\Delta\omega_A/\omega_D$ are 0.5–10%, corresponding to somewhere between 5 and 100 fringes in the mv.

Parameter $\Delta\omega_A$ is also called the ambiguity bandwidth, a term borrowed from radar technology. This term suggests that the frequency spread in the spectrum is associated with an uncertainty in the measurement of the Doppler frequency. In practice, large transit-time bandwidths may, indeed, cause measurement errors because large bandwidths make it more difficult to determine the center frequencies of experimental spectra. In addition, it will be shown that in the case of many particles in the mv, there is an uncertainty in the frequency of the signal that is related to $\Delta\omega_A$ and is caused by the random phase superposition of signals from many particles. There is, however, no random phase superposition when there is only one particle in the mv, and therefore ambiguity broadening imposes no fundamental inaccuracies on the measurement of the frequency of an individual Doppler burst. (The reader will find confusion over this point in some early papers, but this controversy has long been resolved. As a point of interest, it can also be shown that the ambiguity bandwidth can be interpreted in terms of uncertainty in the directions of the light rays in the focused illuminating beams [28, 38].)

The spectrum of the total light flux consists of a Doppler spectrum and a pedestal spectrum. As shown in Fig. 4.17, the pedestal spectrum has the same shape and bandwidth as the Doppler spectrum, but it is centered at zero frequency because each pedestal looks like a Doppler burst with zero Doppler frequency. In general, the Doppler spectrum has lower amplitude than the pedestal spectrum because the visibility of the Doppler signal must always be less than unity. It is clear from Fig. 4.17 that the pedestal component can be removed by high-pass filtering the total signal with a cutoff frequency equal to several times $\Delta\omega_A$.

The shape of the spectrum in Eq. (100) is Gaussian because $a(\mathbf{x})$ was assumed to be Gaussian. For any other intensity distribution the spectral shape is determined by the Fourier transform of $a(\mathbf{x})$.

With suitable approximations, Eq. (98) can be integrated to obtain explicit expressions for $R_{J_D J_D}$ for the case of laminar flow with velocity gradients or turbulent flow with mean velocity gradients and finite turbulence intensity [8, 28, 70]. In all these analyses, C and $\langle D^2 \rangle$ are approximated with constants, $a(\mathbf{x})$ is assumed to be Gaussian, and it is usual to approximate the turbulent probability density function by a normal distribution. As consequences of these assumptions, the envelope of the correlation function and the power spectrum are Gaussian functions.

The correlation function can also be found in the most general case by performing the integrations in Eq. (98) using known functions for $C(\mathbf{x})$, $\langle D^2(\mathbf{x})\rangle$, $a(\mathbf{x})$, $U(\mathbf{x})$, and $f_u(\mathbf{c}, \mathbf{x})$, and the power spectrum can be calculated by Fourier transformation. Explicit expressions cannot, however, be found in general, and numerical integrations are required. Fortunately, only the first and second moments of the spectrum are needed in most applications, and it appears to be possible to obtain analytical expressions for these quantities under fairly weak and nonrestrictive assumptions. The first moment of S_{J_D} can be calculated in terms of the autocorrelation by substituting Eq. (92) into the definition of μ_1 given by Eq. (102a) and manipulating. (The manipulations require the generalized-function Fourier transform of ω, which can be found in [71].) The result is

$$\mu_1 = \frac{\displaystyle\int_{-\infty}^{\infty} \tau^{-2} R_{J_D J_D}(\tau)\, d\tau}{\pi R_{J_D J_D}(0)} \tag{105a}$$

$$= \frac{\displaystyle\iint d^3\mathbf{x}\, d^3\mathbf{c}\; C\langle D^2\rangle f_u(\mathbf{c})a(\mathbf{x}) \int_{-\infty}^{\infty} d\tau\, \tau^{-2}\, a(\mathbf{x}+\mathbf{c}\tau)\cos\,(\omega_S\tau + Kc_1\,\tau)}{\displaystyle\iint C\langle D^2\rangle a^2(\mathbf{x})\, d^3\,\mathbf{x}} \tag{105b}$$

$$= \frac{\displaystyle\iint C(\mathbf{x})\langle D^2(\mathbf{x})\rangle a^2(\mathbf{x})\langle|\omega_S + K_u|\rangle\, d^3\,\mathbf{x}}{\displaystyle\iint C\langle D^2\rangle a^2\, d^3\,\mathbf{x}} + O(N_{FR}^{-2}) \tag{105c}$$

Equation (105c) is an asymptotic result that follows from Eq. (105b) by noting that $a(\mathbf{x}+\mathbf{c}\tau)$ varies slowly with τ in comparison with $\cos(\omega_S + Kc_1)\tau$ when there are many fringes in the mv, and taking the limit as $N_{FR} \to \infty$. In effect, Eq. (105c) ignores transit-time broadening of the spectra associated with individual Doppler bursts. Equation (105c) states that the mean frequency of the spectrum is a volume average of $\langle|\omega_S + Ku(\mathbf{x}, t)|\rangle$ weighted by $C(\mathbf{x})D^2(\mathbf{x})a^2(\mathbf{x}, t)$. It is convenient to denote this averaging operation by

$$(\tilde{\cdot}) = \frac{\int C\langle D^2\rangle a^2(\mathbf{x})(\ )\, d^3\mathbf{x}}{\int C\langle D^2\rangle a^2(\mathbf{x})\, d^3\mathbf{x}} \tag{106a}$$

Thus

$$\mu_1 = \widetilde{\langle|\omega_S + Ku(\mathbf{x}, t)|\rangle} \tag{106b}$$

In a properly designed experiment, ω_S is set at a value large enough to ensure that $\omega_S + Ku > 0$ for all but the most improbable velocity fluctuations. Then Eq. (106b) reduces to

$$\mu_1 = \omega_S + K\widetilde{U} \tag{106c}$$

where $\widetilde{U}$, the filtered mean velocity, is also equal to the mean of the filtered velocity; that is, $\widetilde{U} = \langle \tilde{u}(\mathbf{x}, t) \rangle$.

If the dimensions of the mv are so small that $U(\mathbf{x})$ is constant with negligible error, then $\widetilde{U} \equiv U(0)$, the mean velocity at the center of the mv. This is also true if $C\langle D^2 \rangle$ is constant, $U(\mathbf{x})$ is at most a linear function of position, and $a(\mathbf{x})$ is symmetric about the coordinate planes [28, 70]. In all other cases the relationship between $\widetilde{U}$ and $U(0)$ depends on the details of the weighting factor $C(\mathbf{x})\langle D^2(\mathbf{x})\rangle a^2(\mathbf{x})$. The particle-dependent factor $C\langle D^2 \rangle$ is particularly troublesome because spatial nonuniformities in the scattering particle population are usually not under the control of the experimenter, and worse, they are usually not known. Hence the best procedure by far is to ensure that the scattering population is uniformly distributed in space. Then the volume weighting depends only upon $a^2(\mathbf{x})$, which can be estimated with fair accuracy theoretically. Mean velocity corrections for volume averaging are discussed in [28, 72]. Of course, the ideal situation is achieved when the mv is made so small that mean velocity variations are negligible.

It should be noted that the volume average in Eq. (106a) is consistent with the results found in [28, 70] but differs by a factor $a(\mathbf{x})$ from the average defined in [67]. This has ramifications in the next section, where the statistics of the high-burst-density signal are considered.

The second moment of the power spectrum, defined by Eq. (103a), is a measure of the spectral bandwidth. By differentiating Eq. (93) twice with respect to τ and setting τ equal to zero, it is easy to see that μ_2 is given in terms of $R_{J_D J_D}$ by

$$\mu_2 = -\frac{R''_{J_D J_D}(0)}{R_{J_D J_D}(0)} - \mu_1^2 \tag{107a}$$

Inserting Eq. (98) into Eq. (107a) yields

$$\mu_2 = -[\iint C\langle D^2 \rangle a(\mathbf{x}) f_u(\mathbf{x} + \mathbf{c}\tau)\{[\partial^2 a(\mathbf{x} + \mathbf{c}\tau)/\partial \tau^2] \cos(\omega_S + Kc_1)$$
$$- a(\mathbf{x} + \mathbf{c}\tau)(\omega_S + Kc_1)^2\}_{\tau=0} \, d^3\mathbf{x} \, d^3 \, \mathbf{c}]/[\int C\langle D^2 \rangle a^2(\mathbf{x}) d^3\mathbf{x}] - \mu_1^2 \tag{107b}$$
$$= \Delta\omega_A^2 + \Delta\omega_G^2 + \Delta\omega_T^2 \tag{107c}$$

where $\Delta\omega_A^2$ is the ambiguity broadening given by the first term in braces in Eq. (107b), and $\Delta\omega_G^2 + \Delta\omega_T^2$, the sum of the broadening due to mean velocity gradients and turbulent velocity fluctuations, is given by the last term in braces in Eq. (107b) minus μ_1^2. The expression for $\Delta\omega_A^2$ depends on the spatial structures of $C(\mathbf{x})$, $\langle D^2(\mathbf{x})\rangle$, $a(\mathbf{x})$, and the first two moments of $u(\mathbf{x}, t)$, all of which are accounted for in Eq. (107b). An approximation that is usually satisfactory is obtained by assuming that $C\langle D^2 \rangle$ and f_u are independent of position (locally homogeneous flow), and $a(\mathbf{x})$ is Gaussian. The result is

$$\Delta\omega_A^2 = 8\left[\frac{U^2(0) + \langle u'^2(0)\rangle}{d_m^2} + \frac{V^2(0) + \langle v'^2(0)\rangle}{l_m^2} + \frac{W^2(0) + \langle w'^2(0)\rangle}{h_m^2}\right] \tag{108a}$$

which simplifies further to

$$\Delta\omega_A^2 \cong 8\,\frac{U^2(\mathbf{0})}{d_m^2} \tag{108b}$$

for low-turbulence-intensity flow with mean flow in the x direction.

The equations for $\Delta\omega_G^2$ and $\Delta\omega_T^2$ are obtained from Eq. (107b) by expanding $(\omega_S + Kc_1)^2$, averaging over velocity, and breaking the result into terms associated with mean flow velocity differences in space, and fluctuating velocity differences in space and in time. The equation for the mean gradient broadening is

$$\Delta\omega_G^2 = \frac{\int C\langle D^2\rangle a^2(\mathbf{x})[KU(\mathbf{x}) - (\mu_1 - \omega_S)]^2\,d^3\,\mathbf{x}}{\int C\langle D^2\rangle a^2(\mathbf{x})\,d^3\mathbf{x}} \tag{109a}$$

or, using Eqs. (106a) and (106c),

$$\Delta\omega_G^2 = K^2\,\overline{[U(\mathbf{x}) - \tilde{U}]^2} \tag{109b}$$

Thus $\Delta\omega_G^2$ is the volume-averaged square of the difference between the local velocity and the mean volume-averaged velocity. The turbulent broadening is given by

$$\Delta\omega_T^2 = \frac{K^2 \int C\langle D^2\rangle a^2(\mathbf{x})\langle[u'(\mathbf{x})]^2\rangle\,d^3\mathbf{x}}{\int C\langle D^2\rangle a^2\,(\mathbf{x})\,d^3\,\mathbf{x}} \tag{110a}$$

$$= K^2\,\overline{\langle[u'(\mathbf{x}, t)]^2\rangle} \tag{110b}$$

Its interpretation is obvious.

One of the earliest applications of LDV involved the measurement of turbulence intensity by measurement of the bandwidth of S_{J_D}. According to Eqs. (107c) and 110b), this can be accomplished if $\Delta\omega_A$ and $\Delta\omega_G$ are known. Then

$$\overline{\langle[u'(\mathbf{x}, t)]^2\rangle} = K^{-2}(\mu_2^2 - \Delta\omega_A^2 - \Delta\omega_G^2) \tag{111}$$

can be calculated from measurements of μ_2 for any turbulent probability density function. If the rms is nearly constant across the mv, then $\overline{\langle[u'(\mathbf{x},t)]^2\rangle} = \langle[u'(\mathbf{0}, t)]^2\rangle$. If $\Delta\omega_A$ and $\Delta\omega_G$ can be reduced to about 20% of $\Delta\omega_T$ or less, the correction terms in Eq. (111) are quite small, and $\langle u'^2\rangle$ can be measured with good accuracy. Note, however, that decreasing the size of the mv reduces $\Delta\omega_G$ but increases $\Delta\omega_A$, so there is an optimal size for the mv that depends on the variation of $U(\mathbf{x})$ and $\langle u'^2(\mathbf{x})\rangle$ for any given flow.

For applications in which the precise shape and bandwidth of the power spectrum are not critical, it is satisfactory to model the correlation and spectrum by Gaussian forms that are identical to Eqs. (99) and (100) except that the $\Delta\omega_A^2$ factors in those equations are replaced with μ_2. Further, the bandwidth μ_2 can be approximately generalized to include spectral broadening due to Brownian motion of the scattering particles,

$$\Delta\omega_B = \frac{32\pi KT \sin^2 k}{3\mu_f d_p \lambda^2} \tag{112}$$

and broadening due to random frequency modulation of the laser source,

$$\Delta\omega_0^2 = \langle(\omega_{01} - \omega_{02})^2\rangle - \langle\omega_S\rangle^2 \tag{113}$$

The approximate equation for the total bandwidth becomes

$$\mu_2 = \Delta\omega_B^2 + \Delta\omega_0^2 + \Delta\omega_A^2 + \Delta\omega_G^2 + \Delta\omega_T^2 \tag{114}$$

where the terms are written in rough order of increasing importance.

4.9.5 Burst Density

The light flux is a superposition of Doppler bursts that occur at random times with random amplitudes; i.e., it is a shot-noise process much like the filtered-photoemission process discussed in section 4.6. Hence the random characteristics of the light flux are characterized by a density parameter representing the extent to which the Doppler bursts overlap with one another, on average. In the case of the photoemission process, the appropriate parameter is the mean number of emissions during τ_h, a timescale characteristic of the single-emission pulse. In the case of the light flux, the appropriate parameter is N_e, the effective mean number of particles in the mv. For example, the Doppler bursts from two particles can overlap only if both particles reside in the mv at the same time. Likewise, if N_e particles reside in the mv on average, then N_e Doppler bursts will overlap on average. In analogy with the terminology used in the discussion of the photoemission process, we shall refer to N_e as the burst density.

Upon first consideration it seems reasonable to define N_e to be the mean concentration times the volume of the mv. For example, some investigators [67] use a definition of the form $\int C(\mathbf{x})[a(\mathbf{x})/a(\mathbf{0})]d^3\mathbf{x}$, which reduces to C times the volume defined by $\int [a(\mathbf{x})/a(\mathbf{0})] d^3\mathbf{x}$ when C is constant. This definition is too simple when the particles are polydisperse, for then one good scatterer may produce much more signal than ten poor scatterers, so that the effective number of particles would be one, even though eleven particles were in the mv. It is apparent that some form of amplitude weighting must be used in the definition of N_e. The appropriate choice of amplitude weighting depends somewhat on the uses one intends to make of N_e. That is to say, the definition of N_e should follow naturally from consideration of the statistical properties of J_D. In the following, we develop a particular definition by considering the way in which J_D approaches a joint normal process in the limit of high burst density.

The two-time characteristic function of J_D is defined by

$$\Phi_{J_D}(\Omega_1, \Omega_2) = \langle \exp [j\, \Omega_1 J_D(t) + j\, \Omega_2 J_D(t + \tau)] \rangle \qquad (115)$$

The joint probability density function for $J_D(t)$ and $J_D(t + \tau)$ is the two-dimensional (2-D) Fourier transform of $\Phi_{J_D}(\Omega_1,\Omega_2)$. Hence the characteristic function contains all the statistical information necessary to compute all one-time and two-time moments of J_D. In particular,

$$\langle J_D^2 \rangle = - \left. \frac{\partial^2 \Phi_{J_D}}{\partial \Omega_1^2} \right|_{\Omega_1 = \Omega_2 = 0} \qquad (116)$$

and

$$\langle J_D(t)J_D(t + \tau) \rangle = - \left. \frac{\partial^2 \Phi_{J_D}}{\partial \Omega_1\, \partial \Omega_2} \right|_{\Omega_1 = \Omega_2 = 0} \qquad (117)$$

It can be shown, using the approximation in Eq. (89), that in general, Φ_{J_D} is given by

$$\Phi_{J_D} = \exp [\iiint f_u(\mathbf{c}, \mathbf{x})C(\mathbf{x}, D)(e^{j\beta} - 1)\, d^3\, \mathbf{x}\, dD\, d^3\, \mathbf{c}] \qquad (118)$$

where

$$\beta = \Omega_1\, Da(\mathbf{x}) \cos \Phi(\mathbf{x}, t) + \Omega_2 Da(\mathbf{x} + \mathbf{c}\tau) \cos \Phi(\mathbf{x} + \mathbf{c}\tau, t + \tau) \qquad (119)$$

The interested reader can verify that applications of Eqs. (116) and (117) to this characteristic function yield the same expressions for $R_{J_D J_D}(\tau)$ as Eq. (98).

As the number of particles in the mv becomes large, J_D is a sum of a large number of independent random variables, and we expect from the central limit theorem that J_D will become a joint normal random process. This can be shown by calculating the asymptotic value of Φ_{J_D} in the limit $C \to \infty$, but it must first be recognized that $\langle J_D^2 \rangle \to \infty$ as $C \to \infty$, so that to obtain a useful asymptotic result that has finite mean square value, one must deal with the dimensionless process $J_D^* = J_D(t)/\langle J_D^2 \rangle^{1/2}$. Clearly, $\langle J_D^{*2} \rangle = 1$, independent of C. It follows after some analysis that

$$\Phi_{J_D^*} = \exp \left\{ \iiint f_u(\mathbf{c})C(\mathbf{x}, D) \left[-\frac{\beta^2}{2\langle J_D^2 \rangle} + O\left(\frac{\exp (-\pi^2 N_{FR}^2/24)}{\sqrt{N_e}} \right) \right. \right.$$
$$\left. \left. + O(N_e^{-1}) \right] d^3\, \mathbf{x}\, d^3\, \mathbf{c}\, dD \right\} \qquad (120)$$

where

$$N_e = \frac{[\iint C(\mathbf{x}, D)D^2 a^2(\mathbf{x})\, d^3\mathbf{x}\, dD]^2}{\int C(\mathbf{x}, D)D^4 a^4(\mathbf{x})\, d^3\mathbf{x}\, dD} \tag{121}$$

Ordinarily, one shows that J_D^*, and hence J_D, are asymptotically joint normal by taking the limit of Eq. (120) as $N_e \to \infty$, from which it follows that $\ln \Phi_{J_D}^*$ is quadratic in Ω_1 and Ω_2, hence that the joint probability density for J_D^* (t) and J_D^* $(t + \tau)$ is asymptotically joint normal, and that the rate of convergence to a joint normal distribution is of the order of $N_e^{-1/2}$. (Strictly, J_D^* is a joint normal random process only if its characteristic function is quadratic in the Ωs, but the asymptotic behavior of the characteristic function is exactly the same as for the behavior of the two-time characteristic function, so the more complicated proof is unnecessary.) In the particular case of laser-Doppler signals the rate of convergence is faster, of the order of N_e^{-1}, because the second term in the brackets in Eq. (120) is extremely small for any practical value of N_{FR}.

The purpose of this derivation is to point out two facts. First, the basis for the present definition of N_e is that it is the appropriate parameter for describing the high-particle-density behavior of J_D, and it is a measure of the size of the fourth-order terms in the characteristic function, rather than the third-order term. Second, LDV signals become joint normal at much lower values of N_e than might be expected on the basis of conventional arguments (cf. [12, 67]). For example, if $N_e > 100$ assures joint normal behavior with $N_e^{-1/2}$ convergence, $N_e > 10$ assures an even closer approach to joint normal statistics with N_e^{-1} convergence. This conclusion is supported by experimental experience, which indicates that $N_e > 5 - 10$ is large enough to make the asymptotic formulas for $N_e \to \infty$ valid.

It should be noted that the equation for N_e reduces to

$$N_e = C\left(\frac{\pi^3}{512}\right)^{1/2} d_m l_m h_m \tag{122}$$

when the particle population is monodisperse and uniform in space. Equation (121) clearly weights the strong scatterers more heavily than the weak scatterers. For example, if the particle population consists of only two types, $C(\mathbf{x}, D) = C_1 \delta(D - D_1) + C_2 \delta(D - D_2)$, and

$$N_e = \frac{1 + (C_2 D_2^2/C_1 D_1^2)}{1 + (C_2 D_2^4/C_1 D_1^4)} C_1\left(\frac{\pi^3}{512}\right)^{1/2} d_m l_m h_m \tag{123}$$

which is proportional to C_1 for small values of $C_2 D_2^2/C_1 D_1^2$. Thus natural aerosols usually produce fairly low values of N_e, despite the presence of a great many fine particles. Similarly, when natural aerosols or hydrosols are artificially seeded with good scatters, N_e is approximately the number of good artificial scatters in the mv.

It is now possible to define two asymptotic cases in which $J_D(t)$ exhibits vastly different random properties: high burst density, $N_e \gg 1$, and low burst density, $N_e \ll 1$. These terms will be abbreviated to HBD and LBD, respectively. HBD signals are often referred to as "continuous" signals [68], and the term "individual realizations" is conventionally applied to LBD signals [73]. The intermediate case, $N_e = O(1)$ will be called "$O(1)$ burst density." The reason for devoting so much time to this classification scheme and to the proper definition of N_e is that the choices of signal processing techniques and data processing techniques depend critically upon the type of LDV signal. Specifically, the burst density in combination with the photon density more or less determines the signal processing technique(s) that should be used.

4.9.6 High-Burst-Density Signals

The most desirable characteristic of the HBD signal ($N_e \gg 1$) is that it is nearly continuous; i.e., there are no gaps in it because, according to Eq. (76), the probability of finding zero particles in the mv at any instant is exp $(-N_e)$. Consequently, information on the fluid velocity can be obtained as a continuous function of time. In contrast, the LBD signal yields velocity data only when bursts occur, and the bursts occur at random times, complicating analysis of the data.

This advantage of the HBD signal is offset by additional randomness in the signal that is created by the superposition of many random bursts. A typical HBD signal is illustrated in Fig. 4.18. Its amplitude, being the sum of many random amplitudes, varies randomly in time. More important, the phase of the HBD signal is random also, partly because the phases of the component bursts are random and partly because their amplitudes are random. The instantaneous frequency is defined as the time derivative of the phase, and it also fluctuates randomly about a mean value, even when the frequencies of all the bursts are identical, e.g., steady laminar flow with no velocity gradients. Fluctuations in the phase and instantaneous frequency are referred to as "phase noise" and "ambiguity noise," respectively [67]. They are inherent in the HBD signal, and they place a limitation on the accuracy of the instantaneous Doppler frequency measurement. Thus continuity of the velocity information is accomplished only by sacrificing accuracy.

The equation for the Doppler light flux is

$$J_D(t) = \iint Da(\mathbf{x}) \cos (\omega_S t + Kx) g(\mathbf{x}, t, D) \, d^3 \mathbf{x} \, dD \qquad (124)$$

Following the work of Rice [74] for classical shot-noise processes, it is possible to reduce this equation (which represents a generalized shot-noise process with random points in $\mathbf{x}D$ space) to the form

$$J_D(t) = A(t) \cos \Phi_D(t) \qquad (125)$$

where

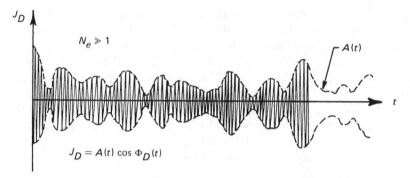

Figure 4.18 High-burst-density Doppler light flux

$$\Phi_D(t) = \omega_s t + K x_m(t) + \phi(t) \tag{126}$$

is the aforementioned random phase of the HBD signal and $A(t)$ is its random amplitude. This reduction is accomplished by decomposing the phase of Eq. (124) into the form $\omega_s t + Kx = \omega_s t + K x_m(t) + K[x - x_m(t)]$, where $x_m(t)$ is any arbitrary function of time. At this point, we do not know how to specify $x_m(t)$, but we expect that it is related to the mean displacement of the velocity field, as determined by $u(\mathbf{x})$. It will be determined shortly by application of certain mathematical requirements that are independent of the steps needed to reduce Eq. (124) to Eq. (125). By inserting this phase decomposition into Eq. (124) and using simple trigonometric identities, it is easy to show that the amplitude and phase are given by

$$A(t) = [F^2(t) + G^2(t)]^{1/2} \tag{127}$$

$$\phi(t) = \tan^{-1} \frac{G}{F} \tag{128}$$

where

$$\begin{bmatrix} F(t) \\ G(t) \end{bmatrix} = \iint Da(\mathbf{x}) \begin{bmatrix} \cos K(x - x_m) \\ \sin K(x - x_m) \end{bmatrix} g(\mathbf{x}, t, D)\, d^3\mathbf{x}\, dD \tag{129}$$

To $O(\exp - N_{FR}^2)$, the correlation functions of F and G are given by

$$R_{FF}\tau = \langle F(t)F(t + \tau) \rangle = \langle G(t)G(t + \tau) \rangle \tag{130a}$$

$$= \frac{1}{2} \iint C(\mathbf{x}) \langle D^2 \rangle a(\mathbf{x}) a(\mathbf{x} + \mathbf{c}\tau) f_u(\mathbf{c})$$

$$\times \cos K[c_1\tau - x_m(t + \tau) + x_m(t)]\, d^3\mathbf{x}\, d^3\mathbf{c} \tag{130b}$$

and

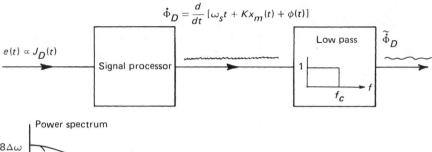

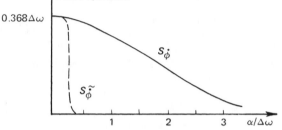

Figure 4.19 Frequency demodulation and the ambiguity noise spectrum

$$\langle F(t)G(t + \tau)\rangle = \langle F(t + \tau)G(t)\rangle = 0 \tag{131}$$

These equations follow from Eq. (130a) by steps that are identical to those leading to Eq. (98) for $R_{J_D J_D}$. Since J_D is a joint normal process for HBD signals, $F(t)$ and $G(t)$ are also joint normal processes. From Eq. (131), they are independent as well.

The Doppler shift information is contained in $\dot{\Phi}_D$, the instantaneous frequency of the HBD signal. Given an input signal of the form $A(t)\cos\Phi_D(t)$, the ideal frequency demodulator (i.e., frequency-to-voltage convertor) would ignore the random amplitude modulation and extract from the input a signal (either a voltage or a digital number) that is directly proportional to the instantaneous frequency,

$$\dot{\Phi}_D = \omega_S + K\dot{x}_m + \dot{\phi}(t) \tag{132}$$

as shown in Fig. 4.19. Ideally the output would be proportional to the fluid velocity, but because of the ambiguity noise $\dot{\phi}(t)$, the real output fluctuates randomly.

In practice, a significant part of this fluctuation is removed by passing the output of the signal processor through a low-pass filter that rejects all frequencies greater than f_c, the low-pass cutoff. The filtered signal $\tilde{\dot{\Phi}}_D$ is the quantity of primary interest. To avoid confusion, the symbols f and $\alpha = 2\pi f$ will be used to represent the frequencies at which $\tilde{\dot{\Phi}}_D$ fluctuates, i.e., the spectral frequency components of the instantaneous frequency. These should not be confused with the Doppler frequency. The reader will find it easiest, perhaps, always to think of $\dot{\Phi}_D$ and $\tilde{\dot{\Phi}}_D$ as voltages proportional to the input signal frequency. It is then obvious that these voltages may also fluctuate in time and that they will have power spectra $S_{\dot{\phi}_D}(\alpha)$ and $S_{\tilde{\dot{\phi}}_D}(\alpha)$.

We expect the mean value of $\tilde{\Phi}_D$ to be simply related to the mean velocity U. This relationship can be determined by using a result derived by Rice [75], which states that the mean value of the instantaneous frequency of a joint normal signal such as J_D in Eq. (124) is equal to the centroid of the power spectrum of the signals. That is, $\langle \dot{\Phi}_D \rangle = \mu_1$, where μ_1 is given by Eq. (102a). Since low-pass filtering does not affect the mean value, $\langle \tilde{\Phi}_D \rangle = \mu_1$ also. Taking the average of Eq. (132) and setting it equal to the value of μ_1 found in Eq. (106c) yields

$$\langle \dot{\Phi}_D \rangle = \omega_S + K\langle \dot{x}_m \rangle + \langle \dot{\phi} \rangle \tag{133a}$$

$$= \omega_S + \frac{K \iint C\langle D^2 \rangle a^2(\mathbf{x})U(\mathbf{x})\, d^3\, \mathbf{x}}{\iint C\langle D^2 \rangle a^2(\mathbf{x})\, d^3\, \mathbf{x}} \tag{133b}$$

Comparison of Eqs. (133a) and (133b) shows that the random variable $\dot{\phi}$ will have zero mean value if and only if $\dot{x}_m(t)$ is defined so that

$$\langle \dot{x}_m \rangle = \frac{\int C\langle D^2 \rangle a^2(\mathbf{x})U(\mathbf{x})\, d^3\, \mathbf{x}}{\int C\langle D^2 \rangle a^2(\mathbf{x})\, d^3\, \mathbf{x}} \tag{134}$$

or in other words,

$$\langle \dot{x}_m \rangle = \tilde{U} \tag{135}$$

where $\tilde{U}$ is the mean velocity volume averaged according to Eq. (106a). Clearly, Eq. (134) can be satisfied by defining

$$x_m(t) = x_m(t_0) + \int_{t_0}^{t} \tilde{u}(t')\, dt' \tag{136}$$

so that

$$\dot{x}_m(t) = \tilde{u}(t) \tag{137}$$

and Eq. (135) follows immediately. Hence the mean displacement is just the displacement due to the volume-averaged Eulerian velocity field. The field is Eulerian rather than Lagrangian because we are interested only in displacements caused by particles that currently reside in the mv. The instantaneous frequency is now

$$\dot{\Phi}_D = \omega_S + K\tilde{u}(t) + \dot{\phi}(t) \tag{138}$$

If we define $\dot{x}_m = \tilde{u}$ so that $\dot{\phi}$ has zero mean value, the signal represented by Eq. (138) is exactly the same as the signal analyzed by Rice [74, 75], even though the present signal is generated by random points in $\mathbf{x}$-D space, whereas in Rice's work the signal is initially generated by random points in time. Hence all the results for $\dot{\phi}$ in [74, 75] can be applied to the present signal without modification. It should be noted that the present definition of $\dot{x}_m$ does not agree with the ad hoc definition of u_0, the "effective velocity seen by the velocimeter" given by George and Lumley in Eq. (2.2.5) of [67]. The difference between these two definitions is quite important because the entire theoretical basis for interpreting HBD LDV signals ultimately rests upon the definition of the "effective" or "measurable" velocity.

Unfortunately, the reasoning needed to establish the correct definition is equally subtle because, as shown here, and as implied in Rice's [74, 75] work, the phase decomposition leading to Eq. (132) is valid for any $x_m(t)$. Hence, up to Eq. (132), $x_m(t)$ could be defined as the time integral of *any* effective velocity, which led Buchhave et al. [68] to suggest that the u_0 decomposition in [67] may not be unique. However, there is only one choice, Eq. (138), that yields $\langle \dot{\phi} \rangle = 0$, and this definition must be used if one intends to make use of Rice's analyses of the statistics of $\dot{\phi}$, in which it is implicit that $\langle \dot{\phi} \rangle = 0$. Moreover, it appears that a correct analysis of the statistics of $\dot{\phi}$ for the case of $\langle \dot{\phi} \rangle \neq 0$ [that is, $\dot{x}_m$ not given by Eq. (138)] would simply lead to results that would depend on $\langle \dot{\phi} \rangle$ in such a way as to ultimately restore $\langle \dot{\phi} \rangle$ to the definition of the effective velocity. Consequently, the velocity "seen by the LDV" is uniquely given by Eq. (138).

Fortunately, it can be shown that all the results derived in [67, 68] could have been derived by starting with the definition

$$u_0 = \frac{\int a(\mathbf{x}) u \, d^3 \, \mathbf{x}}{\int a(\mathbf{x}) \, d^3 \, \mathbf{x}}$$

instead of Eq. (2.2.5) in [67]. This is very close in form to $\tilde{u}$ for the case $C\langle D^2 \rangle =$ const, except that the weighting factor is $a(\mathbf{x})$ instead of $a^2(\mathbf{x})$. Hence the equations in [67] can be corrected by dividing mv dimensions σ_1, σ_2, and σ_3, defined in that reference, by $\sqrt{2}$, thereby converting their $w(\mathbf{x})$ to our $a^2(\mathbf{x})$.

As mentioned above, Rice's results [74, 75] for $\dot{\phi}$ can be applied directly to the present signal. In particular, the power spectrum of the ambiguity noise $S_{\dot{\phi}}(\alpha)$ is very broad, and if S_{J_D} has a Gaussian shape, its value at zero frequency is

$$S_{\dot{\phi}}(\mathbf{0}) \simeq 0.368 \, \Delta\omega \tag{139}$$

where $\Delta\omega$ is the rms bandwidth of $F(t)$ [or $G(t)$], defined by

$$\Delta\omega^2 = \frac{\displaystyle\int_O^\infty \omega^2 S_F(\omega) \, d\omega}{\displaystyle\int_O^\infty S_F(\omega) \, d\omega} \tag{140a}$$

$$= - \frac{R_{FF}''(\mathbf{0})}{R_{FF}(\mathbf{0})} \tag{140b}$$

By inserting R_{FF} from Eq. (130b), into Eq. (140b) and using steps similar to those leading to Eq. (107b), it can be shown that

$$\Delta\omega^2 = \Delta\omega_A^2 + K^2 \, \overbrace{(U - \tilde{U})^2} + K^2 \, \overbrace{\langle (u' - \tilde{u}')^2 \rangle} + \Delta\omega_0^2 + \Delta\omega_B^2 \tag{141}$$

where

$$\widetilde{u}' = \frac{\int CD^2 a^2(\mathbf{x})\widetilde{u}'(\mathbf{x},\, t)\, d^3\, \mathbf{x}}{\int C\langle D^2\rangle a^2(\mathbf{x})\, d^3\, \mathbf{x}} \tag{142}$$

is the volume-averaged velocity fluctuation. The ambiguity bandwidth $\Delta\omega_A^2$ is exactly the same as in Eq. (108a). The second term in Eq. (141) represents the broadening due to mean velocity gradients, also the same as before, and the third term arises from turbulent velocity gradients that cause differences between the local turbulent fluctuations and the volume-averaged turbulent fluctuation, $u'(\mathbf{x}, t) - \widetilde{u}'(\mathbf{x}, t)$. The bandwidths due to source broadening and Brownian motion do not follow from Eqs. (130b) and (140b), but they have been added to the results, as before.

Equation (139) pertains to $\dot{\phi}$, but in practice one is interested in the filtered output of the signal processor $\widetilde{\Phi}_D$, and hence in the properties of the filtered ambiguity noise $\widetilde{\dot{\phi}}$. Rice's [75] result for the spectrum $S_{\dot{\phi}}(\alpha)$ is sketched in Fig. 4.19. Suppose $H(\alpha)$ is the transmission function of the filter, i.e., the ratio of the output to the input when the input is a pure sine wave at frequency α. Then the spectrum of the filtered phase noise is $S_{\dot{\phi}}(\alpha)|H^2(\alpha)|^2$. The cutoff frequency of the filter $\alpha_c = 2\pi f_c$ is usually selected to be greater than the highest turbulent frequencies of interest, and these are somewhat lower than $\Delta\omega$. Then, if $\alpha_c < \Delta\omega$,

$$S_{\widetilde{\dot{\phi}}}(\alpha) = S_{\dot{\phi}}(0)|H(\alpha)|^2 \tag{143a}$$

$$= 0.368\,\Delta\omega|H(\alpha)|^2 \tag{143b}$$

because $S_{\dot{\phi}}$ is nearly constant in the range $0 < \alpha < \Delta\omega$. The filtered ambiguity noise appears to be low-pass-filtered white noise with mean square value

$$\langle(\widetilde{\dot{\phi}})^2\rangle = 0.368\,\Delta\omega\alpha_c \tag{144}$$

which corresponds to an equivalent velocity noise $u_{\widetilde{\dot{\phi}}}$ with rms

$$\langle(u_{\widetilde{\dot{\phi}}}^2)\rangle^{1/2} = K^{-1}\langle(\widetilde{\dot{\phi}})^2\rangle^{1/2} \tag{145}$$

The ambiguity noise and u are statistically independent [67], so the correlation and power spectrum of the measured velocity $K^{-1}\Phi_D = \widetilde{U} + \widetilde{u}'(t) + K^{-1}\widetilde{\dot{\phi}}(t)$ are the sums of the correlations and spectra of $\widetilde{u}(t)$ and $\widetilde{\dot{\phi}}(t)$, respectively. That is,

$$K^{-2}R_{\widetilde{\Phi}_D\widetilde{\Phi}_D}(\tau) = R_{\widetilde{\widetilde{u}}\widetilde{\widetilde{u}}}(\tau) + K^{-2}R_{\widetilde{\dot{\phi}}\widetilde{\dot{\phi}}}(\tau) \tag{146}$$

and

$$K^{-2}S_{\widetilde{\Phi}_D}(\alpha) = S_{\widetilde{\widetilde{u}}}(\alpha) + K^{-2}S_{\widetilde{\dot{\phi}}}(\alpha) \tag{147}$$

where

$$S_{\widetilde{\widetilde{u}}}(\alpha) = |H|^2 S_{\widetilde{u}}(\alpha) \tag{148}$$

is the power spectrum of the filtered values of the measurable (i.e., volume-averaged) velocity fluctuation $\widetilde{u}'(t)$.

Turbulent motions whose scales are of the order of the dimensions of the mv are, of course, attenuated by the volume averaging that is inherent in u'. The finite mv acts as a spatial filter. Detailed calculations of this effect are presented in [67]. As a simple guideline, attenuation is less than 5% for transverse wave numbers of the velocity spectrum in the y direction (that is, the direction of l_m, the largest dimension of the mv) that are less than about $0.6/l_m$, corresponding to about 10 wavelengths in l_m.

4.10 SIGNAL PROCESSORS

The LDV signal processor is designed to measure the Doppler frequency plus any other necessary data from the PMT signal. With certain exceptions, LDV signal processors are special-purpose instruments designed specifically to handle the peculiar characteristics of LDV signals. The primary types of signal processors are correlators, spectrum analyzers, counters, and frequency trackers. Less commonly used techniques include direct computer analysis and filter banks. None of these techniques provides a universally optimal solution to the LDV frequency measurement problem, so the experimentalist must select the type of device that is best suited to the particular application. The main criteria for selection are accuracy, frequency range, ability to extract signal frequency from noise, time resolution, ease of use, and ease of interpreting the output data. The various instruments differ greatly in many of these regards.

Time resolution is defined as the time required to obtain a measurement of the Doppler frequency to within some prescribed accuracy. It is essentially the inverse of the instrument's frequency response. In the following sections, LDV signal processors are identified as either long-time averaging devices or time-resolving devices. The measurement time of the former group of processors is long in comparison with the timescales of the velocity variation, so the output must be interpreted in terms of long-time averages of the velocity. The measurement time of the latter group of processors is of the order of the time for a single Doppler burst, and since this is normally short in comparison with the timescale of the flow, processors in this group are capable of resolving the velocity as a function of time.

Instruments that fall into the long-time averaging class include conventional, general-purpose spectrum analyzers, amplitude correlators, and all but the fastest photon correlators. Devices such as computer-based systems that acquire an individual Doppler burst with a fast analog-to-digital (A-D) converter, but then require a long processing time, must also be included in the class of time-averaging instruments. If the sampling rate needed to resolve the time variation of the flow is sufficiently low, then an instrument normally classed as a time-averaging device may be adequate to follow the signal fluctuations. For example, if a system is able to measure up to 1000 times per second, it would be time resolved for signals whose velocity power spectra extend up to the Nyquist frequency of 500 Hz, but limited to time averaging for signals with frequency spectra higher than that.

Time-resolved LDV signal processors include phase-locked loops, frequency-locked loops, counters, burst correlators, and burst spectrum analyzers. Phase-locked and frequency-locked loops are analog devices that are well suited to the HBD LDV signal, but also capable of processing the LBD signal, within certain limitations. When an instrument can make one or more measurements of the frequency of every burst (up to some maximum burst rate), it is referred to as a "burst processor." The LDV frequency counter is the longest standing instrument of this type. Burst correlators and burst spectrum analyzers were developed later in the evolution of the LDV signal processor because these types of analyses require much more computation and hence very fast electronics if they are to be completed in the span of a single burst. The principal advantage won by the increased computational effort is a more robust measurement of the LDV signal frequency, i.e., ability to measure accurately in the presence of significant noise.

In the following sections, we shall discuss LDV signal processors in order of their combined power to extract signal from noise and to resolve time variations of the velocity. Thus the general forms of correlators and spectrum analyzers are discussed first, with emphasis on time-averaging devices. Time-resolved trackers and counters are discussed next, followed by discussion of the unique aspects of correlators and spectrum analyzers that operate in the burst mode. While none of these instruments is universally optimum, the burst signal processors offer the greatest ability to measure rapidly in the presence of noise.

4.10.1 Amplitude Correlators

The conventional type of correlation analyzer computes the correlation $R_{ee}(\tau) = \langle e(t)e(t + \tau) \rangle$ between the signal amplitudes $e(t)$ and $e(t + \tau)$, where τ is the time delay. These devices are available commercially, but most are intended for general-purpose signal analysis and have not been designed specifically for LDV applications. As a consequence, they are usually too slow for most LDV experiments, although they are useful if the Doppler frequencies are less than about 2–20 kHz. In these cases, the correlation of the Doppler signal over long averaging times can be used to determine the mean flow velocity and the turbulence intensity, with certain corrections as discussed in section 4.9.4. The power spectrum of the Doppler signal can be obtained also by Fourier transforming the measured correlation function. In principle, both the correlation and the spectrum contain information on the full probability density function of the velocity $f_u(c)$, but the broadening effects due to ambiguity and gradients make it difficult to obtain $f_u(c)$ directly, and measurements of moments higher than $\langle (u')^2 \rangle$ require corrections that may be subject to considerable error.

To discuss the applications and limitations of correlation analyzers more fully, it is necessary to describe their operation in more detail. The conventional amplitude correlator converts the signal voltage $e(t)$ into a digital word that is B bits long at intervals of $\Delta\tau$ seconds. An estimate of the correlation function is formed by computing

$$R_{ee}(n\,\Delta\tau) = \frac{1}{(M - |n|)} \sum_{m=1}^{M} e(m\,\Delta\tau)e[(m + n)\,\Delta\tau] \tag{149}$$

wherein the sum over m corresponds to an average over time, and $n = 0, \pm1, \ldots, \pm(N - 1)$ determines the time delay. The maximum time delay is $(N - 1)\,\Delta\tau$, and correlators are usually designated as "N-point" correlators. Typical values of N range from 100 to 1028. It is well known that for most signals, and certainly for LDV signals, this estimate converges to $R_{ee}(n\,\Delta\tau)$ with rms error proportional to $M^{-1/2}$ as M becomes large. Hence the sampling times are usually longer than the timescale of the turbulent flow.

Correlator speed is determined by the rate at which the correlator can perform the multiplications in Eq. (149), an N-point correlator forming $N\,B \times B$ bit products every $\Delta\tau$ seconds. A reasonable number of correlation points is 10 per cycle of the correlation, so if the correlation frequency is f_D, $\Delta\tau$ must be $1/10f_D$, and the correlator product rate must be $10Nf_DB^2$ in bits per second. For example, 10-bit resolution of a 10-MHz signal with 100 lines would require a product rate of 10^{12} bits/s. Clearly, it is helpful to reduce B, and in fact, values of $B = 4$ and 8 bits are typical in general-purposes correlators.

Correlators for general laser velocimetry must work at the highest possible speeds because the Doppler frequencies are normally large. Correlation speed can be improved by reducing the A-D resolution to $B = 1$ bit. Let $e_c(t)$ be the 1-bit digital signal defined by

$$e_c(t) = \begin{cases} 1 & e(t) > 0 \\ -1 & e(t) < 0 \end{cases} \tag{150}$$

Clearly, $e_c(t)$ is just the signal that would be obtained by amplifying $e(t)$ with large gain and clipping it at the ±1 levels. Hence, 1-bit correlation is referred to as clipped correlation. Single-clipped correlators compute $R_{ee_c} = \langle e_c(t)e(t + \tau)\rangle$, while double-clipped correlators compute $R_{e_c e_c} = \langle e_c(t)e_c(t + \tau)\rangle$. Double-clipped correlation is the simplest because it involves only the multiplication of either $1'$ or $-1'$.

A well-known theorem called the *arcsine law* states that the correlation of the double-clipped signal is given by

$$R_{e_c e_c} = \frac{2}{\pi} \arcsin \frac{R_{ee}(\tau)}{R_{ee}(0)} \tag{151}$$

when $e(t)$ is a normal random process (cf. [54], for example), showing that double clipping distorts the correlation on average; that is, $R_{e_c e_c}$ is a biased estimate of R_{ee}. Fortunately, this distortion can be eliminated by adding "auxiliary signals" to $e(t)$ and $e(t + \tau)$ before they are quantized, as shown originally in [76–78]. This procedure is valid for any random process $e(t)$ that has bounded amplitude. The auxiliary signals must have zero mean value, and they must be statistically independent of $e(t)$ and $e(t + \tau)$ and of each other. Finally, the probability distributions of their amplitudes should be uniform. The addition of an auxiliary signal is called linearization, and the function of linearization is to move $e(t)$ [or $e(t + \tau)$] up and down across the 1-bit quantization level, so that each level of $e(t)$ is sampled with uniform probability. This has the same effect as quantizing the signal with a randomly varying quantization level. In this way, the average of the product of the linearized signals correctly converges to $R_{ee}(\tau)$, even though the individual samples of $e(t)$ and $e(t + \tau)$ are quantized very coarsely. In principle, the auxiliary signal may be any random signal that satisfies the conditions stated above, but in practice, a simple triangular wave is used.

Reference [79] describes a 256-point correlator of this type that is capable of correlating every sample pair for $\Delta\tau > 40$ ns. Faster correlation rates ($\Delta\tau > 10$ ns) are achieved by forming time-delayed products at every kth sample, i.e., by ignoring a fraction of the samples. This does not affect the ultimate accuracy of the correlation. Also, it does not significantly lengthen the time required to achieve a certain level of sampling accuracy when the signal is correlated with itself over k $\Delta\tau$, for then the samples in k $\Delta\tau$ are not statistically independent, and therefore they contribute very little to the sampling accuracy. Correlation-time increments such as 10 ns make it possible to measure 10-MHz Doppler signals with 10 correlation points per cycle.

4.10.2 Photon Correlators

The photon correlator correlates the rate of photoemissions, rather than $e(t)$, which is a sum of the filtered photoemission pulses. When using photon correlators, one must also use a fast PMT with no output filtering, so that each photoemission appears as a narrow, discrete pulse. The ideal photon correlator output corresponds to $R_{nn}(\tau) = \langle n(t, t + \Delta\tau)n(t + \tau, t + \tau + \Delta\tau) \rangle$, where $n(t, t + \Delta\tau)$ is the number of photoemission pulses in $(t, t + \Delta\tau)$; that is, $n/\Delta\tau$ is a measure of the instantaneous emission rate. Recall that the conditional average of n given the total light flux J_{tot} is $\dot{\varepsilon}(t)$, given by Eq. (57). It can be shown [55] that

$$R_{nn}(\tau) = \Delta\tau^2 \, R_{\dot{\varepsilon}\dot{\varepsilon}}(\tau) + \langle \dot{\varepsilon} \rangle \, \Delta\tau \, \Lambda\left(\frac{\tau}{\Delta\tau}\right) \tag{152}$$

where

$$\Lambda\left(\frac{\tau}{\Delta\tau}\right) = \begin{cases} 1 - \dfrac{|\tau|}{\Delta\tau} & |\tau| < \Delta\tau \\ 0 & |\tau| > \Delta\tau \end{cases} \tag{153}$$

and

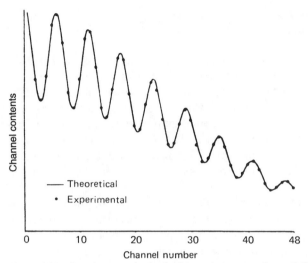

Figure 4.20 Correlogram obtained with a photon correlator. Solid line is a theoretical correlation function fitted through the data for the purpose of determining the Doppler frequency. (From [47])

$$R_{\dot{\varepsilon}\dot{\varepsilon}} = \frac{\eta_q^2}{h^2 v_0^2} \langle J_{tot}(t) J_{tot}(t + \tau) \rangle \qquad (154a)$$

It follows from Eq. (54) that

$$R_{\dot{\varepsilon}\dot{\varepsilon}} = \frac{\eta_q^2}{h^2 v_0^2} (R_{J_B J_B} + R_{J_P J_P} + R_{J_D J_D} + R_{J_R J_R}) \qquad (154b)$$

where the correlations in parentheses are the background, pedestal, Doppler, and reference-beam light-flux correlations, respectively. R_{nn} contains the Doppler signal correlation, but it is superimposed on the pedestal correlation, which has the same shape as the envelope of $R_{J_D J_D}$, plus the background and reference-beam correlations, which will be essentially constants. In addition, there is a narrow spike at $\tau = 0$ caused by the correlation of the photoemission pulses with themselves. In contrast, these additional terms need not appear in the amplitude correlation discussed in the preceding section because J_B, J_P, and J_R are low-frequency signals that can be removed from the analog signal by high-pass analog filters. These terms are not removed in photon correlation because digital high-pass filtering is too time consuming. A typical correlogram obtained using a photon-correlator is shown in Fig. 4.20.

As with amplitude-correlators, photon-correlators employ clipping to improve speed. In photon-correlation the pulse count n is replaced by n_c, which is unity if n exceeds an adjustable clipping level q, and zero otherwise. Single-clipping without linearization is one common mode. In this case, if the light flux is a normal random process,

$$\frac{R_{n_c n}(\tau)}{R_{n_c n}(0)} = \frac{\langle n \rangle - q}{\langle n \rangle + 1} + \frac{1 + q}{1 + \langle n \rangle} R_{nn}(\tau) \tag{155}$$

and $R_{n_c n}$ is nearly equal to R_{nn} if q is set equal to $\langle n \rangle$. When the burst density is small, J_D is not likely to be a normal process, and the single-clipped correlator should use a linearization procedure similar to the one discussed in the preceding section. In the context of photon correlation, this procedure is referred to as scaling, and it is accomplished by leaving a random digital word with uniform probability distribution in the register containing n_c. Detailed theoretical analyses of photon-correlation have been performed by Pike, Jakeman, Ford, and coworkers, and the reader can find references to these works plus extensive bibliographies in [12, 18, 21, 78].

The primary advantage of the correlation technique, either amplitude correlation or photon correlation, is its ability to process very noisy signals and extract relatively accurate frequency information. In principle, the mean photon rate can be less than one photon per Doppler cycle, or even one photon per Doppler burst, and correlations can still be formed, given a sufficiently long averaging time. (Of course, one must correlate between two photons in the same burst to obtain a measure of the frequency, but for any mean number per burst, there is always a finite, albeit perhaps small, probability that two or more photons will occur during a burst.) In this regard, there is very little difference between amplitude correlation and photon correlation. The primary differences are practical. For example, at very low photon densities, the maximum amplitude of $e(t)$ is of the order of the amplitude of a filtered photoemission pulse, which can be small and difficult to detect; in contrast, the photon correlator simply assigns a unit value to each pulse. On the other hand, as the photon density increases, for example, $\dot{\varepsilon}\tau_h = O(1)$, the pulses begin to overlap, and the photon correlator can no longer discriminate between single pulses and groups of pulses. This phenomenon is referred to as "photon pileup," and it seriously distorts the correlation [81]. The amplitude correlator is immune to photon pileup and, in fact, works best in this regime.

At 10-ns speeds, amplitude correlators and photon correlators limit the maximum Doppler shift to about 10 MHz, or perhaps 20 MHz if the number of correlation points is as low as five per Doppler cycle. The maximum frequency is about one order of magnitude lower than the maximum attainable with frequency counters. Consequently, the optical systems used in conjunction with correlator signal processors must employ relatively narrow beam angles when the flow velocities are high.

4.10.3 Spectrum Analysis

The earliest signal-processing technique used in LDV was spectrum analysis, wherein the power spectrum of the PMT signal was measured by long-time averaging. The sweep-type spectrum analyzer is a square-law detector that measures the power of that portion of the input signal that is transmitted through a narrow-band-pass filter whose width is $\Delta\omega_f$ and whose center frequency is ω_f. From the definition of the power spectrum, this power is equal to $S_{J_D}(\omega_f)\Delta\omega_f$ if $\Delta\omega_f$ is sufficiently narrow, and $\Delta\omega_f$ is defined so as to absorb any constant of proportionality. The complete power spectrum is obtained by varying ω_f over the frequency range of interest. In practice, it is more convenient to fix the center frequency of the band-pass filter and translate the spectrum of the signal by mixing it with a local oscillator sine wave whose frequency is swept slowly over the frequency range. This type of analysis is inefficient because the power detector "sees" only a $\Delta\omega_f$-wide band of the signal spectrum, and at any instant it is blind to the signal spectrum outside this band. That is, the swept spectrum analyzer processes $\Delta\omega_f$-wide bands of the spectrum serially. Consequently, the swept spectrum analysis requires about $\Delta\omega/\Delta\omega_f$ times as much time to sample a $\Delta\omega$-wide spectrum as would a device that processed the $\Delta\omega$-wide band in parallel.

Digital spectrum analyzers that operate by continuously fast-Fourier-transforming segments of the sampled input signal are parallel processors, in that they use all the incoming data. These devices can obtain a spectrum in less time than comparable swept-spectrum analyzers. If this range is large enough to contain the maximum Doppler frequency, the fast-Fourier-transform (FFT) analyzer works very well and can perform the same function as a long-time-averaging photon correlator or amplitude correlator followed by FFT analysis. The input can be filtered or, if the A-D converter is replaced by a pulse counter, unfiltered.

Two other types of spectrum analyzers are filter banks and surface acoustic-wave devices. The filter bank consists of a parallel array of band-pass filters; i.e., it is a parallel-processing spectrum analyzer. The signal is input to all of the filters simultaneously, and the instantaneous signal frequency is identified by sampling the output power of each filter to determine the location of the instantaneous spectrum. The Doppler frequency at any instant is set equal to the center frequency of the filter whose output is maximum, so the frequency resolution is determined by the bandwidths of the filters. Baker and Wigley [82] describe a filter bank consisting of 10 filters covering the range 0.63–9.55 MHz in a logarithmic progression. In a certain sense, the filter bank is the ideal LDV signal processor, but it has been used very little in LDV because of the expense of a large number of filters and the difficulty in setting up and maintaining their alignment.

The surface acoustic-wave (SAW) analyzer makes use of the fact that surface acoustic waves are dispersive; i.e., it takes different wave frequencies different times to travel from one end of an SAW crystal to the other. High-speed spectrum analysis is possible (51 μs/spectrum is reported in [83]), but usually at high frequencies.

4.10.4 Frequency Trackers

Frequency trackers measure the instantaneous frequency $\dot{\Phi}_D$ of the LDV signal. The main types of trackers used in LDV are the phase-locked loop (PLL) and the frequency-locked loop (FLL), both of which are analog devices. These instruments work best when the signal is continuous (HBD), but they also work with LBD signals, provided that the time between bursts is not too long. Most commercial frequency trackers hold the value of the velocity measured from the last burst until a new measurement is achieved.

In its simplest form, the PLL consists of an amplitude discriminator that converts the input sinusoidal signal $e(t)$ into a square wave, a mixer that multiplies the square-wave with a reference square-wave, a low-pass amplifier that amplifies the square-wave product and filters the result to produce a voltage that oscillates at the frequency difference, and a voltage-controlled oscillator (VCO) that is controlled by the amplifier output and supplies the reference square-wave. The PLL is "locked" onto the signal if the reference square-wave from the VCO has the same frequency as the input. If the VCO and input signals are 90° out of phase, their product is equally negative and positive, and the low-pass-filtered product (corresponding to a short-time average) is zero. If the input frequency increases, the filtered product becomes positive, and since this is the feedback signal to the VCO, the VCO frequency increases until it again equals the input frequency. In this way, the VCO signal frequency and phase are always locked onto the frequency and phase of the input signal to within feedback errors that are small if the loop gain is large.

The output of the PLL is just the VCO control voltage, which is linearly proportional to the input frequency if the VCO is linear. Successful operation of the PLL is accomplished only when the loop is locked onto the input signal's phase, so the PLL must be capable of following very rapid phase transients to stay in lock. Typical PLL slew rates are of the order of 10^{12} Hz/s. With no signal present, the VCO operates at a constant "free-running" frequency, and it must go through a complicated, nonlinear locking transient to lock onto the signal when it appears. This transient is somewhat random, and it lasts for several periods of the signal frequency. Lock-in will not occur unless the signal frequency is within the "capture bandwidth" of the PLL, which is typically ±10–20% of the free-running frequency. Likewise, the loop will experience loss of lock if the signal frequency falls outside the "lock range," which is typically ±10–30% of the free-running frequency.

When the signal input to a PLL is a HBD LDV signal, the PLL will experience loss of lock for small time periods when the phase noise ϕ changes very rapidly or when the amplitude becomes very small. (Large phase changes are highly correlated with small amplitudes in the HBD signal.) Otherwise, it will track the signal and measure $\dot{\Phi}_D$ accurately. The periods during loss of lock are called "dropouts," and they are usually only a few Doppler cycles long, so that the signal frequency changes very little and remains in the lock range of the PLL during the dropout. During loss of lock the PLL is blind to signals lying outside the capture bandwidth, so if $\dot{\Phi}_D$ changes too much during a dropout, the PLL may not recover lock until such time as $\dot{\Phi}_D$ again passes through the lock range.

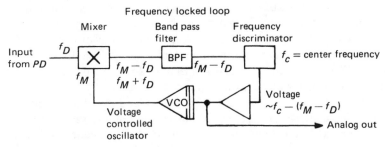

Figure 4.21 Schematic of a frequency-locked loop. (Courtesy TSI, Inc.)

The FLL is a variation of the PLL that possesses a much larger lock range and offers some additional noise rejection. A schematic of a typical FLL is shown in Fig. 4.21. The input signal with frequency $f_D = \dot{\Phi}_D/2\pi$ is mixed (multiplied) with a VCO reference signal at f_M to produce a product consisting of two sine waves, one oscillating at $f_M - f_D$, and the other oscillating at $f_M + f_D$. The product signals are passed through a band-pass filter centered at f_C. Normally f_M is adjusted so that $f_M - f_D \approx f_C$, and the signal frequency $f_M - f_D$ is transmitted, while the signal at $f_M + f_D$ is rejected. The difference between $f_M - f_D$ and f_C is detected by a frequency discriminator (i.e., a PLL) whose voltage output is proportional to $f_C - (f_M - f_D)$. This is the output of the loop, and it is also the feedback signal that controls the VCO. The loop is locked when $f_M - f_D = f_C$. Thus, if f_D increases, $f_C - (f_M - f_D)$ also increases, causing f_M to increase until $f_M = f_D + f_C$. Hence the feedback action of the loop is such as to always center the signal in the middle of the band-pass filter.

In effect, the filter "tracks" the signal and removes noise from the signal prior to frequency measurement by the discriminator. If the filter did not track, the noise bandwidth would necessarily be equal to the range of the Doppler frequency, but with tracking the noise bandwidth determined by the band-pass filter can be made much less. The improvement in the power of the SNR at the frequency discriminator is the ratio of the Doppler frequency range to the band-pass bandwidth. Thus the FLL can process noisier signals than the PLL. The noise immunity of the FLL is further enhanced by the fact that the noise seen by the frequency discriminator always has a mean instantaneous frequency equal to the center frequency of the band-pass filter. In general, the mean frequency of a signal plus noise is "pulled" to a value that is intermediate between the signal frequency and the noise frequency, and this can result in a substantial error if the SNR is low. In the FLL the frequency discriminator sees a signal close to f_C and a noise with mean frequency f_C, so the difference between signal and noise frequencies is small, and the error caused by noise pulling is small. The noise immunity of the FLL and the PLL is discussed in [84].

The disadvantage of a tracking filter is that the frequency discriminator inside the feedback loop does not see the signal unless its frequency is within the pass band. Hence its capture bandwidth at any instant is the bandwidth of the filter, and to provide good noise immunity, this is small, typically ± 1–10%. In processing LBD signals, the Doppler frequency must not change (increase or decrease) by more than half the filter bandwidth during a dropout, or else the FLL will be blind to the new signal frequency when the next burst occurs, and it will not be able to resume tracking. These effects are usually minimized by holding the loop at the last known frequency when a dropout occurs. Even so, the FLL cannot track unsteady flows properly when the data rate is too low. This restriction also applies to the PLL, but it is somewhat less severe because of the wider capture bandwidth.

For example, consider turbulent flow, and suppose we let T_T be the Taylor microscale for the temporal variation of $u(t)$, defined by

$$\left\langle \left(\frac{\partial u'}{\partial t} \right)^2 \right\rangle = \frac{\langle (u')^2 \rangle}{T_T^2} \tag{156}$$

The rms change in velocity during a dropout of mean duration $\langle \Delta t \rangle$ is

$$\Delta u = \frac{\langle (u')^2 \rangle^{1/2}}{T_T} \langle \Delta t \rangle \tag{157}$$

corresponding to an rms Doppler frequency change of

$$K \langle (u')^2 \rangle^{1/2} \frac{\langle \Delta t \rangle}{T_T} \tag{158}$$

in radians per second. If the capture bandwidth $\Delta \omega_C$ of the PLL or the FLL is larger than 3–5 times this value, the tracker will resume tracking after a dropout with high probability. That is, it is highly probable that the frequency of the new Doppler burst will fall within the capture bandwidth, and the tracker measures every burst whose amplitude is large enough to process. Otherwise, if $\Delta \omega_C < K \langle (u')^2 \rangle^{1/2} \langle \Delta t \rangle / T_T$, the tracker will miss many of the rapid, large-amplitude fluctuations in u', and statistical averages computed from the data will be biased. Since $\langle (u')^2 \rangle^{1/2}$ and T_T are determined by the flow, the only remedy is to increase $\Delta \omega_C$, which may not be possible, or to decrease $\langle \Delta t \rangle$, which is equivalent to increasing the data rate. The mean data rate is $\langle \Delta t \rangle^{-1}$, which can be improved by adding more particles or by improving the signal so that more particles can be processed. We shall refer to the ratio $T_T / \langle \Delta t \rangle$ as the mean data density and denote it by

$$\dot{N} T_T \equiv \frac{T_T}{\langle \Delta t \rangle} \tag{159}$$

This is simply the mean number of data points per Taylor microscale.

4.10.5 Frequency Counters

Frequency counters measure the frequency of a signal by accurately timing the duration of an integral number N of cycles of the signal. The timing is performed with respect to the zero crossings of the Doppler component of the signal, so the first step in a counter-processing system is the removal of the pedestals. This is usually accomplished by passing the signal through a fixed high-pass filter, but if the high-pass frequency needed to remove the pedestals of the highest-frequency Doppler bursts exceeds the lowest Doppler frequency that is expected, it may be necessary to use a variable filter or to remove the pedestals optically. For example, suppose $\Delta\omega_A/\omega_D = 5\%$ and $1 \text{ MHz} < f_D < 100 \text{ MHz}$. Then the bandwidths of the pedestals would be 5 MHz when the Doppler frequencies are 100 MHz, and a 10-MHz high-pass filter set to remove these pedestals would also remove the Doppler signals between 1 and 10 MHz. Optical pedestal removal is discussed in [85].

Once the pedestals have been removed, the processor sees only the Doppler signal plus noise. The signal plus noise is converted to a square wave by a Schmidt trigger whose output changes from a low level to a high level whenever the input increases through zero voltage, and vice versa when it decreases through zero. The leading edges of the square wave mark the zero crossings very accurately, and the square wave is compatible with digital logic circuits, which count the number of zero crossings. In the absence of any signal, the zero crossings of the noise would activate the Schmidt trigger, and the counter would measure the noise frequency, even if the noise were weak and the SNR during a typical burst were high. Consequently, it is necessary to discriminate against noise by taking measurements only when a signal is present. This is not a problem with HBD signals because the signal is almost always present, but with LBD signals, the measurements must be limited to times when a burst is present. This is accomplished by setting one or more fixed threshold levels that the signal must exceed to arm the Schmidt trigger. The sketch in Fig. 4.22 shows two threshold levels, one positive and one negative. The Schmidt trigger begins operation when the signal amplitude first exceeds both levels, and terminates at the first zero crossing after the signal fails to cross one of these levels. The time interval in which the signal exceeds the thresholds is called the burst time τ_B, and it is measured by starting a clock at the beginning of the burst detection signal and stopping it at the end. This procedure also makes it impossible to confuse the last few cycles of one burst with the beginning cycles of the next burst.

In section 4.11 the burst detection signal is denoted by $B(t)$. The measurement volume of the LDV is defined by the burst-detection thresholds according to the relations in Eqs. (45) and (46). The minimum light flux J_{min} that appears in these equations can be related to the counter-threshold voltages using the photodetector sensitivity and the value of the gain associated with amplification of the signal between the photodetector and the counter. Since the sensitivity and amplification are often adjustable, the effective size of the mv can be controlled by the experi-

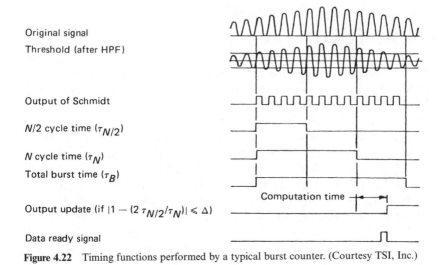

Original signal

Threshold (after HPF)

Output of Schmidt

$N/2$ cycle time $(\tau_{N/2})$

N cycle time (τ_N)

Total burst time (τ_B)

Computation time

Output update (if $|1 - (2\,\tau_{N/2}/\tau_N)| \leqslant \Delta$)

Data ready signal

Figure 4.22 Timing functions performed by a typical burst counter. (Courtesy TSI, Inc.)

menter. In addition, the data rate, i.e., the number of particles per second that produce signals exceeding the thresholds, can also be adjusted. Similar considerations apply to frequency trackers because they also contain inherent threshold voltage levels.

Having defined the burst, the counter can measure its frequency in two ways. First, the N-cycle time τ_N can be measured, and the frequency can be computed from $f_D = N/\tau_N$. The accuracy of this measurement is normally validated by comparing it with the frequency computed from measurements of a smaller number of cycles, say, $N/2$. If the difference in these measurements is less than some prescribed error Δ, for example, if

$$\left|1 - \frac{2\tau_{N/2}}{\tau_N}\right| \leq \Delta \qquad (160)$$

the measurement is accepted; otherwise it is rejected. Comparison of $2\tau_{N/2}$ and τ_N requires a short computation time during which the counter is "dead." The error Δ is normally adjustable. The choice of N and $N/2$ is not unique. For example, many LDV counters use 5/8 comparison, i.e., comparison of the 5-cycle measurement of f_D with the 8-cycle measurement. There is, however, little discernible advantage to 5/8 comparison versus, say, 4/8 comparison.

The mode of measurement described above is called the N-cycle mode. This mode does not guarantee one measurement per burst. For example, if there are 50 cycles in a particular burst, an 8-cycle counter would measure the frequency six times in succession, assuming negligible dead time. Some counters are capable of limiting the measurement to one per burst.

The second method of frequency measurement uses the total burst time τ_B and the number of cycles in the burst N_B to compute the frequency from

$$f_D = \frac{N_B}{\tau_B} \tag{161}$$

This is called the total-burst mode. It provides one measurement per burst, and this measurement clearly represents an average of the particle's velocity during its transit across the mv. For reasons that are explained in section 4.11, the total-burst mode is desirable in processing LBD signals. Furthermore, when the SNR is reasonably high, it is clear that small amounts of noise added to the signal will cause errors in the signal's zero-crossing times at the start and end of the measurement period. These errors result in the smallest frequency error when the measurement time is large. Hence total-burst-mode measurements are more accurate than the N-cycle mode if $N_B > N$, but less accurate if $N_B < N$.

Errors incurred when the SNR is small result from the frequency of the signal plus noise being "pulled" toward the noise frequency. From [84], the counter is essentially immune to this error if the power of the SNR of the burst during the period of measurement is greater than approximately 10. This value is about the same as the minimum SNR needed for PLL, but it is 10–100 times greater than the value required for FLL. Therefore, given the same signal, a counter processor will make fewer successful measurements per second than a FLL, assuming that it is not limited by loss-of-lock problems. The advantage of the counter-type processor is that it is ideally suited to LBD signals, and it has a very large dynamic range and infinite slew rate. For example, given satisfactory pedestal removal, a counter can measure a 1-kHz signal from one burst followed by a 100-MHz signal from the next burst, which may occur a few microseconds later.

The natural output of the counter processor is a digital word corresponding to the measured frequency. In the N-cycle mode it is sufficient to output τ_N, but τ_B is also output for auxiliary use in conjunction with statistical data analysis. In the total-burst mode the outputs are τ_B and N_B. Commercial counters may also provide a voltage proportional to τ_N and/or N/τ_N for the purpose of setting up and monitoring the measurements.

While counters are primarily intended for LBD signals, they work very well with HBD signals, provided the SNR is adequate. Lading and Edwards [86] have shown that counters and frequency trackers produce virtually identical outputs, including the ambiguity noise fluctuations, when the signal is HBD.

4.10.6 Burst-LDV Signal Processors

With increasing speed of digital electronics, it has become possible to construct dedicated LDV "burst processors" that are capable of measuring the frequency of an individual Doppler burst by computation of the correlation or the Fourier transform during the time it takes the burst to occur. These devices have the best ability to extract signal from noise and make accurate measurements of burst frequency in the presence of noise. Successful measurements can be made at SNR = -6 dB, or less, much lower than that required by a counter. Burst correlators and burst spectrum analyzers are essentially similar to the long-time-averaging devices described earlier, except that they operate on a single burst, or a portion thereof.

A key to the successful operation of either device is the detection of the presence of a Doppler burst, especially when the SNR is low, making the unprocessed burst of signal difficult to recognize. Strategies include threshholding on the burst pedestal, which is found by low-pass filtering the total LDV signal, or threshholding on the short-time running average of the mean square signal. With correlation analysis performed continuously, it is also possible to form and estimate the running SNR by monitoring the zero time-delay correlation and the correlation at a time delay greater than the inverse bandwidth of the noise spectrum. In this way, only signals exceeding a prescribed SNR are measured. In order to capture an entire burst without truncation, the PMT signal can be digitized continuously at a high rate and passed through a shift register. After a burst is detected, it can be centered roughly in the middle of the shift register for subsequent analysis. Correlators use one or more high-speed digital signal-processing chips consisting of shift registers, multipliers, and accumulators to sum up the time-lagged products. Burst spectrum analyzers use FFT algorithms to compute the discrete Fourier transform of the digitized signal, whose modulus squared is then used to estimate the power spectrum of the single burst. Typically, length of the digitized signal, in either case, is between 128 and 1024. Longer digital records yield more accurate measurements, but of course, require more computation.

In the case of burst-FFT analysis, the FFT yields an estimate of the Fourier transform of the burst at a discrete set of frequencies. From earlier analysis, the fluid velocity is related to the centroid of the burst spectrum or to its peak. If the sample rate is $1/\Delta t$ samples per second and the digital record consists of N samples, the frequency lines are spaced $1/N\Delta t$ apart, and this defines the basic frequency resolution of the measurements. From the Nyquist criterion, the maximum frequency (i.e., the full-scale Doppler frequency) is one-half of the sample rate, $1/2\Delta t$, so the frequency resolution is just $2/N$ of the full-scale Doppler frequency. If N is not large, or if the actual Doppler frequency is small with respect to the full-scale value, $2/N$ is inadequate for many experimental applications. To improve the resolution, it is common practice to detect the maximum in the burst spectrum and fit a low-order curve to it locally. The curve fit effectively interpolates between the spectral lines to get an improved estimate of the peak frequency of the Doppler signal spectrum. For a pure sinusoidal signal, whose spectra are inherently narrow, the interpolation can improve the resolution to a few tenths of a percent, but for Doppler signals whose bandwidths are inherently large, it is not likely that such accuracy can be achieved for the measurement of a single-burst frequency.

Like the FFT device, the burst correlator must first detect a burst, capture it in the middle of a data input buffer, and then form the convolution of the signal with itself. The time required to compute the correlation can be reduced by restricting the range of time delays to those needed to span the first zero crossing of the correlation, and using the location of the first zero crossing of the correlation to determine the frequency of the Doppler burst. As in the burst-FFT analyzer, the value of Δt determines the resolution of the zero crossing, so the practice is to use adjacent correlation values to more accurately determine the first zero crossing of the correlation by interpolation. Accuracy and dynamic frequency range can be further enhanced by using several digitizers running at different frequencies, and selecting the best digitization on the basis of a first estimate of the Doppler burst frequency found from the highest frequency data.

Burst processors can also determine the transit time of the Doppler burst, for use in burst-time weighting to avoid velocity biasing, and the time of arrival of the burst. Improvements in burst processors will be realized primarily by increasing the record lengths, as improvements on digital computation speeds permit. Longer FFT and correlation records will increase accuracy and/or dynamic range, thereby making the instruments easier to use, as well as more accurate.

4.10.7 Selection of Signal Processors

Guidelines for the selection of a signal processor for a particular measurement problem are necessarily vague because the capabilities of various processors overlap considerably. The SNR is the determining factor in most situations. If the signal is of low photon density, amplitude correlation or photon correlation is the only possibility at present. If the signal is $O(1)$ photon density, photon correlators begin to experience distortions due to photon pileup, but amplitude correlators and spectrum analyzers will work. FLLs and PLLs that incorporate a tracking-filter capability will also work in the high end of this regime, if there are about five photons per Doppler cycle and if the signal is filtered somewhat above the Doppler frequency. High-photon-density signals may be processed with FLLs and PLLs with tracking filters, and counters may also be used if there are enough bursts with SNR > ~10. Of course, amplitude correlators and swept-spectrum analyzers work even better on high-photon-density signals than on low-photon-density signals. Photon correlators can be adapted to high-photon-density signals by the simple artifice of reducing the laser power to produce a low-photon-density signal, but this procedure is clearly wasteful of useful information.

Burst density also influences the choice of instruments. Counters are designed to work best with LBD signals, but they also work very well when the burst density is high. Likewise, trackers are designed to work best with HBD signals, but they will work with LBD signals if the data density, defined as $T_T/\langle\Delta t\rangle$, is not much less than unity.

4.11 DATA PROCESSING

Data processing refers to the procedures used to compute the desired flow properties from the data output by the signal processor. The specific procedure depends, of course, on the general type of signal processor and the type of signal, but it also depends on the type of flow and the type of quantity that is to be computed. Specifically, time-resolved measurements in unsteady flows, especially turbulent flows, require special attention, and most of this section is devoted to this general situation. Steady-flow results are special cases of the more general turbulent-flow results.

4.11.1 Time-Averaging Processors

The amplitude correlator, the photon correlator, and the spectrum analyzer are normally used in the long-time averaging mode; i.e., the correlations and spectra are averaged over many integral timescales of the turbulent motion. The relationship between these quantities and the turbulent-flow field are discussed extensively in section 4.9.4, and they may be used to calculate mean velocity and rms velocity from either correlations or spectra.

In this section we note certain techniques that permit the evaluation of higher-order statistics under special conditions. In particular, suppose that the broadening effects caused by the laser source, Brownian motion, mean velocity gradients, and ambiguity broadening are all negligible in comparison with the broadening caused by turbulent fluctuations, i.e.,

$$\Delta\omega_0, \Delta\omega_B, \Delta\omega_G, \Delta\omega_A \ll \Delta\omega_T$$

Further, suppose that $f_u(\mathbf{c}, \mathbf{x})$ is independent of $\mathbf{x}$. Then, since negligible ambiguity broadening implies that the amplitude factor $a(\mathbf{x})$ is so wide that $a(\mathbf{x} + \mathbf{c}\tau) = a(\mathbf{x})$ for all values of $\mathbf{c}\tau$ that are of interest, we have, from Eq. (98),

$$R_{J_D J_D} \approx \frac{1}{2} \int C\langle D^2\rangle a^2(\mathbf{x}) \, d^3x \int f_u(\mathbf{c}, 0) \cos(\omega_S\tau + K c_1 \tau) \, d^3c \quad (162a)$$

$$= \frac{1}{2} \int C\langle D^2\rangle a^2(\mathbf{x}) \, d^3x \int f_u(c_1, 0) \cos(\omega_S\tau + K c_1 \tau) \, dc_1 \quad (162b)$$

Hence $R_{J_D J_D}$ is proportional to the characteristic function of $f_u(c_1)$ if $\omega_S = 0$. Taking the Fourier transform of Eq. (162b) yields the following power spectrum:

$$S_{J_D}(\omega) = \frac{1}{2K} \int C\langle D^2\rangle a^2(\mathbf{x}) \, d^3\mathbf{x} \, [f_u(\omega - \omega_S) + f_u(\omega + \omega_S)] \quad (163)$$

Thus, if either ω_S or the mean Doppler frequency is large enough in comparison with the spectral bandwidth to make $S_{J_D}(0) \approx 0$, $f_u(\omega + \omega_S)$ will be negligible for $\omega > 0$, and

$$f_u(\omega - \omega_S) = 2KS_{J_D}(\omega) [\int C\langle D^2\rangle a^2(\mathbf{x})\, d^3\mathbf{x}]^{-1} \qquad (164)$$

The complete probability density function for u can be found, and from it, higher-order moments such as $\langle (u')^3\rangle$ can be calculated. When using correlations, it is convenient to take the Fourier cosine transform of the correlogram to obtain the spectrum, so that Eq. (164) can be used. This procedure assumes that the Doppler component of the spectrum can be separated from the other components, such as the pedestal component and the noise component, and it ignores errors associated with the computation of the Fourier transform from a correlogram with a relatively small number of time-delay points.

When the ambiguity bandwidth is not negligible, the measured spectrum is the convolution of the velocity probability density function with a spectrum whose width is determined by the ambiguity bandwidth. There have been a number of efforts, described in [87], to develop algorithms for computing f_u from broadened data, but this procedure is subject to error because, for arbitrary f_u, deconvolution amounts to solving a Fredholm integral equation of the first kind, and this problem requires special conditioning that is not necessarily valid for all forms of f_u.

It should be noted that the effects of burst density appear only in a multiplicative factor in Eq. (164), which essentially normalizes the probability density function. Burst density is irrelevant in correlation and spectral analysis techniques, except as it determines the rates at which the correlation and spectrum can be accumulated statistically.

4.11.2 Time-Resolving Signal Processors

The general-purpose LDV signal processor must be capable of processing both HBD and LBD signals, and the latter type of signal necessitates an ability to make measurements of the velocity within the time span of an individual Doppler burst. The outputs of processors that are designed to perform this function inevitably share certain common characteristics, and it appears to be possible to represent the outputs of all these processors (e.g., counters, trackers, burst correlators, and in the future, burst-spectrum analyzers) with a fairly simple model. The model postulates an ideal signal processor whose output is equal to the instantaneous Doppler frequency, while the signal amplitude exceeds a threshold level J_{min}, and is constant otherwise. The value of the constant may be zero, or more commonly, it may be the value of the last known frequency. The effects of noise in the signal, locking and unlocking transients, and fringe biasing are presumed to have been reduced to negligible levels, and all other idiosyncrasies that may arise in real instruments are ignored.

The output of the ideal processor is simple $\dot{\Phi}_D$ when the burst density is large. We note in section 4.9.6 that the spatially and temporally filtered velocity $\tilde{\tilde{u}}\,(t)$ defined by Eq. (142) represents the measurable velocity in this case.

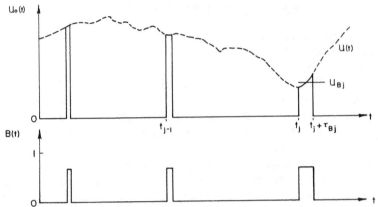

Figure 4.23 Velocity observed with a burst processor and a low-burst-density signal, and the burst indicator function $B(t)$

When the burst density is low, the output of the ideal processor can be represented by an "observed velocity"

$$u_o(t) = \iint w(\mathbf{x}, D)u(\mathbf{x}, t)g(\mathbf{x}, t, D) \, d^3\mathbf{x} \, dD \tag{165}$$

where

$$w(\mathbf{x}, D) = H[a(\mathbf{x})D - J_{min}] \tag{166}$$

indicates the presence of a detectable particle [$H(\ )$ denotes the Heaviside function]. That is, $w(\mathbf{x}, D) = 1$ if a particle is in the mv and its scattering amplitude D is large enough to make the amplitude $a(\mathbf{x})D > J_{min}$, and it is zero otherwise. This definition follows the one used by Buchave et al. [68], but it generalizes the representation to include the effects of random D values, e.g., polydisperse scatterers. For LBDs the occurrence of more than one particle in the mv is negligibly improbable, so at any instant, only one of the terms in $g = \Sigma_j \delta[\mathbf{x} - \mathbf{x}_j(t)]\delta(D - D_j)$ can contribute to the integral. Then,

$$u_o(t) = u[\mathbf{x}_j(t), t] = \mathbf{v}_j(t) \tag{167}$$

during the jth Doppler burst. If no particle is present, $u_o(t) = 0$.

Figure 4.23 indicates the behavior of u_o. In the figure a burst-indicator signal $B(t)$ that is unity when a detectable particle is present, and zero otherwise, is also shown. $B(t)$ can be represented by

$$B(t) = \iint w(\mathbf{x}, D)g(\mathbf{x}, t, D) \, d^3\mathbf{x} \, dD$$

This is a good model of the outputs from trackers, counters, and burst correlators.

If the signal is $O(1)$ burst density, $u_o(t)$ is not a good representation because it sums the individual particle velocities when more than one particle is present. Like-

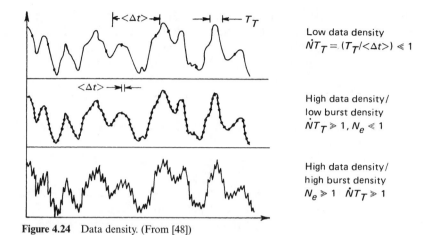

Low data density
$$\dot{N}T_T = (T_T/\langle \Delta t \rangle) \ll 1$$

High data density/
low burst density
$$\dot{N}T_T \gg 1, N_e \ll 1$$

High data density/
high burst density
$$N_e \gg 1 \quad \dot{N}T_T \gg 1$$

Figure 4.24 Data density. (From [48])

wise, $u(t)$ is not a good representation either. It appears that the analysis of the $O(1)$ burst density case is more complex than for the limiting cases, and the author is unaware of any successful attempts to fully analyze the data processing techniques applicable to this type of signal. Consequently, only the high and low limits will be discussed here.

4.11.2.1 Data density. The data density is defined in Eq. (159), in conjunction with the discussion of data-rate effects on frequency trackers. The data density also characterizes the output of a signal processor, and it dictates to some extent the types of data processing methods than can, or must, be used. Data density and burst density are independent concepts because the Taylor microscale varies independently of the size of the measurement volume. Consider, for example, a turbulent flow with mean velocity in the x direction, and suppose that Taylor's time microscale T_T is related to Taylor's spatial microscale in the x direction λ_x by the frozen-field approximation: $\lambda_x = UT_T$. The burst density is $N_e \simeq CA\,d_m$ where A is the projected area of the mv in the x direction, whereas the mean data density is $NT_T \simeq CAUT_T = CA\lambda_x$. Hence, $N/N_e = \lambda_x/d_m$. The ratio λ_x/d_m is rarely smaller than 1, and it is large in a turbulent-flow LDV experiment that has good spatial resolution. Hence the signal can be of high data density (HDD) when it is of LBD (for example, $\lambda_x/d_m = 50$, $\dot{N} = 10$, $N_e = 0.2$), and it is always of HDD when it is of HBD.

The important combinations of data density and burst density are depicted in Fig. 4.24. The low-data-density (LDD) signal is always of LBD if $\lambda_x/d_m = O(1)$ or greater. In this case the mean time between data points $\langle \Delta t \rangle = \dot{N}^{-1}$ is greater than T_T, and the velocity cannot be resolved as a function of time. However, multiple particles hardly ever occur in the mv, so ambiguity noise is avoided, each detectable particle giving a good measurement of its velocity. The HDD, LBD signal is even better in this regard because there is still no ambiguity noise, whereas the mean time between data points is now small in comparison with the T_T, so at least the energy-containing fluctuations can be resolved in time. Finally, at large particle concentrations the data density is very high, but since the burst density is even higher, an ambiguity noise appears in the signal. The HDD, LBD signal is optimal, but good statistical results can be obtained in the other two cases as well.

4.11.2.2 High-burst-density signals. In section 4.9.6 it is shown that the instantaneous frequency of the HBD signal is

$$\tilde{\dot{\Phi}}_D = \omega_S + K\tilde{\tilde{U}} + K\tilde{\tilde{u}}' + \tilde{\dot{\phi}}$$

where the tilde denotes the filtering after the signal processor. It is noted in section 4.10.5 that the outputs of the counter and the frequency tracker are virtually identical for HBD signals. Both outputs will contain small dropout periods that occur when the amplitude envelope of the HBD signal drops below the threshold, coincident with very rapid phase changes. Phase changes may also cause dropouts in the FLL or PLL, owing to loss of lock, and in the counter, owing to $N/2$ or $5/8$ comparison. In either case, the dropouts are a few Doppler cycles long, and this is usually very short in comparison with the T_T, so the signal is still essentially of HDD.

Buchave et al. [68] have analyzed the effects of the dropouts by postulating that the sequence of on (i.e., in lock) and off (i.e., dropout or out of lock) states is a Markov chain. They found that the minimum distortion of the signal statistics is achieved by holding the last known reading during a dropout. With this method, all moments, including the velocity probability density function, are preserved, essentially because the dropout times are uncorrelated. Their model for short dropout times also predicts that the power spectrum of $\dot{\Phi}D$ with dropouts is just $S_{\dot{\phi}_D}$ plus a dropout noise spectrum that is essentially constant out to the cutoff frequency of the low-pass filter. Thus the dropout noise can be treated in the same way as the ambiguity noise.

The single-time statistical moments of $\tilde{\tilde{u}}'(t)$ can be obtained from $\tilde{\dot{\Phi}}'_D = \tilde{\dot{\Phi}}_D(t) - \omega_S - K\tilde{\tilde{U}} = K\tilde{\tilde{u}}'(t) + \tilde{\dot{\phi}}$ by direct computation if $\tilde{\dot{\phi}}$ is small. Otherwise, the statistics can be corrected for $\tilde{\dot{\phi}}$ contamination by using the fact that $\tilde{\dot{\phi}}$ and $\tilde{\tilde{u}}'$ are statistically independent. Thus

$$\langle\tilde{\dot{\Phi}}_D\rangle = \omega_S + K\tilde{\tilde{U}} \tag{168a}$$

$$\sigma_{\dot{\phi}D}^2 = \langle((\tilde{\dot{\Phi}}_D - \langle\tilde{\dot{\Phi}}_D\rangle))^2\rangle = K^2\langle(\tilde{\tilde{u}}')^2\rangle + \langle\tilde{\dot{\phi}}^2\rangle \tag{168b}$$

$$\langle((\tilde{\dot{\Phi}}_D - \langle\tilde{\dot{\Phi}}_D\rangle))^3\rangle = K^3\langle(\tilde{\tilde{u}}')^3\rangle + \langle\tilde{\dot{\phi}}^3\rangle \tag{168c}$$

$$\langle((\tilde{\dot{\Phi}}_D - \langle\tilde{\dot{\Phi}}_D\rangle))^4\rangle = K^4\langle(\tilde{\tilde{u}}')^4\rangle + 2K^2\langle(\tilde{\tilde{u}}')^2\rangle\langle\tilde{\dot{\phi}}^2\rangle + \langle\tilde{\dot{\phi}}^4\rangle \tag{168d}$$

etc., can be corrected with knowledge of $\langle\tilde{\dot{\phi}}^2\rangle$, $\langle\tilde{\dot{\phi}}^3\rangle$, $\langle\tilde{\dot{\phi}}^4\rangle$, and so on. These moments can be calculated using [67] the probability density

$$f_\phi(\mathbf{x}) = \frac{1}{2}\left(1 + \frac{x^2}{\Delta\omega^2}\right)^{-3/2} \tag{169}$$

and taking the filtering into account. Note that Eq. (169) is an even function of its argument, so all odd-order moments of ϕ (and hence Φ_D) vanish. The even-order momts of ϕ can all be calculated in terms of $\Delta\omega^2$ or, for $\dot\phi$, in terms of $\langle\dot\phi^2\rangle$. This last quantity is not readily determined theoretically because the turbulent gradient broadening is unknown in most experiments. Hence $\langle\dot\phi^2\rangle$ is best determined by direct experimental measurement.

Measurements of $\langle\dot\phi^2\rangle$ are, in principle, possible because the spectrum of $\dot\phi$ is white out to f_c. Hence, if f_c is somewhat greater than the maximum turbulent frequency, the $\dot\phi$ spectrum level is readily determined, and $\langle\dot\phi^2\rangle$ can be calculated by extrapolating this level to zero frequency and integrating under the curve. Alternatively, autocorrelation measurements of $\Phi_D - \langle\Phi_D\rangle$ yield the autocorrelation of the velocity plus the autocorrelation of $\dot\phi$. The latter is very narrow, of width f_c^{-1}, so it appears as a spike at zero time delay that extends above the zero-time-delay value of the velocity correlation by the amount $\langle\dot\phi^2\rangle$.

The foregoing procedures are good enough for finding the value $\langle\dot\phi^2\rangle$ needed to correct single-time moments, but other procedures are preferable if the velocity autocorrelation or spectrum is desired. These are based on the fact that the ambiguity noises from two different LDV systems are almost uncorrelated. Morton and Clark [88] report spatial correlation measurements of u using two LDVs with spatially separated measurement volumes. The cross correlation of the LDV signals showed virtually no ambiguity noise, even when the measurement volumes overlapped. Even more surprisingly, van Maanen et al. [89] showed that a single LDV transmitting optical system could be used, and two signals with very weakly correlated ambiguity noises could be contained by collecting light with two photodetectors in different scattering directions.

4.11.2.3 Low-burst density, low-data density signals. Low-burst-density data consist of a series of individual particle-velocity measurements at random times of arrival. These data samples are biased toward high velocities because more particles traverse the mv when the velocity is high than when it is low. Thus the statistical equations that one usually uses to form averages with a set of unbiased samples are invalid when applied to LBD signal data. For example, suppose the data are $u(t_j) = u_j, j = 1, \ldots, J$, where t_j is the arrival time of the jth particle. The customary equation for the sample mean velocity for unbiased data is

$$\frac{1}{J}\sum_{j=1}^{J} u_j$$

Now suppose that the flow is unidirectional and its velocity is a square wave that takes the values V_0 and $2V_0$ for equal times. Then, since the arrival rate of particles is proportional to $u(t)$, there will be twice as many samples when $u = 2V_0$ as when

$u = V_0$. The average calculated from the foregoing equation would be $(1 \times V_0 + 2 \times 2V_0)/(1 + 2) = 5V_0/3$, whereas the true average of the square wave is $3V_0/2$. This discrepancy always occurs, on average, and it is independent of the particle concentration.

The bias effect was first noted by McLaughlin and Tiederman [73], who proposed using the velocity data to statistically weight the samples, a method of correction that is correct for unidirectional flows. George [90], and later Hösel and Rodi [91], proposed using the burst time τ_{B_j} as a statistical weighting factor, on the basis that the burst time is inversely proportional to the magnitude of the three-dimensional velocity vector, and it is readily measured. An alternative method of unbiasing the samples in three-dimensional flow is to weight them with the inverse of the sampling probability [92], but this requires simultaneous measurements of all three velocity components. The present discussion concentrates on the burst time, or "residence time" weighting technique, which is the most widely accepted, albeit not totally proven, technique.

As noted earlier, the observed velocity available from an ideal signal processor with LBD is

$$u_o(t) = \iint w(\mathbf{x}, D)u(\mathbf{x}, t)g(\mathbf{x}, t, D) \, d^3\mathbf{x} \, dD \tag{165'}$$

The relationship between the statistics calculated using burst-time weighting and the statistics of $u(\mathbf{x}, t)$ can be obtained by treating $u_o(t)$ as an ordinary signal and taking time averages. The time average of the nth power of u_o is

$$\overline{u_o^n(t)} = \frac{1}{T} \int_0^T u_o^n(t) \, dt \tag{170a}$$

$$= \frac{1}{T} \sum_j u_j^n(t)\tau_{B_j} \tag{170b}$$

$$= \overline{[u^n]}_o \tag{170c}$$

where u_j, the velocity of the jth particle, is assumed to be constant during the burst time, as usual. Equation (170c) follows from the fact that

$$[u^n(t)]_o \equiv \iint w(\mathbf{x}, D)u^n(\mathbf{x}, t)g(\mathbf{x}, t, D) \, d^3\mathbf{x} \, dD \tag{171a}$$

$$= [u_o(t)]^n \tag{171b}$$

as is apparent from inspection. [Essentially the integral operators in Eqs. (165) and (171a) sample the velocity during a burst. Equation (171a) is the sample of the nth power of u, and Eq. (171b) states that this is equal to the nth power of the sample.] We now assume that the time average of u_o^n equals its ensemble average. Then

$$\overline{u_o^n} = \overline{[u^n]}_o \tag{172a}$$

$$= \langle [u^n]_o \rangle \tag{172b}$$

$$= \langle \iint w(\mathbf{x}, D)u^n(\mathbf{x}, t)g(\mathbf{x}, t, D) \, d^3\mathbf{x} \, dD \rangle \tag{172c}$$

$$= \int C(\mathbf{x})w(\mathbf{x})\langle u^n \rangle \, d^3\mathbf{x} \tag{172d}$$

where

$$w(\mathbf{x}) = \frac{\displaystyle\int_0^\infty w(\mathbf{x}, D)C(\mathbf{x}, D)\, dD}{\displaystyle\int_0^\infty C(\mathbf{x}, D)\, dD} \tag{173}$$

showing that the burst-time-weighted sum in Eq. (170b) equals the volume integral given by Eq. (172d). The former sum is not, however, a sensible average because it averages over the dropout periods when $u_o = 0$. This defect is corrected by dividing Eqs. (170b) and (172d) by the average burst time, given by

$$\frac{1}{T}\int_0^T B(t)\, dt = \frac{1}{T}\sum_j \tau_{B_j} \tag{174a}$$

$$= \langle B(t)\rangle \tag{174b}$$

$$= \int C(\mathbf{x})w(\mathbf{x})\, d^3\mathbf{x} \tag{174c}$$

This yields

$$\frac{\displaystyle\sum_j u_j^n \tau_{B_j}}{\displaystyle\sum_j \tau_{B_j}} = \widetilde{\langle u^n\rangle}$$

where $\widetilde{}$ denotes the volume average:

$$\widetilde{\langle u^n\rangle} = \frac{\int C(\mathbf{x})w(\mathbf{x})\langle u^n(\mathbf{x}, t)\rangle\, d^3\mathbf{x}}{\int C(\mathbf{x})w(\mathbf{x})\, d^3\mathbf{x}} \tag{175}$$

If $\langle u^n(\mathbf{x}, t)\rangle$ is constant within the mv, then $\widetilde{\langle u^n\rangle} = \langle u^n\rangle$.

The mean velocity and mean square fluctuation with respect to the mean that are given by burst-time weighting are

$$\frac{\displaystyle\sum_j u_j \tau_{B_j}}{\displaystyle\sum_j \tau_{B_j}} = \widetilde{U} \tag{176}$$

$$\frac{\displaystyle\sum_j (u_j - U)^2 \tau_{B_j}}{\displaystyle\sum_j \tau_{B_j}} = (U - \widetilde{U})^2 + \widetilde{\langle (u')^2\rangle} \tag{177}$$

respectively, where Eq. (177) follows from Eq. (175) after some manipulation.

The mv in the case of burst-signal processing is determined by the volume average in Eq. (175). This average, with weighting function $C(\mathbf{x})w(\mathbf{x})$, and $w(\mathbf{x})$ given by Eq. (173), is very similar to the average that appears in the case of HBD signals, where the weighting function is $C(\mathbf{x})\langle D^2\rangle a^2(\mathbf{x})$. The integral in Eq. (173) is evaluated by noting that, at fixed $\mathbf{x}$, $w(\mathbf{x}, D)$ is zero unless $a(\mathbf{x})D \geqslant J_{\min}$, that is, unless $D \geqslant D_{\min}(\mathbf{x}) = J_{\min}/a(\mathbf{x})$. For all $D \geqslant D_{\min}$, $w(\mathbf{x}, D) = 1$. Hence, it follows that

$$w(\mathbf{x}) = \int_{J_{\min}/a(\mathbf{x})}^{\infty} \frac{C(\mathbf{x}, D)\, dD}{C(\mathbf{x})} \tag{178}$$

which states that $w(\mathbf{x})$ is just the fraction of the total number of particles for which $D > D_{\min}$, or in other words, $w(\mathbf{x})$ is the probability that a particle at $\mathbf{x}$ has a value of $D > D_{\min}$.

If the scattering population is monodisperse, then $C(\mathbf{x}, D) = \delta(D - D_o)$, and $w(\mathbf{x}) = w(\mathbf{x}, D_o)$. This equals unity if $a(\mathbf{x}) \geqslant J_{\min}/D_o$ and zero otherwise, thereby defining an mv with a sharp boundary. If $a(\mathbf{x})$ is the usual Gaussian ellipsoid, the mv will be an ellipsoid whose major axes are given by l_x, l_y, and l_z from Eq. (46).

In the case of polydisperse particles, the description of the mv is simplified by assuming that $C(\mathbf{x}, D) = C(\mathbf{x})f(D)$, that is, the normalized size distribution of particles $f(D)$ is the same everywhere, but the total number per unit volume may vary in space. Then

$$w(\mathbf{x}) = \int_{J_{\min}/a(\mathbf{x})}^{\infty} f(D)\, dD \tag{179}$$

and the $\mathbf{x}$ dependence enters only through the lower limit of integration. Clearly, $w(\mathbf{x})$ is constant when $a(\mathbf{x})$ is constant, so the mv has an ellipsoidal shape if $a(\mathbf{x})$ is ellipsoidal. The dimensions of the ellipsoid depend on $f(D)$, and this is rather troublesome, since one rarely has this information. Clearly, monodisperse particles are preferable in LBD flows.

It should be noted that the present equations for interpretation of burst-time-weighted statistics differ from those in [68] by the inclusion of the effects of polydispersity. It is suggested in that reference that an on-off scattering volume, that is, $w(\mathbf{x}) = 0$ or 1, is an excellent approximation to real LDV systems, but we see from the present results that $w(\mathbf{x})$ is a 0-1 function only if the scattering particles are monodisperse. Otherwise, $w(\mathbf{x})$ is a continuous function, much like $a^2(\mathbf{x})$.

The first term in Eq. (177) represents the apparent fluctuations caused by mean velocity gradients across the mv. These fluctuations can be significant if the gradients are large, and if they exceed about 10% of the rms turbulent fluctuations, they should be eliminated by reducing the mv size.

The autocorrelation and the power spectrum can be calculated from LDD signals, despite the random times of the data and the gaps between data points. It is shown in [68] that the appropriate unbiased algorithm for the autocorrelation is

$$\langle u'(t)u'(t + \tau)\rangle = \frac{\sum_i \sum_j u'_i u'_j \tau_{B_{ij}}}{\sum_i \sum_j \tau_{B_{ij}}} \tag{180}$$

where $\tau = t_i - t_j$, $i < j$, and $\tau_{B_{ij}}$ is the duration of the overlap between the ith burst and the jth burst, delayed by time lag τ. This equation assumes monodisperse particles, uniform particle concentration, and vanishing mean velocity gradients across the mv. The equation should not be used at zero time lag, because at small values of τ, of the order of the mean burst time, the burst-time-weighted estimate of the autocorrelation contains a large noise spike associated with the correlation of the individual bursts with themselves. The power spectrum can be obtained by taking the Fourier transform of the autocorrelation. If the burst self-correlation spike is retained in the process, the power spectrum will contain a large spectrum that is white out to a frequency that is approximately equal to 1 divided by the mean burst time. Fortunately, the self-correlation spike is easily removed, before Fourier transformation, by replacing it with the zero-time-delay value from Eq. (177). Correlations can be obtained at very short time delays because, for any data rate, there is always a finite probability that the arrival times of two particles will be arbitrarily close. Likewise, spectra can be calculated at frequencies higher than the mean data rate. The penalty associated with low data rates is the long averaging time required to obtain many sample pairs separated by time lags that are smaller than the mean data rate.

4.11.2.4 Low-burst density, high-data density signals. Data processing algorithms for this type of signal are obtained by noting that the data points are so close together that simple interpolation schemes can be used to fill in much of the missing data between bursts. Dimotakis [92] has suggested using linear interpolation, in which case the interpolated curve of some function of the velocity $g(u)$ versus time is a series of trapezoids. Time averaging of this curve yields the following trapezoidal-rule approximation:

$$\overline{g[u(t)]} = \frac{1}{t_N - t_1} \sum_j \frac{1}{2} [g(u_j) + g(u_{j-1})](t_j - t_{j-1}) \qquad (181)$$

In this procedure, biasing is eliminated because the samples are weighted by $t_j - t_{j-1}$, which is inversely proportional to the data rate when the data density is high.

Equation (181) requires digital analysis of the data. Analog analysis is also possible because most signal processors hold the value of the last sample at the output until a new sample is measured. The output signal then looks like a staircase function whose steps are small if the data density is large. The steps add noise, but this can be removed in large part by low-pass filtering. In effect, the low-pass filter performs an analog interpolation. The filtered signal can be analyzed with standard analog instruments such as correlators and mean dc voltmeters, or it can be digitized for subsequent digital analysis. This case certainly offers the simplest analysis procedure and the maximum possible velocity information without contamination by ambiguity noise.

4.11.3 Fringe Biasing

Fringe biasing occurs when the signal processor requires a minimum number of cycles to make a measurement, and certain particle trajectories fail to provide this number. This effect is worst when the velocity is parallel to the fringes so that no particle crosses a fringe, but it is also manifested at oblique angles where particles passing through the center of the mv may cross enough fringes, but those passing near the edges do not. Thus the data rate is greatest when the velocity vector is perpendicular to the fringes, and it decreases as the angle between the velocity and the fringes approaches zero. The resulting data are biased toward samples from perpendicular velocities. Fringe bias is analyzed in [92–94].

There is no satisfactory method of correcting for fringe bias analytically, but it can be reduced to negligible levels in most cases by the simple technique of frequency shifting. Frequency shifting effectively adds cycles to a Doppler burst by moving the fringes with respect to the fluid. Then, if the fringe velocity is large in comparison with the flow velocities, even particles traveling parallel to the fringes will produce an adequate number of cycles as the fringes move over them. Alternatively, for counters, fringe biasing is minimized by using the total-burst mode because this mode requires the least number of cycles for a measurement. Fringe biasing is only important for LBD signals because the signal is almost always present in HBD signals. The data processing procedures discussed in section 4.11.2 assume that fringe biasing has been eliminated.

4.12 FIBER-OPTIC AND LASER DIODE SYSTEMS

Optical fibers and laser diodes offer several possibilities for reducing the size and weight of LDV optical systems. Optical fibers are thin, flexible glass fibers designed to transmit light over long distances with little attenuation. They are used in LDV systems to couple the laser, which is often heavy and bulky, to an optical head and to transmit scattered light signals to a fixed PMT. The optical head contains (1) optics that form pairs of illuminating beams and focus them to a crossing point, (2) scattered-light-collecting optics, and (3) the optics that inject the scattered light into a receiving fiber that takes the light back to the PMT. The use of optical fibers to remove the laser and the PMT from the part of the LDV optics that must be movable makes it much easier to position the LDV measurement volume, and it removes the electronic components of the system from the immediate vicinity of the flow measurement where the environment may not be hospitable to electronics. Fiber-optic-coupled probes are very useful under water, in electrically and/or magnetically noisy environments, or in hot or corrosive environments. Remote location of the laser is particularly advantageous when the laser requires cooling water and a relatively large power supply, as for an Ar^+ laser.

A typical fiber-optic LDV system consists of optics to inject the laser beam into the fiber, the fiber itself, clad in a protective plastic sheath, possibly with sev-

Standard Fiberoptic Probe

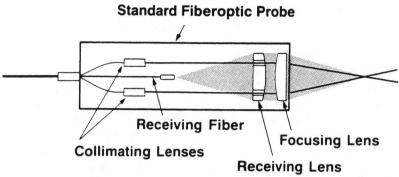

Receiving Fiber

Collimating Lenses

Focusing Lens

Receiving Lens

Figure 4.25 Typical fiber-optic probe. (Courtesy of TSI, Inc.)

eral other fibers, and the optical head (Fig. 4.25), which may be as small as a few centimeters in diameter. The head contains lenses to remove the transmitted light from the end of the fiber and form it into a satisfactory beam, a beam splitter, and a focusing lens. Instead of a single fiber with a beam splitter, it is also possible to use two transmitting fibers, each coupled into their own collimating lens, as shown in Fig. 4.25. To make a single probe, the LDV is operated in the coaxial back-scatter mode so that the focusing lens also collects the scattered light, from whence a set of receiving lenses injects the scattered light into a receiving fiber, which transmits the light to a PMT.

When there are several pairs of beams, as in a multicolor two- or three-dimensional LDV, it is particularly convenient to keep as much of the system as possible in a fixed location. In such systems the beam splitting, color splitting, and frequency shifting can all be done in a fixed unit whose outputs consist of single beams of various colors and frequencies. For example, the Ar⁺ beam can be divided into 514.5-nm pair, with one beam frequency shifted, and a similar 488.0-nm pair. Each pair of beams is transmitted to an optical head, in which the beam splitter is replaced by lenses that expand the outputs of the two fibers and send them through the focusing lenses. Very precise factory alignment is needed to ensure that these optics form a good LDV measurement volume, but once done, optics are all ce-mented into place, and this alignment should not have to be repeated. The princi-pal adjustment left to the experimenter is the positioning of the fibers to inject the laser beams into them efficiently.

The foregoing discussion skirted around certain aspects of optical fiber tech-nology that are, in fact, critical to the successful implementation of fibers in LDV. There are two basic types of fibers: single mode and multimode. Multimode fibers have relatively large diameter cores, i.e., 50–100 μm. The core is clad in a sheath of glass whose refractive index is larger than that of the core, resulting in total internal reflection of light waves that strike the core-cladding interface at shallow angles, and thereby retain the light in the core. The large diameter of the core

permits transverse waves, and hence the output of the fiber contains multiple optical modes, which in turn, create complicated phase fronts and intensity distributions.

Single-mode fibers combine much smaller core diameters, of the order of 5 μm, with a continuously varied profile of refractive index that retains light in the core by gradually bending outward bound light waves back toward the core. They act as wave guides that suppress the formation of transverse oscillation modes. The intensity and phase of the output from a single-mode fiber are more uniform than those from a multimode fiber, and they form a much cleaner LDV fringe pattern. Hence they are preferred for transmitting fibers. This improvement is not, however, achieved without cost. The small diameter of the single-mode core restricts the amount of laser energy that can be transmitted through it and makes it very difficult to inject light into the core with high efficiency. For example, the laser beam to be injected must be focused down to a spot smaller than the core diameter, and the axial location of the focal spot must be within about 2 μm of the end of the fiber and centered on the fiber to within approximately 1/10 μm. To produce coherent outputs, the mode of the injected laser beam should be TEM_{00}. Such stringent requirements require a very fine mechanism for aligning the laser beam with the fiber. With good alignment, 60–80% of the laser energy can be injected successfully. To preserve the polarization of the transmitted light, it is also a good idea to use fibers of the "polarization-preserving" type. Ordinary fibers exhibit stress birefringence, which changes as the fiber is bent and leads to scrambled polarization at the output of the fiber. Polarization-preserving fibers control the polarization by using a strong stress-induced birefringence built into the structure of the fiber. Multimode fibers can be used for transmitting the received light, since that light is already incoherent.

With proper control, laser diodes emit light of sufficient coherence and power to form good LDV signals. They are small and relatively efficient, requiring no cooling water and small power supplies. If such diodes are combined with photodiode light detectors, which are also compact and low powered, it is possible to make an LDV probe that can be held in one hand and is connected to the rest of the system only by electrical wires. An example of such a system is shown in Fig. 4.26. It is a coaxial back-scatter probe similar to the fiber-optic probe in Fig. 4.25, except that it contains both the laser and the detector onboard.

Typical laser diodes operate in the near infrared, 780 and 830 nm, but red emitters are also available at 630 and 670 nm, and they are becoming more powerful. Energies of commercial monolithic laser diodes range from milliwatts to several tens of watts. The less desirable aspects of laser diodes are low beam quality in the form of high divergence, elliptical cross-section and astigmatic wave fronts; low coherence; and strong dependence of the light wavelength on the diode temperature and the supply current. A rule of thumb for the coherence length is that it is given roughly by 1 m divided by the number of milliwatts of output.

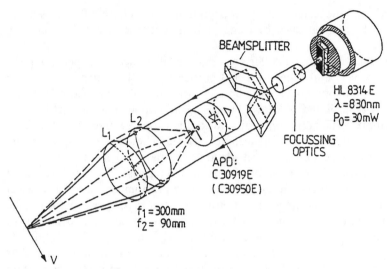

Figure 4.26 Back-scatter laser-diode-based LDV probe. (From [95])

Thus the coherence length of 1-W diode is of the order of 1 mm. Stabilization of the frequency can be achieved by careful control of the temperature and the current.

4.13 PHASE DOPPLER VELOCIMETRY

Phase Doppler velocimetry is a variation of conventional LDV, which makes it possible to measure the diameter of round particles of known refractive index at the same time that the velocity is measured. By using two photodetectors receiving the scattered light at two different directions of observation, one obtains two independent Doppler bursts whose phase difference can be measured, and from this phase difference, it is possible to infer the size of the particle.

The foundations for this technique were laid in a pioneering paper by Durst and Zaré [96] in 1975, in which they showed that the phase of a Doppler signal burst varies with the angle of observation in a manner that depends upon the size of the spherical particle creating that burst. Since the phase of a Doppler burst from a single particle also depends upon the trajectory taken by that particle, it is necessary to establish a reference phase from which a phase difference can be measured. This is done by using two or more photodetectors, one for reference and the others for the signal phase. In practice, the photodetectors collect light scattered in different directions, and Doppler bursts from the same particle are compared to determine the phase differences [97]. This is done by analog phase measurement, by cross correlating the signals, or by cospectral analysis. Phase differences in excess of 2π can be distinguished by using multiple detectors located at properly selected angles.

Phase Doppler instruments offer an important new means of measuring the diameters of droplets and other spherical particles, and they have been applied extensively in liquid sprays, where the droplets are naturally spherical and the refractive index of the liquid is known. By using additional photodetectors, it is possible to remove some of the ambiguity due to the variable refractive index. Also, if the particles are not spherical, the technique is still believed to give approximate measurements whose accuracy may be adequate for many purposes.

Significant advances in the understanding of the phase Doppler instrument have depended upon understanding light-scattering phenomena, in order to select parameters such as the viewing angles in order to optimize the performance of the device. It is known that the light scattered from a particle can be associated with light that has been reflected or refracted or diffracted from the particle. [98] In the case of transparent particles, there are many reflections and refractions created as the light rays penetrate the back of the particle and undergo several reflections on the interior surface of the particle. At each reflection it is possible for some amount of light to be transmitted through the surface. These processes are easily visualized and analyzed for particles that are large enough to be treated using geometric optics. This occurs when the particle diameter is several times the wavelength of light, i.e., around 5–10 μm for visible light. Phase Doppler methods work best in the geometric optics range because the phase difference is insensitive to the refractive index and it is a relatively smoothly increasing function of the particle size. Scattering from particles much smaller than the wavelength of light can also be used, but it is sensitive to the refractive index. The intermediate case, in which the particle diameter is of the order of the optical wavelength, is more difficult than the asymptotic extremes. Measurements of submicron particles are possible [99], while particle diameters as large as several hundred microns are accessible in the other extreme.

4.14 LASER MEASUREMENTS OF VELOCITY FIELDS

From the viewpoint of a fluid dynamicist, the ideal instrument for the experimental study of fluid motion should be able to measure the entire three-dimensional vector field $\mathbf{u}(\mathbf{x}, t)$ as a continuous function of position throughout the volume of the flow domain, and as a continuous function of time. When this measurement is performed at an instant of time, we shall refer to it as an instantaneous velocity field measurement or, more briefly, a field measurement. When sequences of field measurements are made in time, we shall refer to the process as measurement of a field history, although field cinematography would also be descriptive. From such measurements, it would be possible to determine, by direct calculation, fundamental fluid-dynamical quantities such as the vorticity, the deformation tensor, the pressure field, and even the pressure gradient. Each of these quantities appears explicitly in the Navier-Stokes equations, and they afford conceptual means of describing and understanding fluid motion that go beyond the limited vocabulary of pointwise velocity measurements. Such measurements would also provide a form of quantitative flow visualization in which the pattern of the instantaneous velocity field could be seen without the ambiguities and difficulty of interpretation that arise when the flow is visualized by conventional, qualitative techniques. The intent is to combine the accuracy of a single-point method such as LDV with the multipoint nature of flow visualization.

Multipoint measurements can be achieved using LDVs, but this approach is inherently limited due to the wide signal bandwidth that is required for LDV. The simplest approach is to physically scan the LDV measurement volume throughout the fluid domain. The total measurement time, being the product of the number of sample points with the time required to make an LDV measurement at each point, is necessarily long, but for sufficiently slow flows, the field can be treated as being approximately frozen during the scan time. For example, Gollub et al. [100] used a scanning LDV to measure one component of velocity in a horizontal plane through a very slowly evolving thermal convection flow. In faster flows, such as air flow, the scan time becomes so short that measurements are restricted to velocity profiles along single lines, as opposed to 2-D domains, cf. [101]. Multipoint LDV probes with up to six measurement volumes have been described by Nakatani et al. [102, 103]. These systems make use of fiber optics, diode lasers, and diode detectors to provide the miniaturization and cost reduction that are required for feasibility. Unlike scanning LDV systems, measurements are made simultaneously, albeit at a relatively small number of points.

Efforts to make simultaneous measurements of large numbers of vectors in 2-D and 3-D domains have generally taken non-Doppler approaches. Three-dimensional vector field measurements using holographic cinematography have been reported by Weinstein et al. [104], and the results of this work are encouraging. Also, Lee et al. [105] have demonstrated the possibility of using nuclear magnetic resonance to measure fluid fields. However, most of the development in this area to date has concentrated on optical techniques to measure velocity fields in planar, 2-D domains using laser light sheets and conventional 2-D imaging. Various names have been given to these methods, including laser speckle velocimetry, particle image velocimetry, particle image displacement velocimetry, pulsed laser velocimetry, and particle tracking velocimetry. All of these methods can be considered to be special cases of a generic type of instrument that we shall call a pulsed light velocimeter (PLV). This instrument consists of a pulsed light source that illuminates small particles (or other markers) in the fluid for short exposure times, and an optical recording medium that records the locations of the particles at each exposure. Marker velocity is inferred from the displacement of the markers $\Delta \mathbf{x}$ that occurs during the known time between exposure, Δt:

$$\mathbf{u}(\mathbf{x},\, t) \doteq \frac{\Delta \mathbf{x}}{\Delta t} \tag{182}$$

The light source can be incoherent, such as a flash lamp or a strobe light, or coherent, such as a chopped argon laser beam or a pulsed ruby or Nd:YAG laser. The illumination can cover an entire volume, if images are to be recorded in three dimensions using stereoscopic imaging or holography, or a planar 2D light sheet if ordinary 2D imaging is to be used. The recording can be done on a variety of media, but electronic video cameras and photographic film dominate. The markers are usually particles in the 1- to 20-μn range, small enough to follow the fluid motion, but large enough to produce good images. However, progress has been made using molecular markers such as photochromic dyes in liquids [106], photochromic aerosols in gases, and molecular fluorescence in high-speed gas flows. (Molecular fluorescence has also been used to measure the mean fields of velocity, temperature, and pressure in certain gases by virtue of the Doppler shift of the illuminating light relative to the moving molecules [107]. This type of spectroscopic field velocimetry is more closely related to LDV than to the marker displacement techniques.

Measurement of the instantaneous velocity field enjoys its most significant applications in flows that are unsteady. These include transient flows such as explosions, starting transients in ducts and channels, bubble growth and collapse, vortex breakup and unsteady aerodynamics, flows with moving boundaries such as water waves, turbomachinery, internal combustion engines, fluid/structure interactions, and research into the instantaneous coherent structures of turbulent flow.

In the following section we shall consider exclusively the class of instruments that uses a double-pulsed light sheet to illuminate small marker particles and then records their images to store their locations at the time of the first and second light pulses.

More complete discussions of pulsed light velocimetry and other techniques that have been used for measurements of fluid velocity fields can be found in the review articles by Dudderar et al. [108], and Adrian [109, 110] and in the book by Merzkirch [111].

4.15 DOUBLE-PULSED PLANAR LASER VELOCIMETRY

4.15.1 System Parameters

A PLV consists of a pulsed laser source, optics to form the laser output into a thin sheet, marker particles that scatter light in all directions, and a camera that is usually located at 90° to the plane of the light sheet and that records the locations of the particles during each exposure from the light sheet (Fig. 4.27). Position in the fluid is denoted by $\mathbf{x}$, and 2-D position in the image plane of the camera is

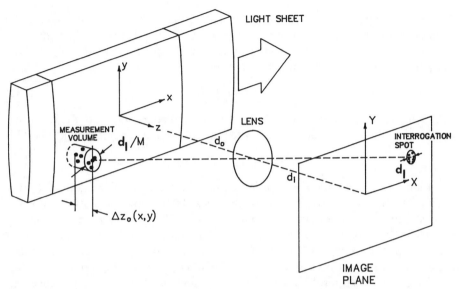

Figure 4.27 Light sheet and image recording system for planar pulsed laser velocimetry

denoted by **X**. Displacements between pulses are denoted by Δx and ΔX, respectively.

The light source need not be a laser if coherent light is not essential, but lasers are generally preferred because of the high brightness of the laser beam, due essentially to its high degree of collimation, and the shortness of the laser pulses. For example, a ruby laser produces 1 Joule (J) of energy in a 20-ns duration pulse with a beam divergence of 3 mradians. The time between pulses Δt is accurate to within better than 10 ns and easily adjustable in steps of 1 μs or less up to values in excess of 1 s; it is usually the most accurate and the most readily changed parameter in the system. The short duration of a laser pulse is capable of freezing the motion of particles moving as fast as 1000 ms^{-1} with less than 20-μm blur.

Typical dimensions of the light sheet are $\Delta z_0 \approx 1$ mm and $\Delta y_0 \approx 100$ mm, depending upon the scale of the flow and the requirements for spatial resolution. Light scattered by each particle in the sheet is imaged by a lens whose magnification will be denoted by $M = d_i / d_0$ (cf. Fig. 4.27). Typically, $0.1 < M < 10$. The observable locations of the particles in the z direction are determined by the combined effects of the thickness of the light sheet, Δz_0, and the depth of the focus of the lens:

$$\delta z = 4(1 + M^{-1})^2 \, (f\#)^2 \lambda \qquad (183)$$

where $f\#$ is the f number of the lens and λ is the wavelength of light.

The character of the recorded image depends critically upon the concentration of the scattering particles C (number per unit volume), their diameters d_p, and the optics of the lens. In the limit of very large particles, the diameter of the image of a particle is simply Md_p. In the limit of very small particles, the diameter of the image is constrained by diffraction to the spot size

$$d_s = 2.44(1 + M)f^\#\lambda \qquad (184)$$

When these effects are coupled with the effect of limited resolution of the recording medium, d_r, an approximate equation for the total diameter of the image is [110]

$$d_\tau = (M^2d_p^2 + d_s^2 + d_r^2)^{1/2} \qquad (185)$$

Film offers much higher resolution than a video camera. For example, Kodak Technical Pan 4415, a favorite film for PLV, has 300 lines/mm resolution (implying at least 2 pixels per line) over a format as large as 100×125 mm, corresponding to $60,000 \times 75,000$ pixels ≈ 4.5 gigapixels. In contrast, the largest video camera array available currently is $2048 \times 2048 = 4$ megapixels, and more common, moderately priced video arrays have 512×512 pixels or 256×256 pixels.

Despite the high resolution of film, video cameras offer such convenience that they are rapidly becoming the medium of choice for PLV. Further, the results obtainable with a good video-based system are comparable to those from 35-mm photographic film, and even 100×125 mm film may not offer substantial improvements in performance if the optics are not optimized to take full advantage of its recording capabilities.

The simplest way to view the difference between photographic PLV and video PLV (sometimes called digital PLV, or DPLV) is to think of the photographic film as an intermediate image storage medium. Without film, the particle images are formed directly in the video camera sensor array. With film, the images, over a possibly wider field of view, are formed and stored on the film, and the video camera subsequently views portions of the film.

4.15.2 Accuracy and Spatial Resolution

The root mean square error of a velocity measurement is given by

$$\left(\frac{\sigma_u}{u^2}\right)^2 = \left(\frac{\sigma_{\Delta x}}{\Delta x^2}\right)^2 + \left(\frac{\sigma_{\Delta t}}{\Delta t^2}\right)^2 \qquad (186)$$

where each σ denotes the rms value of the fluctuation in the quantity denoted by the subscript. Unless the velocity is very large, the uncertainty in time between pulses $\sigma_{\Delta t}$ can be neglected relative to the error caused by uncertainty in the separation between two images, $\sigma_{\Delta x}$. If there are more than 2–4 pixels resolution per particle diameter, the pixelization error is also negligible, and we can assume that

$$\sigma_{\Delta x} = c_\tau d_\tau \qquad (187)$$

where c_τ is a constant that depends upon the procedure used to measure the separation between two images. Then

$$\frac{\sigma_u}{U_{max}} = \frac{c_\tau d_\tau}{\Delta x_{max}} \tag{188}$$

where $\Delta x_{max} = u_{max} \Delta t$ is the maximum displacement caused by u_{max}, the maximum velocity anywhere in the field of view. This relationship can be arranged to look like a statement of Heisenberg's uncertainty principle,

$$\sigma_u \Delta x_{max} = c_\tau u_{max} d_\tau \tag{189}$$

in which the product $c_\tau u_{max} d_\tau$ plays the role of Planck's constant and Δx_{max} bounds the spatial resolution of the instrument. Thus maximum resolution is obtained by minimizing $c_\tau d_\tau$; this can be accomplished by using sophisticated algorithms to minimize the uncertainty in locating the center of each image, e.g., by minimizing c_τ, or by using small particles and high-resolution recording to minimize d_τ.

With current techniques using film, it is possible to achieve $\sigma_u / u_{max} = 0.01$ with $\Delta x_{max} = 0.30$ mm. These values are comparable to the accuracy and spatial resolution afforded by LDV.

4.15.3 Images and Speckle

The concentration of particles determines the type of recorded image and hence the type of image analysis that is required. Fig. 4.28 shows the image plane for concentrations ranging from low to high. In Fig. 4.28a and 4.28b the particles are far enough apart, on average, that their images do not overlap. One then sees isolated images of individual particles. Naturally enough, we call this the particle image mode [112, 113]. When the concentration of particles is large enough, the particle images overlap in the image plane, and if the scattered light is coherent, the overlapping images can interfere with each other. When many images interfere, the resulting pattern is the familiar laser speckle phenomenon (Fig. 4.28c) [111, 114]. This occurs when the source density N_s, defined as the mean number of images per resolution spot size, is large. We find [112]

$$N_s = C\Delta z_0 \frac{\pi d_e^2}{4M^2} \tag{190}$$

where $d_e^2 = d_s^2 + M^2 d_p^2$. The condition for speckle formation is $N_s \gg 1$.

Since the speckle pattern is the interference of light waves scattered from a particular group of particles, the pattern translates with the translation of the group, and the local mean displacement of the group can be inferred from the local mean displacement of the speckle. Thus fluid motion can also be measured when speckle is recorded. We refer to this as the laser speckle mode. Historically, laser speckle velocimetry (LSV) [115–119] for fluids predates the concept of particle

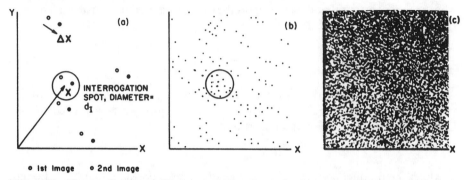

o 1st Image   o 2nd Image

Figure 4.28   Three modes of recording images. (*a*) Low image density occurs when the average number of images per interrogation spot is small. (*b*) High image density occurs when there are many images per interrogation spot. (*c*) Speckle occurs when the concentration of particles is so high that many coherent images of individual particles overlie one another

image velocimetry (PIV) because it originated in studies of solid-surface motion, wherein the light from the surface usually creates speckle. PIV developed in fluid mechanics when it was realized that fluids would seldom contain enough scattering particles to create speckle, and that the most commonly encountered mode would be the particle image mode [112]. Although Meynart [118, 119] referred to the technique as laser speckle in much of his pioneering work, he clearly understood that he was dealing with images.

4.15.4   Image Density in the PIV Mode

Once a field of particle or speckle images is recorded, the challenge is to extract from the photograph the local displacement of the fluid and hence the velocity. With photographs, one way that this is done is to illuminate a small region on the photograph with the beam from a He-Ne laser and to analyze the images within this interrogation spot to determine the displacement ΔX of images within the spot. The procedure is repeated at many different spots until the area of the photograph has been covered. Each spot yields a 2-D vector requiring only two words of memory, which is much more efficient than retaining the entire image of the spot. The dimension of the spot d_I determines the spatial resolution of the measurements. If d_I is 1 mm, a 100 × 125 mm format photo yields 12,500 vectors. The spot may be round, in which case d_I is the diameter, or square, in which d_I is the side dimension. For other shapes, d_I is the characteristic dimension. The specific type of interrogation analysis that is conducted depends upon the mean number of particles within the interrogation spot, a dimensionless quantity that is called the image density [120, 121]:

$$N_{\mathrm{I}} = C\Delta z_0 \frac{\pi d_{\mathrm{I}}^2}{4M^2} \tag{191}$$

(Since d_{I} is the diameter of the spot on the image plane, d_{I}/M is the equivalent diameter in the light sheet.) In general, the probability of finding k particles within the intersection of the interrogation spot with the light sheet is

$$\mathrm{Prob}(k \text{ particles}) = \frac{N_{\mathrm{I}}^k e^{-N_{\mathrm{I}}}}{k!} \tag{192}$$

4.15.5 Interrogation Analysis

4.15.5.1 Low-image-density PIV. In the low-image-density limit of Fig. 4.28a, $N_{\mathrm{I}} \ll 1$. The probabilities of finding one and two particles are N_{I} and $N_{\mathrm{I}}^2/2$, respectively. Hence in this limit, if two images are found in one interrogation spot, the odds are high that both images belong to the same particle. That is, the probability that the two images are the result of two different particles residing in the same spot is small in comparison with the probability that they are the first and second images of one particle lying in the same spot.

These observations make the analysis of low image density very simple. If one finds anything other than two images in an interrogation spot, the spot is ignored, and the interrogation moves to a new spot. If one finds exactly two images in an interrogation spot, one assumes that the images belong to the same particle, and the image displacement $\Delta\mathbf{X}$ is measured directly. From Fig. 4.27 a particle at $\mathbf{x}$ is mapped into an image at -η, where

$$\eta = \frac{d_{\mathrm{i}}}{d_{\mathrm{o}} - z}[x\hat{\mathbf{x}} + y\hat{\mathbf{y}}] \tag{193}$$

Here the origin of the 0x frame lies in the object plane of the camera on the optic axis of the camera lens, $\hat{\mathbf{x}}$ and $\hat{\mathbf{y}}$ are unit vectors, and η is a 2-D vector in the image plane [121, 122]. On the erected image, the particle is located at η, and the image displacement is related to the particle displacement by

$$\Delta\mathbf{X} = M\Delta\mathbf{x} + M\frac{\mathbf{x}}{d_{\mathrm{o}}}\Delta z \tag{194}$$

indicating a dependence of $\Delta\mathbf{X}$ upon the out-of-plane image displacement Δz. This dependence is simply the effect of optical perspective. It can be ignored if the value of $|\mathbf{x}|_{\mathrm{max}}/d_{\mathrm{o}}$ is small, corresponding to paraxial photography. Typically, $|\mathbf{x}|_{\mathrm{max}}/d_{\mathrm{o}}$ is of the order of 0.05–0.1. Perspective effects can be accounted for exactly if one uses stereophotography, or ignored (approximately) if one uses paraxial photography.

Low-image-density PIV has been evaluated as a function of the image density [123]. It was found that bad interrogation data occurred in a small fraction of the measurements because the second image belonged to a different particle than the first image. At small values of N_I, the probability of a bad data vector was small, but then the fraction of interrogation spots that yielded measurements was also small. A compromise figure of $N_I = 0.2$ limited the probability of bad data to approximately 5% of all measurements, while the probability of making any measurement was approximately 0.05 per mm^2 when $d_I = 1$ mm.

Measurements made in a plane passing through the centerline of a submerged water jet are shown in Fig. 4.29. This figure shows the random locations of the data measured by low-image-density PIV, one of its primary disadvantages. The advantage of the low-image-density mode is the simplicity and speed of the interrogation process.

The other main disadvantage of low-image-density PIV is the relatively high incidence of bad data vectors. Bad data can be removed by a postinterrogation process that is done automatically by the computer. The steps in the postinterrogation procedure are shown in Fig. 4.30, wherein the bad vectors in Fig. 4.30a are identified by comparing them to their nearest neighbors and rejecting them if they differ too greatly. The rejected vectors (Fig. 4.30b) are subtracted from the original data (Fig. 4.30b) to produce a field of improved reliability. Note, however, that certain vectors that appear to be erroneous were not rejected because they did not have enough neighbors to make a valid comparison. Manual evaluation by a skilled operator can improve results still further, but the extent of the improvement depends critically upon having a sufficient number of good vectors per unit area.

It is often convenient to interpolate the randomly located vector measurements onto a uniform grid of data. This procedure is essential if derivatives of the field are to be calculated, but it is susceptible to abuse. One should not expect to see more detail in the interpolated results than was available in the random data, and smoothness of the interpolated field is no guarantee of the interpolation's reliability, especially if the interpolation numerically enforces fluid continuity. It is clear from Fig. 4.30 that any substantial increase in the small-scale activity of the flow would be unresolvable by low-image-density techniques and that increasing the data density is a high priority.

4.15.5.2 Particle tracking velocimetry. One method of measuring particle displacements in the PIV technique is particle tracking velocimetry (PTV), in which each particle image is analyzed to locate its center, and the centers of (possibly multiple pulse) image tracks are connected to determine the particle's trajectory and velocity [125–127]. PTV is usually applied to low-image-density fields in which the number of images is not so large as to make image-by-image analysis prohibitively time consuming, or to introduce ambiguities due to crossing trajectories. If the image density increases, simple tracking methods fail because they are unable to successfully associate images with the correct particles. Hence, PTV and low-image-

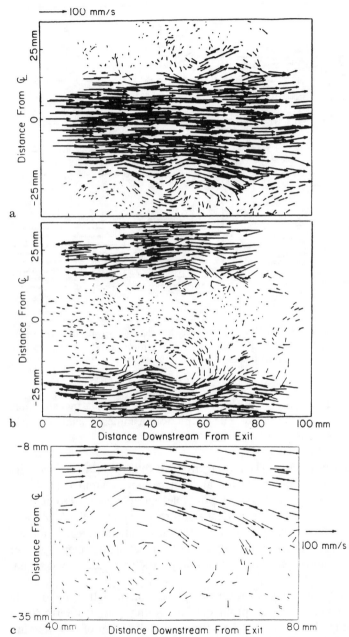

Figure 4.29 Vector maps of a 36-mm-diameter axisymmetric water jet measured by low-image-density PIV. (*a*) Stationary reference frame. (*b*) Reference frame moving at 92% of the centerline velocity. (*c*) Enlargement of the shear layer vortex in Fig. 4.29*b*. (From [123])

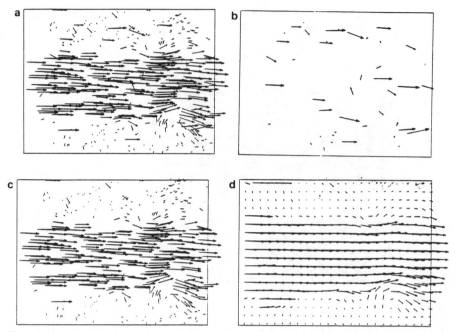

Figure 4.30 Postinterrogation processing of low-image-density PIV data: (*a*) original data, (*b*) errone-ous vectors identified by computer, (*c*) vector field after removal of erroneous vectors, (*d*) vector field interpolated onto a uniform grid. (From [124])

density PIV are often used synonymously, whereas in fact, one is a measurement method and the other is a type of image. The distinction becomes crucial in certain advanced analysis techniques.

4.15.5.3 High-image-density PIV. The aforementioned difficulties associated with low image density can be overcome in the high-image-density limit (Fig. 4.28*b*) because each interrogation spot contains, in principle, enough particles to produce a valid velocity measurement. Both experience and theoretical studies [128] indi-cate that $N_I > 10$–20 provides excellent results if the maximum displacement is less than $0.25d_I$ and the fluid velocity gradients do not create displacements greater than d_I. Smaller values are satisfactory if multiple pulses are used, provided that gradients are very small.

The price paid for the richness of the image information in high-image-density PIV is complexity of its analysis. The problem of analyzing a single interrogation spot is illustrated in Fig. 4.31, in which the pairings and direction of flow that are made obvious in Fig. 4.31*a* are not at all obvious in Fig. 4.31*b*. Clearly, given data in the form portrayed in Fig. 4.31*b,* some form of statistical analysis is needed to determine the correct pairings.

○ **1st image**

● **2nd image**

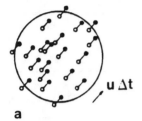 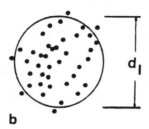

a   b

Figure 4.31 Analysis of images in a high-image-density interrogation spot. (*a*) Pairs of images are identified by a connecting line, and first images are marked. (*b*) The pattern from Fig. 4.31*a* as it appears on a double-exposure photograph. (From [109])

Statistical analysis can be performed in the image plane of the interrogation spot or in its Fourier transform plane (Fig. 4.32). The latter can be obtained from the former by digitized Fourier transform or by optical Fourier transform using a lens as the transform element, as in Fig. 4.32. In the image-plane approach, the objective of the analysis is to find a vector displacement that translates the first images of the particles such that they coincide with the second images. This can be accomplished by performing a two-dimensional spatial correlation of $I(\mathbf{X})$, the image intensity of the interrogation spot [120, 121, 129]. Figure 4.33*a* illustrates such a spot. The correlation is defined by

$$R(\mathbf{s}) = \int_{\text{spot}} I(\mathbf{X})I(\mathbf{X} + \mathbf{s})\,d\mathbf{X} \qquad (195)$$

where $\mathbf{s}$ is the separation vector. The spatial correlation of the image field in Fig. 4.33*a* is given by Fig. 4.33*c*. It contains a large peak centered at $\mathbf{s} = 0$ that corresponds to each image correlating with itself, and two smaller peaks centered at the mean image displacements $\pm\Delta\mathbf{X}$. They correspond to the first images correlating with the second image, and vice versa.

The image displacement can be found by locating the centroid μ_{D+} of the positive displacement correlation peaks [121]. (The correlation is reflectionally symmetric, $R(\mathbf{s}) = R(-\mathbf{s})$, so the other peak at $\mu_{D-} = -\mu_{D+}$ is redundant). The centroidal velocity is given by

$$\boldsymbol{\mu}_1 = \frac{\boldsymbol{\mu}_{D+}}{M\Delta t} \qquad (196)$$

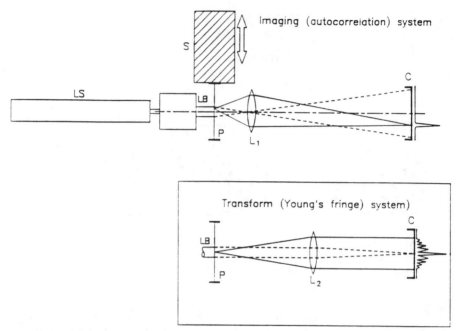

Figure 4.32 Two-dimensional analysis of an interrogation spot in the image plane, and in the Fourier transform plane. (From [130])

assuming paraxial recording with magnification M [cf. Eq. (194)]. The symbol μ_1 is introduced to emphasize that it is an *estimate* of $\mathbf{u}(\mathbf{x}, \mathbf{t})$.

It has been shown that the expected value of the centroidal velocity given by Eq. (196) corresponds to a volume average of the velocity in the intersection of the interrogation spot and the laser light sheet [121]. The weighting function is determined by the intensity distribution of the interrogation beam, the intensity distribution of the light in the illuminating sheets, and the displacement of the images relative to the dimensions of the measurement volume. The dependence upon velocity implies a statistical bias toward low velocities. Fortunately, this bias is small (~1%) unless the velocity differences within the measurement volume are very large [126]. This function defines the measurement volume. If I_I and I_{0I} were top-hat functions, the measurement volume would simply be their geometric intersection. When I_I tapers smoothly to zero, the mv tapers similarly, much like the LDV mv.

An expanded view of a sample correlation function shows that it also contains many low-level maxima corresponding to randomly overlapping particle images at various separations $\mathbf{s}$ (see Fig. 4.33). Successful measurement of the vector displacement is achieved if the peaks associated with the first and second image correlations are taller than any of the noise peaks. Statistically, the noise peaks can occasionally exceed the signal peaks, in which case a spurious measurement is

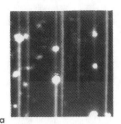

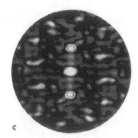

a   b   c

Figure 4.33 (*a*) Interrogation spot image field 1 mm × 1 mm. (*b*) Modulus of the 2-D Fourier transform of the image field in Fig. 4.33*a*. (*c*) Fourier transform of Fig. 4.33*b*, corresponding to the 2-D correlation function of Fig. 4.33*a*

made, unless error correction procedures are employed. A numerical simulation of these effects shows how the probability that the interrogation gives a valid vector and the accuracy of the vector depend upon image density, velocity gradients, and the maximum allowed displacement [128].

The alternative to 2-D spatial correlation is analysis in the Fourier transform plane [115–119]. Two-dimensional Fourier transformation of $I(\mathbf{X})$ in Fig. 4.33*a* yields the pattern of bright and dark Young's fringes in Fig. 4.33*b*. The fringes are perpendicular to the mean vector displacement, and their spacing is inversely proportional to the displacement magnitude. Thus $\Delta\mathbf{X}$ can be inferred from the fringes by measuring their orientation and their wavelength. Numerous methods have been proposed for this purpose [131–135].

Instead of calculating Young's fringes by 2-D FFT, it is possible to form them optically by placing a lens one focal length away from the photograph. Then each particle image on the photograph acts as a small source whose light is collimated by the lens into a nearly parallel beam. Each pair of images produces two light waves that interfere with each other to produce a fringe pattern, exactly as in LDV. If N particle images exist in the interrogation spot, the number of fringe patterns will be the combination of N things taken two at-a-time, or $N^2/2$. Most of these fringe patterns are oriented randomly, with random wavelength, but N_p of them will be the same, where N_p represents the number of pairs of images that belong to the same particle. If there are N_{p_0} particles in the interrogation spot at the time of the first exposure, the number that remain in the spot after the second exposure will be N_p, where N_p is necessarily less than N_{p_0} because some particles will be convected out of the measurement volume.

Historically, the first form of PIV/LSV analysis Fourier transformed the image optically, digitized the intensity of Young's fringes, and analyzed the fringes numerically. Numerous methods have been proposed for this purpose [131–135]. The most natural and the most powerful method of analyzing Young's fringe pattern is to Fourier transform it. This results in a pattern that is very similar to the correlation function [128] (Fig. 4.34).

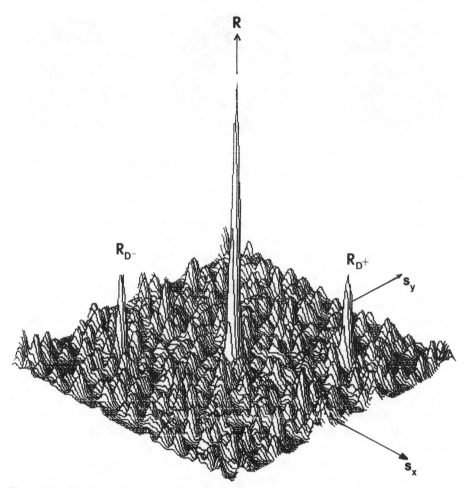

Figure 4.34 Correlation function for the transmitted intensity. (From [128])

4.15.6 Automated Analysis

Figure 4.35 shows details of a typical computer-controlled system for automatically analyzing double-pulsed PIV photographs to obtain 2-D vector fields. The photograph is mounted in a digitally controlled X-Y table, and it is illuminated by light from a He-Ne laser, which defines the interrogation spot. If the system were video based instead of photographic, everything would remain the same except that the X-Y table and the light source would be removed and the video camera would view the flow directly. The interrogation spots would be determined in software and they would frequently correspond to rectangles with top-hat resolution.

Depending upon the application, the interrogation spots are either Fourier

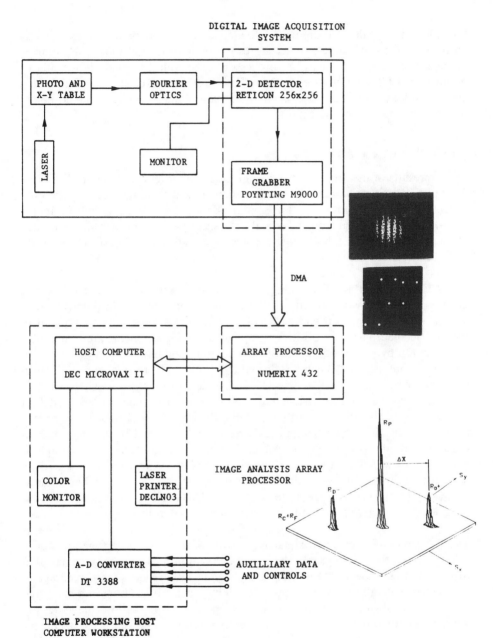

Figure 4.35 Computer-controlled interrogation system

transformed optically and digitized by video camera, or they are imaged directly onto the camera. A frame grabber stores the image and transmits it to an array processor for analysis. The array processor makes full 2-D analysis feasible by performing FFT computations quickly. The host computer stores, manipulates, and scans the vector data. After each interrogation, it also controls the movement of the X-Y table to a new location.

4.15.7 Results and Applications

Sample results of measurements in a submerged water jet are shown in Fig. 4.36. A 27-mm-diameter nozzle is located four diameters above a horizontal plate onto which a 28.8-cm/s jet impinges at normal incidence and spreads radially along the wall. Data are obtained from a vertical light sheet passing through the centerline of the jet. The photograph was interrogated using the system described above and 2-D autocorrelation analysis.

4.15.7.1 Image shifting. The ambiguity in the direction of flow that exists in a double-pulsed PIV was resolved in this experiment by an image-shifting technique, in which the image field is shifted between exposures, so that one component of the displacement is always positive [136]. Negative displacements are found after interrogation by subtracting the artificial shift. The technique is analogous to frequency shifting in LDV. It can be effected by photography through a rotating mirror [123] or by electro-optic means [137]. The electro-optic system shown in Fig. 4.37 uses a specially cut birefringent crystal in front of the camera lens to deflect the light rays scattered from particles, so that the image of a particle is not shifted if the scattered light is horizontally polarized, and is shifted by an amount $\mathbf{X}_s$ if the scattered light is vertically polarized. The polarization of the scattered light is controlled by switching the linear polarization of the light from the laser from horizontal to vertical between the first and second pulses. Since switching is done by a pockel cell, very fast operation is possible. This device is applicable when the particles do not cause a loss of polarization upon scattering.

4.15.7.2 Validation. In approximately 5–10% of the first interrogations of a double-pulsed PIV photograph, the tallest correlation peak is a noise peak rather than a true signal peak. (Multiple pulsing significantly reduces the percentage of bad data.) Generally, the noise peak is randomly located, resulting in a measured vector that is obviously different from its neighbors. As in the low-image-density analysis, these vectors can be removed, but with high-image-density data the removal is much easier, and it can be done with greater confidence. We refer to this as post interrogation validation. It should be noted that the invalid vectors can also be removed during interrogation by applying a validation criterion that calls for the tallest autocorrelation peak to be larger than the next tallest peak by a prescribed ratio. If this ratio, called the detectability, is high, only the interrogation

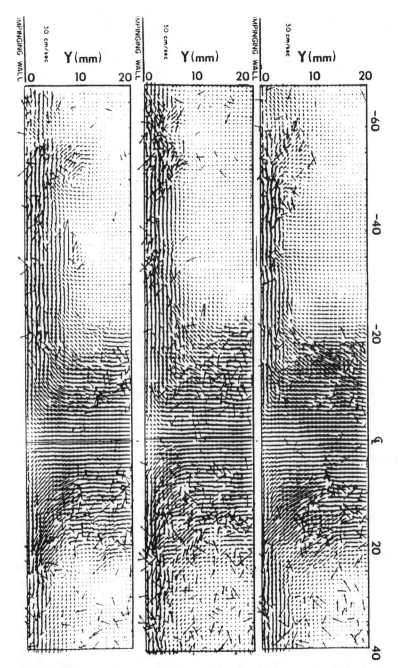

Figure 4.36 Vector maps of impinging jet flow obtained from correlation analysis. The spacing is 0.7 mm on a 143 × 26 node grid. Approximately 5% of the raw measurements are bad. (From [130])

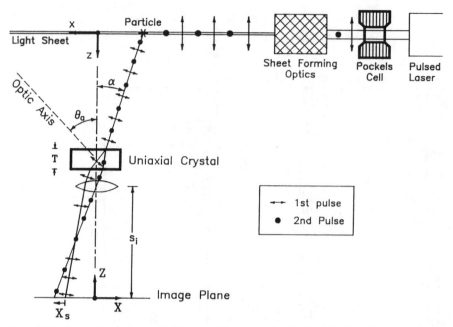

Figure 4.37   System for electro-optical image shifting to resolve the directional ambiguity. (From [137])

spots with very strong signals and weak noise peaks will be accepted. A concomitant side effect of setting a strong criterion for data validation is a reduction in the probability of successful interrogation. However, with a proper choice of parameters, excellent results can be achieved [128]. Depending upon the characteristics of the flow, it is useful to have the capability to adjust the parameters in a validation method and to be able to perform validation both during and after the interrogation. Some manual intervention by an operator is sometimes called for as a final step in perfecting the data.

The data in Fig. 4.38 were found from the raw interrogation data by postinterrogation data removal, followed by interpolation to replace deleted vectors and a small amount of low-pass spatial filtering to reject high-frequency noise in the field, in preparation for differentiation of the data. Vortices can be readily seen in the shear layer surrounding the core of the jet and in the region above the wall jet boundary layer. The latter vortices are responsible for an unsteady flow separation that occurs at a radius of approximately 55 mm. The accuracy of the vector measurements is estimated to be better than 1% of full-scale velocity.

The data in Fig. 4.38 are sufficiently dense and accurate to permit spatial differentiation to extract fundamental flow information. Three vorticity fields corresponding to the total velocity fields in Fig. 4.38 are presented in Fig. 4.39. The grey-tone map is scaled according to the tone bar under the plots. Figure 4.39*d*

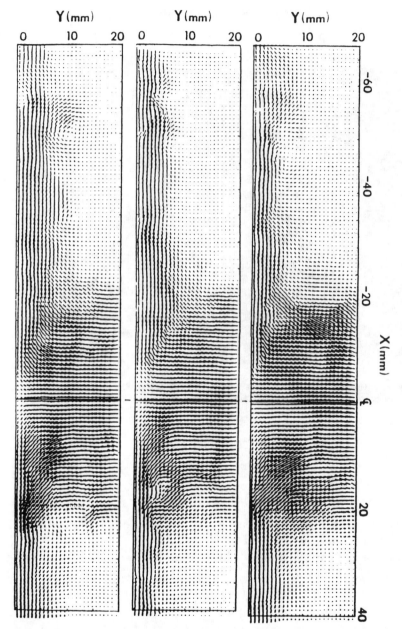

Figure 4.38 Vector maps obtained from the raw data in Fig. 4.36 by postinterrogation data validation, interpolation over missing data, and spatial filtering. (From [130])

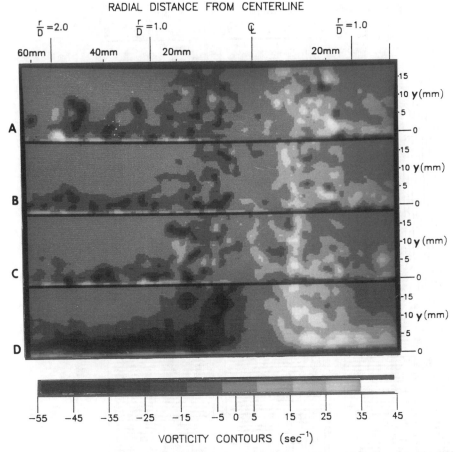

RADIAL DISTANCE FROM CENTERLINE

$\frac{r}{D}=2.0$ $\frac{r}{D}=1.0$ ₵ $\frac{r}{D}=1.0$

60mm   40mm   20mm   20mm

−55   −45   −35   −25   −15   −5 0 5   15   25   35   45

VORTICITY CONTOURS (sec⁻¹)

Figure 4.39 Impinging jet. (*a–c*) Contours of the vorticity of the total vector fields shown in Fig. 4.38. (*d*) Contours of the mean vorticity

is the average vorticity found by ensemble averaging the vorticity fields from 11 photographs taken at widely separated times.

The deformation tensor e_{ij} is defined by

$$e_{ij} = \frac{1}{2}\left(\frac{\partial u_i}{\partial x_j} + \frac{\partial u_j}{\partial x_i}\right) \tag{197}$$

From the u_1 and u_2 data in Fig. 4.38, it is possible to evaluate the mean values of e_{11}, e_{22}, e_{33}, and e_{12}. The results in Fig. 4.40 were calculated from the ensemble-averaged mean velocity field described above [138]. The e_{11} and e_{22} contours indicate high strain rates in the region around the stagnation point, as expected, and

RADIAL DISTANCE FROM CENTERLINE

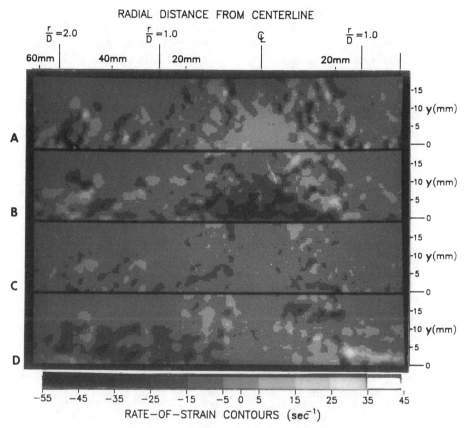

Figure 4.40 Contours of the rate-of-strain components for the instantaneous total velocity field whose vorticity field is shown in Fig. 4.39a: (a) e_{11}, (b) e_{22}, (c) e_{33}, and (d) e_{12}

the e_{33} component indicates some out-of-plane motion. The e_{12} shear component is generally high where the vorticity is high.

4.16 APPLICATIONS OF LDV AND PIV

PIV has been applied to a number of flows including wind-tunnel flows [139], gas flow in an internal combustion engine cylinder [140], laminar flames, low-Re wakes behind cylinders in water [141], and waves and turbulent channel flows. Several other applications are reviewed in the literature [110, 142, 143]. At the present time, most applications involve relatively small scale laboratory systems, but as with LDV, one can expect the range of application to broaden substantially as the technique matures.

Laser-Doppler velocimetry has been applied to a truly impressive range of fluid flows, covering virtually the entire field of fluid mechanics, far too numerous to list in detail. The interested reader can find many examples in the proceedings of international symposia and workshops on the applications of laser velocimetry [144–151], and in a collection of milestone papers on LDV [152].

NOMENCLATURE

$a(\mathbf{x})$	amplitude distribution of Doppler burst, Eq. (29)
A	cross-sectional area of measurement volume
$\langle A \mid B \rangle$	conditional average of A, given the value of B
$A(t)$	amplitude envelope of high-burst-density signal, Eq. (127)
B	number of analog-to-digital conversion bits
$B(t)$	burst indicator
c	total velocity variable in the velocity probability density function, Eq. (88)
C_a	anode capacitance
$C(\mathbf{x}, t)$	number density of all particles per unit volume, Eq. (96)
$C(\mathbf{x}, t, D)$	number density of D-type particles per unit volume
d_e	diameter of image light field
$d_{e^{-2}}$	e^{-2} diameter of focused Gaussian illuminating beam, Eq. (34)
d_f	fringe spacing, Eq. (49)
d_i	image distance
d_o	object distance
d_p	particle diameter
d_r	resolution of film or video camera
d_s	diameter of diffraction-limited spot, Eq. (184)
d_m, l_m, h_m	e^{-2} dimensions of the laser-Doppler velocimetry measurement volume, Eqs. (41)–(43)
D	Doppler signal-amplitude factor, Eq. (26b)
D_a	aperture diameter
D_c	coherence diameter
$D_{e^{-2}}$	e^{-2} diameter of unfocused illuminating beam
D_p	aerodynamic diameter of a particle
δD	range of Doppler signal-amplitude factor
e	photodetector output voltage
e_c	clipped photodetector voltage
e_n	shot-noise voltage
e_N	total-noise voltage signal
$\mathbf{E}$	electric vector
f	focal length; frequency
f_c	collecting-aperture distance from measurement volume; low-pass cutoff frequency

f_C	center frequency
f_D	Doppler frequency
f_M	frequency of voltage-controlled oscillator
f_u	probability density for **u**, Eq. (88)
$f(\mathbf{x'}, t', \mathbf{x}, t)$	probability density for particle position at t', given it was at **x** at t, Eq. (88)
f_{3-dB}	frequency response of particles
f_ϕ	probability density for $\dot{\phi}$
$f\#$	focal length/aperature diameter
Δf	bandwidth of the photodetector/filter system
$F(t), G(t)$	amplitude functions, Eq. (129)
g	internal gain of a photomultiplier tube
$g(\mathbf{x}, t, D)$	particle presence indicates function, Eq. (68)
G	scattering gain, Eq. (65)
h	Planck's constant
$h(t)$	impulse response function of photomultiplier tube plus filters
$h_f(t)$	impulse response of the filter system after the photomultiplier tube
$h_p(t)$	impulse response of the combined photomultiplier tube/filter system
H	transfer function of filter, Eq. (143a); Heaviside function
I	intensity
I_{00}	fringe intensity, Eq. (48)
j	$= \sqrt{-1}$
J	light flux
k	wave number, $= 2\pi / \lambda$; integer number; number of photoemissions; Boltzmann's constant, Eq. (112)
K	laser-Doppler velocimeter sensitivity vector, Eq. (14)
K_c	coherence factor
l_x, l_y, l_z	dimension of measurement volume defined by threshold criterion, Eq. (46)
m	refractive index ratio; integer
M	number of samples, Eq. (149); magnification of camera lens
n	integer; refractive index of fluid
n_p	refractive index of particle; total number of particles of fluid
$n(t, t + \Delta t)$	number of photon pulses in $(t, t + \Delta t)$, Eq. (152)
$n(V, t, D)$	number of D-type particles in volume V, Eq. (71)
N	mean number of particles, Eq. (77); integer value; number of cycles; random number of images in an interrogation spot
$\dot{N}$	data rate, Eq. (159)
N_B	number of cycles in a burst
N_D	mean number of particles in e^{-2} measurement volume, Eq. (52)
N_e	mean effective number of particles in measurement volume, Eq. (121)

N_{FR}	number of fringes in measurement volume, Eq. (51)
N_I	image density, Eq. (191)
N_s	source density, Eq. (190)
$N(V, t, D)$	mean number of D-type particles in V at t, Eq. (74a)
$\mathfrak{n}$	total number of particles in the entire fluid volume
$\hat{\mathbf{p}}$	polarization direction (unit vector)
p_0	probability, Eq. (72)
P	laser beam power; pedestal amplitude factor, Eq. (26b)
q	clipping level, Eq. (155); quantum of electron charge
r	$= \mid \mathbf{r} \mid$
$\mathbf{r}$	position in far field
R	autocorrelation of interrogation spot image intensity, Eq. (195)
R_L	load resistance
R_{xx}	autocorrelation of variable $\mathbf{x}$, cf. Eq. (91)
$\mathbf{s}$	position in correlation plane
$\hat{\mathbf{s}}$	propagation direction
s_0, s_1	distance in direction of propagation
S_x	power spectrum of variable $\mathbf{x}$, cf. Eq. (92); photodetector sensitivity
SNR	signal-to-noise power ratio, Eq. (64a)
t	time
t_j	random emission time
t_1	particle time constant, Eq. (66)
Δt	time between events; time between light pulses
T	averaging time; absolute fluid temperature, Eq. (112)
T_T	Taylor microscale for time, Eq. (156)
u_o	observed velocity
$\mathbf{u}'$	fluctuating velocity vector, $= (u', v', w')$
$\mathbf{u}(\mathbf{x}, t)$	total velocity vector, $= (u, v, w)$
$\mathbf{U}$	mean velocity vector, $= (U, V, W)$
$\mathbf{v}(t)$	total Lagrangian velocity, $= \mathbf{u}[\mathbf{x}(t), t]$
V	visibility, Eq. (30); volume
$\overline{V}$	peak visibility, Eq. (31)
V_D	volume of e^{-2} measurement volume
V_m	material volume
V_0	constant velocity
ΔV	overlap volume
$w(\mathbf{x})$	measurement volume indicator function for all particles, Eq. (173)
$w(\mathbf{x}, D)$	measurement volume indicator function for D-type particles, Eq. (166)
$\mathbf{x}$	position
$\mathbf{x}(t)$	particle trajectory
$\mathbf{x}_m(t)$	mean effective displacement in high-burst-density signal, Eq. (126)

$\Delta \mathbf{x}$	displacement of particle between light pulses
$\mathbf{X}$	position in the image plane
$\mathbf{X}_s$	displacement caused by image shifting
$\Delta \mathbf{X}$	displacement of particle image between light pulses, $= \Delta X, \Delta Y$
$\Delta y_0, \Delta z_0$	height and thickness of laser light sheet
α	circular frequency of ambiguity noise spectrum
α_c	$= 2\pi f_c$
β	factor in Eq. (119)
$\delta(\ )$	Dirac delta function
δz	depth of focus, Eq. (183)
Δ	comparison accuracy tolerance in a counter
$\dot{\varepsilon}$	mean emission rate, Eq. (55)
$\dot{\varepsilon}(t)$	conditional mean emission rate given $J_{\text{tot}}(t)$, Eq. (57)
ζ	radial distance from centerline of laser beam
η	location of particle image in the image plane, Eq. (193)
η_q	quantum efficiency, Eq. (55)
$\theta_{e^{-2}}$	e^{-2} divergence angle of a Gaussian beam
θ_{12}	angle between two sources
κ	half-angle between illuminating beams
λ	wavelength of light
λ_x	Taylor microscale in x direction
$\lambda(\mathbf{x}, t, D)$	particle density function in x-D space, Eq. (72)
Λ	triangle function, Eq. (153)
μ_{D+}	centroidal location of the positive displacement peak in the correlation plane
μ_f	dynamic viscosity of fluid
μ_1	mean frequency of power spectrum, Eq. (102a)
μ_2	bandwidth of power spectrum, Eq. (103a)
ν	frequency (Hz)
ν_0	laser frequency (Hz)
ξ	dummy variable
π	constant, $= 3.1416$
ρ_p	particle density
σ	scattering coefficient, Eq. (3)
σ_u	root-mean-square uncertainty in measurement of $\mathbf{u}$
τ	time delay
τ_B	burst time
τ_f	time constant of filter system after the photomultiplier tube
τ_h	time constant of combined photomultiplier tube/filter system, Eq. (59)
τ_N	time for N cycles
$\tau_{N/2}$	time for $N/2$ cycles
τ_p	time constant of photomultiplier tube anode pulses

$\Delta\tau$	time-delay increment
ϕ	phase noise, Eq. (128)
$\dot{\phi}$	ambiguity noise
Φ	phase (radians)
$\dot{\Phi}$	instantaneous frequency
Φ_{J_D}	characteristic function of J_D, Eq. (115)
Ψ	phase change
ω	circular frequency (radians/s)
ω_c	circular cutoff frequency, $= 2\pi f_c$, Eq. (144)
$\Delta\omega$	frequency bandwidth, Eq. (140a)
$\Delta\omega_A$	ambiguity bandwidth, Eqs. (100), (101), (108a)
$\Delta\omega_B$	bandwidth due to Brownian motion, Eq. (112)
$\Delta\omega_C$	captive bandwidth of a phase-locked loop or a frequency-locked loop
$\Delta\omega_f$	spectrum-analyzer filter bandwidth
$\Delta\omega_G$	bandwidth due to mean gradient broadening, Eq. (109a)
$\Delta\omega_T$	bandwidth due to turbulent fluctuations, Eq. (110a)
$\Delta\omega_0$	bandwidth due to laser source, Eq. (113)
Ω	solid angle subtended by collecting aperture
Ω_1, Ω_2	argument variables in characteristic function

Subscripts

B	background radiation
D	Doppler signal
e^{-2}	quantity evaluated at e^{-2} points
E	electronics noise
H	heterodyne
i, j	particles, emission times, arrival times, or other random events
J_D	Doppler light flux
J_P	pedestal light flux
li	wave scattered from lth illuminating beam by ith particle
m	measured quantity (high-burst-density signals); measurement volume
max	maximum value
min	minimum detectable light flux
n	photon count
o	observed quantity
O	illuminating beam
P	pedestal quantity
peak	peak value, occurs when $\mathbf{x} = \mathbf{0}$
R	reference beam
RF	radio frequency
S	frequency shift; source

tot	total light flux
x, y, z	direction
$0l$	lth illuminating beam

Superscripts

~	filtered
∧	unit vector
•	time derivative
—	time averaged
~	volume averaged
*	complex conjugate; normalized value

REFERENCES

1. H. Z. Cummins, N. Knable, and Y. Yeh, Observation of Diffusion Broadening of Rayleigh Scattered Light, *Phys. Rev. Lett.,* vol. 12, pp. 150–153, 1964.
2. Y. Yeh and H. Z. Cummins, Localized Fluid Flow Measurements with an He-Ne Laser Spectrometer, *Appl. Phys. Lett.,* vol. 4, pp. 176–178, 1964.
3. D. K. Kreid, Measurements of the Developing Laminar Flow in a Square Duct: An Application of the Laser-Doppler Flow Meter, M.S. thesis, University of Minnesota, Minneapolis, 1966.
4. R. J. Goldstein and D. K. Kreid, Measurement of Laminar Flow Development in a Square Duct Using a Laser Doppler Flowmeter, *J. Appl. Mech.,* vol. 34, pp. 813–817, 1967.
5. J. W. Foreman Jr., R. D. Lewis, J. R. Thornton, and H. J. Watson, Laser Doppler Velocimeter for Measurement of Localized Flow Velocities in Liquids, *IEEE Proc.,* vol. 54, pp. 424–425, 1966.
6. J. W. Foreman Jr., E. W. George, and R. D. Lewis, Measurement of Localized Flow Velocities in Gases with a Laser-Doppler Flowmeter, *Appl. Phys. Lett.,* vol. 7, pp. 77–80, 1965.
7. R. N. James, Application of a Laser-Doppler Technique to the Measurement of Particle Velocity in Gas-Particle Two-Phase Flow, Ph.D. thesis, Stanford University, Stanford, Calif., 1966.
8. R. J. Goldstein and W. F. Hagen, Turbulent Flow Measurements Utilizing the Doppler Shift of Scattered Laser Radiation, *Phys. Fluids,* vol. 10, pp. 1349–1352, 1967.
9. E. Rolfe and R. M. Huffaker, Laser Doppler Velocity Instrument for Wind Tunnel Turbulence and Velocity Measurements, NASA Rep. N68-18099, 1967.
10. F. Durst and M. Zaré, Bibliography of Laser-Doppler Anemometry Literature, Sonderforschungsberich 80, University of Karlsruhe, 1974.
11. F. Durst, A. Melling, and J. H. Whitelaw, *Principles and Practice of Laser-Doppler Anemometry,* Academic, New York, 1976.
12. C. A. Greated and T. S. Durrani, *Laser Systems in Flow Measurement,* Plenum, New York, 1977.
13. B. Chu, *Laser Light Scattering,* pp. 271–290, Academic, New York, 1974.
14. B. S. Rinkevichius, *Laser Anemometry,* Energy Publishing House, Moscow, 1979 (in Russian).
15. B. M. Watrasiewicz and M. J. Rudd, *Laser Doppler Measurements,* Butterworth, London, 1976.
16. H. D. Thompson and W. H. Stevenson (eds.), *The Use of the Laser Doppler Velocimeter for Flow Measurements, Proc. of the First Int. Workshop on Laser Velocimetry, 1972,* Project Squid Headquarters, Jet Propulsion Center, Purdue University, West Lafayette, Ind., 1972.
17. H. D. Thompson and W. H. Stevenson (eds.), *Proc. of the Second Int. Workshop on Laser Velocimetry, 1974,* Engineering Experiment Station, Purdue University, West Lafayette, Ind., 1974.
18. H. Z. Cummins and E. R. Pike (eds.), *Photon Correlation and Light Beating Spectroscopy, Proc. of NATO Advanced Study Inst., 1973,* Plenum, New York, 1974.
19. E. R. G. Eckert (ed.), *Minnesota Symp. on Laser Anemometry Proc., 1975,* University of Minnesota, Department of Conferences, Minneapolis, 1976.

20. P. Buchave, J. M. Delhaye, F. Durst, W. K. George Jr., K. Refslund, and J. H. Whitelaw (eds.), *The Accuracy of Flow Measurements by Laser Doppler Methods, Proc. of the LDA-Symposium, 1974,* Hemisphere, Washington, D.C., 1977.
21. H. Z. Cummins and E. R. Pike (eds.), *Photon Correlation Spectroscopy and Velocimetry, Proc. of NATO Advanced Study Inst., 1976,* Plenum, New York, 1977.
22. H. J. Pfeifer and J. Haertig (eds.), *Applications of Non-Intrusive Instrumentation in Flow Research,* AGARD CP-193, 1976.
23. L. S. G. Kovasznay, A. Favre, P. Buchave, L. Fulachier, and B. W. Hansen (eds.), *Proc. of the Dynamic Flow Conf. 1978 on Dynamic Measurements in Unsteady Flows,* Skovlunde, Denmark, 1978.
24. H. D. Thompson and W. H. Stevenson (eds.), *Laser Velocimetry and Particle Sizing, Proc. of the Third Int. Workshop on Laser Velocimetry, 1978,* Hemisphere, Washington, D.C., 1979.
25. M. Kerker, *The Scattering of Light,* chap. 3, Academic, New York, 1969.
26. D. A. Jackson and D. M. Paul, Measurement of Hypersonic Velocities and Turbulence by Direct Spectral Analysis of Doppler Shifted Laser Light, *Phys. Lett.,* vol. 32, p. 77, 1970.
27. D. A. Jackson and D. M. Paul, Measurement of Supersonic Velocity and Turbulence by Laser Anemometry, *J. Phys. E,* vol. 4, pp. 173–176, 1970.
28. R. V. Edwards, J. C. Angus, M. J. French, and J. W. Dunning Jr., Spectral Analysis of the Signal from the Laser Doppler Flowmeter: Time-Independent Systems, *J. Appl. Phys.,* vol. 42, pp. 837–850, 1971.
29. L. E. Drain and B. C. Moss, The Frequency Shifting of Light by Electro-Optic Techniques, *Opto-electronics J.,* vol. 4, pp. 429–436, 1972.
30. E. I. Gordon, A Review of Acoustooptical Deflection and Modulation Devices, *Proc. IEEE,* vol. 54, pp. 1391–1401, 1966.
31. T. Suzuki and R. Hioki, Translation of Light Frequency by a Moving Grating, *J. Opt. Soc. Am.,* vol. 57, pp. 1551–1552, 1967.
32. M. Born and E. Wolf, *Principles of Optics,* Pergamon, New York, 1959.
33. R. J. Adrian and W. L. Earley, Evaluation of LDV Performance Using Mie Scattering Theory, pp. 426–454 in [19], 1976.
34. R. J. Adrian and K. L. Orloff, Laser Anemometer Signals: Visibility Characteristics and Application to Particle Sizing, *Appl. Opt.,* vol. 16, pp. 677–684, 1977.
35. H. Kogelnik and T. Li, Laser Beams and Resonators, *Appl. Opt.,* vol. 5, pp. 1550–1567, 1966.
36. H. Weichel and L. S. Pedrotti, A Summary of Useful Laser Equation—An LIA Report, *Electro-Opt. Sys. Des.,* pp. 22–36, July 1976.
37. F. Durst and W. H. Stevenson, Properties of Focused Laser Beams and the Influence on Optical Anemometer Signals, pp. 371–388 in [19], 1976.
38. R. J. Adrian and R. J. Goldstein, Analysis of a Laser Doppler Anemometer, *J. Phys. E,* vol. 4, pp. 505–511, 1971.
39. M. J. Rudd, A New Theoretical Model for the Laser Doppler Meter, *J. Phys. E,* vol. 2, pp. 723–726, 1969.
40. D. B. Brayton, Small Particle Signal Characteristics of a Dual Scatter Laser Velocimeter, *Appl. Opt.,* vol. 13, pp. 2346–2351, 1974.
41. W. M. Farmer, Measurement of Particle Size, Number Density and Velocity Using a Laser Interferometer, *Appl. Opt.,* vol. 11, pp. 2603–2609, 1972.
42. D. M. Robinson and W. P. Chu, Diffraction Analysis of Doppler Signal Characteristics for a Cross Beam Laser Doppler Velocimeter, *Appl. Opt.,* vol. 14, pp. 2177–2181, 1975.
43. D. W. Roberds, Particle Sizing Using Laser Interferometry, *Appl. Opt.,* vol. 16, pp. 1861–1865, 1977.
44. L. E. Drain, Coherent and Non-Coherent Methods in Doppler Optical Beat Velocity Measurement, *J. Phys. D,* vol. 5, pp. 481–495, 1972.
45. V. J. Corcoran, Directional Characteristics in Optical Heterodyne Detection Processes, *J. Appl. Phys.,* vol. 36, pp. 1819–1825, 1965.

46. R. J. Adrian, Turbulent Convection in Water over Ice, *J. Fluid Mech.,* vol. 69, pp. 753–781, 1975.
47. J. W. Foreman Jr., Optical Path Length Difference Effects in Photomixing with Multimode Gas Laser Radiation, *Appl. Opt.,* vol. 6, pp. 821–829, 1967.
48. L. M. Fingerson, R. J. Adrian, R. K. Menon, S. L. Kaufman, and A. A. Naqwi, *Data Analysis, Laser Doppler Velocimetry and Particle Image Velocimetry,* TSI, Inc., St. Paul, Minn., 1993.
49. G. R. Grant and K. L. Orloff, Two Color Dual-Beam Backscatter Laser Doppler Velocimeter, *Appl. Opt.,* vol. 12, pp. 2913–2916, 1973.
50. F. L. Crossway, J. O. Hornkohl, and A. E. Lennert, Signal Characteristics and Signal Conditioning Electronics for a Vector Velocity Laser Velocimeter, pp. 396–444 in [16], 1972.
51. R. J. Adrian, A Bi-Polar Two Component Laser Velocimeter, *J. Phys. E,* vol. 4, pp. 72–75, 1975.
52. C. Greated, Measurement of Reynolds Stresses with an Improved Laser Flow Meter, *J. Phys. E,* vol. 3, pp. 753–756, 1970.
53. W. Yanta and G. J. Crapo, Applications of the Laser Doppler Velocimeter to Measure Subsonic and Supersonic Flows, AGARD Pre-Print 193, pp. 2.1–2.8, 1976.
54. A. Papoulis, *Probability, Random Variables and Stochastic Processes,* chap. 16, McGraw-Hill, New York, 1965.
55. W. T. Mayo, Modeling Laser Velocimeter Signals as Triply Stochastic Poisson Processes, pp. 455–484 in [19], 1976.
56. W. H. Stevenson, R. Dos Santos, and S. C. Mettler, Fringe Model Fluorescence Velocimetry, AGARD Pre-Print 193, pp. 21.1–21.9, 1976.
57. Dow Diagnostics, Indianapolis, Ind.
58. Duke Standards, Palo Alto, Calif.
59. K. T. Whitby, The Physical Characteristics of Sulphur Aerosols, *Atmos. Environ.,* vol. 12, pp. 135–159, 1978.
60. W. V. Feller and J. F. Meyers, Development of a Controllable Particle Generator for LDV Seeding in Hypersonic Wind Tunnels, pp. 342–357 in [19], 1976.
61. J. K. Agarwal and L. M. Fingerson, Evaluation of Various Particles for Their Suitability as Seeds in Laser Velocimetry, pp. 50–66 in [24], 1979.
62. N. G. Jerlov and E. S. Nielsen (eds.), *Optical Aspects of Oceanography,* Academic, New York, 1974.
63. O. B. Brown and H. R. Gordon, Size-Refractive Index Distribution of Clear Coastal Water Particulates from Light Scattering, *Appl. Opt.,* vol. 13, pp. 2874–2880, 1974.
64. J. K. Agarawal and P. Keady, Theoretical Calculation and Experimental Observation of Laser Velocimeter Signal Quality, *TSI Q.,* vol. 6, pp. 3–10, 1980.
65. E. Brockmann, Computer Simulation of Laser Velocimeter Signals, pp. 328–331 in [24], 1979.
66. R. J. Adrian, Estimation of LDA Signal Strength and Signal-to-Noise Ratio, *TSI Q.,* vol. 5, pp. 3–8, 1979.
67. W. K. George and J. L. Lumley, The Laser-Doppler Velocimeter and Its Application to the Measurement of Turbulence, *J. Fluid Mech.,* vol. 60, pp. 321–362, 1973.
68. P. Buchave, W. K. George Jr., and J. L. Lumley, The Measurement of Turbulence with the Laser-Doppler Anemometer, *Annu. Rev. Fluid Mech.,* vol. 11, pp. 443–503, 1979.
69. S. Chandrasekhar, Stochastic Problems in Physics and Astronomy, *Rev. Mod. Phys.,* vol. 15, pp. 1–89, 1943.
70. R. J. Goldstein and R. J. Adrian, Measurement of Fluid Velocity Gradients Using Laser Doppler Techniques, *Rev. Sci. Instrum.,* vol. 42, pp. 1317–1320, 1971.
71. M. J. Lighthill, *Fourier Analysis and Generalized Functions,* p. 43, Cambridge University Press, Cambridge, 1970.
72. D. K. Kreid, Laser-Doppler Velocity Measurements in Non-Uniform Flow: Error Estimates, *Appl. Opt.,* vol. 8, pp. 1872–1881, 1974.
73. D. K. McLaughlin and W. G. Tiederman, Biasing Correcting for Individual Realization of Laser Anemometer Measurements in Turbulent Flows, *Phys. Fluids,* vol. 16, pp. 2082–2088, 1973.

74. S. O. Rice, Mathematical Analysis of Random Noise, *Bell Syst. Tech. J.,* vol. 23, pp. 282–332, 1944; vol. 24, pp. 46–156, 1945.
75. S. O. Rice, Statistical Properties of a Sine Wave Plus Random Noise, *Bell Syst. Tech. J.,* vol. 27, pp. 109–156, 1948.
76. P. Jespers, P. T. Chu, and A. A. Fettweis, New Method for Computing Correlation Functions, presented at *Int. Symp. on Information Theory,* Brussels, 1962.
77. J. Ikebe and T. Sato, A New Integrator Using Random Voltage, *Electrotech. J.,* vol. 7, pp. 43–47, 1962.
78. B. P. T. Veltman and H. Kwakernaak, Theorie und Technik der Polaritats-Korrelation für die Dynamische Analyse nieder Frequentor Signale und Systeme, *Regelungstechnik,* vol. 9, pp. 357–364, 1961.
79. K. H. Norsworthy, A New High Product Rate 10 Nanosecond, 256 Point Correlator, *Phys. Scr.,* vol. 19, pp. 369–378, 1978.
80. E. R. Pike, How Many Signal Photons Determine a Velocity?, pp. 285–289 in [24], 1979.
81. A. E. Smart and W. T. Mayo Jr., Applications of Laser Anemometry to High Reynolds Number Flows, *Phys. Scr.,* vol. 19, pp. 426–440, 1978.
82. R. J. Baker and G. Wigley, Design, Evaluation and Application of a Filter Bank Signal Processor, pp. 350–363 in [20], 1977.
83. M. Alldritt, R. Jones, C. J. Oliver, and J. M. Vaughan, The Processing of Digital Signals by a Surface Acoustic Wave Spectrum Analyzer, *J. Phys. E,* vol. 11, pp. 116–119, 1978.
84. R. J. Adrian, J. A. C. Humphrey, and J. H. Whitelaw, Frequency Measurement Errors Due to Noise in LDV Signals, pp. 287–311 in [20], 1977.
85. H. H. Bossel, W. J. Hiller, and G. E. A. Meier, Noise Cancelling Signal Difference Method for Optical Velocity Measurements, *J. Phys. E,* vol. 5, pp. 893–896, 1972.
86. L. Lading and R. V. Edwards, The Effect of Measurement Volume on Laser Doppler Anemometer Measurements as Measured on Simulated Signals, pp. 64–80 in [20], 1977.
87. J. R. Abiss, The Structure of the Doppler-Difference Signal and the Analysis of Its Autocorrelation Function, *Phys. Scr.,* vol. 19, pp. 388–395, 1978.
88. J. B. Morton and W. H. Clark, Measurements of Two-Point Velocity Correlations in Pipe Flow Using Laser Anemometers, *J. Phys. E,* vol. 4, pp. 809–814, 1971.
89. H. R. E. van Maanen, K. van der Molen, and J. Blom, Reduction of Ambiguity Noise in Laser-Velocimetry by a Cross Correlation Technique, pp. 81–88 in [20], 1977.
90. W. K. George Jr., Limitations to Measuring Accuracy Inherent in the Laser Doppler Signal, pp. 20–63 in [20], 1977.
91. W. Hösel and W. Rodi, New Biasing Elimination Method for Laser-Doppler Velocimeter Counter Processing, *Rev. Sci. Instrum.,* vol. 48, pp. 910–919, 1977.
92. P. E. Dimotakis, Single Scattering Particle Laser Doppler Measurements of Turbulence, pp. 10.1–10.14 in [22], 1976.
93. P. Buchhave, Biasing Errors in Individual Particle Measurements with LDA-Counter Signal Processor, pp. 258–278 in [20], 1977.
94. M. C. Whiffen, Polar Response of an LV Measurement Volume, pp. 591–592 in [19], 1976.
95. D. Dopheide, M. Faber, G. Reim, G. Taux, Laser and Avalanche Diodes for Velocity Measurement by Laser Doppler Anemometry, *Exp. Fluids,* vol. 6, pp. 289–297, 1988.
96. F. Durst and M. Zaré, Laser Doppler Measurements in Two-Phase Flows, pp. 403–429, in [20], 1977.
97. W. P. Bachalo, Method for Measuring the Size and Velocity of Spheres by Dual-Beam Light Scatter Interferometry, *Appl. Opt.,* vol. 19, pp. 363–370, 1980.
98. H. C. van de Hulst, *Light Scattering by Small Particles,* Dover, New York, 1981.
99. A. Naqwi, F. Durst, and G. Kraft, Sizing Submicrometer Particles Using a Phase Doppler System, *Appl. Opt.,* vol. 130, pp. 4903–4913, 1991.
100. J. P. Gollub, A. R. McCarriar, and J. F. Steinman, Convective Pattern Evolution and Secondary Instabilities, *J. Fluid Mech.,* vol. 125, pp. 259–281, 1982.

101. D. R. Williams and M. Economou, Scanning Laser Anemometer Measurements of a Forced Cylinder Wake, *Phys. Fluids,* vol. 30, pp. 2283–2285, 1987.
102. N. Nakatani, M. Tokita, T. Izumi, and T. Yamada, LDV Using Polarization-Preserving Optical Fibers for Simultaneous Measurement of Multidimensional Velocity Components, *Rev. Sci. Instrum.,* vol. 56, p. 2025, 1985.
103. N. Nakatani, A. Maegawa, T. Izumi, T. Yamada, and T. Sakabe, Advancing Multipoint Optical Fiber LDV's—Vorticity Measurement and Some New Optical Systems, in *Laser Anemometry in Fluid Mechanics,* vol. III, pp. 3–18, LADOAN, Instituto Superior Tecnico, Lisbon, Portugal, 1988.
104. L. M. Weinstein, G. B. Beeler, and M. Lindemann, High-Speed Holocinematographic Velocimeter for Studying Turbulent Flow Control Physics, AIAA Paper 85-0526, 1985.
105. S. J. Lee, M. K. Chung, C. W. Mun, and Z. H. Cho, Experimental Study of Thermally Stratified Unsteady Flow by NMR-CT, *Exp. Fluids,* vol. 5, pp. 273–281, 1987.
106. A. T. Popovich and R. L. Hummel, A New Method of Nondisturbing Turbulent Flow Measurements Very Close to a Wall, *Chem. Eng. Sci.,* vol. 22, pp. 21–25, 1967.
107. B. Hiller and R. K. Hanson, Simultaneous Planar Measurements of Velocity and Pressure Fields in Gas Flows Using Laser-Induced Flourescence, *Appl. Opt.,* vol. 27, pp. 33–48, 1988.
108. T. D. Dudderar, R. Meynart, and P. G. Simpkins, Full-Field Laser Metrology for Fluid Velocity Measurement, *Opt. Lasers Eng.,* vol. 9, pp. 163–200, 1988.
109. R. J. Adrian, Multipoint Optical Measurements of Simultaneous Vectors in Unsteady Flow—A Review, *Int. J. Heat Fluid Flow,* vol. 7, pp. 127–145, 1986.
110. R. J. Adrian, Particle-Imaging Techniques for Experimental Fluid Mechanics, *Annu. Rev. Fluid Mech.,* vol. 23, pp. 261–304, 1991.
111. W. Merzkirch, *Flow Visualization,* 2nd ed., Academic, New York, 1987.
112. R. J. Adrian, Scattering Particle Characteristics and Their Effect on Pulsed Laser Measurements of Fluid Flow: Speckle Velocimetry vs. Particle Image Velocimetry, *Appl. Opt.,* vol. 23, pp. 1690–1691, 1984.
113. C. J. D. Pickering and N. A. Halliwell, Speckle Photography in Fluid Flows: Signal Recovery with Two-Step Processing, *Appl. Opt.,* vol. 23, pp. 1128–1129, 1984.
114. J. C. Dainty (ed.), *Laser Speckle and Related Phenomena,* Springer-Verlag, New York, 1975.
115. T. D. Dudderar and P. G. Simpkins, Laser Speckle Photography in a Fluid Medium, *Nature,* vol. 270, pp. 45–47, 1977.
116. D. B. Barker and M. E. Fourney, Measuring Fluid Velocities with Speckle Patterns, *Opt. Lett.,* vol. 1, pp. 135–137, 1977.
117. R. Grousson and S. Mallick, Study of Flow Pattern in a Fluid by Scattered Laser Light, *Appl. Opt.,* vol. 16, pp. 2334–2336, 1977.
118. R. Meynart, Instantaneous Velocity Field Measurement in Unsteady Gas Flow by Speckle Velocimetry, *Appl. Opt.,* vol. 22, pp. 535–540, 1983.
119. R. Meynart, Speckle Velocimetry Study of Vortex Pairing in a Low R Unexcited Jet, *Phys. Fluids,* vol. 26, pp. 2074–2079, 1983.
120. R. J. Adrian and C. S. Yao, Development of Pulsed Laser Velocimetry of Measurement of Fluid Flow, in G. Patterson and J. L. Zakin (eds.), *Proc. 8th Symp. on Turbulence,* pp. 170–186, University of Missouri–Rolla, 1983.
121. R. J. Adrian, Statistical Properties of Particle Image Velocimetry Measurements in Turbulent Flow, in R. J. Adrian, D. F. G. Durao, F. Durst, and J. H. Whitelaw (eds.), *Laser Anemometry in Fluid Mechanics,* vol. III, pp. 115–130, LADOAN, Instituto Superior Tecnico, Lisbon, 1988.
122. L. Lourenco and M. C. Whiffen, Laser Speckle Methods in Fluid Dynamics Applications, in R. J. Adrian et al. (eds.), *Laser Anemometry in Fluid Mech.,* vol. II, pp. 51–68, LADOAN, Instituto Superior Tecnico, Lisbon, 1986.
123. C. C. Landreth, R. J. Adrian, and C. S. Yao, Double Pulsed Particle Image Velocimeter with Directional Resolution for Complex Flows, *Exp. Fluids,* vol. 6, pp. 119–128, 1988.
124. J. A. Ochs, Generation of Vorticity Contour Maps from 2-D Randomly Located Velocity Fields with Errors, Senior Research Project Rep., TAM Dep., Univ. of Illinois, Urbana, 1987.

125. M. Gharib, M. A. Hernan, A. H. Yavrouian, and V. Sarohia, Flow Velocity Measurement by Image Processing of Optically Activated Tracers. AIAA Paper 85-0172, 1985.

126. T. P. K. Chang, A. T. Watson, and G. B. Tatterson, Image Processing of Tracer Particle Motions as Applied to Mixing and Turbulent Flow, Parts I and II, *Chem. Eng. Sci.,* vol. 40, pp. 269–285, 1985.

127. J. C. Kent and A. R. Eaton, Stereo Photography of Neutral Density He-Filled Bubbles for 3-D Fluid Motion Studies in an Engine Cylinder, *Appl. Opt.,* vol. 21, pp. 904–912, 1982.

128. R. D. Keane and R. J. Adrian, Optimization of Particle Image Velocimeters, Part I: Double Pulsed Systems, *Meas. Sci. Tech.,* vol. 1, pp. 1202–1215, 1990.

129. W. H. Peters and W. F. Ranson, Digital Imaging Techniques in Experimental Stress Analysis, *Opt. Eng.,* vol. 21, pp. 427–431, 1982.

130. C. C. Landreth and R. J. Adrian, *Proc. 4th Int. Symp. on Applications of Laser Anemometry to Fluid Mech.,* Lisbon, July 11–14, 1988, Instituto Superior Tecnico, Lisbon, 1988.

131. G. H. Kaufman, A. E. Ennos, B. Gale, and D. J. Pugh, An Electro-Optical Read-Out System for Analysis of Speckle Photographs, *J. Phys. E Sci. Instrum.,* vol. 13, pp. 579–584, 1980.

132. B. Ineichen, P. Eglin, and R. Dandliker, Hybrid Optical and Electronic Image Processing for Strain Measurements by Speckle Photography, *Appl. Opt.,* vol. 19, pp. 2191–2195, 1980.

133. R. Meynart, Digital Image Processing for Speckle Flow Velocimetry, *Rev. Sci. Instrum.,* vol. 53, pp. 110–111, 1982.

134. G. A. Reynolds, M. Short, and M. C. Whiffen, Automated Reduction of Instantaneous Flow Field Images, *Opt. Eng.,* vol. 24, pp. 475–479, 1985.

135. S. A. Isacson and G. H. Kaufman, Two-Dimensional Digital Processing of Speckle Photography Fringes, 2: Diffraction Halo Influence for the Noisy Case, *Appl. Opt.,* vol. 24, pp. 1444–1447, 1985.

136. R. J. Adrian, Image Shifting Technique to Resolve Directional Ambiguity in Double-Pulsed Velocimetry, *Appl. Opt.,* vol. 26, pp. 3855–3858, 1986.

137. C. C. Landreth and R. J. Adrian, Electro-Optical Image Shifting for Particle Image Velocimetry, *Appl. Opt.,* vol. 27, pp. 4216–4220, 1988.

138. C. C. Landreth and R. J. Adrian, Impingement of a Low Reynolds Number Turbulent Jet on a Flat Surface, *Exp. Fluids,* vol. 9, pp. 74–84, 1990.

139. J. Kompenhans and J. Reichmuth, 2-D Flow Field Measurements in Wind Tunnels by Means of Particle Image Velocimetry, presented at 6th Int. Congress on Appl. of Lasers and Electro-Optics, San Diego, Calif., Nov. 8–12, 1987.

140. D. L. Reuss, R. J. Adrian, C. C. Landreth, D. T. French, and T. Fansler, Instantaneous Planar Measurements of Velocity and Large-Scale Vorticity and Strain Rate in an Engine Using Particle Image Velocimetry, SAE Paper 89061.

141. L. M. Lourenco and A. Krothapalli, Application of PIDV to the Study of the Temporal Evolution of the Flow Past a Circular Cylinder, in R. J. Adrian et al. (eds.), *Laser Anemometry in Fluid Mechanics,* vol. III, pp. 161–178, Ladoan, Instituto Superior Technico, Lisbon, 1988.

142. N. A. Halliwell (ed.), Special Issue on Progress in Particle Image Velocimetry, *Opt. Lasers Eng.,* vol. 9, 1988.

143. I. Grant (ed.), Selected Papers on Particle Image Velocimetry, *SPIE Milestone Series,* vol. MS99, SPIE Opt. Engr. Press, Bellingham, Wash., 1994.

144. R. J. Adrian, D. F. G. Durao, F. Durst, H. Mishina, and J. H. Whitelaw (eds.), *Laser Anemometry in Fluid Mechanics,* 404 pp., Ladoan, Lisbon, 1984.

145. R. J. Adrian, D. F. G. Durao, F. Durst, H. Mishina, and J. H. Whitelaw (eds.), *Laser Anemometry in Fluid Mechanics—II,* 518 pp., Ladoan, Lisbon, 1986.

146. R. J. Adrian, T. Asanuma, D. F. G. Durao, F. Durst, and J. H. Whitelaw (eds.), *Laser Anemometry in Fluid Mechanics—III,* 541 pp., Ladoan, Lisbon, 1988.

147. R. J. Adrian, T. Asanuma, D. F. G. Durao, F. Durst, and J. H. Whitelaw (eds.), *Laser Anemometry in Fluid Mechanics,* 534 pp., Springer-Verlag, Berlin, 1989.

148. R. J. Adrian, D. F. G. Durao, F. Durst, M. Maeda, and J. H. Whitelaw (eds.), *Applications of Laser Techniques to Fluid Mechanics,* 467 pp., Springer-Verlag, Berlin, 1991.

149. R. J. Adrian, D. F. G. Durao, F. Durst, M. Maeda, and J. H. Whitelaw (eds.), *Laser Techniques and Applications in Fluid Mechanics,* Springer-Verlag, Berlin, 1993.
150. A. Dybbs and P. A. Pfund (eds.), *Intl. Symp. on Laser Anemometry,* ASME FED-vol. 33, 305 pp., ASME, New York, 1985.
151. A. Dybbs and B. Ghorashi (eds.), *Laser Anemometry Advances and Applications,* 845 pp., ASME, New York, 1991.
152. R. J. Adrian (ed.), Selected Papers on Laser Doppler Velocimetry, *SPIE Milestone Series,* vol. MS78, SPIE Opt. Eng. Press, Bellingham, Wash., 1991.

VOLUME FLOW MEASUREMENTS

G. E. Mattingly

5.1 INTRODUCTION

One of the most common types of material measurements in American industry is the determination of fluid quantity, or the measurement of fluid flow rate. Untold sums of money are spent daily on fluid custody transfers that are based upon the "read-out" information from fluid metering devices. The following examples are based upon information available in 1978–1979:

1. Amounts
   a. *Petroleum industry.* Worldwide production is 22 billion barrels of crude petroleum per year. In the United States the metering of crude is performed at 650,000 measurement sites by 50,000 employees.
   b. *Gas.* In the United States, 20 trillion ft^3 of natural gas are metered each year, through 42 million fluid quantity devices.
   c. *Wastewater discharge.* The EPA has issued 90,000 discharge permits to municipalities and industries. New treatment plants cost $6 to $10 billion per year.
2. Costs
   a. *Fluids.* The petroleum-refining, chemical, and beverage industries account for $1 trillion per year. The interbasin transfer of water from Mississippi to Texas alone amounts to $1.3 billion per year.
   b. *Meters.* Fluid flow devices cost $843 million per year.

This chapter is reprinted from the previous edition.

Note, particularly, the amount of natural gas that is metered in the United States each year. At an estimated cost of $2/ft^3, one can quickly quantify the material and dollar value that is associated with each 0.01% of systematic uncertainty that might exist in this measurement system.

The purposes of volume flow measurements are

1. Effective regulation
2. Energy distribution
3. Custody transfer/equity in marketplace
4. International competition

A wide variety of fluid meters maintain and/or control material quality and quantity in practically every continuous industrial process. Because of the cost of the fluids and the measurements, it is essential that fluid quantity and flow-rate measurements are made as precisely and as accurately as required by the parties involved. It is also incumbent upon those involved in fluid custody transfer to establish the traceability chains that link their measurements to the appropriate standards that are involved. Only in this manner can fluid volume measurements be performed equitably, with the confidence of seller and buyer alike.

5.2 CLASSIFICATION OF METERING DEVICES

Every meter consists of two distinct parts, each of which performs a specific function. The first part—the primary element—is in contact with the fluid and produces an interaction with the fluid. Examples are

1. Orifice plate
2. Turbine wheel
3. Vortex-shedding strut

The second part—the associated secondary element—converts one or more reactions received from the primary element into an observable quantity. Examples of secondary elements that would be associated with the above primary elements are

1. A manometer to exhibit the differential pressure generated in the fluid by the orifice plate
2. An electrical system that magnetically senses the revolving turbine blade tips
3. A system that detects the vortices being shed behind the strut

When a meter is to be selected, both the primary and the secondary elements must be considered, as they must operate together, as a unit, in the particular environment where the measurements are to be made. Furthermore, any other factors that may introduce variation into the measurement results should also be considered in the evaluation of the fluid metering system. Such factors might include a person who manually reads the manometer or a computer or microprocessor that calculates a final result from analog or digital data received from the secondary element. These factors and their influence are discussed further in sections 5.4 and 5.5.

Meters can be further classified into those that determine fluid quantity and those that indicate fluid flow rate [1]. The various types of fluid measuring devices within each of these classes are as follows:

1. Quantity meters
   a. Weighing (for liquids)
      (1) Weighers
      (2) Tilting traps
   b. Volumetric (for liquids)
      (1) Calibrated tank
      (2) Reciprocating piston
      (3) Rotating piston or ring piston
      (4) Rotating disk
      (5) Sliding and rotating vane
      (6) Gear and lobed impeller (rotary)
   c. Volumetric (for gases)
      (1) Bellows
      (2) Liquid sealed drum
      (3) Gear and lobed impeller (rotary)
2. Rate meters
   a. Differential pressure
      (1) Orifice
      (2) Venturi
      (3) Nozzle
      (4) Centrifugal
      (5) Pitot tube
      (6) Linear resistance
   b. Momentum
      (1) Turbine
      (2) Propeller
      (3) Cup anemometer
   c. Variable area
      (1) Gate
      (2) Cones, floats in tubes
      (3) Slotted cylinder and piston
   d. Force
      (1) Target
      (2) Hydrometric pendulum

  e. Thermal
   (1) Hot wire, hot film
   (2) Total heating
  f. Fluid surface height or "head" type
   (1) Weirs
   (2) Flumes
  g. Other
   (1) Electromagnetic
   (2) Tracers
   (3) Acoustic
   (4) Vortex shedding
   (5) Laser
   (6) Coriolis

For the most part, the quantity meters listed above are used where spatial requirements and temporal response characteristics are not as important as precision and accuracy. For example, liquid weighing devices, calibrated volumetric tanks, reciprocating-piston devices, and liquid sealed-drum systems for gases are among the devices used most as "primary" standards for the calibration of fluid flow-rate devices. The simplicity of these devices and their readily accessible and "checkable" components produce precision and accuracy and, in turn, confidence in their results. A major conclusion of sections 5.4 and 5.5 is that several levels of checks can and should be applied to these systems; however, here we note only that these types of systems have special status as very precise and accurate devices.

The remaining devices in the list are more commonly used today in industrial installations for measuring the rate of fluid flow; they are subdivided according to the physical principles that form the basis of their operation. In the section that follows, selected meter types are described briefly, with emphasis on principles of operation and general performance characteristics.

5.3 SELECTED METER PERFORMANCE CHARACTERISTICS

In this section, the characteristics of the more commonly used types of fluid metering devices are described, as well as several new and innovative types of metering devices that hold considerable promise for future applications.

5.3.1 Orifice Meters

Owing to their simplicity, cost, ruggedness, and widespread acceptance, differential-pressure-type meters—and notably orifice meters—deserve special at-

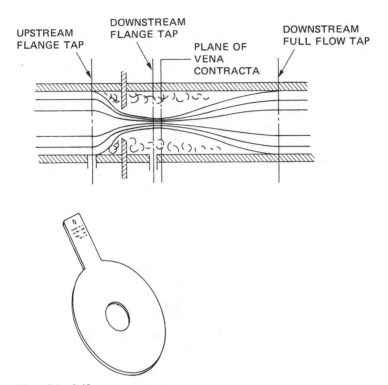

UPSTREAM
FLANGE TAP

DOWNSTREAM
FLANGE TAP

PLANE OF
VENA
CONTRACTA

DOWNSTREAM
FULL FLOW TAP

Figure 5.1 Orifice meter

tention [2]. An orifice-meter "run" is shown in Fig. 5.1. The primary element of an orifice meter consists of

1. The upstream meter-tube section, including any of several types of devices intended to "condition" the flow
2. The orifice fixture and plate assembly
3. The downstream meter-tube section
4. The secondary element, not shown in Fig. 5.1, which might consist of a U-tube manometer in conjunction with fluid temperature measurement instrumentation

The ideal performance characteristics for orifice meters are given in App. 5.A for incompressible fluids. Appendix 5.B contains ideal performance characteristics for compressible fluids, and App. 5.C contains real-gas characteristics.

The features that affect sustained and accurate orifice-meter performance are as follows:

1. Orifice plate
   a. Dimensions
      (1) Hole diameter
      (2) Beveling
      (3) Roughness
      (4) Flatness
      (5) Edge shape and sharpness
      (6) Thickness
   b. Centeredness in pipe
2. Upstream and downstream meter-tube conditions
   a. Round
   b. Straight
   c. Smooth
   d. Length
3. Tube-bundle straightening vanes
   a. Size, number, and configuration
   b. Parallelism
   c. Ruggedness
4. Orifice flanges and fittings
   a. Pressure rating
   b. No weld seams or grooves
5. Pressure taps (see App. 5.F)
   a. Flange
   b. Pipe
   c. D-D/2 taps
   d. Corner
   e. Other (pipe, Duzenburg)
   f. Burrs, tolerances, and peripheral locations
6. Thermometer wells
   a. Downstream of meters
   b. Located to measure average temperature (see App. 5.G)
7. Flow and fluid effects
   a. Profile
   b. Swirl
   c. Pulsations
   d. Contaminants
   e. Plate bending

A paddle-type orifice plate is shown at the lower left of Fig. 5.1. In use, this plate is sandwiched between pipe flanges so that the orifice-plate hole is concentric with the pipe centerline. The differential pressure (upstream minus downstream) across this plate is then related to the fluid flow rate, according to

$$\dot{Q} = A_2 C_D \left(\frac{2\,\Delta p}{\rho(1-\beta^4)} \right)^{1/2} \tag{1}$$

where $\dot{Q}$ = volumetric flow rate
$\quad A_2$ = throat area
$\quad C_D$ = discharge coefficient
$\quad \Delta p$ = upstream-to-downstream pressure differential
$\quad \rho$ = fluid density
$\quad \beta$ = ratio of throat diameter to internal diameter of upstream pipe

Equation (1) is based upon a series of assumptions, such as those governing Bernoulli phenomena, i.e., steady flow of an incompressible fluid in which there is no phase change.

Equivalently, Eq. (1) can be written

$$\dot{Q} = KA_2 \left(\frac{2\,\Delta p}{\rho} \right)^{1/2} \tag{2}$$

where K is termed the flow coefficient:

$$K = \frac{C_D}{(1-\beta^4)^{1/2}} \tag{3}$$

Both the discharge coefficient and, correspondingly, the flow coefficient depend on the exact locations of the pressure tappings upstream and downstream of the orifice plate. For a given meter and pressure-tapping arrangement, a conventional calibration procedure, as described in section 5.3, can determine the dependence of C_D upon the flow Reynolds number Re:

$$\text{Re} = \frac{DV\rho}{\mu} \tag{4}$$

where D = pipe internal diameter
$\quad V$ = average fluid velocity
$\quad \mu$ = fluid dynamic viscosity

If conventional pressure-tapping configurations are used, it may be satisfactory to calculate the discharge or flow coefficients via such formulas as are available in the metering literature (see App. 5.E). Figure 5.2 depicts the American pressure-tapping convention, which uses the so-called flange taps. These have the advantage that (1) they are conveniently drilled radially straight through the flanges that hold the plate, and (2) they are located close to the plate, where they are capable of sensing a large differential pressure; this, in turn, is desirable because it can increase metering sensitivity. Flange taps have the disadvantage of being located 1 inch on either side of the orifice plate. Thus, flange-tapped meters of different sizes do not conform to geometric scaling laws.

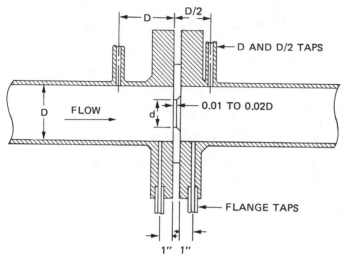

Figure 5.2 Thin-plate orifice meter

Also shown in Fig. 5.2 are the so-called D-$D/2$ taps. These taps do conform to geometric scaling laws. Because of their locations, these taps sense fluid pressures that are "smoother," in the sense that turbulent pressure fluctuations at these tap locations are generally smaller than those that occur nearer the orifice plate. Nearer the plate, where the flow is stagnating against the plate (upstream and downstream surfaces) and turning so as to flow radially inward toward the orifice hole, the turbulence generally affects pressure-measuring instrumentation more significantly.

U-tube or single-leg differential manometers can tend to "bounce" markedly when sensing pressures near the orifice plate. In addition, a variety of other factors influence manometer bounce. For example, the diameter of the tapping hole; the diameter, length, and configuration of the hose or tubing connecting the manometer to the meter; and the inertial characteristics of the manometer fluid all exert some influence (see App. 5.F). This bouncing phenomenon can cause difficulty in obtaining precise metering performance from an orifice meter.

A variety of manometer reading schemes have been devised to reduce the detrimental effects of bouncing. A frequently used method is to simply "eyeball" an average value of the manometer column height after observing several cycles of the bounce. The average value of the square roots of several such values is taken to be the true value for the particular flow. A second scheme that is used on single-leg differential manometers is to read consecutive maximum and minimum values of the bouncing column of manometer fluid. After "max-min" pairs of readings have been recorded some 10 to 20 times, the average of all these pairs is taken to approximate the true average value. Another scheme is to install needle valves in the tubes connecting the manometer to the orifice meter. Partially closing these valves tends to smooth the bouncing effect. The partially closed valves also limit the time response of the combined system, so that if the flow is changing with time, it may not be detected, or a lag will exist that will impede timely meter performance. Still another scheme is an extension of this "damping with valves" technique. This consists of installing identical, rapidly acting valves with small actuation displacements in the manometer lines. At random intervals, the valves are quickly and simultaneously closed, either manually or electrically, so that the manometer is isolated from the meter. The then-stationary manometer reading is recorded, and the valves are opened. After a random interval, the cycle is repeated. When this has been done 10 to 20 times, the average value of the square root of the pressure difference is taken as an approximation of the true value.

Recent efforts to analyze and optimize the performance of orifice meters at the National Bureau of Standards (NBS) have involved computer modeling techniques [5]. From the results, a variety of well-known effects on orifice-meter performance can be quantified. For example, the effects of β ratio, inlet-velocity distribution, orifice-plate thickness, turbulent intensity, and swirl can be assessed.

5.3.2 Venturi Tubes and Flow Nozzles

Figure 5.3 shows a Venturi tube. These differential-pressure-type devices have a shape that closely approximates the streamline patterns of the flow through a reduced cross-sectional area; as a result, their pressure-loss characteristics are generally lower than those of orifice plates. Hence, the higher cost of a Venturi tube is offset in time by reduced pumping costs incurred in producing the flow through the meter. Correspondingly, the discharge coefficients of these devices tend to approximate unity more closely, and there is less variation in the performance of Venturi tubes than orifice meters.

Just as with orifice meters, Venturi tubes can be of the concentric or eccentric type. Both types are shown in Fig. 5.3. The eccentric type provides more "self-cleaning" than the concentric type. Thus, it performs more satisfactorily on flows containing sediments or particulates that might, in horizontally installed meters, deposit in corners and thereby change the effective geometry of the meter and, in turn, its performance characteristics.

5.3.3 Elbow Meters

Figure 5.4 shows the primary and secondary components of the elbow meter. While this type of device does not generally match the precision and uncertainties of other differential-pressure-type devices, it does not increase the pressure-loss characteristic of the flow system in which it is installed. Also, the cost of the meter is simply the cost of the manometer and the connection to the existing elbow, which

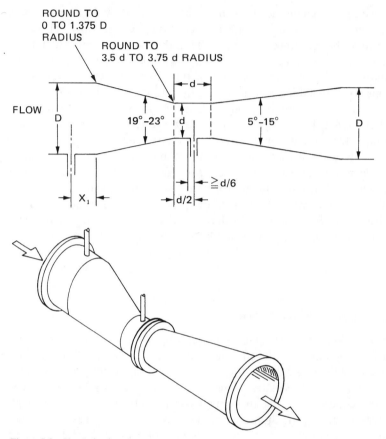

ROUND TO
0 TO 1.375 D
RADIUS

ROUND TO
3.5 d TO 3.75 d RADIUS

FLOW

D

d

19°–23°

d

5°–15°

D

$\geqq$ d/6

X_1

d/2

Figure 5.3 Venturi tube

is quite small in comparison to the cost of an orifice or Venturi primary device. The elbow meter is considered to be nonintrusive, since it does not change the pattern of flow in the pipe and does not introduce structural elements into the flow.

5.3.4 Pitot Tubes

Figure 5.5 shows a sketch of a primary device that is based upon differential pressure. This differential is established, separately, between the high pressure of the flow stagnating near the free tip of a cantilevered tube and the low static pressure of the flow around the side of the same tube. By displacing the sensing taps axially along the tube, any detrimental interaction between them is avoided. Because the distribution of flow in the center region of the pipe is generally quite uniform in the cross-stream directions, a good estimate is obtained of the maximum flow ve-

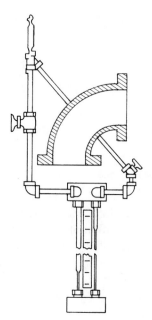

Figure 5.4 Elbow meter

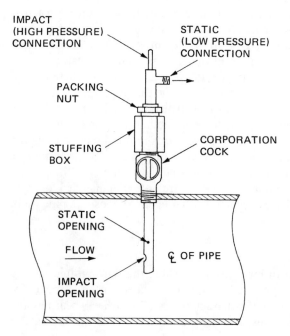

IMPACT
(HIGH PRESSURE)
CONNECTION

STATIC
(LOW PRESSURE)
CONNECTION

PACKING
NUT

STUFFING
BOX

CORPORATION
COCK

STATIC
OPENING

FLOW

¢ OF PIPE

IMPACT
OPENING

Figure 5.5 Pitot tube

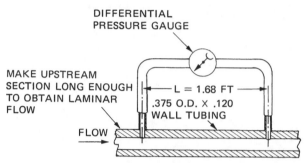

Figure 5.6 One possible arrangement of the components of a laminar flowmeter

locity near the centerline. By relating this value to the cross-sectional average velocity through (1) a calculation, assuming a known radial distribution of velocity in
the conduit, or (2) a calibration of the device, one can generally obtain an accuracy
in the range quoted in Fig. 5.5.

5.3.5 Laminar Flowmeters

Figure 5.6 shows a typical arrangement of the primary and secondary components
of a laminar flowmeter. The basis of the device is the establishment of laminar flow
between the pressure taps, and use of the well-known relationship between laminar
flow rate through a tube of known cross-sectional area and the pressure drop across
its known length. These devices generally have the advantage that they operate
bidirectionally and indicate, via the sign of the differential pressure, the direction
of the flow. As shown in Fig. 5.6, laminar flowmeters can be nonintrusive.

Figure 5.7 shows other arrangements for the primary element. These can involve a single capillary tube that is a constriction in the larger pipeline, or an
intrusive bundle of smaller capillary tubes in which the flow is arranged to be
laminar. Other intrusive elements, such as honeycomblike structures, can also be
used. Because these generally are composed of noncircular cross-sectional shapes,
it is recommended that each device be calibrated to obtain assured performance.
It is essential with these types of primary elements that the geometry of the conduit
remain unchanged. Therefore, laminar flowmeters are not recommended for measuring flows containing particulates or materials that tend to deposit on pipe walls.

5.3.6 Turbine Meters

Figure 5.8 shows a sketch of a turbine meter, in which the flowing fluid spins a
propeller wheel whose angular speed is in some way related to the average flow
rate in the conduit. Depending upon the characteristics of the wheel's bearings and
their longevity, this device often gives accuracies up to ±0.25% of the flow rate.

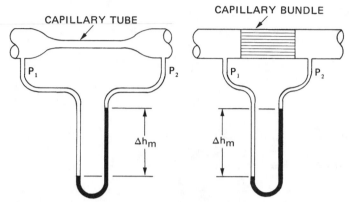

Figure 5.7 Two additional types of primary elements for laminar flowmeters

The angular speed of the wheel is generally detected by the passage of the blade tips past a coil pickup on the pipe. Increased resolution can be obtained by increasing the number of pickups and summing the pulses received, by increasing the number of blades, or by shrouding the blade tips and placing magnetic buttons on the shroud so as to generate more pulses per revolution of the wheel. With proper arrangement of the pertinent geometry of the wheel and blades and with good bearings, turbine meters can produce a frequency output that is proportional to flow rate within a few tenths of a percent over a flow-rate range of 30 to 1. At

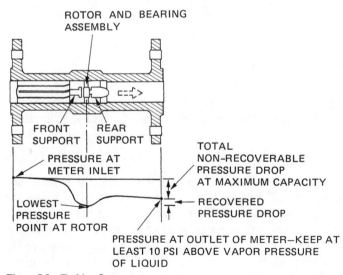

Figure 5.8 Turbine flowmeter

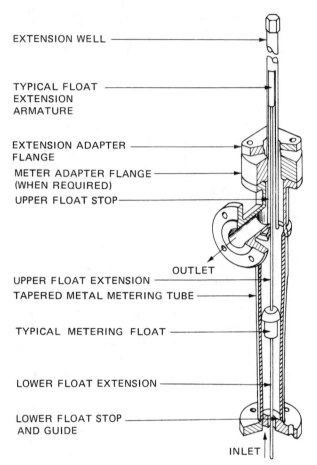

EXTENSION WELL

TYPICAL FLOAT
EXTENSION
ARMATURE

EXTENSION ADAPTER
FLANGE

METER ADAPTER FLANGE
(WHEN REQUIRED)

UPPER FLOAT STOP

OUTLET

UPPER FLOAT EXTENSION

TAPERED METAL METERING TUBE

TYPICAL METERING FLOAT

LOWER FLOAT EXTENSION

LOWER FLOAT STOP
AND GUIDE

INLET

Figure 5.9   Typical rotameter configuration

the low end of this range, fluid viscous effects are a limiting factor. At the high end, the interaction between blade tips or magnetic buttons and the pickup is limiting.

5.3.7 Rotameters

A typical rotameter configuration is shown in Fig. 5.9. The vertically installed device operates when the upward fluid drag on the float is balanced in the upwardly diverging tube by the weight of the float. The vertical elevation of the float is then an indication of the flow rate, and this is generally read manually if the metering tube (and the fluid) is transparent. If the pressure within the fluid stream to be measured is too high to allow glass or plastic tapered tubes, metal can be used; then the float position can be sensed magnetically or electrically.

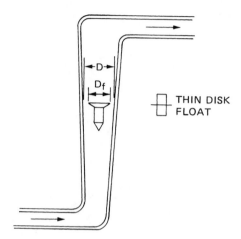

Figure 5.10 Rotameter tube and floats

By appropriate arrangement of the divergence of the metering tube, the vertical position of the float may be made linearly proportional to the flow rate. Other arrangements are also feasible, such as one in which the float position is logarithmically dependent upon flow rate. The floats used with rotameters can have a variety of shapes and sizes; Fig. 5.10 shows two popular shapes. These are interchangeable, so that a wide range of flow rates can be measured in the same tapered tube.

5.3.8 Target Meters

Figure 5.11 shows a target meter. These devices operate on the principle that the fluid drag on a disk supported in a pipe flow is related to the average flow rate. This fluid drag can be sensed without significant movement or rotation of the disk,

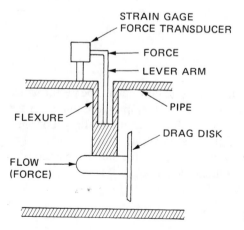

Figure 5.11 Target meter

(a)

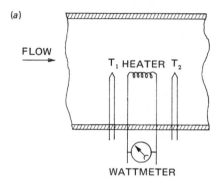

(b)

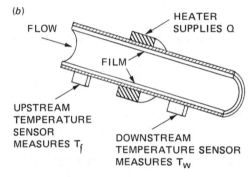

Figure 5.12 Thermal flowmeter. (a) $\Delta T = T_2 - T_1$. (b) $\Delta T = T_w - T_f$

which would alter its drag characteristics; various types of secondary devices, such as strain gages and fluid-activated bellows, can be used to detect the disk drag.

Target meters can be concentric or eccentric, as shown in Fig. 5.11. The calibration curves vary, owing to the effects of the strut supporting the disk. These devices can also be used as bidirectional meters, with the sense of the force indicating the flow direction. Because of their simple structure, they can be used to measure the flow rate of dirty fluids, which may have entrained sediments, as long as these do not alter the critical geometrical arrangement shown in Fig. 5.11.

5.3.9 Thermal Flowmeters

Figure 5.12 shows a typical arrangement for a thermal flowmeter. The performance of this type of meter is based upon the increase in fluid temperature that is sensed between two thermometers when heat is added to the fluid between the thermometers. For a given fluid heating rate q, the fluid mass flow rate $\dot{M}$ is proportional to the rate of heat addition and inversely proportional to the downstream-to-upstream temperature difference ΔT. In compatible units,

$$\dot{M} = \frac{q}{c_p \, \Delta T} \tag{5}$$

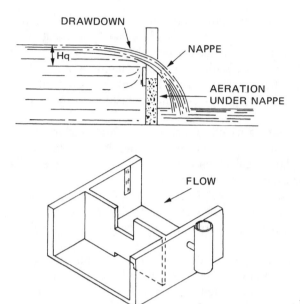

Figure 5.13 Sharp-crested weir

where c_p is the fluid specific heat at constant pressure. If sufficiently large distances separate the temperature sensors so that complete cross-sectional mixing occurs, and if no heat is lost to or through the pipe wall, the device should not require calibration. On the other hand, if the heat addition occurs locally on the centerline of the pipe or only in the fluid layers along the pipe wall, flow calibration is required to achieve best performance.

Thermal flowmeters can be arranged to be nonintrusive, as shown in the lower portion of Fig. 5.12. They can also operate through cooling, instead of heating.

5.3.10 Weirs and Flumes

Figure 5.13 shows a sharp-crested weir and the flow pattern that might exist when the weir is installed in a much wider channel. The upstream depth of water above the weir crest is related to the flow rate in the channel through the rating curve for the device. This upstream depth is generally measured with a pressure transducer sensing fluid depth via the static pressure or with a depth measurement made with a surface height gage installed in an auxiliary chamber attached to the channel, as shown in the lower portion of Fig. 5.13. In this chamber, surface heights can be measured on the quiescent surface more precisely than they can in the flow channel, where waves or other surface irregularities can impede good depth measurements.

For proper performance of the device, the cavity under the nappe must be aerated. Otherwise, the pressure distribution through the nappe will not be atmospheric, as is assumed in the construction of the rating curve. This aeration is virtually guaranteed if the channel width is much larger than the width of the weir crest. It should be noted that a weir is essentially an open-channel version of an orifice-plate meter, wherein the differential pressure is the depth measurement. The flume (not sketched in Fig. 5.13) corresponds in a similar fashion to the venturi meter.

Weirs and flumes are frequently selected to measure the flow rates in canals, streams, and rivers. It is then essential that sediment and other debris be kept from clogging the primary (or secondary) element, and to this end, self-cleaning features are essential. The weir can be arranged to be self-cleaning via a small slit in the partition under the crest. This slit should be sized so as to pass any silt buildup that might deposit on the upstream side of the partition. In this manner the original geometry, on which the calculated or calibrated rating curve is based, will be preserved.

The calibration of open-channel metering devices can be very difficult—especially for the very large devices that are installed in rivers. It is necessary to determine the flow rate as accurately as is required over as wide a flow-rate range as occurs naturally, and to use these data to produce a rating curve. The process is considerably easier for prefabricated or portable units, which can be installed in laboratory facilities, where precise determinations of the flow rate can be routinely made.

A calibration method that has recently been used at NBS to study the performance characteristics of Parshall flumes is computer modeling [6, 7]. With this technique, the meter geometry and inlet flow distribution can be treated as boundary conditions for the problem of determining the entire flow field within the meter, together with the free-surface position. The flow field can be integrated to obtain the flow rate, and the free-surface distribution produces the depth to be measured via the secondary device, thus giving a computed rating curve. Such a technique could allow the quantitative assessment of unusual inlet flow distributions and changes in critical metering geometry, such as tilt, convergence angles, and crest heights.

5.3.11 Magnetic Flowmeters

Figure 5.14 shows the arrangement of the coils and electrodes that make up the typical magnetic flowmeter. This meter operates on the principle that a conducting fluid will generate a voltage proportional to the flow rate as it passes through a magnetic field. The typical device is nonintrusive, with magnetic coils insulated from the flowing fluid and the electrodes maintaining electrical contact with the fluid. This type of device is available commercially in a wide range of pipe sizes.

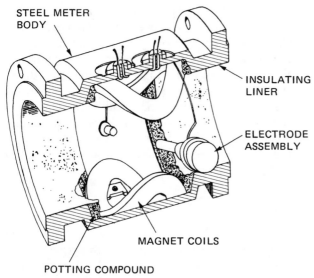

STEEL METER
BODY

INSULATING
LINER

ELECTRODE
ASSEMBLY

MAGNET COILS

POTTING COMPOUND

Figure 5.14   Magnetic flowmeter

5.3.12   Acoustic Flowmeters

A variety of acoustic flowmeters are commercially available. These use several
different physical principles; the most prevalently used acoustic flowmeter is based
upon a Doppler principle. Figure 5.15 shows one arrangement, in which a pair of
ultrasonic transmitters beam to receivers placed across a flow conduit. One of the
transmitters beams downstream, while the other beams upstream. The detected
differences in the times of travel are then related to the average velocity of the flow
in the conduit.

These devices have been installed in closed and open conduits in a range of
sizes that includes rivers. While the transmitters and receivers must maintain good

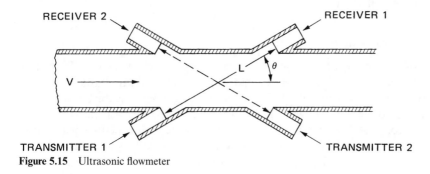

RECEIVER 2

RECEIVER 1

L

θ

V

TRANSMITTER 1

TRANSMITTER 2

Figure 5.15   Ultrasonic flowmeter

acoustic contact with the fluid, this type of device is considered to be nonintrusive. Recently, at NBS a new acoustic-type flowmeter was developed. This meter, which is also nonintrusive, uses a controlled source to produce sound in a closed pipe containing a flow of variable-temperature air. Via feedback circuitry, the controlled source maintains the sound wavelength at four to five times the internal pipe diameter. The sound is sensed by two microphones spaced four to five pipe diameters apart along the pipe and mounted flush with the inner pipe wall. In this configuration the difference in phase angles sensed by the two microphones allows calculation of (1) the fluid density via the sound speed-temperature relationship and (2) the average fluid velocity and thus the volumetric flow rate. The product of density and this flow rate gives the mass flow rate [8].

5.3.13 Vortex-Shedding Meters

A vortex-shedding meter is composed of a diametral strut or a series of chordal struts placed across the cross-sectional area of a pipe. Annular arrangements have also been used. For Reynolds numbers above certain threshold values, the flow about these shapes produces vortices in the near-wake regions behind the struts. These vortices are shed into the more distant wake regions at frequencies that are proportional to the flow velocity toward the strut. Any of a variety of schemes is used to sense the vortex-shedding frequency, from which the flow rate may be computed.

Sensors for vortex detection include pressure transducers mounted on one or more of the downstream surfaces of the strut, thermal sensors that detect transients produced by the vortex effect of altering the heat flux from some portion of the strut surface, and strain sensors that detect vortex-produced cross-stream oscillations of the strut itself. Since vortex shedding constitutes the basis of operation, it is important to use a strut shape that produces strong, frequent vortices. The complex nature of vortex shedding requires that the meters be calibrated to characterize their performance. Because vortex shedding takes place over a wide range of Reynolds numbers, these types of meters are specified to operate over flow-rate ranges of 100 to 1.

5.3.14 Laser Flowmeters

Because of the many advantages of laser-Doppler velocimetry (LDV), this velocity-measuring principle offers considerable promise as the basis of a flowmeter. By focusing two beams of the same laser light (one beam phase-shifted relative to the other) in a flowing fluid so that an interference fringe pattern is produced across the flow path in the volume of intersection of the two beams, one can determine the local fluid velocity from light scattered by particulates transported with the fluid velocity. This can be done nonintrusively through transparent sections of the flow conduit and, from measurements of the wavelength of the laser light and the relative angular displacement of the intersecting beams, the local fluid velocity can be determined without calibration. Thus one can generate a laser-based flowmeter in the following way: by orienting the fringe pattern in the sensing volume so as to measure the velocity component parallel to the pipe centerline, and by either traversing this so-called sensing volume over the cross-sectional area of the flow and integrating results to obtain volumetric flow rate, or by determining the relationship between a particular local velocity, say the centerline value, and the average velocity. Critical in all this is maintenance of the optical alignment and preservation of clean windows through which the laser light may pass into and out of the flow field.

5.3.15 Coriolis-Acceleration Flowmeters

Besides the gravitational and centrifugal acceleration meters described above, there are meters that use Coriolis acceleration as their basis of operation [9]. These devices are most readily (although, in principle, not solely) adapted to liquids because of their high density. To meter fluid in a closed conduit, a U-shaped tube containing the fluid flow to be metered is vibrated with the tips of the U fixed and the motion of the bend normal to the plane of the U. Within this tube the fluid flow rate produces a torquing motion about the axis of symmetry of the U. The amplitude of this torsional motion gives the fluid flow rate, and the period provides the fluid density. The meter is thus capable of mass flow-rate and density measurements in a nonintrusive fashion, since no structural element enters the conduit.

5.3.16 Flow-Conditioning Devices

A significant factor that affects, to some degree, the performance of every type of fluid meter is the distribution of the flow into the meter. Because of the infinitude of piping arrangements that can precede and follow an installed flowmeter, a very wide range of flow anomalies can exist and can radically alter the meter's performance. Whether a meter is designed for a particular flow profile or is flow calibrated in a laboratory, factors that cause the actual flow to differ from the expected flow will also cause a difference in meter operation. One way to ensure that the flow profile at the meter inlet is the one expected is to install a suitable flow conditioner at an appropriate distance upstream of the flowmeter.

It would seem that flow-profile anomalies could be minimized by designing the inlet piping to allow viscous and turbulent diffusion to produce the desired flow distribution. This desired distribution is produced by long (i.e., many pipe diameters), straight lengths of piping of a constant diameter that matches that of the meter. Similar piping downstream of the meter prevents any flow peculiarities from propagating upstream and affecting the meter. Several difficulties arise with this approach to establishing designed flow conditions:

1. Different types of meters are affected differently by different types of flow anomalies.
2. Because of the nonlinear nature of the physical laws that govern the flow fields within meters, it is difficult to predict the effects of flow anomalies on different types of meters.
3. Experiments to test all types of meters with all types of anomalous flow profiles would be extremely time consuming and expensive.
4. Many meter users cannot spare the space needed to install long, straight lengths of upstream and downstream piping to achieve designed meter performance.

Because of such difficulties, a variety of conditioning devices have been devised to produce a desired flow condition. The device may be designed to produce the flow that is expected to occur in the particular pipe for the particular Reynolds number, after passing down a length equal to many pipe diameters. Or, it may be designed to produce the same specific flow distribution, regardless of the anomalous pattern that enters the flow conditioner. The latter is often regarded to be the easier of the two alternatives.

It is of paramount importance to satisfactorily condition the flow without producing a large static-pressure loss as the fluid flows through the conditioner. Such considerations led to the use of flow conditioners that included one or more screens, or metal plates having many holes supported across the flow cross section. The principle here was that the (generally turbulent) mixing occurring just downstream of the screen would tend to eliminate abnormal distributions of fluid kinetic energy.

Because the effects of fluid swirl are widely known to radically alter the expected performance of many types of meters, these are of prime concern to users of fluid meters. Severe secondary motions or swirl can be introduced into pipe flow by elbows in the piping. Perhaps the elbow arrangement most likely to introduce the greatest swirl is two elbows installed close together and oriented so that they are "out of plane." This generally produces a vortex in the pipe flow, where the angular vorticity vector lies along the pipe centerline. This type of swirl persists for many, many diameters of pipe length before the effects of fluid viscosity can dissipate it. Even a single elbow can set up vortices in pipe flow. These secondary motions can be pictured as having vorticity vectors that are oppositely directed, lie parallel to but off the pipe centerline, and are symmetric with respect to the plane containing the centerlines of the pipes joined by the elbow. Again a considerable length of pipe is required for viscous diffusion to dissipate such motions.

To remove such anomalous motions from pipe flow via flow conditioning, the principles of waveguiding are widely used. Thus a wide variety of straightening tubes, baffles, and radial panels have been arranged to obstruct the swirling velocity components and (preferably with low pressure loss) reduce it to an acceptable level or eliminate it. In American metering practice, various arrangements for bundling numerous small tubes have been used [1, 2]. Efforts to produce an optimal flow conditioner (i.e., one that has maximum capability for removing anomalies and minimal pressure loss) are continuing [10, 11].

5.4 PROVING—PRIMARY AND SECONDARY STANDARDS

Calibration requires flow measurement with maximum absolute accuracy and precision, usually with an apparatus that collects the total flux of fluid during a measured time interval. This flux is conventionally measured gravimetrically or volumetrically, with fluid density measurements usually required in either case. Standards of mass, length, time, and temperature are required for these flow-rate measurements, but the uncertainty in flow-rate calibrations is greater than can be attributed to uncertainty in the basic standards.

Uncertainties involved in flow-rate measurements result from the kinds of calibration facilities and procedures used. Uncertainties common to all facilities and procedures result from the lack of ability to achieve the following.

- Set and maintain a steady flow rate.
- Measure flow rate without error.
- Separate imprecision in flow stability from imprecision of the calibration flow standard. A meter may be differentiated from a flow standard, as the latter is built upon other, more basic standards directly, with minor corrections based on momentum and energy principles.
- Establish and determine the pertinent fluid properties.
- Completely remove spurious or systemic flow disturbances.

Uncertainties that are pertinent to specific facilities and procedures vary considerably; thus, only generic examples are discussed here.

Depending on the fluid involved, various procedures and equipment are used to measure the bulk mass flow rate. For liquid flow, direct weighing and timing techniques are commonly used:

1. In a static-weigh mode, before and after the flow is collected, via a timed diverter valve in a weigh tank
2. In a dynamic-weigh mode, with the weighing and timing operations performed while the flow is being collected in a weigh tank.

5.4.1 Liquid Flow: Static Weighing Procedure

Figure 5.16 is a schematic diagram of apparatus used for liquid flow-rate measurement. Rotation of the diverter valve allows fluid to be collected in the weigh tank and actuates a timer that measures the collection interval. This time interval, in conjunction with the collected mass as determined by weighings before and after

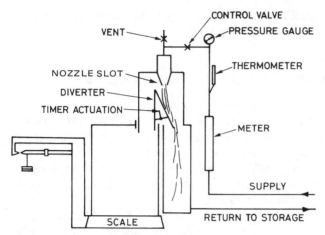

Figure 5.16   Static-weighing calibrator

collection of the liquid, gives the flow rate. The apparatus and procedure used should receive careful attention, to ensure accuracy of measurement.

It is desirable that the diverter cut through the liquid stream as rapidly as possible (in 30 ms or less) to help reduce the possibility of a significant diverter error. This is accomplished by rapid diverter travel through a thin liquid sheet formed by the nozzle slot. Generally, this liquid sheet has a length of 25 to 50 times its thickness. Also, the pressure drop across the nozzle slot should be kept small to avoid excessive splashing and turbulence in (and on) the emitted jet and in the weigh tank.

Experience has shown that, for a well-designed system, the switching error for one start-stop cycle of the diverter may correspond to an error of 0 to 10 ms. This error is dependent upon the flow rate, the velocities of traverse (in each direction) of the diverter tip through the liquid sheet, and the exact location of timer actuation with respect to the liquid sheet emerging from the nozzle slot. Appendix 5.D describes a convenient procedure for determining the switching error introduced by the diverter start-stop cycle.

Atmospheric buoyancy should be considered in a precision weighing procedure. Its magnitude may be in the range 0.1–0.2% of the actual weight, depending upon the densities of the ambient air and the liquid, and the weight of the tank.

In those calibration applications requiring a conversion between volumetric and gravimetric units, the measurement of liquid density is of extreme importance. Density error may range from 0.002% to 0.1%, depending on the particular technique employed and the accuracy of temperature measurements at the flowmeter during the density determination and in the collection tank.

Because loss by evaporation is a source of error, the volatility of the liquid should be considered whenever a liquid-flow-calibration system is vented to the atmosphere. However, for liquids of high vapor pressure such as gasoline, refrigerants, and liquefied gases, considerable refinement of the techniques described here is necessary to eliminate both loss by evaporation and vapor formation in the flowmeter and the meter discharge lines.

Typical systematic errors for current state-of-the-art liquid-flowmeter, static-weigh calibration systems are as follows:

Source	Error, %
Diverter switching	0.025
Time	0.005
Mass from weighing scale	0.025
Mass-to-volume conversion	0.03

These errors combine to form a possible overall systematic error in the range 0.04–0.08%, depending upon the procedure used to combine the individual errors, and on whether or not conversion from mass to volume units is required. The precision of such a system is best evaluated by performing repeated observations to obtain a measure of the closeness of repeated observations, or a value for the standard deviation. Generally, the value of the standard deviation for one observation is in the range 0.03–0.15%. This includes the imprecision (sensitivity and repeatability) of the flowmeter under test, which, of course, varies considerably among different types of meters.

5.4.2 Liquid Flow: Dynamic Weighing Procedure

The static-weighing method of flowmeter calibration is time consuming and thus not well suited for those applications in which convenience and speed of operation are important. Therefore dynamic weighing is utilized frequently. In this procedure, the time interval required to collect a preselected mass of liquid is measured; the weighing is performed while liquid is flowing into the weigh tank.

A diverter is not used in dynamic-weighing procedures. Rather, under a condition of steady flow, the weigh-tank dump valve is closed. As the weight of liquid in the tank increases, it overcomes the resistance of a tare-value counterpoise mass on the end of the weigh beam, which then rises, starting the timer (Fig. 5.17). An additional preselected mass is added to the pan, depressing the weigh beam. When it rises again, the timer is stopped. This procedure requires acceleration of the weighing scale just prior to both the start and stop actuations of the timer.

Four important dynamic phenomena take place during the dynamic weighing cycle. They are

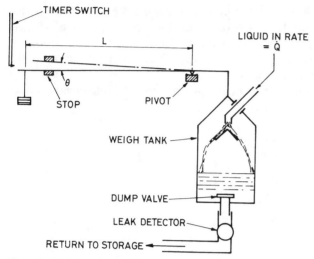

Figure 5.17 Dynamic-weighing calibrator

- A change in the impact force of the falling liquid between the initial and final weigh points
- Collection of an extra amount of liquid from the falling column by the rising level in the tank
- A change in the inertia of the scale and liquid in the weigh tank, with a resultant change in the time required to accelerate the weigh beam to the timer actuation point
- Forces due to waves in the tank

Generally, the decrease in impact forces is equal and opposite to the additional weight of liquid collected from the vertically falling column. Thus these two effects cancel each other. The change in inertia between the initial and final weight points can affect the indicated flow rate by as much as 0.5% if this inertial error is not accounted for [12]. On smaller dynamic weighing systems, the inertia effect can be eliminated by using a substitution weighing technique. Surging or oscillations of liquid within the weigh tank may have a strong influence on the precision of the weighing procedure. Baffles and other arrangements within the tank can reduce, but not eliminate completely, this undesirable phenomenon, which is always most pronounced at higher rates of flow.

From this discussion, it may be seen that dynamic-weighing procedures may introduce unknown systematic errors and weighing imprecision into the calibration procedure. Thus a static check and calibration of the instrumentation is not sufficient to prove absolute accuracy because the response time of the instrumentation under transient conditions is very important.

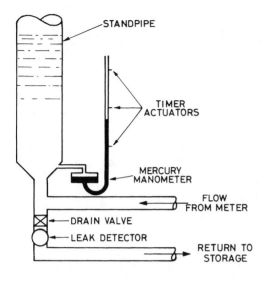

Figure 5.18 Standpipe type of dynamic weighing calibrator

Another type of dynamic procedure for determining liquid flow rate is the so-called standpipe system. The principle of operation involves determining the liquid quantity that dynamically fills the left leg of the manometerlike apparatus sketched in Fig. 5.18. The dynamic elevations of the mercury column in the right leg of the device can then be related to the volume of liquid in the left leg. This type of device offers the obvious advantage of speed in determining liquid flow rate. It is important that the relationship between the liquid quantity in the left leg of the device and the mercury column height in the right leg be very well characterized. A disadvantage of this system is that the collection of fluid in the standpipe increases the pressure in the test pipeline. This is transmitted to the output of the pump (or head tank) producing the desired flow rate and tends to decrease it. Thus the calibration of meters in such systems can be affected by the decreasing flow rate. Of course, the magnitude of this effect depends upon the geometric features of the standpipe, the amount of fluid being collected, and the characteristics of the pump and reservoir system.

At times, load cells are used on flowmeter calibration systems instead of mechanical-weigh scales. These have the advantages of simplicity of installation and digital readout. Like scale systems, load-cell systems have to be properly calibrated at frequent intervals to assure accuracy.

5.4.3 Gas Flow: Static Procedure

Figure 5.19 shows the layout of an NBS facility that can be used to measure gas flow via a static procedure similar to that described above for liquids. Figure 5.20 shows the vacuum collection tank depicted in Fig. 5.19. The measurements are

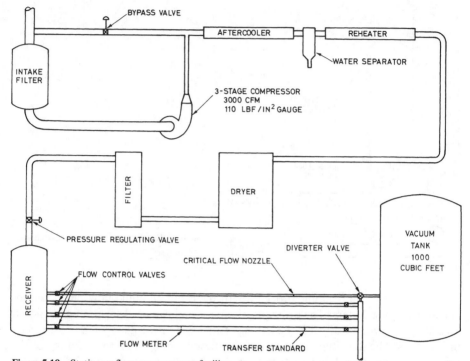

Figure 5.19   Static gas-flow measurement facility

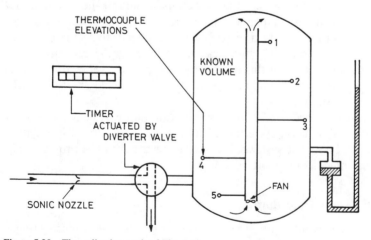

Figure 5.20   The collection tank of Fig. 5.19

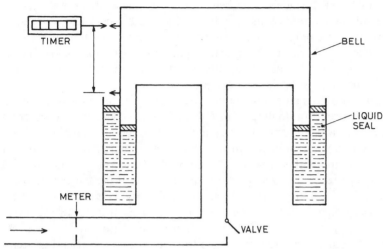

Figure 5.21   Bell-type dynamic gas-flow prover

called "static," as both the collected gas volume and the density change over the collection interval can be measured at presumably stationary conditions. Dynamics are involved only in the timed opening and closing of a diverter valve. The collection volume is found from the weight and density of water and/or gaseous nitrogen and argon fillings of the tank. (The density is determined from temperature and pressure measurements.) Results of multiple filings of the tank indicate that a volume uncertainty as small as 0.02% can be achieved. The temperature of gas in the larger tank is measured with 25 copper-constantan thermocouples, all made from a single batch of wire, located over the interior of the tank. A sonic nozzle prevents the tank pressure (which is practically zero at the initiation of collection) from affecting the back pressure on the meter under test. It is estimated that facilities of this type are capable of determining gas flow rate with an uncertainty of ±0.26% [12].

5.4.4 Gas Flow: Dynamic Procedure

Gas flow can also be measured with bell-type provers, as illustrated in Fig. 5.21. Similarly, gas can be collected via the motion of mercury-sealed pistons in vertical glass tubes; the collection process is timed, after an initial acceleration period, by a timer actuated by two photosensors when their light beams are interrupted by the piston. The light beams that traverse the glass tube are placed a known vertical distance apart. Figure 5.21 shows that the vertical motion of the bell in its annular bath of sealing liquid is similarly timed by a switch that is actuated successively by two arms mounted on the side of the bell. Internal-diameter measurements are used to derive volume per unit of distance traveled for both types of provers; additional measurements are required to account for displacement motion of the sealing liquid in the bell-type prover. The transfer of air between a standard volumetric measure and a bell prover is sometimes used for its calibration, but dimensional calibration is recommended for large provers.

Careful attention to numerous details is necessary to avoid measurement difficulties. These arise from small rates of flow and small collected volumes, from the dynamics of the measurement process, and from the difficulty of making meaningful gas-temperature measurements in small gas-flow systems. A nearly constant laboratory temperature, both spatially and temporally, is used in conjunction with sufficient piping upstream from the meter and calibrator to bring the gas temperature to that of the calibration system, meter, and laboratory. This not only reduces temperature measurement problems but also prevents heat transfer in the meter and prover, a very important requirement. Thermal insulation and/or heat exchangers are also used and are recommended to ensure equal meter and gas temperatures in difficult environments.

It is estimated that, with facilities such as these, gaseous flow rates may be determined to within 0.26% [12].

5.4.5 Ballistic Calibrators

A type of piston prover that is quite similar in operation to that described above for gas flow is the ballistic calibrator, which can be used on either gases or liquids. This system has several advantages, including simplicity, bidirectionality, compactness, and ruggedness, which permit it to be installed in the field on pipelines that transport large quantities of costly fluids such as petroleum. A typical arrangement is shown in Fig. 5.22.

The ballistic calibrator, sometimes referred to as the "ball prover," determines fluid flow rate via the following sequence:

1. The desired flow rate is produced by the flow-control valve, downstream of the meter. The fluid flows from the reservoir, through the pump to the meter, through valve A, and back to the reservoir. All other valves designated by letters in Fig. 5.22 are closed. The entire piping system is completely filled with the test fluid. All valving is electrically operated.

2. At the instant at which fluid collection is to begin, valves B, D, E, and C are opened, and valve A is closed. This initiates the motion of the spheroid (ball) from behind the detector switch, which, when actuated, starts a timer. The flow from the meter now passes through valves B and D and into the region behind the ball. The fluid in front of the moving ball passes through valves E and C and back into the reservoir.

FLOW REVERSING VALVES

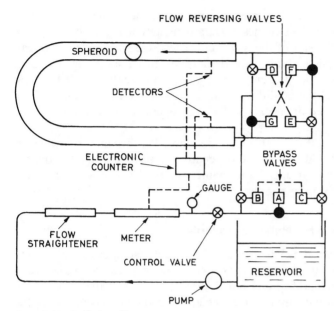

Figure 5.22 Ballistic calibrator

3. When the ball reaches the second detector switch, the timer is stopped, valves D and E close, and valve A opens, so that the flow from the meter is returned to the reservoir.
4. Quickly, before the desired flow rate changes, valve A is closed as valves G and F are opened. This initiates ball motion in the opposite direction. When the ball passes the first detector switch, the timer is again started. Fluid from the meter passes through valves B and G and into the region behind the ball. The fluid in front of the ball passes through valves F and C and into the reservoir.
5. When the ball reaches the other detector switch, the timer is stopped, valves G and E close, and valve A opens to return the flow to the reservoir.

Each of the two "ball travels" displaces a volume of fluid between the detector switches that start and stop the timer. Thus, each pass enables one to calculate the volumetric flow rate. Any "diverterlike" error in the system (see App. 5.D) that would cause a systematic error to be associated with the displaced volume in one direction and its corresponding time interval tends to be canceled when the two displaced volumes and the two respective time intervals are summed and used to determine the flow rate.

It is, of course, necessary to assume that there is no leakage past the ball during its travel. Also, one must accurately calculate or measure the volume displaced by the ball as it moves between the switches. By approximately sizing the ball-pipeline diameter in relation to the size of the meter under test, the speed of the ball can be kept low. This adds plausibility to the assumption that the displaced volume can be measured by slowly filling or emptying the ball pipeline through the motion of the ball and measuring the fluid displaced. It allows the critical volume of the ball prover to be determined or checked in the same dynamic manner in which the device is actually used.

The integrity of the ball must also be maintained. If it becomes scratched so that leakage occurs, it should be replaced or repaired immediately. In addition, the switching must be carefully arranged and maintained so that any switching error is either canceled out or kept negligible.

5.4.6 NBS Facilities and Secondary Standards

The conventional calibration capabilities at NBS include facilities for calibrating flowmeters and wind-velocity sensors. Flowmeters of many types and line sizes can be calibrated in water, air, and hydrocarbon fluids. A summary of these facilities is given in Table 5.1. The usual calibration procedure at NBS consists in testing the device repeatedly at each of five flow rates spanning the normal range of usage of the meter, in the fluid actually metered by the instrument. If NBS is unable to calibrate a device in the fluid actually metered, sometimes a surrogate fluid is used. How well this works is dependent upon specific details, and each case is treated individually.

In the repeated testing of a meter, the fluid flow rate is determined five consecutive times by gravimetric or volumetric means, together with the meter response. The 25 points generated provide a database with which the performance of the device can be predicted statistically, as a result of a single day's testing. The entire process is repeated on another day to provide data concerning "day-to-day" repeatability when the device is not removed from the pipeline. The calibration procedure and the results are described in a Report of Calibration, which allows the requestor to claim "conventional traceability" to NBS for fluid flow measurements made with the calibrated device.

A flow-metering device that has been satisfactorily calibrated can be used as a secondary, or transfer, standard to calibrate other meters. There are advantages and disadvantages to such a scheme. The main advantage is speed and correspondingly lower cost. The quality of such a "secondary" calibration depends upon the characteristics of the original device and how these are maintained. A critical disadvantage of such a scheme involves the satisfactory interruption of any "communication" between two meters in the same pipeline via the common flow rate. This is, of course, essentially the problem one faces when installing a flowmeter downstream of anything that produces "flow disturbances," or variations of the inlet flow profile expected by virtue of the pipe flow physics. For example, two turbine

Table 5.1 NBS flow-calibration facilities and performance characteristics

Facility	Capabilities	Flow determination system, features
	Gas and Liquid Metering in Closed Conduits	
Low air flow	1.5 m³/min (50 scfm) max.; 3440 kPg (500 psig) max.; ambient temperature (20°C) ± 0.25% uncertainty	Bell-type provers, mercury-sealed piston devices, critical nozzles
High air flow	83 m³/min (3000 scfm) max.; 861 kPg (125 psig) max.; ambient temperature (20°C) ± 0.25% uncertainty	Constant-volume collection tank, critical nozzles
Low water flow*	2.5 kg/s (40 gpm) max.; 345 kPg (50 psig) max.; ambient temperature (20°C) ± 0.13% uncertainty	Dynamic weighing
High water flow	6.25 kg/s (10,000 gpm) max.; 517 kPg (75 psig) max.; ambient temperature (20°C) ± 0.13% uncertainty	Static weighing
Low-liquid-hydrocarbon flow	10 kg/s (200 gpm) max.; 207 kPg (30 psig) max.; ambient temperature (20°C) ± 0.13% uncertainty	Dynamic weighing
High liquid-hydrocarbon flow	100 kg/s (2000 gpm) max.; 348 kPg (50 psig) max.; ambient temperature (20°C) ± 0.13% uncertainty	Static weighing
	Wind Sensors	
Wind tunnel	0.9- × 0.9-m test section; 4.5 cm/s to 9 m/s ± 1.0% uncertainty	Velocities measured using laser velocimetry, hot-wire anemometry
Wind tunnel (dual test section)	1.5- × 2.1-m test section; 46 m/s max. ± 0.3% uncertainty	Low stream turbulence, adjustable pressure gradient; velocities measured using hot-wire anemometry and pitot tube
	1.2- × 1.5-m test section; 82 m/s max. ± 0.3% uncertainty	Low stream turbulence; velocities measured using hot-wire anemometry and pitot tube
Wind tunnel	1.4- × 1.4-m test section; 27 m/s max.	Can be operated in steady or fluctuating mode; gusts range from 0.1 to 25 Hz; amplitudes range to 50% of mean flow

*Antifreeze additives may be used.

meters in the same pipeline can interact in a significant way via the swirl imparted to the flowing stream by the upstream meter.

Communication between two meters in the same pipeline can be satisfactorily interrupted in several ways. For example, if long lengths of straight, constant-diameter piping separate the two meters, then viscous diffusion of anomalous effects can occur, resulting in the "proper" inlet flow entering the downstream meter. The specific length of piping required depends critically upon

1. How susceptible the downstream meter is to particular disturbances in the inlet flow
2. The nature of the disturbances imparted to the flow by the upstream meter
3. The nature of the flow in the pipe connecting the two meters

It is generally found, however, that testing laboratories and laboratories operated by meter manufacturers do not contain enough space to accommodate the "long, straight lengths" required by the installation specifications for the meter in question. Thus, one confronts the problem of artificially "conditioning" the flow so that either (1) the effect of long, straight lengths of piping on the pipeflow is somehow duplicated, or (2) the anomalous flow effects are reduced (or eliminated) without incurring excessive fluid pressure loss, so that the flow entering the downstream meter does not contain disturbances or anomalies that would impair meter performance. The latter is the purpose of flow conditioners. Then, the downstream unit, operating in conjunction with its upstream conditioner, becomes a reliable product for sale, or a satisfactory secondary, or transfer, flow standard.

Flow conditioners have in the past and will in the future play a critical role in fluid metering. They are available in a wide variety of shapes and sizes, as the result of the very wide range of flow disturbances that are known to impair meter performance and the very diverse range of intuition and imagination that have produced them [10, 11].

5.5 TRACEABILITY TO NATIONAL FLOW STANDARDS— MEASUREMENT ASSURANCE PROGRAMS FOR FLOW

Conventional practice for establishing the traceability of a flow measurement includes many activities. For example, the owner of a newly purchased and properly installed flowmeter might define traceability through the flowmeter manufacturer's quoted performance for the device. The manufacturer, in turn, might define traceability as the spot-checking of finished products against a secondary standard meter that is installed in the pipeline of the calibration facility. Calibration against an in-line standard meter has the significant benefits mentioned earlier. Of course, from time to time the manufacturer should check the performance of the secondary standard against a "master" standard. The master standard probably is on a shelf most of the time and thus is not degraded through normal use, as the in-line standard may be. Again, from time to time the manufacturer should check the performance of the master meter.

The master meter can be calibrated at NBS, or it might be checked in house if facilities for gravimetric or volumetric calibration as described above are available. Such a calibration can, however, be a rather slow process that may impair the normal production routine. The traceability that the manufacturer might cite for its gravimetric or volumetric test facility would probably be a calibration of its system by the state office of weights and measures. For a gravimetric system this generally means that state weights were brought to the producer's laboratory and used to calibrate its scale system. To check its timing system, the manufacturer would either calibrate its timers against a timing standard or use the WWV time signal from NBS. Further, the manufacturer would check its diverter system (or inertial corrections if the system were dynamic) and pertinent fluid properties in some appropriate manner. Since the weights and timing standards would all be solidly traceable to NBS, the new meter owner would be assured of the traceability of flow measurements to NBS.

However, this traceability chain can have some very weak links, as indicated by the following questions:

1. What statistics justify the manufacturer's reliance on spot checks of its products?
2. For what period of time should one rely on manufacturer's specifications for the performance of a new meter?
3. Are the flow and fluid parameters properly bracketed by the spot-checking calibration procedure?
4. Is the meter purchaser using the secondary element in the manner in which it was used in the spot-check calibration?
5. Is the fluid flow profile at the manufacturer's lab the same as that at the user's installation?

The link between the user and the manufacturer may not be the only weak one. For example, the manufacturer might ask

1. How do I justify my spot-checking arrangement?
2. When I check two flowmeters in the same pipeline, is there any interaction that affects either meter?
3. When I check my flow determination system with a static (or nearly static) process, how well do I account for the dynamics involved in its actual use?
4. What is the flow profile of the fluid entering my meters in the calibration facility?
5. Is the calibration I receive from NBS for my master meter indicative of the performance of this device when I use it as I do in my lab?

Some questions can be asked at NBS as well:

1. When a manufacturer sends a master meter for calibration, how much of the manufacturer's auxiliary pipework and secondary element should be included to get the calibration that is wanted?
2. When NBS personnel operate and read the secondary element, is the technique the same as that used by the operators in the manufacturer's laboratory?
3. Is the flow profile entering the meter at NBS indicative of that prevailing in the manufacturer's lab?
4. How good is the NBS Flow Determination System?
5. How often should the NBS Flow Determination System be checked?

Thus it seems apparent that the traceability chain can have some weak links, if not some broken sections. Several of these questions could perhaps be avoided if users sent their units directly to NBS for calibration. However, this might not solve all the problems.

So the question remains: What should be done, and by whom, to establish traceability? To answer this question, it is necessary to decide on what is meant by traceability and what constitutes traceability for flow measurements. Following this, it should be decided how flow traceability should be established and maintained.

The following definitions of traceability have been proposed [13]:

1. Traceability is the ability to demonstrate conclusively that a particular instrument or artifact standard has either been calibrated by NBS at accepted intervals, or has been calibrated against another standard in a chain or echelon of calibrations, ultimately leading to a calibration performed by NBS.
2. Traceability to designated standards (national, international, or well-characterized reference standards based upon fundamental concepts of nature) is an attribute of some measurements. Measurements have traceability to the designated standards if and only if scientifically rigorous evidence is produced on a continuing basis to show that the measurement process is producing measurement results (data) for which the total measurement uncertainty relative to national or other designated standards is quantified.
3. Traceability means the ability to relate individual measurement results to national standards or nationally accepted measurement systems through an unbroken chain of comparisons.
4. Traceability implies a capability to quantitatively express the results of a measurement in terms of units that are realized on the basis of accepted reference standards, usually national standards.

Although there are similarities among these four definitions, the differences among them are more interesting. The salient difference between the first two is the involvement with calibrations of instruments. The second definition does not include the word "instrument" or "calibration," whereas the first definition is intrinsically based upon instruments and their calibrations. The second definition stresses the results of measurement processes.

This contrast—between instruments and calibrations on the one hand and the results of measurement processes on the other—is considered by this author to be the basis for distinguishing between "static" and "dynamic" traceability.

5.5.1 Static Traceability

By static traceability is meant the collection of activities that metrologists must perform to quantify the performance of the various components of their measurement facilities. For example, in the United States, in the case of flow measurement, this might mean checking a weighing facility (or volumetric prover) with weights or volumetric test measures from a state office of weights and measures (OWM) that is itself traceable to NBS. Also, it would mean checking the laboratory's timing mechanisms against NBS traceable timing standards. Furthermore, if liquid density were required to convert mass measurements to volume, then static traceability would necessitate establishing the relationship between density-measuring instruments and NBS standards.

However, if the diverter mechanism in the metrologist's flow facility were faulty, if a peculiar flow profile entered a flowmeter under test, if a significant human error recurred during flowmeter tests, or if the processing of the raw data to final results were wrong, then evidently static traceability would not be sufficient to establish the resulting flow measurements in the laboratory. Something more would be required, namely, quantitative proof that the entire flow measurement process in the laboratory continuously produces results for which the uncertainty relative to specified standards is quantified. It is this extra sufficiency condition that constitutes dynamic traceability.

5.5.2 Dynamic Traceability

Properly established dynamic traceability is intended to assess, in a realistic manner, all the pertinent factors influencing the flow measurement results produced in a facility. These factors include

1. The particular scheme or instruments arranged for determining mass or volumetric flow rate
2. The environmental conditions prevailing during the test period
3. Operator idiosyncrasies, if present
4. The calculation procedures used for determining final test results from raw data

To establish the dynamic traceability of flow measurements to national standards (and between the standards of nations on an international basis), measurement assurance programs (MAPs) for fluid flow are required [14–16].

5.5.3 Measurement Assurance Programs

The purpose of a flow MAP is to

1. Quantitatively establish the total uncertainty of flow measurement processes
2. Provide proof that flow measurements are as good as specified
3. Evaluate the entire flow measurement system in question (operators, environment, methods, instruments, etc.)

The advantages of a flow MAP are that it

1. Provides, clearly and quantitatively, the limitations of a particular method of measurement
2. Gives a clear description of the factors affecting the uncertainty of the measurement
3. Provides guidance for obtaining satisfactory results where small uncertainties are required
4. Provides the rationale for simplifying existing procedures
5. Provides a means for cyclically monitoring the performance of measurement processes

A flow MAP can be arranged within individual countries in which the national laboratory is a participant and, usually, the organizer of the activity. A MAP is composed of two essential elements. The first is an artifact (defined as "any object made by man, especially with a view to subsequent use"); the second is an algorithm (defined as "a particular rule or procedure devised for solving a specific problem"). Artifacts consist of a flowmeter(s) adjacent pipework, and the auxiliary instrumentation that is required to establish the dynamic traceability of flow measurements. Algorithms consist of the testing procedure devised to produce the desired data, the analyses of these data, and the scheduling through which the measurements traced are continually checked. The process is sketched in Fig. 5.23 alongside a conventional calibration procedure.

To initiate the flow MAP, the originating laboratory (possibly the national standards organization) generally selects a simple, widely used, reliable type of transfer standard, perhaps a flowmeter that is capable of performance adequate to the intended task, i.e., making good, repeatable measurements for a particular fluid, flow-rate range, and pipeline size. Confidence can be added to the test results by using two flowmeters connected in series in the flow facility. To ensure that the particular flowmeters are performing as expected, it is recommended that the performance of these meters be cyclically checked at a particular laboratory, such as the originating laboratory. Control-chart records should be maintained and analyzed to quantify any changes observed in performance. A typical chart for an NBS turbine meter is shown in Fig. 5.24.

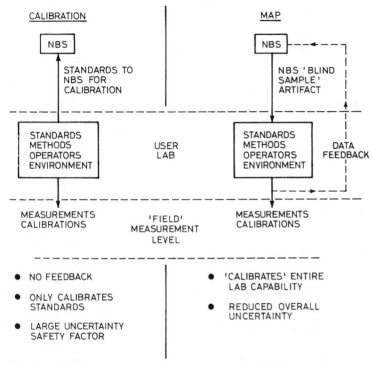

Figure 5.23 Comparison of conventional calibration and MAP procedures

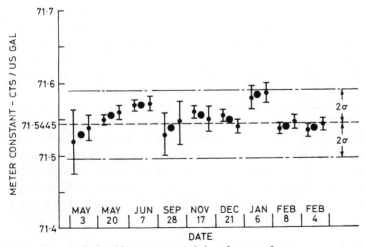

Figure 5.24 Typical turbine-meter control chart for meter factor

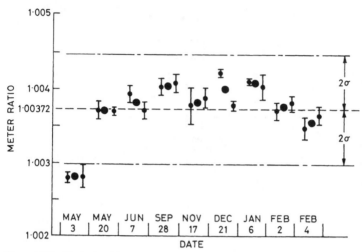

Figure 5.25 Typical turbine-meter control chart for ratio

Tandem meter arrangements, composed of different types of meters or meters of the same type but of different sizes connected with appropriate fittings, should also be capable of rendering realistic data to assure flow measurements. The use of two meters in the same pipeline, so that the same flow passes through both, produces data in which considerable credibility can be placed [17]. This tandem meter-testing configuration has the obvious advantage of providing redundancy for the data generated. The tandem arrangement also provides the opportunity to quickly and efficiently monitor the relative responses of the two meters to the flow through both of them. For example, if the two meters were identical turbine meters, their relative responses to the flow could be the ratio of their respective frequencies. In such a case, when the ratio meter monitoring the two signals continually displays a value that lies within an acceptably small tolerance of an expected value, increased confidence can be placed in the data being taken. This expected value (which should approximate unity for identical meters) should be obtained from previous testing conducted in the originating laboratory.

Experience at NBS has shown that two flowmeters that are identical in make, manufacturer, and model can perform quite differently [17, 18]. A typical control chart for the ratio performance of a particular pair of NBS turbine meters is shown in Fig. 5.25. Deviation of the monitored ratio of frequencies from the expected value by more than the preset tolerance would be interpreted as indicating that one of the two meters had been changed somehow—an event that would obviously impair the credibility of the test results. The testing should then not proceed until the ratio could be made to conform to the expected value within the specified tolerance.

It is possible that identical malfunctions could occur simultaneously to both

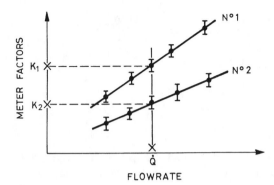

Figure 5.26 Normal meter performance

meters in such a way as to shift their performance by the same amount and in the same direction; however, the probability of this event is assumed to be negligibly small.

Figure 5.26 shows a chart of flowmeter performance. For a flow rate $\dot{Q}$ the respective flowmeter factors are K_1 and K_2. Figure 5.27 shows the ratio performance R_{12} for these two meters at the flow rate $\dot{Q}$ and the tolerance $\pm\delta$ for the expected ratio value $R_{12,\dot{Q}}$. When, as shown in Fig. 5.28, the ratio display indicates a value R such that

$$R_{12,\dot{Q}} - \delta \leq R \leq R_{12,\dot{Q}} + \delta \tag{6}$$

The testing of the two meters continues, and the meter factors are calculated via the flow determination system of the laboratory under test. These meter factors then contribute to the database for this laboratory.

Figure 5.29 illustrates circumstances in which testing should not be continued. Here the ratio display indicates a value R such that

$$R \gg R_{12,\dot{Q}} \tag{7}$$

and exceeding the specified tolerance. The reason for the unacceptably high value of R is illustrated as a "partially damaged" meter 2. A particle of dirt may have

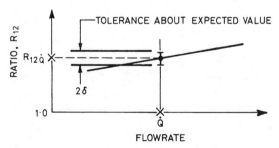

Figure 5.27 Expected ratio for flow rate $\dot{Q}$

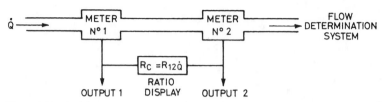

Figure 5.28 Satisfactory condition for conducting the test

gotten into the bearing of this turbine meter, so that the rotational frequency of its impeller was decreased. This would decrease the denominator of the ratio R, producing the observed high value. In this case, a decision to continue the test on the basis that the upstream meter 1 is probably functioning properly is detrimental to (1) the performance of the laboratory under test, and (2) confidence in the database that will be used to establish dynamic traceability among the participating laboratories.

There is an additional advantage to testing flowmeters in tandem. At each flow rate, one obtains repetitive and simultaneous determinations of the performance of the two meters, i.e., the respective meter constants or flow factors. Correlation of these results enables one to decompose the total variation observed for each meter into components [19]. These component variations are, for each meter, those attributable (1) to the meters themselves, and (2) to the flow facility in which the meters are tested. Figures 5.30–5.33 illustrate how the total variation observed in the meter-factor results for each meter can be decomposed into these two components.

In Fig. 5.30 are shown the actual meter performance and expected value of meter factor for the two meters. Figure 5.31 shows the piping configuration under test; the flow rate $\dot{Q}(t)$ is taken to be a function of time. The time-averaged value of this unsteady flow rate is $\dot{\bar{Q}}(t)$, as shown in Fig. 5.30. Figure 5.32 shows a possible flow rate $\dot{Q}(t)$ as varying sinusoidally with mean value $\dot{\bar{Q}}(t)$. The figure also shows the time interval over which the laboratory under test determines flow rate. For these conditions, it is apparent that this laboratory's meter-factor results will be markedly scattered, in spite of the fact that the meters are very repeatable,

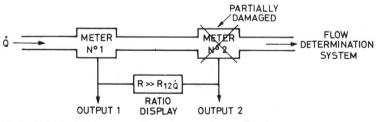

Figure 5.29 Unsatisfactory conditions for continuing the test

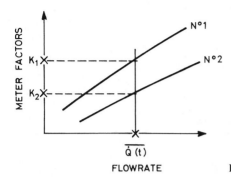

FLOWRATE

Figure 5.30   Normal meter performance

high-resolution devices, as shown in Fig. 5.30. This variation is displayed in Fig. 5.33, which shows the scattering of the meter-factor data about mean values of K_1 and K_2. The standard deviations $\sigma_{1,T}$ and $\sigma_{2,T}$ represent the quantified scattering about the mean values.

The correlation coefficient r_{12} is computed for these simultaneously determined meter factors at a particular flow rate such as $\dot{Q}(t)$. Under the assumption that the sources of variation are statistically independent, one can interpret the square of this correlation coefficient as that fraction of the total observed meter-factor variation that can be associated with both meters. The two meters are assumed statistically independent from each other except for their physical connection via piping and the (unsteady) flow passing through them. These two connections are taken together and referred to as the facility component of the total meter-factor variation observed for each meter; that is,

$$\sigma_{1,T}^2 = \sigma_{1,M}^2 + \sigma_{1,F}^2$$
$$\sigma_{2,T}^2 = \sigma_{2,M}^2 + \sigma_{2,F}^2 \tag{8}$$

where $\sigma_{1,T}$ and $\sigma_{2,T}$ are, as before, the standard deviations for the total meter-factor variation. The quantities $\sigma_{1,M}^2$ and $\sigma_{2,M}^2$ are the portions of the total variation in meter factor that are due to the meters themselves. The quantities $\sigma_{1,F}$ and $\sigma_{2,F}$ are those portions of the total observed meter-factor variations attributable to the facility. According to the assumptions described above,

$$r_{12} = 1 \tag{9}$$

and

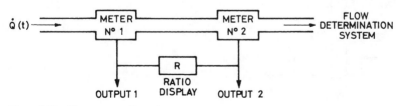

Figure 5.31   Flow-test configuration

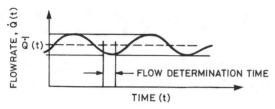

Figure 5.32 Assumed flow-rate variation

$$r_{12}^2 = \frac{\sigma_{1,F}^2}{\sigma_{1,T}^2} = \frac{\sigma_{2,F}^2}{\sigma_{2,T}^2} = 1 \tag{10}$$

Therefore, it is concluded that

$$\sigma_{1,M} = \sigma_{2,M} = 0 \tag{11}$$

so that the variation observed in the meter-factor results for each meter is due entirely to the facility, and that none of the total variation can be attributed to the meters (see also App. 5.H).

In real situations, the circumstances are not so simple. However, this analysis is one more way in which increased confidence can be placed in the resulting database. When, through the course of testing the same set of meters in the prescribed manner in a series of laboratories, the same (or similar) results are continually obtained for meter variation, all participants are more confident of the implications of the tests, that is, the traceability sought.

It should be noted that the temporal variation of $\dot{Q}(t)$ assumed via Fig. 5.32 should not necessarily violate the ratio criteria of Fig. 5.27; this is because of the nature of the meter characteristics in Fig. 5.30. In fact, for the situation indicated by Figs. 5.30–5.33, the laboratory under test appears to obtain the "correct" values for K_1 and K_2 in spite of the undesirable fact that the flow cannot be steadied. However, for different meter characteristics, or for different flow-rate variations, a wide range of alternative results is feasible. An important example is the situation wherein the inability to produce a steady flow results in both excessive variations in, and systematically wrong values for the meter factor.

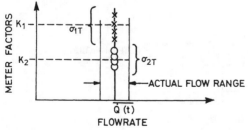

Figure 5.33 Meter-factor results for single flow rate

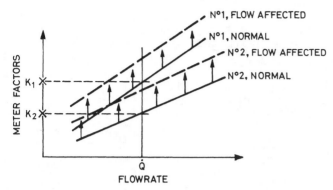

Figure 5.34 Assumed meter performance and flow-profile effects

5.5.4 The Role of Flow Conditioning in the Artifact Package

The flow velocity profile is widely known to have a significant effect on the performance of flowmeters. For this reason, the incorporation of flow conditioners into the MAP algorithm has been found to be very effective [17, 18]. Furthermore, when present development efforts have finally evolved optimal flow conditioners, it is expected that entirely new arrangements of devices and meters will become available for inclusion in flow MAPs.*

The role of flow-conditioning devices in flow MAPs will be to aid in determining whether participating laboratories have anomalous flow profiles in their facilities. Figures 5.34–5.36 indicate how flow conditioners could quantitatively assess flow-profile problems in laboratories. For the assumed meter performance charac-

---

*At NBS, computer modeling efforts have produced preliminary flow-conditioner configurations that appear to have very good conditioning capabilities. Laser-based experiments are planned to validate these results. Dr. F. Kinghorn at the National Engineering Laboratory (NEL) in the United Kingdom, and Dr. M. Sens at Société Nationale des Gaz du Sud-Ouest in France are actively developing other flow-conditioning ideas.

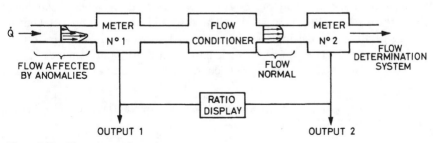

Figure 5.35 Flow-test configuration

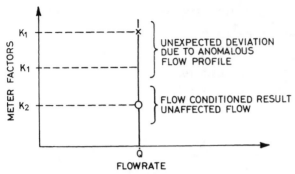

Figure 5.36 Meter-factor results for single flow rate

teristics shown in Fig. 5.34, the meter factor for a flow affected with "anomalies" is increased from the value measured in "normal" profiles. When the proper flow-conditioning element is bolted between* the tandemly piped meters, only the upstream meter (meter 1) is affected by flow anomalies (Fig. 5.35). The downstream meter, being screened from these anomalies (and any introduced into the pipe flow by the upstream meter), exhibits normal performance. The results of the tests of Fig. 5.35 are shown in Fig. 5.36. The unexpected deviation represents, quantitatively, the effect of the flow anomaly on meter 1.

5.5.5 Test Program

The test program must be devised so that (1) high confidence can be placed upon the artifact package, and (2) the database produced is adequate to the task of clearly evaluating the significant components of the system in question. To establish and maintain high confidence in the artifact package, an adequate testing program is carried out in the originating laboratory, with close monitoring of meter performance. The results, in control-chart form, are updated in timely fashion and disseminated to the participants[†] in the MAP. The tandem-meter configuration is used with the ratio criteria (or some alternative). The flow-conditioning device is used between the metering units (meters plus matched adjacent meter tubes).

It is recommended that only one or two flow rates be used to test the artifact in each of the participating laboratories. These should be uniquely specified in terms of the pertinent dimensionless parameter, such as Reynolds number. In this manner, attention is focused on each participant laboratory's performance in determining meter factors at specific, closely controlled flow conditions, rather than

---

*Suggestion attributed to Dr. E. A. Spencer of the Flow Measurement Division, National Engineering Laboratory, United Kingdom.

†The participants in the program have to agree on the level of anonymity they collectively wish to maintain through the operation of the MAP.

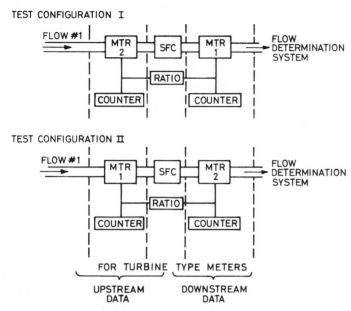

Figure 5.37 Schematic test procedure for a single flow rate

diffused by attempting to characterize overall meter performance (i.e., a calibration curve) over a wide range of flows.

Each of the two flowmeters is tested in the upstream and downstream positions. This generates two statistically independent sets of data for each participant in the MAP, for each flow rate specified for testing. More than one flow rate is recommended to allow a laboratory facility with, say, a low-flow-rate capability, to participate at least partially. A sketch of the piping configurations is shown in Fig. 5.37. In the upper portion of the figure, meter 2 occupies the upstream position, and meter 1 is downstream. With this arrangement, all flows are tested, and sufficient repeat testing is performed to evaluate the "switch-off, switch-on" repeatability and/or the "day-to-day" or "running" repeatability. When testing in this configuration is completed, the tests are duplicated for the arrangement shown in the lower portion of Fig. 5.37.

5.5.6 Data Analysis

The method of analysis used to assess the variance in flow measurement results among participating laboratories was formulated by W. J. Youden [20] and has been used at NBS for years in interlaboratory testing programs. In this procedure, the statistically independent results for each flow condition for the respective meters in the same position are plotted, one meter factor on the ordinate and the

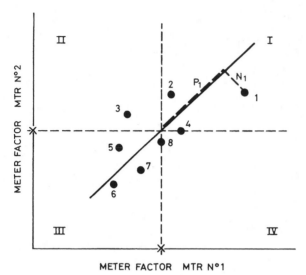

Figure 5.38 Graphic representation of the Youden analysis

other on the abscissa (Fig. 5.38). Thus, each laboratory is represented by a single point that is labeled with a coded number, letter, or symbol to maintain the anonymity of the laboratories. Usually, the originating laboratory is custodian of such information and of the data used to perform this analysis. Plots such as that shown in Fig. 5.38 have been referred to as Youden plots, in honor of Youden's significant contribution to NBS efforts to develop round-robin testing programs [20, 21].

It is recommended that, before round-robin testing programs begin among participating laboratories, the originating laboratory thoroughly test all equipment and instrumentation according to the test procedures devised. This serves to "run in" the artifact and establish its performance, stability, etc., according to the originating laboratory's measurement processes. In addition, it allows the procedural steps in the algorithm to be evaluated for effectiveness and efficiency. The resulting database enables a Youden plot to be generated and analyzed to determine how well the whole set of MAP ingredients performed in the originating laboratory. In a sense, these results indicate, quantitatively, the degree of resolution that one might expect in using this MAP to examine the flow measurement processes of the participating laboratories.

The Youden analysis is performed by drawing vertical and horizontal lines, respectively, through the medians of the abscissa and ordinate data (see Fig. 5.38). The intersecting median lines then divide the data into four quadrants. In Cartesian notation, data lying in the first of these quadrants is, to some extent, systematically inaccurate, since each point is "high" relative to the best available estimate of the true values of the turbine-meter constants, namely, the coordinates of the intersection of the medians. Similarly, points lying in the third quadrant are systematically low. Data in the second and fourth quadrants are termed inconsistent or random, since points in these areas are "low-high" and "high-low" relative to the intersection of the medians. Thus, the degree to which the data lie in a circular pattern about this intersection or are distributed elliptically with the major axis at a slope of $+1$ or -1 quantifies the nature of the variation in the flow-facility and turbine-meter systems.

The total variation in the data can be categorized by calculating standard deviations based upon the parallel and perpendicular projections of all the data points onto a line drawn through the median intersection with a slope of $+1$. These are

Systematic

$$\sigma_s = \left(\frac{1}{N-1} \sum_{i=1}^{N} P_i^2 \right)^{1/2} \tag{12}$$

Random

$$\sigma_r = \left(\frac{1}{N-1} \sum_{i=1}^{N} N_i^2 \right)^{1/2} \tag{13}$$

The ratio of these parallel and perpendicular standard deviations gives the degree and orientation of the ellipticity of the data:

$$e = \frac{\sigma_s}{\sigma_r} \tag{14}$$

If this ratio is larger than unity, the variation can be interpreted as predominantly systematic. Depending on its magnitude, possible causes for this variation can be sought in either the meter's performance, the flow facility, or both. A ratio e that is very much less than unity could indicate that the transfer standard package is not capable of sufficient resolution, and other metering devices need to be selected. Alternatively, there could be insufficient resolution in the flow facility, or operator inconsistencies. This pattern has not been observed in NBS studies.

If the ratio is close to unity, then systematic and random variations are similar. Decisions can then be made as to whether the radius of the data spread is acceptable or whether improvement is desirable.

The typical results shown in Fig. 5.39 are simulated data for the test procedure of Fig. 5.37. The various participants are indicated by symbols. The two points for each laboratory indicate the repeatability of that facility. Generally, repeatability is tested by switching the flow off and then on again. The solid symbols in Fig. 5.39 denote results obtained in the originating laboratory after the artifact package was

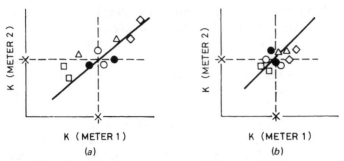

Figure 5.39 Typical Youden results for a single flow rate. (*a*) Upstream meters. (*b*) Downstream meters

returned from the participating laboratories. The repeatability of the results indicates that the artifact did not change during the round robin testing program.

It is not possible to prescribe specific responses to the results of such an analysis. At present, the individual participants are entirely on their own. Their positions on the Youden plots can provide the impetus for reexamining some (or all) of the ingredients of the flow measurement processes.

When other participants are the competitors of a particular laboratory whose results lie outside the main body of results, the impetus for improvement can be immense. On the other hand, when a facility's performance is comfortably close to that of the national standards laboratory, a very satisfied feeling develops in that facility. In fact, this situation can, in some senses, be more of a detriment than an outlier, for example, if excellent agreement is used as the excuse to postpone or eliminate the periodic checking of a laboratory's measurement processes. It is highly erroneous to believe that, because good (or excellent) agreement was obtained once via a flow MAP, the system should "forever after" be left alone.*

To avoid such faulty thinking, the flow MAP should be designed so that a repeat test is performed at a later date. The scheduling should be based upon meter performance as shown by the control charts (Figs. 5.24 and 5.25). That is, the artifact package should exhibit stable performance over the time interval required to test a whole round of laboratories.

If such testing intervals are excessively long, then the set of participating laboratories should be subdivided such that the testing interval is commensurate with the stability shown by the control charts. When a group of flow laboratories is subdivided into smaller groups, the testing efforts of the originating laboratory are increased, owing to the fact that the "before-and-after" testing must be done for each subround. To reduce the amount of extra testing, MAPs can be arranged regionally. In such a scheme, local originating laboratories are chosen by geograph-

---

*This is similar to the fluid-metering adage that, "if one plans to make a very accurate, conventional calibration of a very accurate flowmeter, one should plan to do this only once!"

ical location. To distribute the work load, this duty can rotate among the participants. Cross checks among regional originating laboratories can be used to unite the results of all participants.

To satisfactorily and realistically demonstrate that one's future fluid flow measurements will be good as specified, "high-confidence," realistic, dynamic-traceability chains are required. These dynamic-traceability chains can be set up with the establishment of MAPs for flow. In these, generic flow measurement conditions (i.e., flowmeter types, fluids, flow-rate ranges, and pipeline sizes) are assessed, and MAPs are designed around these conditions, so that confident links of dynamic traceability connect national laboratories to the nation's high-quality flow measurement laboratories. Similar programs are designed to establish the assurance of flow measurements down the traceability chain to the field level. Participation in such programs remains the responsibility of those making fluid flow measurements. Ultimately, conventional flowmeter calibration practice should give way to the enhanced assurance attainable via flow MAPs.

The national laboratories of countries that depend upon fluid resources would be well advised to establish flow MAPs within their borders and set up dynamic-traceability links with other national laboratories. In this way, the increasingly critical and costly transfer of fluid resources can occur in the world's marketplaces with satisfactory and demonstrable equity for all concerned.

APPENDIX 5.A IDEAL PERFORMANCE CHARACTERISTICS FOR DIFFERENTIAL-PRESSURE-TYPE METERS: INCOMPRESSIBLE FLUIDS

For the steady, incompressible (i.e., isentropic) flow of an ideal fluid through two cross-sectional areas $A_1 > A_2$ in a closed conduit, the continuity equation is

$$\dot{m}_1 = \rho_1 A_1 V_1 = \rho_2 A_2 V_2 = \dot{m}_2 \tag{A1}$$

where $\dot{m}_1$, $\dot{m}_2$ = constant mass flow rates
  ρ_1, ρ_2 = fluid densities
  V_1, V_2 = average fluid velocities through A_1, A_2

Conservation of energy in the absence of mechanical work and elevation change gives

$$0 = \frac{dp}{\xi} + \frac{V\,dV}{g} \tag{A2}$$

where p = absolute pressure
  ξ = specific weight
  V = average fluid velocity normal to cross-sectional area
  g = acceleration of gravity

Integrating Eq. (A2) between A_1 and A_2 gives

$$\frac{V_2^2 - V_1^2}{2g} = \frac{p_1 - p_2}{\xi} \tag{A3}$$

where, if $\beta = D_2/D_1$, and D_1 and D_2 are the respective diameters at A_1 and A_2,

$$V_2 = \left[\frac{2g(p_1 - p_2)}{\xi(1 - \beta^4)}\right]^{1/2} \tag{A4}$$

Thus, the ideal mass flow rate $\dot{M}_I$ is

$$\dot{M}_I = A_2\left[\frac{2\rho(p_1 - p_2)}{1 - \beta^4}\right]^{1/2} \tag{A5}$$

APPENDIX 5.B IDEAL PERFORMANCE CHARACTERISTICS FOR DIFFERENTIAL-PRESSURE-TYPE METERS: COMPRESSIBLE FLUIDS

For an ideal gas, the equation of state is

$$p = \rho RT \tag{B1}$$

where p = absolute pressure
ρ = fluid density
R = gas constant
T = absolute temperature

The isentropic relationship between pressures and densities at the respective stream locations is

$$\rho_2 = \rho_1\left(\frac{p_2}{p_1}\right)^{1/\gamma} \tag{B2}$$

where $\gamma = c_p/c_v$
c_p = specific heat at constant pressure
c_v = specific heat at constant volume

Thus continuity and energy considerations give

$$V_2 = \left[\frac{2\gamma p_1(1 - r^{1 - 1/\gamma})}{(\gamma - 1)\rho_1(1 - \beta^4 r^{2/\gamma})}\right]^{1/2} \tag{B3}$$

where $r = p_2/p_1$. Thus the ideal mass flow rate $\dot{M}_I$ for compressible fluids is

$$\dot{M}_I = \frac{A_2 p_1}{T_1^{1/2}}\left[\frac{r^{2/\gamma}(r^{2/\gamma} - r^{1 + 1/\gamma})}{1 - \beta^4 r^{2/\gamma}}\right]^{1/2}\left(\frac{g}{R}\frac{2\gamma}{\gamma - 1}\right)^{1/2} \tag{B4}$$

where the bracketed term is called the fluid meter function X, and the term in parentheses is called the fluid meter constant K_X. Then Eq. (B4) can be written

$$\dot{M}_1 = \frac{A_2 p_1}{T_1^{1/2}} XK_X \tag{B5}$$

APPENDIX 5.C REAL, COMPRESSIBLE ORIFICE-FLOW CALCULATION

The equation characterizing the orifice metering of gases is [2]

$$\dot{Q}_h = C'\sqrt{h_w p_f} \tag{C1}$$

where $\dot{Q}_h$ = volumetric rate of flow at reference conditions, L^3/T
  C' = orifice flow constant*
  h_w = differential pressure referred to reference temperature, L
  p_f = absolute static pressure, F/L^2

The orifice flow constant* is computed via

$$C' = F_b F_r YF_{pg} F_{tb} F_{tf} F_g F_{pv} F_m F_a F_l \tag{C2}$$

where F_b = basic orifice factor
  F_r = Reynolds number factor
  Y = expansion factor
  F_{pb} = pressure-base factor
  F_{tb} = temperature-base factor
  F_{tf} = flowing-temperature factor
  F_g = specific-gravity factor
  F_{pv} = supercompressibility factor
  F_m = mercury-manometer factor
  F_a = orifice thermal expansion factor
  F_l = gage-location factor

APPENDIX 5.D DIVERTER EVALUATION AND CORRECTION

Experience has shown that diverter systems can contribute significantly to errors in flow measurement processes where static-weighing schemes are used to determine liquid flow rates. In a well-designed diverter system, the timing error (the error in starting and stopping the liquid collection interval) may be reduced to less than 10 ms. On the other hand, if such systems are not properly evaluated and periodically checked, significant errors can result.

*Not to be confused with the discharge coefficient.

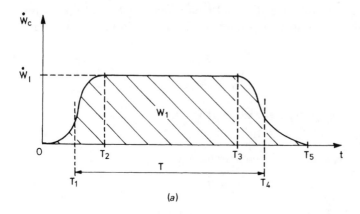

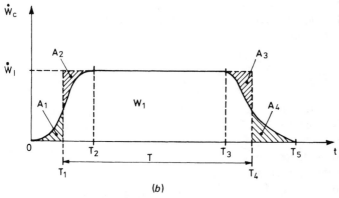

Figure 5.D1 (*a*) Graph of $\dot{W}_c$ versus time. (*b*) The areas of interest are shaded

This error is dependent upon

1. The liquid flow rate and its distribution within the diverter system
2. The motion of the diverter through the liquid jet
3. The location of the timing switches activated by the moving diverter

To evaluate the diverter system and quantify its performance characteristics, a flow sensor such as a turbine meter is first installed in the facility to monitor flow rate. This is used to vary the flow rate in some reasonable manner over the range of operation of the diverter. If W_1 is the liquid weight collected during time interval t, the collection process may be plotted in terms of the collected flow rate $\dot{W}_c$, as shown in Fig. 5.D1*a*.

The time $t = 0$ denotes the instant at which the diverter mechanism is activated. The time T_1 is the starting point for the timer recording the diversion interval. Time T_2 is the instant at which the collected flow rate $\dot{W}_c$ is $\dot{W}_1$, which is being monitored by the turbine meter, and T_3 is the instant at which the diverter system is activated to stop the collection. Time T_4 is the instant at which the timer stops, and T_5 denotes the end of the collection, including dripping into the collection tank.

The area under the collection curve is then

$$W_1 = \int_0^{T_5} \dot{W}_c(t)\, dt$$

This weight will be the correct one for the diversion interval $T = T_4 - T_2$, where $W_1 = \dot{W}_1 T$, when

$$\int_0^{T_1} \dot{W}_c(t)\, dt + \int_{T_4}^{T_5} \dot{W}_c(t)\, dt = \int_{T_1}^{T_2} [\dot{W}_1 - \dot{W}_c(t)]\, dt + \int_{T_3}^{T_4} [\dot{W}_1 - \dot{W}_c(t)]\, dt$$

This equality is shown graphically in Fig. 5.D1*b* as

$$A_1 + A_4 = A_2 + A_3$$

which is satisfied if

$$A_1 = A_2 \qquad \text{and} \qquad A_3 = A_4$$

or if

$$A_1 = A_3 \qquad \text{and} \qquad A_2 = A_4$$

Given the collection characteristics of the diverter, i.e., the function $\dot{W}_c(t)$, these areas can be made equal by proper adjustment of the times T_1 and T_4 or, equivalently, the positions of the switches that start and stop the timer. Alternatively, the error incurred by associating the time $T = T_4 - T_1$ and the area under the curve W_1 can be determined as a function of flow rate. With this determined, a correction can be made to obtain the true weight corresponding to time interval T.

Several schemes are available for producing diverter correction as a function of flow rate. One such scheme, described here, is based upon amplifying the diverter error by repetitive diversions that, when totaled and compared to a single diversion, permit simple evaluation of the error. With the desired flow rate through the diverter system, the turbine-meter generating frequency f gives the volumetric flow rate $\dot{Q}$ via

$$\dot{Q} = \frac{f}{K}$$

where K is the turbine constant. The weight flow rate is

$$\dot{W} = \frac{f\xi}{K}$$

where ξ is the fluid specific weight. If the error associated with the timed interval T is denoted by ε, then a single diversion for a collected weight W gives

$$f_1 = \frac{K_1}{\xi_1} \frac{W_1}{T + \varepsilon} \tag{D1}$$

Where the subscript 1 denotes a single diversion.

When N repetitive diversions (as many as 25) collect a total weight $\sum_{i=1}^{N} W_i$ in the total time interval $\sum_{i=1}^{N} (T_i + \varepsilon)$, then

$$f_2 = \frac{K_2}{\xi_2} \frac{\displaystyle\sum_{i=1}^{N} W_i}{\displaystyle\sum_{i=1}^{N}(T_i + \varepsilon)}$$

where the subscript 2 denotes values pertaining to repetitive diversions. If the N repetitive diversions are equal in duration, then

$$T = \sum_{i=1}^{N} T_i = NT_i$$

$$f_2 = \frac{K_2}{\xi_2} \frac{\displaystyle\sum_{i=1}^{N} W_i}{\displaystyle\sum_{i=1}^{N} T_i} \frac{1}{1 + N\varepsilon/T} \tag{D2}$$

The ratio of Eqs. (D1) and (D2) produces

$$\frac{f_1 \xi_1 / K_1}{f_2 \xi_2 / K_2} = \frac{W_1/(T + \varepsilon)}{\sum_{i=1}^{N} W_i / \sum_{i=1}^{N} T_i[1/(1 + N\varepsilon/T)]}$$

$$= \frac{(W_1/T)[1/(1 + \varepsilon/T)]}{\sum_{i=1}^{N} W_i / \sum_{i=1}^{N} T_i[1/(1 + N\varepsilon/T)]}$$

For convenience, let

$$\dot{W}_1 = \frac{W_1}{T} \qquad \dot{Q}_1 = \frac{f_1 \xi_1}{K_1}$$

$$\dot{W}_2 = \frac{\displaystyle\sum_{i=1}^{N} W_i}{\displaystyle\sum_{i=1}^{N} T_i} \qquad \dot{Q}_2 = \frac{f_2 \xi_2}{K_2}$$

where $\dot{Q}_1$ and $\dot{Q}_2$ are the respective volumetric flow rates. Then

$$\frac{\dot{Q}_1}{\dot{Q}_2} = \frac{\dot{W}_1[1/(1 + \varepsilon/T)]}{\dot{W}_2[1/(1 + N\varepsilon/T)]}$$

Solving for ε/T produces the fractional correction for the time T, or

$$\frac{\varepsilon}{T} = \frac{\dot{Q}_1 \dot{W}_2/\dot{Q}_2 \dot{W}_1 - 1}{N - \dot{Q}_1 \dot{W}_2/\dot{Q}_2 \dot{W}_1}$$

When the dependency of ε/T upon flow rate is determined, corrections can be made if the error is significant. Alternatively, the switch position can be changed until subsequent evaluations of this diverter error indicate it is negligibly small for all flow rates, in which case, it can be justifiably excluded from the calculations performed on calibration data.

APPENDIX 5.E EMPIRICAL FORMULAS FOR ORIFICE DISCHARGE COEFFICIENTS

Conventional practice in the United States for locating pressure taps in an orifice meter includes formulas for calculating discharge coefficients [1]. Because these formulas are quite involved, we present instead a simpler relationship produced by J. Stolz [3]. This empirical result can be used to calculate discharge coefficients for any tapping arrangements. It is

$$C_d = 0.5959 + 0.0312\beta^{2.1} - 0.184\beta^8 + 0.0029\beta^{2.5} \left(\frac{10^6}{Re_{D_1}}\right)^{0.75}$$

$$+ 0.09 \frac{L_1 \beta^4}{1 - \beta^4} - 0.0337L_2'\beta^3$$

where L_1 and L_2' are the distances of the centerlines of the pressure-tap holes from the upstream surface of the orifice plate, nondimensionalized by the internal diameter of the upstream pipe. When

$$L_1 \geqslant 0.4333$$

it is recommended that the value 0.039 be used for the coefficient of the $\beta^4/(1 - \beta^4)$ term.

This relationship produced a fit to a database of orifice discharge coefficients to within 0.2%.

APPENDIX 5.F PRESSURE MEASUREMENTS

The measurement of fluid pressure consists in determining one or more of the three quantities P_0, P_s, and P_v, where, from energy considerations,

$$P_0 = P_s + P_v \tag{F1}$$

where P_0 = fluid stagnation pressure when stagnation is ideal
P_s = fluid static pressure
P_v = fluid dynamic pressure

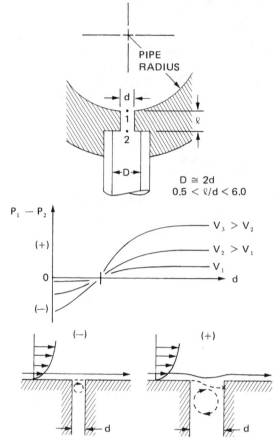

Figure 5.F1   Pressure characteristics for different tapping hole diameters

The interrelationships among these three pressures depend [4] upon

1. Fluid characteristics and properties
2. The nature of the real or ideal stagnation process
3. Pertinent laminar or turbulent flow characteristics

5.F.1 Sensing Static Pressure

Since static-pressure sensing schemes often constitute the critical link between the primary and secondary elements of a flowmeter, it is essential to good metering that they allow optimal overall performance. Conventionally, it is assumed that very small, square-edged holes drilled through flow conduits normal to parallel flow give the correct static pressure. Should this assumption not be completely

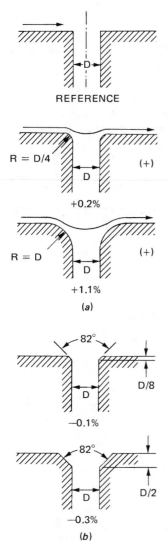

Figure 5.F2 Pressure characteristics for different tapping hole geometries. (*a*) Rounding. (*b*) Countersinking

true, it is further assumed that meter calibrations will account for discrepancies. Figures 5.F1 and 5.F2 illustrate qualitatively the kinds of dynamics that can occur and the corresponding effect on static-pressure measurements.

5.F.2 Sensing Total Pressure

Ideally, total pressures can be sensed only by stagnating the flow isentropically. Thus, one may define the Pitot coefficient C_p as

$$C_p = \frac{P_{tI} - P_s}{P_v} \tag{F2}$$

where P_{tI} is the indicated total pressure. Real effects that influence the stagnation process for total-pressure sensors are

1. Geometry. Examples are probe shape and angle of attack.
2. Viscosity including fluid properties and flow characteristics. For Reynolds numbers based on the outside radius of Pitot tubing, it is found that for $Re > 1000$, $C_p \approx 1$; for $50 < Re < 1000$, $0.99 < C_p < 1$; and for $Re < 10$, $C_p > 1$ and is asymptotic to 5.6/Re.
3. Transverse gradients. Complex flow patterns can occur, with the general result that indicated total pressures are high.
4. Turbulence. Both total-pressure and static-pressure sensors are affected by turbulence via

$$P_{tI} = P_s + \rho \frac{V^2}{2} + \frac{\rho(v')^2}{2} \tag{F3}$$

where ρ = fluid density
V = mean axial fluid velocity
v' = root-mean-square magnitude of the turbulence velocity

Thus the indicated static pressure P_{Is} becomes

$$P_{Is} = P_s + \frac{\rho(v')^2}{2} \tag{F4}$$

It should be noted that v' must be 20% of the mean velocity V before the turbulence terms amount to 4% of the mean dynamic pressure P_v.

APPENDIX 5.G TEMPERATURE MEASUREMENT AND RECOVERY FACTOR

The measurement of the temperature of a real fluid generally consists in reading the output signal of a thermometer as the fluid stagnates against the sensor surface and equilibrates thermally with this sensor surface. Because of the viscous and thermal diffusion properties of real fluids, the thermometer generally does not indicate the thermodynamically ideal stagnation temperature

$$T_0 = T_s + T_v \tag{G1}$$

where T_0 = thermodynamically ideal stagnation temperature
T_s = fluid static temperature
T_v = ideal dynamic temperature = $V^2/2Jgc_p$
V = initial local fluid velocity of stagnating fluid
J = mechanical equivalent of heat

The Prandtl number Pr expresses the ratio of fluid viscous to thermal diffusion effects; that is,

$$\text{Pr} = \frac{\text{viscous effects}}{\text{thermal diffusion}} = \frac{\mu c_p}{k} \tag{G2}$$

where μ = dynamic viscosity of fluid
$\quad k$ = thermal conductivity

Thus when, as in air, with

$$0.65 < \text{Pr} < 0.7 \tag{G3}$$

the thermal diffusion effects are greater than the viscous effects, the real dynamic temperature is less than the ideal. Similarly, when, as in water, with

$$1 < \text{Pr} < 13 \tag{G4}$$

the viscous effects are greater than the thermal diffusion effects, the real dynamic temperature is greater than the ideal. Of course, for

$$\text{Pr} = 1 \tag{G5}$$

the viscous heating effects generated by the stagnating real fluid are exactly diffused thermally, so that the real dynamic temperature is the ideal.

Thus [4]

$$T_{pi} = T_s + rT_v \tag{G6}$$

where T_{pi} = ideal probe temperature
$\quad r$ = recovery factor

The recovery factor can then be written

$$r = \frac{T_{pi} - T_s}{T_0 - T_s} \tag{G7}$$

It has been found that

$$r = \text{Pr}^{1/2} \tag{G8}$$

Therefore the temperature sensed by a sensor in a fluid stream depends upon the following:

1. The viscous and thermal diffusion properties
2. The characteristics of the sensor, i.e., shape and orientation to the flow
3. The nature of the fluid stagnation process, including any pertinent fluid motion effects, such as turbulence, that might alter the molecular viscous and thermal diffusion phenomena described above
4. Any other pertinent thermal losses from the temperature sensor

APPENDIX 5.H ANALYSIS OF VARIANCE
WITH TWO FLOWMETERS IN SERIES*

When two flowmeters are calibrated in the same line, analysis of variance techniques may be applied for the purpose of assessing the portion of the variation that is due to the flow facility and each meter.

Suppose two meters that respond similarly to flow phenomena are installed in series on a pipeline. They need not be identical meters or even meters of the same type, but for present purposes let us consider that they are identical turbine meters. For N determinations of the flow rate for a single flow setting via, say, primary standards, we can compute the respective average meter constants as

$$\overline{K}_x = \frac{1}{N}\sum_{i=1}^{N} K_{xi} \qquad \overline{K}_y = \frac{1}{N}\sum_{i=1}^{N} K_{yi} \qquad (H1)$$

where K_{xi} and K_{yi} are the turbine-meter constants determined N times via calibration techniques at a flow rate, and overbars denote averages. The standard deviations are

$$\sigma_x = \left[\frac{1}{N-1}\sum_{i=1}^{N}(K_{xi} - \overline{K}_x)^2\right]^{1/2} \qquad (H2)$$

$$\sigma_y = \left[\frac{1}{N-1}\sum_{i=1}^{N}(K_{yi} - \overline{K}_y)^2\right]^{1/2} \qquad (H3)$$

The correlation coefficient is

$$r = \frac{\sum_{i=1}^{N}(K_{xi} - \overline{K}_x)(K_{yi} - \overline{K}_y)}{(N-1)\sigma_x\sigma_y} \qquad (H4)$$

Now assume that the variances σ_x^2 and σ_y^2 are composed of portions σ_f^2 due to the facility and portions σ_{my}^2 and σ_{mx}^2 due to the meters:

$$\sigma_x^2 = \sigma_f^2 + \sigma_{mx}^2 \qquad (H5)$$

$$\sigma_y^2 = \sigma_f^2 + \sigma_{my}^2 \qquad (H6)$$

That fraction of the total variance for each meter that correlates with the other is assumed to be due to the variation attributable to the facility, or

$$r^2 = \frac{\sigma_f^2}{\sigma_y^2} = \frac{\sigma_f^2}{\sigma_x^2} \qquad (H7)$$

*See [19].

so

$$1 = r^2 + \frac{\sigma_{my}^2}{\sigma_y^2} \tag{H8}$$

or

$$\frac{\sigma_{my}^2}{\sigma_y^2} = 1 - r^2 \tag{H9}$$

is that fraction of the total variance that remains after the correlated portion is removed. Thus the standard deviations of the meters themselves are

$$\sigma_{my} = \sigma_y \sqrt{1 - r^2}$$
$$\sigma_{mx} = \sigma_x \sqrt{1 - r^2} \tag{H10}$$

In this manner, the meter performance can be quantified in any such test, and the performance of the facility can be assessed as viewed through the response of the two meters. For example, if the variations in the meter constants correlate perfectly with each other (that is, $r = 1$), then the interpretation would be that

$$\sigma_{my} = 0 \qquad \sigma_{mx} = 0 \tag{H11}$$

The total variation observed in the meter constants would be due to the facility.

On the other hand, if no correlation were observed between the respective meter constants, then $r = 0$ and

$$\sigma_{my} = \sigma_y \qquad \sigma_{mx} = \sigma_x \tag{H12}$$

The interpretation here would be that all the observed variation in the meter constants is due to the meters themselves or to factors that were completely uncorrelated.

NOMENCLATURE

A	area
c_p, c_v	specific heat at constant pressure, constant volume
C'	flow constant
C_D, C_d	dimensionless discharge coefficients
D	inside pipe diameter
e	ellipticity
F	meter factor
f	frequency (Hz)
g	gravitational acceleration
h	height, differential pressure
J	Joule's constant

k	thermal conductivity
K	flow coefficient, flowmeter constant
L	characteristic length
$\dot{m}, \dot{M}$	mass rate of flow
M	molecular weight
N	number of samples
p	pressure
Pr	Prandtl number
q	heating rate
$\dot{Q}$	volume flow rate, volumetric flow rate
r	recovery factor, correlation coefficient, pressure ratio
R	ratio of meter constants, gas constant
Re	Reynolds number
St	Strouhal number
t	time
T	temperature, absolute temperature, time
V	average fluid velocity in conduit, point fluid velocity
$\dot{W}$	weight flow rate
β	ratio of orifice hole to pipe diameter
γ	specific-heat ratio
δ	allowable tolerance
ε	expected error
μ	viscosity
ν	kinematic viscosity
ρ	density
σ	standard deviation
ξ	specific weight

Subscripts

a	orifice thermal expansion
b	basic
c	collected
f	fluid, facility in which meter is tested
F	facility in which meter is tested
g	gravity
i	sample number
I	ideal
l	gage location
m	manometer, specific flowmeter
M	specific flowmeter
p	probe
pb	pressure base
pv	supercompressibility

Q̇	specific value of flow rate
r	random, specific Reynolds number
s	systematic, static
t	total
T	total
tb	temperature base
tf	flowing temperature
v	dynamic
x,y	collected particular meters
0	stagnation
1,2	positions along conduit, specific meters in same pipe

Superscripts

−	average of a set of values
—	time averaged
′	deviation from time-smoothed value
·	time derivative

REFERENCES

1. H. S. Bean (ed.), *Fluid Meters—Their Theory and Applications,* 6th ed., ASME, New York, 1971.
2. American Gas Association, *Orifice Metering of Natural Gas,* Rep. 3, Arlington, Va., 1972.
3. R. W. Miller, National and International Orifice Coefficient Equations Compared to Laboratory Data, *Proc. of the ASME Winter Annu. Meeting,* New York, 1979.
4. R. P. Benedict, *Fundamentals of Temperature, Pressure and Flow Measurement.* Wiley, New York, 1969.
5. R. W. Davis and G. E. Mattingly, Numerical Modeling of Turbulent Flow Through Thin Orifice Plates, *Proc. of Symp. on Flow on Open Channels and Closed Conduits,* U.S. National Bureau of Standards Spec. Publ. 484, 1977.
6. R. W. Davis, Numerical Modeling of Two-Dimensional Flumes, *Proc. of Symp. on Flow in Open Channels and Closed Conduits,* U.S. National Bureau of Standards Spec. Publ. 484, 1977.
7. R. W. Davis and S. Deutsch, A Numerical-Experimental Study of Parshall Flumes, to appear in *J. Hydraul. Res.*
8. B. Robertson and J. E. Potzick, A Long Wavelength Acoustic Flow Meter (in preparation).
9. K. Plache, Coriolis/Gyroscopic Flow Meter, *Proc. of the ASME Winter Annu. Meeting,* Atlanta, 1977.
10. F. C. Kinghorn and K. A. Blake, The Design of Flow Straightener-Nozzle Packages for Discharge Side Testing of Compressors, *Proc. of the Conf. on Design and Oper. of Ind. Compressors, Glasgow,* Institute of Mechanical Engineering, New York, 1978.
11. F. C. Kinghorn, A. McHugh, and W. D. Dyet, The Use of Etoile Flow Straighteners with Orifice Plates in Swirling Flow, *Proc. of the ASME Winter Annu. Meeting,* New York, 1979.
12. M. R. Shafer and F. W. Ruegg, Liquid-Flowmeter Calibration Techniques, *Proc. of the ASME Winter Annu. Meeting,* New York, 1957.
13. B. C. Belanger, Traceability—An Evolving Concept, *Am. Soc. Test. Mater. Standardization News,* February 1979.
14. J. M. Cameron, Measurement Assurance, U.S. National Bureau of Standards Intern. Rep. 77-1240, 1977.

15. P. E. Pontius and J. M. Cameron, Realistic Uncertainties and the Mass Measurement Process—An Illustrated Review, U.S. National Bureau of Standards Monograph 103, 1967.
16. P. E. Pontius, Measurement Assurance Programs—A Case Study: Length Measurements, Pt. I, Long Gage Blocks, U.S. National Bureau of Standards Monograph 149, 1975.
17. G. E. Mattingly, P. E. Pontius, H. H. Allion, and E. F. Moore, A Laboratory Study of Turbine Meter Uncertainty, *Proc. of Symp. on Flow in Open Channels and Closed Conduits,* U.S. National Bureau of Standards Spec. Publ. 484, 1977.
18. G. E. Mattingly, W. C. Pursley, R. Paton, and E. A. Spencer, Steps Toward an Ideal Transfer Standard for Flow Measurement, *FLOMEKO Symp. on Flow,* Groningen, The Netherlands, 1978.
19. W. Strohmeier, *Notes on Turbine Meter Performance,* Rep. TN 17, Fischer and Porter Co., Warminster, Pa., 1971.
20. W. J. Youden, Graphical Diagnosis of Interlaboratory Test Results, *J. Ind. Qual. Control,* vol. 15, no. 11, pp. 133–137, 1959.
21. C. Eisenhart and W. J. Youden, *Dictionary of Scientific Biography,* vol. XIV, Scribners, New York, 1976.

FLOW VISUALIZATION BY DIRECT INJECTION

Thomas J. Mueller

A man is not a dog to smell out each individual track;
he is a man to see, and seeing, to analyze.

F. N. M. Brown

6.1 INTRODUCTION

Throughout the history of aerodynamics and hydrodynamics there has been a great interest in making flow patterns visible. The visualization of complex flows has played a unique role in the improvement of our understanding of fluid dynamic phenomena. Flow visualization has been used to verify existing physical principles and, in the process, has led to the discovery of numerous flow phenomena. In

It is a pleasure to thank H. Werlé (ONERA), R. W. Hale (Sage Action, Inc.), W. C. Wells and J. M. Hample (USAF/WAL), C. R. Smith and T. Wei (Lehigh University), W. J. McCroskey, K. W. McAlister, and L. W. Carr (U. S. Army Aeroflightdynamics Directorate-ARTA), and G. N. Malcolm (Eidetics International, Inc.) for generously providing photographs and written descriptions of their research. My sincere thanks go to my colleagues at the University of Notre Dame, R. C. Nelson, S. M. Batill, J. T. Kegelman, and B. J. Jansen Jr., for sharing their research in flow visualization as well as their helpful comments during the writing of this chapter.

addition to obtaining qualitative global pictures of the flow, the possibility of acquiring quantitative measurements without introducing probes, which invariably disturb the flow, has provided the necessary incentive for the development of a large number of visualization techniques. Although clean air and water are transparent, smoke or other particles in air, and dye or other particles in water, provide the necessary contamination for flow visualization. For very practical reasons, the study of fluid mechanics was concerned with the flow of water and other liquids until relatively recent times. Man's interest in flight, however, pointed out the necessity of visualizing air flows to understand the mechanics of objects moving through the air. Many substances have been used to visualize the flow of air and water. In air, smoke, helium bubbles, dust particles, and even glowing iron particles, have been used; in water, a variety of dyes, particles, neutrally buoyant spheres, and both air and hydrogen bubbles have been employed.

An important consideration for any method of flow visualization is, What does the picture show about the fluid motion and how can the flow patterns in the picture be interpreted? Does the injected substance (e.g., smoke, dye, or bubbles) trace out a streamline, streak line, or path line? In boundary layers, separated regions, and other regions of relatively high vorticity gradients, why does the injected substance accumulate in some places and not in others? To begin to answer these questions, one must review several fundamental notions of fluid mechanics, namely, the definitions of streamline, streak line, and path line.

Streamlines, streak lines, and path lines are three curves that have been defined to help describe the flow of a fluid. A streamline is a curve that is everywhere tangent to the instantaneous velocity vectors or, in other words, everywhere parallel to the instantaneous direction of the flow. A streak line is the locus of all fluid particles that have passed through a prescribed fixed point during a specified interval of time. A path line is the curve traversed by a particular fluid particle during a specified interval of time. When the flow is steady (i.e., not dependent upon time), the streamline, streak line, and path line that pass through the same point are identical. Conversely, when the flow is unsteady (i.e., dependent upon time), these three lines are, in general, different.

If the identity of individual particles or bubbles can be followed for a given length of time, then the trajectory of these particles or bubbles may be recorded and a path line obtained. Smoke lines or dye lines emanating continuously from each opening of the smoke or dye injector are streak lines. If the smoke or dye line can be interrupted for a specific time interval, then a time–streak line can be created. The type of fluid curve that may be obtained, and/or how the visual flow pattern may be used to gain insight into the problem under study, will be discussed for each method presented.

The present discussion of flow visualization by direct injection will only consider the use of smoke and helium bubbles in air, and dye and hydrogen bubbles in water. Because of the author's experience with the use of smoke, a somewhat more extensive treatment of this subject will be presented.

The purpose of this presentation is to acquaint the reader with the types of equipment and procedures necessary to use these flow visualization techniques, as well as the advantages and disadvantages of each technique.

6.2 AERODYNAMIC FLOW VISUALIZATION

It was recognized as early as 1759 by John Smeaton that the natural wind was variable in direction and speed and, therefore, unreliable for aerodynamic research. An artificial and controllable wind was required. Although many early investigations used the whirling machine, invented in 1746 by Ellicott and Robins, or some variation of this basic idea, it was not until the wind tunnel was invented by F. H. Wenham in Great Britain about 1871 that systematic aerodynamic experiments were possible. This first wind tunnel had a 260-mm^2 cross section with air blown into the inlet by a steam-driven fan at speeds of up to 18 m/s [1]. Flow visualization in wind tunnels closely followed the development of these facilities. In this section, the use of smoke and helium bubbles as visualizing agents will be discussed. Two methods of introducing smoke will be presented, namely, the smoke-tube and the smoke-wire methods.

6.2.1 Smoke-Tube Method

The use of smoke to visualize the flow in wind tunnels began around the turn of the century and was due to L. Mach (Vienna, 1893) and E. J. Marey (Paris, 1899). Some years later, Prandtl and his associates (Gottingen, 1923), L. F. G. Simons and N. S. Dewey (Teddington, 1930) and W. S. Farren (Cambridge, 1932) experimented with smoke visualization. The important advances toward the eventual use of smoke visualization as a research tool began in the 1930s with the work of A. M. Lippisch (Darmstadt, 1937) and F. N. M. Brown (Notre Dame, 1937). A more complete history of the use of smoke in wind tunnels is given in [2] and [3].

Although Brown and Lippisch established subsonic smoke-tunnel techniques, which still represent the state-of-the-art, Brown's equipment had important advantages not possible with that built by Lippisch. For example Brown's three-dimensional tunnels had much larger inlet contraction ratios (24:1 compared to 12:1) and more antiturbulence screens, which produced lower turbulence levels at the high velocities. In fact, by making smaller test sections, while increasing the contraction to 48:1 and 96:1, Brown was able to produce low turbulence levels at speeds up to about 60 m/s. Furthermore, by introducing the smoke upstream of the screens instead of inside the tunnel, as Lippisch and most of the others did, the smoke injection rake did not add to the disturbances in the test section. Expanding the techniques of Brown, V. P. Goddard (Notre Dame, 1959) was able to produce the world's first supersonic smoke tunnel. With this equipment, smoke photographs were taken at speeds up to 404 m/s. It is not surprising that these

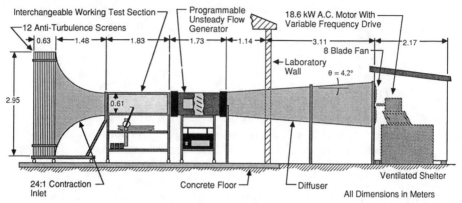

Figure 6.1 Low-turbulence subsonic smoke tunnel with unsteady flow generator (all dimensions in millimeters). (Courtesy of T. J. Mueller, University of Notre Dame)

wind tunnels, smoke generators, and photographic techniques have been copied extensively.

The most important requirement for a smoke flow visualization tunnel is that it should have a low turbulence level in the test section. This is readily accomplished by using several screens followed by a large contraction in area ahead of the test section. In the following section, a description of the equipment and techniques developed by Brown and Goddard will be presented. These facilities have been used continuously by faculty, students, and sponsored research projects over the past four decades. The equipment will be described as it now stands in the Aerospace Laboratory at the University of Notre Dame.

6.2.1.1 Low-turbulence subsonic wind tunnels. There are four nonreturn or indraft subsonic wind tunnels in the Notre Dame Aerospace Laboratory. All were designed with the same basic concepts developed by Brown. A summary of the original specifications of all of these tunnels is given in [2]. The two largest identical tunnels that recently received new motors and controllers will be described in detail.

Each of the two large indraft tunnels is powered by an 18.6-kW (25 hp) ac motor, with a solid-state speed control (see Fig. 6.1). The motor drives an eight-bladed fan that is 1220 mm in diameter. The motor and fan assembly is mounted on a concrete base and is isolated from the diffuser. To eliminate mechanical vibrations, the diffuser is isolated from the test section by a 101-mm section of sponge rubber. The motor is enclosed in a ventilated shelter. The open area of the shelter walls may be varied by raising or lowering canvas curtains. This shelter protects the motor from the elements and also provides a wind break for the tunnel exhaust. In order to produce an oscillating free stream, a special unsteady-flow generator was constructed and adapted to the existing subsonic wind tunnel [4, 5]. An unsteady velocity was produced by the pitching motion of four $610 \times 150 \times 6$ mm

(24 × 6 × 0.25 in.) louvers mounted horizontally across the flow channel. The louvers were driven by a dc servomotor that was controlled by a microcomputer. In the first series of experiments, a nonreversing streamwise periodic velocity was produced by oscillating the louvers about a fixed position. This method was used, since it resulted in a velocity waveform with low harmonic distortion relative to rotating the louvers at constant speed. Nonetheless, a technique to further reduce harmonic distortion of velocity variation was developed and employed throughout these experiments. Through the use of computer control, the velocity waveform in the test section had excellent repeatability and less than 4% total harmonic distortion in amplitude (0.2% spectral energy).

Several interchangeable working sections are available for these tunnels. Both constant-area and tapered working sections are on hand. These tunnels can be adapted quickly to use square test sections of 610 × 610, 48:1, and 96:1. All of these test sections are 1828 mm in length and have 6.35-mm plate glass on at least one side, with a flat black or black velvet back. However, the 431 × 431 and 305 × 305 mm test sections have additional contractions and diffuser sections attached to them. An externally mounted two-component strain gage balance may be used in conjunction with any desired working section. Antiturbulence screening is used upstream of the contraction cone. There are five bronze screens measuring 5.51 × 7.09 meshes/cm (14 × 18 meshes/in.), followed by seven nylon screens measuring 7.87 × 7.87 meshes/cm (20 × 20 meshes/in.). The bronze screens are made of wire 0.305 mm (0.012 in.) in diameter, and the nylon screens have a thread diameter of 0.076 mm (0.003 in.). The main contraction cone (see Fig. 6.1) has a square cross section and a shape given by Smith and Wang [6]. The measured turbulence (disturbance) levels, using a single-sensor hot wire, were less than 0.08% for a 1- to 2500-Hz bandwidth and less than 0.025% for a 25- to 2500-Hz bandwidth. Spectra did not reflect a turbulence character; hence, acoustic influences probably dominated at these low levels. Several peaks in the spectra could be correlated with fan blade passage frequencies and electronic noise, although these were several orders of magnitude below the major constituents of the spectrum [7].

6.2.1.2 Smoke. The word "smoke" is used in a very broad sense in flow visualization techniques and includes a variety of smokelike materials such as vapors, fumes, and mists. The smoke used must be generated in a safe manner and must possess the necessary light-scattering qualities so that it can be readily photographed. It is also important that the smoke not adversely affect the wind tunnel into which it is introduced nor the model being studied. Another desirable but not absolutely necessary qualification is that the smoke be nontoxic in the unlikely event the experimenters are exposed to it. Finding a smokelike substance that meets all these criteria is not an easy task.

A large number of materials have been used to generate smoke, e.g., the combustion of tobacco, rotten wood, and wheat straw, the products of reaction of various chemical substances such as titanium tetrachloride and water vapor, and the vaporization of hydrocarbon oils, to name just a few. The smokelike materials used may be referred to as aerosols, since aerosols are composed of colloided particles suspended in a gas. A great deal of interest has been focused on aerosol generation and its properties because of its close relationship to meteorology, air pollution, cloud chambers, smokes, combustion of fuels, colloid chemistry, etc.

Two very practical items must be carefully examined before the choice of smokelike material is made for flow visualization. The smoke or aerosol particles must be as small as possible, so they will closely follow the flow pattern being studied. These smoke particles must be large enough to scatter a sufficient amount of light, so that photographs of the smoke pattern can be obtained. Although many materials and substances have particle sizes below 1 micron (μm), the most practical ones for flow visualization are tobacco smoke, rosin smoke, carbon black, and oil smoke [8,9]. Tobacco smoke and carbon-black particles range from 0.01 to about 0.20 μm, while rosin smoke particles range from about 0.01 to about 1.0 μm and oil smoke particles range from 0.30 to about 1.0 μm. There is no doubt that all these particles are small enough to follow the flow. It should be noted that the flow particles in water vapor (fog) are generally much larger than 1 μm (i.e., 1 to about 50 μm). If one now considers the light-scattering ability of the particles used, another constraint becomes apparent. Particles should be larger than about 0.15 μm to scatter a sufficient amount of light to be readily seen. This light-scattering criterion indicates that tobacco smoke and carbon-black particles are mostly lower than 0.15 μm and therefore would be more difficult to photograph.

Resin is a semisolid, organic substance exuded from various plants and trees or prepared synthetically, whereas rosin is the hard, brittle resin remaining after oil of turpentine has been distilled. Maltby and Keating [10] also mention that some ammonium chlorate is present in the resin canisters, which must be stored away from heat, since this substance is unstable. The device is simply a smoke bomb adapted for wind-tunnel use.

There are, of course, many possible hydrocarbon mixtures, i.e., oils, which undoubtedly could be used to produce smoke by combustion or vaporization. From the point of view of laboratory safety, it would be desirable to use vaporization rather than the combustion technique. Furthermore, it would also be safer to use an oil that would vaporize at the lowest possible temperature and be the least flammable. Considering the oils most commonly used to produce smoke, mineral oil requires the highest temperature for vaporization, while charcoal lighter fluid requires the lowest. The second lowest temperature for vaporization is for kerosene [8,9]. Since kerosene is less flammable than charcoal lighter fluid, it is the obvious choice. Kerosene seems to offer the best compromise when particle size, light-scattering ability, low vaporization temperature, and low flammability are considered.

6.2.1.3 Kerosene smoke generation and rake system. Although the first oil smoke generator was developed by Preston and Sweeting [11], one of the most successful

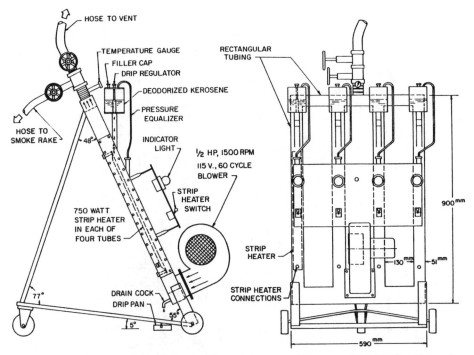

Figure 6.2 Schematic of Brown's four-tube oil smoke generator. (Courtesy of T. J. Mueller, University of Notre Dame)

oil smoke generators was designed by F. N. M. Brown in 1961. Large quantities of dense kerosene smoke were produced quickly and safely with this generator. This four-tube kerosene generator is shown schematically in Fig. 6.2.

A flat electric strip heater is located inside a 51 mm (2 in.) square thin wall conduit tube. The entire unit is set at a convenient angle (about 60°), and a sight-feed oiler is mounted on the unit at the upper end of each tube, so the oil drips on the upper end of the strip heater. It has been determined that a drip rate of approximately two drops per second is more than sufficient to produce the desired amount of smoke. Faster rates result in inefficient and wasteful operation. Further-more, an extremely fast drip rate can result in back-firing of the unit. A squirrel cage blower mounted at the low end of the unit is used to force the smoke through the system. The squirrel cage blower is more or less mandatory; in the event of back-firing, the sudden increase in pressure is easily transmitted through the rotor.

Before entering the smoke rake, the smoke is allowed to pass through a heat exchanger made of 42-mm-diameter pipe, as shown in Fig. 6.3. The prime function of this heat exchange is to cool the smoke down to room temperature. The entire system has drain cocks conveniently located, one at the bottom of each tube of the

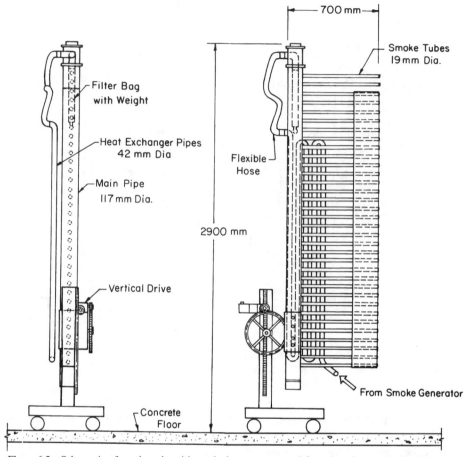

Figure 6.3 Schematic of smoke rake with vertical movement used for subsonic tunnels. (Courtesy of T. J. Mueller, University of Notre Dame)

generator itself to remove excess oil not converted into smoke, and others at the bottom of the heat exchanger to remove whatever oil might have been condensed. After passing through the heat exchanger condenser system, the smoke flows into a 117-mm manifold and is passed through an absorbent cloth filter. This filter serves a dual purpose; it removes most of the remaining lighter tars and aids in distributing the smoke uniformly into the evenly spaced 19-mm tubes that extend from the manifold. These evenly spaced tubes determine the initial smoke-line spacing. Such an array of tubes with the manifold has a rakelike appearance, and the assembly is referred to as a smoke rake. Appropriate measures should be taken to guard against leaks in the system and to provide an additional outside exhaust for the smoke generator, since the excessive and prolonged inhalation of oil smoke could be a health hazard.

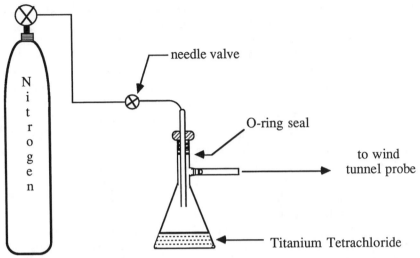

Figure 6.4 Flow visualization apparatus using titanium tetrachloride. (Courtesy of R. C. Nelson, University of Notre Dame)

Oil smoke generators of this type have been constructed in single, double, and quadruple units. This grouping arrangement of units is basically a method of increasing the volumetric output of smoke. The smoke rake shown in Fig. 6.3 can be moved up or down, as desired, from the subsonic smoke-tunnel control panel. A fixed-position smoke rake of similar design is used in conjunction with the supersonic tunnels.

6.2.1.4 Titanium tetrachloride. Although the use of titanium tetrachloride ($TiCl_4$) for flow visualization dates back to Simmons and Dewey in 1931 [12], it has recently been used to great advantage for a variety of separated and vortex flows [13–15]. The direct application of $TiCl_4$ is somewhat hazardous, and therefore its use is usually limited to short time periods. Titanium tetrachloride is a colorless liquid that, when exposed to the moisture in air, forms hydrochloric acid vapor and very small titanium dioxide particles. Recently, Visser et al. [14] at Notre Dame developed a safer and convenient system for inert pressurized gas. This system, shown in Fig. 6.4, allows for the continued production of $TiCl_4$ smoke, which can be introduced at any point in the wind tunnel. An open-return wind tunnel that exhausts outside of the laboratory is necessary for the prolonged and safe use of $TiCl_4$.

6.2.1.5 Steam. The only reason for trying to replace the kerosene smoke generation technique of Brown would be to produce a nontoxic and nonflammable smokelike substance. All products of combustion, reactions of chemical substances, oil and paraffin, vapors, and aerosols are toxic to some degree. Many of these substances are flammable, and some are corrosive or chemically active. The only technique available that does not have one or more of these undesirable properties appears to be the steam–liquid nitrogen method. However, generating large quantities of steam in some type of boiler presents a different safety hazard. After generation, the steam is mixed with liquid nitrogen and introduced into the wind tunnel. Although the apparatus used to generate the steam is somewhat different in [16] and [17], the end product is the same. Both the Massachusetts Institute of Technology and Iowa State University laboratories have used steam with reasonable success. The stream of steam–liquid nitrogen mixtures have been used in both indraft and closed-circuit subsonic wind tunnels.

Although the use of steam for flow visualization has the advantages of being clean and nontoxic, it does have disadvantages. For example, the system temperature must be controlled carefully if a neutrally buoyant fog is to be obtained. As pointed out earlier, water vapor is composed of much larger particles on the average than oil smoke. It also has the disadvantage that the steam condenses to water on cold model surfaces, walls, and other protrusions in the test section. The steam photographs reasonably well and can produce useable visualization records.

6.2.1.6 Photographic techniques. Photographs can be taken with a variety of cameras: still, stereo, cinematic, and stereocinematic. Photographs may be made with steady light but are most commonly made with short-interval (20 μs) strobolumes, so placed as to illuminate the flow field under investigation. To obtain sufficient intensity and to provide a uniform illumination, three to five lamps are used. The strobolumes are triggered by the camera shutter, or other means, as dictated by the problem.

Still photographs are made with a 101.6 × 127 mm (4 × 5 in.) view camera using a $f_{4.5}$ or $f_{6.3}$ coated lens. Polaroid type 57 film or Kodak Royal-X Pan film, thrice overdeveloped in D-11, have been used to obtain maximum contrast. The Royal-X Pan negatives are projection printed on contrast 5 bromide paper. If linear measurements are to be made directly from prints, a dimensionally stable paper is used. An SLR 35-mm camera and Tri-X film have also been used successfully.

Three-dimensional pictures may be taken with a stereographic camera designed and constructed by Brown and his colleagues. This apparatus consists of two 101.6 × 127 mm (4 × 5 in.) cameras mounted on a base casting supported by a heavy but mobile base. The lens separation can be varied from 203 to 610 mm. A stereo comparator is used to extract data from the photographs obtained. This apparatus has been used also with two 16-mm movie cameras for stereocinematic investigation. High-speed movies are taken with a Wollensak WF-3 Fastex 16-mm camera, capable of speeds from 1000 to 8000 frames/s. Using Eastman 4-X negative film 7224, camera speeds of up to 4000 frames/s can be used with good results. Various combinations of high-intensity lights (i.e., 1000 and 2000 W) are used for taking high-speed movies. It is important for the experimenter to have access to a dark room, so that film may be developed quickly. If unsatisfactory results are obtained, a new series of photographs may be taken with little time lost. Since photography is a fine art, a great deal of experimentation is necessary in both taking and processing photographs and movie film.

One of the most significant developments in the field of smoke visualization in the past 15 years has been the introduction of laser light to illuminate smoke. Although other light sources have been used for many years to produce narrow sheets of light, the minimal spreading of laser sheets together with the variety of power levels available have had a dramatic effect on smoke visualization. The sheet of laser light is usually produced by passing the laser beam through a cylindrical lens or a glass rod [18, 19].

A computer-compatible video camera system offers a great deal of potential for extracting quantitative data from flow visualization experiments. These are limitations based upon the speed, which suggest that such a system be used in conjunction with the proven methods of obtaining still and high-speed cinematic photographs.

6.2.1.7 Digital image processing. The interest in obtaining quantitative information from smoke photographs has led to use of digital computers to analyze the global and local flow patterns [20, 21]. The use of digital image processing to analyze flow visualization images from photographs or video frames is increasing very rapidly [22–27]. The recent review of digital image processing in flow visualization by Hesselink [21] is an excellent starting point for those interested in this rapidly developing field. Examples of the results obtained from a study of the interaction of turbulent scales in boundary layer flows are described by Corke [25, 28].

A relatively new technique that has the potential to examine complex three-dimensional flow structure is the scanning laser light sheet used together with a high-speed video system and an image processing technique [29]. A pulsed 40-W copper-vapor laser has been used in a system developed by Kegelman and Roos [29–32] to study the coherent structures in reattaching laminar and turbulent shear layers. The pulsed laser beam, at about 8.5 mJ per pulse for 6000 pulses/s, is spread into a sheet by reflection from a cylindrical mirror. The mirror is mounted on a galvanometer that rotates about an axis perpendicular to the cylindrical mirror axis. The cylindrical mirror is positioned at the focal point of a parabolic mirror, so that the light sheet is always reflected parallel to the axis parabolic mirror. Applying electrical current to the galvanometer produces a light sheet that scans the flow field seeded with smoke. A high-speed video camera is synchronized with both the pulsed laser and the motion of the light sheet, so that one image is obtained for each position of the light sheet. Each image obtained is digitized and stored on a VAX 780 computer system. The raw video images of the smoke visualizations are processed to remove video noise and to correct for nonuniform light intensity as well as perspective distortion. This new technique allows for three-dimensional images to be reconstructed and transferred to a graphics processor and examined closeup [29]. Although this technique is in the development stage, it is clear that it has great potential for the study of shear layer transition and turbulence because it can produce quantitative data from flow visualization studies. There are, of course, many other recent developments in the use of digital image processing to study both laminar and turbulent flows [33–37].

6.2.1.8 Low-turbulence supersonic and transonic wind tunnels. The Aerospace Laboratory at Notre Dame also houses three supersonic wind tunnels and one transonic nonreturn tunnel, developed by extending the successful low-speed smoke-tunnel concepts of Brown to much higher speeds.

Notre Dame's supersonic installation consists of three separate diffusers, connected to a common manifold and three Allis-Chalmers rotary vacuum pumps. Each of these positive displacement vane vacuum pumps is driven by a 125-hp induction motor at a speed of 435 rpm, delivering 94 m³/min at 457 mm of mercury vacuum. By suitable arrangement of the tunnel exhaust valves, any one of the three tunnel stations may be operated in a continuous mode. The pressure difference essential for operating a given tunnel is produced by turning on one, two, or all three of the rotary vacuum pumps. Operation of any of these tunnels is limited by the moisture content of the ambient air. Dryers are not feasible because they would interfere with the introduction of smoke into the tunnel. It has been determined that good data can be obtained with ambient dew points as high as about −4°C.

The world's first supersonic smoke tunnel [38], called the pilot tunnel, was designed using the method of Foelsch for a Mach number of 1.38 and is shown in Fig. 6.5. The aluminum block test section is about 101 mm long and has a square cross-sectional area of 4032 mm². The upper nozzle block has a piece of 12.7-mm lucite sandwiched in the center, so that smoke lines may be lighted from above when necessary. The side walls consist of a 19-mm-thick Parallel-O-Plate glass. The glass extends over almost the entire nozzle length (i.e., from upstream of the throat to the end of the test section). The converging portion of the supersonic nozzle and the inlet contraction cone were designed using the method of Smith and Wang [6]. The overall contraction ratio from the beginning of the inlet to the nozzle throat is about 93:1. Seven screens are located at the upstream side of the contraction cone: a single brass screen with 5.51 × 7.09 meshes/cm (14 × 18 meshes/in.) and six nylon screens with 7.87 × 7.87 meshes/cm (20 × 20 meshes/in.). The pilot tunnel can be operated continuously using a single rotary vacuum pump. Several other supersonic nozzle configurations and sizes are also available. A transonic smoke tunnel with slotted upper and lower walls and clear plastic side walls has recently been fabricated and calibrated [8].

6.2.1.9 Visualization techniques for supersonic flow. The same oil smoke generator, basic smoke rake, and photographic techniques have been used with the supersonic tunnels as with the subsonic tunnels for direct smoke line visualization. The most critical item is, of course, lighting.

When taking pictures in the pilot tunnel, three lamps are arranged, as shown

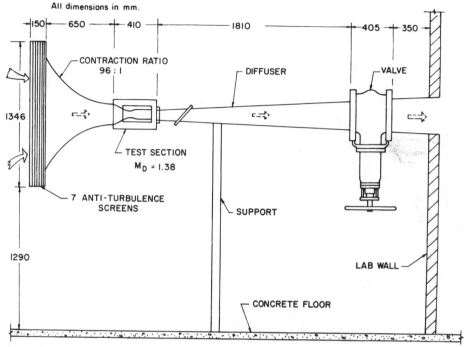

Figure 6.5 Pilot supersonic smoke tunnel. (Courtesy of T. J. Mueller, University of Notre Dame)

in Fig. 6.6. When using the other larger nozzles, four and sometimes five lamps are used. Any variation in the basic lighting technique can be used when taking an ordinary direct photograph of the smoke lines, e.g., front or back lighting. On some occasions, the back-lighting technique is most useful, as maximum scattering occurs when the light is brought in from behind at about 130° from the direct line of sight. However, it must be noted that this will result in a loss of contrast because the black background is sacrificed with this technique. The laser light sheet should also be considered for high-speed applications.

As the oil smoke is not visible through the schlieren system (i.e., its index of refraction is not much different from that of air), the use of other materials was studied by Goddard [39]. Although other gases might be used, nitrous oxide was found to be convenient because of its availability and relatively low cost. The nitrous oxide can be admitted indirectly into the smoke rake, although it would be better to pass the gas through some sort of piping system to return the gas to room temperature. The cloth bag inside the rake is allowed to remain as an aid to the uniform dispersion of the nitrous oxide. The gas flow is regulated during tunnel operation until the minimum flow to produce satisfactory streak lines (as observed through the schlieren) is obtained. This method of supersonic streak-line visualiza-

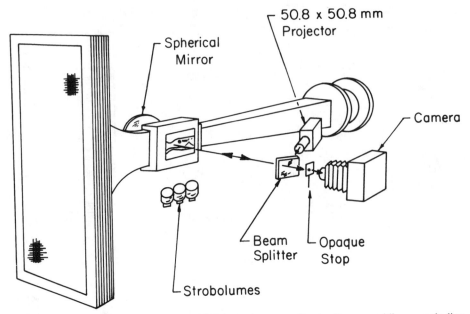

Figure 6.6 Modified schlieren system and lighting arrangement for simultaneous schlieren-smoke line photographs. (Courtesy of T. J. Mueller, University of Notre Dame)

tion is quite attractive because of its simplicity, low cost, and adaptability to any existing installation where the proper inlet reduction ratio and antiturbulence screening can be provided.

Goddard [39] developed a modified nonparallel, double-pass schlieren system for use with the supersonic smoke tunnel. It is fabricated from a single spherical mirror, where the light source is located at the center of curvature. This type of schlieren system can also be further classified as to whether it is truly coincident or noncoincident (i.e., whether the source and knife edge are located on or slightly off the optical axis). To accomplish the desired results (i.e., to take a combined smoke and schlieren photograph), it becomes apparent that the true coincident type is a requisite, from the standpoint of light-gathering capabilities.

The ordinary coincident schlieren system is designed to use a slit source and knife edge stop. On occasion, the design may even use a circular source and some sort of iris mechanism to function as a circular stop. Because a dark background is necessary for the photographing of smoke lines, a small circular opaque stop is used in place of an iris mechanism. Such modification results in a dark field with bright lines appearing wherever the light rays are deviated. A schematic of the optical arrangement is shown in Fig. 6.6.

Figure 6.6 indicates how the system was used in conjunction with the photographing of the streamlines. In use, the schlieren light source is steady; the desired contrast of the wave patterns recorded on the film negative is controlled by the camera shutter speed. As the flash duration of the strobolume lights that illuminate the streak lines is of the order of 20 μs, the shutter speed has little effect on the photographing of the smoke lines. With such an arrangement, the streamline flow and wave patterns are recorded simultaneously on the same photographic film.

Originally, a red light source and a green stop were used, with the idea that the green stop would cause little interference with the photographing of the smoke lines. It was found that a small opaque stop, 2.381 mm in diameter, did not interfere with the light-gathering capabilities of the camera. Further, it is far easier to produce various sizes of opaque stops than it is to cut small circular green stops from a gelatin filter. Opaque stops are easily produced by photographing a white circle on a black background at the desired magnification. The negative so produced is then mounted in a 50.8 × 50.8 mm glass slide and, as a unit, is ready for use as a stop. The red filter was not removed even after determining that use as a stop was satisfactory. The red light gives some semblance of monochromatic light. Even more important is the fact that, as an absolutely dark field cannot be fully realized, the red light provides a better background for the smoke lines; the photographs produced then have a good contrast between the smoke lines and the background.

6.2.1.10 Application of smoke-tube method. The smoke tubes or smoke filaments or, as many prefer, smoke lines, that emanate from each opening of the smoke rake are streak lines. Since a continuous tube or streak of smoke is usually used, the resulting patterns are the result of all previous motion and thus produce an integrated effect and must be considered as such. The smoke can hinder as well as serve the experimenter in defining flow phenomena. For example, as will be pointed out later in this section, there are very shallow waves that appear in the transition region. To make these waves visible, a very thin layer of smoke must be used. If the smoke is too thick, the waves cannot be seen, and it is very easy to conclude that no disturbances exist. Yet, if the very limited quantity of smoke necessary to define the shallow waves is used, phenomena that take place at a larger distance above the surface of the model will not be defined, and erroneous conclusions can likewise be reached. It is imperative therefore to use various quantities of smoke and to take a sufficient quantity of photographic data necessary to completely define the flow phenomena. It is also important to remember that the patterns represent a material deformation, or Lagrangian representation of the motion; therefore care must be taken in interpreting smoke streak line photographs. Furthermore, it is always desirable to compare the data from smoke and dye techniques with pressure, velocity, and force measurements as well as with theoretical or numerical analysis.

Transition in attached shear layers The transition process in the boundary layer of an axisymmetric body is an important factor that determines the magnitude and direction of the aerodynamic forces acting on the body. These forces are closely

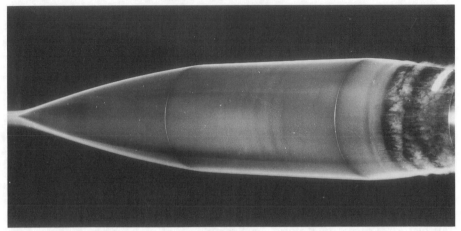

Figure 6.7 Smoke photograph of nonspinning axisymmetric body for $\alpha = 0°$ and $Re_L = 815,000$. (Courtesy of T. J. Mueller, R. C. Nelson, and J. T. Kegelman, University of Notre Dame)

related to how rapidly the boundary layer grows and whether or not it separates from the body surface. Both of these factors have a significant effect on the aerodynamic forces. The discovery of large-scale coherent structures in turbulent flows, mostly through flow visualization experiments, has drastically changed our approach to studying these complex problems. The crucial ingredients of turbulent flows appear to be vortex interactions. The three-dimensional vortex structures in turbulent flows usually originate in the transition region. Recent experimental and theoretical studies have helped to provide a detailed understanding of some of the modes of transition [40–49].

It is generally agreed that transition from laminar to turbulent flow may be described as a series of events that take place more or less continuously, depending on the flow problems studied. Since turbulence is essentially an unsteady three-dimensional phenomenon, the breakdown of a two-dimensional laminar flow may be viewed as the process whereby small-amplitude velocity fluctuations, or traveling wave disturbances, acquire significant amplitude and three-dimensionality. In one path to turbulence, the velocity fluctuation or traveling wave front that is initially straight (for the flow over a flat plate) develops spanwise undulations that are enhanced by secondary instabilities. For the flow over a nonspinning axisymmetric body, the traveling wave front is axisymmetric.

A study of the boundary layers on spinning and nonspinning axisymmetric bodies was performed by J. T. Kegelman, R. C. Nelson, and the author at the University of Notre Dame in 1978. The baseline model for these flow visualization, pressure, and force studies is an axisymmetric model consisting of 3-caliber secant-ogive-nose, a 2-caliber cylindrical midsection, and a 1-caliber 7° conical boattail. Two smooth baseline models were designed, one to be used in the flow visualization and force tests and the other for measuring the pressure distribution around the

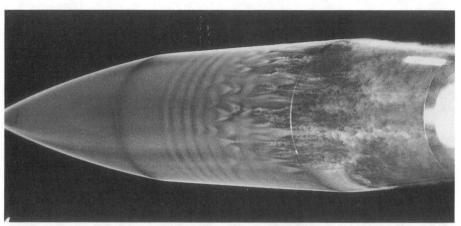

Figure 6.8   Smoke photograph of nonspinning axisymmetric body for $\alpha = 0°$ and $Re_L = 1,030,000$. (Courtesy of T. J. Mueller, R. C. Nelson, and J. T. Kegelman, University of Notre Dame)

body. The flow visualization model was designed to be constructed in three parts. The cylindrical midsection contains the bearings, mounting supports, and drive system. The flow visualization models were anodized black to improve photographic contrast between the model and smoke filaments. For zero spin and zero angle of attack, a single smoke tube was positioned to impinge on the sharp nose of the baseline model in a symmetrical fashion. These experiments covered the range of Reynolds numbers based upon body length (Re_L) between 315,000 and 1,030,000.

The high-speed smoke photographs and pressure data indicated at least three different modes of transition over the range of conditions used during the study of spontaneous transition at zero angle of attack [40–43]. The three types that occurred over the cylindrical part of the body started with two-dimensional Tollmien-Schlichting waves. Depending upon the Tollmien-Schlichting amplitudes reached for between 4 and 5 calibers along the body, the following may occur:

1. The waves may stop growing in the locally favorable pressure gradient and thus constitute part of the initial disturbances for the locally or fully separated boundary layers over the boattail, e.g., Fig. 6.7.
2. For somewhat larger amplitudes of the waves, a secondary instability takes place, leading to staggered Λ vortex formations seen in Fig. 6.8.
3. When still larger amplitudes of the waves are reached, in-line, or aligned, Λ vortices are formed.

Recently, Kegelman [41, 43] was able to change the staggered patterns to aligned patterns by acoustic enhancement of the amplitude, all other conditions remaining unchanged. A more complete description of these results is presented in [49].

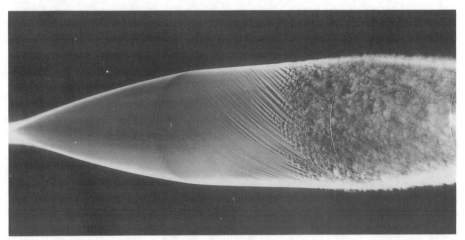

Figure 6.9 Typical striations in the smoke resulting from cross-flow vortices for $\alpha = 0°$, $V/U_\infty = 0.848$ (1250 rpm), and $Re_L = 315,000$. (Courtesy of T. J. Mueller, R. C. Nelson, and J. T. Kegelman, University of Notre Dame)

For a spinning axisymmetric body, vortices originate in the cross flow and spiral around the body [50]. These cross-flow vortices eventually break down into turbulence but do so in a distinctly different manner from the axisymmetric waves. Depending on the Re_L and spin ratio, the transition process is initiated by either the breakdown of axisymmetric waves (i.e., Tollmien-Schlichting waves) or the breakdown of the vortices generated in the cross flow. Furthermore, for certain combinations of these parameters, both the axisymmetric waves and the cross-flow vortices occur. The simultaneous occurrence of these two phenomena was first discovered at the University of Notre Dame using smoke visualization. Because of the complex nature of the transitional process and the sensitivity of the individual events in this process, experiments are very difficult. The most important recent contributions to understanding the physics of the transition process have come from flow visualization experiments. The nonintrusive nature of flow visualization plus its global view have been important factors in its success.

The baseline model was studied at spin rates from zero to 4500 rpm at angles of attack from zero to 10° for Re_L between 315,000 and 1,030,000. For the spinning model at zero angle of attack, the phenomenon was primarily related to the ratio of the peripheral velocity to the free stream velocity V/U, and relatively independent of Re (i.e., it was not significantly affected by changes in Re for a given V/U) [50]. Tests were conducted for a range of V/U between zero and 1.67. There were no notable changes in the boundary layer characteristics for $V/U < 0.4$, with the exception of a slight skewness in the tips of the vortex trusses. When vortex trusses were present, this skewness could be seen for V/U values as low as 0.1. As V/U increased, striations in the smoke (manifestations of the cross-flow vortices) ap-

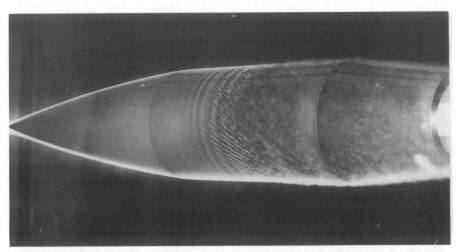

Figure 6.10 The simultaneous appearance of Tollmien-Schlichting waves and cross-flow vortices for $\alpha = 0°$, $V/U_\infty = 0.61$ (2900 rpm), and $Re_L = 1,030,000$. (Courtesy of T. J. Mueller, R. C. Nelson, and J. T. Kegelman, University of Notre Dame)

peared at an angle approximately equal to $\tan^{-1}$ of V/U, as shown in Fig. 6.9. The wavelength of the striations λ/D was approximately 3.8×10^{-2} and remained constant regardless of spin ratio or Re. As Re was increased, the transition process took place over a shorter distance. The striations broke down into a corkscrew shape just before becoming turbulent (see Fig. 6.9). The transition zone moved forward with increasing spin rate, and the transition process took place over a shorter distance as Re was increased. Furthermore, when the striations appear toward the end of the midsection, they are superimposed (shown in Fig. 6.10) on the two-dimensional (axisymmetric) Tollmien-Schlichting waves, which are similar to those that appear on the nonspinning body. At high values of V/U (> 1.0), the boundary layer was fully turbulent along the entire midsection, regardless of Re.

Magnus force and flow visualization data were also obtained for the baseline model for $Re_L = 315,000$ and $1,030,000$. The angle of attack of the model was varied from zero to $10°$ in $2°$ increments, and the nondimensional spin rate V/U was varied from zero to 1.67. The data exhibited both positive and negative side forces over the spin rates tested. Several trends were observed, which agreed with both the visual data and force measurements of the spinning model at angle of attack [50]. At $2°$ and $4°$ angles of attack, transition took place via the formation and breakdown of the striations, closely resembling the transition process at the zero angle of attack. These striations occurred symmetrically about the axis of rotation; however, they were axisymmetric about the plane of angle of attack. The striations moved forward and became shorter as spin was increased. They occurred at an angle $\theta = \tan^{-1} V/U$. At a moderate angle of attack $\alpha = 6°$, the striations

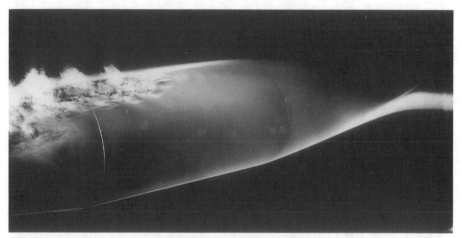

Figure 6.11 Starboard view at $\alpha = 10°$, $V/U_\infty = 0$ (0 rpm), and $Re_L = 315,000$. (Courtesy of T. J. Mueller, R. C. Nelson, and J. T. Kegelman, University of Notre Dame)

appeared superimposed over the large-amplitude waves. At a high angle of attack, $\alpha = 10°$, a patch region only formed at zero spin (Fig. 6.11) and passed over the starboard side of the model as the spin ratio increased (Figs. 6.12–6.14), coincident with the magnitude of the negative Magnus force. No striations were formed. At high spin rates and high angles of attack, the side force became negative, corresponding to a positive Magnus force.

It is clear that with this flow visualization technique, it is possible to conduct a detailed investigation of the three-dimensional deformations that initiate the breakdown of both Tollmien-Schlichting waves and the vortex tubes resulting from the cross-flow instability.

Influence of sound on a laminar wake Brown and Goddard [51] investigated the effect of sound on the laminar wake behind airfoils and flat plates with sharp leading and trailing edges. In the visual study of the flow past an airfoil or flat plate, the three basic characteristics of the wake flow with or without sound are the frequency of vortex pair formation, the geometry of the vortices (i.e., the spacing ratio h/λ, where h is the vertical distance between rows and λ is the horizontal distance between vortices in the same row), and V_w, the wake speed, as compared to the free stream velocity. Since the wake becomes visible with the introduction of smoke, the frequency of vortex formation can be determined using standard stroboscopic techniques. The geometry of the wake can be determined by linear measurements made from photographic prints on dimensionally stable photographic paper. Knowing both the free stream velocity and the frequency of vortex formation, the wake speed can be determined. The introduction of sound had a direct influence upon the frequency of vortex formation. The wake frequency fol-

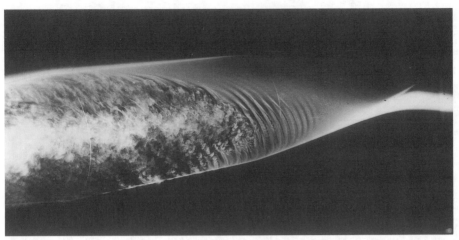

Figure 6.12 Starboard view at $\alpha = 10°$, $V/U_\infty = 0.65$ (1000 rpm), and $\mathrm{Re}_L = 315,000$. (Courtesy of T. J. Mueller, R. C. Nelson, and J. T. Kegelman, University of Notre Dame)

lows the sound frequency for a limited range both above and below the natural frequency of formation. The vortex spacing ratio varied with the sound frequency. However, the wake speed was invariant with change in sound frequency. The effects of sound on the wake flow behind a typical flat plate are summarized in Fig. 6.15, where results for two different free stream speeds are shown. The sound control limits, upper and lower, obtained from experiments are indicated in Fig. 6.15. The-oretical curves of the spacing ratio variation with change in sound frequency are

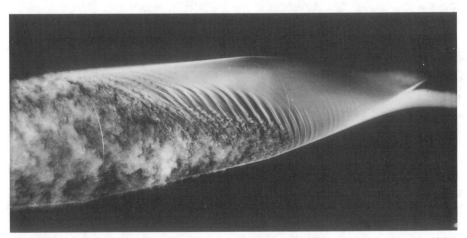

Figure 6.13 Starboard view at $\alpha = 10°$, $V/U_\infty = 1.0$ (1500 rpm), and $\mathrm{Re}_L = 315,000$. (Courtesy of T. J. Mueller, R. C. Nelson, and J. T. Kegelman, University of Notre Dame)

Figure 6.14 Starboard view at $\alpha = 10°$, $V/U_\infty = 1.65$ (2500 rpm), and $Re_L = 315,000$. (Courtesy of T. J. Mueller, R. C. Nelson, and J. T. Kegelman, University of Notre Dame)

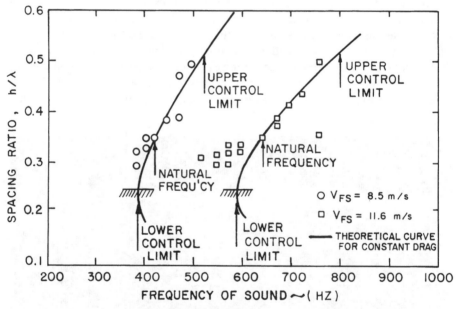

Figure 6.15 Variation of spacing ratio with sound frequency for a flat plate. (Courtesy of University of Notre Dame)

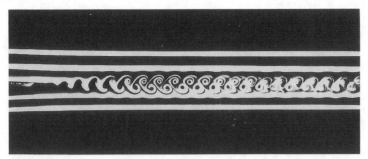

Figure 6.16   Flat plate of Fig. 6.15, U_∞ = 8.5 m/s and sound frequency of 0 Hz. (Courtesy of University of Notre Dame)

also shown. These curves were generated by making use of von Karman's work, where he showed that the drag of an object can be determined from wake characteristics. Since the drag is unchanged when sound is introduced and knowing the wake characteristics without sound, the theoretical curve can be generated from the change in wake characteristics with sound. Figures 6.16 and 6.17 are smoke photographs of the flat plate wakes used in Fig. 6.15, and show the difference in spacing ratio when sound is introduced.

Vortex breakdown on a delta wing   The breakdown of the vortices emanating from the leading edge of delta wings has been of interest since the first delta wing aircraft were designed about four decades ago. In recent years this interest has expanded as a possible method of increasing the maneuverability of high-speed aircraft. Different types of vortex breakdowns have been identified in vortex-tube experiments. The two most common types on wings are the bubble and spiral modes of breakdown. R. C. Nelson and his colleagues at Notre Dame have made numerous contributions to the understanding of vortex breakdown phenomena using the smoke-tube method and a laser light sheet [52–57]. A sketch of the experimen-

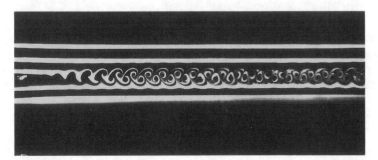

Figure 6.17   Flat plate of Fig. 6.15, U_∞ = 8.5 m/s and sound frequency of 470 Hz. (Courtesy of University of Notre Dame)

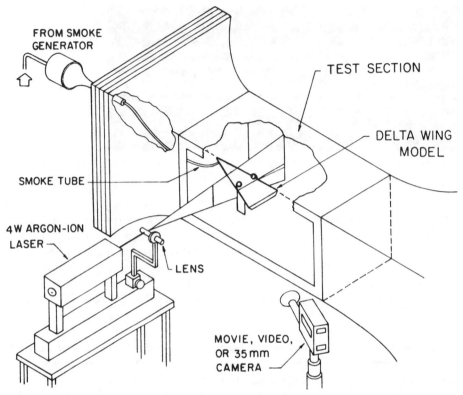

FROM SMOKE
GENERATOR

TEST SECTION

DELTA WING
MODEL

SMOKE TUBE

4W ARGON-ION
LASER

LENS

MOVIE, VIDEO,
OR 35mm
CAMERA

Figure 6.18 Experimental arrangement delta wing and laser light sheet. (Courtesy of R. C. Nelson, University of Notre Dame)

tal arrangement is shown in Fig. 6.18. At low free stream velocities, i.e., down to 3 m/s, details of the flow can be clearly seen that are often imperceptible at higher speeds. Results for a thin, sharp-edged delta wing with a sweep angle of 85° are shown in Figs. 6.19 and 6.20. In Fig. 6.19 the spiral nature of the vortices is emphasized by the appearance of striations in the smoke. The striations become visible when the flow is accelerated around the leading edge and the smoke mixes with entrained flow in the vortex. In this photograph, the vortex on the right is breaking down at approximately the midchord position, while the left vortex does not break down until somewhere in the wake. The combination of high sweep and low free stream velocity usually resulted in axisymmetric vortex breakdown as previously discussed. Which vortex would break down first could not be predicted and was observed to change back and forth at irregular intervals. This was probably the result of small changes in the free stream conditions due to gusts or changes in the direction of the wind at the tunnel exit.

In the photograph shown in Fig. 6.20, it is possible to actually see the roll-up of

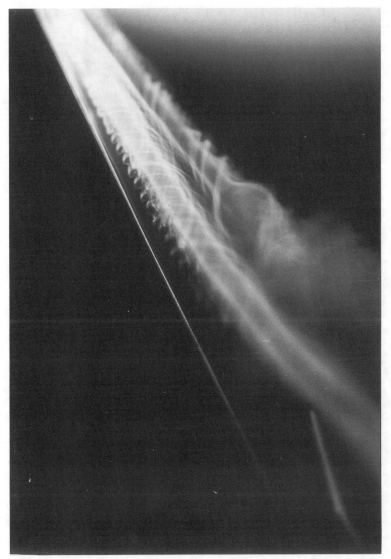

Figure 6.19 Smoke flow visualization with flood lamp illumination, sweep $= 85°$, $\alpha = 40°$, and $V = 3$ m/s. (Courtesy of R. C. Nelson, University of Notre Dame)

the shear layer that forms the primary vortices and the development of secondary vortical-like structures in the shear layer. The growth of these secondary structures is similar to the evolution of the classic Kelvin-Helmholtz instability.

Titanium tetrachloride was used to tag the cores of the vortices originating from the leading-edge extension (LEX) and the delta wing of a generic fighter

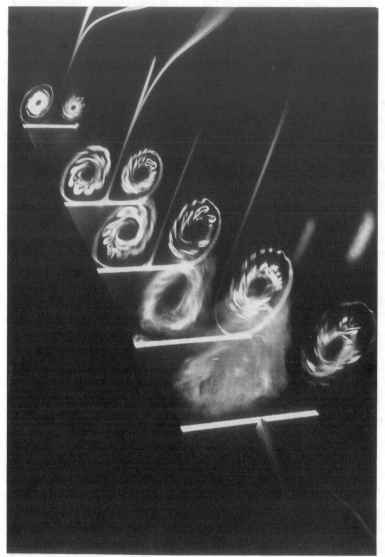

Figure 6.20 Lateral laser sheet cross sections, sweep = 85°, α = 40°, and V = 3 m/s. (Courtesy of R. C. Nelson, University of Notre Dame)

model [14, 15]. The TiCl$_4$ was introduced through a hypodermic tube connected to the system shown in Fig. 6.21 at the apex of the LEX and at the delta wing LEX juncture. A typical photograph shown in Fig. 6.21 indicates how clearly the individual vortex structures as well as their interactions can be identified. This smoke-tube type of method is referred to as "vortex core tagging" and has produced several important results [15, 57].

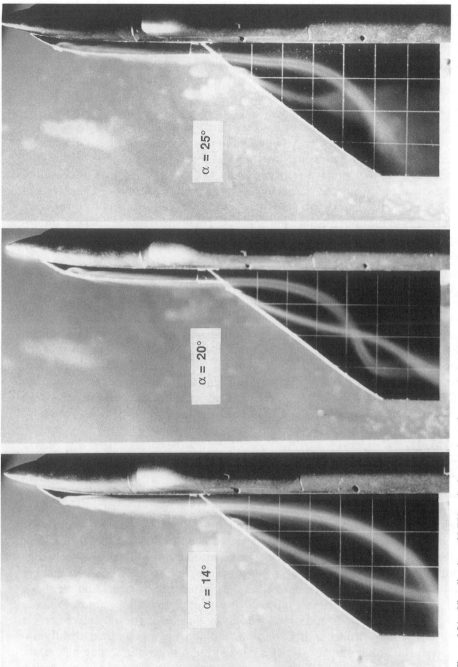

α = 25°

α = 20°

α = 14°

Figure 6.21 Visualization of LEX and wing vortices of a generic fighter model using TiCl₄. (Courtesy of R. C. Nelson, University of Notre Dame, and G. Erickson, NASA LRC)

Dynamic stall on an oscillating airfoil In unsteady-flow experiments, useful information concerning the gross behavior of a complicated flow field can often be
obtained even when individual streak lines cannot be identified. An example of
such a situation is dynamic stall on an oscillating airfoil.

The main obstacle to the continuing development and improvement of the
conventional helicopter configuration has been the dynamic retreating blade stall
that accompanies high loading and advance ratios [58]. The most important characteristics of dynamic stall are the unusually high lift coefficients, angles of attack
α and pitching moments C_m. A large amount of research has been directed at
understanding and documenting this complex unsteady-flow problem in the past
decade (e.g., [58–60]). In the research under the direction of W. J. McCroskey,
smoke visualization has been very helpful in studying this problem. Thin layers of
smoke were emitted through slits in the airfoil's leading edge and downstream
through slits in the upper surface of the airfoil. High-speed movies and still photographs were taken from directly above the airfoil using high-intensity quartz lamps
and high-intensity strobe lights, respectively [58, 59].

It has been found that the predominant feature of dynamic stall is the shedding
of a strong vortexlike disturbance from the leading-edge region. An airfoil whose
incidence is much higher than the static stall angle, is shown in Fig. 6.22. The
vortex shed from the leading-edge region moves downstream over the upper surface of the profile, as shown in Fig. 6.23, for Re based upon chord c and free stream
velocity U_∞ of 10^6. The passage of this vortex distorts the chordwise pressure distribution and produces transient forces and moments that are basically different from
their static counterparts. The pressure coefficient C_p versus the position along the
airfoil chord *(X/c)* is also shown in Fig. 6.23, where ω is the angular velocity, t the
time, and k the reduced frequency, defined as angular velocity times the chord
divided by 2 times the free stream velocity.

Supersonic flow visualization Figure 6.24 shows the supersonic flow past a wedge.
This photograph was made using a modified schlieren system to be able to obtain
simultaneous schlieren and smoke lines [38]. This photograph shows the two fundamental ways in which supersonic flow tends to follow parallel surfaces: flow
through a shock wave and expansion around a corner. In the flow through the
shock wave, the abrupt deflection of the streamline to flow parallel to the front
surface of the wedge can be observed. The expansion at the shoulder of the wedge
to follow parallel to the main body of the wedge is clearly seen. The Mach number
(Ma) of the flow can be determined from the photograph by measuring the shock
angle and the streamline deflection. In a similar way, streamlines can be followed
throughout the flow field, and so map the entire flow field. To study wake flows, a
lucite wedge-shaped plug with a rounded leading edge was inserted slightly upstream of the nozzle throat [61]. This configuration resembles a plug nozzle, referred to as the expansion-deflection nozzle, or may be thought of as representing
a strut, a flame holder, a Scramjet fuel injector, etc.

$$\alpha = 15° + 10° \sin \omega t \qquad k = \frac{\omega c}{2U_\infty} = 0.15$$

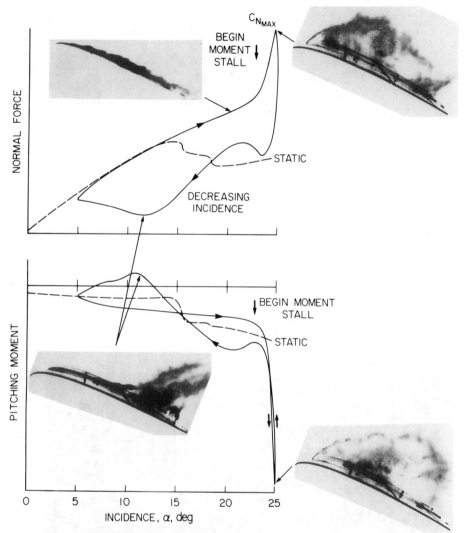

Figure 6.22 Normal force and pitching moment on NACA 0012 airfoil during dynamic stall at $\alpha = 15° + 10° \sin \omega t$, $k = \omega c/2U_\infty = 0.15$, Re $= 2.5 \times 10^6$ [59]. [Courtesy of W. J. McCroskey, U.S. Army Aeromechanics Laboratory (AVRADCOM)]

DYNAMIC STALL

$$\alpha = 15° + 14° \sin \omega t \qquad Re = 10^6 \qquad k = 0.1$$

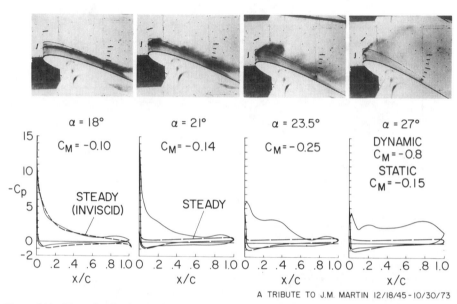

A TRIBUTE TO J.M. MARTIN 12/18/45 - 10/30/73

Figure 6.23 Flow visualization and pressure measurements of the vortex-shedding phase of dynamic stall on an oscillating airfoil [60], pitch axis at $X/c = 0.25$. [Courtesy of W. J. McCroskey, U.S. Army Aeromechanics Laboratory (AVRADCOM)]

Figure 6.24 Supersonic flow past a 5° half-angle wedge, Ma = 1.38. (Courtesy of University of Notre Dame)

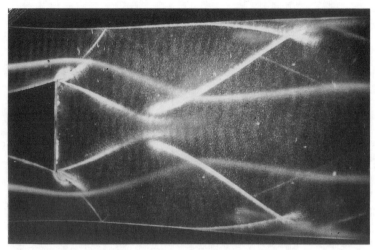

Figure 6.25 Simultaneous smoke line and opaque-stop schlieren photograph of wake flow with laser light source. (Courtesy of T. J. Mueller, University of Notre Dame)

A simultaneous smoke line and opaque-stop schlieren photograph of the wake flow using a laser light source is presented in Fig. 6.25. Smoke is not visible in a schlieren system, since it has approximately the same index of refraction as air. By measuring the local deflection of the smoke lines passing through the recompression shock wave and wave angles, the Ma immediately ahead of and behind the recompression shock can be determined. Another series of experiments was run with an actual expansion-deflection nozzle. A comparison of the Ma obtained from smoke line–shock wave patterns and from total- and static-pressure measurements is shown in Fig. 6.26. This correlation is for the Ma immediately downstream of the recompression shock versus the distance measured along the recompression shock from the nozzle centerline.

These high-speed flow visualization techniques have also been used to study a transonic cascade problem [62]. More recently, these techniques have been used to develop design criteria for improved high-speed smoke wind tunnels [8].

6.2.2 Smoke-Wire Method

Although the smoke-tube method has been used successfully to study many complex flow problems, there are fundamental flow phenomena that require the ability to produce small but discrete smoke filaments (streak lines) and to be able to locate the filaments accurately within the flow field so that small-scale details may be studied.

The smoke-wire technique, developed by Raspet and Moore in the early 1950s and subsequently improved and extended [63–67], is capable of producing very fine

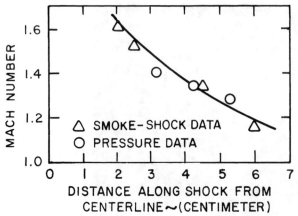

Figure 6.26 Correlation of smoke line–shock data and pressure data for Mach number immediately downstream of recompression shock for E-D nozzle. (Courtesy of T. J. Mueller, University of Notre Dame)

smoke filaments and can be used to study the detailed structure of complex flow phenomena. The "smoke" is produced by vaporizing oil from a fine (~0.1 mm) wire by the use of resistive heating. The technique was initially applied to the measurement of velocity profiles in a boundary layer. Applications have included investigations of transition [44–47] and the large eddy structure in turbulent shear flows [67–69]. Corke et al. [66] indicate a number of benefits associated with the smoke-wire technique, but for the applications documented here, the method's primary benefit is the fine structure of the smoke streak lines. A study of the separation bubble near the leading edge of an airfoil at low Re including transition will be presented. This particular phenomenon presents some unique challenges that indicate the need for flow visualization, and the smoke-wire is an ideal candidate. It can be used to study the physics of this complex flow phenomenon in regions of slow, recirculating flow where other measurement techniques such as hot wire and pressure transducers are very difficult to use. The wake behind a bluff body in a linear shear flow will also be presented.

The smoke-wire technique is limited and, fortunately, ideally suited to applications where the Re based on wire diameter is small (~20.0). For practical applications, this requires wind tunnel speeds of the order of 4–6 m/s. The method consists of a fine wire positioned in the flow field, coated with oil, and heated by passing an electrical current through the wire [70]. These experiments included stainless steel and tungsten wires measuring 0.025, 0.076, and 0.152 mm (0.001, 0.003, and 0.006 in.) in diameter. The strength, resistive heating characteristics, and size are all important factors in choosing the "best" wire for a given application. To minimize the disturbance produced by the wire that must be close to the model in the test section, the Re based upon the wire diameter was maintained at less than 20.0.

As the wire is heated, it expands and sags, which is not desirable if accurate placement of the smoke streak lines is required. Therefore the wire was prestressed, so that when heated, there was no noticeable sagging. This required a prestress level of approximately 1.03×10^9 Pa (1.5×10^5 psi) for the stainless wire. Since this is quite near the yield stress for the wire, it was important that once it was stressed, it was handled carefully.

6.2.2.1 Applying oil to the wire. A number of different liquids have been used to coat the smoke wire to produce the smoke filaments. These included several types of lubricating and mineral oils and a commercially available product, Life-Like Model Train Smoke, produced by Life-Like Products, Baltimore, Maryland. The results were similar to those achieved in [71], which indicated that the model train smoke produced the best smoke filaments. This product is composed of a commercial-grade mineral oil to which a small amount of oil of anise and blue dye has been added. The oil is very easy to work with, and such small quantities are used that the commercial product is ideal and inexpensive. There would be obvious safety problems associated with large quantities of the smoke, but since there are such small amounts produced and in this application the open-circuit tunnel exits to the outside of the laboratory, there were no safety problems. Reference [66] even indicates that since the amount of smoke is so small, the method is quite suitable for limited use in closed-circuit tunnels.

There appear to be a number of methods that can be used to coat the wire, each with its own advantages and disadvantages. Reference [66] documents a pressurized gravity feed method that might be suitable for a vertical wire. A method similar to this was tried, but there were problems with the large droplets running down the wire and coating the surface. These droplets would periodically be blown off the wire and would wet the surface of the airfoil model. Nagib [67] has developed a "windshield wiper" device that automatically coats the wire by wiping oil along the wire, but this apparatus would have to be located within the test section, which was unsuitable for the types of tests to be conducted. The technique used for these applications was manual coating. The rear wall of the test was fitted with an easily removable section. Between each use of the wire, the section was removed and the wire carefully wiped with a cotton applicator soaked with oil. This provided a uniform coating with no fouling of the model. The method is somewhat cumbersome, but since each coating could be accomplished within a few seconds, it was adequate.

6.2.2.2 Timing circuit. As the coated wire is heated, fine smoke streak lines are formed at each droplet on the wire (~8 lines/cm for the 0.076-mm wire). Depending upon the current through the wire, the beads can be vaporized very rapidly or, if a lower current is used, continuous filaments of adequate density will emanate from the wire for as long as 2s. With even lower current, the streaklines become fainter and cannot be photographed. For example, the 0.076-mm-diameter 302 stainless steel wire of 0.04 m length was heated using a power supply setting of approximately 0.7 amp (A). Both ac and dc power supplies were tried. However, the dc power supply was selected due to ease of control and the fact that the steady state current results in smoother and better defined smoke lines. Because of the relatively short duration of the smoke generation for a single wire coating, it is important that the event being photographed, the lighting, the camera, and the smoke, be properly controlled and synchronized. To accomplish this, a timing circuit was designed. The design is a modification of those in [66] and [70], and is included here in Fig. 6.27.

The "power set" potentiometer controls the amount of current desired to pass through the wire, and the "pulse length" potentiometer controls the length of time this current is applied. The "burst" potentiometer can be switched into the circuit to enable the operator to control the frequency of current pulses during the burn, resulting in an intermittent pattern of smoke. The wire is heated and cooled due to the pulsed voltage, the smoke density varies in a similar manner, and streak time lines can be formed. An additional potentiometer in this "switched" circuit controls the on/off period of these bursts. The "camera delay" potentiometer controls the time delay before the camera is triggered.

All the above controls can be preset, reducing the photographing of a given event to a single-stop operation. The circuit applies power to the smoke wire, and with the appropriate user set delay, it will activate a camera and lights using a solenoid attached to a camera trigger. The controls are set, the wire oiled, and the start button momentarily depressed. This causes the smoke to be generated and the camera triggered when the smoke has reached the desired intensity and location. A similar circuit could be designed so the camera could be triggered by some event within the tunnel, such as an oscillating flap or airfoil. This would allow for a conditional photographic sampling. The timing circuit is invaluable in the practical application of the smoke-wire technique.

6.2.2.3 Application of the smoke-wire method. The smoke tube makes use of rather large quantities of smoke, which are produced in a smoke generator outside of the wind tunnel and introduced into the tunnel just upstream of the inlet. Visualization data acquired using this method are characterized by a few (from one to about six) rather thick smoke streak lines (i.e., about 16 mm in diameter). The smoke-wire method, however, is capable of producing very thin smoke streak lines (i.e., about 1 mm in diameter). These very thin streak lines are, of course, subject to the same limitations as the thick ones discussed earlier. Because the smoke wire is usually located in the test section near the model under study, care must be taken to minimize the disturbances produced by the pressure of the wire. Therefore Re based upon the wire diameter must be kept low enough that the wire wake is steady and as small as possible (i.e., $R_{ed} < 40$).

The performance of airfoils operating in low-Re incompressible flows has been of increasing interest in the past decade. This interest has been a result of the desire

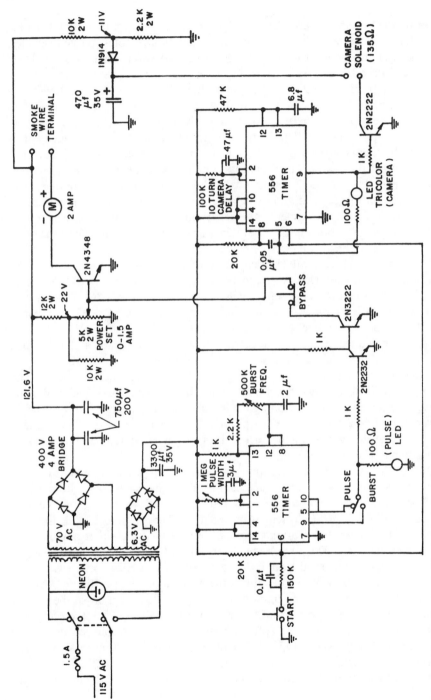

Figure 6.27 Smoke-wire timing circuit design. (Courtesy of University of Notre Dame)

to improve the low-speed performance of general aviation aircraft and high aspect ratio sailplane wings, as well as to improve the design of remotely piloted vehicles, jet engine fan blades, and propellers at high altitudes. Wind turbine rotors and free flying model aircraft also represent applications where low-Re performance is very important. Many significant aerodynamic problems appear to occur below chord Reynolds numbers Re_c of about 200,000. Although recent advances have been made, there are problems that require more study if further improvements in performance are to be realized. These problems are all related to the management of the airfoil boundary layer. A very important area of concern is the occurrence and behavior of the leading-edge separation bubble and the associated transition phenomena in the free shear layer. It is well known that the development and characteristics of this separation bubble are highly sensitive to Re, pressure gradient, and disturbance environment. The separation bubble plays a critical role in determining the development of the boundary layer, which in turn, affects the overall performance of the airfoil [4, 5, 7, 72–74].

An airfoil model with a 0.25-m chord and a 0.40-m span with a NACA 66_3-018 airfoil section was fitted with end plates and mounted in the 0.6×0.6 m square cross-section test section of the wind tunnel shown in Fig. 6.1. A schematic representation of the model and end plates is shown in Fig. 6.28. Also shown in Fig. 6.28 are the two smoke-wire locations used [72]. The horizontal wire (A-A) was used to introduce a sheet of fine smoke streak lines in a plane along the span of the airfoil model. The wire was located 65 mm forward of the leading edge of the airfoil and was parallel to the leading edge. The vertical location of the wire was adjustable using a screw-track device attached to the outside of each end plate. This allowed for accurate positioning of the sheet of smoke relative to the airfoil. The vertical wire (B-B) was used to produce a sheet of streak lines in a plane normal to the leading edge of the airfoil. This wire was located 430 mm forward of the leading edge of the airfoil.

Different lighting and photographic procedures were used for each of the two wire orientations (Fig. 6.29). In both cases, still and high-speed movie photography were used. The still photographs were taken using a Graflex Graphic View camera with an ACU-Tessar 210 ($f_{6.3}$) lens using Polaroid Type 57 and Kodak Royal-X Pan films. The high-speed motion pictures were taken using either a Wollensak WF-3 Fastex camera (1000–3000 frames/s) or a DBM-5 Milliken camera (64–500 frames/s). Both cameras used Kodak 4-X negative, 16-mm 7224 film.

For the planform views of the model, the horizontal wire position (A-A) was used, and the cameras were mounted below the glass floor section of the model. The camera was positioned normal to the chord of the airfoil for each angle of attack studied. Lighting for the still photographs was accomplished using two high-intensity General Radio Type 1532 strobolumes having a 20-ms flash duration and triggered by the camera shutter. For the planform views, the lights were directed along the span of the airfoil, as shown in Fig. 6.29. To help reduce the light

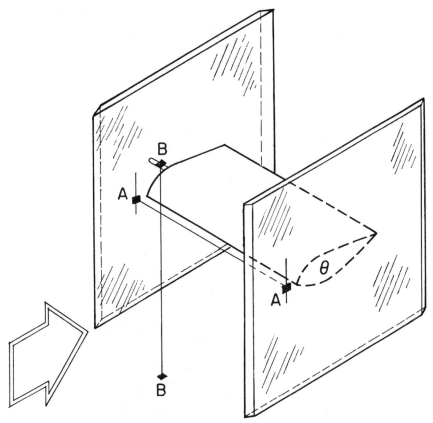

Figure 6.28 Schematic of wind-tunnel model end plates and wire location. (Courtesy of T. J. Mueller and S. M. Batill, University of Notre Dame)

intensity falloff across the span of the wing, a mirror was placed on the back wall of the tunnel; this helped provide uniform illumination across the span. For the vertical wire position, the camera was aimed along the spanwise axis of the wing. The model was illuminated from above and below the wing, normal to the wing, through 25.0-mm slits in the top and bottom of the test section. For the Fastex camera, continuous lighting was supplied by a single 1000-W quartz lamp. The Milliken camera was synchronized with two GENRAD Type 1540 strobolumes triggered by the camera. Each of the methods mentioned provide adequate illumination and contrast for photographing the streak lines.

The leading-edge separation bubble, shown in Fig. 6.30, is formed when the laminar boundary layer separates from the surface as a result of the strong adverse pressure gradient downstream of the point of minimum pressure [72]. This sepa-

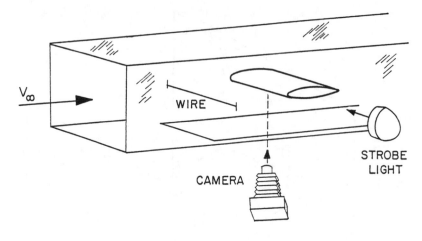

Horizontal Wire Position

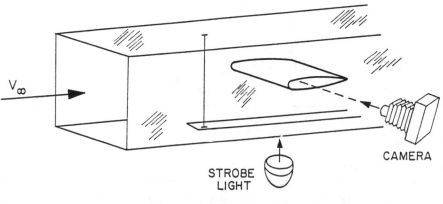

Vertical Wire Position

Figure 6.29 Wire, camera, and light locations for planform and profile views. (Courtesy of T. J. Mueller and S. M. Batill, University of Notre Dame)

rated shear layer is very unstable, and transition usually begins a short distance downstream of separation as a result of the amplification of velocity disturbances present immediately after separation. Reattachment can occur while the shear layer is undergoing transition or after the transitional process is complete and the flow is turbulent. The region between separation and reattachment is referred to as the

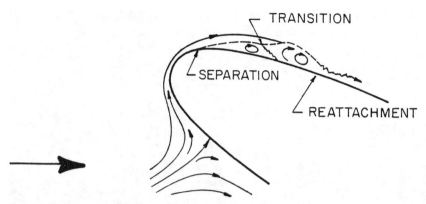

Figure 6.30 Schematic of leading-edge separation bubble. (Courtesy of University of Notre Dame)

separation bubble. Figure 6.31 shows the leading-edge separation bubble on the NACA 66_3-018 airfoil at $Re_c = 40,000$ and $a = 12°$. The vertical smoke-wire configuration (B-B) was used to obtain this photograph [70].

It is generally agreed that transition from laminar to turbulent flow may be described as a series of events that take place more or less continuously, depending on the flow problem being studied. Since turbulence is essentially a three-dimensional phenomenon, the breakdown of a two-dimensional laminar flow may be viewed as the process whereby finite amplitude velocity fluctuations, or traveling wave disturbances, acquire significant three-dimensionality [75]. Transition has been very graphically described as the process by which the straight and parallel vortex lines of the two-dimensional flow deform into a constantly changing and twisting three-dimensional mess called "turbulence" [76].

The flow visualization study of this phenomenon demonstrates the usefulness of the smoke-wire technique. Using both still and high-speed motion picture photography, detailed visual records of the physics of the separation bubble are acquired. The actual photographic records yielded detailed streak line dynamics, which provide significant insight into the dynamics of the flow field.

A series of experiments was conducted in which the horizontal (A-A) wire was used for both profile and planform views of the airfoil. A sketch of the airfoil and relative wire positions used is shown in Fig. 6.32. For the profile views, only a small center section of the smoke wire was coated. The angle of attack was 12° and $Re_c = 55,000$. In Fig. 6.33 the wire was located so that the smoke sheet was close to the stagnation streamline (called wire position 1). In Fig. 6.34, the smoke wire has been raised approximately 1 cm to wire position A2 and the smoke sheet lies above the top surface of the airfoil. The contamination of the laminar flow above the airfoil is evident as the effect of the transitioning and reattaching turbulent boundary layer propagates away from the airfoil surface. These photographs represent only two of a series that graphically illustrate the streak line dynamics in this com-

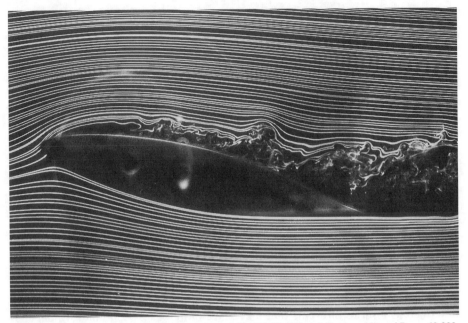

Figure 6.31   Smoke-wire visualization for a smooth NACA 66_3-018 airfoil at $\alpha = 12°$ and $Re_c = 40,000$. (Courtesy of T. J. Mueller and S. M. Batill, University of Notre Dame)

plex flow field and that were possible due to the accuracy in positioning of the smoke filaments.

An examination of the smoke photographs substantiates the notions of a highly unstable two-dimensional flow that breaks down in a very definite manner to a three-dimensional chaotic turbulent flow. These smoke photographs represent the most definitive visual description of separated shear layer transition available.

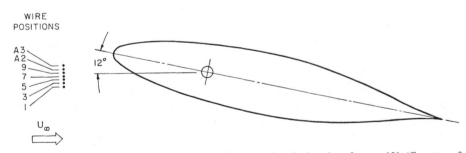

Figure 6.32   Profile view of NACA 66_3-018 airfoil and smoke-wire locations for $\alpha = 12°$. (Courtesy of T. J. Mueller and S. M. Batill, University of Notre Dame)

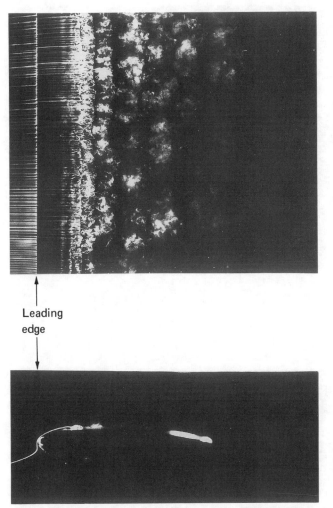

Leading
edge

Figure 6.33 Smoke-wire position 1 for $\alpha = 12°$ and $Re_c = 55,000$. (Courtesy of T. J. Mueller, S. M. Batill, and B. J. Jansen Jr., University of Notre Dame)

Some structure is also visible in the developing turbulent flow. Although this visual technique has only been applied for $Re_c < 120,000$, the basic transition process should follow the same series of events at higher Re. For example, using the same airfoil, the beginning of the transition process moves toward the separation location as the free stream velocity is increased. The length of the transition region also decreases with higher free stream velocities. Thus the understanding gained at low Re can definitely help develop a physical model of transition that will be useful at high Re.

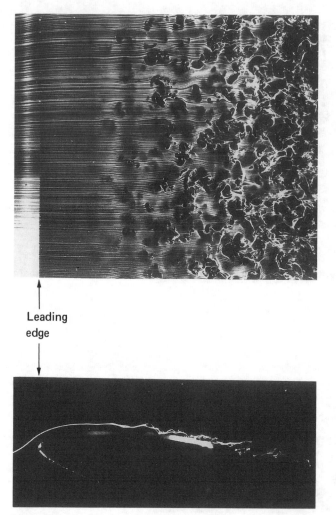

Leading
edge

Figure 6.34 Smoke-wire position A2 for $\alpha = 12°$ and $Re_c = 55,000$. (Courtesy of T. J. Mueller, S. M. Batill, and B. J. Jansen Jr., University of Notre Dame)

6.2.3 Helium Bubble Method

The most popular agent for airflow visualization in wind tunnels has been smoke. As discussed earlier, the use of smoke requires a low turbulence level in the wind-tunnel test section to minimize diffusion. Furthermore, the smoke particles are so small that they cannot be observed or photographed individually. A larger, neutrally buoyant tracer agent is needed to follow the path of individual particles. Another desirable quality of such a technique is the fact that particles can be used in existing wind tunnels with moderate turbulence.

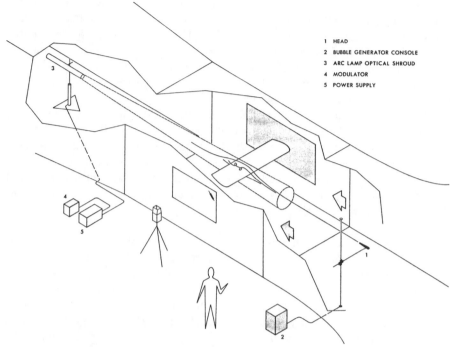

1 HEAD
2 BUBBLE GENERATOR CONSOLE
3 ARC LAMP OPTICAL SHROUD
4 MODULATOR
5 POWER SUPPLY

Figure 6.35 General arrangement of the helium bubble visualization system in a wind tunnel [77–81]. (Courtesy of Sage Action, Inc.)

The soap bubble is an ideal particle, since its size and buoyancy can be controlled. The scientific interest in soap bubbles dates back to Robert Hooke and Isaac Newton [77]. The first use of soap bubbles for flow visualization in wind tunnels appears to be by Redon and Vinsonneau [78] in 1936 at Marseille, France. In 1938, Kampè de Fériet [79] reported the use of bubbles to make statistical measurements of turbulence in a wind tunnel. Interest in this method of flow visualization seems to have disappeared until B. V. Johnson [80] generated small bubbles to study the flow within a cylindrical vortex tube in 1961. The most significant and practical bubble-generation system for wind-tunnel use developed from this latter work without the knowledge of the earlier work of Redon. The development of this modern system was begun in 1967 by Hale et al. [81]. The complete system consists of a bubble generator and lighting and optical components for illuminating the bubbles (all available from Sage Action, Inc., P.O. Box 416, Ithaca, NY 14850) and the photographic equipment to record the paths of the bubbles. A typical arrangement of this system in a large wind tunnel is shown in Fig. 6.35. The bubble generator consists of a head in which the bubbles are actually formed, and a console that supplies the constituents to the head. Neutral buoyancy is achieved by filling the bubbles with helium. The console meters the helium, a bubble film solution (BFS), and air to control bubble size, mean specific weight, and generation rate. Using Sage Action BFS 1035, bubbles from approximately 1 mm (0.03937 in.) to 5 mm (0.19685 in.) in diameter can be generated at rates up to 500/s. Two basic heads are presently available. A small, low-speed head can be used in airflows up to 15 m/s (50 ft/s), while a larger, high-speed head can be used in airflows up to approximately 60 m/s (200 ft/s).

For certain applications, a vortex filter is employed to remove any nonneutrally buoyant bubbles and liquid droplets. A high-speed head injects the bubbles tangentially into the cylindrical vortex filter, and only the neutrally buoyant bubbles exit axially from the center tube. These bubbles can then be ducted through flexible plastic tubing to the flow field under study without impinging on the interior wall of the tubing. This arrangement has been particularly useful for internal flow studies such as air cooling of electronic systems.

6.2.3.1 Lighting and photography. Because the bubbles reflect about only 5% of the incident light, careful selection and placement of the light sources are necessary. The principal aim is to shine as much light as possible on the bubbles while keeping the background dark. This may require painting the models and background areas with low-reflectivity paint (i.e., flat black), even when they are not in the direct beam of the source. Usually, a well-defined light beam is directed along the mean airflow direction and the line of sight of the observer or camera is essentially perpendicular to the light beam axis. Since each application is somewhat different, an all-purpose lighting system cannot be chosen. For small illumination, however, Sage Action, Inc. [81] suggests the Eimac 150-W Xenon arc lamp; an optical shroud and arc lamp modulator were specifically designed for this lamp. The advantages of this lighting system are the lamp's compactness and its adaptability to electronic modulation or chopping for quantitative measurements. This modulation makes the system very useful for a wide range of air velocities, since a continuous variation of the chopping frequency is possible. Illumination of a larger field is possible with a variety of incandescent light sources that are available at reasonably low cost. Mechanical modulation can be used with these light sources if necessary. A detailed description of the various illumination arrangements and modulation capabilities is presented in [81].

Bubble trace photography, in common with smoke photography, is quite different from conventional photography. For streak photographs it is desirable to obtain a high trace intensity and a low background exposure. In addition to using increased illumination, it is necessary to use highly sensitive film. Two types of film have been used [81]: Kodak 2475 with ASA 1000 and Kodak Royal-X Pan with ASA 1250. As discussed in the section on smoke photography, the sensitivity of such film may be further increased through special processing. The image intensity of a bubble streak is independent of camera shutter speed but dependent, instead, on the bubble velocity. Furthermore, the shutter speed determines the streak length and number of bubbles in a given photograph. If the bubble generation rate is R and the shutter speed is T, the number of bubbles in the photograph is RT. The number of bubbles that pass all the way through the camera's field of view is $(RT - RL/V)$, where L is the field dimension and V is the air speed. Hence a large value of R allows high shutter speeds, which in turn, reduce exposure of the model and background [81].

Bubble traces in a number of tests have been recorded by using closed-circuit television. There are several important advantages to this method. Certain television cameras provide higher sensitivity for bubble trace recordings than the fastest available photographic emulsions, and such cameras are readily available. It is possible to record sequentially with closed-circuit television or motion picture photography, and this is especially vital in unsteady flows, where the events are often not repeated. An advantage of using television is that the video-scanning process records the complete path of every bubble that passes through the camera's field of vision; a disadvantage of using a movie camera is that about one-half of the bubble path information is lost because of the mechanical shutter. As television also permits immediate replay, on-the-spot adjustments can be made on the bubble generator, the camera, and/or the lighting. This advantage, as well as the advantage of being able to monitor the live camera view, results in effective use of the researcher's time.

6.2.3.2 Application of helium bubble techniques. According to the manufacturer, well over 320 bubble-generator systems have been sold to universities, government facilities, and industry. Almost 100 of these systems have been delivered outside the United States, primarily to Europe and Japan. This equipment and the associated techniques have been used for aerodynamic wind-tunnel testing, the design and evaluation of ventilation systems, wind effects on structures, airflow through blowers and fans, natural convection systems, and many other practical flow situations.

Early in the development and application of the bubble visualization techniques, a comparison was made between the potential flow solution and streak photography for the two-dimensional Karman-Trefftz airfoil [81]. The potential flow solution and the visualization experiment were obtained for $\alpha = -5°$. The tunnel speed used was 30.48 m/s (100 ft/s) to minimize boundary layer effects including separation. The results of this study are shown in Fig. 6.36. Except for the slight separation near the trailing edge, the agreement between the two flow fields is very good. The crossing of the bubble paths indicates that the flow was not completely steady. The effect of using different shutter speeds for the flow over the two-dimensional Karman-Trefftz airfoil is shown in Fig. 6.37. The angle of attack was zero, and the free stream velocity was 15.24 m/s (50 ft/s) for both photographs. The shutter speed was 0.50 s in Fig. 6.37a, while it was 0.10 s in Fig. 6.37b. While the overall flow pattern is visible in Fig. 6.37a, it is difficult to follow individual streaks. Although individual bubble streaks are evident in Fig. 6.37b, some of the streaks are a little out of focus as a result of the small depth-of-field obtained by

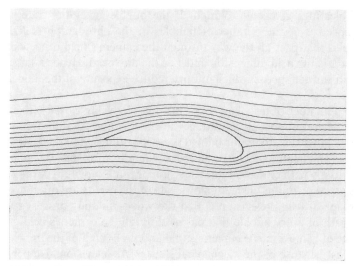

(a) Streamlines of Potential Flow Field

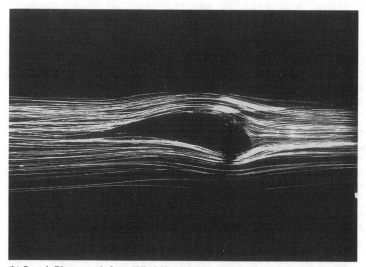

(b) Streak Photograph from Wind-Tunnel Test (f$_2$)

Figure 6.36 Comparison of the potential flow streamline pattern with a helium bubble streak photograph for a two-dimensional Karman-Trefftz airfoil [81]. (Courtesy of Sage Action, Inc.)

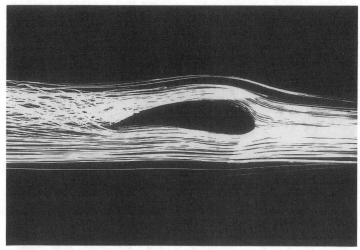

(a) Photograph taken at 1/2 s (f$_4$)

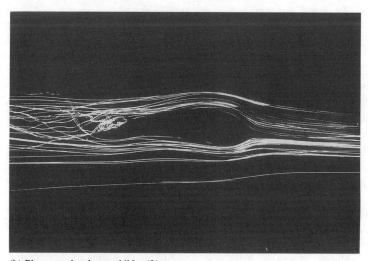

(b) Photograph taken at 1/10 s (f$_4$)

Figure 6.37   Helium bubble photograph of the flow over a two-dimensional airfoil for different exposures [81]. (Courtesy of Sage Action, Inc.)

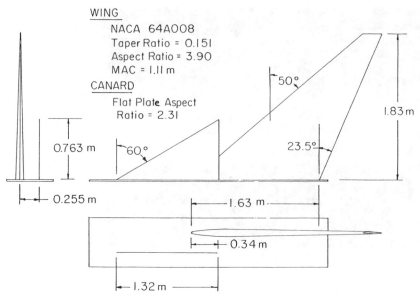

Figure 6.38   Three-view drawing of semi-span close-coupled canard model for wind-tunnel tests, scale 1:24 [81,82]. (Courtesy of Sage Action, Inc.)

using a low stop. This depth-of-field problem is common to most flow visualization techniques using direct injection.

The helium bubble technique was used together with force and pressure measurements to study a close-coupled canard (CCC) configuration [82, 83]. The CCC configuration has the principal advantage of high total lift at large α, combined with reduced trim drag. This is an important capability for advanced air-superiority fighter aircraft. The objective of this study was to obtain a good physical understanding of the canard wing flow field at high angles of attack, so that a simplified theory could be formulated.

A drawing of the CCC half-span model is shown in Fig. 6.38. The wing section was a NACA 66A008 airfoil laid out in a direction inclined 25% outward relative to the chordal direction or nearly normal to the trailing edge. The canard was a flat plate, 0.48 cm thick at the root with a semicircular leading edge and a blunt trailing edge. For the flow visualization studies, videotape recordings were made to supplement the still photography to identify unsteady effects. Surface oil flow visualization using titanium dioxide was also employed.

Figure 6.39 shows two representative streak photographs for $\alpha = 25°$. This angle was chosen because it was found from force data to be in the regime of favorable interaction. Figure 6.39a shows the wing without the canard, and Fig. 6.39b with the canard. The wing in badly stalled without the canard, and a turbulent separation region covers the whole upper surface. The remainder of the flow

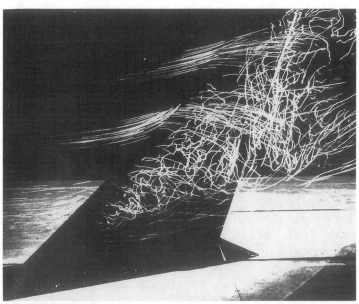

(a) Without Canard

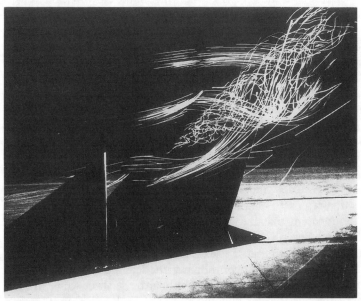

(b) With Canard

Figure 6.39 Plan view of flow pattern over upper surface of wing, with and without canard. Reynolds number $= 0.73 \times 10^6$ based upon wing mean aerodynamic chord [81,82]. (Courtesy of Sage Action, Inc.)

follows smoothly around the separation region, which is fed by only a small stream tube close to the wing root. No evidence of an organized leading-edge vortex or a tip vortex is apparent for this case.

There seem to be two important differences in the wing flow field when the canard is attached. A tight leading-edge vortex, extending out to a distance about equal to the canard semi-span, is formed inboard. As this vortex is in a narrow shadow region, it is not evident in Fig. 6.39b. The bursting of this vortex, however, is shown dramatically as the core becomes turbulent, and enlarges as it passes over the wing and off the trailing edge. The flow is laminar outside this turbulent "funnel," but it has appreciable rotation. The other notable occurrence is the unusual spanwise flow sandwiched between the burst vortex system and the upper surface of the wing. This flow originates inboard as flow that has gone over the leading-edge vortex and then reattached behind. After traveling underneath the burst vortex, it almost reaches the tip before turning back again into the free stream direction. Again, there is no evidence of a tip vortex.

The leading-edge vortex on the canard bursts near the apex, although this is not shown in Fig. 6.39b. This is the same behavior the canard alone would exhibit at this angle of attack. The core of this burst vortex is very turbulent and increases considerably as it moves rearward to the trailing edge and into the wake above the wing. It appears that the presence of the wing holds this burst trailing vortex system down near the canard and thus closer to the wing plane. However, the flow is quite smooth between the canard and the wing, which suggests a "channeling" effect that preserves the leading-edge vortex on the inboard portion of the wing.

The separation line ahead of the leading-edge vortex on the inboard portion of the wing was shown quite distinctly by surface flow visualization. However, in a number of other locations, a comparison of the surface patterns with those of the outer flow revealed significant differences in flow direction, indicating a drastic change in flow direction through the boundary layer.

An average or quasi-steady flow field was constructed from the superposition of streaks from a number of photographs [82, 83]. Considerable effort was needed to establish the three-dimensional relationship between the streaks. The result of this superposition for $\alpha = 25°$ is shown in Fig. 6.40. The broad arrows describe the overall flow, which is steady and well ordered; the narrow arrows are actual bubble trajectories that represent the random motion in the burst vortex core.

It is reasonably apparent that the flow separates all along the wing leading edge, creating a leading-edge vortex sheet. This implies that there is no pressure difference or load across the wing at the leading edge, equivalent to the Kutta-Joukowski condition at the trailing edge. Inboard, the leading-edge vortex sheet rolls up tightly to form a concentrated vortex. Outboard, the sheet wraps around the turbulent core of the burst vortex as the core seems to detach from the wing surface, causing the strong, visible spanwise flow *outside* the boundary layer from the low-velocity spanwise flow *inside* the boundary layer on yawed wings. This

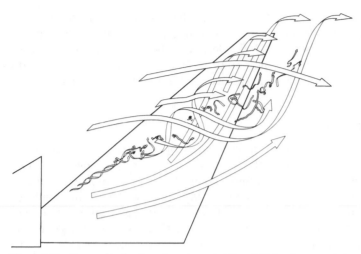

Figure 6.40   Composite drawing of average wind flow field for close-coupled canard [82]. (Courtesy of Sage Action, Inc.)

difference has been recognized for some time. Consequently, lower pressures are expected on the upper wing surface outboard and so more lift from the tip region at such incidence.

By comparison, the inside of the burst vortex is nearly a dead-air region. The velocities are much lower and almost completely random, implying that the pressure is practically constant. The vortex sheet around this stagnant region is roughly conical in shape from the point of bursting to the wing trailing edge. These two properties of the burst vortex system make theoretical formulation much easier. Furthermore, aerodynamic measurements on a full-span model, including pressure contours on the upper surface of the wing, correlate with the burst vortex observed.

The helium bubble method has been used to study a variety of complex flow problems. For example, the flow field surrounding a parachute during the opening process has been investigated using this method [84]. Velocity profiles were obtained from bubble streak photographs for a case late in the inflation process and for the fully inflated case. These data suggested that a potential flow mathematical model would suffice to provide a reasonable description of the opening process.

An extensive study was performed to determine the feasibility of using the neutrally buoyant bubbles at transonic speeds [85]. The bubbles were successfully injected and photographed in a transonic flow with Ma = 0.9. New lighting techniques and methods of strengthening the bubble film were examined.

The complex flow fields related to rotary wing aerodynamics have also been studied using the helium bubble method [86, 87]. The flow visualization portion of the research documented in [86] revealed a small, well-defined ground vortex at moderate and high wind velocities. The force data obtained showed that this ground vortex and the trailing vortex systems of the main-rotor flow field produced tail-rotor thrust perturbations. In [87] the flow field about an isolated rotor blade in rectilinear flow was examined. The results of these experiments revealed details of the formation of the tip vortex and the overall rollup of the vortex sheet.

6.2.4 Concluding Remarks

Smoke visualization in wind tunnels has been developed to the point where it can produce a global view of very complex subsonic and supersonic flow fields. This global picture is helpful in understanding complicated fluid dynamic phenomena, as well as indicating specific regions where quantitative measurements should be taken. When the flow is steady, the smoke streak lines are identical to streamlines that can be readily compared with analytical or numerical results for the same flow. When the flow under investigation is periodic, as the vortex shedding from a sharp flat plate, the vortex spacing obtained from smoke photographs agrees well with theoretical predictions. Shedding frequencies for this type of flow may also be determined by using a mechanical or electronic strobe during the wind-tunnel experiments. Even when the flow is unsteady, the streak line pattern obtained is of considerable value in helping to understand complicated flow phenomena. The recent development of digital image processing techniques to obtain quantitative data suggests almost unlimited promise for the future.

The use of helium bubbles in wind-tunnel experiments is increasing. The ability to follow helium bubbles photographically and obtain particle paths and velocities has provided the motivation for the development of this visualization method. Preliminary studies indicate that this method can also be used at transonic speeds. It should be clear that there are many fluid dynamic and heat transfer problems that can be studied using the smoke-tube, smoke-wire, and helium bubble techniques described above.

6.3 HYDRODYNAMIC FLOW VISUALIZATION

Flow visualization in water has played an important role in the understanding of fluid motions. It is clear that visual observations of flow phenomena were the first and, for a long time, the only experimental techniques available. It is difficult to imagine that da Vinci, Galileo, Newton, Bernoulli, Euler, and many others did not take advantage of flow visualization in their studies of fluid mechanics.

One of the most important discoveries in the history of fluid mechanics was a result of using aniline dye to produce colored water. This was the Osborne Reynolds experiment (1883) with a small filament of colored water in the center of a tube filled with clear water. As the velocity was increased through the tube, the transition from laminar to turbulent flow was observed by watching the formation of eddies and the subsequent diffusion of the colored water [88]. Many other experiments with a variety of visualizing agents have been developed and performed since 1883. Excellent reviews of the many available techniques have been presented by Clayton and Massey [89], Merzkirch [90], Werle [91], and Gad-el-Hak [92]. A brief description of many of the existing water-tunnel facilities in the world has been given by Erickson [93]. A description of the use of a water towing tank as an experimental tool is given by Gad-el-Hak [94]. In this section, the use of dye and hydrogen bubbles as visualizing agents will be discussed.

6.3.1 Dye Method

A large variety of experimental equipment and procedures has been used in connection with the dye injection method of flow visualization. From the famous experiment of Reynolds to the present, the use of dye injection as a visualization agent has steadily increased. The most significant developments in this field, however, began at the Office National d'Etudes et de Recherches Aerospatiales (ONERA) in France with the work of Roy and Werle [91] in the early 1950s. The equipment and techniques developed at ONERA established the state of the art and have been copied by many laboratories throughout the world. For this reason, one of the ONERA vertical water tunnels and visualization techniques, and a somewhat similar, though smaller, facility at the U.S. Air Force Wright Aeronautical Laboratories (USAF/WAL), will be discussed.

6.3.1.1 Vertical water tunnel at ONERA. As with the wind tunnels used for smoke visualization, the most desirable requirements for a water tunnel to be used for visualization purposes are uniform flow and a low turbulence level in the test section. The TH1 ONERA water tunnel [95, 96] is vertical, of the open-circuit type, and operates under the action of gravity, as shown in Figs. 6.41 and 6.42. It consists of a constant-head reservoir, flow straighteners, contraction section, test section (220 × 220 × 700 mm), test section extension (200 mm long), and discharge section. The capacity of the reservoir provides for a useful testing time of 2 min at a speed of 10 cm/s. Additional water capacity is available, permitting speeds of up to 50 cm/s for up to 12 min. A variety of model mounting and dye injection methods are available. Two larger vertical water tunnels of similar design are also in use in the ONERA Laboratory [96].

6.3.1.2 Vertical water tunnel at USAF/WAL. The facility at USAF/WAL was built as a pilot tunnel to develop components for a much larger tunnel. The USAF/WAL pilot water tunnel (shown in Fig. 6.43) is vertical and of the open-circuit type; as in the ONERA setup, the tunnel operates intermittently under the action of gravity. It is made up of a reservoir, turbulence damper, contraction section, test section (146 × 146 × 457 mm), and discharge section. The reservoir capacity allows for a run time of 2 min at 15 cm/s and an extended run time with the

Figure 6.41   Photograph of the ONERA vertical water tunnel. (Courtesy of H. Werlé, ONERA)

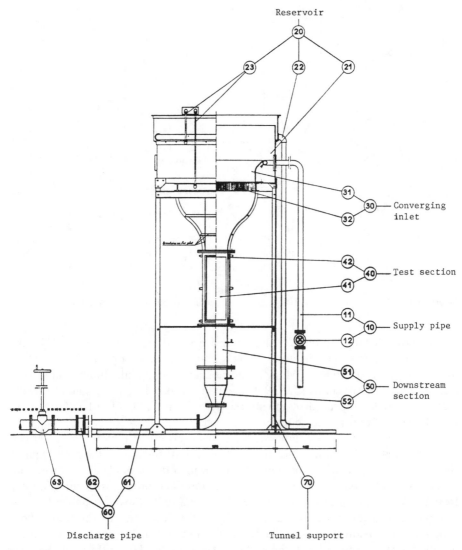

Figure 6.42 Schematic of the ONERA vertical water tunnel. (Courtesy of H. Werlé, ONERA)

addition of water from the spray bar. The open-pore foam used as the flow straight-ener appears to adequately damp any turbulence from the spray bar. Currently, models are mounted from the side walls, and up to eight dye (four colors) tubes are available. The four dye reservoirs are pressurized from a nitrogen bottle through a pressure regulator. Each dye reservoir holds 0.12 liter (4 oz.) and has two valved tubes available.

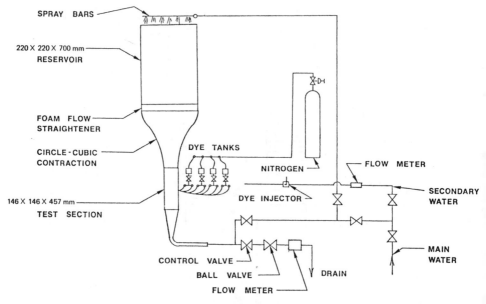

Figure 6.43   Schematic of the USAF/WAL pilot water tunnel. (Courtesy of USAF/WAL)

6.3.1.3 Dye. A large number of dyes and dye solutions have been used for the marking of filament lines in water tunnels [90]. The most popular dyes used are milk, food coloring, and ink. Many other substances, including fluorescent materials, have also been used [18, 19, 97]. It has been found that the stability of the filaments may be improved by mixing the dye with milk [90, 91]. The fat content of milk has been presumed to retard its diffusion.

The injection of dye without significantly altering the flow under study is a primary concern. It is important that the velocity of the injected dye be equal to the velocity of the surrounding flow. This helps maintain a stable dye filament and reduces the disturbance to the surrounding flow. The problems of injecting dye from a rake upstream of the model are the same as with the early smoke tunnels. The presence of the injection rake or tube in the flow disturbs the tunnel flow, which in turn, disturbs the dye filaments. To circumvent this problem, the dye is usually injected from holes in the surface of the model. Using this method, care must be taken so that the dye has as small a velocity component perpendicular to the surface as possible, or the effect will be equivalent to blowing into the boundary layer. It is difficult to accurately pulse dye filaments for direct velocity measurements, and thus, continuous dye filaments are generally used.

6.3.1.4 Photographic techniques. Because the velocities involved in most water-tunnel experiments are quite low and there is great contrast between the colored dyes and the surrounding water flow, no exotic photographic techniques are necessary. Photo flood lights and still or movie cameras with color film are usually used. Readily available and relatively inexpensive color or black and white video systems may also be used.

6.3.1.5 Application of dye method. The use of dye filaments in a water tunnel is analogous to the use of smoke filaments in a wind tunnel and thus is subject to the same limitations. The dye method has been used to study a very large number of two- and three-dimensional steady and unsteady flows [91]. These flows may be further classified according to type of phenomenon studied: jets, wakes, boundary layer structure, vortices, etc. Only a few representative flows will be discussed.

Vortices from a swept-back wing (ONERA) Because of their importance in a large number of fluid dynamic situations, vortices have been studied for many years. The relatively recent introduction of supersonic aircraft configurations has resulted in the use of water tunnels to investigate the complicated three-dimensional flow field and vortex interactions produced. In particular, vortex-induced lift resulting from controlled leading edge separation provided the motivation for many studies of slender wings [93]. A variety of water-tunnel experiments has been performed in attempts to understand the development and breakdown of vortex cores [91, 93]. Vortices formed on the upper surface of delta wings at incidence are shown in Fig. 6.44. The continuous injection of dye through holes in the surface of these delta wings clearly shows the shape and structure of these vortices. Flow visualization results of this type have been useful in identifying the effects of delta wing geometry and incidence angle on the position and structure of the vortices. Intermittent dye injection has also been used to determine the velocity along the vortex axis [91]. The phenomenon of vortex breakdown on highly swept-back-wing aircraft at high incidence is shown in Fig. 6.45. This phenomenon has received a great deal of attention, both theoretical and experimental [98], and has been verified in flight. For delta wings, flow visualization results have been successfully compared with data obtained at supersonic speeds. An extensive correlation of vortex flows was performed by Erikson [93]. This is not a surprising conclusion, since vortices of this type have been treated theoretically as inviscid, with reasonable success. Erikson [93] has also been able to correlate the location of vortex bursting obtained from dye studies in water tunnels with results obtained from flight tests. A recent comparison of the vortices produced from the LEX and wing of a generic fighter aircraft model in both a wind tunnel using T_1Cl_4 and a water tunnel using dye show excellent agreement [15]. An example of this comparison is shown in Fig. 6.46.

Forward-swept wing with canard (USAF/WAL) Interest in forward-swept-wing aircraft is growing rapidly. This type of configuration, made possible by lightweight, nonmetallic composite materials, would have greater range and fuel economy, lower stall speeds, spin resistance, and improved low-speed control for easier landings and takeoffs than the present swept-back-wing design. Some of the present

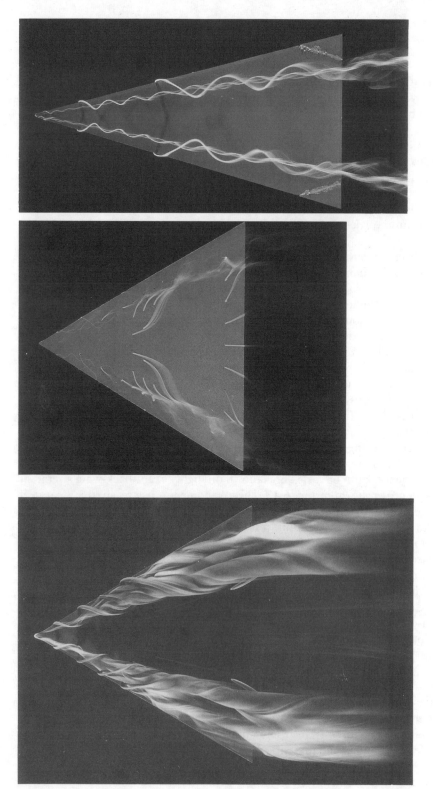

Figure 6.44 Slender delta wing configurations at incidence. (Courtesy of H. Werlé, ONERA)

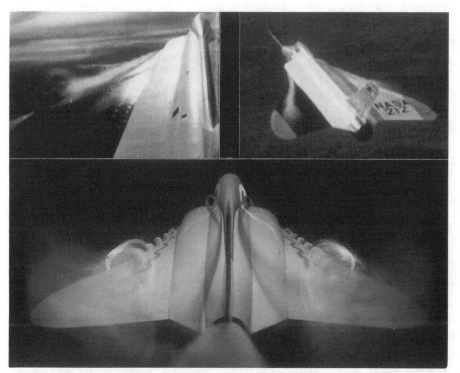

Figure 6.45 Vortex flow on a highly swept-back-wing aircraft (*top*) from flight tests and (*bottom*) from a 1/72-scale model in the water tunnel. (Courtesy of H. Werlé, ONERA)

aircraft designs use canards to improve maneuverability. The forward-swept-wing design concept is relatively new, and therefore a large number of analytical and experimental studies are necessary to understand the complex flow over such a geometry.

Water-tunnel experiments, using dye injection, were performed at the Wright Aeronautical Laboratories to study the influence of the canard on the main wing flow. For these experiments, a half-span model of a Grumman Aerospace Corporation design was placed in the USAF/WAL vertical water tunnel, as shown in Fig. 6.47. The model was studied at pitch angles (α) from $-2°$ to $+17°$, and at each model pitch angle the canard was pitched at angles (β) from $-5°$ to $+25°$. The influence of the canard on the flow over the main wing is shown in the series of photographs in Fig. 6.48, where $\alpha = 5°$ and $\beta = 2.5°$, $5°$, $10°$, and $20°$. The water-tunnel velocity was 6 cm/s (0.2 ft/s), and the Reynolds number was 52,493/m (16,000/ft). Experiments of this type will undoubtedly continue, together with analytical and numerical studies, as this new configuration evolves.

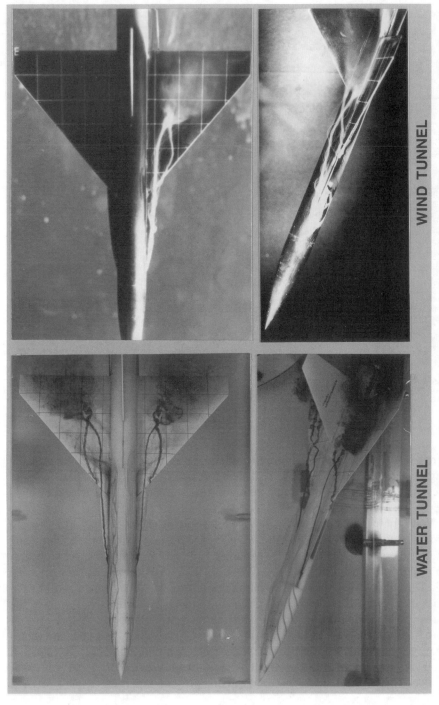

WATER TUNNEL

WIND TUNNEL

Figure 6.46 Comparison of water-tunnel and wind-tunnel flow visualization for a generic fighter with 50° swept wing and 21° angle of attack. (Courtesy of G. N. Malcolm, Eidetics International Inc.)

Figure 6.47 Forward-swept wing with canard in the USAF/WAL vertical water tunnel. (Courtesy of USAF/WAL)

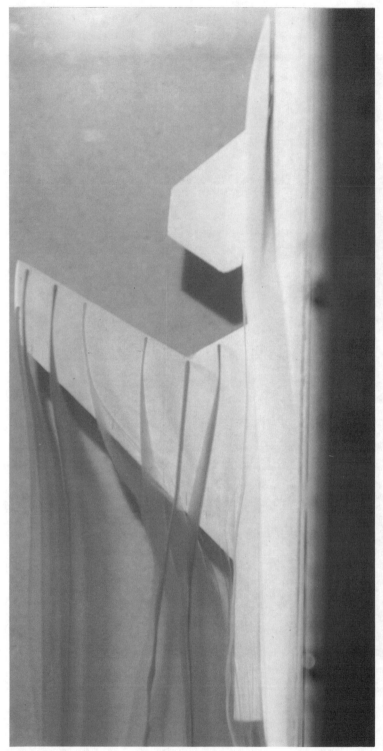

Figure 6.48a Forward-swept wing model with canard at $\alpha = 5°$ and $\beta = 2.5°$. (Courtesy of USAF/WAL)

428

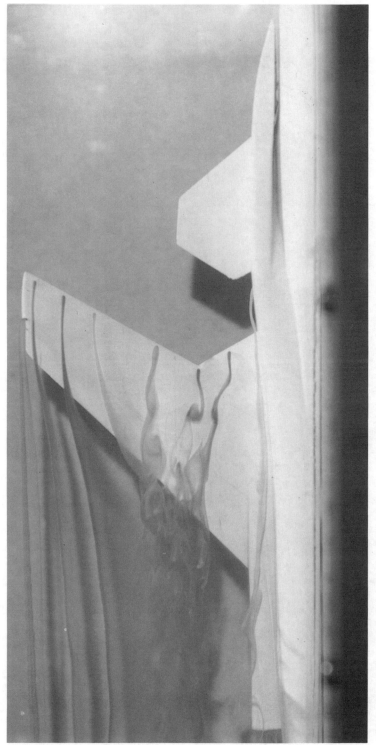

Figure 6.48b Forward-swept wing model with canard at $\alpha = 5°$ and $\beta = 5°$. (Courtesy of USAF/WAL)

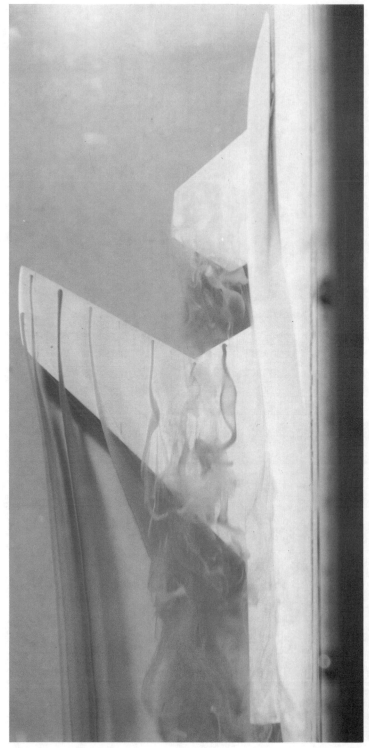

Figure 6.48c Forward-swept wing model with canard at $\alpha = 5°$ and $\beta = 10°$. (Courtesy of USAF/WAL)

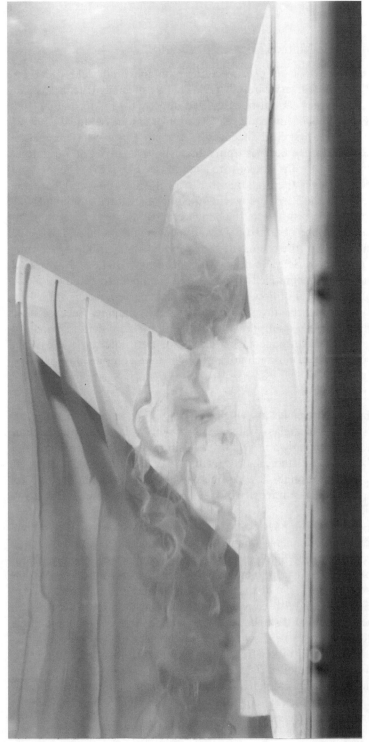

Figure 6.48*d*   Forward-swept wing model with canard at $\alpha = 5°$ and $\beta = 20°$. (Courtesy of USAF/WAL)

6.3.2 Hydrogen Bubble Method

Visualization of flow fields using dye injection techniques that yield streak line patterns must be interpreted very carefully [99]. This is especially true in unsteady flows. Although streamlines cannot be visualized directly by any technique [100], quantitative measurements of unsteady velocity profiles can be obtained using the hydrogen bubble method first introduced by Geller [101] in 1954. This method generates very small hydrogen bubbles from a fine wire that is part of a dc circuit, similar to the circuit used for the smoke-wire method (see Fig. 6.27). The predominant force on the bubbles is the drag due to the local water motion. Decreasing the bubble diameter has the effect of lowering the buoyancy force at a greater rate than the drag force. For this situation, the bubbles follow the local water motions, and velocity profiles may be obtained for two-dimensional low-speed flows [102, 103]. An excellent review of the use of the hydrogen bubble method to determine velocity profiles is given by Schraub et al. [102], Mattingly [104], and Merzkirch [18]. The hydrogen bubble technique is also useful in the qualitative study of global flow patterns around bodies. For flow problems of this type, separation, transition, and unsteady wakes have been investigated using this technique. Because of its usefulness in obtaining both quantitative and qualitative flow field information, this technique has seen steadily increasing usage, for both internal and external flows. Examples of the use of this technique to understand complex flow phenomena will be presented.

6.3.2.1 Free-surface water channel at Lehigh University. The Lehigh University facility has a free-surface plexiglass water channel with a 5-m working section, 0.9 m wide by 0.3 m deep (shown in Fig. 6.49). Using a speed control unit with a shaft speed feedback circuit, stable speeds from 0.01 to 0.60 m/s may be attained. By combining a settling sponge, flow straightener–screen combination, and 5:1 inlet contraction, a turbulence intensity of 0.4% and spanwise flow uniformity of ±2% can be achieved.

The viewing and recording system used in this facility consists of two high-speed closed-circuit video cameras (manufactured by the Video Logic Corporation) and synchronized strobe lights to provide 120 frames/s with effective frame exposure times of 10^{-5} s. An exceptionally clear picture can be obtained from the high-resolution screen using 250 horizontal, direct overlay rasters with a sweeping frequency of 25.2 kHz. With conventional lenses, fields of view as small as 6×6 mm can be obtained at distances of 0.5 m. A split-screen capability allows two different fields of view to be displayed simultaneously and recorded. All recorded data can be played in flicker-free slow motion (both forward and reverse) as well as single frame, for detailed data analysis. Once a video sequence has been recorded, still photographs of individual stop-action frames can be taken directly from the video screen using conventional photography or via a videographic copier that interfaces directly with the video recorder.

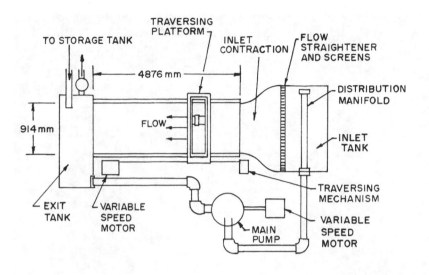

a) Schematic of Flow Facility

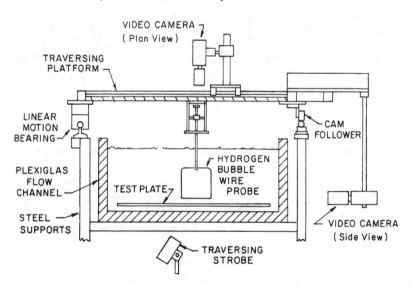

b) End-on View of Channel (Looking Downstream) and Traversing Platform

Figure 6.49 Lehigh University free-surface water channel. (Courtesy of C. R. Smith and T. Wei, Lehigh University)

433

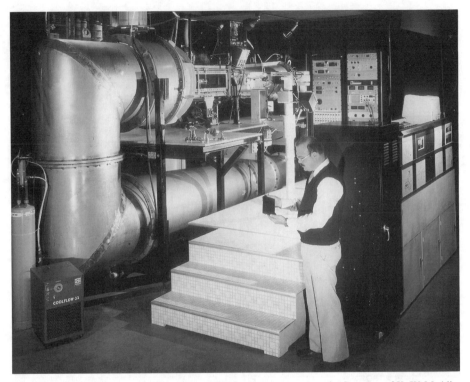

Figure 6.50 Aeroflightdynamics Directorate's 21- by 31-cm water tunnel. (Courtesy of K. W. McAlister, U.S. Army Aeroflightdynamics Directorate ARTA)

6.3.2.2 Horizontal water tunnel at the U.S. Army aeroflightdynamics directorate.
This tunnel is of a closed-circuit design [105] and is operated under the U.S. Army Aviation Research and Technology Activity to obtain load and flow visualization data (Fig. 6.50). The circuit contains approximately 4000 L of water and, with the exception of the fiberglass contraction section and the plexiglass windows, is constructed of stainless steel. The tunnel passages are predominantly circular, except for the rectangular test section, which measures 0.2 m wide by 0.3 m high by 1.0 m long. The transition from circular to rectangular cross section takes place in the contraction and diffuser sections adjoining the test section.

Prior to entering the 10:1 contraction, the flow is straightened by two sets of honeycombs, and the turbulence is reduced by four sections of screening. Two large storage tanks are used for dissolving fresh chemicals before filling the tunnel and for holding that portion of the water withdrawn from the tunnel when model changes are made. Whenever water is pumped back into the tunnel, it must first pass through a filter system designed to remove contaminates down to 5 μm. The velocity that can be obtained in the test section is continuously variable from nearly 0 to over 6 m/s. The presence of cavitation-induced air bubbles represents a serious limitation on the maximum usable speed in a test program, since air bubbles can severely interfere with the viewing of smaller bubbles intentionally generated during flow visualization studies. Under atmospheric conditions, the amount of air dissolved in water is generally found to be about 2% by volume. This air remains in solution until drawn out by cavitation to form a lingering air bubble of detectable size. Typical low-pressure sources responsible for cavitation include the impeller, turning vanes, points of tunnel-wall separation, and the leading edge of a lifting airfoil. The problem can be alleviated by subjecting the water in the tunnel to a vacuum and extracting a majority of the dissolved air. Degassing is normally only necessary with each initial filling of the tunnel. Once the air has been removed, the water can be transferred later to the holding tanks (while model changes are made) and returned to the tunnel with negligible reingestion of air. During a test, the tunnel can be pressurized up to 15 psig in order to postpone the development of cavitation bubbles caused by the small amount of air that inevitably remains in solution.

6.3.2.3 Bubble generation. Flow visualization by the hydrogen bubble method involves the placement of a fine metallic wire within the flow to serve as the cathode of a dc circuit. An anode (also submerged), consisting of any suitable conductive object, is placed nearby. Supplying voltage to the circuit causes the liberation of hydrogen at the cathode through electrolysis, with oxygen being released at the anode. Although the composition of ions produced depends on the particular electrolyte used, the significant reaction is the decomposition of the water according to $2H_2O \rightarrow 2H_2 + O_2$. Since the hydrogen bubbles formed are more numerous than the oxygen bubbles, better flow visualization is obtained using the cathode. Hydrogen bubbles forming on the electrode are swept off by the flow to form a continuous stream.

If the electrode is a straight wire and a voltage is supplied in the form of square-wave pulses, discrete lines of bubbles are produced. If the electrode is formed by the kinking of the wire in a zigzag fashion, the bubbles will shed at each downstream kink to produce a series of streak lines in the flow (Fig. 6.51). Since the size of the bubbles released from the wire is of the order of one-half the wire diameter; the use of an extremely fine wire (25–50 μm) usually renders buoyancy effects negligible [102]. Because the hydrogen bubbles are generated in a region where there is a velocity defect due to the wire, some distance is required before the bubbles reach the velocity of the fluid motion under investigation. Although common water supplies generally contain enough impurities to provide a current path between the two electrodes, the addition of a small amount of electrolyte such as sodium sulfate (0.15 g of Na_2SO_4 per liter of water works well) greatly enhances bubble generation, producing denser bubble lines with a consequent increase in clarity and contrast of the visual data obtained.

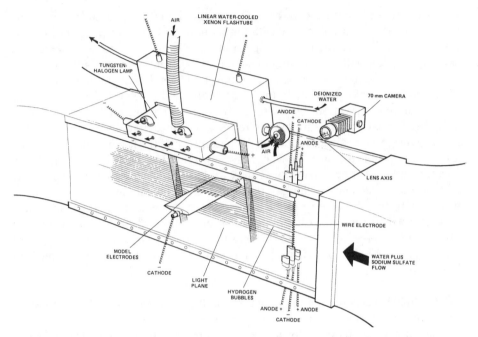

Figure 6.51 Electrodes and lighting used for hydrogen bubble visualization. (Courtesy of K. W. McAlister, U.S. Army Aeroflightdynamics Directorate, ARTA)

6.3.2.4 Lighting and photographic techniques. The technique used to obtain still and high-speed movies in the water tunnel at the Army Aeroflightdynamics Directorate is typical of those used in many facilities of this type and similar to those described in the smoke-tube method. The bubbles are illuminated by a narrow sheet of light directed through the upper test section window and extending over a distance of 30 cm in the free stream direction (Fig. 6.51). Baffles are used to control the width of the light passing into the test section. The path of the light is canted 10° from the plane of the bubbles to provide a backlighting component relative to the camera [105]. Both continuous and flash sources of light are produced over the 30-cm length. Continuous lighting is provided by a single 1000-W tungsten-halogen lamp. This lamp is placed behind heat-absorbing glass plates to protect the test-section windows, and the heat absorbed is then convected away from these plates by a high-volume stream of air. Once the light has entered the main optical housing, it is turned 90° by a mirror that can be retracted when this mode of lighting is no longer required. Continuous lighting is used for general viewing, for low-speed movies, and for long exposures (normally longer than 15 ms) with a single-frame camera.

When an instantaneous visualization of the flow field is needed, the mirror is

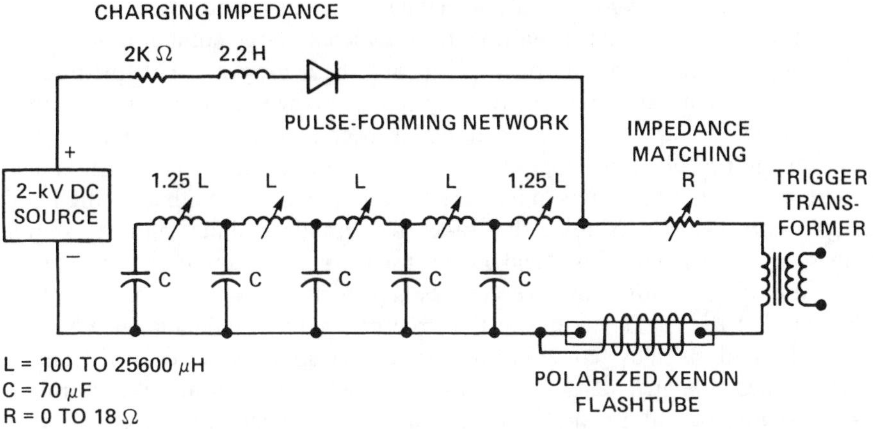

CHARGING IMPEDANCE

Figure 6.52 Circuit for shaping and protracting flashtube discharges. (Courtesy of K. W. McAlister, U.S. Army Aeroflightdynamics Directorate, ARTA)

retracted, and a 10,000-W xenon flashtube is activated. The strobe can either be synchronized to the shutter of a high-speed camera or operated in a single-flash mode and synchronized with a particular phase of an unsteady event. There are occasions (especially in rotational flows) when a slight amount of streaking is desirable in order to visualize the motion of the particles in a single exposure. Streaking is provided by a pulse-forming network with variable inductance (Fig. 6.52) to protract the photoemission of the flashtube [106]. This permits particle-path visualization to be obtained over a Reynolds number range requiring high levels of bubble illumination that are normally beyond the practical limits of continuous light sources. The values shown in the circuit diagram are used to produce well-shaped light pulses from 1- to 16-ms duration.

The single-frame photography system consists of an automatic 70-mm film magazine, a bellows type of focusing body, and a 240-mm lens that is coupled with an electronically controlled aperture and shutter (15 ms minimum). This composite camera is mounted on a rigid post, so that the film plane is located a nominal distance of 152 cm from the center of the test section. This combination of lens and film plane-to-subject distance was found to offer the best compromise of image size, maximum depth of field, and minimum perspective distortion. The cinematic system consists of a Milliken high-speed camera, capable of indexing 16-mm film up to 500 frames/s. This camera is mounted on a rotatable arm that allows the lens axis to be coaxially positioned in front of the lens of the single-frame camera system. Since most continuous sources of light do not provide adequate illumination for short-duration exposures, a specially tailored charge-discharge circuit for the flashtube is used to generate strong bubble illuminations at camera speeds up to 250 frames/s.

6.3.2.5 Application of hydrogen bubble method. Since its introduction in 1954, the hydrogen bubble method has been used to study a very large number of low-speed fluid dynamic phenomena. Although, with care, the hydrogen bubble method can be used to obtain measurements of velocity, it has also been used in many instances to obtain global flow patterns. This method has been especially useful in the area of transition and turbulent boundary layer structure, and unsteady flow phenomena.

The objective of this study, under the direction of C. R. Smith at Lehigh University, was to visually examine the wake of a circular cylinder and to examine spanwise variations in the flow. Evidence of a spanwise variation in the wake region had been previously noted by several investigators using dye injection (e.g., [107] and [108]), but the characteristics of the structure and its Re dependence had not been established, since dye could not be injected in a spanwise sheet away from the surface of the cylinder. The use of a hydrogen bubble wire alleviates the shortcomings of dye injection and reveals an ordered spanwise structure of the near wake of a cylinder.

A 1.2-cm-diameter stainless steel circular cylinder, 0.9 m in length, was mounted spanwise across the channel at the half depth of the channel. The cylinder was polished to eliminate roughness and any spanwise irregularities that might give rise to spanwise flow irregularities. To examine the effect of end conditions, the cylinder was mounted alternately using both rigid end supports and a guy wire support arrangement. Identical results were obtained using either support technique.

For the present investigation, platinum wire, 0.025 mm in diameter, was used. Bubble lines were made visible by illuminating the flow with a light source of high intensity directed at an oblique angle to the line of sight. Placement and orientation of the bubble wire within a flow was accomplished by a bubble-wire probe consisting of two conductive, insulated metal prongs between which the platinum wire was stretched, with the ends secured by soldering. The probe was constructed of brass and insulated with heat-shrink tubing and red glyptol. The probe was mounted to a calibrated traversing mechanism that could be adjusted in 0.05-mm increments.

The bubble wire was energized by square-wave voltage pulses supplied by a voltage generator with a range of 0–90 V and a maximum output of 2.5 A. The pulse duration was continuously adjustable, and the pulsing frequency could be varied from 0 to 340 Hz. The pulsing frequency was monitored by a multifunction counter that could measure frequencies from 5 Hz to 100 MHz.

After about a minute of operation, the quality of bubble lines shed from the wire began to degenerate. This was due to a buildup on the wire of positively charged ions, which may be removed by reversing the polarity of the voltage supply for a few seconds, then switching it back. However, abrupt changes in the voltage output can damage the pulse generator and/or cause a breakage of the bubble wire. Also, the residual capacitance present in the generator causes the square-wave signal to persist for a few seconds after the generator has been shut down. Therefore,

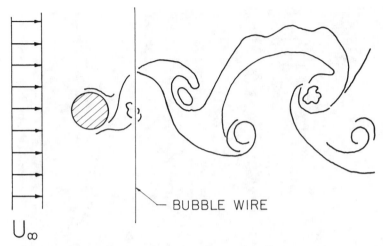

BUBBLE WIRE

U_∞

Figure 6.53 Schematic of flow over a circular cylinder—side view. (Courtesy of C. R. Smith and T. Wei, Lehigh University)

to avoid potential damage to the equipment, it is necessary to wait for the residual pulses to fade away before switching the polarity in either direction.

Using the flow visualization system, both vertical and horizontal visualizations were performed over a Reynolds number range of $50 < Re_d < 2500$. Figure 6.53 shows a typical side view of the vortex shedding pattern for $Re_d = 300$ with a pulse rate of 30 Hz. The picture in Fig. 6.54 is a single video frame obtained directly from the recorded video signal using a videographic copier. This figure illustrates the very organized, periodic vortex shedding pattern that has become a well-known fluid dynamic phenomenon, and gives just a hint of a spanwise behavior (as indicated by the folded edges of the vortex wake region).

To visualize the spanwise structure, the hydrogen bubble wire was located parallel to the cylinder, as shown in Fig. 6.55. As indicated, the wire could be moved relative to the cylinder to facilitate visualization of the flow structure. With the wire positioned essentially in-line with both the trailing edge and the top of the cylinder, Fig. 6.56 shows schematically the top view pattern that can be observed. The corresponding side view vortex pattern is also shown for reference. What came as a total surprise was the appearance of spanwise "pockets" of inflow normal to the plane of the bubbles, which occur between the transverse vortices. In the top view of Fig. 6.56, these pockets appear just behind the cylinder. As the pockets convect downstream, the bubble sheet reveals the presence of axially oriented vortices of relatively uniform spacing. These vortices (shown schematically between transverse vortices of like rotation) were of small core size and strong vorticity relative to the transverse vortices, and were observed to occur in counterrotating pairs.

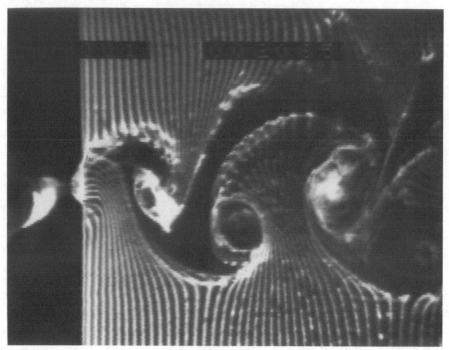

Figure 6.54 Photograph using a videographic copier of the periodic vortex shedding pattern behind a circular cylinder. (Courtesy of C. R. Smith and T. Wei, Lehigh University)

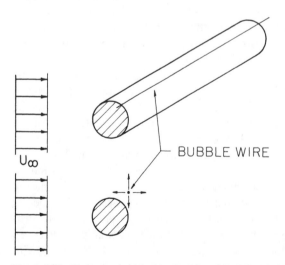

Figure 6.55 Hydrogen bubble wire location for study of the spanwise flow structure. (Courtesy of C. R. Smith and T. Wei, Lehigh University)

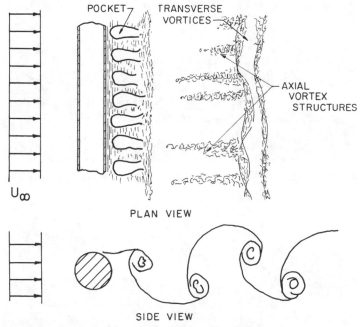

Figure 6.56 Sketch of the hydrogen bubble patterns from the top and side of the circular cylinder. (Courtesy of C. R. Smith and T. Wei, Lehigh University)

Figure 6.57 is a four-picture top view sequence illustrating the formation of the spanwise structure. These pictures are single video frames obtained by taking a Polaroid photograph directly from the video monitor. The orientation of each picture is the same as the top view in Fig. 6.56, with the cylinder just out of the field of view, to the left. For these pictures the bubble wire was running in a dc mode such that a continuous sheet of bubbles was generated. The formation of spanwise pockets is illustrated in Figs. 6.57a and 6.57b, with the evolution to axial vortices shown in Fig. 6.57c as a result of the penetration of the bubble sheet by the axial vortices. In Fig. 6.57d the remnants of the visualization bubbles depicting the spanwise structure have dissipated at the downstream side of the transverse vortex (although their presence is still in evidence), and the formation of a new set of spanwise pockets has just started.

The phenomenon illustrated in Figs. 6.56 and 6.57 is by no means unique to one flow condition. The spanwise structure was determined to be directly related to the transition of the transverse vortices to turbulence, appearing for all Re examined from $Re_d = 150$ (the accepted transition value) up to $Re_d = 2500$. The phenomenon was found not to be a function of the cylinder end condition, nor peculiar to free stream conditions in the channel. This latter point was confirmed by

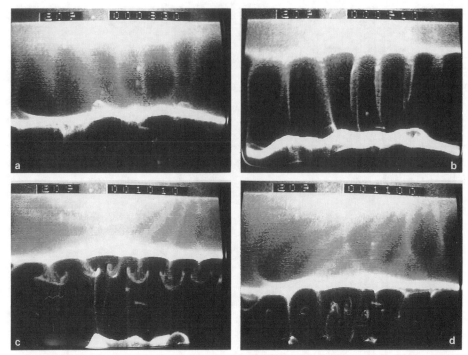

Figure 6.57 Hydrogen bubble photographs from video monitor of the spanwise structure. (Courtesy of C. R. Smith and T. Wei, Lehigh University)

towing the cylinder through a quiescent channel, with identical structures appearing in the wake.

From these visualizations, it appears that the formation of counterrotating axial structures in the wake of a circular cylinder is a necessary part of the process of transition to turbulence of the vortices shed from the cylinder. Further results from this study are contained in [109].

6.3.2.6 Flow over an oscillating airfoil. Dynamic stall occurs in many practical situations when a lifting airfoil must undergo a change in incidence beyond its static-stall angle. The hydrogen bubble experiments of McAlister and Carr [105, 109] at the U.S. Army Aeroflightdynamics Directorate, together with the smoke visualization experiments of McCroskey (of the same laboratory), discussed earlier, have helped in the understanding of the phenomenon of dynamic stall. Although certain details of the stall process may be dependent on the airfoil geometry and the particular flow environment (e.g., pitch rate and Reynolds number), the general characteristics of the stall vortex are believed to be qualitatively invariant. An important step toward a better understanding of dynamic stall therefore would be to examine the collapse of the boundary layer and the initial development of this vortex under conditions amenable to more definitive flow visualizations than are normally possible in air. The closed-circuit water tunnel at the U.S. Army Aeroflightdynamics Directorate was used to examine this problem. To study the flow pattern around the airfoil, a wire cathode was stretched across the test section and oriented normal to the direction of flow. Although both platinum and stainless steel are good noncorrosive materials, stainless steel was selected because of its superior tensile strength. A nonconductive model of a NACA 0012 airfoil with a modified leading edge, with a chord of 10 cm and a span of 21 cm, was used for these experiments. Electrodes were also placed at nine chordwise locations along the upper surface of the model during its construction. The sinusoidal pitching motion of the airfoil is accomplished by a flywheel, connecting rod, and rack and gear mechanism, which function to transform a circular motion, first, to reciprocating motion and then to airfoil pitch oscillations.

Typical results using both the fine-wire electrode upstream of the airfoil and the surface electrodes built into the airfoil's upper surface are shown in Fig. 6.58. This figure shows the location of flow separation, the wake vortex patterns, the shear layer vortices, and the dynamic stall vortex. For this airfoil at Re = 21,000, a reduced frequency of $k = 0.25$ and $\alpha = 10° + 10° \sin \omega t$, the hydrogen bubble experiments were important factors in arriving at the following conclusions.

The onset of dynamic stall was found to begin with a rapid movement of flow reversal toward the leading edge of the airfoil. At one point, a thin layer of reversed flow was observed to momentarily exist throughout the entire upper surface boundary layer without causing any appreciable disturbance to the viscous-inviscid interface.

The free shear layer that was created between the region of reversed flow and the inviscid stream was not stable. This instability resulted in a transformation of the free shear layer into a multitude of discrete clockwise vortices, out of which emerged a dominant "shear layer vortex."

Once the flow had reversed up to the leading edge of the airfoil, a protuberance grew and eventually developed into the "dynamic stall vortex" that has been observed in the high-Re experiments.

6.3.2.7 Digital image processing. With recent developments in high-speed digital video equipment, it is now possible to obtain quantitative information from dye and hydrogen bubble visualization experiments [37, 110–114]. Because many of the visualization experiments in water are performed at much lower velocities than those using smoke in air, currently available digital video equipment, computers, and light sources are adequate to obtain good results. The use of colored dyes in water enhances the ability to obtain meaningful results.

Automated image processing of hydrogen bubble flow visualization has been used to establish local, instantaneous velocity profile information in turbulent

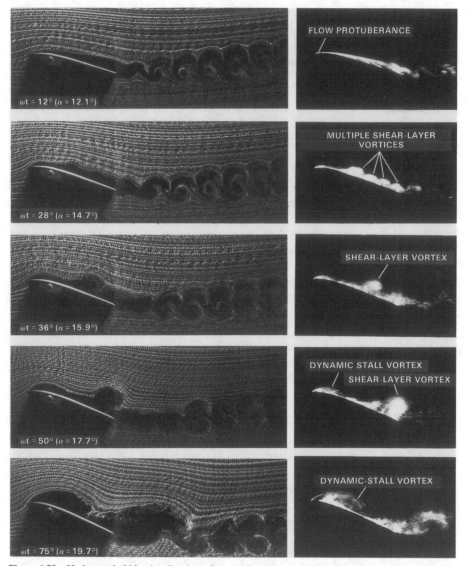

Figure 6.58 Hydrogen bubble visualization of the critical stages of dynamic stall. (Courtesy of K. W. McAlister and L. W. Carr, U.S. Army Aeroflightdynamics Directorate, ARTA)

boundary layers [111, 114]. Using special smoothing and gradient detection algo-rithms, individual bubble lines were computer identified by Lu and Smith [111]. They were subsequently able to determine local velocity behavior as a function of position and time, evaluate time-averaged turbulence properties, and correlate probe-type turbulent burst detection techniques with the corresponding visualization data. More recently, a technique for quantitative identification and detection of hairpin-type vortices within a turbulent boundary layer was developed by Smith and Lu [112]. As a result of this type of research, the new field of computer-aided flow visualization has emerged.

6.3.3 Concluding Remarks

Although restricted to comparatively low speed flows, the dye and hydrogen bubble visualization methods have been responsible for much of our understanding of many basic steady and unsteady flow phenomena. These techniques have been widely used for almost three decades, but new ways of using them continue to emerge. For example, the use of cross-line wires to visualize quadrilateral flow elements whose deformation and rotation can be measured directly has been reported [113].

6.4 CONCLUSIONS

The smoke-tube and helium bubble methods in air have been found to be useful over the widest range of speeds, from low subsonic to transonic and even supersonic velocities. Although the smoke-wire method in air and the dye and hydrogen bubble methods in water are usually used at low and moderate velocities, they provide greater detail of the flow than the other methods.

While the usefulness of the direct injection methods of flow visualization is apparent, each flow problem studied should be approached with care. The importance of the results obtained for a given problem will be directly related to the time and effort invested by the experimenters. The interpretation of flow visualization data should be made in conjunction with data from other measurement techniques as well as with theoretical or numerical analyses.

REFERENCES

1. N. H. Randers-Pehrson, Power Wind Tunnels, *Smithson. Misc. Collect.*, vol. 93, no. 4, 1935.
2. T. J. Mueller, Smoke Visualization of Subsonic and Supersonic Flows (The Legacy of F. N. M. Brown), University of Notre Dame Report UNDAS TN-3412–1, or AFOSR TR-78-1262, June 1978.
3. T. J. Mueller, On the Historical Development of Apparatus and Techniques for Smoke Visualization of Subsonic and Supersonic Flows, AIAA Paper 80-0420-CP, presented at the AIAA 11th Aerodynamic Testing Conference, March 18–20, 1980.
4. M. Brendel, Experimental Study of the Boundary Layer on a Low Reynolds Number Airfoil in Steady and Unsteady Flow, Ph.D. thesis, University of Notre Dame, Notre Dame, Ind., May 1986.
5. M. Brendel and T. J. Mueller, Boundary Layer Measurements on an Airfoil at a Low Reynolds Number in an Oscillating Freestream, *AIAA J.*, vol. 26, no. 3, pp. 257–263, 1988.
6. R. H. Smith and C. T. Wang, Contracting Cones Giving Uniform Throat Speeds, *J. Aeronaut. Sci.*, vol. 11, no. 4, pp. 356–360, 1944.

7. M. Brendel and T. J. Mueller, Boundary-Layer Measurements on an Airfoil at Low Reynolds Numbers, *J. Aircraft,* vol. 25, no. 7, pp. 612–617, 1988.

8. S. M. Batill, R. C. Nelson, and T. J. Mueller, High Speed Smoke Flow Visualization, Air Force Wright Aeronautical Laboratories Report AFWAL/TR-3002, 1981.

9. T. J. Mueller, Gases: Smokes, chapt. 5, in W. J. Yang (ed.), *Handbook of Flow Visualization,* pp. 45–63, Hemisphere, New York, 1989.

10. R. L. Maltby and R. F. A. Keating, Smoke Techniques for Use in Low Speed Wind Tunnels, AGARD-ograph 70, pp. 87–109, 1962.

11. J. H. Preston and N. E. Sweeting, An Improved Smoke Generator for Use in the Visualization of Airflow, Particularly Boundary Layer Flow at High Reynolds Numbers, ARCR&M 2023, ARC 7111, October 1943.

12. L. F. G. Simmons and N. S. Dewey, Photographic Records of Flow in the Boundary Layer, Reports and Memoranda of the Aeronautical Research Council 1335, London, 1931.

13. P. Freymuth, W. Bank, and M. Palmer, Use of Titanium Tetrachloride for Visualization of Accelerating Flows Around Airfoils, in W. J. Yang (ed.), *Flow Visualization III,* pp. 99–105, Hemisphere, New York, 1985.

14. K. D. Visser, R. C. Nelson, and T. T. Ng, Method of Cold Smoke Generation for Vortex Core Tagging, *J. Aircraft,* vol. 25, no. 11, pp. 1069–1071, 1988.

15. G. N. Malcolm and R. C. Nelson, Comparison of Water and Wind Tunnel Flow Visualization Results on a Generic Fighter Configuration at High Angles of Attack, AIAA Paper 87-2423, presented at AIAA Atmospheric Flight Mechanics Conference, August 17–19, 1987.

16. R. L. Bisplinghoff, J. B. Coffin, and C. W. Haldeman, Water Fog Generation System for Subsonic Flow Visualization, *AIAA J.,* vol. 14, no. 8, pp. 1133–1135, 1976.

17. S. A. Brandt and J. D. Iversen, Merging of Aircraft Trailing Vortices, *J. Aircraft,* vol. 14, no. 12, pp. 1212–1220, 1977.

18. W. Merzkirch, *Flow Visualization,* 2nd ed., Academic, San Diego, Calif., 1987.

19. W. Merzkirch, Techniques of Flow Visualization, NATO AGARD Report 302, December 1987.

20. H. M. Nagib, T. C. Corke, and J. L. Way, Computer Analysis of Flow Visualization Records Obtained by the Smoke-Wire Technique, presented at the Dynamic Flow Conference, Baltimore, Md., September 1978.

21. L. Hesselink, Digital Image Processing in Flow Visualization, *Annu. Rev. Fluid Mech.,* vol. 20, pp. 421–485, 1988.

22. M. A. Hernan and J. Jiminez, The Use of Digital Image Analysis in Optical Flow Measurements, *Proceedings of 2nd Symposium on Turbulent Shear Flows,* London, Paper 702, pp. 7.7–7.14, July 1979.

23. C. C. Giamati, Application of Image Processing Techniques to Fluid Flow Data Analysis, NASA Tech. Memo 82760, February 1981.

24. J. P. Borleteau, Concentration Measurement with Digital Image Processing, IEEE, ICIASF 83 Record, pp. 37–42, 1983.

25. T. C. Corke, Two-Dimensional Match Filtering as a Means of Image Enhancement in Visualized Turbulent Boundary Layers, AIAA Paper 84-534, January 1984.

26. T. C. Corke and Y. Guezennec, Discrimination of Coherent Feature in Turbulent Boundary Layers by the Entropy Method, AIAA Paper 84-534, January 1984.

27. T. Utami and T. Ueno, Visualization and Picture Processing of Turbulent Flow, *Exp. Fluids,* vol. 2, pp. 25–32, 1984.

28. T. C. Corke, A New View on Origin Role and Manipulation of Large Scales in Turbulent Boundary Layers, Ph.D. thesis, Illinois Institute of Technology, Chicago, 1981.

29. J. T. Kegelman, A Flow-Visualization Technique for Examining Complex Three-Dimensional Flow Structures, *10th Symposium on Turbulence,* University of Missouri-Rolla, September 22–24, 1986.

30. F. W. Roos and J. T. Kegelman, Control of Coherent Structures in Reattaching Laminar and Turbulent Shear Layers, *AIAA J.,* vol. 24, no. 12, pp. 1956–1963, 1986.

31. F. W. Roos and J. T. Kegelman, Influence of Excitation on Coherent Structures in Reattaching Turbulent Shear Layers, AIAA Paper 86-0112, presented at AIAA 24th Aerospace Science Meeting, Reno, Nev., January 6–8, 1986.

32. F. W. Roos and J. T. Kegelman, Structure and Control of Flow over a Backward-Facing Step, presented at Forum on Unsteady Flow Separation, ASME Fluids Engineering Spring Conference, Cincinnati, Ohio, June 15–17, 1987.

33. I. Kimura and T. Takamori, Image Processing of Flow Around a Circular Cylinder by Using Correlation Technique, in C. Véret (ed.), *Flow Visualization IV,* pp. 221–226, Hemisphere, New York, 1987.

34. M. Geron, D. Khelif, and J. -L. Bousgarbies, Digital Image Processing Applied to the Statistical Study of the Mixing Zone Structure Constraining Walls Effect, in C. Véret (ed.), *Flow Visualization IV,* pp. 227–228, Hemisphere, New York, 1987.

35. J. C. A. Garcia and L. Hesselink, 3-D Reconstruction of Fluid Flow Visualization Images, in C. Véret (ed.), *Flow Visualization IV,* pp. 235–240, Hemisphere, New York, 1987.

36. T. Kobayashi, T. Saga, and S. Segawa, Some Considerations on Automated Image Processing of Pathline Photographs, in C. Véret (ed.), *Flow Visualization IV,* pp. 241–246, Hemisphere, New York, 1987.

37. L. J. Lu and C. R. Smith, Application of Image Processing of Hydrogen Bubble Flow Visualization for Evaluation of Turbulence Characteristics and Flow Structure, in C. Véret (ed.), *Flow Visualization IV,* pp. 247–261, Hemisphere, New York, 1987.

38. V. P. Goddard, J. A. McLaughlin, and F. N. M. Brown, Visual Supersonic Flow Patterns by Means of Smoke Lines, *J. Aerospace Sci.,* vol. 26, no. 11, pp. 761–762, 1959.

39. V. P. Goddard, Development of Supersonic Streamline Visualization, Report to the National Science Foundation on Grant 12488, March 1962.

40. J. T. Kegelman, R. C. Nelson, and T. J. Mueller, Smoke Visualization of the Boundary Layer on an Axisymmetric Body, AIAA Paper 79–1635, presented at the AIAA/Atmospheric Flight Mechanics Conference for Future Space Systems, Boulder, Colo., August 6–8, 1979.

41. J. T. Kegelman, Experimental Studies of Boundary Layer Transition on a Spinning and Nonspinning Axisymmetric Body, Ph.D. dissertation, Department of Aerospace and Mechanical Engineering, University of Notre Dame, Notre Dame, Ind., September 1982.

42. J. T. Kegelman, R. C. Nelson, and T. J. Mueller, The Boundary Layer on an Axisymmetric Body with and Without Spin, *AIAA J.,* vol. 21, no. 11, pp. 1485–1491, 1983.

43. J. T. Kegelman and T. J. Mueller, Experimental Studies of Spontaneous and Forced Transition on an Axisymmetric Body, *AIAA J.,* vol. 24, no. 3, pp. 397–403, 1986.

44. W. S. Saric and A. S. W. Thomas, Experiments on the Subharmonic Route to Turbulence in Boundary Layers, in T. Tatsumi (ed.), *Turbulence and Chaotic Phenomena in Fluids,* pp. 117–122, North-Holland, Amsterdam, 1984.

45. W. S. Saric, V. W. Kozlov, and V. Ya. Levchenko, Forced and Unforced Subharmonic Resonance in Boundary-Layer Transition, AIAA Paper 84-0007, presented at the AIAA 22nd Aerospace Sciences Meeting, Reno, Nev., January 9–12, 1984.

46. W. S. Saric, Visualization of Different Transition Mechanisms, in H. L. Reed (ed.), Gallery of Fluid Motion, *Phys. Fluids,* vol. 29, no. 9, p. 2770, 1986.

47. A. S. W. Thomas, Experiments on Secondary Instabilities in Boundary Layers, in J. P. Lamb (ed.), *Proceedings of the Tenth U.S. National Congress of Applied Mechanics,* pp. 436–444, ASME, New York, 1987.

48. T. Herbert, Analysis of Secondary Instabilities in Boundary Layers, in J. P. Lamb (ed.), *Proceedings of the Tenth U.S. National Congress of Applied Mechanics,* pp. 445–456, ASME, New York, 1987.

49. T. J. Mueller, The Role of Smoke Visualization and Hot-Wire Anemometry in the Study of Transition, in W. George and R. Arndt (eds.), *Advances in Turbulence,* pp. 195–227, Hemisphere, New York, 1989.

50. J. T. Kegelman, R. C. Nelson, and T. J. Mueller, Boundary Layer and Side Force Characteristics

of a Spinning Axisymmetric Body, AIAA Paper 80-1584-CP, presented at the AIAA Atmospheric Flight Mechanics Conference, August 11–13, 1980.

51. F. N. M. Brown and V. P. Goddard, The Effect of Sound on the Separated Laminar Boundary Layer, Final Report, NSF Grant G11712, pp. 1–67, University of Notre Dame, Notre Dame, Ind., 1963.

52. R. C. Nelson, Flow Visualization of High Angle of Attack Vortex Wake Structures, AIAA Paper 85-0102, presented at AIAA 23rd Aerospace Sciences Meeting, Reno, Nev., January 14–17, 1985.

53. R. C. Nelson, The Role of Flow Visualization in the Study of High-Angle-of-Attack Aerodynamics, in M. J. Hemsch and J. M. Nielsen (eds.), *Tactical Missile Aerodynamics,* pp. 43–88, AIAA, New York, 1986.

54. F. M. Payne, T. T. Ng, R. C. Nelson, and L. B. Schiff, Visualization and Wake Surveys of Vortical Flow over a Delta Wing, *AIAA J.,* vol. 26, no. 2, pp. 137–143, 1988.

55. F. M. Payne, T. T. Ng, and R. C. Nelson, Experimental Study of the Velocity Field on a Delta Wing, AIAA Paper 87-1231, presented at the AIAA 19th Fluid Dynamics, Plasma Dynamics and Lasers Conference, Honolulu, Hawaii, June 8–10, 1987.

56. K. D. Visser, K. P. Iwanski, R. C. Nelson, and T. T. Ng, Control of Leading Edge Vortex Breakdown by Blowing, AIAA Paper 88-0504, presented at the AIAA 26th Aerospace Sciences Meeting, Reno, Nev., January 11–14, 1988.

57. P. E. Olsen and R. C. Nelson, Vortex Interaction over Double Delta Wings at High Angles of Attack, AIAA Paper 89-2191, presented at AIAA 7th Applied Aerodynamics Conference, Seattle, Wash., July 31 to August 2, 1989.

58. J. M. Martin, R. W. Empey, W. J. McCroskey, and F. X. Caradonna, An Experimental Analysis of Dynamic Stall on an Oscillating Airfoil, *J. Am. Helicopter Soc.,* January 1974.

59. W. J. McCroskey, L. W. Carr, and K. W. McAlister, Dynamic Stall Experiments on Oscillating Airfoils, *AIAA J.,* vol. 14, no. 1, pp. 57–63, 1976.

60. W. J. McCroskey, Three-Dimensional and Unsteady Separation at High Reynolds Numbers, in *Notes of Unsteady Aerodynamics,* AGARD Lecture Series 94, Von Karman Institute for Fluid Dynamics, Brussels, 1969.

61. T. J. Mueller, C. R. Hall Jr., and W. P. Sule, Supersonic Wake Flow Visualization, *AIAA J.,* vol. 7, no. 11, pp. 2151–2153, 1969.

62. W. B. Roberts and J. A. Slovisky, Location and Magnitude of Cascade Shock Loss by High-Speed Smoke Visualization, *AIAA J.,* vol. 17, no. 11, pp. 1270–1272, 1979.

63. J. J. Cornish, A Device for the Direct Measurements of Unsteady Air Flows and Some Characteristics of Boundary Layer Transition, Mississippi State University, Aerophysics Research Note 24, p. 1, 1964.

64. C. J. Sanders and J. F. Thompson, An Evaluation of the Smoke-Wire Technique of Measuring Velocities in Air, Mississippi State University, Aerophysics Research Report 70, October 1966.

65. H. Yamada, Instantaneous Measurements of Air Flows by Smoke-Wire Technique, *Trans. Jpn. Mech. Eng.,* vol. 39, p. 726, 1973.

66. T. Corke, D. Koga, R. Drubka, and H. Nagib, A New Technique for Introducing Controlled Sheets of Smoke Streaklines in Wind Tunnels, Proceedings of International Congress on Instrumentation in Aerospace Simulation Facilities, IEEE Publication 77 CH 1251-8 AES, p. 74, 1974.

67. H. M. Nagib, Visualization of Turbulent and Complex Flows Using Controlled Sheets of Smoke Streaklines, in T. Asanuma (ed.), *Flow Visualization,* pp. 257–263, Hemisphere, New York, 1979.

68. N. Kasagi, M. Hirata, and S. Yokobori, Visual Studies of Large Eddy Structures in Turbulent Shear Flows by Means of Smoke-Wire Method, in T. Asanuma (ed.), *Flow Visualization,* pp. 245–250, Hemisphere, New York, 1979.

69. T. Torii, Flow Visualization by Smoke-Wire Technique, in T. Asanuma (ed.), *Flow Visualization,* pp. 251–256, Hemisphere, New York, 1979.

70. S. M. Batill and T. J. Mueller, Visualization of Transition in the Flow over an Airfoil Using the Smoke-Wire Technique, *AIAA J.,* vol. 19, no. 3, pp. 340–345, 1981.

71. D. Cornell, Smoke Generation for Flow Visualization, Mississippi State University, Aerophysics Research Report 54, November 1964.
72. T. J. Mueller and S. M. Batill, Experimental Studies of Separation on a Two-Dimensional Airfoil at Low Reynolds Numbers, *AIAA J.,* vol. 20, no. 4, pp. 457–463, 1982.
73. G. S. Schmidt and T. J. Mueller, A Study of the Laminar Separation Bubble on an Airfoil at Low Reynolds Numbers Using Flow Visualization Techniques, AIAA Paper 87-0242, presented at AIAA 25th Aerospace Sciences Meeting, Reno, Nev., January 12–15, 1987.
74. T. J. Mueller, The Visualization of Low Speed Separated and Wake Flows, AIAA Paper 87-2422, presented at AIAA Atmospheric Flight Mechanics Conference, Monterey, Calif., August 17–19, 1987.
75. T. Cebeci and P. Bradshaw, *Momentum Transfer in Boundary Layer,* Hemisphere, New York, 1977.
76. R. Betchov, Transition, in W. Frost and T. H. Moulden (eds.), *Handbook of Turbulence,* vol. 1, pp. 147–164, Plenum, New York, 1977.
77. R. W. Hale, P. Tan, and D. E. Ordway, Experimental Investigation of Several Neutrally Buoyant Bubble Generators for Aerodynamic Flow Visualization, Naval Research Review, pp. 19–24, June 1971.
78. M. H. Redon and M. F. Vinsonneau, Etude de l'Ecoulement de l'Air Autour d'une Maquette, l'Aeronautique, *Bull. Aerotech.,* vol. 18, no. 204, pp. 60–66, 1936.
79. J. Kampè de Fériet, Some Recent Research on Turbulence, *Proceedings of the Fifth International Congress for Applied Mechanics,* pp. 352–355, John Wiley, New York, 1939.
80. F. S. Owen, R. W. Hale, B. V. Johnson, and A. Travers, Experimental Investigation of Characteristics of Confined Jet-Driven Vortex Flows, United Aircraft Research Laboratory Report R-2494-2, AD-328 502, November 1961.
81. R. W. Hale, P. Tan, R. C. Stowell, and D. E. Ordway, Development of an Integrated System for Flow Visualization in Air Using Neutrally Buoyant Bubbles, Sage Action, Inc., Report SAI-RR 7107, 62 pp., December 1971.
82. R. W. Hale and D. E. Ordway, High-Lift Capabilities from Favorable Flow Interaction with Close-Coupled Canards, Sage Action, Inc., Report SAI-RR 7501, 38 pp., December 1975.
83. R. W. Hale, P. Tan, and D. E. Ordway, Prediction of Aerodynamic Loads on Close-Coupled Canard Configurations—Theory and Experiment, Sage Action, Inc., Report SAI-RR 7702 (ONR-CR215-194-3F), 28 pp., July 1977.
84. P. C. Klimas, Helium Bubble Survey of an Opening Parachute Flow Field, *J. Aircraft,* vol. 10, no. 9, pp. 567–569, 1973.
85. L. S. Iwan, R. W. Hale, P. Tan, and R. C. Stowell, Transonic Flow Visualization with Neutrally Buoyant Bubbles, Sage Action, Inc., Report SAI-RR 7304, 25 pp., December 1973.
86. R. W. Empey and R. A. Ormiston, Tail-Rotor Thrust on a 5.5-Foot Helicopter Model in Ground Effect, presented at the 30th Annual National Forum of the American Helicopter Society, May 1974.
87. R. W. Hale, P. Tan, R. C. Stowell, L. S. Iwan, and D. E. Ordway, Preliminary Investigation of the Role of the Tip Vortex in Rotary Wing Aerodynamics Through Flow Visualization, Sage Action, Inc., Report SAI-RR 7402, 32 pp., December 1974.
88. G. A. Tokaty, *A History and Philosophy of Fluid Mechanics,* G. A. Foulis and Co., Ltd., Henley-on-Thames, Oxfordshire, 1971.
89. B. R. Clayton and B. S. Massey, Flow Visualization in Water: A Review of Techniques, *J. Sci. Instrum.,* vol. 44, pp. 2–11, 1967.
90. W. Merzkirch, *Flow Visualization,* Academic, New York, 1974.
91. H. Werlé, Hydrodynamic Flow Visualization, *Annu. Rev. Fluid Mech.,* vol. 5, pp. 361–382, 1973.
92. M. Gad-el-Hak, Visualization Techniques for Unsteady Flows: An Overview, *J. Fluids Eng.,* vol. 110, pp. 231–243, 1988.
93. G. E. Erickson, Vortex Flow Correlation, Air Force Wright Aeronautical Laboratories Report AFWAL TR 80-3143, January 1981.

94. M. Gad-el-Hak, The Water Towing Tank as an Experimental Facility, *Exp. Fluids*, vol. 5, no. 5, pp. 289–297, 1987.
95. H. Werlé, Methodes d'Etude par Analogie Hydraulique des Ecoulements Subsonique, Supersonique et Hypersonique, (NATO) AGARD Report 399, 1960.
96. H. Werlé and H. Gallon, The New Hydrodynamic Visualization Laboratory of the Aerodynamics Division, Rech. Aerosp. Report 1982-5, pp. 1–23, 1982.
97. D. R. Campbell, Flow Visualization Using a Selectively Sensitive Fluorescent Dye, Aerospace Research Laboratories Report ARL 73-005, May 1973.
98. P. Poisson-Quinton and H. Werlé, Water Tunnel Visualization of Vortex Flows, *Astronaut. Aeronaut.*, June 1967.
99. F. R. Hama, Streaklines in a Perturbed Shear Flow, *Physi. Fluids*, vol. 5, no. 6, 1962.
100. K. W. McAlister and L. W. Carr, Water Tunnel Visualization of Dynamic Stall, *ASME J. Fluids Eng.*, vol. 101, pp. 376–380, 1979.
101. E. W. Geller, An Electrochemical Method of Visualizing the Boundary Layer, *J. Aeronaut. Sci.*, vol. 22, no. 12, pp. 869–870, 1955.
102. F. A. Schraub, S. J. Kline, J. Henry, P. W. Runstadler Jr., and A. Littel, Use of Hydrogen Bubbles for Quantitative Determination of Time-Dependent Velocity Fields in Low-Speed Flows, *ASME J. Basic Eng.*, vol. 87, pp. 429–444, 1965.
103. E. Kato, M. Suita, and M. Kawamata, Visualization of Unsteady Pipe Flows Using Hydrogen Bubble Technique, *Proceedings of the International Symposium on Flow Visualization*, pp. 342–346, Bochum, West Germany, September 9–12, 1980.
104. G. E. Mattingly, The Hydrogen-Bubble Flow-Visualization Technique, DTMB Report 2146 (AD 630 468), February 1966.
105. K. W. McAlister and L. W. Carr, Water Tunnel Experiments on an Oscillating Airfoil at $R = 21,000$, NASA TM 78446, March 1978.
106. K. W. McAlister, A Pulse Forming Network for Particle Path Visualization, NASA TM 81311, 1981.
107. F. R. Hama, Three-Dimenstional Vortex Pattern Behind a Circular Cylinder, *J. Aeronaut. Sci.*, vol. 24, p. 156, 1957.
108. J. H. Gerrard, The Wakes of Cylindrical Bluff Bodies at Low Reynolds Numbers, *Philos., Trans. R. Soc. Ser. A*, vol. 288, p. 351, 1978.
109. T. Wei and C. R. Smith, Secondary Vortices in the Wake of Circular Cylinders, *J. Fluid Mech.*, vol. 169, pp. 513–533, 1986.
110. D. Rockwell, R. Atta, L. Kramer, R. Lawson, C. Lusseyran, C. Magness, D. Sohn, and T. Staubli, Flow Visualization and Its Interpretation, AGARD Symposium on Aerodynamics and Related Hydrodynamic Studies of Water Facilities, AGARD-CPP-413, 1986.
111. L. J. Lu and C. R. Smith, Image Processing of Hydrogen Bubble Flow Visualization for Determination of Turbulence Statistics and Bursting Characteristics, *Exp. Fluids*, vol. 3, pp. 349–356, 1985.
112. C. R. Smith and L. J. Lu, The Use of Template-Matching Technique to Identify Hairpin Vortex Flow Structures in Turbulent Boundary Layers, in S. J. Kline (ed.), *Proceedings of Zaric International Seminar on Wall Turbulence*, 20 pp., 1989.
113. D. Rockwell, R. Atta, C.-H. Kuo, C. Hefele, C. Magness, and T. Utsch, On Unsteady Flow Structure from Swept Edges Subjected to Controlled Motion, *Proceedings of Workshop II on Unsteady Separated Flow*, pp. 299–312, 1988.
114. C. R. Smith, Computer-Aided Flow Visualization, chap. 24, in W. J. Yang (ed.), *Handbook of Flow Visualization*, pp. 375–391, Hemisphere, New York, 1989.
115. T. Matsui, H. Nagata, and H. Yasuda, Some Remarks on Hydrogen Bubble Technique for Low Speed Water Flows, in T. Asanuma (ed.), *Flow Visualization*, pp. 215–220, Hemisphere, New York, 1979.
116. R. Reznicek (ed.), *Flow Visualization V*, Hemisphere, New York, 1990.
117. Y. Tanida and H. Miyashiro (eds.), *Flow Visualization VI*, Springer-Verlag, Berlin, 1992.

OPTICAL SYSTEMS FOR FLOW MEASUREMENT: SHADOWGRAPH, SCHLIEREN, AND INTERFEROMETRIC TECHNIQUES

Richard J. Goldstein and T. H. Kuehn

7.1 INTRODUCTION

Optical techniques that have been used in the measurement of flow include (1) direct visualization, where some type of marker (e.g., dye, bubbles, solid particles) is followed along with the fluid motion; (2) laser-Doppler systems, in which the frequency shift of scattered illumination from such a marker—usually a particle— is measured; and (3) measurement of what might be called index of refraction of a medium, from which some property or properties of the flow are determined.

Only the methods in the last category are examined in this chapter.* These include schlieren, shadowgraph, and interferometric techniques** that are used to study density fields in transparent media, usually gases or liquids. References [1–10] have considerable information and extensive bibliographies on these methods.† Although all three methods depend on variation of the index of refraction in a trans-

---

Mr. H. D. Chiang was of great assistance during the preparation of this material—in particular, with the flow-visualization photographs and the references. E. R. G. Eckert, A. G. Hayener, A. Roshko, and A. B. Witte kindly supplied photographs from their research.

*Techniques falling into categories 1 and 2 are described in Chaps. 6 and 4, respectively.

**Birefringent fluids, which are doubly refractive when subject to shear stresses, are discussed in Chap. 9.

†Much of the material in this chapter is taken from [8].

parent medium and the resulting effects on a light beam passing through the test region, quite different quantities are measured with each one. Shadowgraph systems are used to indicate the variation of the second derivative (normal to the light beam) of the index of refraction. With a schlieren system, the first derivative of the index of refraction (in a direction normal to the light beam) determines the light pattern. Interferometers respond directly to differences in optical path length, essentially giving the index-of-refraction field within the flow.

Optical measurements have many advantages over other techniques. Perhaps the major one is the absence of an instrument probe that could influence the flow field. The light beam can also be considered essentially inertialess, so that very rapid transients can be studied. The sensitivities of the three optical methods are quite different, so that they can often be used to study a variety of systems. Thus interferometers are often used to study flows in which density gradients are small, while schlieren and shadowgraph systems are often employed in studying shock and flame phenomena, in which very large density gradients are present.

All three techniques are valuable when visualizing flows in which density differences occur naturally or are artificially induced. When used quantitatively, these techniques can determine density, pressure, and/or temperature variations in the flow. From these, other properties of the flow field (e.g., laminar versus turbulent nature, boundary layer thickness, shock angles, points of separation, and reattachment) can often be inferred.

Shadowgraph, schlieren, and interferometric measurements are essentially integral; they integrate the quantity measured over the length of the light beam. For this reason they are well-suited to measurements in two-dimensional fields, where there is no index of refraction or density variation in the field along the light beam, except at the beam's entrance to and exit from the test (disturbed) region. These latter variations can be considered as sharp discontinuities, or appropriate end corrections can be made. Axisymmetric fields can also be studied [8]. If the field is three-dimensional, an average (along the light beam) of the measured quantity can still be determined. Since both schlieren and shadowgraph systems are primarily used for qualitative studies, this is often acceptable; and even in interferometric studies, the averaging done by the light beam can sometimes be advantageous. Observations of local three-dimensional variations in the flow have been made of such phenomena as the rise of thermal plumes and turbulent bursts [11].

In the three methods to be studied, the index of refraction (or one of its spatial derivatives) determines the resulting illumination or light pattern. The index of refraction of a homogeneous medium is a function of the thermodynamic state, often only the density. According to the Lorenz-Lorentz relation, the index of refraction of a homogeneous transparent medium can be obtained from

$$\frac{1}{\rho} \frac{n^2 - 1}{n^2 + 2} = \text{const} \tag{1}$$

When n $\cong$ 1, this reduces to the Gladstone-Dale equation,

$$\frac{n-1}{\rho} = C \tag{2}$$

or

$$\rho = \frac{n-1}{C} \tag{3}$$

which holds quite well for gases. The constant C, called the Gladstone-Dale constant, is a function of the particular gas and varies slightly with wavelength. Usually, instead of using C directly, the index of refraction at a standard condition n_0 is given:

$$n-1 = \frac{\rho}{\rho_0} (n_0 - 1) \tag{4}$$

or

$$\rho = \rho_0 \frac{n-1}{n_0-1} \tag{5}$$

When the first or second derivative (say, with respect to y) is determined as in a schlieren or shadowgraph apparatus, then, using from Eqs. (3) and (4),

$$\frac{\partial \rho}{\partial y} = \frac{1}{C} \frac{\partial n}{\partial y} = \frac{\rho_0}{n_0 - 1} \frac{\partial n}{\partial y} \tag{6}$$

$$\frac{\partial^2 \rho}{\partial y^2} = \frac{1}{C} \frac{\partial^2 n}{\partial y^2} = \frac{\rho_0}{n_0 - 1} \frac{\partial^2 n}{\partial y^2} \tag{7}$$

The question arises as to what phenomena in a flowing fluid cause the density variations that are observed. Naturally occurring variations are present in compressible flows, for example, high-speed flows involving shock waves. Natural convection flow fields are nonuniform in density, as are the flows in flames and combustion systems. The mixing of fluids of different density can be studied, as can forced convection flow fields, when heat or mass transfer produces nonuniform densities. An artificial density distribution can be introduced into a flow through local heating—often of a transient nature—from a heating wire or spark. This "tracer" can then be followed.

Many of the measurements using index-of-refraction methods involve temperature variation in a two-dimensional flow field. It is of interest to see how the temperature affects the index of refraction and its derivatives if the flowing medium is a gas.

If the pressure can be assumed constant and the ideal gas equation of state ($\rho = P/RT$) holds, then

$$\frac{\partial n}{\partial y} = -\frac{CP}{RT^2} \frac{\partial T}{\partial y} = -\frac{n_0 - 1}{T} \frac{\rho}{\rho_0} \frac{\partial T}{\partial y} \tag{8}$$

or

$$\frac{\partial T}{\partial y} = -\frac{T}{n_0 - 1} \frac{\rho_0}{\rho} \frac{\partial n}{\partial y} \tag{9}$$

and

$$\frac{\partial^2 n}{\partial y^2} = C\left[-\frac{\rho}{T}\frac{\partial^2 T}{\partial y^2} + \frac{2\rho}{T^2}\left(\frac{\partial T}{\partial y}\right)^2\right] \tag{10}$$

Note that Eq. (9), which applies to a schlieren study, shows a relatively simple relationship between the gradient of the temperature and the gradient of the index of refraction. For a shadowgraph, the equivalent relation, Eq. (10), is more complicated, although under many conditions, the second term may be small.

The index of refraction of a gas as measured in an interferometer can indicate the temperature directly, from Eqs. (3) and (4), assuming constant pressure and the prefect-gas equation of state,

$$T = \frac{C}{n-1}\frac{P}{R} = \frac{n_0 - 1}{n-1}\frac{P}{P_0}T_0 \tag{11}$$

Low-speed flows in which heat transfer occurs can often be approximated as constant-pressure systems in which Eqs. (8)–(11) apply.

For a reversible, adiabatic (isentropic) process in an ideal gas, the pressure, temperature, and density all vary with

$$P\left(\frac{1}{\rho}\right)^k = \text{const}$$

or

$$\frac{P}{P_0} = \left(\frac{\rho}{\rho_0}\right)^k \tag{12}$$

where k is the ratio of specific heats (constant pressure to constant volume) and ρ_0 refers to some reference condition. If the pressure variation is derived from the optical measurements, then from Eqs. (5) and (12),

$$\frac{P}{P_0} = \left(\frac{n-1}{n_0-1}\right)^k \tag{13}$$

and

$$\frac{\partial P}{\partial y} = P\frac{k}{n-1}\frac{\partial n}{\partial y} \tag{14}$$

or

Table 7.1 Index of refraction for air and water at 20°C and 1 atm

λ,nm	$n_{air} - 1$	n_{H_2O}	$(dn/dT)_{air}$ (°C)$^{-1}$	$(dn/dT)_{H_2O}$ (°C)$^{-1}$
546.1	2.733×10^{-4}	1.3345	-0.932×10^{-6}	-0.895×10^{-4}
632.8	2.719×10^{-4}	1.3317	-0.927×10^{-6}	-0.880×10^{-4}

$$\frac{\partial n}{\partial y} = \frac{1}{P}\frac{\partial P}{\partial y}\frac{n-1}{k} \tag{15}$$

Similarly, the expression relating the second derivatives of the pressure and index of refraction could be obtained for an isentropic flow.

The index of refraction of a liquid is primarily a function of temperature and, for accurate results, should be obtained from direct measurement rather than from Eq. (1). Many tabulated results and empirical expressions for the index of refraction of liquids are available. For comparison,Table 7.1, derived from [12, 13], cites values at 20°C and 1 atm for air and water. The two wavelengths chosen are a commonly used mercury line (546.1 nm) and the visible line from a CW He-Ne laser (632.8 nm).

Of the three systems to be discussed, two of them, namely, schlieren and shadowgraph, can be described by geometric or ray optics, although under certain conditions, diffraction effects can be significant. Interferometers, as the name implies, depend on the interference of coherent light beams, and some discussion of physical (wave) optics is required.

7.2 SCHLIEREN SYSTEM

7.2.1 Analysis by Geometric or Ray Optics

To study both schlieren and shadowgraph systems, it is necessary to analyze the path of a light beam in a medium whose index of refraction is a function of position. Consider Fig. 7.1, which shows a light beam, traveling initially in the z direction, passing through a medium whose index of refraction varies (for simplicity) only in the y direction. At time τ the beam is at position z, and the wave front (surface normal to the path of the light) is as shown. After a time interval $\Delta\tau$, the light has moved a distance of $\Delta\tau$ times the velocity of light, which in general, is a function of y, and the wave front or light beam has turned an angle $\Delta\alpha'$. The local value of the speed of light is c_0/n. With reference to Fig. 7.1, and assuming that only small deviations occur, the distance Δz that the light beam travels during time interval $\Delta\tau$ is

$$\Delta z = \frac{c_0}{n}\Delta\tau \tag{16}$$

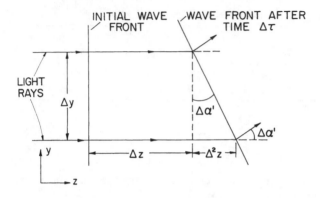

$$\Delta z = (c_0/n)\Delta\tau$$

$$\Delta^2 z = \Delta z_y - \Delta z_{y+\Delta y}$$

$$= -c_0\left[\Delta(1/n)/\Delta y\right]\Delta\tau\,\Delta y$$

$$\Delta\alpha' \cong \Delta^2 z/\Delta y = -n\left[\Delta(1/n)/\Delta y\right]\Delta z$$

$$d\alpha' = 1/n\,(\partial n/\partial y)dz = \left[\partial(\ln n)/\partial y\right]dz$$

Figure 7.1 Bending of light rays in an inhomogeneous medium

Now

$$\Delta^2 z = \Delta z_y - \Delta z_{y+\Delta y} = -\frac{\Delta(\Delta z)}{\Delta y}\Delta y \qquad (17)$$

or

$$\Delta^2 z = -c_0\frac{\Delta(1/n)}{\Delta y}\Delta\tau\Delta y \qquad (18)$$

The angular deflection of the ray is

$$\Delta\alpha' \approx \frac{\Delta^2 z}{\Delta y} = -n\frac{\Delta(1/n)}{\Delta y}\Delta z \qquad (19)$$

In the limit, Δy and Δz are considered to be very small,

$$d\alpha' = \frac{1}{n}\frac{\partial n}{\partial y}dz = \frac{\partial(\ln n)}{\partial y}dz \qquad (20)$$

For small deflections, the angle of the light beam is the slope $\partial y/\partial z$ of the light beam, and thus

$$\frac{\partial^2 y}{\partial z^2} = \frac{1}{n}\frac{\partial n}{\partial y} \tag{21}$$

If the angle remains small, this expression holds over the light path through the disturbed region. If the entering angle is zero, the angle at the exit of the test region is

$$\alpha' = \int \frac{1}{n}\frac{\partial n}{\partial y}dz = \int \frac{\partial(\ln n)}{\partial y}dz \tag{22}$$

where the integration is performed over the entire length of the light beam in the test region.

If the test region is enclosed by glass walls and the index of refraction within the test section is different from that of the ambient air n_a, then from Snell's law, an additional angular deflection is present. If α is the angle of the light beam after it has passed through the test section and emerged into the surrounding air,

$$n_a \sin \alpha = n \sin \alpha' \tag{23}$$

If the test-section windows are plane and of uniform thickness, then for small values of α and α',

$$\alpha = \frac{n}{n_a}\alpha' \tag{24}$$

Equation (22) gives

$$\alpha = \frac{n}{n_a}\int \frac{1}{n}\frac{\partial n}{\partial y}dz \tag{25}$$

If the $1/n$ within the integrand does not change greatly through the test section, then

$$\alpha = \frac{1}{n_a}\int \frac{\partial n}{\partial y}dz \tag{26}$$

or, since $n_a \approx 1$,

$$\alpha = \int \frac{\partial n}{\partial y}dz \tag{27}$$

Note that if a gas (not at extremely high density) is the test fluid, $\alpha' \cong \alpha$.

The angle α in Eq. (27) is in the yz plane. If there is a variation of index of refraction in the x direction, then again, assuming small angular deviation, a similar expression would give the deflection in the xz plane proportional to $\partial n/\partial x$.

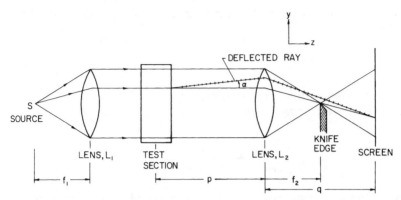

Figure 7.2 Typical schlieren system using lenses

If variations of n in the x and y directions as well as the effect of significant angular deflection are included, the resulting equations for the path of the light beam, equivalent to Eq. (21), are [2]

$$y'' = \frac{1}{n}[1 + (x')^2 + (y')^2]\left(\frac{\partial n}{\partial y} - y'\frac{\partial n}{\partial z}\right) \tag{28a}$$

$$x'' = \frac{1}{n}[1 + (x')^2 + (y')^2]\left(\frac{\partial n}{\partial x} - x'\frac{\partial n}{\partial z}\right) \tag{28b}$$

where primes refer to differentiation with respect to z.

Note that the light beam is turned in the direction of increasing index of refraction. In most media this means that the light is bent toward the region of higher density.

A schlieren system is basically a device to measure or indicate the angle α, typically of the order of 10^{-6}–10^{-3} rad, as a function of position in the xy plane normal to the light beam. Consider the system shown in Fig. 7.2. A light source, which we assume to be rectangular (of dimensions a_s by b_s) and at the focus of lens L_1 provides a parallel light beam entering the field of disturbance in the test section. The deflected rays, when disturbance is present, are indicated by cross-hatched lines. The light is collected by a second lens L_2, at whose focus a knife-edge is placed, and then passes onto a screen located at the conjugate focus of the test section. As is shown below, if the screen is not at the focus of the disturbance, shadowgraph effects will be superimposed on the schlieren pattern.

If no disturbance is present, then ideally, the light beam at the focus of L_2 would be as shown in Fig. 7.3, with dimensions a_0 by b_0, which are related to the initial dimensions by

$$\frac{a_0}{a_s} = \frac{b_0}{b_s} = \frac{f_2}{f_1} \tag{29}$$

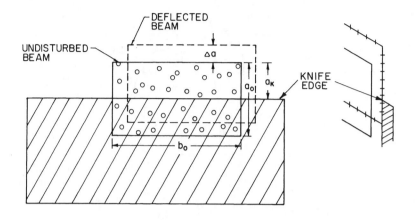

Figure 7.3 View of deflected and undisturbed beams at the knife-edge of a schlieren system

where f_1 and f_2 are the focal lengths of L_1 and L_2, respectively.

As shown in Fig. 7.3, the source is usually adjusted so that the shorter dimension a_0 is at right angles to the knife-edge to maximize sensitivity. The knife-edge (typically a razor blade) is adjusted, when no disturbance is present, to cut off all but an amount a_K (typically, $a_K = a_0/2$) of the height a_0. When the knife-edge is moved across the beam exactly at the focus, the illumination at the screen should decrease uniformly; if the knife-edge is not in the focal plane, the image at the screen will not darken uniformly. The illumination at the screen when no knife-edge is present is I_0, and with the knife-edge inserted in the focal plane the illumination is

$$I_K = \frac{a_K}{a_0} I_0 \tag{30}$$

The light passing through each section of the test region comes from all parts of the source. Thus at the focus, not only is the image of the source composed of light coming from the whole field of view, but light passing through every point in the field of view gives an image of the source at the knife-edge. If the light beam at a position x,y in the test region is deflected by an angle α, then from Fig. 7.4, the image of the source coming from that position will be shifted at the knife-edge by an amount

$$\Delta a = \pm f_2 \alpha \tag{31}$$

where the sign is determined by the orientation of the knife-edge; it is positive when (as in Fig. 7.4) $\alpha > 0$ gives $\Delta a > 0$, and negative if the knife-edge is reversed,

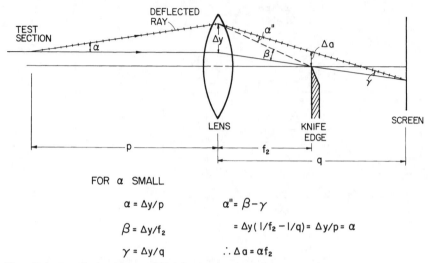

Figure 7.4 Ray displacement at knife-edge for a given angular deflection

so that $\alpha > 0$ leads to $\Delta a < 0$. The illumination at the image of position x, y on the screen will be (see Fig. 7.3)

$$I_d = I_K \frac{a_K + \Delta a}{a_K} = I_K\left(1 + \frac{\Delta a}{a_K}\right) \tag{32}$$

where Δa is positive if the light is deflected away from the knife-edge, and negative if the light is deflected toward the knife-edge. The relative intensity or contrast is

$$\text{contrast} = \frac{\Delta I}{I_K} = \frac{I_d - I_K}{I_K} = \frac{\Delta a}{a_K} = \pm \frac{\alpha f_2}{a_K} \tag{33}$$

using Eq. (31).

Note that the sensitivity of the schlieren system for measuring the deflection is

$$\frac{d(\text{contrast})}{d\alpha} = \frac{f_2}{a_K} \tag{34}$$

or proportional to f_2 and inversely proportional to a_K. For a given optical system, minimizing a_K by movement of the razor blade would maximize the contrast. However, this would limit the range for deflection of the beam toward the knife-edge to

$$\alpha_{\text{max,neg}} = \frac{a_K}{f_2} \tag{35}$$

as all deflections this large or larger would give (neglecting diffraction) no illumination. The maximum angle of deflection away from the knife-edge that could be measured is

$$\alpha_{max} = \frac{a_0 - a_K}{f_2} \tag{36}$$

as a deflection of this magnitude would permit all the source illumination to pass to the screen for equal range in both directions, $a_K = a_0/2$ and

$$\alpha_{max,neg} = \alpha_{max} = \frac{a_0}{2f_2} = \frac{a_s}{2f_1} \tag{37}$$

Note from Fig. 7.3 that deflections in the x direction are parallel to the knife-edge and will not affect the illumination at the screen; so, if density gradients in the x direction within the test region are to be studied, the knife-edge must be turned at right angles. For maximum sensitivity (since $a_s < b_s$) the source should also be rotated 90°.

Combining Eqs. (26) and (33) gives

$$\text{contrast} = \frac{\Delta I}{I_K} = \pm \frac{f_2}{a_K n_a} \int \frac{\partial n}{\partial y} dz \tag{38}$$

Assuming a two-dimensional field with $\partial n/\partial y$ constant at a given x, y position over the length L in the z direction,

$$\text{contrast} = \pm \frac{f_2}{a_K} \frac{1}{n_a} \frac{\partial n}{\partial y} L \tag{39}$$

This equation holds for every x, y position in the test section and gives the contrast at the equivalent position in the image on the screen.

If the deflection is toward the knife-edge, the field will darken and the contrast will be negative. Using the coordinate system of Fig. 7.3, if the knife-edge covers up the region $y < 0$ (that is, knife-edge pointing upward) at the focus, then

$$\frac{\Delta I}{I_K} = + \frac{f_2}{a_K} \frac{1}{n_a} \frac{\partial n}{\partial y} L \tag{40}$$

If the knife-edge is reversed and covers the region $y > 0$, then

$$\frac{\Delta I}{I_K} = - \frac{f_2}{a_K} \frac{1}{n_a} \frac{\partial n}{\partial y} L \tag{41}$$

Changing the knife-edge reverses the dark and light images on the screen. The brighter areas of the image represent regions in the test section where the index of refraction (and thus usually the density) increases in the direction away from the knife-edge (Fig. 7.5). For a gas, Eq. (38) can be rewritten using Eq. (6).

$$\frac{\Delta I}{I_K} = \pm \frac{f_2}{a_K n_a} \frac{n_0 - 1}{\rho_0} \int \frac{\partial \rho}{\partial y} dz \tag{42}$$

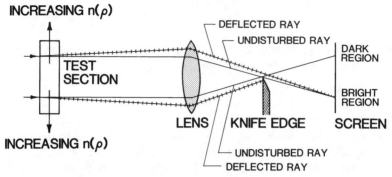

Figure 7.5 Effect of index-of-refraction gradient on illumination at screen

and equivalent to Eq. (39),

$$\frac{\Delta I}{I_K} \cong \pm \frac{f_2}{a_K} \frac{n_0 - 1}{\rho_0} \frac{\partial \rho}{\partial y} L \tag{43}$$

taking $n_a \cong 1$.

For a gas at a constant pressure,

$$\frac{\Delta I}{I_K} = \mp \frac{f_2}{a_K n_a} \frac{n_0 - 1}{\rho_0} \int \frac{\rho}{T} \frac{\partial T}{\partial y} dz \tag{44}$$

and, as in Eq. (43),

$$\frac{\Delta I}{I_K} \cong \mp \frac{f_2}{a_K} \frac{n_0 - 1}{\rho_0} \frac{\rho}{RT^2} \frac{\partial T}{\partial y} L \tag{45}$$

For a liquid, where n is a function only of T,

$$\frac{\Delta I}{I_K} = \pm \frac{f_2}{a_K n_a} \int \frac{\partial T}{\partial y} \frac{dn}{dT} dz \tag{46}$$

If the field is two-dimensional and n does not change greatly, then

$$\frac{\Delta I}{I_K} = \pm \frac{f_2}{a_K n_a} \frac{\partial T}{\partial y} \frac{dn}{dT} L \tag{47}$$

$$\frac{\Delta I}{I_K} \cong \pm \frac{f_2}{a_K} \frac{\partial T}{\partial y} \frac{dn}{dT} L \tag{48}$$

In a quantitative study, measurements must be made of the illumination or contrast, usually of the image on a photographic negative. These are quite time consuming, and the resulting accuracy has not usually warranted the effort. Standard schlieren systems are employed for qualitative studies of density or tempera-

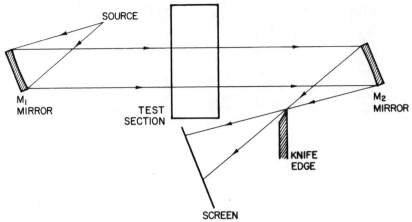

Figure 7.6   Typical schlieren system using converging mirrors

ture fields, although quantitative measurements of such items as shockwave angles or positions can be made. The minimum value of the contrast that can easily be observed is of the order of 0.05, a value that can be used in determining the sensitivity of the system. Since n and its derivatives vary somewhat with wavelength, it is preferable that the light source be relatively monochromatic, although for both schlieren and shadowgraph systems, this is usually not of major importance.

7.2.2 Applications and Special Systems

The high cost of large aberration-free lenses usually precludes construction of the system shown in Fig. 7.2, but the optically similar system using concave mirrors shown in Fig. 7.6 is widely used. The source and knife-edge should be in the same plane and on opposite sides of the axes of the two mirrors in the "Z" arrangement shown. This eliminates the aberration coma, although astigmatism is still present in the off-axis system. The legs of the Z should each be at the same angle to the line between the two mirrors; this angle should be as small as possible to minimize astigmatism [14]. Schlieren photographs taken with a system similar to that of Fig. 7.6 are shown in Figs. 7.7–7.9.

Figure 7.7 is the image of a helium jet entering a still atmosphere of air. Different images of the flow field in the jet are shown. Although the Reynolds number is essentially constant for the different photographs, the difference in knife-edge orientation, shown for each view, dramatically changes the image. As mentioned above, the schlieren image lightens when the index of refraction increases in the direction away from the knife-edge. Thus the lighter regions on the photo show that the local density gradient is positive in the direction toward the knife-edge, while the reverse is true for the darker regions.

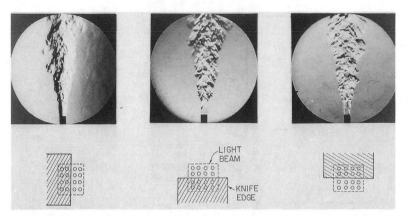

Figure 7.7 Schlieren images of a helium jet entering an atmosphere of air: The effect of knife-edge orientation (Re = 630)

Figure 7.8 presents schlieren images of the same jet geometry as shown in Fig. 7.7 for four different Reynolds numbers. The change is apparent in the character of the jet from a relatively steady flow to flow with large-scale eddies as the Reynolds number increases.

Figure 7.9 contains a series of schlieren photographs taken of a Mach 3 flow of air into which air is injected through a tangential slot [15]. The relative amount of air injected is given by M, the ratio of the mass velocity of the injected flow to that of the mainstream. Various flow phenomena and their dependence on M can be observed in Figs. 7.7–7.9. These phenomena include the expansion fan at the top edge of the splitter plate (between the main and injected flows), a separation region, lip shock, recompression shock, and alternative expansions and compressions. Other applications include measurements of acoustic excitation of a subsonic jet [16], ultrasonic wave pressure distribution [17], boundary layer growth along a shock tube wall [18], shock fronts generated by explosive events [19], and natural convection boundary layers adjacent to an array of vertical flat plates [20].

A number of variations on the schlieren systems shown in Figs. 7.2 and 7.6 have been used, and two of these are shown in Fig. 7.10. In Fig. 7.10a, one plane and one converging mirror are used. Since plane mirrors are easier to make than converging mirrors, this apparatus would be somewhat less expensive than the one in Fig. 7.6. The main advantage of this system, however, is the amplification of the angle representative of the disturbance. As the beam passes through the test section twice, the deflection angle is doubled and, all other parameters being the same, so is the sensitivity. However, this double passage causes, in general, a slight blurring of the image, as the beam does not go through exactly the same part of the test section on each passage. In addition, since the screen cannot be at the focus of both views of the test section, some shadowgraph effects are present.

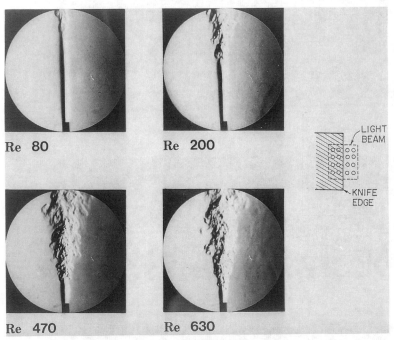

Re 80

Re 200

LIGHT
BEAM

KNIFE
EDGE

Re 470

Re 630

Figure 7.8 Schlieren images of the flow structure of a helium jet entering air at different Reynolds numbers

A single-mirror system, shown in Fig. 7.10b, is still simpler, although the blurring of the image is still more serious than in Fig. 7.10a because the light is not parallel. The source and knife-edge are at conjugate foci, and for convenience they are often kept close together and thus at a distance twice the focal length f from the mirror. If α is the deflection of the beam after a single transit of the test section, the sensitivity [cf. Eq. (34)] is $4f/a_K$. Since the source and knife-edge in the two systems shown in Fig. 7.10 are in such close proximity, a splitter plate is sometimes used to give them a larger physical separation. It should be apparent that, in all these systems, a camera placed in the beam after the knife-edge and focused on the test section can be used in place of the screen.

Other schlieren systems have used one of the optical arrangements shown above, usually that of Fig. 7.6, but without the knife-edge. In a color schlieren [21] the knife-edge is replaced by colored filters held at the focus, and the deflection of the light beam (necessarily nonmonochromatic) gives rise to different colors. Other variations use colored filters near the light source and an aperture at the focus [22]. In another system, an aperture is placed at the focus, giving a darker image for any deflection or density gradient, irrespective of the direction of the gradient. An opaque disk at the focus gives a brightening of the image for a light deflection in any direction.

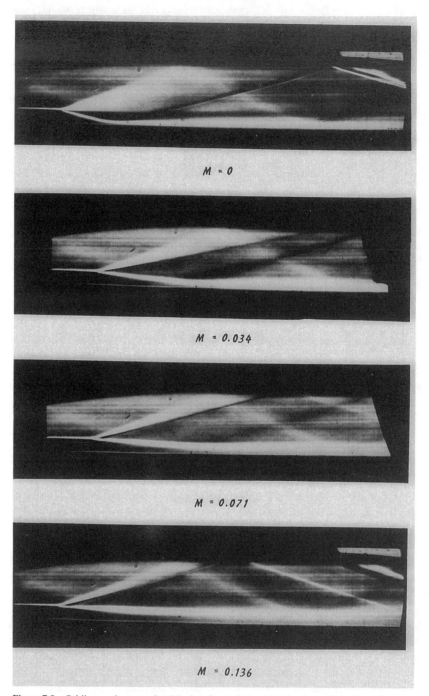

Figure 7.9 Schlieren photographs: Mach-3 flow of air with injection of air through a tangential slot (M = ratio of mass velocity of injected flow to mainstream flow). $(Ma)_1 = 3.01$; $h = 0.239$ in; $s = 0.182$ in.

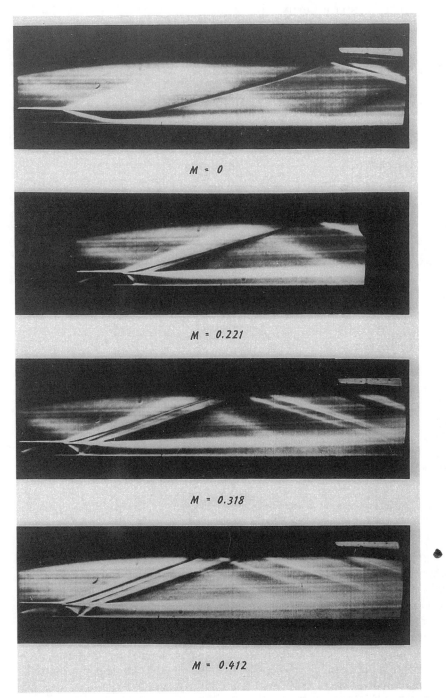

$M = 0$

$M = 0.221$

$M = 0.318$

$M = 0.412$

Figure 7.9 (*Continued*) Schlieren photographs: Mach-3 flow of air with injection of air through a tangential slot (M = ratio of mass velocity of injected flow to mainstream flow). $(Ma)_1 = 3.01$; $h = 0.239$ in; $s = 0.182$ in.

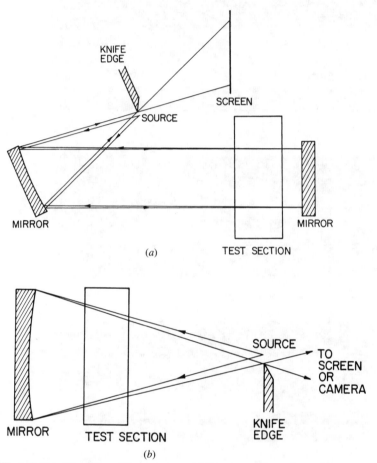

Figure 7.10 Alternative schlieren systems. (*a*) One converging and one plane mirror. (*b*) One converging mirror. (*c*) Holographic schlieren system

A holographic schlieren system records the image on a holographic plate as shown in Fig. 7.10*c*. The knife-edge and screen are not introduced until the hologram is reconstructed. This allows a range of knife-edge positions and angles to be used to obtain the best contrast, even though the original event recorded on the hologram may have had a very short duration.

Quantitative studies with standard schlieren systems require measurement of the local light intensity. This has been done with a photomultiplier to study density fluctuations in a supersonic turbulent jet [23] and to measure density profiles in natural and forced convection thermal boundary layers [24]. Experiments similar to those of [23] have been performed in a convecting liquid with a grating rather than a single knife-edge [25]. These systems for studying the fluctuations of the

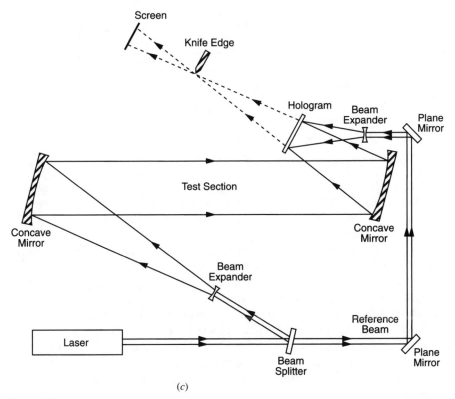

(c)

Figure 7.10   Continued

index of refraction are closely related to crossed-beam measurements to study turbulence in shear flow.

Crossed-beam systems [26, 27] depend on fluctuations in the absorption or scattering coefficients in the flow. The local statistical properties of the flow are determined from cross correlation of the fluctuating outputs of two photodetectors measuring the intensity of two orthogonal light beams that cross the flow. In an early study the light beam fluctuation was produced by a water mist in a subsonic airflow [26]. Other studies used what are essentially crossed schlieren beams to study flow in subsonic jets [28] and supersonic jets [29]. These systems relate closely to the single-beam device described in [23, 25]. A parallel-beam system has been used to study the coherent structures in axisymmetric jets [30]. A laser beam is split into five equally spaced parallel beams, and the deflection of each is measured by an optical detector array.

Standard schlieren systems can also be used to derive quantitative information

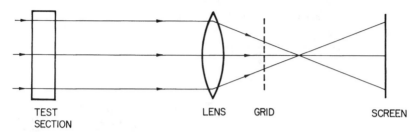

TEST   LENS   GRID   SCREEN
SECTION

Figure 7.11   Grid-schlieren system

on the flow field. For example, a scanning microdensitometer has been used to determine the local light intensity across the image of a flow field [24]. In general, measurements of this type require a calibration relating the light intensity to the light ray displacement at the knife-edge.

A special apparatus for quantitative studies is the Ronchi or grid-schlieren system [31–33]. The grid, shown as part of a focusing-lens schlieren system in Fig. 7.11, could be used in any of the optical arrangements previously described. In general, however, quantitative studies are best performed using parallel light that passes through the test section only once. The grid has equally spaced opaque lines (whose widths are usually set equal to the spacing) on a transparent sheet. It is placed before the focus, as shown in Fig. 7.11. The resulting schlieren image is a series of lines that are parallel if no disturbance is present. When a disturbance is present, the displacement of these fringes is directly proportional to the local angular deflection α.

The grid could also be placed right at the focus where the knife-edge is located in a standard schlieren system; then, the beam just passes through one of the gaps between two opaque lines when no disturbance is present. Thus the spacing between the lines equals a_0 (Fig. 7.3), and the screen is uniformly illuminated. In the presence of a disturbance, the deflection of the light beam at the focus Δa causes the beam to traverse the grid, producing a light or dark image, depending on the magnitude of Δa. The resulting image is a series of fringes, called isophates, representing regions of constant angular deflection (often approximating constant-density gradient). Placement of the grid at the focus of the second lens is often impractical. The required line spacing on the grid may be so small as to cause significant diffraction effects.

Other systems that may be of interest in flow studies include a self-illuminated schlieren system [35] and a stereoscopic schlieren [36]. The latter two, though cumbersome to use, have application to the study of three-dimensional fields. A modification of the sharp-focusing system using holography permits local measurements all along the light beam by using reconstruction from a single hologram [37].

Most of these special schlieren systems, including the sharp-focusing systems, the Ronchi system, and other quantitative schlieren systems described, have not

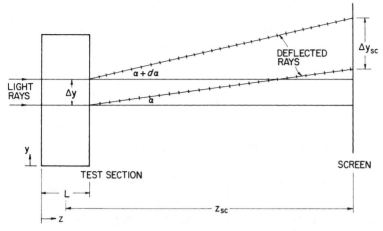

Figure 7.12 Displacement of light beam for shadowgraph evaluation

been widely used. Most common are systems similar to the one shown in Fig. 7.6, which provide qualitative flow visualization, although measurement of shock angles and locations, flow-separation positions, and even the region of transition to turbulence can be made. Schlieren interferometers are described in section 7.4.

7.3 SHADOWGRAPH SYSTEM

In a shadowgraph system the linear displacement of the perturbed light is measured, rather than the angular deflection as in a schlieren system. The shadowgraph image can be understood by referring to Fig. 7.12, which shows a parallel light beam entering a nonuniform test section. To simplify the derivation, index of refraction variations in the test section are assumed to exist only in the y direction.

Consider the illumination at the exit of the test section. The linear displacement of the light beam is probably not large there because of the relatively short distance the light has traveled. If the illumination is uniform entering the test section, it should still be closely uniform there. The beam, however, is no longer parallel, having been deflected by an angle α, which is a function of y. The illumination within the region defined by Δy at this position is within the region defined by Δy_{sc} at the screen. If the initial intensity is I_T, then at the screen,

$$I_0 = \frac{\Delta y}{\Delta y_{sc}} I_T \tag{49}$$

If z_{sc} is the distance to the screen, then

$$\Delta y_{sc} = \Delta y + z_{sc} d\alpha \tag{50}$$

The contrast is

$$\frac{\Delta I}{I_T} = \frac{I_0 - I_T}{I_T} = \frac{\Delta y}{\Delta y_{sc}} - 1 \cong -z_{sc}\frac{\partial \alpha}{\partial y} \qquad (51)$$

Combining this with Eq. (26) gives

$$\frac{\Delta I}{I_T} = -\frac{z_{sc}}{n_a}\int\frac{\partial^2 n}{\partial y^2}dz \qquad (52)$$

If the index of refraction is only a function of density, then

$$\frac{\Delta I}{I_T} = -\frac{z_{sc}}{n_a}\int\frac{\partial^2 \rho}{\partial x^2}\frac{\partial n}{\partial \rho}dz \qquad (53)$$

assuming $dn/d\rho$ is constant. For a gas, Eq. (10) could be substituted into Eq. (52). If there is also a variation of n in the x direction, then, equivalent to Eq. (52),

$$\frac{\Delta I}{I_T} = -\frac{z_{sc}}{n_a}\int\left(\frac{\partial^2 n}{\partial x^2} + \frac{\partial^2 n}{\partial y^2}\right)dz \qquad (54)$$

Shadowgraphs, like schlieren and interferometer photographs, are often taken of phenomena that can be approximated as two-dimensional. As with all integrating optical systems, good qualitative flow visualization can be obtained, even of three-dimensional phenomena. Note that variations of the index of refraction in both the x and y directions are obtained from a single shadowgraph image, while schlieren systems usually only indicate variations normal to the knife-edge.

Different optical geometries are possible for shadowgraph systems, as shown in Fig. 7.13. Parallel light systems, as in Fig. 7.13a, are easiest to understand, although the lensless and mirrorless system of Fig. 7.13b is also usable if the distance from the source to the test region is large. A variation of the system shown in Fig. 13b has been used to study the flow near the surface of a cone with an opening angle equal to the spreading angle of the light beam [38]. Other combinations of mirrors and lenses analogous to the schlieren systems of Figs. 7.2, 7.6, and 7.10 have been used.

It should be noted that if a mirror or lens is used after the test region in a shadowgraph, it should be placed so that the conjugate focus of the test section is not in the plane of the screen. At the conjugate focus the parts of the beam deflected at different angles in the test region are all brought back together, so there is no linear displacement there and thus no shadowgraph effect. In fact, schlieren systems are usually focused on the test region to eliminate shadowgraph effects.

Standard shadowgraph systems are rarely used for quantitative density measurements. The contrast would have to be measured accurately, and Eq. (53), for example, integrated twice to determine the density distribution. Even the density or temperature gradient, which is of interest in heat transfer studies, would require

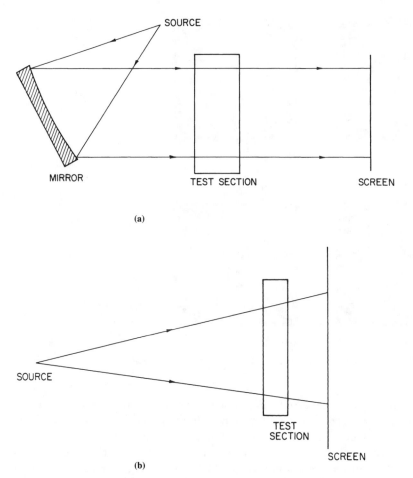

Figure 7.13 Alternative shadowgraph systems. (*a*) One converging mirror. (*b*) No lens or mirror

one integration. If, however, large gradients of density are present, as in a shock wave or a flame, shadowgraph pictures can be very useful. Quantitative measurements of such things as shock angles and the location of boundary layer transition can be made (see page 28 of [3]).

Figure 7.14 shows shadowgraphs of the helium jets whose schlieren photos are shown in Fig. 7.8. The same optical system was used as is described in reference to Figs. 7.7 and 7.8 (i.e., a system similar in layout to that in Fig. 7.6), with the knife-edge removed and the camera set out of focus to obtain a shadowgraph image.

Figure 7.15 is a shadowgraph showing the large-scale structure in the mixing

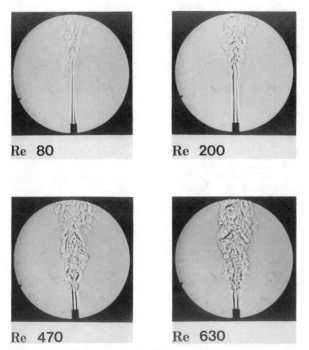

Figure 7.14 Shadowgraphs of a helium jet entering an atmosphere of air

layer between parallel streams of helium (upper region) and nitrogen (lower region) [39]. The helium velocity is about 3 times the velocity in the nitrogen layer.

Numerous applications have included natural convection in double-diffusive plumes [40], swirling flames [41], subsonic and transonic flow past a circular cylinder [42], tip vortices from helicopter rotors [43], shock wave/turbulent boundary layer interactions [44], gas dynamics [45], gun muzzle shock [46, 47], a comparison between hot film and optical measurements of shock wave motion [48], and solidification of an aqueous solution [49].

7.4 INTERFEROMETERS

7.4.1 Basic Principles

Interferometers are often used in quantitative studies. Unlike the schlieren and shadowgraph systems, an interferometer does not depend upon the deflection of a light beam to determine density or index of refraction variation. In fact, refraction effects are usually of second order and undesirable in interferometry, as they introduce deviations or errors in the evaluating equations. To understand interferomet-

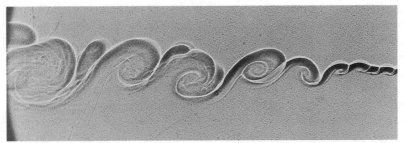

Figure 7.15 Shadowgraph of mixing of parallel-flowing streams of helium (above) and nitrogen (below). (From [28])

ric measurements, one must consider the wave nature of light, and this is perhaps best done by first examining a particular system that is widely used.

The Mach-Zehnder interferometer is often employed in aerodynamic studies. One of the main advantages of this system over other interferometers is the large displacement of the reference beam from the test beam. In this way, the reference beam can pass through a uniform field. In addition, since the test beam passes through the disturbed region only once, the image is sharp, and optical paths can be clearly defined. References [7, 50–56] discuss some of the details of the optics of Mach-Zehnder interferometers.

Figure 7.16 is a schematic diagram of a Mach-Zehnder interferometer. A monochromatic light source is used in conjunction with a lens to obtain a parallel

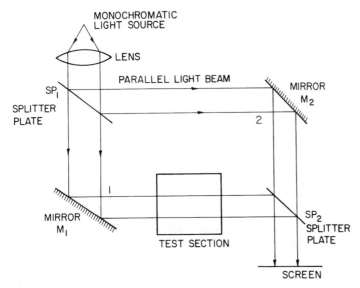

Figure 7.16 Mach-Zehnder interferometer

beam of light. The requirement of a very narrow spectral width for the light source is critical when using an interferometer. The parallel light beam strikes the first splitter plate SP_1, which is a partially silvered mirror permitting approximately half the impinging light to pass directly through it. This transmitted light follows path 1 to mirror M_1, where it is reflected toward the second splitter plate SP_2. The light reflected by SP_1 follows path 2 to mirror M_2, where it is also reflected toward SP_2. The second splitter plate also transmits about half the impinging light and reflects most of the rest. The recombined beams from paths 1 and 2 pass on to the screen. Note that there would also be another recombined beam leaving SP_2, but in general, only one beam from the final splitter is used. The mirrors and splitter plates are usually set at corners of a rectangle [54]. They should then all be closely parallel and at an angle of $\pi/4$ to the initial parallel beam.

Let us assume that in Fig. 7.16 the mirrors are perfectly parallel and that in both paths 1 and 2 there are no variations in optical properties normal to either beam. This requires uniform properties, not only between the mirrors but also in the splitter plates. (Note that the effect of a variation in the thickness of a splitter plate, i.e., a slight wedge, can be corrected by a rotation of one of the mirrors if both surfaces of the plate are flat.) The two beams, from paths 1 and 2, emerging from SP_2 are then parallel.

The amplitude of a plane light wave in a homogeneous medium can be represented by

$$A = A_0 \sin \frac{2\pi}{\lambda}(c\tau - z) \tag{55}$$

where A_0 = peak amplitude
c = speed of light
τ = time
z = distance
λ = wavelength

Consider the amplitude of beam 1 at a fixed position past SP_2 to be

$$A_1 = A_{01} \sin \frac{2\pi c\tau}{\lambda} \tag{56}$$

The other beam could be represented at the same position by

$$A_2 = A_{02} \sin \left(\frac{2\pi c\tau}{\lambda} - \Delta \right) \tag{57}$$

where the phase difference Δ appears because the two beams will probably not be exactly in phase because of a difference in their path lengths. Before the first splitter plate, the two beams (1 and 2) were, of course, one beam and in phase.

Since beams 1 and 2 come from the same source, they are coherent and can interfere with each other. This is implicit in Eqs. (56) and (57) if Δ is not a function of time. Adding Eqs. (56) and (57), and assuming that $A_{01} = A_{02} = A_0$,

$$A_T = A_1 + A_2 = A_0 \left[\sin \left(\frac{2\pi c\tau}{\lambda} - \Delta \right) + \sin \frac{2\pi c\tau}{\lambda} \right] \qquad (58)$$

which can be rewritten

$$A_T = 2A_0 \cos \frac{\Delta}{2} \sin \left(\frac{2\pi c\tau}{\lambda} - \theta \right) \qquad (59)$$

where θ is a new phase difference. Thus the sum of the two waves is a new wave of the same frequency and wavelength.

The intensity of the combined beam is the quantity observed visually or measured on a photographic plate. The intensity I is proportional to the square of the peak amplitude, or

$$I \sim 4A_0^2 \cos^2 \frac{\Delta}{2} \qquad (60)$$

Note that when $\Delta/2\pi$ is an integer (say, i), the peak intensity is 4 times that of either of the two beams, but when $\Delta/2\pi$ is a half-integer $(i + 1/2)$, the intensity is zero. The interesting yet not too difficult paradox in this latter case is, Where did the energy go?

The optical path along a light beam is defined by

$$PL = \int n \, dz \qquad (61)$$

or

$$PL = \int \frac{c_0}{c} dz = \lambda_0 \int \frac{dz}{\lambda} \qquad (62)$$

Thus the optical path length is the vacuum wavelength times the real light path in wavelengths (which can vary along the path). In Fig. 7.16, the difference between paths 1 and 2 is

$$\overline{\Delta PL} = \int_1 n \, dz - \int_2 n \, dz \qquad (63)$$

$$\overline{\Delta PL} = PL_1 - PL_2 = \lambda_0 \left(\int_1 \frac{dz}{\lambda} - \int_2 \frac{dz}{\lambda} \right) \qquad (64)$$

The phase difference between two points a distance of z apart [cf. Eq. (55)] is $2\pi \, dz/\lambda$. The difference in phase of the two beams upon recombination is

$$\Delta = 2\pi \left(\int_1 \frac{dz}{\lambda} - \int_2 \frac{dz}{\lambda} \right)$$

or

$$\frac{\Delta}{2\pi} = \frac{\overline{\Delta PL}}{\lambda_0} \tag{65}$$

If $\overline{\Delta PL}/\lambda_0$ is zero or an integer, then from Eqs. (60) and (65), there is constructive interference, and the field on the screen in Fig. 7.16 is uniformly bright.

7.4.2 Fringe Pattern with Mach-Zehnder Interferometer

Consider a Mach-Zehnder interferometer with beams 1 and 2 passing through homogeneous media, so that initially, the recombined beam is uniformly bright ($\overline{\Delta PL}$ is assumed to be zero). If a disturbance (inhomogeneity) were put in part of the field of light beam 1, the path difference ΔPL would no longer be zero, nor would the field be uniform. At any position on the cross section of the beam (neglecting refraction), Eq. (63) could be used to obtain ε, the path-length difference in terms of vacuum wavelengths:

$$\varepsilon = \frac{\overline{\Delta PL}}{\lambda_0} = \frac{1}{\lambda_0} \int (n - n_{\text{ref}}) \, dz \tag{66}$$

where n_{ref} is the reference value of the index of refraction in reference beam 2. If $\overline{\Delta PL}/\lambda_0$ is an integer, the field will be bright, while if $\overline{\Delta PL}/\lambda_0$ is a half-integer, the field will be dark. Thus the initially uniformly bright field will have a series of bright and dark regions (fringes), each one representative of a specific value of $\overline{\Delta PL}/\lambda_0$ and differing in magnitude from the adjacent fringe of the same intensity by a value $\varepsilon = \overline{\Delta PL}/\lambda_0 = 1$. If there is a gas in light beam 1, the Gladstone-Dale relation, Eq. (2), can be used in Eq. (66):

$$\varepsilon = \frac{c}{\lambda_0} \int (\rho - \rho_{\text{ref}}) \, dz \tag{67}$$

If the field is two-dimensional, in that the only variations in the index of refraction along the light beam (i.e., in the z direction) are the sharp discontinuities at the entrance and exit of the test section (see [8] for a treatment of axisymmetric density fields), and ρ only varies over a length L, the fringe shift is given by

$$\varepsilon = \frac{n - n_{\text{ref}}}{\lambda_0} L \tag{68}$$

For a gas,

$$\varepsilon = \frac{c}{\lambda_0} (\rho - \rho_{\text{ref}}) L \tag{69}$$

or

$$\rho - \rho_{\text{ref}} = \frac{\lambda_0 \varepsilon}{CL} = \frac{\lambda_0 \varepsilon}{n_0 - 1} \rho_0 \tag{70}$$

If the pressure is constant and the ideal gas law is used, then

$$\frac{1}{T} = \frac{\lambda_0 R}{PCL} \varepsilon + \frac{1}{T_{ref}} \tag{71}$$

$$T = \frac{PCLT_{ref}}{PCL + \lambda_0 R\varepsilon T_{ref}} \tag{72}$$

or

$$T - T_{ref} = \left(\frac{-\varepsilon}{PCL/(\lambda_0 RT_{ref}) + \varepsilon}\right)T_{ref} \tag{73}$$

For a two-dimensional field in a liquid, Eq. (68) could be written

$$n = \frac{\lambda_0 \varepsilon}{L} + n_{ref} \tag{74}$$

where n and n_{ref} would have to be known as functions of temperature. For small temperature differences,

$$\varepsilon = \frac{L}{\lambda_0} \frac{dn}{dT} (T - T_{ref}) \tag{75}$$

and

$$T - T_{ref} = \frac{\varepsilon\lambda_0}{L} \frac{1}{dn \, / \, dT} \tag{76}$$

If $\lambda_0 = 546.1$ nm and L is 30 cm, each fringe (that is, $\varepsilon = 1$) represents a temperature difference of about 2°C in air at 20°C and 1 atm. In water under the same conditions, each fringe represents a temperature difference of about 0.02°C.

If the initial optical path lengths of beams 1 and 2 are not exactly equal (but their difference is less than the coherence length of the light source), the above equations can still be used to determine density and temperature differences between different parts of the cross section of the test beam 1, assuming uniform optical path length over the cross section of the reference beam. Then n_{ref}, ρ_{ref}, and T_{ref} refer to a uniform portion of the test region, and all properties at other locations in the test section are measured relative to the properties there. This is how interferograms—the fringe patterns obtained with an interferometer—are normally evaluated.

When the interferometer beams are recombined parallel to each other (called infinite fringe setting), each fringe is the locus of points in a two-dimensional field where the optical path length or density is constant. These fringes are useful for qualitative flow-field visualization as well as for quantitative studies; they can delineate boundary layers, mixing regions, etc. Similarities exist between the infinite fringe setting of a Mach-Zehnder interferometer where optical path-length con-

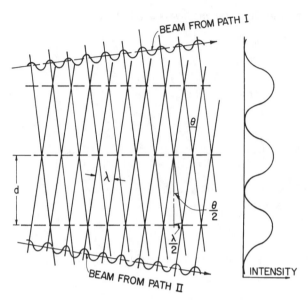

Figure 7.17 Intensity pattern from two intersecting plane light beams

tours are observed and color contours of a color schlieren system where index of refraction gradient contours are visualized [57].

A Mach-Zehnder interferometer is not always used with the beams parallel upon recombination, as in the discussion above. Consider two beams, each of which is uniform (in phase) normal to the direction of its propagation, although diverging slightly, at a small angle θ, from the other beam, as represented by the two wave trains in Fig. 7.17. Lines are shown drawn through the crests (maxima if amplitude) for each wave train to represent the planes (wave fronts) normal to the direction of propagation. Constructive interference occurs where the maxima of the two beams coincide; dashed lines representing the loci of these positions are shown. If a screen is placed approximately normal to the two beams, the intensity distribution on the screen follows a cosine-squared law [Eq. (60)], as shown in Fig. 7.17. Thus parallel, equally spaced, alternately dark and bright fringes (called wedge fringes) appear on the screen when there are no disturbances in either field. The difference in optical path length between the two beams varies linearly across the field of view with wedge fringes, so that only one fringe in the field (the zero-order fringe) can represent equal path lengths for the two beams. From Fig. 7.17, the spacing between the fringes is

$$d = \frac{\lambda / 2}{\sin \theta / 2} \tag{77}$$

or

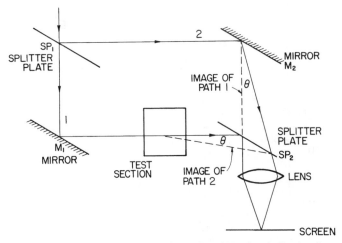

Figure 7.18 Light rays for Mach-Zehnder interferometer, indicating the preferred position of focus

$$d \sim \frac{\lambda}{\theta}$$

for small θ. For fringes to be observable, θ must be very small. For example, if d is about 5 mm, θ (using the green mercury line) is about 10^{-4} rad. As θ is decreased to zero, the fringes get further and further apart, approaching the "infinite fringe pattern" found when the two beams are parallel.

When a Mach-Zehnder interferometer is adjusted to give wedge fringes, the fringes are localized as shown in Fig. 7.18. Only a pair of rays is shown; after diverging, they are brought to a focus on the screen (or film in a camera) by a focusing lens or mirror. The angular separation of the beams is greatly exaggerated on the figure. The dashed lines represent the paths from the virtual object of the beams in paths 1 and 2 as they would appear along the other path. The fringes are localized where, tracing backward along the real and imaginary paths, the rays intersect. The plane of localization can be adjusted (once the beams are close enough to parallel produce fringes, that is, d not too small) most conveniently by rotation of SP_2 (and of M_2, to keep the fringes in view) about two orthogonal axes in the plane of SP_2. If the fringes are localized at M_2, the rotation of M_2 will not affect the plane of localization and will only change the fringe spacing and orientation. To get both the fringes (localized at M_2) and the test section (actually, the center of the test section, as discussed below) in focus on the screen or in the camera, the interferometer mirrors can be placed on the corners of a 2-1 rectangle, the distance from SP_2 to M_2 then being the same as the distance from SP_2 to the middle of the test section.

When a disturbance is present within the test section (which is in beam 1 in Fig. 7.18), the optical path is no longer uniform in this beam. The fringes then are

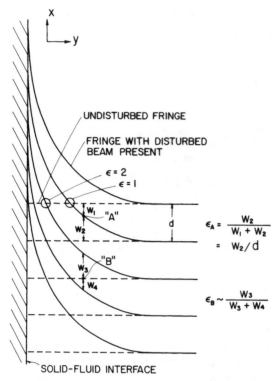

$$\epsilon_A = \frac{W_2}{W_1 + W_2}$$
$$= W_2/d$$

$$\epsilon_B \sim \frac{W_3}{W_3 + W_4}$$

Figure 7.19   Fringe shift pattern with wedge fringes

no longer straight, but rather are curved as in Fig. 7.19. In this figure, the original (undisturbed) positions of the fringes are shown by dashed lines. The undisturbed fringes are normally aligned in a direction in which the expected index of refraction (density) gradient will be large. The difference in optical path length from the original value or from the reference positions in the field of view where the fringes have not changed is shown in the figure in terms of the fringe shift ε. If the total fringe shift is large, only integral values of ε are usually measured; for small differences in optical path length, fractional values of ε can be measured as shown.

The possibility of measuring fractional values of ε is one advantage of wedge fringes. In addition, it is difficult to be certain with infinite fringes that the undisturbed (or reference) field is at the maximum brightness, so there is an uncertainty in the reference position. This problem does not occur with wedge fringes as long as there is a region of known uniform properties in the field of view. Contours of constant optical path length can also be obtained with wedge fringes by superposing the disturbed interference pattern over the undisturbed pattern. The resulting moiré fringe pattern gives grey lines that can represent constant-density lines in a two-dimensional test region. If there are irregularities in the undisturbed image

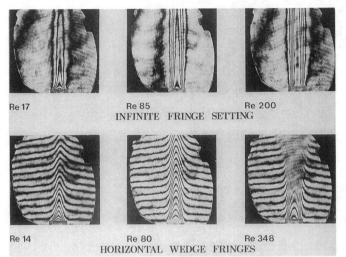

Figure 7.20   Interferograms of a low-Reynolds-number helium jet entering a still atmosphere of air

due to faulty optical parts, this superposition method can still be used to obtain quantitative results.

Interferograms are often evaluated by using a microscope with a moving bed to determine the locations of individual fringes. The fringe positions can be put in digital form by using a digitizing tablet or its equivalent [58, 59] on which the image of the interferogram is projected and a cursor is used to record the fringe locations, or by digitizing the image obtained from a video camera. Optical noise often limits the accuracy and useability of fringe pattern data. One-dimensional and two-dimensional Fourier transform techniques have been devised to reduce the errors associated with random noise, which improves the accuracy of the retrieved fringe shift data [60, 61]. Other techniques are given in [62, 63]. No matter how the fringe locations are evaluated, some reference marks are normally required in the field to clearly indicate the position of test surfaces and to evaluate the scale of the image.

7.4.3 Examples of Interferograms

Interferograms are often used for quantitative studies of heat transfer [8]. In direct-flow measurements they have been used to study the mixing of fluids of different densities (often of different compositions), of compressible (often supersonic) flows, and of flows in which some effective tracer, perhaps a hot spot from a local heat source, can be used to change the index of refraction or density of the flowing field.

The flow of a jet of helium at low Reynolds number into a still atmosphere of air is shown in Fig. 7.20. The upper set of figures represents the infinite fringe

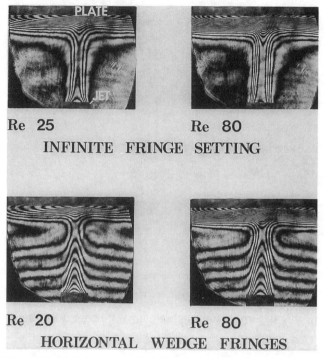

Figure 7.21 Interferograms of a helium jet impinging on a flat plate

condition. The lower set of figures shows wedge fringes, which are often more convenient for quantitative evaluation, particularly when the total fringe shift in the field is small. The jet used in these photos is axisymmetric, and a transformation [8] would have to be made to convert the fringe position into the axisymmetric density field.

Figure 7.21 shows the impingement of a helium jet on a flat plate. Figure 7.22 also shows impingement of a jet, but near the edge of the plate, where the flow field of the helium around to the opposite side of the plate can be observed.

Figure 7.23 contains interferograms of a high-speed projectile (sphere) and a cone with a cylindrical projection, taken with a holographic interferometer (see below) [64, 65]. Figures similar to these have been used to determine the turbulent fluctuations in the wake of objects in a ballistic range [64].

Figure 7.24 compares a shadowgraph photo and two holographic interferograms of a sharp-tipped spike with conical flare in a Mach 3 flow [66]. Figure 7.25 shows the transition from laminar to turbulent flow in the boundary layer on a slender sharp-tipped cone [67]. The interferograms of these figures were taken with holographic interferometers in a fashion that yields results similar to those obtained with a Mach-Zehnder interferometer.

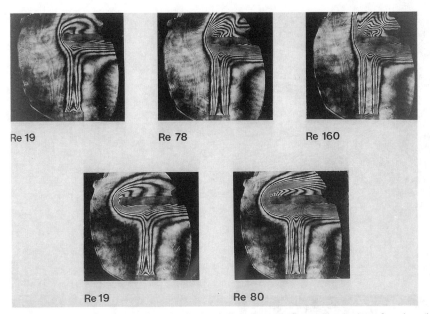

Figure 7.22 Interferograms of an impinging helium jet with flow around edge of a plate (infinite-fringe setting)

Figure 7.26 contains two interferograms of the flow through a cascade of gas turbine blades (E. R. G. Eckert, personal communication). In each photo, one of the blades is heated (note the heating wire connection) for better visualization of the flow around the blades. The incoming flow is parallel to the heating wire connection. Note the difference in the boundary layer growth on the concave and convex sides of the blades. Also, there is a marked difference in the boundary layer, depending on the direction of the incoming flow relative to the blade orientation. Close examination of the flow shows boundary layer thickening, and even separation. Additional interferograms can be found in the published literature. Recent examples include natural convection flow from an open cavity [68] and combined natural and forced convection in a rotating enclosure [69]. Both of these studies used the infinite fringe setting.

7.4.4 Design and Adjustment

Figures 7.16 and 7.18 are schematics of a typical Mach-Zehnder interferometer. Instead of lenses to form the parallel light beam and to bring the wide beam down into a smaller region for focusing on a screen or through the aperture of a camera, mirrors are often used to decrease the cost of the system.

(a)

(b)

Figure 7.23 Holographic interferograms of high-speed flow over an object. (a) One-half-inch-diameter sphere; 1 atm pressure; Ma≃ 6. (b) Shock interaction of 60° cone cylinder projected at Ma≃ 3.5; spark-produced blast wave. (From [38, 39])

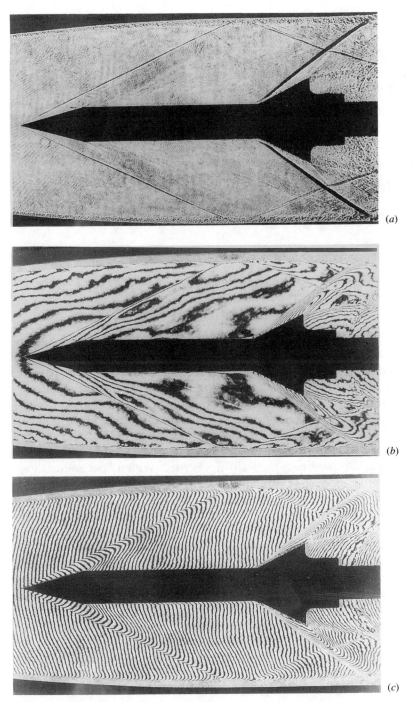

Figure 7.24   Flow over sharp-tipped spike with conical flare; pressure 100 psia; Ma = 2.98. (*a*) Shadowgraph. (*b*) Infinite-fringe interferogram. (*c*) Wedge-fringe interferogram. (From [40])

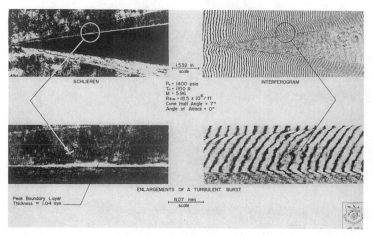

Figure 7.25 Holographic schlieren and interferometric depiction of boundary-layer transition on a slender, sharp-tipped cone. (From [40])

For a sufficiently long coherence length (see below), a relatively monochromatic light source must be used. In the past, low-pressure lamps with filters were often used. Today, lasers are used most often as the light source for interferometers. In fact, they are necessary for some special types. Lasers offer a quite large coherence length, which can be of great value. With other light sources, if thick windows must be used on the test section, or if the fluid in the test section has a refractive index very different from that of air, it may be necessary to have a compensating tank to ensure that the path lengths of the beams are not too different. With a laser light source, such tanks are normally not needed. With a laser source, however, spurious fringes may be present, owing to multiple reflections in some of the optical components. Note that the test section should be placed in path 1, as shown in Figs. 7.16 and 7.18, rather than in the other light beam. In this way, the beam representing the shadow image of the test section does not pass through the last splitter plate, which could cause considerable astigmatism [70].

The alignment of an interferometer, such as is shown in Figs. 7.16 and 7.18, is somewhat complicated and time consuming but is not so horrendous a task as is often perceived. A number of methods that greatly simplify the task have been described [50, 54, 71–73]. The chief concern is to align the reference and test beams so that they are closely parallel. If they are not, the fringe spacing is so small that the fringes cannot be detected.

The parallelism of the beam leaving the paraboloidal mirror F can be determined by measuring its dimensions at various positions along its path or, more accurately, by reflecting it back with a plane mirror and observing the focus of this returned beam. The two path lengths are then usually set approximately equal. The

Figure 7.26 Wedge-fringe interferograms used for visualizing flow over a heated gas-turbine blade held in a cascade; oncoming flow direction is parallel to the visible wire carrying the heating current. (From [42])

difference in length must be smaller than the coherence length of the source-filter combination, which increases as the light becomes more monochromatic. If the light-intensity variation with wavelength is Gaussian with a bandwidth of $\Delta\lambda$, the coherence length or optical path difference over which fringes can still be observed is approximately $\lambda^2/\Delta\lambda$ [74].

To obtain fringes, beams 1 and 2 (Fig. 7.18) must be nearly parallel to each other following SP_2. This can be accomplished by aligning the images of two objects observed by looking back through SP_2, preferably with a small telescope. The two objects examined should be far apart and must both be located before the first splitter SP_1, so that images are obtained for paths 1 and 2. Rotation of M_2 and SP_2 about two orthogonal axes moves each pair of virtual images of the two objects. When both images of each object are superimposed, the two beams leaving SP_2 are closely parallel, and fringes should appear in the field of view. While focusing on the center of the test section and M_2 with the telescope, camera, or screen, fringes are made as sharp as possible by further rotation of SP_2 while rotating M_2 to keep the fringe spacing from getting too small.

When the plane of focus of the fringes is in M_2 (and the center of the test section), the final adjustment of optical path length can be made. It is advantageous to keep the interferometer set close to the zero-order fringe (zero optical path length between the two light beams), since that is the position for sharpest fringes, even for filtered light. In addition, white light and the zero-order fringe are useful for measuring the index of refraction for fluids whose optical properties are unknown or in tracing fringes through regions of large density gradients. In many interferometers the path length of one of the beams can be altered by translation of a mirror (M_1). If white light (an incandescent bulb suffices) replaces the monochromatic light on half the field of view, the mirror can be translated until the zero-order white-light fringe is observed. In practice, it is often helpful, particularly if the initial setting is far from the zero-order fringe, to use light of varying coherence lengths, from the most monochromatic to white, to make each set of fringes as sharp as possible while translating the mirror.

Once adjusted, the interferometer, if properly constructed, usually needs only minor adjustment. Placing the unit in a vibration-free constant-temperature area helps to maintain alignment.

7.4.5 Errors in a Two-Dimensional Field

Since the Mach-Zehnder interferometer is of great value for quantitative studies in two-dimensional fields, considerable attention [8, 75–85] has been directed toward the corrections that must be applied when the idealizations assumed in the derivation of Eqs. (70), (73), and (76) are not strictly met. The two most significant errors usually encountered are due to refraction and end effects.

Refraction occurs when there is a density (really, index of refraction) gradient normal to the light beam that causes the beam to "bend." The resulting error increases with increasing density gradient and with increasing path length in the disturbed region L. It is refraction that usually hinders accurate interferometric measurements in thin forced-convection boundary layers.

End effects are caused by deviation from two-dimensionality in the density field, particularly where the light beam enters and leaves the disturbed region. The end effects are usually large when the disturbed region is large normal to the light beam direction and the test-section length along the light beam is relatively short. End effects are often significant with thick thermal boundary layers. Thus, if an experimental apparatus is designed to minimize refraction error, the end effect may be large, and vice versa. Further details and a simplified analysis of the errors are contained in [8].

Errors associated with misalignment of the test surface with the light beam also exist. In addition to the error in determining surface position and blockage of the light beam near the surface, an error in measuring the index of refraction gradient at the surface can be quantified in terms of the misalignment angle [85].

7.4.6 Other Interferometers

Interferometric systems other than the Mach-Zehnder have been used in flow studies. These, in general, produce interferograms that can be evaluated in a manner similar to that for the Mach-Zehnder patterns. In some systems, a grating divides the initial beam into two coherent beams, one of which traverses the test region. The beams are recombined on another grating, yielding an interference pattern. One system [86, 87] uses two gratings, with the reference and test beams close to each other. In a four-grating apparatus [88], the beams are further apart, but residual fringes are often superimposed on the pattern. A laser light source has also been used with a grating interferometer [89].

Several special interferometers with a laser light source have been used or suggested for flow measurement [90–93]. Systems that are used with laser sources include schlieren or shearing interferometers. In one such system [94–96], a very small wire or stop is placed at the focus of a standard schlieren system to block the central maximum of the Fraunhofer diffraction pattern. Then the phase distribution produced when the beam passes through a disturbance in the test region can be observed in reference to an undisturbed part of the test beam. The result is a fringe pattern similar to that observed with a Mach-Zehnder interferometer set for infinite fringe spacing. A similar system uses a spot that has an optical thickness of one-quarter wavelength at the focus. This gives the undiffracted light a phase shift with respect to the diffracted light that passes around the spot. The phase-shifted and unphase-shifted portions of the beam interfere to form a fringe pattern similar to the infinite fringes produced with a Mach-Zehnder interferometer [97]. Another system [98] uses a small glass shearing plate in place of the knife-edge. When the angle between the incoming light beam and the normal to the plate is approximately 50°, the first two reflected beams interfere, resulting in fairly straight parallel fringes. A Wollaston prism [99–101] and a grating [102] have also been used to produce finite fringe interference patterns. Two Ronchi gratings have been used in the shadowgraph system shown in Fig. 7.13a between the test section and the screen [103, 104]. Moiré finite fringes are produced, which have been used to study compressible flows.

Shearing interferometers in which the reference beams are sheared slightly [100, 105] in the lateral direction have been used, with conventional and laser light sources. A polarization interferometer [7, 106] is a wave-shearing interferometer in which two coherent beams, polarized at right angles to one another, are produced from a single incoming beam. The two beams often diverge at a finite angle, and when they are recombined after passage through the test section, a fringe pattern results. Either a finite fringe or an infinite fringe field can be obtained using three Wollaston prisms [107, 108]. Errors caused by misalignment of a test surface are discussed in [109].

In a shearing interferometer, the test and reference beams are often separated by only a small amount δ. Let us assume this separation is in the y direction and

$$\Delta y' = \delta \qquad (78)$$

When the beams are recombined, the interference field is representative of the local difference in optical path length $\overline{\Delta PL}$ between positions y and $y + \delta$ (or $y + \Delta y'$). Equivalent to Eq. (68), the corresponding fringe shift is

$$\varepsilon = \frac{\overline{\Delta PL}}{\lambda_0} = \frac{n_{y + \Delta y'} - n_y}{\lambda_0} L \qquad (79)$$

$$\varepsilon = \frac{\Delta n}{\Delta y} \Delta y' \frac{L}{\lambda_0} = \frac{\Delta n}{\Delta y} \delta \frac{L}{\lambda_0} \qquad (80)$$

If the beam separation is small,

$$\varepsilon = \frac{\partial n}{\partial y} \frac{1}{\lambda_0} L\delta \qquad (81)$$

or

$$\frac{\partial n}{\partial y} = \frac{\lambda_0 \varepsilon}{L\delta} \qquad (82)$$

For a gas,

$$\frac{\partial \rho}{\partial y} = \frac{\lambda_0 \varepsilon}{CL\delta} = \frac{\lambda_0 \rho_0 \varepsilon}{(n_0 - 1)L\delta} \qquad (83)$$

Assuming constant pressure, the perfect gas law gives

$$\frac{\partial T}{\partial y} = -\frac{T\lambda_0}{n_0 - 1} \frac{\rho_0}{\rho} \frac{\varepsilon}{L\delta} \qquad (84)$$

Note that δ is a constant (often adjustable) of the sharing interferometer. As long as it is small, the local derivative of the index of refraction (or density or temperature) in a two-dimensional field is proportional to the fringe shift.

An advantage of a shearing interferometer over a Mach-Zehnder interferometer is that the density or temperature gradient near the surface of an object can be measured directly. The sensitivity is proportional to the shear spacing as well as to the length of the test object in the light beam direction. With a small shear spacing, the errors due to refraction and thick-window effects tend to vanish. Among the disadvantages is the need to integrate the fringe displacement to obtain density distributions. Also, only the density gradient in the direction of the beam shear can be measured, although this direction may be changed. The closest to a surface that an accurate measurement of the density gradient can be made is of the order of one-half the shear displacement.

7.4.7 Holography

Holography can also be used in flow measurement [110–121]. In a holographic interferometer, the light beam (in this case, almost necessarily a highly uniform beam from a laser source) is split into two coherent beams, one of which passes through the test section, while the other (reference beam) bypasses the test section. The two beams are recombined on a photographic plate. The resulting hologram is a diffraction grating formed by the emulsion on the plate. If the hologram on the developed plate is viewed via the reference beam, the interference pattern can be interpreted in a manner analogous to a Mach-Zehnder interferogram.

An example of a holographic system is shown in Fig. 7.27 [122]. Although holographic interferometry is normally done using a double exposure, real-time holographic interferograms can be obtained for flow studies using a precise and adjustable hologram mount. With this system, either infinite fringes or, for more quantitative measurement, wedge fringes can be obtained. Another holographic system design is shown in Fig. 7.28 [123]. This system is useful for the study of very rapid phenomena such as the passage of shock waves and explosive events. The Q-switched ruby laser has a wavelength of 694 nm and pulse widths typically between 10^{-8} and 10^{-7} s. A holographic system with two reference beams has been developed [124]. Fringe spacing or orientation can be changed by tilting or rotating the hologram during reconstruction.

Although somewhat more inconvenient to use than a Mach-Zehnder interferometer, a relatively low-cost system can be set up for holographic interferometry. This is because most of the optical components of the holographic apparatus need not be extremely uniform or of very high quality to produce good interferograms. In holographic interferometry, two wave fronts passing along essentially the same path but at different times are compared, whereas in conventional interferometry, two wave fronts passing along different paths are compared at the same time.

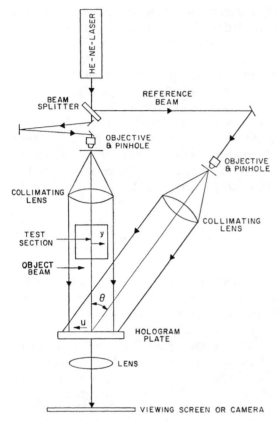

Figure 7.27 Holographic interferometer

Therefore imperfections in optical components do not cause spurious fringes as in conventional interferometry.

Another advantage of holography is the possibility of total separation of the flow-field experiment from the data analysis. In conventional schlieren analysis, the location of an optical filter, such as a knife-edge, must be determined while the experiment is in progress. Holography freezes the wave front information on the hologram during the experiment. The position of the knife-edge can then be determined during the hologram reconstruction. Many positions may be necessary for optimum analysis of the flow field. Other types of filters can also be used during the reconstruction phase if desired.

A third advantage of using a holographic system is that a single photograph (hologram) can give information on the interference patterns that would otherwise have to be observed by looking in different directions through the test section. This

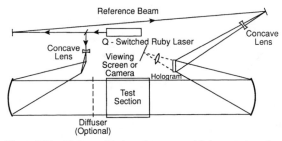

Figure 7.28 Holographic interferometer with large test section suitable for measurement of rapid phenomena

follows from the large amount of information that is available on the hologram when a diffuser plate is located before the test section to distribute the object beam in different directions. The hologram contains the record of very complex wave fronts produced in passing through the test section. Viewing the hologram from a specific direction provides essentially the equivalent of the interference pattern that would be obtained by a beam passing through the test section in that direction.

7.4.8 Three-Dimensional Measurements

The interferometric measurement techniques described in the previous sections apply only to fluids in which the index of refraction is invariant in the direction of the light beam. Many applications exist in which the index-of-refraction field varies in all three coordinate directions, $n(x, y, z)$. A single interference pattern does not contain enough information to determine the nature of the three-dimensional field. The most common method employed in the measurement of three-dimensional fields is to pass a collimated beam through the field in different directions to obtain several interference patterns. Typically, the light beams remain parallel to the x-y plane. Each beam produces an image function that is equal to the instantaneous integral of the light beam through the fluid in that direction:

$$\text{PL}_{\theta_1} = \int_{S_l} n(x, y, z) \, ds \tag{85}$$

For reconstructing the field, individual planes parallel to the x-y plane at specified values of z are considered separately. These two-dimensional fields are then assembled later to reconstruct the entire three-dimensional field. Therefore our attention will be focused on the reconstruction process that occurs in a single z plane. With the value of z constant at $z = z_k$, Eq. (85) becomes

$$\text{PL}(\rho, \theta_l, z_k) = \int_{S_l} n(x, y, z_k) \, ds \tag{86}$$

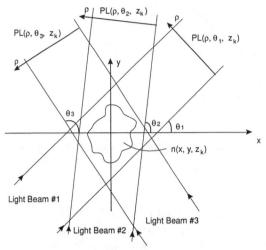

Figure 7.29 Optical path-length measurements obtained by passing three collimated light beams through an index-of-refraction field in the z_k plane

where ρ is the coordinate perpendicular to the light beam direction in the z_k plane and θ_l is the angle of the ray measured from the x axis. This integration process is shown for three beams in Fig. 7.29. The difference in path length between a beam passing through the test fluid and a reference beam or beam passing through the fluid at an earlier time becomes

$$\Delta PL(\rho,\, \theta_l,\, z_k) = \int_{S_l} [n(x,\, y,\, z_k) - n_{\text{ref}}(x,\, y,\, z_k)]\, ds \qquad (87)$$

This optical path length difference can produce a fringe pattern as in Eq. (66).

The general form of equation (87) can be written as

$$\phi_l = \int_{S_l} f(x,y)\, ds \qquad (88)$$

The optical path-length function ϕ_l is known at discrete angles θ_l and at discrete values of ρ. The task is to determine the original index-of-refraction field $f(x, y)$ given the path-length data ϕ_l. Several inversion schemes have been developed in the refractionless limit when no opaque objects block the light beam [125].

The first method developed was based on a two-dimensional Fourier transform technique [126, 127]. A series representation using orthogonal functions was refined and applied to the study of jet flows [128]. A direct inversion technique was also developed [129]. These methods require path-length data over an 180° angle of view for asymmetric index of refraction fields. This is possible for optical configurations in which the measurement volume contains a steady field that can be rotated over a 180° angle. However, this is not possible for unsteady flows and other flow configurations that cannot be rotated.

Series expansion methods do not require a complete 180° angle of view and are therefore more practical in many applications. A sample method or sine method was proposed [129]. A grid method was also demonstrated by the same authors. This is conceptually the simplest inversion technique and can be explained as follows. The function $f(x, y)$ is approximated by a grid of $M \times N$ rectangular elements or dimensions l_x and l_y each having a constant value of the index of refraction n. The total optical path length of a ray passing through this grid is given by the sum of the geometrical path length of the ray through each element in which it passes multiplied by the corresponding element index of refraction. This can be written as

$$\phi(\rho_j, \theta_l) = \sum_{m=0}^{M-1} \sum_{n=0}^{N-1} W_{mn}(\rho_j, \theta_l) n(l_x m, l_y n)$$

where the weighting function W is determined strictly by geometry. One then solves for the values of n that best satisfy this set of equations for all θ_l and ρ_j. The best results are obtained, i.e., minimum error, when the system has a redundancy factor larger than 3 [130]. This implies that the total number of path-length data, $\phi(\rho_j, \theta_l)$, is at least 3 times as large as the number of grid elements, $M \times N$. These series methods can be solved using matrix inversion techniques. Another solution method that has become widely used due to its minimal use of computer time is the algebraic reconstruction technique or ART method [131]. Although this iterative method was originally developed for underdetermined systems of equations, its applicability for an overdetermined set of equations was demonstrated using the direct additive technique [129]. The inversion techniques reviewed in this paragraph are all capable of providing reconstructions with as little as a 30° angle of view. However, the larger the angle of view, the better the accuracy.

The entire Fourier transform inversion process can be performed optically rather than using a computer to perform the inversion numerically [132]. A collimated beam is passed through the test section three times in three different directions in succession with a beam rotation between the passes. The resulting interference pattern is recorded on a hologram, which contains an infinite fringe pattern that is proportional to the index-of-refraction distribution in the x-y plane. The technique has been demonstrated using a vertical heated wire in air. However, the method is limited to index-of-refraction fields that are invariant or only slightly changing in the z direction.

Several investigations have been performed using a single light beam and rotating the model. Transonic flow near an aerodynamic model in a blowdown wind tunnel was investigated in one study [133]. The model was constructed of lucite near the testing region to eliminate light blockage and allow a 180° field of view. A holographic interferometer was used in the finite fringe mode with a Q-switched double-pulsed ruby laser that gave an exposure time of 20 ns. Another study, which

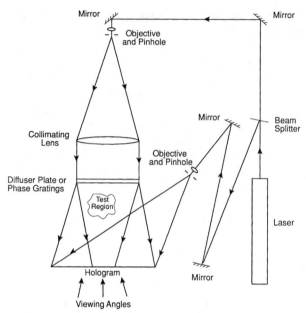

Figure 7.30 Optical apparatus for recording multidirectional holographic interferograms

examined transonic flow from a nozzle, used a Q-switched ruby laser in a holographic interferometer to obtain holograms at seven different angles, covering an 180° field of view [58]. One group used a shearing interferometer to measure the three-dimensional temperature field in a finite-sized Bénard cavity in natural convection [134]. The ART solution method was used in the data analysis and produced results that agreed well with three-dimensional numerical calculations. The density field in air around a rotating helicopter blade was measured in another study [135], using a pulsed ruby laser in a holographic interferometer to obtain holograms of the flow field near the rotor tip over an 180° field of view. Corrections were made for refraction and shocks.

Another method used to obtain multiple angles of view is to record all the information on a single hologram. A diffuser plate or phase grating is placed before the test region to provide the different angles of light beam propagation through the test fluid. A typical optical setup is shown in Fig. 7.30. The range of viewing angle depends on the angle of scattering or deflection before the test region and the width of the hologram with respect to the test region. Most series reconstruction methods provide reasonable accuracy with 30°–45° angle of view. This type of setup was used [136] with an opal glass diffusing plate behind the test section. Infinite fringe holographic interferograms were obtained to measure the natural convective temperature field above a heated horizontal rectangular plate in water. The range of viewing angle was 30°, which was doubled to 60° by assuming a plane of symmetry in the center of the plate. A hologram reconstruction method using a thin-beam technique was demonstrated in [137]. This method reduces the problem of fringe localization with a diffuse light source and makes the data aquisition process more convenient. Another method to improve data analysis is to replace the diffuser plate with a phase grating. The phase grating diffracts the light beam in a few predetermined directions through the test region and improves the fringe contrast in the reconstructed images. In [138] the collimated beam passed through two holographic phase gratings in series. These gratings generated 17 light beams with a total angle of view of 62°. Diffuser plates were used on two sides of the test region, and three holographic plates on the other two sides in [128]. Each hologram provided a field of view of about 15°, and the total angle of view was approximately 100°. The jet studied was planar symmetric, so that the 90° angle of view was effectively 180°.

Distinct collimated beams can also be passed through the test region simultaneously at different angles. Four collimated beams have been used spaced 45° apart to give a total angle of view of 135° [139, 140]. Two beams were sent to one holographic plate and camera, and the other two beams sent to another plate and camera. Wedge fringe patterns were produced in real time on both holographic plates. Using this configuration, unstationary thermal mixing was measured in a small tank with a stirrer by photographing all four fringe patterns simultaneously in real time. The sine inversion method with the ART solution scheme was used to reduce the data.

All the inversion techniques outlined earlier assume that the light rays pass through the object undeflected by refraction. When refraction is significant, corrections should be made as in conventional interferometry. An iterative correction procedure has been described [141] in which 20 iterations were sufficient to reduce the average error between actual and computed distributions to less than 1% for all test cases considered. A number of iterative ray tracing correction procedures and linearized inverse scattering techniques are outlined in [142]. The convergence and reliability characteristics of these techniques are not well defined.

An opaque object often blocks a portion of the light beam. Examples include aerodynamic bodies, projectiles, heated surfaces, and supports. Several algorithms have been devised to overcome the problem of missing data in the reconstruction process. One group [143] used an interpolation method to estimate the values of the missing fringe data. They achieved satisfactory results for supersonic flow around a cone. A unified iterative approach using a priori information was devised by another group [144]. The method was employed on a commercial computerized tomography scanner and could be readily adapted to optical imaging systems. An iterative convolution method that does not use a priori information was shown to be superior to series expansion methods for the test cases considered [145]. Another method [146] uses a modified grid technique to solve for the index of refraction in the outermost portions of the test region first and then proceeds in toward the blockage.

7.5 CONCLUSION

The properties of shadowgraph, schlieren, and interferometric systems have been described. The basic operating equations for all three systems have been derived, with emphasis on their use in determining density and temperature distributions in two-dimensional fields.

Shadowgraph and schlieren systems are used principally for qualitative descriptions of a density field. Because schlieren and shadowgraph photographs yield information on the first and second derivatives of density, their widest application can be found in systems with steep gradients of density and temperature, such as flame fronts and shock waves. Interferometers can be used in quantitative studies of two-dimensional (including axisymmetric) density and temperature fields.

The advent of lasers permits novel and useful interferometer designs. Lasers can be used in schlieren interferometers and in holography for temperature and density measurements. Holographic interferometers using laser light sources can be used in the evaluation of three-dimensional density fields.

Computer-aided image analysis and data processing are becoming more widely used [147]. These include optical methods such as tomographic reconstruction. Other more traditional methods are improved with the addition of pattern recognition, digital filtering, and image subtraction.

NOMENCLATURE

a_s, a_0, a_K	dimension of schlieren beam normal to knife-edge at source, at knife-edge, and above knife-edge when no disturbance is present, respectively
A	amplitude of light beam
A_T	amplitude of recombined light beams from paths 1 and 2
A_0	maximum of amplitude
A_1	amplitude of light beam from path 1
A_2	amplitude of light beam from path 2
A_{01}	maximum of amplitude from path 1
A_{02}	maximum of amplitude from path 2
b_0, b_s	dimension of schlieren beam parallel to knife-edge at source and knife-edge, respectively
c	speed of light
c_0	speed of light in a vacuum
C	Gladstone-Dale constant
d	fringe spacing
f	focal length of lens or mirror, index-of-refraction field
i	integer
I	light intensity
I_d	illumination at screen of disturbed field when knife-edge is present

I_K	illumination at screen of undisturbed field when knife-edge is present
I_T	illumination at exit of test section (and at screen if no deflection) in shadowgraph system
I_0	illumination at screen with no knife-edge
j	integer
k	ratio of constant-pressure to constant-volume specific heats
K	knife-edge
l_x	length of reconstruction elements in x direction
l_y	length of reconstruction elements in y direction
L	length of test section in light beam direction z
m	integer in x direction on reconstruction plane
M	ratio of mass velocity of injected flow to mainstream flow; number of reconstruction elements in x direction
M_1, M_2	mirror
n	index of refraction; integer in y direction on reconstruction plane
n_a	index of refraction of air outside test section
n_{ref}	index of refraction in reference region
n_y	index of refraction at x, y
$n_{y+\Delta y'}$	index of refraction at $x, y + \Delta y'$
n_0	index of refraction at standard conditions
N	number of reconstruction elements in y direction
p	object distance from lens or mirror
P	pressure
P_0	pressure at standard conditions
$\overline{PL}$	optical path length
$\overline{\Delta PL}$	difference in optical path length
q	image distance from lens or mirror
R	gas constant in terms of mass
s	distance along light beam
S	source
Sl	length of beam passing through field to be reconstructed in θ_i direction
SP	splitter plate
T	temperature
T_{ref}	temperature in reference region
W	distances defined in Fig. 7.19; weighting function in grid reconstruction technique
x	direction normal to y and z
y	direction perpendicular to z, usually the direction in which the gradient of density or temperature lies
y_{sc}	height of light ray at screen
z	direction along light beam
z_k	direction perpendicular to reconstruction plane

z_{sc}	distance from test section to screen
α	angular deflection of light ray as measured in air outside test section; same as α' if $n \cong n_a$
$\alpha_{max,neg}$	maximum deflection angle toward knife-edge that can be measured with schlieren system
α_{max}	maximum deflection angle away from knife-edge that can be measured with schlieren system
α'	angular deflections of light ray within test fluid
α''	angle defined in Fig. 7.4
β	angle defined in Fig. 7.4
γ	angle defined in Fig. 7.4
δ	characteristic length in shearing interferometers; displacement of beams $\Delta y'$
Δ	phase difference; change of
$\Delta \alpha$	deflection of light beam away from schlieren knife-edge
Δy	width of light beam
$\Delta y'$	displacement of beams in shearing interferometer δ
Δz	distance that the light beam travels during time $\Delta \tau$
$\Delta \tau$	time interval
ε	interferometer fringe shift; optical path-length difference in vacuum wavelengths
θ	angle between interferometer beams when recombined
θ_l	angle of beam through field to be reconstructed
λ	wavelength of light
λ_0	vacuum wavelength
ρ	density, perpendicular distance to light beam from origin in reconstruction plane
ρ_{ref}	density in reference region
ρ_0	density at standard conditions
τ	time
ϕ_l	optical path-length function generated from reconstruction plane in θ_i direction

REFERENCES

1. F. J. Weinburg, *Optics of Flames,* Butterworth, London, 1963.
2. R. W. Ladenburg, B. Lewis, R. N. Pease, and H. S. Taylor, *Physical Measurements in Gas Dynamics and Combustion,* chaps. A1–A3, Princeton University Press, Princeton, N.J., 1954.
3. D. W. Holder, R. J. North, and G. P. Wood, Optical Methods for Examining the Flow in High-Speed Wind Tunnels, pts. I and II, AGARD, 1965. (Also, Schlieren Methods, Notes in Applied Science 31, National Physics Laboratory, London, 1963.)
4. N. F. Barnes, Optical Techniques for Fluid Flows, *J. Soc. Motion Pict. Telev. Eng.,* vol. 61, pp. 487–511, 1953.

5. R. C. Dean Jr., *Aerodynamic Measurements,* MIT Gas Turbine Laboratory, Cambridge, Mass., 1953.
6. H. Shardin, Toepeler's Schlieren Method: Basic Principles for Its Use and Quantitative Evaluation, Navy Translation 156, 1947.
7. W. Hauf and U. Grigull, Optical Methods in Heat Transfer, in J. P. Hartnett and T. F. Irvine Jr. (eds.), *Advances in Heat Transfer,* vol. 6, p. 133, Academic, New York, 1970.
8. R. J. Goldstein, Optical Techniques for Temperature Measurement, in E. R. G. Eckert and R. J. Goldstein (eds.), *Measurements in Heat Transfer,* 2d ed., p. 241, Hemisphere, Washington, D.C., 1976.
9. G. S. Settles, Modern Developments in Flow Visualization, *AIAA J.,* vol. 24, no. 8, pp. 1313–1323, 1986.
10. T. P. Davies, Schlieren Photography: Review, Applications and Bibliography, *Proc. SPIE Int. Soc. Opt. Eng.,* vol. 348, pp. 878–887, 1982.
11. T. Y. Chu and R. J. Goldstein, Turbulent Convection in a Horizontal Layer of Water, *J. Fluid Mech.,* vol. 60, pp. 141–159, 1973.
12. Landolt-Bornstein, *Physikalisch-Chemische Tabellen,* Suppl. 3, p. 1677, 1935.
13. L. Tilton and J. Taylor, Refractive Index and Dispersion of Distilled Water for Visible Radiation at Temperatures 0 to 60°C, *J. Res. Nat. Bur. Stand.,* vol. 20, pp. 419–477, 1938.
14. G. S. Speak and D. J. Walters, Optical Considerations and Limitations of the Schlieren Method, ARC Tech. Rep. 2859, London, 1954.
15. R. J. Goldstein, E. R. G. Eckert, F. K. Tsou, and A. Haji-Sheikh, Film Cooling with Air and Helium Injection Through a Rearward-Facing Slot into a Supersonic Flow, *AIAA J.,* vol. 4, pp. 981–985, 1966.
16. S. N. Heavens, Visualization of the Acoustic Excitation of a Subsonic Jet, *J. Fluid Mech.,* vol. 100, pp. 185–192, 1980.
17. B. Porter, Quantitative Real-Time Schlieren System for Ultrasound Visualization, *Rev. Sci. Instrum.,* vol. 55, no. 2, 1984.
18. B. E. L. Deckker, Boundary Layer on a Shock Tube Wall and at a Leading Edge Using Schlieren, *Proc. 2d Int. Symp. Flow Visualization,* W. Germany, pp. 413–417, 1980.
19. P. L. Lu, R. Landini, and A. Lacson, Application of Stroboscopic Laser Schlieren Photographic Technique to Explosion Events, *Proc. SPIE Int. Soc. Opt. Eng.,* vol. 348, pp. 410–417, 1982.
20. G. Tanda, Natural Convection Heat Transfer from a Staggered Vertical Plate Array, *J. Heat Transfer,* vol. 115, pp. 938–945, 1993.
21. W. L. Howes, Rainbow Schlieren and Its Applications, *Appl. Opt.,* vol. 23, pp. 2449–2460, 1984.
22. G. S. Settles, Color Schlieren Optics: A Review of Techniques and Applications, *Proc. 2d Int. Symp. Flow Visualization,* W. Germany, pp. 749–759, 1980.
23. M. R. Davis, Quantitative Schlieren Measurements in a Supersonic Turbulent Jet, *J. Fluid Mech.,* vol. 51, pp. 435–447, 1972.
24. B. Hannah, Quantitative Schlieren Measurements of Boundary Layer Phenomena, *Proc. 11th Int. Congr. High Speed Photography,* pp. 539–545, 1974.
25. G. E. Roe, An Optical Study of Turbulence, *J. Fluid Mech.,* vol. 43, pp. 607–635, 1970.
26. M. J. Fisher and F. R. Krause, The Crossed-Beam Correlation Technique, *J. Fluid Mech.,* vol. 28, pp. 705–717, 1967.
27. M. Y. Su and F. R. Krause, Optical Crossed-Beam Measurements of Turbulence Intensities in a Subsonic Jet Shear Layer, *AIAA J.,* vol. 9, pp. 2113–2114, 1971.
28. L. N. Wilson and R. J. Damkevala, Statistical Properties of Turbulent Density Fluctuations, *J. Fluid Mech.,* vol. 43, pp. 291–303, 1970.
29. B. H. Funk and K. D. Johnston, Laser Schlieren Crossed-Beam Measurements in a Supersonic Jet Shear Layer, *AIAA J.,* vol. 8, pp. 2074–2075, 1970.
30. G. T. Kaghatgi, Study of Coherent Structures in Axisymmetric Jets Using an Optical Technique, *AIAA J.,* vol. 18, pp. 225–226, 1980.

31. V. Ronchi, Due Nuovi Metodi per lo Studio delle Superficie e dei Sistemi Ottici, *Ann. Regia Sci. Normale Super Pisa,* vol. 15, 1923 (bound 1927).

32. P. F. Darby, The Ronchi Method of Evaluating Schlieren Photographs, *Tech. Conf. Opt. Phenomenon Supersonic Flow,* NAVORD Rep., pp. 74–76, 1946.

33. D. A. Didion and Y. H. Oh, A Quantitative Schlieren-Grid Method for Temperature Measurement in a Free Convection Field, Mechanical Engineering Department Tech. Rep. 1, Catholic University of America, Washington, D.C., 1966.

34. L. A. Watermeier, Self-Illuminated Schlieren System, *Rev. Sci. Instrum.,* vol. 37, pp. 1139–1141, 1966.

35. A. Kantrowitz and R. L. Trimpi, A Sharp Focusing Schlieren System, *J. Aerospace Sci.,* vol. 17, pp. 311–314, 1950.

36. J. H. Hett, A High Speed Stereoscopic Schlieren System, *J. Soc. Motion Pict. Telev. Eng.,* vol. 56, pp. 214–218, 1951.

37. R. D. Buzzard, Description of Three-Dimensional Schlieren System, *Proc. 8th Int. Congr. High Speed Photography,* pp. 335–340, 1968.

38. M. C. Schmidt and G. S. Settles, Alignment and Application of the Conical Shadowgraph Flow Visualization Technique, *Exp. Fluids,* vol. 4, pp. 93–96, 1986.

39. A. Roshko, Structure of Turbulent Shear Flows: A New Look, *AIAA J.,* vol. 14, pp. 1349–1357, 1976.

40. T. J. McDougall, Double-Diffusive Plumes in Unconfined and Confined Environments, *J. Fluid Mech.,* vol. 133, pp. 321–343, 1983.

41. V. M. Domkundwar, V. Sriramulu, and M. C. Gupta, Flow Visualization Studies on Swirling Flames, *Proc. 2d Int. Symp. Flow Visualization,* W. Germany, pp. 51–55, 1980.

42. O. Rodriguez, The Circular Cylinder in Subsonic and Transonic Flow, *AIAA J.,* vol. 22, pp. 1713–1718, 1984.

43. S. P. Parthasarathy, Y. I. Cho, and L. H. Back, Wide-Field Shadowgraphy of Tip Vortices from a Helicopter Rotor, *AIAA J.,* vol. 25, pp. 64–70, 1987.

44. G. S. Settles and H.-Y. Teng, Flow Visualization Methods for Separated Three-Dimensional Shock Wave/Turbulent Boundary-Layer Interactions, *AIAA J.,* vol. 20, pp. 390–397, 1982.

45. L. A. Cross and E. A. Strader, High Repetition Rate, High Resolution Shutterless Photography for Gas Dynamic Studies, *Proc. SPIE Int. Soc. Opt. Eng.,* vol. 348, pp. 380–387, 1982.

46. E. M. Schmidt, R. E. Gordnier, and K. S. Fansler, Interaction of Gun Exhaust Flowfields, *AIAA J.,* vol. 22, pp. 516–517, 1983.

47. E. M. Schmidt and D. D. Shear, Optical Measurements of Muzzle Blast, *AIAA J.,* vol. 13, pp. 1086–1091, 1975.

48. F. W. Roos and T. J. Bogar, Comparison of Hot-Film Probe and Optical Techniques for Sensing Shock Motion, *AIAA J.,* vol. 20, pp. 1071–1076, 1982.

49. C. S. Magirl and F. P. Incropera, Flow and Morphological Conditions Associated with Unidirectional Solidification of Aqueous Ammonium Chloride, *J. Heat Transfer,* vol. 115, pp. 1036–1043, 1993.

50. E. R. G. Eckert, R. M. Drake Jr. and E. Soehngen, Manufacture of a Zehnder-Mach Interferometer, Wright-Patterson AFB, Tech. Rep. 5721, ATI-34235, 1948.

51. F. D. Bennett and G. D. Kahl, A Generalized Vector Theory of the Mach-Zehnder Interferometer, *J. Opt. Soc. Am.,* vol. 43, pp. 71–78, 1953.

52. T. Zobel, The Development and Construction of an Interferometer for Optical Measurement of Density Fields, NACA Tech. Note 1184, 1947.

53. H. Shardin, Theorie und Anwendung des Mach-Zehnderschen Interferenz-Refraktometers, *Z. Instrumentenkd.,* vol. 53, pp. 396, 424, 1933. (DRL Trans. 3, University of Texas.)

54. L. H. Tanner, The Optics of the Mach-Zehnder Interferometer, ARC Tech. Rep. 3069, London, 1959.

55. L. H. Tanner, The Design and Use of Interferometers in Aerodynamics, ARC Tech. Rep. 3131, London, 1957.

56. D. Wilkie and S. A. Fisher, Measurement of Temperature by Mach-Zehnder Interferometry, *Proc. Inst. Mech. Eng.,* vol. 178, pp. 461–470, 1963.
57. W. L. Howes, Rainbow Schlieren vs. Mach-Zehnder Interferometer: A Comparison, *Appl. Opt.,* vol. 24, pp. 816–822, 1985.
58. L. T. Clark, D. C. Koepp, and J. J. Thykkuttathil, A Three-Dimensional Density Field Measurement of Transonic Flow from a Square Nozzle Using Holographic Interferometry, *J. Fluid Eng.,* vol. 99, pp. 737–744, 1977.
59. G. Ben-Dor, B. T. Whitten, and I. I. Glass, Evaluation of Perfect and Imperfect Gas Interferograms by Computer, *Int. J. Heat Fluid Flow,* vol. 1, pp. 77–91, 1979.
60. D. J. Bone, H.-A. Bachor, and R. J. Sandeman, Fringe-Pattern Analysis Using a 2-D Fourier Transform, *Appl. Opt.,* vol. 25, pp. 1653–1660, 1986.
61. C. Roddier and F. Roddier, Interferogram Analysis Using Fourier Transform Techniques, *Appl. Opt.,* vol. 26, pp. 1668–1673, 1987.
62. J. J. Snyder, Algorithm for Fast Digital Analysis of Interference Fringes, *Appl. Opt.,* vol. 19, pp. 1223–1225, 1980.
63. W. J. McKeen and J. D. Tarasuk, Accurate Method for Locating Fringes on an Interferogram, *Rev. Sci. Instrum.,* vol. 52, pp. 1223–1225, 1981.
64. A. B. Witte, J. Fox, and H. Rungaldier, Localized Measurements of Wake Density Fluctuations Using Pulsed Laser Holographic Interferometry, *AIAA J.,* vol. 10, pp. 481–487, 1972.
65. A. B. Witte and R. F. Wuerker, Laser Holographic Interferometry Study of High-Speed Flow Fields, *AIAA J.,* vol. 8, pp. 581–583, 1970.
66. A. G. Havener and R. J. Radley Jr., Supersonic Wind Tunnel Investigations Using Pulsed Laser Holography, ARL 73-0148, 1973.
67. A. G. Havener, Detection of Boundary-Layer Transition Using Holography, *AIAA J.,* vol. 15, pp. 592–593, 1977.
68. R. A. Showole and J. D. Tarasuk, Experimental and Numerical Studies of Natural Convection with Flow Separation in Upward-Facing Inclined Open Cavities, *J. Heat Transfer,* vol. 115, pp. 592–605, 1993.
69. F. J. Hamady, J. R. Lloyd, K. T. Yang, and H. Q. Yang, A Study of Natural Convection in a Rotating Enclosure, *J. Heat Transfer,* vol. 116, pp. 136–143, 1993.
70. D. B. Prowse, Astigmatism in the Mach-Zehnder Interferometer, *Appl. Opt.,* vol. 6, p. 773, 1967.
71. E. W. Price, Initial Adjustment of the Mach-Zehnder Interferometer, *Rev. Sci. Instrum.,* vol. 23, p. 162, 1952.
72. D. B. Prowse, A Rapid Method of Aligning the Mach-Zehnder Interferometer, *Aust. Def. Sci. Serv.,* Tech. Note 100, 1967.
73. J. M. Cuadrado, M. V. Perez, and C. Gomez-Reino, Equilateral Hyperbolic Zone Plates: Their Use in the Alignment of a Mach-Zehnder Interferometer, *Appl. Opt.,* vol. 26, pp. 1527–1529, 1987.
74. M. Born and E. Wolf, *Principles of Optics,* 3d ed., p. 319, Pergamon, New York, 1965.
75. E. R. G. Eckert and E. E. Soehngen, Studies on Heat Transfer in Laminar Free Convection with the Mach-Zehnder Interferometer, Air Force Tech. Rep. 5747, ATI-44580, 1948.
76. G. P. Wachtell, Refraction Effect in Interferometry of Boundary Layer of Supersonic Flow Along Flat Rate, *Phys. Rev.,* vol. 78, p. 333, 1950.
77. R. E. Blue, Interferometer Corrections and Measurements of Laminar Boundary Layers in Supersonic Stream, NACA Tech. Note 2110, 1950.
78. E. R. G. Eckert and E. E. Soehngen, Distribution of Heat Transfer Coefficients Around Circular Cylinders in Crossflow at Reynolds Numbers from 20 to 500, *Trans. ASME,* vol. 74, pp. 343–347, 1952.
79. M. R. Kinsler, Influence of Refraction on the Applicability of the Zehnder-Mach Interferometer to Studies of Cooled Boundary Layers, NACA Tech. Note 2462, 1951.
80. W. L. Howes and D. R. Buchele, A Theory and Method for Applying Interferometry to the Measurement of Certain Two-Dimensional Gaseous Density Fields, NACA Tech. Note 2693, 1952.

81. W. L. Howes and D. R. Buchele, Generalization of Gas-Flow Interferometry Theory and Interferogram Evaluation Equations for One-Dimensional Density Fields, NACA Tech. Note 3340, 1955.

82. W. L. Howes and D. R. Buchele, Practical Considerations in Specific Applications of Gas-Flow Interferometry, NACA Tech. Note 3507, 1955.

83. W. L. Howes and D. R. Buchele, Optical Interferometry of Inhomogeneous Gases, *J. Opt. Soc. Am.*, vol. 56, pp. 1517–1528, 1966.

84. J. M. Mehta and W. M. Worek, Analysis of Refraction Errors for Interferometric Measurements in Multicomponent Systems, *Appl. Opt.*, vol. 23, pp. 928–933, 1984.

85. R. D. Flack Jr., Mach-Zehnder Interferometer Errors Resulting from Test Section Misalignment, *Appl. Opt.*, vol. 17, pp. 985–987, 1978.

86. R. Kraushaar, A Diffraction Grating Interferometer, *J. Opt. Soc. Am.*, vol. 40, pp. 480–481, 1950.

87. J. R. Sterrett and J. R. Erwin, Investigation of a Diffraction Grating Interferometer for Use in Aerodynamic Research, NACA Tech. Note 2827, 1952.

88. F. J. Weinberg and N. B. Wood, Interferometer Based on Four Diffraction Gratings, *J. Sci. Instrum.*, vol. 36, pp. 227–230, 1959.

89. J. R. Sterrett, J. C. Emery, and J. B. Barber, A Laser Grating Interferometer, *AIAA J.*, vol. 3, pp. 963–964, 1965.

90. R. J. Goldstein, Interferometer for Aerodynamic and Heat Transfer Measurements, *Rev. Sci. Instrum.*, vol. 36, pp. 1408–1410, 1965.

91. A. K. Oppenheim, P. A. Urtiew, and F. J. Weinberg, On the Use of Laser Light Sources in Schlieren-Interferometer Systems, *Proc. R. Soc. London Ser. A*, vol. 291, pp. 279–290, 1966.

92. L. H. Tanner, The Design of Laser Interferometers for Use in Fluid Mechanics, *J. Sci. Instrum.*, vol. 43, pp. 878–886, 1966.

93. G. E. A. Meier and M. Willms, On the Use of Fabry-Perot-Interferometer for Flow Visualization and Flow Measurement, *Proc. 2d Int. Symp. Flow Visualization*, W. Germany, pp. 725–739, 1980.

94. E. L. Gayhart and R. Prescott, Interference Phenomenon in the Schlieren System, *J. Opt. Soc. Am.*, vol. 39, pp. 546–550, 1949.

95. E. B. Temple, Quantitative Measurement of Gas Density by Means of Light Interference in Schlieren System, *J. Opt. Soc. Am.*, vol. 47, pp. 91–100, 1957.

96. J. B. Brackenridge and W. P. Gilbert, Schlieren Interferometry: An Optical Method for Determining Temperature and Velocity Distributions in Liquids, *Appl. Opt.*, vol. 4, pp. 819–821, 1965.

97. R. C. Anderson and M. W. Taylor, Phase Contrast Flow Visualization, *Appl. Opt.*, vol. 21, pp. 528–536, 1982.

98. C. J. Wick and S. Winnikow, Reflection Plate Interferometer, *Appl. Opt.*, vol. 12, pp. 841–844, 1973.

99. R. D. Small, V. A. Sernas, and R. H. Page, Single Beam Schlieren Interferometer Using a Wollaston Prism, *Appl. Opt.*, vol. 11, pp. 858–862, 1972.

100. W. Merzkirch, Generalized Analysis of Shearing Interferometers as Applied to Gas Dynamic Studies, *Appl. Opt.*, vol. 13, pp. 409–413, 1974.

101. G. Smeets, Observational Techniques Related to Differential Interferometry, *Proc. 11th Int. Congr. High Speed Photography*, pp. 283–288, 1974.

102. S. Yokozeki and T. Suzuki, Shearing Interferometer Using the Grating as the Beam Splitter, *Appl. Opt.*, vol. 10, pp. 1575–1580, 1971.

103. J. Stricker and O. Kafri, A New Method for Density Gradient Measurements in Compressible Flows, *AIAA J.*, vol. 20, pp. 820–823, 1982.

104. J. Stricker, Axisymmetric Density Field Measurements by Moire Deflectometry, *AIAA J.*, pp. 1767–1769, 1983.

105. O. Bryngdahl, Applications of Shearing Interferometry, in E. Wold (ed.), *Progress in Optics*, vol. 4, Wiley, New York, 1965.

106. R. Chevalerias, Y. Latron, and C. Veret, Methods of Interferometry Applied to the Visualization of Flows in Wind Tunnels, *J. Opt. Soc. Am.*, vol. 47, pp. 703–706, 1957.

107. W. Z. Black and W. W. Carr, Application of a Differential Interferometer to the Measurement of Heat Transfer Coefficients, *Rev. Sci. Instrum.*, vol. 42, pp. 337–340, 1971.
108. W. Z. Black and J. K. Norris, Interferometric Measurement of Fully Turbulent Free Convective Heat Transfer Coefficients, *Rev. Sci. Instrum.*, vol. 45, pp. 216–218, 1974.
109. R. D. Flack Jr., Shearing Interferometer Inaccuracies Due to a Misaligned Test Section, *Appl. Opt.*, vol. 17, pp. 2873–2875, 1978.
110. M. H. Horman, An Application of Wavefront Reconstruction to Interferometry, *Appl. Opt.*, vol. 4, pp. 333–336, 1965.
111. L. O. Heflinger, R. F. Wuerker, and R. E. Brooks, Holographic Interferometry, *J. Appl. Phys.*, vol. 37, pp. 642–649, 1966.
112. L. H. Tanner, Some Applications of Holography in Fluid Mechanics, *J. Sci. Instrum.*, vol. 43, pp. 81–83, 1966.
113. O. Bryngdahl, Shearing Interferometry by Wavefront Reconstruction, *J. Opt. Soc. Am.*, vol. 58, pp. 865–871, 1968.
114. F. P. Kupper and C. A. Dijk, A Method for Measuring the Spatial Dependence of the Index of Refraction with Double Exposure Holograms, *Rev. Sci. Instrum.*, vol. 43, pp. 1492–1497, 1972.
115. Franz Mayinger and Walter Panknin, Holography in Heat and Mass Transfer, *Heat Transfer 1974 (Proc. 5th Int. Heat Trans. Conf.)*, vol. VI, pp. 28–43, Japan Society of Mechanical Engineers/ Society of Chemical Engineers, Tokyo, 1974.
116. C. M. Veret, Applications of Flow Visualization Techniques in Aerodynamics, *Proc. SPIE Int. Soc. Opt. Eng.*, vol. 348, pp. 114–119, 1982.
117. J. Surget, Holographic Interferometer for Aerodynamic Flow Analysis, *Proc. 2d Int. Symp. Flow Visualization*, W. Germany, pp. 743–747, 1980.
118. G. Lee, D. A. Buell, and J. P. Licursi, Laser Holographic Interferometry for an Unsteady Airfoil Undergoing Dynamic Stall, *AIAA J.*, vol. 22, pp. 504–511, 1983.
119. W. D. Bachalo and M. J. Houser, Optical Interferometry in Fluid Dynamics Research, NACA Tech. Note 2693, 1952.
120. J.-J. Hwang and T.-M. Liou, Augmented Heat Transfer in a Rectangular Channel with Permeable Ribs Mounted on the Wall, *J. Heat Transfer*, vol. 116, pp. 912–920, 1993.
121. T.-M. Liou, W.-B. Wang, and Y.-J. Chang, Holographic Interferometry Study of Spatially Periodic Heat Transfer in a Channel with Ribs Detached from One Wall, *J. Heat Transfer*, vol. 117, pp. 32–39, 1993.
122. W. Aung and R. O'Regan, Precise Measurement of Heat Transfer Using Holographic Interferometry, *Rev. Sci. Instrum.*, vol. 42, pp. 1755–1759, 1971.
123. J. D. Trolinger, Flow Visualization Holography, *Opt. Eng.*, vol. 14, pp. 470–481, 1985.
124. J. D. Trolinger, Application of Generalized Phase Control During Reconstruction to Flow Visualization Holography, *Appl. Opt.*, vol. 18, pp. 766–774, 1979.
125. D. Mewes and W. Ostendorf, Application of Tomographic Measurement Techniques for Process Engineering Studies, *Int. Chem. Eng.*, vol. 26, pp. 11–21, 1986.
126. R. N. Bracewell and A. C. Riddle, Inversion of Fan-Beam Scans in Radio Astronomy, *Astrophys. J.*, vol. 150, pp. 427–434, 1967.
127. P. D. Rowley, Quantitative Interpretation of Three-Dimensional Weakly Refractive Phase Objects Using Holographic Interferometry, *J. Opt. Soc. Am.*, vol. 59, pp. 1496–1498, 1969.
128. R. D. Matulka and D. J. Collins, Determination of Three-Dimensional Density Fields from Holographic Interferograms, *J. Appl. Phys.*, vol. 42, pp. 1109–1119, 1971.
129. D. W. Sweeney and C. M. Vest, Reconstruction of Three-Dimensional Refractive Index Fields from Multidirectional Interferometric Data, *Appl. Opt.*, vol. 12, pp. 2649–2664, 1973.
130. A. K. Tolpadi, *Coupled Three-Dimensional Conduction and Natural Convection Heat Transfer*, Ph.D. thesis, University of Minnesota, 1987.
131. R. Gordon, R. Bender, and G. T. Herman, Algebraic Reconstruction Techniques (ART) for Three-Dimensional Electron Microscopy and X-Ray Photography, *J. Theor. Biol.*, vol. 29, pp. 471–481, 1970.

132. G. N. Vishnyakov, G. G. Levin, B. M. Stepanov, and V. N. Filinov, Optical Tomography at a Limited Number of Projections, *Proc. SPIE Int. Soc. Opt. Eng.,* vol. 348, pp. 596–600, 1982.
133. R. A. Kosakoski and D. J. Collins, Application of Holographic Interferometry to Density Field Determination in Transonic Corner Flow, *AIAA J.,* vol. 12, pp. 767–770, 1974.
134. H. H. Oertel Jr., Three-Dimensional Convection Within Rectangular Boxes, in *Natural Convection in Enclosures,* ASME Publ. HTD.-vol. 8, pp. 11–16, 1980.
135. R. Snyder and L. Hesselink, Optical Tomography for Flow Visualization of the Density Field Around a Revolving Helicopter Rotor Blade, *Appl. Opt.,* vol. 23, pp. 3650–3656, 1984.
136. D. W. Sweeney and C. M. Vest, Measurement of Three-Dimensional Temperature Fields Above Heated Surfaces by Holographic Interferometry, *Int. J. Heat Mass Transfer,* vol. 17, pp. 1443–1453, 1974.
137. H. H. Chau and O. S. F. Zucker, Holographic Thin-Beam Reconstruction Technique for the Study of 3-D Refractive-Index Field, *Opt. Commun.,* vol. 8, pp. 336–339, 1973.
138. C. M. Vest and P. T. Radulovic, Measurement of Three-Dimensional Temperature Fields by Holographic Interferometry, in *Applications of Holography and Data Processing,* pp. 241–249, Pergamon, London, 1977.
139. D. Lubbe, *Ein Messverfahren fur Instationare, Dreidimensionale Verteilungen und Seine Anwendung auf Mischvorgange,* Universitat Hannover, 1952.
140. F. Mayinger and D. Lubbe, Ein Tomographisches mebverfahren and Seine Anwendung auf Mischvorgange und Stoffaustausch, *Warme Stoffubertragung,* vol. 18, pp. 49–59, 1984.
141. S. Cha and C. M. Vest, Tomographic Reconstruction of Strongly Refracting Fields and Its Application to Interferometric Measurement of Boundary Layers, *Appl. Opt.,* vol. 20, pp. 2787–2794, 1981.
142. C. M. Vest, Tomography for Properties of Materials that Bend Rays: A Tutorial, *Appl. Opt.,* vol. 24, pp. 4089–4094, 1985.
143. T. F. Zien, W. C. Ragsdale, and W. C. Spring III, Quantitive Determination of Three-Dimensional Density Field by Holographic Interferometry, *AIAA J.,* vol. 13, pp. 841–842, 1975.
144. B. P. Medoff, W. R. Brody, M. Nassi, and A. Macovski, Iterative Convolution Backprojection Algorithms for Image Reconstruction from Limited Data, *J. Opt. Soc. Am.,* vol. 73, pp. 1493–1500, 1983.
145. C. M. Vest and I. Prikryl, Tomography by Iterative Convolution: Empirical Study and Application to Interferometry, *Appl. Opt.,* vol. 23, pp. 2433–2440, 1984.
146. G. Schwarz and H. Knauss, Quantitative Experimental Investigation of Three-Dimensional Flow Fields Around Bodies of Arbitrary Shapes in Supersonic Flow with Optical Methods, *Proc. 2d Int. Symp. Flow Visualization,* W. Germany, pp. 737–741, 1980.
147. F. Mayinger, Image-Forming Optical Techniques in Heat Transfer: Revival by Computer-Aided Data Processing, *J. Heat Transfer,* vol. 115, pp. 824–834, 1993.

FLUID MECHANICS MEASUREMENTS IN NON-NEWTONIAN FLUIDS

Christopher W. Macosko and Paulo R. Souza Mendes

8.1 INTRODUCTION

Many liquid flow problems of interest today deal with non-Newtonian fluids. The main types of non-Newtonian liquids are polymer solutions, molten polymers, and concentrated dispersions such as slurries, food products, pastes, sealants, inks, and paints.

The departures of non-Newtonian fluids from typical Newtonian behavior can be severe. The viscosity of a typical non-Newtonian fluid is strongly shear thinning. If viscosity data of low shear rate are used for pump selection and pressure calculations, considerable error can result. Non-Newtonian fluids can even behave opposite to Newtonian fluids in certain flows. In the presence of inertia ($Re_D > 16$), Newtonian jets contract after emerging from a tube, while polymer solutions can expand. Both droplet breakup and the onset of turbulence can be suppressed by the addition of very small amounts of high-molecular-weight polymer. Lubrication, flow instabilities, mixing, and tank draining become considerably more complicated with non-Newtonian fluids. Some of these and other examples are reviewed by Tanner [1], Bird et al. [2], and Middleman [3].

Why do non-Newtonian fluids present such a problem for fluid mechanical analysis? For incompressible Newtonian liquids there is just one constant, the viscosity μ, which relates stress to rate of deformation:

$$\mathbf{T} = -p\mathbf{1} + 2\mu\mathbf{D} \qquad (1)$$

where $\mathbf{T}$ is the total-stress tensor, p is the isotropic pressure, and $\mathbf{D}$ is the rate-of-deformation tensor. We will assume in this chapter that $\mathbf{T}$ is always symmetric. This is a good assumption for amorphous materials, including most liquids of engineering interest. The discussion will also be restricted to isotropic materials.

Since the Newtonian viscosity is a function only of temperature (and a weak function of pressure), generally if we measure the velocity distribution, we can determine the stresses. However, for non-Newtonian fluids, stress can have a complex dependence on the strain history:

$$\mathbf{T} = -p\mathbf{1} + \underset{-\infty}{\overset{t}{\mathcal{F}}}[\mathbf{C}_t(t')] \tag{2}$$

where $\mathcal{F}$ is a tensor-valued functional that describes how the stress relates to the flow kinematics and $\mathbf{C}_t$ is defined below. This relation, the rheological constitutive equation for the fluid, is not easy to obtain, and even if it is known, it can be extremely difficult to apply to anything but the simplest flows. This means that usually both stress and velocity measurements are needed to understand a non-Newtonian flow problem. The experimental problem is compounded further by the fact that many non-Newtonian materials of interest are hard to work with. They are often very viscous, opaque, require high temperatures, or contain volatile solvents.

The goal of this chapter is to show how some of these experimental difficulties can be overcome. We point out the special problems encountered with non-Newtonian fluids in trying to utilize the methods described in the previous chapters. We also describe some flow measurement methods that are unique to non-Newtonian materials, particularly flow birefringence.

Before describing the measurement methods, however, we summarize some results from the field of rheology. We first look at some of the basic types of flow behavior we can expect from non-Newtonian fluids. Then we examine some of the more useful constitutive relations and discuss how these relations can be verified. The reader familiar with rheology or more interested in problems with measurements in complex flows can skip ahead to section 8.5.

8.2 MATERIAL FUNCTIONS

A Newtonian fluid requires only a single material constant to relate stress to deformation. For a non-Newtonian fluid one or more material functions are required. To determine these material functions can be an impossible experimental problem. Fortunately, a limited number of material functions are required to completely characterize a given simple flow. To help determine which rheological measurements should be made, it is valuable to see which functions can possibly arise and what they look like for typical materials. This is an important aspect of rheology. Here we are only able to summarize the main points without proof. Most advanced rheology texts [2, 4–7] discuss this problem in more detail.

8.2.1 Some Mechanics Concepts

To assist in the following discussion, we now introduce a few important concepts. Let us consider a flow described by the velocity field $\mathbf{v}(\mathbf{x}, t)$. At some fixed instant of time t, the velocity gradient tensor $\mathbf{L}$ is defined by the relation $d\mathbf{v} = \mathbf{L}d\mathbf{x}$. Here, $d\mathbf{v}$ is the difference between the velocities of the particles that occupy $\mathbf{x} + d\mathbf{x}$ and $\mathbf{x}$, at time t. The ijth component of $\mathbf{L}$ is therefore $L_{ij} = \partial v_i/\partial x_j$, so that $dv_i = (\partial v_i/\partial x_j)dx_j$ (repeated indices imply summation from 1 to 3).

The rate-of-deformation tensor $\mathbf{D}$ is the symmetric part of $\mathbf{L}$, while the antisymmetric part is the vorticity tensor $\mathbf{W}$:

$$\mathbf{D} \equiv \frac{1}{2}(\mathbf{L} + \mathbf{L}^\dagger) \qquad \mathbf{W} \equiv \frac{1}{2}(\mathbf{L} - \mathbf{L}^\dagger) \tag{3a}$$

or

$$D_{ij} \equiv \frac{1}{2}(L_{ij} + L_{ji}) \qquad W_{ij} \equiv \frac{1}{2}(L_{ij} - L_{ji}) \tag{3b}$$

The principal values of $\mathbf{D}$ give the rates of stretch of material filaments that are instantaneously aligned with the principal directions. The vorticity tensor is related to the angular velocity of the fluid particles.

The rate of deformation $\mathbf{D}$ is the only kinematic field needed to determine the stress field in a Newtonian fluid flow. In flows of (non-Newtonian) materials with memory, the present state of stress depends on the material's past history of deformation. To deal with this dependence, we define the deformation gradient $\mathbf{F}_t(t')$ by the relation $d\mathbf{x}' = \mathbf{F}_t(t')d\mathbf{x}$, where $d\mathbf{x}$ is an infinitesimal material filament at the present time t, while $d\mathbf{x}'$ is the same filament at a previous time t'. Clearly, $dx'_i = (\partial x'_i/\partial x_j)\,dx_j$, so that $F_{ij} \equiv \partial x'_i/\partial x_j$.

The notation convention we have adopted here for deformation is such that the subscript indicates the reference instant of time of the unstrained material configuration. Thus, in the above discussion we implicitly chose the present instant of time t as the no-strain reference time, so that $\mathbf{F}_t(t) = \mathbf{1}$. For the sake of simplicity, wherever there are no grounds for confusion, we will drop both the subscript and the argument of $\mathbf{F}$. We will also adopt the same procedure with other strain-related quantities to be introduced later in this text.

A comparison between the magnitudes of $d\mathbf{x}'$ and $d\mathbf{x}$ is a measure of the strain of the filament during the time interval $(t - t')$. One way to evaluate the strain is through the ratio

$$\frac{d\mathbf{x}' \cdot d\mathbf{x}'}{d\mathbf{x} \cdot d\mathbf{x}} = \frac{\mathbf{F}d\mathbf{x} \cdot \mathbf{F}d\mathbf{x}}{|d\mathbf{x}|^2} = \hat{e} \cdot \mathbf{F}^\dagger\mathbf{F}\hat{e} \tag{4}$$

where $\hat{e} \equiv d\mathbf{x}/|d\mathbf{x}|$ is a unit vector aligned with the filament at the present time t. The tensor quantity that naturally arises in the above ratio, $\mathbf{C}_t(t') \equiv \mathbf{F}_t^\dagger(t')\mathbf{F}_t(t')$, is the right Cauchy-Green strain tensor, or just the strain tensor. The Finger tensor is just the inverse of $\mathbf{C}$, that is, $\mathbf{B} \equiv \mathbf{C}^{-1} = \mathbf{F}^{-1}(\mathbf{F}^\dagger)^{-1}$. Note that these two strain tensors are both symmetric, and according to their definitions, it is clear that $\mathbf{C}_t(t) = \mathbf{B}_t(t) = 1$.

It is worth noting that the choice of the present time as the reference (zero strain) time is arbitrary, but it is the usual choice for motions of fluids when there is no clear or known rest state. When there is a well-defined rest configuration, it is often chosen as the reference configuration, as we will do in section 8.2.3 (see also chap. 1 of [7]).

The rate-of-strain tensor $\mathbf{D}$ is related to the strain tensor $\mathbf{C}$ as follows (d/dt is the material time derivative):

$$\mathbf{D} = \lim_{t' \to t} \frac{d\mathbf{C}}{dt'} \tag{5}$$

8.2.2 Steady Shear Flows

A quantity that frequently arises in rheology is the deformation rate, defined as $\dot{\gamma} \equiv \sqrt{2\,\mathrm{tr}\,\mathbf{D}^2}$. If $\dot{\gamma}$ (or $\mathbf{D}$) is not a function of position, the flow is said to be homogeneous. For homogeneous simple shear the only nonzero velocity component is in the flow direction, and it is only a function of the cross direction $v_1 = f(x_2)$. For steady simple shear,

$$v_1 = \dot{\gamma} x_2 \qquad v_2 = 0 \qquad v_3 = 0 \tag{6}$$

where $\dot{\gamma}$ is a constant, the shear rate. For such a simple flow, the rate-of-deformation tensor reduces to

$$2\mathbf{D} = \begin{bmatrix} 0 & \dot{\gamma} & 0 \\ \dot{\gamma} & 0 & 0 \\ 0 & 0 & 0 \end{bmatrix} \tag{7}$$

Note that the previously defined deformation rate reduces to the constant $\dot{\gamma}$ in Eq. (6), and that is why we used the same symbol to denote both quantities. Because of the symmetry of a shear deformation, the stress tensor can only have four different components,

$$\mathbf{T} = \begin{bmatrix} T_{11} & T_{12} & 0 \\ T_{12} & T_{22} & 0 \\ 0 & 0 & T_{33} \end{bmatrix} \tag{8}$$

For an incompressible material, we can determine the normal stresses only to within an arbitrary pressure p. Thus the deformation can affect only the normal stress differences. It follows then that there are only three independent stress quantities in shear:

$$T_{12}, \ T_{11} - T_{22}, \ T_{22} - T_{33} \tag{9}$$

These stresses will be, in general, a function of the history of the shear deformation. The simplest history is that of steady shearing. In this case the three stress quantities can only be a function of the shear rate, $2D_{12}$ or $\dot{\gamma}$, and are independent of time:

$$T_{12} = \tau(\dot{\gamma}) \tag{10}$$

$$T_{11} - T_{22} = N_1(\dot{\gamma}) \tag{11}$$

$$T_{22} - T_{33} = N_2(\dot{\gamma}) \tag{12}$$

Since in the limit of low shear rates these stresses go with $\dot{\gamma}$ and $\dot{\gamma}^2$ it is most common to define viscosity and normal-stress coefficients, as follows.

Viscosity

$$\eta(\dot{\gamma}) = \frac{T_{12}}{\dot{\gamma}} \tag{13}$$

First normal-stress coefficient

$$\Psi_1(\dot{\gamma}) = \frac{T_{11} - T_{22}}{\dot{\gamma}^2} \tag{14}$$

Second normal-stress coefficient

$$\Psi_2(\dot{\gamma}) = \frac{T_{22} - T_{33}}{\dot{\gamma}^2} \tag{15}$$

These three functions are called the viscometric material functions. They have been fairly well studied for polymeric fluids. Figure 8.1 illustrates all three functions for (1) a polyisobutylene solution in oil and (2) polyethylene at typical extrusion or molding temperature. Note that for both materials the second normal-stress coefficient is negative and considerably smaller than the first. Also note that at low shear rates, particularly for the polymer solution, all three material functions approach constant limiting values:

$$\lim_{\dot{\gamma} \to 0} \eta(\dot{\gamma}) = \eta_0 \tag{16}$$

$$\lim_{\dot{\gamma} \to 0} \Psi_1(\dot{\gamma}) = \Psi_{10} \tag{17}$$

$$\lim_{\dot{\gamma} \to 0} \Psi_2(\dot{\gamma}) = \Psi_{20} \tag{18}$$

These limiting values are often used in phenomenological and molecular constitutive equations. The trends in Fig. 8.1 seem to be generally true for polymeric liquids. For dispersions, very few normal-stress data are available, although they are predicted theoretically [8]. For dilute suspensions the viscosity function does look

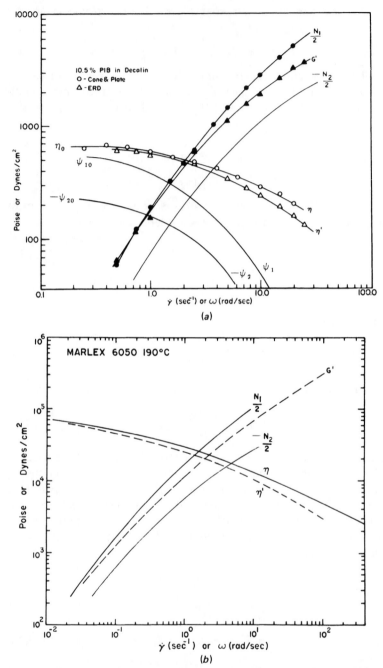

Figure 8.1 Examples of the three steady-shear (viscometric) material functions: (a) 10.5% polyisobutylene in decalin and (b) a high-density polyethylene melt, Marlex 6050, made by the Phillips Petroleum Co. (From [9].) Also shown are the two linear viscoelastic material functions G' and η', Eqs. (26) and (29)

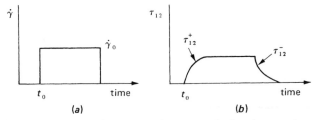

Figure 8.2   Start-up for constant shear rate and relaxation experiment. (*a*) Step increase in shear rate at time t_0. (*b*) Shear-stress response. For a Newtonian fluid the stress would also be a step function

like Fig. 8.1, but concentrated dispersions often do not show a constant limiting viscosity. Instead, viscosity seems to increase continuously at low rates, indicating a yield stress. This "plastic" behavior is discussed further in section 8.3.2. Some concentrated suspensions (corn starch is one example) have a shear-thickening viscosity.

8.2.3 Transient Shear

It is often in transient response that non-Newtonian fluids behave so differently from Newtonian fluids. There are an infinite number of possible time histories of deformation; however, a few types that occur frequently in process flow problems can be generated fairly easily in various rheometers.

The flows discussed in this section are homogeneous, although $\dot{\gamma}$ is in general a function of time. Moreover, a well-defined rest configuration is available, and therefore here it is convenient to choose this rest state as the unstrained configuration. Accordingly, we define the total strain, $\gamma_{t_0}(t)$, between times t_0 (when the motion starts) and t as $\gamma_{t_0}(t) \equiv \int_{t_0}^{t} \dot{\gamma}(t')dt'$. For other flows in which the present configuration is chosen to be the undeformed one, the total strain is $\gamma_t(t') \equiv \int_{t}^{t'} \dot{\gamma}(t'')dt''$.

One of the most important responses is start-up for steady shear flow. Shear rate is set essentially instantaneously at a fixed value, and the shear and normal stresses are recorded as they come to equilibrium. This is illustrated in Fig. 8.2. In Fig. 8.3 the data have been normalized by their equilibrium viscosity values. Note the large overshoot that occurs on start-up and the faster relaxation at high shear rates. This is characteristic of most polymeric materials. We can define start-up and relaxation shear-stress material functions η^+ and η^- as

$$\eta^+(t,\dot{\gamma}_0) = \frac{\tau_{12}^+(t,\dot{\gamma}_0)}{\dot{\gamma}_0} \tag{19}$$

$$\eta^-(t,\dot{\gamma}_0) = \frac{\tau_{12}^-(t,\dot{\gamma}_0)}{\dot{\gamma}_0} \tag{20}$$

Similar functions can be defined for the normal stresses.

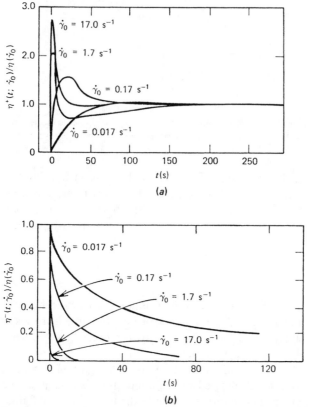

Figure 8.3 Normalized transient viscosity. (*a*) Step start-up. (*b*) Stopping of flow at various constant shear rates $\dot{\gamma}_0$. Data are for a 2.0% polyisobutylene solution. (From [2])

Another important transient test is a step application of strain, $\gamma_{t_0}(t) = \gamma_0 H(t - t_0)$, illustrated in Fig. 8.4. Here $H(t - t_0)$ is the Heaviside unit step function, which is zero for negative arguments and unity otherwise. This shows clearly the viscoelastic character typical of many polymeric solutions. At short times the stress is large, typical of a rubbery solid. At long times the stress relaxes back to zero. For a Newtonian liquid, the stress would be zero at all times, except for a pulse at $t = t_0$. To describe the data, we define a stress-relaxation modulus

$$G(t - t_0, \gamma_0) = \frac{\tau_{12}(t, \gamma_0)}{\gamma_0} \qquad (21)$$

This material function is illustrated in Fig. 8.5 for a 20% polystyrene solution. We note that for small strains, $G(t - t_0)$ is independent of strain, and at large strains it has roughly the same shape but is just shifted vertically. This behavior is typical

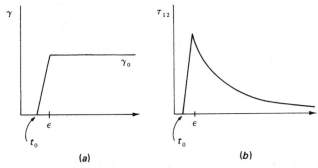

(a)   **(b)**

Figure 8.4 Stress-relaxation experiment. (*a*) It takes a small but finite time ε to accomplish a step strain γ_0. (*b*) Stress response. For a Hookean solid, the stress would remain constant at its maximum value; for a Newtonian liquid, relaxation would be instantaneous

of other polymeric liquids. We find these observations particularly important in development of viscoelastic constitutive relations in the next section.

It is also possible to subject a sample to a step increase in stress. This is known as creep and is frequently used for solids testing. Figure 8.6 shows typical data. Initially, the creep is rapid, but then it slows to a steady viscous flow characterized

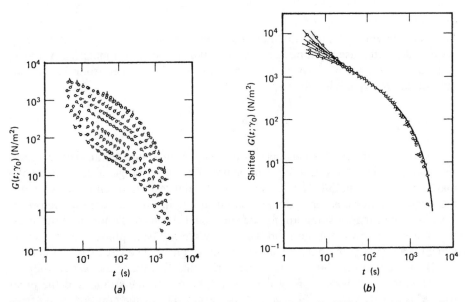

Figure 8.5 Log relaxation modulus $G(t, \gamma_0) = \tau_{12}(t)/\gamma_0$ versus log time for 20% polystyrene in Aroclor. Shear magnitudes range from 0.41 (uppermost curve) to 25.4 (bottom curve). In Fig. 8.5*b* the data of Fig. 8.5*a* are shifted vertically. (From [2])

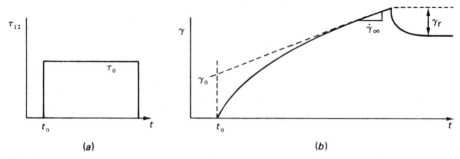

Figure 8.6 Creep experiment. (*a*) Step increase in shear stress. (*b*) Resultant shear strain. The equation for the asymptote is $\gamma_0 + \dot{\gamma}_\infty t$

by a constant shear rate $\dot{\gamma}_\infty$. The data are usually described in terms of the creep compliance $J(t, \tau_0)$:

$$\gamma_{t_0}(t) = \int_{t_0}^{t} \dot{\gamma}(t')dt' = J(t,\tau_0)\tau_0 \tag{22}$$

The γ_0 intercept at t_0 in Fig. 8.6*b* is used to define the equilibrium creep compliance

$$J_e = \frac{\gamma_0}{\tau_0} \tag{23}$$

For many polymeric liquids the limit of J_e at very small stresses approaches a constant that is often used as a basic measure of the elasticity of a material, just as η_0 [Eq. 16] is a measure of viscosity [10]. The same limit is obtained from the recoverable strain after removing the stress, Fig. 6*b*.

$$J_e^0 = \lim_{\tau_0 \to 0} \frac{\gamma_0}{\tau_0} = \lim_{\tau_0 \to 0} \frac{\gamma_r}{\tau_0} \tag{24}$$

Another transient testing method adapted from solids testing is the use of sinusoidal oscillations. Because it can be used for a wide range of liquid and solid materials and can be highly automated, sinusoidal testing, often called dynamic mechanical analysis, is very popular. As illustrated in Fig. 8.7, the sample is subjected to a continuous small-amplitude sinusoidal strain oscillation, and the stress is monitored. For a Hookean solid the stress wave will be in phase with the strain, while for a Newtonian liquid it will be 90° out of phase. The viscoelastic liquid (or solid) will lie between the two. To characterize this "betweenness," we decompose the stress into two waves; one wave τ' is in phase with the strain, and the other wave τ'' is exactly 90° out of phase with the strain. The prime notation arises from the fact that the decomposition of the τ wave can be conveniently expressed as a complex number:

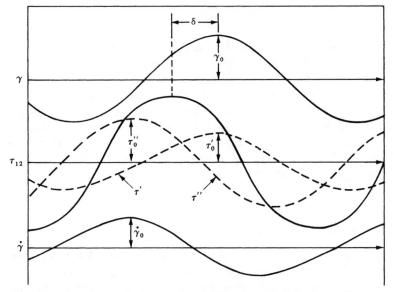

Figure 8.7 Sinusoidal oscillation experiment. The stress wave is decomposed into two waves: τ', which is in phase with the strain, and τ'', which is 90° out of phase with the strain. Note that τ'' is in phase with the rate of strain $\dot{\gamma}$

$$\tau_{12}(t) = \tau^* = \tau' + i\tau'' \qquad (25)$$

The magnitudes of the τ' and τ'' waves are used to define two dynamic moduli, as follows.

Elastic or in-phase modulus

$$G'(\omega) = \frac{\tau_0'}{\gamma_0} \qquad (26)$$

Viscous or out-of-phase modulus

$$G''(\omega) = \frac{\tau_0''}{\gamma_0} \qquad (27)$$

The phase shift δ between the strain and stress waves is

$$\tan \delta = \frac{G''}{G''} \qquad (28)$$

Note that we can also define a dynamic viscosity $\eta'(\omega)$ by looking at the part of the stress that is in phase with the rate of strain $\dot{\gamma}$. As can be seen in Fig. 8.7, τ'' is in phase with the $\dot{\gamma}$ wave; thus

$$\eta'(\omega) = \frac{\tau_0''}{\dot{\gamma}_0} \tag{29}$$

The G' and G'' and η' material functions are then sufficient to characterize a material's response to small strain oscillation. At larger strains, however, higher-order terms and harmonics can enter. The linear moduli are usually measured as a function of oscillation frequency and temperature. Typical data for several polymeric liquids are shown in Figs. 8.1 and 8.8. The shapes are quite characteristic for most polymers of high molecular weight: G' has a limiting low-frequency slope of 2, which decreases to nearly a plateau and then rises again. G'' has a limiting slope of 1, goes through a local maximum, and then rises again. The blends in Fig. 8.8 show that the high-molecular-weight component has the greatest influence on the rheology.

8.2.4 Material Functions in Extension

Besides shear, the other major classification of flows is extension. Many processes involve extensional flow, for example, fiber spinning, foaming, and flow in any converging or diverging channel. All real flows can be made up of some combination of shear and extension. To a non-Newtonian fluid, the stresses in extension are fundamentally different from the ones measured in shear, and in general, it is not possible to predict one from the other.

In pure extension (irrotational flow), velocities can vary only in the flow direction:

$$v_1 = a_1 x_1 \qquad v_2 = a_2 x_2 \qquad v_3 = a_3 x_3 \tag{30}$$

For incompressible fluids, by the continuity relation, only two velocities can be independent, i.e.,

$$\sum_{i=1}^{3} a_i = 0 \tag{31}$$

The rate-of-deformation tensor has only diagonal components,

$$\mathbf{D} = \begin{bmatrix} a_1 & 0 & 0 \\ 0 & a_2 & 0 \\ 0 & 0 & -(a_1 + a_2) \end{bmatrix} \tag{32}$$

as does the stress tensor,

$$\mathbf{T} = \begin{bmatrix} T_{11} & 0 & 0 \\ 0 & T_{22} & 0 \\ 0 & 0 & T_{33} \end{bmatrix} \tag{33}$$

As discussed with regard to shear flow, for incompressible fluids, we can measure normal stresses only to within an arbitrary constant. Thus there are only two

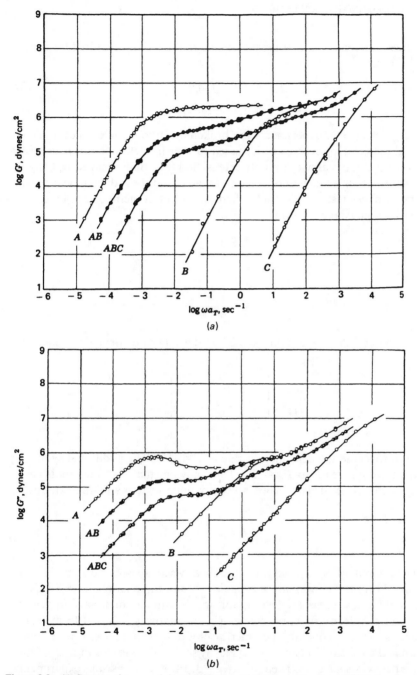

Figure 8.8 (*a*) Storage shear modulus and (*b*) loss modulus reduced to 160°C, for three narrow-distribution polystyrenes and two blends. Viscosity-average molecular weights $\times 10^{-4}$ are *A*, 58; *B*, 5.9; *C*, 0.89. Sample *AB* is a blend of equal parts of *A* and *B;* sample *ABC* is an equal blend of all three. Here, a_T is an empirical factor used to shift data from other temperatures to 160°C. (From [11])

normal-stress differences, and at most two material functions, required for general extensional flow:

$$T_{11} - T_{22} = f_1(a_p, t) \tag{34}$$

$$T_{22} - T_{33} = f_2(a_p, t) \tag{35}$$

Usually, extensional experiments are broken down into three types: uniaxial, biaxial, and planar. However, as is discussed further in section 8.4, it is generally much more difficult to carry out purely extensional tests than shear tests. Only for uniaxial deformation is much information available, and those data appear reliable only, so far, for very viscous materials where solidlike test methods can be used.

For steady uniaxial extension,

$$v_1 = \dot{\varepsilon} x_1$$

$$v_2 = -\frac{1}{2}\dot{\varepsilon} x_2 \tag{36}$$

$$v_3 = -\frac{1}{2}\dot{\varepsilon} x_3$$

where $\dot{\varepsilon}$ is the constant extension rate. Thus the rate-of-deformation tensor becomes

$$\mathbf{D} = \begin{bmatrix} \dot{\varepsilon} & 0 & 0 \\ 0 & -\frac{1}{2}\dot{\varepsilon} & 0 \\ 0 & 0 & -\frac{1}{2}\dot{\varepsilon} \end{bmatrix} \tag{37}$$

Note that in this case the deformation rate is $\dot{\gamma} = \sqrt{3}\dot{\varepsilon}$. The additional symmetry of uniaxial extension leads to $T_{22} = T_{33}$. Thus only one material function is necessary to describe uniaxial data. This is usually done in terms of the extensional viscosity

$$\eta_e(\dot{\varepsilon}) = \frac{T_{11} - T_{22}}{\dot{\varepsilon}} \tag{38}$$

For a Newtonian fluid, η_e is just 3 times the shear viscosity because $T_{11} = 2\mu\dot{\varepsilon}$ and $T_{22} = -\mu\dot{\varepsilon}$. From the continuum mechanics theory, the relation $\eta_e = 3\eta_0$ is expected to hold at low stress levels, even for non-Newtonian materials. In general, the extensional viscosity function behavior is qualitatively different from that of the shear viscosity. Frequently, polymeric solutions exhibit extensional "thickening," while their shear viscosities are typically shear thinning (see Fig. 8.9). Other qualitative behaviors have also been reported [13, 14], and the various experimental difficulties make it nearly impossible to draw definitive conclusions about the true behavior of η_e. For example, results are confounded by the difficulty in determining whether steady extension has, in fact, been achieved during the experiment. Figure 8.10 shows results at very high extensions for a low-density polyethylene melt. We

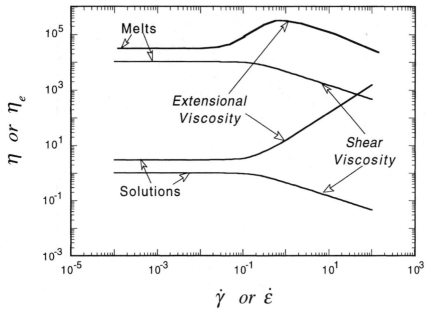

Figure 8.9 Typical behavior of steady shear and steady extensional viscosity for polymer melts and polymeric solutions

see that even at total strain of over 7 (extension ratios of 10^3), the steady state has not been attained.

In light of the data in Fig. 8.10 and the fact that in most real flows the extensional strain seldom exceeds 2, it may be more appropriate to focus on transient characterization in extension. The deformation histories defined for shear in the previous section can be applied to extension. For small strains the same interrelationships hold as given for shear, and the time-dependent extensional modulus is just $3G(t)$. Thus small strain extension can be predicted from shear. For example, $E' = 3G'$ and $E'' = 3G''$, and extensional stress growth can be predicted from shear data for strains up to 1. At larger strains these linear viscoelastic results no longer hold, and actual extensional measurements are needed to predict behavior [15]. Some of the nonlinear constitutive equations given in the next section are fairly successful in describing this nonlinear behavior.

8.3 CONSTITUTIVE RELATIONS

The goal of fluid mechanics is to understand particular flow problems. To analyze a problem and to compare results against experimental measurements, the fluid mechanicist uses the balance equations, boundary conditions, and a constitutive equation. The rheologist's goal is to determine that constitutive relation.

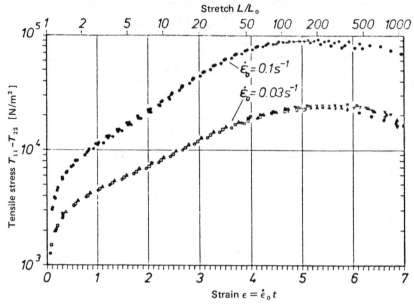

Figure 8.10 Tensile stress versus strain at constant rate of extension for a low-density polyethylene melt. (From [12])

What type of constitutive equations can we expect? Although this area continues to be the subject of much research and there is much that we do not know, we can make some generalizations. First, since they are liquids, non-Newtonian fluids usually can be treated as incompressible (the major exception is foam). The assumption of incompressibility simplifies constitutive relations. Second, non-Newtonian liquids can be grouped by the nature of their time dependence or the memory of their past state. Viscous materials are time independent; they have no memory. Plastic materials have perfect memory or solidlike behavior up to a yield stress. After the yield stress, they usually can be considered as viscous materials. Materials with a fading memory are called viscoelastic. Below, we indicate some of the important models in each of these three categories. Entire books are available on constitutive relations [6, 16, 17]; here we examine only a few of the more elementary equations. Since the problems of interest generally involve complex flows, the simplest constitutive equations are often the only ones for which we can hope to obtain a solution to the flow problem.

8.3.1 General Viscous (Reiner-Rivlin) Fluid

A viscous fluid has no memory of its past. At a given time instant t, only the instantaneous kinematics at t determines the extra-stress $\boldsymbol{\tau}$ at that same time instant. The extra stress is defined as $\tau \equiv \mathbf{T} + p\mathbf{1}$. Thus, if the flow kinematics is uniquely determined by $\mathbf{D}$,

$$\tau = \mathbf{f}(\mathbf{D}) \tag{39}$$

Most fluids are isotropic, and in this case, $\mathbf{f}$ must be isotropic. The most general isotropic form of $\mathbf{f}$ is [18–20]

$$\tau = \alpha_0 \mathbf{1} + \alpha_1 \mathbf{D} + \alpha_2 \mathbf{D}^2 \tag{40}$$

where α_i are scalar functions of the principal invariants of $\mathbf{D}$, which are defined as

$$I_\mathbf{D} = \text{tr } \mathbf{D} \qquad II_\mathbf{D} = \frac{1}{2}(I_\mathbf{D}^2 - \text{tr } \mathbf{D}^2) \qquad III_\mathbf{D} = \det \mathbf{D} \tag{41}$$

For an incompressible fluid, α_0 can be lumped in with the arbitrary pressure p:

$$\mathbf{T} = -p\mathbf{1} + \alpha_1 \mathbf{D} + \alpha_2 \mathbf{D}^2$$

The Newtonian fluid is just a special case of the Reiner-Rivlin fluid, with $\alpha_1 = 2\mu$ and $\alpha_2 = 0$. As mentioned above, in general, $\alpha_1 = \alpha_1(I_\mathbf{D}, II_\mathbf{D}, III_\mathbf{D})$. It is this feature that is most useful about the general viscous model. Often the most important rheological effect is the shear-rate dependence of viscosity. In polymeric liquids and concentrated suspensions, the viscosity can change by 10^3 over the accessible shear-rate range. In polymer processing operations, particularly flow through channels and dies, simple equations that accurately describe the shear-rate-dependent viscosity are essential for process modeling.

Normally, α_2 is neglected. In steady shear flow it predicts only a second normal stress coefficient and zero for the first coefficient, which is essentially opposite to all experimental results. There are several useful forms for α_1. As indicated above, α_1 should be a function of all the invariants; however, for an incompressible fluid, $I_\mathbf{D} = 0$, and for simple shear $III_\mathbf{D} = 0$. Thus α_1 is usually assumed to be a function $\eta(II_\mathbf{D})$ of $II_\mathbf{D}$ only, but in general, it can be a function of $II_\mathbf{D}$ and $III_\mathbf{D}$. Therefore we are interested in constitutive equations of the form

$$\mathbf{T} = -p\mathbf{1} + \eta(II_\mathbf{D})2\mathbf{D} \qquad \tau = \eta(II_\mathbf{D})2\mathbf{D} \tag{42}$$

This model is often called the generalized Newtonian fluid (GNF) model. We note that, since $II_\mathbf{D}$ is directly related to the deformation rate $\dot{\gamma}$, we can equivalently write that $\eta = \eta(\dot{\gamma})$, which is a more usual form. Below, several common expressions for $\eta(\dot{\gamma})$ are given.

8.3.1.1 Power law model. The most widely used form of the viscosity function is the power law model

$$\eta = m\dot{\gamma}^{n-1} \qquad \tau_{ij} = m(\dot{\gamma})^{n-1}(2D_{ij}) \qquad \tau = 2m\dot{\gamma}^{n-1}\mathbf{D} \tag{43}$$

This equation is most often applied to steady, simple shear flows, where, since $D_{12} = D_{21} = \dot{\gamma}$, the extra stress becomes

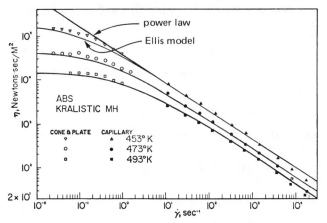

Figure 8.11 Viscosity versus shear rate for an ABS polymer melt. (From [27])

$$\tau_{12} = \tau_{21} = m\dot{\gamma}^n \tag{44}$$

with no other stress components. Equation (44) is often how the power law appears in the literature, but it is important to remember that this is only for simple shear. For other flows the full three-dimensional equation, Eq. (43), must be used, but since most materials behave differently in shear and extension, generally Eq. (43) is only a good approximation in steady shear flows. If the dependence of η on the third invariant is kept, however, the applicability of the GNF can be extended to axisymmetric complex flows [21]. A GNF model with a wider range of applicability is obtained by assuming in Eq. (39) that the extra stress depends on kinematic tensors other than **D**[23, 23].

In the processing range of many polymeric liquids and dispersions, the power law is a good approximation to viscosity shear-rate data. Figure 8.11 shows viscosity versus shear rate for an acrylonitrile-butadiene-styrene (ABS) polymer melt. At high shear rate, $\dot{\gamma} > 1$, the power law fits the data well, with m a function of temperature. The power law has been used extensively in polymer process models [3, 24, 25].

One of the obvious disadvantages of the power law is its failure to describe the low-shear-rate region. Since n is usually less than 1, at low shear rate, η goes to infinity rather than to a constant η_0, which is predicted theoretically and usually observed experimentally (Fig. 8.11).

8.3.1.2 Ellis model. Three-parameter models, like the Ellis model below, have been proposed to provide a Newtonian region at low shear rate and power law dependence at high rates:

$$\frac{\eta}{\eta_0} = \frac{1}{1 + m(\dot{\gamma})^{n-1}} \tag{45}$$

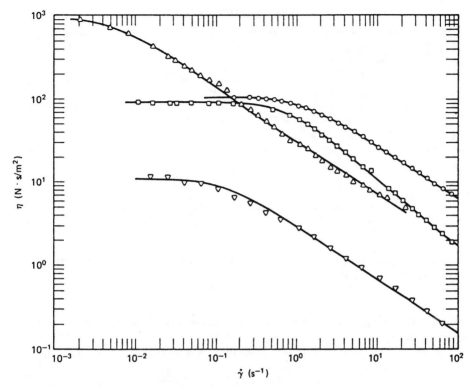

Figure 8.12 Fit of Carreau model to viscosity versus shear rate data: (△) 2.0% polyisobutylene in Primol; (□) 7% aluminum laurate in a mixture of decalin and *m*-cresol; (▽) 0.75% polyacrylamide in 95/5 mixture of water and glycerin; (○) 5% polystyrene in Aroclor. (From [2])

Though it is somewhat more complex than the power law, the Ellis model has been shown to fit a much wider range of viscosity versus shear-rate data. This can be seen in Fig. 8.11. The Ellis model has been used in a number of complex flow problems [26].

8.3.1.3 Other models. A number of other empirical expressions for $\eta(\dot{\gamma})$ are available. Some non-Newtonian liquids show a limiting viscosity η_∞ at high shear rates. This can be treated with the modified Ellis model of Bird and Carreau [28]:

$$\frac{\eta - \eta_\infty}{\eta_0 - \eta_\infty} = \frac{1}{[1 + (\lambda\dot{\gamma})^2]^{(1-n)/2}} \tag{46}$$

where λ has units of time and can be considered a time constant of the fluid. Figure 8.12 shows viscosity data for a soap and three polymer solutions compared with the Carreau model. A slight modification of this model has been proposed by

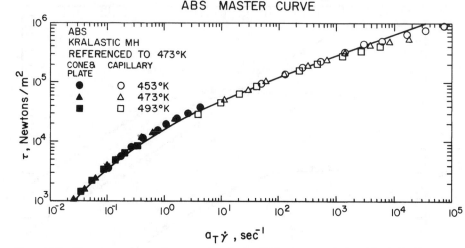

Figure 8.13 Shear stress versus shear-rate data for an ABS polymer melt from Fig. 8.11. The master curve was made using Eq. (49). (From [33])

Yasuda et al. [29], which has a remarkable capability of fitting well most shear viscosity data.

To fit experimental data more accurately, a power series is frequently used [30]:

$$\eta(\dot\gamma) = a_0 + a_1\dot\gamma + a_2\dot\gamma^2 + \dots \tag{47}$$

This equation seems to have its greatest use in numerical modeling. Bird et al. [2] give a number of other empirical models for $\eta(\dot\gamma)$.

The temperature and pressure dependence of viscosity can also be critical to understanding some processing problems. Van Krevelen [31] reviews both these areas, and Goldblatt and Porter [32] have reviewed the problem of pressure dependence. Exponential relations have been found useful to correlate both temperature and pressure dependence:

$$\eta_0 = Ke^{E_\eta/RT}e^{bp} \tag{48}$$

where E_η is the temperature dependence of the zero-shear-rate viscosity and b is its pressure dependence.

Another approach [11] is the WLF equation, based on free-volume changes. It indicates a shift for viscosity with respect to a reference state:

$$\log a_T = \log \frac{\eta(T)}{\eta(T_{ref})} = \frac{C_1(T - T_{ref})}{C_2 + T - T_{ref}} \tag{49}$$

Figure 8.13 shows that the data of Fig. 8.11 can be shifted quite well with this approach. Van Krevelen [31] shows that the WLF equation is most useful for amor-

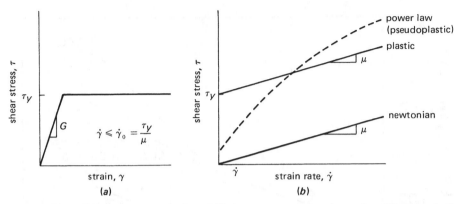

Figure 8.14 (*a*) Shear stress versus strain and (*b*) shear stress versus strain rate for a Bingham plastic material

phous polymers close to their glass transition temperature T_g, while the exponential form appears better for $T \geq T_g + 100$. Temperature dependence for suspensions has not been studied extensively.

8.3.2 Plastic Behavior

A plastic material is one that shows little or no deformation up to a certain level of stress. Above this yield stress, the material flows like a liquid. Plasticity is common to widely different materials. Many metals yield at strains of less than 1%. Concentrated suspensions of solid particles in Newtonian liquids also show a yield stress followed by nearly Newtonian flow. Such materials are often called Bingham plastics, after E. C. Bingham, who first described paint in this way in 1919. House paint and food substances like margarine, mayonnaise, and ketchup are good examples of Bingham plastics.

A simple model for a plastic material is Hookean behavior at stresses below yield, and Newtonian behavior above. For one-dimensional deformations,

$$
\begin{aligned}
\tau &= G\gamma_{t_0}(t) & \tau \leq \tau_y \\
&\quad \mu\dot{\gamma} + \tau_y & \tau > \tau_y
\end{aligned}
\tag{50}
$$

Figure 8.14 illustrates this behavior and compares the Bingham plastic flow to Newtonian and power law fluids. We can see why power law or a strong shear-thinning behavior is frequently called pseudoplastic.

An important feature of plastic behavior is that, if the stress is not constant over a body, parts of it may flow while the rest acts like a solid. Fredrickson [34] illustrates this for flow in a tube. The shear stress changes linearly from zero at the center of the tube to a maximum at the wall. Thus the central portion of the material flows like a solid plug. Neck formation during uniaxial extension of a solid at

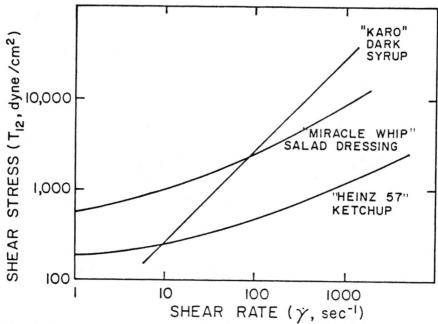

Figure 8.15 Flow data for several food products. (Courtesy Graco Co., Minneapolis, Minnesota)

constant strain rate is another example of this type of behavior. At the smallest sample cross section or at an inhomogeneity, the stress during the test just exceeds the yield stress, and large deformation can occur.

Example: The Ketchup Bottle

We have all been frustrated by that malevolent Bingham plastic, ketchup. To exceed its yield stress in the neck of the bottle, one must frequently tap the bottle, and then, when the shear stress at the wall exceeds τ_y, flow is rapid. Figure 8.15 shows shear stress versus shear rate data for ketchup and several other food products; $\tau_y \approx 200$ dyn/cm^2 for this ketchup sample. (Note that, in contrast to Fig. 8.14, this is a log-log plot.) Will ketchup empty under gravity from a typical bottle?

In the neck of the bottle the wall shear stress τ_w will be balanced by the pressure head of ketchup in the bottle. If we can approximate the bottle as a tube of length L and diameter D and a cylindrical reservoir of height H, then [note Eq. (98)]

$$\tau_w(\pi DL) = p\frac{\pi D^2}{4} \quad \text{or} \quad \tau_w = \frac{pgHD}{4L} \tag{51}$$

The density is $\rho \approx 1$ g/cm^3, the gravitational acceleration is $g = 980$ cm/s^2, and a bottle with a "standard" neck has $D \approx 1.5$ cm and $L \approx 6$ cm. If the bottle is partially full, $H \approx 4$ cm. Then $\tau_w \approx 200$ dyn/cm^2, and the ketchup should not flow without some thumping. Note that the situation is probably worse, since we have assumed atmospheric pressure above the ketchup in the bottle.

It is typically less, owing to the partial vacuum created as the bottle is inverted. For a "wide mouth" bottle, $D \simeq 3$ cm and $\tau_w \simeq 400$ dyn/cm^2, which may make mealtimes flow more smoothly.

To handle deformations occurring in more than one direction, Eq. (43) should be put into three-dimensional form. The only significant change is to replace the one-dimensional τ in the yield criterion with some scalar function of the invariants of $\boldsymbol{\tau}$. There are a number of yield criteria in the literature [20]. The von Mises criterion, which uses the second invariant of $\boldsymbol{\tau}$, is the most common:

$$\boldsymbol{\tau} = G\mathbf{B}_{t_0}(t) \qquad |\mathrm{II}_\tau| \leq \tau_y^2 \tag{52}$$

where $\mathbf{B}_{t_0}(t)$ is the Finger tensor, as defined earlier, and

$$\boldsymbol{\tau} = 2\left(\mu + \frac{\tau_y}{\dot{\gamma}}\right)\mathbf{D} \qquad |\mathrm{II}_\tau| > \tau_y^2 \tag{53}$$

Note that postyield behavior other than Newtonian flow can readily be substituted. The most common is power law viscosity.

Other constitutive equations for plastic materials are described by Fredrickson [34], Argon [16], Larson [6], and Macosko [7]. However, these relations are less well developed than the viscoelastic constitutive equations described in the next sections. This state of affairs is partially due to the difficulty in obtaining accurate data, independent of rheometer geometry, on plastic materials.

8.3.3 Linear Viscoelasticity

Most polymeric liquids show the phenomenon called "fading memory." At short times they behave like a rubbery solid, and at long times like a viscous liquid. This can perhaps be best seen in the simple stress-relaxation experiment that was discussed in section 8.2.3. Recall that when a polymeric liquid is subjected to a sudden, small step strain, the stress rises quickly and then relaxes with time. This is illustrated in Fig. 8.16.

Maxwell, over a century ago, suggested that this type of behavior could be modeled (for simple shear) by a linear combination of the ideal elastic or Hookean solid and the Newtonian liquid:

$$\tau + \left(\frac{\eta_0}{G_0}\right)\frac{\partial \tau}{\partial t} = \eta_0 \dot{\gamma} \tag{54}$$

The Maxwell model is often represented as a series combination of springs (elastic elements) and dashpots (viscous) elements as in Fig. 8.17. From Eq. (54) and the spring-and-dashpot representation, we see that for slow motions the dashpot or Newtonian behavior dominates. For rapidly changing stresses the derivative term dominates, and thus at short times the model approaches elastic behavior.

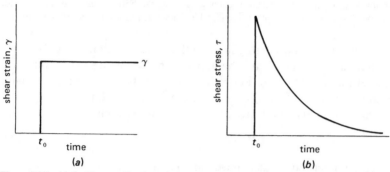

(a)   (b)

Figure 8.16   (a) Sudden application of shear strain to a sample and (b) resultant stress response

The Maxwell model is a simple linear combination of viscous and elastic effects and is called the linear viscoelastic model.

The term η_0/G_0 in Eq. (54) has units of time and is the characteristic relaxation time λ_0 of the system. This can be seen more clearly if we put Eq. (54) into an integral form:

$$\tau = \int_{-\infty}^{t} \left[\frac{G_0}{\lambda_0} e^{-(t-t')/\lambda_0} \right] \dot{\gamma}(t') dt' \tag{55}$$

The function inside the brackets is called a relaxation modulus $G(t - t_0)$ [note Eq. (21)]. It is multiplied by the strain rate, which can be a function of time $\dot{\gamma}(t)$.

An equivalent integral representation is in terms of a memory function $M(t - t')$ times the strain $\gamma_t(t')$:

$$\tau = -\int_{-\infty}^{t} \frac{G_0}{\lambda_0} e^{-(t-t')/\gamma_0} \gamma_t(t') = -\int_{-\infty}^{t} M(t - t') \gamma_t(t') dt' \tag{56}$$

Note that the memory function is just the derivative of the modulus

$$M(t - t') = \frac{dG(t - t')}{dt'} \quad \text{or} \quad M(s) = -\frac{dG(s)}{ds} \tag{57}$$

To better fit the data, several relaxation times are generally used:

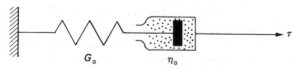

Figure 8.17   Spring-and-dashpot representation of a Maxwell model

Table 8.1 Relaxation times for low-density polyethylene at 150°C

k	λ_k,s	G_k,Pa
1	10^3	1.00
2	10^2	1.80×10^2
3	10^1	1.89×10^3
4	10^0	9.80×10^3
5	10^{-1}	2.67×10^4
6	10^{-2}	5.86×10^4
7	10^{-3}	9.48×10^4
8	10^{-4}	1.29×10^5

From [35].

$$G(s) = \sum_{k=1}^{N} G_k e^{-s/\lambda_k} \tag{58}$$

Usually 5–10 relaxation times are adequate to fit the typical experimental data range. This is illustrated in Fig. 8.18, in which Laun [35] uses eight relaxation times to fit the shear relaxation modulus for a low-density polyethylene. We clearly see the contribution of each relaxation time. The values of λ_k and G_k are given in Table 8.1.

Polymer molecular theories like the Rouse model [2, 6, 11] suggest an infinite series form for the relaxation spectra:

$$G_k = G_0 \frac{\lambda_k}{\sum_k \lambda_k} \qquad \lambda_k = \frac{\lambda_0}{k^2} \tag{59}$$

Clearly, one could also construct a continuous relaxation-modulus function.

These integral models can be made three-dimensional by simply substituting the stress tensor $\boldsymbol{\tau}$ for the shear stress, and the rate-of-deformation tensor $\mathbf{D}$ for the shear rate. Thus, in general, we can write

$$\boldsymbol{\tau} = \int_{-\infty}^{t} G(t - t')2\mathbf{D}(t')dt' \tag{60}$$

or the small-strain tensor $\mathbf{E}$ for the shear strain $\gamma_t(t')$

$$\boldsymbol{\tau} = -\int_{-\infty}^{t} M(t - t')\mathbf{E}_t(t')dt' \tag{61}$$

The small-strain tensor $\mathbf{E}_t(t')$ is related to $\mathbf{D}$ by $\mathbf{E}_t(t') = 2\int_t^{t'} \mathbf{D}(t'')dt''$, similar to the definition of $\gamma_t(t')$ in the one-dimensional case. Actually, the Finger tensor $\mathbf{B}$, which is a measure of strain regardless of how large the strain is, tends to the small-strain

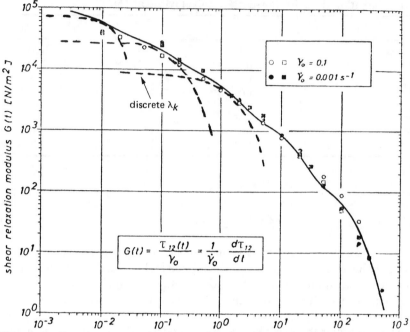

Figure 8.18 Shear-relaxation modulus versus time for a low-density polyethylene at 150°C. The solid line is the sum of eight exponential relaxation times given in Table 8.1. (From [35])

tensor when the total deformation is small. More precisely, if the total strain is small, then $\mathbf{E}_t(t') \simeq 1 - \mathbf{B}_t(t')$.

Equations (60) and (61) should be applicable to any type of flow, shear or extension, at small strains or strain rates as long as the material has a fading memory.

The general linear viscoelastic model (generalized Maxwell model) has been quite successful in fitting small-strain data for polymeric liquids. Figures 8.18 and 8.19 illustrate this. The distribution of relaxation times given in Table 8.1 was used to calculate all three curves. For sinusoidal oscillations, Eq. (60) gives

$$G'(\omega) = \Sigma_k G_k \frac{\omega^2 \lambda_k^2}{1 + \omega^2 \lambda_k^2} \tag{62}$$

$$G''(\omega) = \Sigma_k G_k \frac{\omega \lambda_k}{1 + \omega^2 \lambda_k^2} \tag{63}$$

For small strains or strain rates, all the various transient shear tests can be interrelated using linear viscoelasticity. The interrelations are summarized in terms of J_e^0 and λ_0 in Table 8.2.

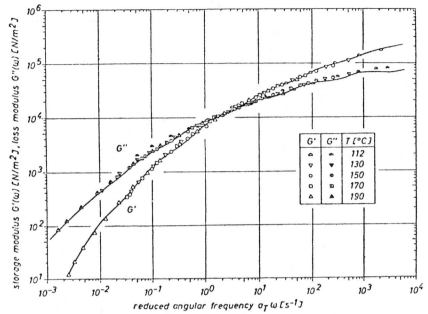

Figure 8.19 Dynamic shear moduli for the same low-density polyethylene as in Fig. 8.18. The lines were calculated from the same relaxation spectra. (From [35])

8.3.4 Nonlinear Viscoelasticity

A major drawback of the Maxwell model is that although it is reasonable for small-strain transients, it predicts a Newtonian viscosity in shear and does not predict normal stresses. Hundreds of papers have been written on how to improve the model or write new models to show this type of nonlinear behavior. Many of these models become quite complex and really unusable with complex flows. Here we present two of the simplest nonlinear models and indicate how a better fit to rheological data can be achieved with greater complexity.

As discussed earlier, the linear viscoelasticity theory is valid in the limit of small strains and strain rates. Some of the quantities that appear in this linear theory ($\mathbf{E}$, $\partial \boldsymbol{\tau}/\partial t$, for example) are not frame indifferent. This means that constitutive relations involving these quantities will predict that the material mechanical behavior is a function of the observer! To fix that, the partial time derivatives that appear in the differential models are replaced with another time derivative that is frame indifferent [6, 17], namely

$$\frac{\delta \boldsymbol{\tau}}{\delta t} \equiv \frac{d\boldsymbol{\tau}}{dt} - \mathbf{W}\boldsymbol{\tau} + \boldsymbol{\tau}\mathbf{W} - a(\mathbf{D}\boldsymbol{\tau} + \boldsymbol{\tau}\mathbf{D}) \qquad (64)$$

Table 8.2 Limiting relations for linear viscoelasticity

Transient shear test		Equilibrium creep compliance J_e^0		Longest relaxation time λ_0
Steady shear	$\lim\limits_{\dot\gamma \to 0}$	$\dfrac{\psi_1}{2\eta^2}$	$\lim\limits_{\dot\gamma \to 0}$	$\dfrac{\psi_1}{2\eta}$
Sinusoidal oscillations	$\lim\limits_{\omega \to 0}$	$\dfrac{G'}{(G'')^2}$	$\lim\limits_{\omega \to 0}$	$\dfrac{G'}{G''\omega}$
Creep	$\lim\limits_{\tau_0 \to 0}$	$\dfrac{\gamma_0}{\tau_0}$	$\lim\limits_{\tau_0 \to 0}$	$\dfrac{\gamma_0}{\dot\gamma_\infty}$
Constrained recoil	$\lim\limits_{\tau_0 \to 0}$	$\dfrac{\gamma_r}{\tau_0}$	$\lim\limits_{\tau_0 \to 0}$	$\dfrac{\gamma_r}{\dot\gamma_\infty}$
Stress relaxation	$\lim\limits_{\dot\gamma_0 \to 0}$	$\displaystyle\int_0^\infty \dfrac{\eta^-}{\eta_0^2}\,dt$	$\lim\limits_{\dot\gamma_0 \to 0}$	$\displaystyle\int_0^\infty \dfrac{\eta^-}{\eta_0}\,dt$
Stress growth	$\lim\limits_{\dot\gamma \to 0}$	$\displaystyle\int_0^\infty \dfrac{\eta_0 - \eta^+}{\eta_0^2}\,dt$	$\lim\limits_{\dot\gamma_0 \to 0}$	$\displaystyle\int_0^\infty \left(1 - \dfrac{\eta^+}{\eta_0}\right)dt$

The rate $\delta\tau/\delta t$ is the interpolated convected time derivative of τ. When $a = 1$, we get the upper convected time derivative; when $a = -1$, $\delta\tau/\delta t$ becomes the lower convected time derivative; and for $a = 0$, the corotational time derivative is recovered. Using it, we obtain the interpolated Maxwell model:

$$\tau + \lambda_0 \frac{\delta\tau}{\delta t} = 2\mu\mathbf{D} \tag{65}$$

The upper convected Maxwell (UCM) model ($a = 1$) predicts a first normal-stress difference, but it still gives a constant viscosity and no second normal-stress difference. Using the corotational or Jaumann derivative ($a = 0$) [2] gives shear thinning that is much too strong to be realistic. To achieve more realistic shear thinning, one can replace μ in Eq. (65) with a shear-rate-dependent viscosity [36]. It is also possible to choose different values for the interpolating parameter a, depending on the fluid to be modeled, such as to fit better experimental data [37].

Many other differential constitutive models have been developed. One approach is to formally expand the stress in upper convected time derivatives of the rate of strain ($a = 1$). The result is similar to that for the general viscous fluid except that, owing to the time dependence of the fluid, the series is infinite. To solve actual problems, the series is truncated, usually at second order [1]:

$$\mathbf{T} = -p\mathbf{1} + \beta_1(2\mathbf{D}) + \beta_2\frac{\delta(2\mathbf{D})}{\delta t} + \beta_3(2\mathbf{D})^2 \tag{66}$$

where the β_i are material constants. Clearly, β_1 gives the shear viscosity, and β_2 and β_3 are elasticity parameters. For steady shear, the limiting values of the material functions are simply related to the β_i:

$$\eta_0 = \beta_1 \qquad \Psi_{10} = -2\beta_2 \qquad \Psi_{20} = 2\beta_2 + \beta_3 \tag{67}$$

The steady extensional viscosity is

$$\eta_e = 3\beta_1 + 3(\beta_2 + \beta_3)\dot{\varepsilon} \tag{68}$$

The second-order fluid model, as this model is called, is strictly valid for small departures from Newtonian behavior, flows that have both low shear rates and slow transients. Because it is simple and valid in this limit, it has frequently been used to get a qualitative idea of how elasticity affects a steady flow, such as the direction of a secondary flow [26] and the magnitude of the rod climbing effect [38]. A slight modification of the second-order model is the Criminale-Ericksen-Filbey (CEF) equation [39], in which the coefficients are allowed to be functions of the invariants of **D**.

The other approach to improving upon the Maxwell model is to work with the integral equations. Although these are more difficult to apply to flow problems, they have been more successful in fitting rheological data.

If the (negative of the) small-strain tensor **E** in Eq. (61) is replaced with the Finger tensor **B**, time-dependent response can be modeled to larger strains* [40–42]:

$$\mathbf{\tau} = \int_{-\infty}^{t} M(t - t')\mathbf{B}_t(t')dt' \tag{69}$$

However, this rubberlike liquid model still does not have a shear-thinning viscosity. This requires some strain dependence in the memory function. Many forms have been suggested for this strain dependence. One of the simplest is the factorized memory function proposed by Wagner [15], who noticed that for many polymeric materials, as illustrated in Fig. 8.8, the shape of the relaxation modulus (or memory function) does not change significantly with strain. This suggests that for large strains it can be factored into the time-dependent, small-strain modulus and a strain-dependent damping function h. When this idea is combined with the rubberlike liquid, we obtain

---

*Some authors (e.g., [2]) prefer to replace $(-\mathbf{E})$ with $(\mathbf{B} - \mathbf{1})$ instead, which gives $\mathbf{\tau} = \int_{-\infty}^{t} M(t - t')[\mathbf{B}_t(t') - \mathbf{1}]dt'$. Both approaches are equivalent as far as the total stress **T** is concerned, since the extra term $(-\mathbf{1})\int_{-\infty}^{t} M(t - t')dt'$ can be absorbed into the isotropic pressure term of **T**.

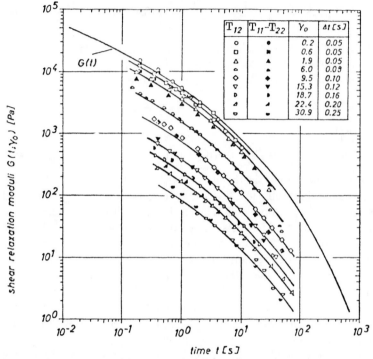

Figure 8.20 Time and strain dependence of the stress-relaxation modulus for a low-density polyethylene melt at 150°C. (From [35])

$$\boldsymbol{\tau} = \int_{-\infty}^{t} M(t - t')h(I_{\mathbf{B}}, II_{\mathbf{B}})\mathbf{B}_t(t')dt' \tag{70}$$

where $I_{\mathbf{B}}$ and $II_{\mathbf{B}}$ are the first and second invariants of **B**.

The K-BKZ model has been found to fit a wide variety of shear and extensional data for polymer melts. Laun [35] used the strain dependence of the stress-relaxation modulus for low-density polyethylene (shown in Fig. 8.20) to determine the damping function shown in Fig. 8.21. With these data and the linear viscoelastic relaxation times on the same material given in Table 8.1, Wagner was able to use Eq. (70) to predict the start-up and steady state viscosity and first normal-stress coefficient data given in Figs. 8.22 and 8.23. For the start-up viscosity, Eq. (70) becomes

$$\eta^+(t, \dot{\gamma}_0) = \sum_k \frac{G_k\lambda_k}{(1 + n\dot{\gamma}_0\lambda_k)^2}[1 - e^{-t_{r,k}}(1 - n\dot{\gamma}_0\lambda_k t_{r,k})] \tag{71}$$

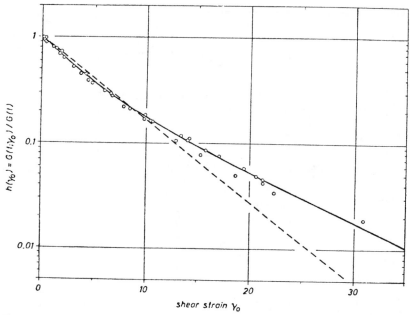

Figure 8.21 Damping function versus strain calculated from the data of Fig. 8.20. Dashed line: $h = \exp(-0.18\gamma_0)$. Solid line: $h = 0.57 \exp(-0.310\gamma_0) + 0.43 \exp(-0.106\gamma_0)$. (From [35])

where $t_{r,k} = t/\lambda_k + n\dot{\gamma}_0 t$ and $n = 0.18$. This success is encouraging. The factorized memory-function model is of the same form as that of Doi and Edwards [43], developed from molecular theory.

A problem with Wagner's model is that it is not very accurate in small strains, in the linear viscoelastic regime, or in very large strains. An improvement can be achieved by substituting a sigmoidal damping function for the exponential one, as done by Papanastasiou et al. [44]:

$$h(I_B, II_B) = \frac{\alpha}{(\alpha - 3) + \beta I_B + (1 - \beta)II_B} \tag{72}$$

The damping function given in Eq. (72) has been frequently used in conjunction with Eq. (70) [45].

8.3.5 Discussion

A difficult task for the experimental fluid mechanicist is to determine which constitutive relation is applicable to a particular problem and to obtain the appropriate parameters for the model. If the fluid is Newtonian, the latter is quite simple. Given the temperature and pressure of interest, the Newtonian viscosity can be deter-

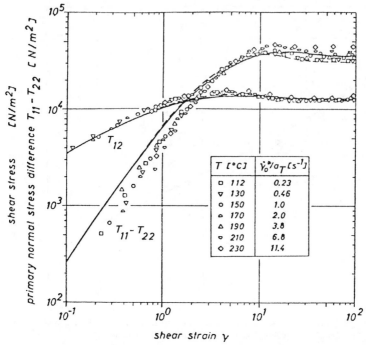

Figure 8.22 Start-up of shear stress and first normal-stress difference versus strain for a low-density polyethylene melt. (From [35])

mined with a single simple experiment such as a falling ball or efflux from a tube. For non-Newtonian fluids, rheological characterization can require a considerable amount of experimental work. Even the simple falling-ball and tube-flow experiments can give erroneous "viscosities" for non-Newtonian materials. The measurement of non-Newtonian material functions is discussed in section 8.4.

To help simplify the fluid characterization problem, it is important to identify the type and range of deformations that are anticipated in the flow of interest. Is the flow primarily shear or extensional? Are the timescales short or long? Are the total strains greater or less than 1–2? By answering these questions first, the amount of rheological measurement needed can be considerably reduced. For example, if the basic problem is the determination of pressure drop in steady conduit flows, viscosity versus shear-rate data may be adequate. If the problem is surface instability in a coating flow, the time dependence of the stresses to changing deformations may be more important than fitting the shear-rate dependence.

A useful dimensionless group in this regard is the Deborah number [46], which is the ratio of the fluid relaxation time λ_0 to some appropriate flow time t_r:

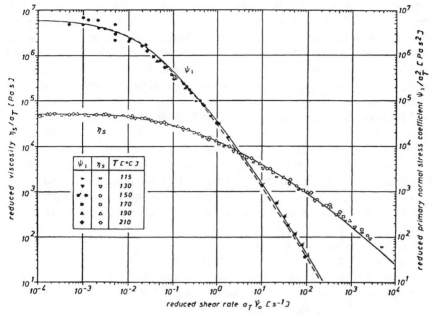

Figure 8.23 Steady state shear viscosity and first normal-stress coefficient versus shear rate for a low-density polyethylene melt. (From [35])

$$De = \frac{\lambda_0}{t_r} \qquad (73)$$

If $t_r < \lambda_0$, transient, often solidlike response may dominate the flow behavior. Steady state data may have little relevance, and a purely viscous model will not be likely to explain the phenomenon. Another important dimensionless group is the Weissenberg number, which is the product of a relaxation time and a characteristic deformation rate:

$$Ws = \lambda_0 \dot{\gamma}_c \qquad (74)$$

When Ws is large, significant departure from the Newtonian fluid behavior is expected.

8.4 RHEOMETRY

In section 8.2 we discussed the stresses that need to be measured to completely characterize a fluid in shear or extensional flow. The real job of the experimental rheologist is to actually achieve the kinematics of these simple flows and then measure the stresses. In some sense this task is the reverse of that of the experimental fluid mechanicist. In principle, the latter studies a known fluid in a complex flow. The rheologist begins with a very simple flow, one in which the kinematics are completely determined regardless of the type of fluid, and then uses it to characterize a complex fluid. Clearly, in any study of non-Newtonian fluids these roles must merge. The rheologist must use fluid mechanics methods to verify that a new rheometer does indeed produce the kinematics assumed. The fluid mechanicist must get involved with rheometry to do serious work with non-Newtonian flows.

A rheometer, then, is a flow device in which the stresses can be measured and the kinematics are known or can be determined from a few simple measurements, regardless of the fluid's constitutive equation. Below, we briefly survey the most common and useful rheometers. We give but do not derive the working equations for data analysis, and point out the advantages and disadvantages of each. However, there are several potential systematic errors in making measurements with these flow devices. For these and for further information on rheometers, the reader is referred to Macosko [7], to the several texts that focus on rheometry [47–49], and to the specific references given below.

As with our discussion of the material functions, the treatment of rheometers can be broken into the two basic flows: shear and extension. Shear measurements have constituted most of rheometry; however, one of the most active areas of research today is in developing new extensional devices.

8.4.1 Shear Rheometers

The shear geometries in common use as rheometers are shown with their coordinate systems in Fig. 8.24. The working equations for some of these geometries are given in the following subsections.

8.4.1.1 Concentric-cylinder rheometer.

Working equations Refer to Fig. 8.25 for definitions of the parameters.

Shear stress

$$T_{21} = \tau_{r0}(R_i) = \frac{M_i}{2\pi R_i^2 L} \tag{75}$$

Shear strain

$$\gamma = \frac{\theta \overline{R}}{R_0 - R_i} \tag{76a}$$

or for a narrow gap,

$$\gamma = \frac{\Omega t \overline{R}}{R_0 - R_i} \tag{76b}$$

Drag Flows

Coordinates
x_1 x_2 x_3

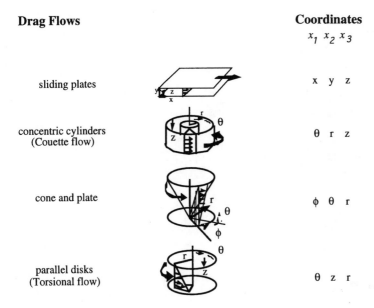

sliding plates   x   y   z

concentric cylinders
(Couette flow) θ r   z

cone and plate ϕ θ r

parallel disks
(Torsional flow) θ z   r

Pressure Flows

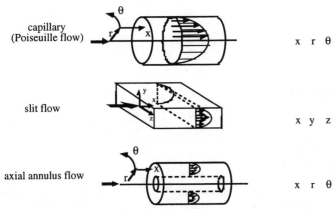

capillary
(Poiseuille flow)   x   r θ

slit flow   x   y   z

axial annulus flow   x   r θ

Figure 8.24 Common shear-flow geometries. In each sketch, x_1 is the flow direction, x_2 is the direction of the shear gradient, and x_3 is the neutral direction. (Adapted from [44])

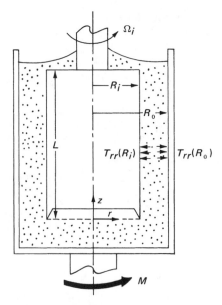

Figure 8.25 Schematic of concentric-cylinder rheometer

where $\theta = \Omega t$ is the angular displacement for steady rotation and $\overline{R} = (R_0 + R_i)/2$ is the mean radius.

Shear rate

$$\dot{\gamma}(R_i) = \dot{\gamma}(R_0) = \frac{\Omega_i \overline{R}}{R_0 - R_i} = \frac{2\Omega_i}{1 - k^2} \qquad \kappa = \frac{R_i}{R_0} > 0.99 \tag{77}$$

(narrow gap, homogeneous), or

$$\dot{\gamma}(R_i) = \frac{2\Omega_i}{n(1 - k^{2/n})} \qquad n = \frac{d(\ln M_i)}{d(\ln \Omega_i)} \qquad \kappa < 0.99 \tag{78}$$

(wide gap, nonhomogeneous), or

$$\dot{\gamma}(R_0) = \frac{2\Omega_0}{n(1 - k^{2/n})} \qquad n = \frac{d(\ln M_i)}{d(\ln \Omega_i)} \qquad \kappa < 0.99 \tag{79}$$

Normal stress (narrow gap)

$$T_{11} - T_{22} = T_{\theta\theta} - T_{rr} = \frac{[T_{rr}(R_0) - T_{rr}(R_i)]\overline{R}}{R_0 - R_i} \tag{80}$$

Corrections The most common corrections are due to end effects and fluid inertia problems. The contribution to torque from the bottom can be greatly reduced by using a large gap or by trapping a gas pocket, as indicated in Fig. 8.25. Conical bottoms with the cone angle to match the shear rate between the cylinders are also used [7]. Fluid inertia can be a problem in transient studies of low-viscosity fluids [50, 51]. It also leads to secondary flows (Taylor vortices) if the inner cylinder rotates [51, 52].

Criteria for secondary flow

$$\frac{\rho^2\Omega^2(R_0 - R_i)^3 R_i}{\eta_0} < 1700 \tag{81}$$

Shear heating

$$\frac{M}{M_0} = 1 - b\frac{Br}{12} \tag{82}$$

where

$$\eta(T) = \eta_0(T_0)e^{bT} \qquad Br = \frac{\eta_0 R^2\Omega}{k_T T_0} \tag{83}$$

Utility This is the best geometry for lower-viscosity systems ($\eta_0 < 100$ Pa s), but it is hard to load and clean out high-viscosity materials. The device is good for high shear rates. The gravity settling of suspensions has less effect than in cone-and-plate rheometers (section 8.4.1.2). Normal stresses are hard to measure, owing to the curvature and the need to transmit the signal through a rotating shaft. Rod climbing with a large gap can also be used to measure normal stresses [38, 52].

8.4.1.2 Cone-and-plate rheometer.

Working equations Refer to Fig. 8.26 for definitions of the symbols.

Shear stress

$$T_{12} = T_{\phi\theta} = \frac{3M}{2\pi R^3} \tag{84}$$

Shear strain (homogeneous)

$$\gamma = \frac{\phi}{\beta} \tag{85}$$

Shear rate

$$\dot{\gamma} = \frac{\Omega}{\beta} \tag{86}$$

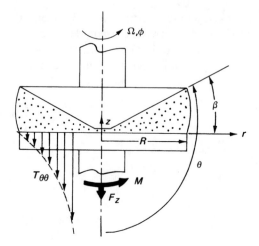

Figure 8.26 Schematic of cone-and-plate rheometer

Normal stress

$$N_1 = T_{\phi\phi} - T_{\theta\theta} = \frac{2F_z}{\pi R^2} \tag{87}$$

$$N_1 + 2N_2 = -\frac{\partial T_{\theta\theta}}{\partial(\ln r)} \tag{88}$$

Corrections

Inertia and secondary flow

$$N_1 = \frac{2F_z}{\pi R^2} - 0.15\rho\Omega^2 R^2 \tag{89}$$

$$\frac{M}{M_0} = 1 + 6 \times 10^{-4} \, Re^2 \tag{90}$$

$$Re = \frac{\rho\Omega^2\beta^2 R^2}{\eta_0} \tag{91}$$

Edge failure for high-viscosity samples occurs around $\dot\gamma = 1 \text{ s}^{-1}$.

Gap opening, oscillations

$$\frac{6\pi R\eta_0}{K\beta^3} < \text{material relaxation time} \tag{92}$$

Shear heating

$$\frac{M}{M_0} = 1 - b\frac{Br}{20} \tag{93}$$

where *Br* is given in Eq. (83).

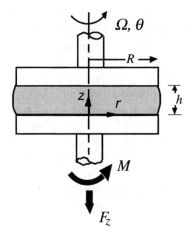

Figure 8.27 Schematic of parallel-plate rheometer

Utility The cone-and-plate rheometer is the most common instrument for normal-stress measurements. It has the simplest working equations for homogeneous deformation. It requires a stiff, well-aligned instrument and is useful for low- and high-viscosity materials. High-viscosity use is limited by elastic edge failure; low-viscosity use, by inertia corrections, secondary flow, and loss of sample at edges.

8.4.1.3 Parallel-plate rheometer.

Working equations Refer to Fig. 8.27 for definitions of the symbols.

Shear strain (nonhomogeneous, depends on position)

$$\gamma = \frac{\theta r}{h} \tag{94}$$

Shear rate (at perimeter)

$$\dot{\gamma}_R = \frac{\Omega R}{h} \tag{95}$$

Shear stress

$$T_{12} = T_{\theta z} = \frac{M}{2\pi R^3}\left[3 + \frac{d(\ln M)}{d(\ln \dot{\gamma}_R)}\right] \tag{96}$$

Normal stress

$$N_1 - N_2 = \frac{F_z}{2\pi R^2}\left[2 + \frac{d(\ln F_z)}{d(\ln \dot{\gamma}_R)}\right] \tag{97}$$

Corrections For inertia and secondary flow, use Eqs. (89) and (90) with $h/R = \beta$. For edge failure, the corrections are the same as for the cone-and-plate rheometer. For shear heating, they are similar to those for cone-and-plate rheometers [2, 7].

Utility The key advantage over the cone-and-plate rheometer is the ability to independently vary shear rate (and shear strain) by the rotation rate Ω or by changing the gap h. This permits an increased range with a given experimental setup. For very viscous materials and soft solids, sample preparation and loading are simpler. Edge failure can be delayed to higher shear rates by decreasing the gap during an experiment. This same effect requires a change of cone angle in cone-and-plate rheometers.

The main disadvantage is the nonhomogeneous strain field, but if only small strain or a steady rate of strain material functions is required, this is not a problem.

8.4.1.4 Capillary rheometer.

Working equations Refer to Figs. 8.28 and 8.29 for definitions of the parameters. A key assumption in the analysis is that there is fully developed steady flow.

Wall shear stress

$$\tau_w = \frac{R\Delta P}{2L} \qquad \Delta P = P_4 - P_5 \tag{98}$$

Wall shear rate

$$\dot{\gamma}_w = \frac{4Q}{\pi R^3}\left[\frac{3}{4} + \frac{1}{4}\frac{d(\ln Q)}{d(\ln \Delta P)}\right] \tag{99}$$

for nonhomogeneous deformation and

$$\dot{\gamma}_w = \left(\frac{3n+1}{n}\right)\frac{Q}{\pi R^3} \tag{100}$$

for the power law.

Viscosity

$$\eta = \frac{\tau_w}{\dot{\gamma}_w} \tag{101}$$

First normal-stress difference

$$(T_{11} - T_{22})^2 = 8\tau_w^2(B^6 - 1) \tag{102}$$

which requires constitutive assumptions [7], and

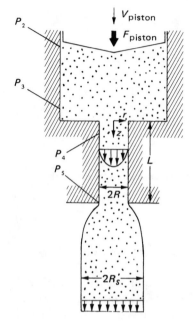

Figure 8.28 Cross section of a typical capillary rheometer. The extrudate swell is shown. The pressure transducer readings at P_2, P_3, etc., are indicated in Fig. 8.29

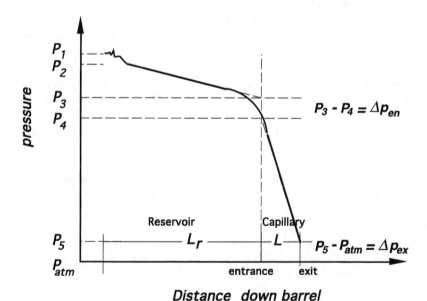

Figure 8.29 Pressure profile in a capillary or slit rheometer. See Fig. 8.28 for the locations of the readings. $P_1 - P_{atm}$ is the total force per unit area acting on the piston; $P_1 - P_2$ is due to piston friction; $P_2 - P_3$ to reservoir losses; $P_3 - P_4 \equiv \Delta p_{en}$ to entrance loss; $P_4 - P_5$ is the pressure drop due to steady flow; $P_5 - P_{atm} \equiv \Delta p_{ex}$ is the exit pressure drop

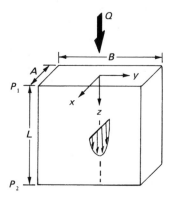

Figure 8.30   Coordinates and dimensions for the slit rheometer

$$B = \frac{R_S}{R} - 1.10 \tag{103}$$

the extrudate swell ratio, corrected for swell observed in slow Newtonian flow.

Corrections   The measurement of ΔP from piston force requires correction for friction, reservoir loss, and entrance loss (note Fig. 8.29). The measurement of Q from piston velocity requires correction for compressibility [7]. Leakage around the piston leads to errors. The value of Q from extrudate weight requires melt-density data. Viscous heating may be a problem [24, 27, 33, 53, 54]. Finally, $2R_s$ is very difficult to measure accurately, owing to density changes and the long time it takes to reach equilibrium [7].

Utility   The capillary rheometer is frequently chosen over drag-flow rheometers because it can achieve higher shear rates with high-viscosity systems. The shape is similar to many dies and pipes in process flows, and it is often used as a process simulator. It is inexpensive, and high accuracy can often be achieved with long capillaries.

   The main disadvantages include variation of shear rate and residence time across the flow, melt fracture and shear heating with high-viscosity samples, and the other corrections given above.

8.4.1.5 Slit rheometer.

Working equations   Refer to Fig. 8.30 for definitions of the parameters. Key assumptions in the analysis are fully developed steady flow and $A/B \leq 0.1$.

Wall shear stress

$$\tau_w = \frac{A\,\Delta P}{2L} \tag{104}$$

Wall shear rate

$$\dot{\gamma} = \frac{6Q}{A^2B} \left[\frac{2}{3} + \frac{1}{3} \frac{d(\ln Q)}{d(\ln \Delta P)} \right] \qquad (105)$$

Viscosity

$$\eta = \frac{\tau_w}{\dot{\gamma}_w} \qquad (106)$$

$$\eta = \frac{A^3B \, \Delta P}{4QL} \frac{n}{2n + 1} \qquad (107)$$

(power law).

Corrections These are the same as for the capillary rheometer.

Utility The main advantage of this device over the capillary rheometer is the flat side wall. This permits the use of flush-mounted pressure transducers and provides for flow visualization. Disadvantages include more difficult construction and possible errors due to side walls.

8.4.1.6 Closing comments. Before closing section 8.4.1, it is worth noting that in all of the geometries discussed above, the shear stress is inferred from total torque, force, or pressure drop measurements. A different setup, the sliding plate rheometer, has recently been proposed [55] in which the shear stress is measured directly and locally with the aid of a shear-stress transducer developed by Dealy and co-workers [56]. This geometry has a number of advantages, including errors introduced by edge effects are eliminated; the actual wetted area does not need to be known; there are no extraneous friction torques or forces affecting the measurements; and large strains are easily achieved [56].

8.4.2 Extensional Rheometry

As indicated in section 8.2.4 for non-Newtonian fluids, extension is fundamentally different from shear; extensional viscosity is a different material function from shear viscosity. Constitutive equations that are similar in shear can predict quite different results in extension. Furthermore, many important flows are highly extensional—fiber spinning, film blowing, and bubble growth, for example. Thus there has been great interest during the last two decades in making extensional rheological measurements.

Another reason for this research activity is that it is so difficult to generate homogeneous extensional flow, especially for low-viscosity liquids, which represent a particular challenge in extensional rheometry. Often different measuring techniques give altogether different results [57]. The basic problem is that flow over stationary boundaries results in shear stresses, but without such boundaries, it is difficult to control the deformation of a low-viscosity fluid. Surface tension, gravity, and flow instabilities all conspire to change the streamlines. A further problem arises from the large strains that are often required for stresses in memory fluids to reach their steady straining limit. In shear flows, streamlines are parallel so that large strains can be achieved with long residence times. The streamlines in extensional flow diverge (or converge), meaning that to achieve infinitely large strain, a sample must become infinitely thin in one direction. It may not be possible to attain a steady rate of extension in some materials, since they may rupture or deform unstably at high strains.

Many different methods have been tried to circumvent these problems and generate purely extensional flows. These are described by Laun [35], Macosko [7], Petrie [58], Dealy [59], James [60], and Whorlow [48]. Here we summarize those extensional rheometers that appear most promising. They are shown schematically in Fig. 8.31. At present, only the first geometry, simple tension, is accepted as an extensional rheometer, and then only for high-viscosity liquids. Below, we give the working equations for two methods of generating uniaxial extension using the tensile geometry. Following this is a short discussion of the other geometries, with pertinent references.

8.4.2.1 Tension-translating clamp.

Working equations See Fig. 8.32. Key assumptions are uniform drawdown, no end effects, and no surface-tension effects.

Strain

$$\varepsilon = \ln \frac{L}{L_0} \tag{108}$$

Strain rate

$$\dot{\varepsilon} = \frac{d(\ln L)}{dt} \tag{109}$$

Stress

$$T_{11} - T_{22} = \frac{F}{A} = \frac{FL}{A_0 L_0} \tag{110}$$

$$T_{11} - T_{22} = \frac{FL}{\pi R_0^2 L_0} \tag{111}$$

(cylinder). Equations (110) and (111) assume an incompressible material with initial cross-sectional area A_0 and length L_0. Cross sections other than cylindrical may be used.

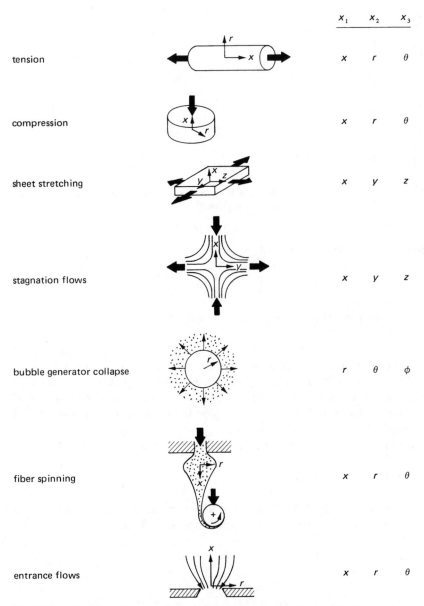

	x_1	x_2	x_3
tension	x	r	θ
compression	x	r	θ
sheet stretching	x	y	z
stagnation flows	x	y	z
bubble generator collapse	r	θ	ϕ
fiber spinning	x	r	θ
entrance flows	x	r	θ

Figure 8.31 Extensional flow geometries. Only the first three geometries give homogeneous deformations

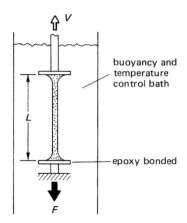

Figure 8.32 Schematic of an extensional rheometer with a translating clamp

Corrections For surface tension [61],

$$T_{11} - T_{22} = \frac{F}{A} - \frac{\gamma}{R}\left(1 - \frac{2R}{L}\right)$$ (112)

For $L/R \geqslant 10$, end effects are believed to be negligible.

Utility Homogeneous deformation and servo control of the clamp permit a wide range of steady and transient tests: start-up, recovery creep, and stress relaxation [62, 63]. The constant strain rate requires an exponentially increasing velocity. Sample preparation requires very high viscosity ($\geqslant 10^6$ Pa s) or a solid sample at ambient temperature in order to clamp or glue the ends. Surface-tension effects and the density difference between the sample and the bath appear to require viscosity $\geqslant 10^4$ Pa s. Limitations on bath length restrict the total strain $\varepsilon \leqslant 4$.

8.4.2.2 Tension-rotating clamp.

Working equations See Fig. 8.33. Key assumptions are uniform drawdown, no end effects, no surface-tension effects, and an incompressible fluid.

Strain

$$\varepsilon = \frac{R\theta}{L}$$ (113)

where θ is the angular displacement.

Strain rate

$$\dot{\varepsilon} = \frac{\Omega R}{L}$$ (114)

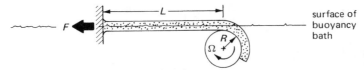

Figure 8.33 Schematic of a rotating-clamp extensional rheometer. Other setups employ pairs of rotating clamps at each end [64–66]

Stress

$$T_{11} - T_{22} = \frac{F}{A} = \frac{F}{\pi R_0^2 e^{-\varepsilon}} \qquad (115)$$

Corrections For surface tension [61],

$$T_{11} - T_{22} = \frac{F}{A} - \frac{\gamma}{R} \qquad (116)$$

Utility The advantages of this device over the translating clamp are (1) a simpler apparatus is required to generate constant strain rate [67–69], (2) a much shorter bath is required to achieve high strains [12, 61, 64–66], (3) higher strain rates may be possible, and (4) density match may not be so critical, since the sample floats. However, tests other than constant rate are generally more difficult. For recoverable strain the sample must be cut and measured. The sample size is generally larger, and sample preparation is more critical. End and thermal effects may reduce the stability of the flow [68] unless pairs of rotating clamps are used at each end and special temperature control is provided [12, 66].

8.4.2.3 Other extensional rheometers. The other geometries shown in Fig. 8.31 are either less well established or are not able to generate steady extension [Eq. (30)]. However, because the simple-tension rheometers are apparently limited to high-viscosity samples, there is continued interest in these other test geometries.

Compression is the opposite of tension, but in practice, the test requires a short sample to prevent buckling. If the ends are clamped as in the tension tests, considerable shear deformation will develop. Thus the ends must be maintained parallel but allowed to slip. This can be achieved by lubricating the end plates with a low-viscosity liquid, as Chatraei et al. [70] have demonstrated. This technique can be simplified by just coating the sample ends [70–72].

One of the problems of this method is the loss of lubricant, which typically occurs at a Henky strain ε [$\equiv -\ln(L/L_0)/2$] of about unity [73]. The viscosity of the lubricant plays an important role [74]; if it is too small, the lubricant will be rapidly squeezed out, while if it is too large, there will be shear as well as extension in the sample. Planar squeezing can also be obtained with an adaptation of the squeezing technique, as recently done by Khan and Larson [75].

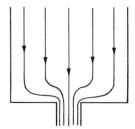 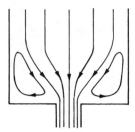

Figure 8.34 Streamlines for entrance flow of two different fluids. (*a*) No recirculation, typical of high-density polyethylene. (*b*) With recirculation, typical of low-density polyethylene and polystyrene melts. (From [49])

Equal biaxial stretching of a sheet can also be used to generate steady compression. Stephenson and Meissner [76] have built a sophisticated device for sheet stretching, with eight pairs of rotating clamps arranged in a circle and eight automated scissors to cut the sheet between the rollers. Controlling the rollers at the same speed generates equal biaxial stretching, while other programs can give any combination of extension rates. Again, this method appears limited to high-viscosity samples with $\eta \geq 10^5$ Pa s.

Sudden contraction flows (flows in tubes or channels with an abrupt variation of cross-sectional area) have a strong extensional component and are not limited to high-viscosity materials. Cogswell [77] has reviewed attempts to use converging flows to determine extensional viscosity. However, for non-Newtonian fluids the entrance flow pattern can change dramatically with different materials, as shown in Fig. 8.34 [78–83]. Cogswell [84] and Binding [85, 86] reported simplified analyses that give the extensional viscosity from measurements of entrance pressure drop (Δp_{en}; see Fig. 8.29). Binding's analysis is more complete, but the working equations are not as easy to use as the ones given by Cogswell's analysis [87]. Steady extension can be achieved by a properly shaped die and the use of a thin lubricant layer at the wall [88–90]. Macosko et al. [89] measured steady planar extensional viscosities on a polystyrene melt with such a lubricated die. However, these experiments are difficult and messy and, given the results reported so far, do not seem to be worth the trouble.

Planar extension can also be achieved by pressurizing a hollow cylinder of melt [13, 91]. By pulling on the tube, uniaxial extension can be superposed on the planar extension. Fabrication of the hollow cylinder samples is typically difficult and only possible for high-viscosity materials.

Bubble growth, shown in Fig. 8.31, gives uniaxial compression, while bubble collapse gives extension. This method is attractive for low-viscosity liquids. The deformation is not homogeneous throughout the sample, and the determination of extensional viscosity requires the assumption of a constitutive equation [92–96].

With lower-viscosity liquids ($\eta < 10^3$ Pa s), extensional rheometry is much

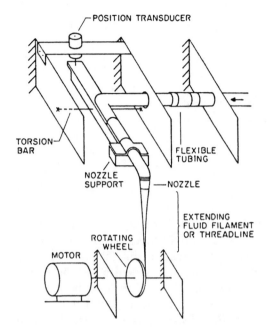

Figure 8.35 Typical fiber-spinning apparatus. (From [97])

more problematic. Yet, the effects of extensional flow can be much greater for polymer solutions than for polymer melts. Because an experimental method for generating a purely extensional flow of low-viscosity fluids is not known, we have to content ourselves with indexing methods. These indexing techniques rely on flows that have a strong extensional component but that also have shear. Moreover, the strain history is typically poorly defined. Nonetheless, because the extensional effects are in general very strong, these indexing techniques are useful in the characterization of processing materials.

The most common indexing method is perhaps fiber spinning, owing to its applicability to industrial processes [58]. Figure 8.35 shows a typical experimental setup [97]. In this technique the material is continuously extruded from a tube, and the outcoming fiber is collected and stretched by a rotating wheel or vacuum suction [7]. Ideally, the fiber radius should decrease with the square root of the distance from the exit, but photographic results show that it usually decreases with a different power [58, 97, 98]. Surface tension, gravity, inertia, and the elastic memory of the fluid can all cause deviation from ideality.

A variation of this method that has the advantage of a well-defined strain history is the so-called tubeless syphon [60]. A nozzle is dipped in a bath of the test fluid, a vacuum applied, and the fluid sucked out of the bath. The nozzle is slowly raised, and a free standing, rising column of fluid develops.

A newer measuring method for extensional viscosity is the opposed-nozzles

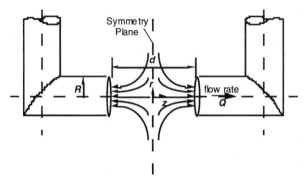

Figure 8.36 Flow between opposed nozzles

technique developed by Fuller et al. [99]. Two tube inlets are positioned opposing each other (Fig. 8.36), and when fluid is sucked into the tubes, a stagnation flow is created in the region between the nozzles. The fluid tends to pull the tubes together, and the reaction force is measured. If we assume pure extension in the region between the nozzles, and no flow far away, we can get a single relation between the measured force and the stress difference $T_{11} - T_{22}$. Conservation of mass yields directly that the extension rate is $\dot{\varepsilon} = 2Q/(\pi R^2 d)$, where Q is the volume flow rate at each nozzle, R is the nozzle inner radius, and d is the nozzle spacing.

It should be pointed out that the flow is neither purely extensional nor homogeneous, as assumed in the analysis [22], so that the strain history of all fluid particles is not the same. An assessment of the effects of shear and flow inhomogeneities on the measurements is given by Souza Mendes et al. [100]. The advantages of this technique are that it allows measurements for a wide range of extension rates; fluids can be of viscosities as low as 10^{-2} Pa s; and it is a simple and clean experiment. A commercial instrument using this technique is available by Rheometrics, Inc. Cai et al. [101] compared extensional viscosities obtained using the opposed-nozzles device with other measurement techniques, and considerable disparities were observed.

James et al. [102, 103] recently proposed a measurement technique using a converging channel flow at high Reynolds numbers. It is based on the idea that shearing is confined to the thin boundary layers near the channel walls. Thus a shear-free extensional flow is attained in the core. In the convergent section a nearly constant $\dot{\varepsilon}$ is produced, provided no vortices appear. The absence of vortices remains to be checked via flow visualization. Limitations of the technique include the presence of a preshear history and the difficulty in validating it by measurements with Newtonian fluids [60].

Sridhar et al. [104] discuss a filament-stretching method that is also applicable to polymer solutions. The sample is positioned between coaxial discs, which start moving relative to each other with an exponentially increasing velocity. This gives a constant $\dot{\varepsilon}$, and the technique has the advantage of a well-defined deformation history. One deficiency of this method is that it seems to be limited to rather low extension rates.

8.5 MEASUREMENTS IN COMPLEX FLOWS

As discussed above, in a rheometer the kinematics of the flow is determined from some simple external measurement such as flow rate or angular velocity. Now we turn our attention to complex flows in which the kinematics are unknown. As pointed out in the introduction, experimental analysis of a complex non-Newtonian flow usually requires measurement of both the velocity field and the stresses. In this section we discuss special problems encountered in making these measurements in non-Newtonian fluids. The basic techniques, such as pressure transducers and hot-wire or laser-Doppler anemometry, are discussed in the other chapters of this book. Here we concentrate on the differences in using these methods with non-Newtonian fluids. Before looking at each method, we make some general comments on working with non-Newtonian materials.

Essentially all non-Newtonian liquids are either dispersions of solid or liquid particles or polymer solutions and melts. Since non-Newtonian fluids tend to be of high viscosity, most flow problems are of low Reynolds number, and turbulence is usually not encountered. In fact, the addition of small amounts of axisymmetric particles or polymers tends to suppress turbulence. The high viscosity of many non-Newtonian materials can, however, cause other problems not usually encountered in flow measurements.

Two common assumptions in Newtonian studies, namely, isothermal flow and no slip at the wall, need to be checked carefully. Because of their high viscosity and relatively low thermal conductivity, non-Newtonian fluids, and in particular, polymer melts, can generate significant heat through viscous dissipation. Temperature increases of over 50°C have been measured for polymer melts flowing through a slit or tube [27, 33]. Temperature profiles can be highly nonuniform, since generation is proportional to the square of the velocity gradient. Winter [53] has done an extensive review of this problem area.

With high-viscosity polymer melts and concentrated suspensions, particularly those that show a yield stress, it appears that the no-slip wall boundary condition can fail to hold. Chauffoureaux et al. [105] reported direct evidence for wall slip with polyvinyl chloride melts. Snelling and Lontz [106] and Uhland [107] indicated that slip occurs for other polymer melts in die flows. Kraynik and Schowalter [108] have shown evidence for wall slip with aqueous solutions of polyvinyl alcohol and sodium borate using a hot-film anemometer at the wall.

Careful experiments in a capillary rheometer were conducted by Ramamurthy [109, 110], who reported finite slip velocities for HDPE and LLDPE resins. Hatzikiriakos and Dealy [111] used the slide plate rheometer described in section 8.4 to obtain clear experimental evidence of slip in HDPE.

Denn [112] recently discussed the implications of slip in extrudate surface defects such as sharkskin and melt fracture. Piau and El Kissi [113] present another point of view regarding these same defects, which illustrates that little is known about the physical mechanisms that govern the appearance of these phenomena.

Although the nonslip assumption is valid in most cases, these counterexamples suggest that it must be checked in each case where high-viscosity, non-Newtonian materials are being studied.

Another problem with non-Newtonian fluids is composition instability. Since they are generally prepared from two or more components, it is possible for relative concentrations to change during experiments, owing to solvent evaporation, degradation of polymer, or settling out of suspended particles. Water-soluble polymer molecules are particularly subject to bacteriological attack. Fungicides should be added to samples that must be used for more than several days. The high temperatures necessary to process molten polymers can lead to thermal and oxidative degradation. High-molecular-weight polymer chains are also susceptible to mechanical degradation in strong flow fields [114].

With concentrated suspensions, settling and sample uniformity are major problems. In fact, the effect of changing sample homogeneity during a complex flow may be as important as the non-Newtonian properties. As is discussed below, particles migrate during flow. Suspended particles can settle out completely in bends and expansions. In flows of rodlike particles, such as glass-fiber suspensions, it can be particularly difficult to obtain reproducible results.

8.5.1 Pressure Measurements

Special problems arise in pressure measurements on non-Newtonian fluids, due to high viscosity, high temperature, and elastic effects.

High viscosity means high pressures. Over 1000 atm is not uncommon in polymer processing equipment. Sensitivity is typically adequate down to one-hundredth or one-thousandth of full scale; thus several transducers with different ranges are often needed for a given study. With molten polymers, temperatures can be as high as 350°C. Zero shift due to temperature changes can be particularly troublesome and should be checked carefully in setting up an experiment. The calibration constant is less sensitive to temperature, but this should also be verified periodically during service. A number of pressure transducers that have been designed for the plastics industry can operate accurately under high-temperature and wide pressure-range conditions (for example, those manufactured by Dynisco, Kulite and Sensotec).

One of the surprising discoveries in non-Newtonian fluid mechanics concerns the influence of mounting holes on pressure readings. Typically, pressure transducers are mounted with their sensing diaphragms not in direct contact with the flow. Often, these diaphragms are rather large. Typically, a small fluid-filled hole is used to connect the transducer to the wall of the flow field of interest. In using manome-

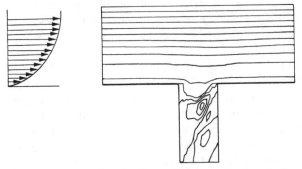

Figure 8.37 Streamlines for Couette flow near a wall cavity from numerical calculations of Crochet and Bezy [116]

ters, such a connection is a necessity. It had been established that the size and shape of such holes had no effect with Newtonian fluids, and it was assumed for many years that this was also true for non-Newtonian fluids. However, in 1968, by comparing various normal-stress measurements in rheometers with pressure holes and by total thrust, Lodge and coworkers [115] discovered a significant influence of these pressure holes.

Figure 8.37 illustrates the problem. We see that the streamlines for shear flow near a wall are bent as the flow passes over a cavity in the wall. Non-Newtonian fluids can have a normal tension along the streamline. As the flow is deflected, this normal tension tends to "pull" fluid out of the hole, reducing the pressure read by the transducer. Experimental and theoretical studies [115, 117] show that this pressure error, the difference between the pressure at the bottom of the cavity and at a flush transducer, is about -20% of the first normal-stress difference:

$$p_{\mathrm{H}} \simeq -0.2(T_{11} - T_{22}) \tag{117}$$

The hole error can be eliminated by using a flush-mounted transducer. Or, since the error will be approximately the same for each hole, pressure differences can be correctly measured. The pressure-hole error seems to be less significant for high-viscosity liquids, probably because the shear stresses are so much larger than the normal stresses; i.e., the flow has small Weissenberg number [118, 119].

One advantage of the pressure-hole error is that it can be used to measure normal stresses. This has been applied to the design of an on-line rheometer [117, 120].

8.5.2 Velocity Measurements

Most velocity measurement methods rely on optics. This means that the fluid must be transparent to the wavelength of interest. This can be a problem, since many non-Newtonian fluids are opaque or contain too many small particles. If gas bubbles become entrained in a high-viscosity, non-Newtonian fluid, they can be very difficult to remove. A few can be tolerated or even used for tracers, but a large number will scatter too much light.

Hydrogen-bubble generation by water hydrolysis from a wire stretched across the flow can be used with water-based fluids. The same techniques as with non-Newtonian fluids can be applied [121]. However, with polymer solutions, polymer degrades on the wire, fouling it and reducing the uniformity of the generation pattern. This is shown by the streaks in Fig. 8.38. If the wire is carefully cleaned periodically, the method can be utilized. Another problem is wire bending and breakage, owing to the high stresses generated by viscous liquids. The wire curvature is apparent in Fig. 8.38.

Similar problems are encountered in using hot-film or hot-wire anemometry, i.e., fouling due to thermal degradation of the sample and mechanical failure due to high viscosity. As discussed above, high-viscosity fluids can lead to viscous dissipation around the probe, altering the heat transfer character and thus the interpretation of the results. The nature of laminar heat transfer from non-Newtonian fluids has not been well studied. Kraynik and Schowalter [108] discuss some of these problems with hot-film anemometry.

Particle tracer methods for flow visualization have been used extensively to measure velocity profiles in non-Newtonian fluids. Most materials already have many (often too many) impurity particles that can serve as tracers. An important problem in these studies is that of particle migration. Leal [122] has reviewed this subject. Due to inertia or deformability, particles in Newtonian fluids can migrate. If the particles are small enough, Brownian motion and viscosity counterbalance the migration, and the tracers follow fluid path lines. However, in non-Newtonian fluids, additional stresses can act on particles. Karnis and Mason [123] report that particles in non-Newtonian Couette flow actually migrated opposite to the direction expected for Newtonian fluids.

Laser-Doppler anemometry is subject to the same contamination and tracer problems as discussed above. Nonisothermal conditions can cause distortion, owing to the temperature dependence of the index of refraction. Since most non-Newtonian fluids are strongly shear thinning, velocity profiles are steeper near walls than for Newtonian fluids. This can present special measurement problems, both in focusing the crossed beams near the wall and in dealing with the finite scattering volume. The problems can be compounded by the relatively low velocities that are often of interest in these more viscous materials. Special low-frequency techniques need to be employed. Kraemer and Meissner [124] discuss these problems in a laser-Doppler study on polymer melts. Boger and Mackay [125] have also reviewed these and other aspects of flow visualization in polymer solutions.

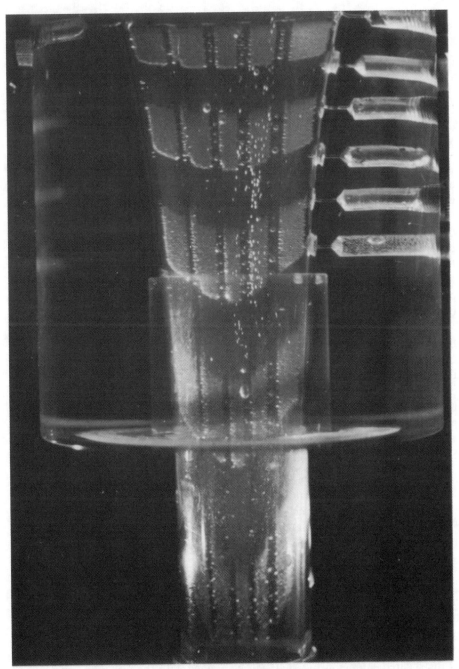

Figure 8.38 A 3.5% polyacrylamide solution in water flowing over a 25-μm-diameter tungsten wire. A voltage of 100 V was used to generate the hydrogen bubbles shown. (From [88])

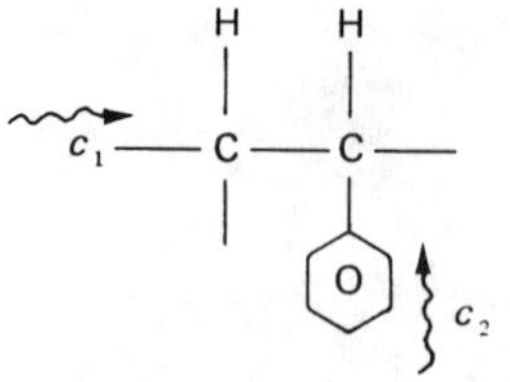

Figure 8.39 Light travels faster across a polystyrene chain than along it: $c_1 \gg c_1$

8.5.3 Flow Birefringence

To fully characterize a complex flow of a non-Newtonian fluid, the stress as well as velocity distribution are needed. Flow birefringence can be used to obtain stress distributions in transparent polymeric liquids [126].

Nearly all asymmetric molecules also show optical anisotropy; i.e., their index of refraction is different in different directions along the molecule. For example, in polystyrene, light travels faster across the chains, through the large benzene rings, than along them (Fig. 8.39). Thus, if polarized light is transmitted through an oriented polystyrene sample, it will be separated into two mutually perpendicular components that are out of phase with each other and rotated with respect to the incident beam. This effect is called double refraction or birefringence.

Birefringence has been used extensively to study stress patterns in solids. This field is often referred to as photoelasticity [127]. To see how the effect is used, consider a solid rectangle subjected to simple shear (Fig. 8.40). The stress distribution can be represented by an ellipsoid with principal axes p_I and p_{II}. The stress ellipsoid can be related to the stresses measured on the boundaries of the solid by the difference between the principal stresses $\Delta p = p_I - p_{II}$, and the angle χ_p that p_I makes with respect to the shear direction χ_1.

Shear stress

$$T_{12} = \frac{1}{2}\,\Delta p \, \sin(2\chi_p) \tag{118}$$

Normal-stress difference

$$T_{11} - T_{22} = \Delta p \cos (2\chi_p) \tag{119}$$

Since this stress field will orient the molecules of the solid, we will see birefringence (if it is transparent). In principle, since stress and the index of refraction are both second-rank tensors, we need a fourth-rank tensor (81 components) to relate them. However, for many homogeneous, amorphous materials such as glass and many polymers, the two tensors are related by a simple constant called the stress optical coefficient C. Thus the two principal axes of the polarized light in the plane of the sheared sample in Fig. 8.40 will be aligned with the principal stresses, and their difference will be proportional to Δp:

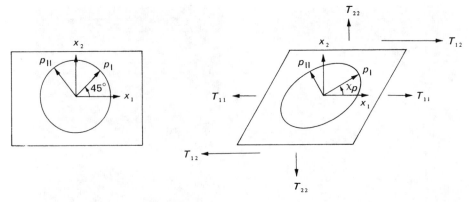

Figure 8.40 Stress ellipsoid in simple shear

$$\chi_n = \chi_p \qquad (120)$$

$$\Delta n = C\Delta p \qquad (121)$$

These same relations should hold for homogeneous liquids. They can be tested by measuring both the principal stresses and birefringence in the same flow. This has been done for a number of Newtonian and some polymeric liquids [126].

Figure 8.41 shows χ_n and Δn versus shear stress for decalin, a Newtonian oil, and polyisobutylene in decalin. For the decalin, we see that $\chi_n = 45°$ and is independent of shear stress, which by Eq. (119) means there are no normal shear stresses, as expected for a Newtonian fluid. For the polymer solution, we see that χ_n decreases, indicating normal stresses. The χ_p from cone-and-plate thrust mea-

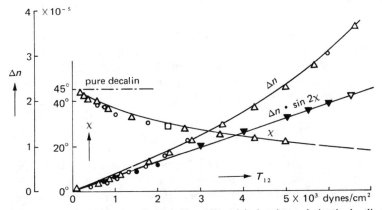

Figure 8.41 Test of stress-optical law for a 15% polyisobutylene solution in decalin: (□) χ_p from total thrust in cone and plate and χ_n, from birefringence data; (○) at 30°C; and (△) at 50°C. (From [126])

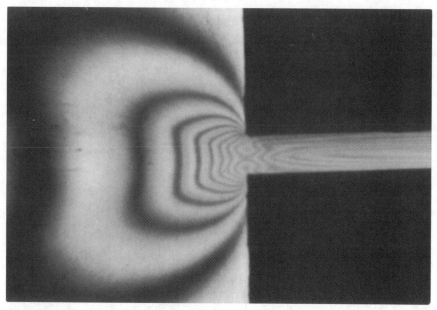

Figure 8.42 Photograph through crossed polarizers of high-density polyethylene at 200°C entering a slit die at 17cm³/min. (From [128])

surements [see Eq. (88)] is in good agreement with χ_n. As expected from combining Eqs. (121) and (118), $\Delta n \sin (2\chi)$ gives a straight line versus T_{12}. The slope is C, the stress optical coefficient. The only exceptions to $C = $ const appear to be suspensions or dilute polymer solutions, in which the solvent and polymer indexes of refraction are not closely matched. In these cases the shape of the particle or of the polymer will result in an additional "form" birefringence that cannot readily be removed.

The great advantage of flow birefringence is that it can provide the needed stress distribution in a non-Newtonian flow. The technique can also be sensitive to low stress levels. For example, in Fig. 8.41, we see that 10^3 dyn/cm² (~0.02 psi) gives $\Delta n \simeq 3 \times 10^{-7}$. Such changes in Δn are readily measurable. This can be seen from the basic relation between Δn and the measured phase shift δ:

$$\Delta n = \frac{\delta \lambda}{2\pi L} \tag{122}$$

where λ is the wavelength of the light used and L is its path length through the sample. Thus for a mercury lamp (560 nm) and a 1-cm sample thickness, $\Delta n = 3 \times 10^{-7}$ means $\delta \simeq 2°$, which can readily be detected with a suitable compensator. Another advantage of birefringence is that it can be combined with laser-Doppler anemometry. Some of the same optical systems can be used for both techniques [129].

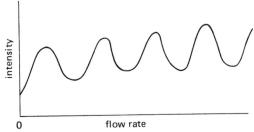

Figure 8.43 Representative output from a photocell during a steady increase in flow rate for a birefringence experiment

The birefringence effect can be demonstrated by simply placing crossed polars over a section of the flow with transparent boundaries. Figure 8.42 shows the birefringence around the entrance to a slit die. If white light is used, the fringes are colored; with monochromatic light, they are black and white. From Fig. 8.42 we can estimate Δn by counting the fringes. If there are no more fringes outside the field of the photo, then for the small circular fringe at the entrance to the slit, $\delta = 9(2\pi)$. Thus, by Eq. (111), assuming 560-nm light and $L \simeq 8$ mm gives $\Delta n \simeq 4 \times 10^{-4}$. Using $C \simeq 2 \times 10^{-9}$ Pa^{-1} for polyethylene, the difference in principal stresses at the entrance is $\Delta p \simeq 2 \times 10^5$ Pa $\simeq 2$ atm.

To measure δ more precisely, a photomultiplier can be used. If the flow rate is slowly increased from zero, the trace of intensity from the photomultiplier will look like Fig. 8.43, an oscillation with each 2π change in δ. Calibration of the oscillation pattern can give δ fairly accurately; however, at the extrema the resolution is not as good. Janeschitz-Kreigl [126] and Gortemacher [130] discuss a modulation method for improving resolution with the photomultiplier technique.

The classical method for precise birefringence measurements is to use a slit of light and a compensator that can accurately measure χ_n and δ. Such an optical system is shown in Fig. 8.44. With this arrangement the light is in the x_3 direction, and the birefringence is proportional to the principal stresses in the $x_1 x_2$ plane, the plane of Fig. 8.40. Other planes can be studied if the light can be introduced in another direction. However, great care must be taken against distortion of the beam. The two major problems are temperature gradients and "parasitic" birefringence from the windows and from end effects in the flow near them. The index of refraction is a function of temperature; typically, $dn/dT \simeq 10^{-3}$ (°C)$^{-1}$. Thus a relatively small temperature gradient, $dT/dx \simeq 0.1$ to 1°C/mm, can cause the incident light to bend and reflect off one of the walls. Janeschitz-Kriegl [126] analyzes this problem for the Couette rheometer. He shows that reflections from metallic walls are particularly bad, since they have high adsorption coefficients. The use of glass walls provides nearly perfect reflection and allows one to obtain data despite some temperature gradients. Of course, better temperature control and shorter path

Ocular

lens

analyzer

quarter wave plate

sample

diaphragm

polarizer

lens

slit

lens
lens

source monochromatic

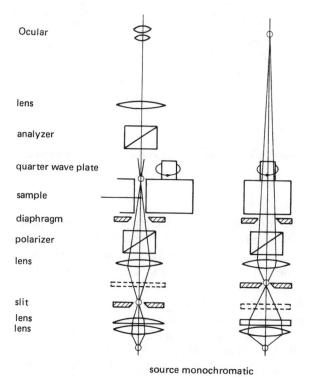

Figure 8.44 Optical alignment of birefringence apparatus for Couette flow. (From [89])

lengths reduce the problem. Janeschitz-Kriegl used a slow crossflow in his Couette apparatus to improve temperature control.

Parasitic birefringence can come from stresses inherent in the glass windows and from thermal stresses and pressure on the glass that can arise during operation. The shear flow over the windows will also generate birefringence. The main approach is to match each window and the flow field over it as closely as possible in an attempt to cancel out the birefringence from both. Stress-free glass can be obtained, and the uses of rubber gaskets can reduce the thermal stress problem. A fairly recent review on optical rheometry is available by Fuller [131], where recent advances in flow birefringence are discussed. The subject is presented in detail by Lodge in Chapter 9 of Macosko [7].

NOMENCLATURE

a_T temperature-shift factor in Eq. (49)
a_1, a_2, a_3 constants

b	constant
$\mathbf{B}$	Finger strain tensor
C	stress optical coefficient
$\mathbf{C}$	right Cauchy-Green strain tensor
C_1	constant
C_2	constant
$d\Phi/dt$	material time derivative of scalar Φ, $= \partial\Phi/\partial t + \mathrm{grad}\,\Phi \cdot \mathbf{v}$
$d\mathbf{A}/dt$	material time derivative of vector or tensor $\mathbf{A}$; $= (\partial\mathbf{A}/\partial t) + \mathbf{LA}$
$d\mathbf{x}$	length of a material filament at time t
$d\mathbf{x}'$	length of a material filament at time t'
$\mathbf{D}$	rate-of-deformation tensor
De	Deborah number
D_{ij}	ijth component of $\mathbf{D}$
$\hat{\mathbf{e}}$	unit vector
$\mathbf{E}$	small-strain tensor, $= \lim_{\gamma \to 0}(\mathbf{1} - \mathbf{B})$
E_η	viscosity activation energy
$\mathbf{F}$	deformation gradient tensor
F_{ij}	ijth component of $\mathbf{F}$
G	elastic modulus
G'	elastic or in-phase modulus
G''	viscous or out-of-phase modulus
$G(t - t_0, \gamma_0)$	stress-relaxation modulus
h	strain-dependent damping function
$H(t)$	Heaviside unit step function
$\mathbf{I_D}, \mathbf{II_D}, \mathbf{III_D}$	invariants of the rate-of-deformation tensor
J_e	creep compliance
J_e^0	limit of J_e as $\tau_0 \to 0$
k	relaxation times
K	constant
$\mathbf{L}$	velocity gradient tensor
L_{ij}	ijth component of $\mathbf{L}$
m	preexponential constant in Eq. (43)
M	memory function
n	power law exponent
$N_1(\dot{\gamma})$	first normal-stress difference
$N_2(\dot{\gamma})$	second normal-stress difference
p	pressure
Re	Reynolds number
t	time
t_{ref}	reference configuration time
$\mathbf{T}$	stress tensor
T_{ij}	ijth component of $\mathbf{T}$
$\mathbf{v}$	velocity vector

$\mathbf{W}$	vorticity tensor
W_{ij}	ijth component of $\mathbf{W}$
Ws	Weissenberg number
$\mathbf{x}$	position vector
$\mathbf{1}$	unit tensor
$\alpha_0, \alpha_1, \alpha_2$	coefficients
$\beta_1, \beta_2, \beta_3$	coefficients
$\gamma_{t_{\text{ref}}}(t)$	total shear strain during time interval $(t - t_{\text{ref}}) = \int_{t_{\text{ref}}}^{t} \dot{\gamma}(t')dt'$
$\dot{\gamma}$	deformation rate, $= \sqrt{2 \text{ tr } \mathbf{D}^2}$, or shear rate
$\dot{\varepsilon}$	extension rate
η	non-Newtonian shear viscosity
$\eta_e(\dot{\varepsilon})$	elongational viscosity
η^*	complex viscosity
η'	real part of complex viscosity
η''	imaginary part of complex viscosity
η_0, η_∞	low-shear-rate and high-shear-rate limiting viscosities
$\eta^+(t, \dot{\gamma}_0)$	start-up shear-stress material function
$\eta^-(t, \dot{\gamma}_0)$	relaxation shear-stress material function
λ_k	relaxation time
μ	Newtonian viscosity
$\boldsymbol{\tau}$	extra-stress tensor
τ_{ij}	components of $\boldsymbol{\tau}$
τ^*	complex shear stress
τ'	real part of complex shear stress
τ''	imaginary part of complex shear stress
χ	angle of principal stress with respect to shear direction x_1
Ψ_1	first normal-stress coefficient
Ψ_2	second normal-stress coefficient

REFERENCES

1. R. I. Tanner, *Engineering Rheology,* Oxford University Press, 1985.
2. R. B. Bird, R. C. Armstrong, and O. Hassager, *Dynamics of Polymeric Liquids,* vol. 1, 2d ed., Wiley, New York, 1987.
3. S. Middleman, *Fundamentals of Polymer Processing,* McGraw-Hill, New York, 1977.
4. G. Astarita and G. Marrucci, *Principles of Non-Newtonian Fluid Mechanics,* McGraw-Hill, New York, 1974.
5. W. R. Schowalter, *Fundamentals of Non-Newtonian Fluid Mechanics,* Wiley, New York, 1979.
6. R. G. Larson, *Constitutive Equations for Polymer Melts and Solutions,* Butterworths, Boston, Mass., 1988.
7. C. W. Macosko, *Rheology: Principles, Measurements and Applications,* VCH, New York, 1994.
8. H. Brenner, *Int. J. Multiphase Flow,* vol. 1, p. 195, 1974.
9. C. W. Macosko and W. M. Davis, *Rheol. Acta,* vol. 13, p. 814, 1974.
10. W. W. Graessley, *Adv. Polym. Sci.,* vol. 16, p. 1, 1974.

11. J. D. Ferry, *Viscoelastic Properties of Polymers,* 3d ed., Wiley, New York, 1980.
12. T. Raible, A. Demarmels, and J. Meissner, *Polym. Bull.,* vol. 1, p. 397, 1979.
13. H. M. Laun and H. Schuch, *J. Rheol.,* vol. 33, p. 119, 1989.
14. H. A. Barnes, J. F. Hutton, and K. Walters, *An Introduction to Rheology,* Elsevier, Amsterdam, 1989.
15. M. H. Wagner, *Rheol. Acta,* vol. 15, p. 136, 1976.
16. A. S. Argon (ed.), *Constitutive Equations in Plasticity,* MIT Press, Cambridge, Mass., 1975.
17. D. D. Joseph, *Fluid Dynamics of Viscoelastic Liquids,* Springer-Verlag, New York, 1990.
18. C. Truesdell and W. Noll, *The Non-Linear Field Theories of Mechanics,* Springer-Verlag, Berlin, 1965.
19. P. Chadwick, *Continuum Mechanics,* Halsted-Wiley, New York, 1976.
20. L. E. Malvern, *Introduction to the Mechanics of Continuous Media,* p. 37, Prentice-Hall, Englewood Cliffs, N.J., 1969.
21. B. Debbaut and M. J. Crochet, *J. Non-Newtonian Fluid Mech.,* vol. 30, p. 169, 1988.
22. P. R. Souza Mendes, M. Padmanabhan, L. E. Scriven, and C. W. Macosko, *Rheol. Acta,* vol. 34, p. 209, 1995.
23. R. Schunk and L. E. Scriven, *J. Rheol.,* vol. 34, p. 1085, 1990.
24. J.-F. Agassant, P. Avenas, J.-Ph. Sergent, and P. J. Carreau, *Polymer Processing,* Oxford University Press, New York, 1991.
25. Z. Tadmor and C. G. Gogos, *Principles of Polymer Processing,* Wiley, New York, 1979.
26. R. B. Bird, *Annu. Rev. Fluid Mech.,* vol. 8, p. 13, 1976.
27. H. W. Cox and C. W. Macosko, *AIChE J.,* vol. 20, p. 785, 1974a.
28. R. B. Bird and P. J. Carreau, *Chem. Eng. Sci.,* vol. 23, p. 427, 1968.
29. K. Yasuda, R. C. Armstrong, and R. E. Cohen, *Rheol. Acta,* vol. 20, p. 163, 1981.
30. Z. Tadmor and I. Klein, *Engineering Principles of Plasticating Extrusion,* Reinhold, New York, 1970.
31. D. W. Van Krevelen, *Properties of Polymers,* 2d ed., Elsevier, Amsterdam, 1977.
32. P. H. Goldblatt and A. Porter, *J. Appl. Polym. Sci.,* vol. 20, p. 1199, 1976.
33. H. W. Cox and C. W. Macosko, *Soc. Plast. Eng. Tech. Pap.,* vol. 20, p. 28, 1974.
34. A. G. Fredrickson, *Principles and Applications of Rheology,* Prentice-Hall, Englewood Cliffs, N.J., 1964.
35. H. M. Laun, *Rheol. Acta,* vol. 17, p. 1, 1978.
36. J. L. White and A. B. Metzner, *J. Appl. Polym. Sci.,* vol. 8, p. 1367, 1963.
37. N. Phan-Thien and R. I. Tanner, *J. Non-Newtonian Fluid Mech.,* vol. 2, p. 353, 1977.
38. D. D. Joseph and R. Fosdick, *Arch. Ratl. Mech. Anal.,* vol. 49, p. 321, 1972.
39. W. O. Criminale Jr., J. L. Ericksen, and G. L. Filbey Jr., *Arch. Ratl. Mech. Anal.,* vol. 1, p. 410, 1958.
40. S. Rivlin, in F. R. Eirich (ed.), *Rheology,* vol. 2, p. 351, Academic, New York, 1956.
41. L. R. G. Treloar, *The Physics of Rubber Elasticity,* 3d ed., Oxford, New York, 1974.
42. A. S. Lodge, *Elastic Liquids,* Academic, New York, 1964.
43. M. Doi and S. F. Edwards, *J. Chem. Soc. Faraday Trans. II,* vol. 74, pp. 1789, 1802, 1818, 1978; vol. 75, p. 32, 1979.
44. A. C. Papanastasiou, L. E. Scriven, and C. W. Macosko, *J. Rheol.,* vol. 27, p. 387, 1983.
45. R. I. Tanner, in P. Moldenaers and R. Keunings (eds.), *Theoretical and Applied Rheology,* vol. 1, p. 12, Elsevier, Amsterdam, 1992.
46. S. Middleman, *The Flow of High Polymers,* Wiley, New York, 1968.
47. K. Walters, *Rheometry,* Halsted-Wiley, London, 1975.
48. R. W. Whorlow, *Rheological Techniques,* 2d ed., Ellis Horwood, London, 1992.
49. J. M. Dealy, *Rheology for Molten Plastics,* Van Nostrand Reinhold, New York, 1982.
50. J. M. Dealy and K. F. Wissbrun, *Melt Rheology and Its Role in Plastics Processing,* Van Nostrand Reinhold, New York, 1990.

51. J. L. Schrag, *Trans. Soc. Rheol.*, vol. 21, p. 399, 1977.
52. M. M. Denn and J. Roisman, *AIChE J.*, vol. 15, p. 545, 1969.
53. H. H. Winter, *Adv. Heat Transfer*, vol. 13, p. 205, 1977.
54. M. S. Carvalho and P. R. Souza Mendes, *ASME J. Heat Transfer*, vol. 114, p. 582, 1992.
55. A. J. Giacomin, T. Samurkas, and J. M. Dealy, *Polym. Eng. Sci.*, vol. 29, p. 499, 1989.
56. J. M. Dealy, in P. Moldenaers and R. Keunings (eds.), *Theoretical and Applied Rheology*, vol. 1, p. 39, Elsevier, Amsterdam, 1992.
57. K. Walters, in P. Moldenaers and R. Keunings (eds.), *Theoretical and Applied Rheology*, vol. 1, p. 16, Elsevier, Amsterdam, 1992.
58. C. J. S. Petrie, *Extensional Flows*, Pittman, London, 1979.
59. J. M. Dealy, *J. Non-Newtonian Fluid Mech.*, vol. 4, p. 9, 1978.
60. D. F. James, in D. deKee, and P. N. Kaloni (eds.), *Recent Developments in Structural Continua*, vol. 2, p. 217, John Wiley, New York, 1990.
61. H. M. Laun and H. Münstedt, *Rheol. Acta*, vol. 17, p. 415, 1978.
62. H. Münstedt, *J. Rheol.*, vol. 23, p. 421, 1979.
63. V. S. Au-Yeung and C. W. Macosko, in G. Astarita, G. Marrucci, and L. Nicolais (eds.), *Rheology*, vol. 2, *Applications*, p. 717, Plenum, New York, 1980.
64. J. Meissner, *Rheol. Acta*, vol. 8, p. 78, 1969.
65. J. Meissner, *J. Appl. Polym. Sci.*, vol. 16, p. 2877, 1972.
66. T. Raible and J. Meissner, in G. Astarita, G. Marrucci, and L. Nicolais (eds.), *Rheology*, vol. 2, *Applications*, p. 425, Plenum, New York, 1980.
67. C. W. Macosko and J. M. Lorntson, *Soc. Plast. Eng. Tech. Pap.*, vol. 19, p. 461, 1973.
68. R. W. Connelly, L. J. Garfield, and G. H. Pearson, *J. Rheol.*, vol. 23, p. 651, 1979.
69. Y. Ide and J. L. White, *J. Appl. Polym. Sci.*, vol. 20, p. 2511, 1976; vol. 22, p. 1061, 1978.
70. S. Chatraei, C. W. Macosko, and W. W. Winter, *J. Rheol.*, vol. 25, p. 433, 1981.
71 P. R. Soskey and H. H. Winter, *J. Rheol.*, vol. 29, p. 493, 1985.
72. S. A. Khan, R. K. Prud'homme, and R. G. Larson, *Rheol. Acta*, vol. 26, p. 144, 1987.
73. A. C. Papanastasiou, C. W. Macosko, and L. E. Scriven, *J. Numer. Methods Fluids*, vol. 6, p. 819, 1986.
74. R. B. Secor, Ph.D. thesis, University of Minnesota, 1988.
75. S. A. Khan and R. G. Larson, *Rheol. Acta*, vol. 30, p. 1, 1991.
76. S. E. Stephenson and J. Meissner, in G. Astarita, G. Marrucci, and L. Nicolais (eds.), *Rheology*, vol. 2, *Applications*, p. 431, Plenum, New York, 1980.
77. F. N. Cogswell, *J. Non-Newtonian Fluid Mech.*, vol. 4, p. 23, 1978.
78. A. E. Everage and R. L. Balman, *Nature*, vol. 273, p. 213, 1978.
79. D. V. Boger and H. Nguyen, *Polym. Eng. Sci.*, vol. 18, p. 1037, 1978.
80. S. A. White, A. D. Gotsis, and D. G. Baird, *J. Non-Newtonian Fluid Mech.*, vol. 24, p. 121, 1987.
81. D. V. Boger, *Annu. Rev. Fluid Mech.*, vol. 19, p. 157, 1987.
82. D. V. Boger and R. J. Binnington, *J. Non-Newtonian Fluid Mech.*, vol. 35, p. 339, 1990.
83. D. M. Binding and K. Walters, *J. Non-Newtonian Fluid Mech.*, vol. 30, p. 233, 1988.
84. F. N. Cogswell, *Polym. Eng. Sci.*, vol. 12, p. 64, 1972.
85. D. M. Binding, *J. Non-Newtonian Fluid Mech.*, vol. 27, p. 173, 1988.
86. D. M. Binding, *J. Non-Newtonian Fluid Mech.*, vol. 41, p. 27, 1991.
87. M. Padmanabhan, Ph.D. thesis, University of Minnesota, 1992.
88. H. H. Winter, C. W. Macosko, and K. E. Bennett, *Rheol. Acta*, vol. 18, p. 323, 1979.
89. C. W. Macosko, M. A. Ocansey, and H. H. Winter, in G. Astarita, G. Marrucci, and L. Nicolais (eds.), *Rheology*, vol. 2, *Applications*, p. 723, Plenum, New York, 1980.
90. R. B. Secor, C. W. Macosko, and L. E. Scriven, *J. Non-Newtonian Fluid Mech.*, vol. 23, p. 355, 1987.
91 J. F. Stevenson, S. C.-K. Chung, and J. T. Jenkins, *Trans. Soc. Rheol.*, vol. 19, p. 397, 1975.

92. G. H. Pearson and S. Middleman, *AIChE J.,* vol. 23, pp. 714–725, 1977.
93. E. D. Johnson and S. Middleman, *Polym. Eng. Sci.,* vol. 18, p. 963, 1978.
94. R. Y. Ting and D. L. Hunston, *J. Appl. Polym. Sci.,* vol. 21, p. 1825, 1977.
95. H. Münstedt and S. Middleman, *J. Rheol.,* vol. 25, p. 24, 1981.
96. A. C. Papanastasiou, L. E. Scriven, and C. W. Macosko, *J. Non-Newtonian Fluid Mech.,* vol. 16, p. 53, 1984.
97. K. Baid and A. B. Metzner, *Trans. Soc. Rheol.,* vol. 21, p. 237, 1977.
98. C. B. Weinberger and J. D. Goddard, *Int. J. Multiphase Flow,* vol. 1, p. 465, 1974.
99. G. G. Fuller, C. A. Cathey, B. Hubbard, and B. E. Zebrowski, *J. Rheol.,* vol. 31, p. 235, 1987.
100. P. R. Souza Mendes, M. Padmanabhan, L. E. Scriven, C. W. Macosko, in *Proc. 3d World Conf. on Exp. Heat Transfer, Fluid Mechanics & Thermodynamics,* Honolulu, Hawaii, Elsevier, New York, 1993.
101. J. J. Cai, P. R. Souza Mendes, C. W. Macosko, L. E. Scriven, and R. B. Secor, in P. Moldenaers and R. Keunings (eds.), *Theoretical and Applied Rheology,* vol. 2, p. 1012, Elsevier, Amsterdam, 1992.
102. D. F. James, G. M. Chandler, and S. J. Armour, *J. Non-Newtonian Fluid Mech.,* vol. 35, p. 421, 1990.
103. D. F. James, G. M. Chandler, and S. J. Armour, *J. Non-Newtonian Fluid Mech.,* vol. 35, p. 445, 1990.
104. T. Sridhar, V. Tirtaatmadja, D. A. Nguyen, and R. K. Gupta, *J. Non-Newtonian Fluid Mech.,* vol. 40, p. 271, 1991.
105. J. C. Chaufforeaux, C. Dehennau, and J. Vanrijckevorsel, *J. Rheol.,* vol. 23, p. 1, 1979.
106. G. R. Snelling and J. F. Lontz, *J. Appl. Polym. Sci.,* vol. 3, p. 257, 1960.
107. E. Uhland, *Rheol. Acta,* vol. 18, p. 1, 1979.
108. A. M. Kraynik and W. R. Schowalter, *J. Rheol.,* vol. 25, p. 95, 1981.
109. A. V. Ramamurthy, *Adv. Polym. Technol.,* vol. 6, p. 489, 1986.
110. A. V. Ramamurthy, *J. Rheol.,* vol. 30, p. 337, 1986.
111. S. G. Hatzikiriakos and J. M. Dealy, *J. Rheol.,* vol. 35, p. 497, 1991.
112. M. M. Denn, in P. Moldenaers and R. Keunings (eds.), *Theoretical and Applied Rheology,* vol. 1, p. 45, Elsevier, Amsterdam, 1992.
113. J. M. Piau and N. El Kissi, in P. Moldenaers and R. Keunings (eds)., *Theoretical and Applied Rheology,* vol. 1, p. 70, Elsevier, Amsterdam, 1992.
114. A. Casale and R. S. Porter, *Polymer Stress Reactions,* Academic, New York, 1978.
115. J. M. Broadbent, A. Kaye, A. S. Lodge, and D. G. Vale, *Nature,* vol. 217, p. 55, 1968.
116. M. J. Crochet and M. J. Bezy, *J. Non-Newtonian Fluid Mech.,* vol. 5, p. 201, 1979.
117. K. Higashitani and A. S. Lodge, *Trans. Soc. Rheol.,* vol. 19, p. 307, 1975.
118. C. D. Han, *Rheology in Polymer Processing,* Academic, New York, 1976.
119. G. Ehrmann and H. H. Winter, *Kunststofftechnik,* vol. 12, p. 156, 1973.
120. A. S. Lodge, U.S. Patent 3,777,549, Seiscor Corp., Tulsa, Okla., 1973.
121. F. A. Schraub, S. J. Kline, J. Henry, P. W. Runstadler, and A. Littel, *Trans. ASME J. Basic Eng.,* vol. 87, p. 429, 1965.
122. L. G. Leal, *Annu. Rev. Fluid Mech.,* vol. 12, p. 435, 1979.
123. S. Karnis and S. G. Mason, *Trans. Soc. Rheol.,* vol. 10, p. 571, 1966.
124. H. Kraemer and J. Meissner, in G. Astarita, G. Marrucci, and L. Nicolais (eds.), *Rheology,* vol. 2, *Applications,* p. 463, Plenum, New York, 1980.
125. D. V. Boger and D. Mackay, in A. Collyer and D. Clegg (eds.), *Rheological Measurements,* chap. 14, Elsevier, London, 1988.
126. H. Janeschitz-Kriegl, *Adv. Polym. Sci.,* vol. 6, p. 170, 1969; *Polymer Melt Rheology and Flow Birefringence,* Springer, Berlin, 1983.
127. A. W. Hendry, *Photoelastic Analysis,* Pergamon, New York, 1966.

128. C. D. Han, *Trans. Soc. Rheol.,* vol. 18, p. 163, 1974.
129. G. G. Fuller and L. G. Leal, *Rheol. Acta.,* vol. 19, p. 580, 1980.
130. F. H. Gortemacher, Ph.D. thesis, T. H., Delft, 1976.
131. G. G. Fuller, in P. Moldenaers and R. Keunings (eds.), *Theoretical and Applied Rheology,* vol. 1, p. 55, Elsevier, Amsterdam, 1992.

MEASUREMENT OF WALL SHEAR STRESS

Thomas J. Hanratty and Jay A. Campbell

9.1 INTRODUCTION

A fluid flowing past a boundary exerts normal and tangential stresses on it. Normal stresses or pressures are readily measured by connecting a small hole on the surface to a manometer. Consequently, numerous studies have been made of the pressure distribution on solid surfaces in contact with a flowing fluid. The measurement of the tangential or shear stresses is much more difficult. However, the extra effort needed to obtain such data is rewarding, since information about the variation of the wall shear stress is, often, quite useful in analyzing a flow field. The six principal methods developed for measuring the local wall shear stress, aside from the extrapolation of direct velocity measurements, are

The Stanton tube [1]
Direct measurement [2]
Thermal method [3]
The Preston tube [4]
The sublayer fence [5, 6]
The electrochemical technique [7, 8]

The first such measurements were reported in the historic paper by Stanton et al. [1], who wanted to determine the fluid velocity close to a boundary in order to find out whether slip existed at a wall for a turbulent flow. To do this, they used a rectangular shaped Pitot tube that had the wall of the pipe as one of its sides. The difference of the pressure measured with this Pitot tube from the static pressure was used to determine the velocity at the center point of the tube. By carrying out experiments in a fully developed laminar flow, Stanton and his coworkers found that the relation between the measured pressure difference and the velocity was not given by the usual equation for a Pitot tube in a free flow. They established a calibration curve for the "effective center point" of their instrument, now called a Stanton tube, and used it to measure velocities in a fully developed turbulent flow. From these experiments it was shown, for the first time, that for a turbulent flow close to a wall the variation of the time-averaged velocity is given by

$$\overline{U} = \frac{\overline{\tau}_w}{\mu} y \tag{1}$$

It was established that, if the "effective center" of the Stanton tube was located close enough to the wall, the velocity calculated from the pressure measurement could be related to the wall shear stress by using Eq. (1). The Stanton tube thereby provides an indirect method for measuring $\overline{\tau}_w$.

The local tangential force on a surface can be determined directly by allowing some portion of the surface to be movable against a restoring force. Measurements are made of the displacement of the element or of the force required to keep it in a null position. The choice of a size for the element is a balance between the need to have a sufficiently large force acting upon it and the desire to obtain local measurements of the wall shear stress. The first local measurements with such a device were made by Kempf [2], who used panels 308 × 1010 mm to measure the drag force at several stations along the bottom of a 77-m-long pontoon. In recent years the technique has been developed sufficiently that a movable element as small as 9 mm can be used. However, Settles [9] describes these gages as being, "delicate, expensive, cranky and undependable." He points out that "major errors are likely to occur in strong pressure-gradient flows, and balance survival during intermittent wind tunnel starting and stopping transients is a full concern."

Fage and Falkner [10] studied the relation between the local wall shear stress and the rate of heat transfer from small thermal elements mounted flush with the surface. They used a nickel strip, 0.262 cm long and 26.2 cm wide, embedded 0.107 cm in an ebonite block. A heating current was passed through the nickel strip, and its temperature was determined by measuring the resistance. The current was controlled manually so that the temperature of the element was kept constant. The heating current was related to the local wall shear stress at different locations on a circular cylinder in a flow stream. Ludwieg [3] developed an instrument, using this concept, that consisted of a heated copper block that was movable and therefore could be located at different positions over a surface. With Tillmann, Ludwieg used this instrument to establish the law of the wall for turbulent boundary layers [11].

Much simpler and more compact versions of this gage were developed by Liepmann and Skinner [12], who used as the heating element a 12.7-μm platinum wire buried in a groove in the surface of bakelite, by Bellhouse and Schultz [13, 14], who used a platinum film deposited on a glass substrate, and by McCroskey and Durbin [15], who used photoetching techniques. The latter two probes have been developed into commercially available instruments that use standard hot-wire and hot-film equipment. At present, the usual mode of operation of wall film gages is to control the average temperature of the heated element using feedback circuitry and to determine how the electric heating current varies with wall shear stress. One of the chief difficulties in using these various heat transfer probes is that heat is lost to the substrate as well as to the fluid, so that the effective length of the probe can be much larger than that of the heating element. Works by Rubesin et al. [16] with embedded heated wires and by Sandborn [17, 18] with 0.001-cm wires lying on top of the surface offer methods for greatly reducing this effect.

Preston [4] suggested what is probably the simplest method for determining wall shear stress. He measured the impact pressure on round Pitot tubes resting on a surface. The tubes had outside diameters varying from 0.74 to 3.08 mm and a ratio of inside diameter to outside diameter of 0.6. For turbulent flows the tubes were too large to be entirely in the region where Eq. (1) is valid. However, on the basis of measurements made by Ludwieg and Tillmann [11], Preston argued that, for a given fluid, the velocity variation close to a wall is a universal function that depends only on the local wall shear stress and not on the geometry of the system or on the previous history of the flow field:

$$\overline{U} = (\overline{\tau}_w/\rho)^{1/2} f\left[\frac{y\rho(\overline{\tau}_w/\rho)^{1/2}}{\mu}\right] \tag{2}$$

In the immediate vicinity of the wall,

$$f\left[\frac{y\rho(\overline{\tau}_w/\rho)^{1/2}}{\mu}\right] = \frac{y\rho(\overline{\tau}_w/\rho)^{1/2}}{\mu} \tag{3}$$

in order to be consistent with Eq. (1).

Thus the theoretical basis for using Preston's instrument in turbulent flows is the assumption that $\overline{U} = f(\overline{\tau}_w, y)$ over a greater range of wall distances than the range of validity of Eq. (1). If this assumption is correct, much larger Pitot tubes can be used to measure $\overline{\tau}_w$ than originally suspected. Therefore the difficulties associated with the manufacture of a Stanton tube would not justify its choice over a conventional Pitot tube. Preston carried out experiments for fully developed turbulent flow in a pipe for which $\overline{\tau}_w$ can be calculated from pressure drop measurements as

$$\overline{\tau}_w = \frac{d}{4}\left|\frac{dP}{dx}\right| \tag{4}$$

to establish a universal calibration curve for a round Pitot tube resting on a wall.

There has been some doubt about the accuracy of the law of the wall, and therefore a number of investigators have attempted to develop other devices, similar to the Stanton tube, that consist of a wall obstruction completely immersed in the viscous sublayer. One of the more interesting of these is the sublayer fence invented by Konstantinov and Dragnysh [5, 6]. The difference in pressure before and behind a sharp edge projecting through the surface, normal to the flow, is related to the wall shear stress. The advantages of this instrument over the Stanton gage are that it gives an almost doubled pressure reading, it eliminates the necessity of a separate static pressure tapping, and it gives readings both in forward and in reversed flows.

A mass transfer analog of the heated surface film was developed by Reiss and Hanratty [7, 8]. An electrochemical reaction is carried out on the surface of an electrode mounted flush with the wall. The voltage on the electrode is kept large enough that the reaction rate is so fast that it is mass transfer controlled, yet small enough that no side reactions are occurring. Under these conditions the concentration of the reacting species is maintained at approximately zero at the electrode surface. The current flowing through the electrode circuit, which is proportional to the rate of mass transfer, is related to the velocity gradient at the surface. This instrument holds advantages over the thermal technique in that it avoids problems associated with substrate heat losses. In principle, it can be calibrated analytically and can be easily applied to situations requiring complicated sensor configurations. The mass transfer process occurring at the electrode is characterized by large Schmidt numbers so that these probes have the desirable feature that there is a quite large range of flow rates over which the concentration boundary layer is within a region where Eq. (1) describes the velocity field. These electrochemical probes have disadvantages, in that they can be used only in liquid flows and in equipment that is compatible with the chemicals.

Mitchell and Hanratty [19] showed how electrochemical probes can be used to study the time-averaged and fluctuating velocity gradient for a turbulent flow and pointed out the need to account for spatial averaging and for the time response of the concentration boundary layer. Fortuna and Hanratty [20] later improved the analysis for frequency response presented by Mitchell and Hanratty. Dimopoulos and Hanratty [21] and Son and Hanratty [22] showed how electrochemical techniques can be used for studying laminar boundary layers. Son and Hanratty [22] developed a sandwich electrode in which a pair of rectangular electrodes separated by a thin layer of insulation is oriented with the long sides perpendicular to the flow. By comparing mass transfer rates to the front and back electrodes, the direction of flow, and therefore the separation position of a boundary layer, can be determined. Tournier and Py [23] and LeBouche and Martin [24] further developed this sandwich electrode for application to boundary layer flows. Sirkar and Hanratty [25, 26] demonstrated how a pair of rectangular electrodes in a chevron arrangement can be used to measure both components of the velocity gradient at the wall. Py and Gosse [27] suggested the use of a pair of semicircular electrodes for this purpose. However, because of the difficulty in fabricating these semicircular electrodes, Py [28] used pairs of rectangles with their long dimension parallel to the flow to measure both components of the fluctuating velocity gradient in a turbulent flow. Most of the early work with these electrochemical techniques used the ferricyanide-ferrocyanide reaction in an excess of sodium hydroxide. Py [29] suggested the use of the iodine reaction in an excess of potassium iodide. This system has been found to be more stable and to cause less problems with respect to electrode contamination. Mao and Hanratty [30, 31] have used inverse mass transfer techniques to analyze time-varying flows. The application of these numerical methods to a sandwich electrode offers the possibility to study flows that are reversing their direction [32].

Any of the six techniques can be used to measure time-averaged wall shear stresses in fully developed flows and in boundary layers with zero pressure gradient. The choice depends on the system in which they are to be used and on whether the measuring technique will interfere with the flow. For other flow situations, it appears that the heat transfer or mass transfer probes have the widest applicability. The possibility of measuring the fluctuations in the wall shear stress, as well as the time average, makes them particularly attractive choices for time-varying flows. A description of all six techniques will be presented. However, much attention will be given to the heat and mass transfer techniques because of their greater potential and because the theoretical problems associated with their use are particularly intricate.

The chapter is closed with a discussion of the use of velocity measurements at distances from the wall where Eq. (1) is not applicable, to determine $\bar{\tau}_w$ in turbulent flows. Also described are newly developed techniques that measure the thinning of oil films, the Doppler shift of light scattered by particles, and the time of flight of a heated marker.

9.2 DIRECT MEASUREMENTS

The need of aerodynamicists to measure shear stresses in high-speed flows (see [33]) has led to the development of a number of ingenious designs for compact gages to measure directly the local wall shear stress. Winter [34] and Archarya [35] give descriptions of these instruments and review progress made in solving the following problems, which Winter cites to be associated with their use:

1. Provision of a transducer for measuring small forces or deflections, and the compromise between the requirement to measure local properties and the necessity of having an element of sufficient size that the force on it can be measured accurately.
2. The effect of the necessary gaps around the floating element.
3. The effects of misalignment of the floating element.
4. Forces arising from pressure gradients.

5. The effect of gravity or of acceleration if the balance is to be used in a moving vehicle.
6. Effects of temperature changes.
7. Effects of heat transfer.
8. Use with boundary layer injection or suction.
9. Effects of leaks.
10. Protection of the measuring system against transient normal forces during starting and stopping in a supersonic tunnel.

These instruments have been used to establish the turbulent skin friction law for a flat plate for incompressible [33, 36–38] and for compressible flows [39, 40]. They have also been used for measurements on an airplane at Mach numbers up to 4.9 [41].

The gage is, usually, calibrated by a static method. A force is applied to the element by suspending weights from a thread with the instrument mounted vertically [42] or with a pulley arrangement and the instrument mounted horizontally [33]. As pointed out by Mabey and Gaudet [42], very thin threads have to be used with small gages. In fact, these authors found it necessary to use a human hair.

Since this technique requires a portion of the wall to be movable in a direction parallel to the boundary, the sensing element must have a gap around its perimeter. Because of the presence of this gap, effects appear under flow conditions that are not taken into account in the calibration. The presence of the discontinuity at the surface alters the wall shear stress in regions close to the gap. If the element is not aligned almost perfectly with the contour of the wall, depressions or protrusions can give rise to additional forces because of flow disturbances and of pressure forces acting on the protruding surfaces. If pressure variation associated with the flow is significant over the element surface, fluid circulation can occur through the gaps and a force can exist on the head because of the difference in pressures of the fluid in the upstream and downstream gap. This latter difficulty appears to be a fundamental problem with this type gage that cannot be eliminated and may severely limit its application in flows with pressure gradients.

The use of large elements and large gaps reduces errors due to misalignment. However, too large a gap should be avoided because of the change of shear stress across the gap. Winter [34] recommends that gu^*/ν should not exceed 100, where g is the gap size. A discussion of errors associated with misalignment at the gap has been presented by Allen [43–45].

For flows with small pressure gradients and small wall shear stress gradients, large balances may be used. Such balances are relatively easy to calibrate and require small buoyancy corrections arising from nonuniformities of the pressure in the gap around the floating plate. An example is the 368-mm-diameter gage used by Winter and Gaudet [39] to study airflows for $16 \times 10^6 < \mathrm{Re}_x < 200 \times 10^6$ at $0.2 < \mathrm{Ma} < 2.8$. In large gradients the use of as small a gap as possible is desired so as to reduce flow through the gap. This requirement, in addition to the require-

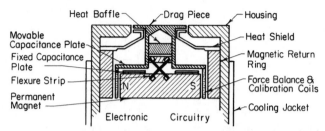

Figure 9.1 Construction of Kistler skin-friction gage as presented in the literature from Kistler Instrument Co.

ment of a small element, magnifies the alignment problem. The calibration is more difficult, and the correction for buoyancy may be uncertain.

A paper describing the use of one of these small skin friction gages [42] describes tests performed with the Kistler gage shown in Fig. 9.1. This instrument, which has a head with a diameter of 9.4 mm and which is designed to be insensitive to linear acceleration, was at one time available commercially (sold by Sundstrand Data, Inc., Redmond, Washington). It has a feedback circuit that ensures that the floating element remains in a fixed position relative to its housing. Consequently, the gap around the element is kept uniform and can be much smaller than that required if the element is allowed to deflect, as is the case in the conventional skin friction balance. Mabey and Gaudet mounted the balances in a smooth steel plug, which could be located in holes drilled at different locations in the flat plate with which tests were made. By doing this, it was possible to align the instrument even with or slightly below the surface surrounding it before mounting in the test section.

An unevenness of pressure makes misalignment problems more severe. As shown by Acharya [35], a pressure force is introduced that can be significant in comparison with the force associated with shear stresses. Frei [46] has shown that this problem can be overcome in airflows by sealing the gap between the floating element and the wall with a liquid, held in place by surface tension. If this instrument is not carefully designed, the force due to the surface tension of the liquid can introduce new errors. An analysis of this effect is given by Frei and Thomann [47], who present measurements with a ring-shaped element located in a cylindrical flow section. The element had a length of 10 mm and a circumference equal to that of the tube, 200 mm. A circular element with a diameter of 60 mm and a gap of 0.1 mm has also been used in Thomann's laboratory.

Most floating head devices have a feedback mechanism that keeps the sensing element stationary [42, 45, 47]. Petri [48], however, has used a determination of the displacement to derive the surface force. Since the amount of displacement is restricted, a very sensitive measuring technique had to be developed. The floating element was attached to a thin cantilever plate that is sensitive to forces only in

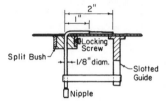

Figure 9.2 Construction of Preston tube as given by Preston [4]

one direction. A deflection of the cantilever plate caused by a movement of the element produces a change in the capacitance with respect to a second stationary parallel plate that is mounted a short distance from the cantilever. This change in capacitance, which is proportional to the displacement, is picked up by an electric circuit. An advantage of this measurement technique is that it is compatible with small floating elements. The one used by Petri was only 4 mm square. This method for measuring small displacement has been exploited by Schmidt et al. [49], who used a 0.5 × 0.5 mm element with gaps of 10 μm.

9.3 PRESTON TUBE

The complexity of the design of the surface force gages and the great care that must be taken in their use make the Preston tube a particularly attractive alternative. The arrangement by Preston is shown in Fig. 9.2 and is described by him as follows: "The brass stem is 1/8 in. diameter and the mouth of the Pitot is 2 in. from the stem. . . . The front part of the tube was made of stainless steel. It was bent so that the first contact with the wall was at the mouth, and was soldered into the brass stem 1.0 in. from the mouth" [4]. In using the Preston tube, care must be taken in locating the static pressure tap so that its reading is not being affected by blockage effects due to the presence of the Pitot tube at the wall. Although most Preston tubes used in tests have followed Preston's specification of a ratio of inside to outside diameter, results do not appear to be too sensitive to this parameter. For example, Patel [50] and Rechenberg [51] find no effect provided the ratio is greater than 0.2.

The principle of operation is that the instrument obeys the equation for a Pitot tube that has an effective center at $y = \frac{1}{2}K_t d_t$, where d_t is the outside tube diameter. Suppose the Preston tube lies entirely in a region close to the wall where $U = (\tau_w / \mu)y$. Then a Reynolds number can be defined as

$$d_t^+ = \left[\frac{1}{2} \left(\frac{1}{2} K_t \right)^2 \frac{d_t^2 \tau_w}{\rho \nu^2} \right]^{1/2} \tag{5}$$

For large enough values of d_t^{+2} the difference of the impact pressure at the effective center of the tube from the static pressure is given by the following relation, provided $U = (\tau_w / \mu)y$:

$$\frac{\Delta P}{\tau_w} = \frac{1}{2}\left(\frac{1}{2}K_t\right)^2 \frac{d_\tau^2 \tau_w}{\rho v^2} \tag{6}$$

Young and Mass [52] have shown that for Pitot tubes in a shear flow at large Reynolds numbers, $K_t = 1.30$ for a tube with a ratio of inside to outside diameter of 0.60. For very small Reynolds numbers it can be expected that viscous effects will be important and therefore that K_t will be a function of the Reynolds number, d_t^+. Equation (6) can be applied to a turbulent flow if it is assumed that the flow is uniform in the spanwise direction at all instances. Then

$$\frac{\overline{\Delta P}}{\overline{\tau}_w} = \frac{1}{2}\left(\frac{1}{2}K_t\right)^2 \frac{d_t^2 \overline{\tau}_w}{\rho v^2}\left[1 + \left(\frac{\overline{\tau_w'^2}}{\overline{\tau}_w^2}\right)\right] \tag{7}$$

here $\overline{\tau}_w$ is now the time-averaged wall stress and $\overline{\tau_w'^2}$ is the mean square of the fluctuations in the wall stress. Measurements carried out in a number of laboratories would indicate that $\overline{\tau_w'^2}/\overline{\tau}_w^2 \cong 0.15$.

The most convenient system for calibrating a Preston tube is a fully developed pipe flow for which the pressure gradient is linearly related to the wall shear stress by Eq. (4). It is usually better to make the pressure gradient measurements without the Preston tube in place, since flow blockage could affect the results. When using this calibration procedure, the pressure taps used for measuring pressure gradient should not be located in regions where the pipe walls are tapered, and tests should be carried out to ensure that the flow is symmetric [53]. Calibrations of Preston tubes in this manner in laminar flows give values of K_t close to that obtained by Young and Maas [52]. This supports the physical interpretation that a Preston tube behaves the same as a Pitot tube located in a free shear layer.

The application of Eqs. (6) and (7) to interpret Preston tubes in most turbulent flows is not possible because the tube cannot be made small enough to sample only a region for which $\overline{U} = \overline{\tau}_w y/\mu$. The use of the Preston tube under these circumstances requires that the region of the flow it sees be described by the law of the wall, Eq. (2). If this is true, then it can be shown by dimensional reasoning that

$$\frac{\overline{\Delta P}}{\overline{\tau}_w} = f\left(\frac{d_t^2 \overline{\tau}_w}{\rho v^2}\right) \tag{8}$$

where $f(d_t^2\overline{\tau}_w/\rho v^2)$ is given by Eq. (6) for very small values of the argument. An alternative form of Eq. (8), used by Preston, is

$$\frac{\overline{\Delta P}d^2}{\rho v^2} = f\left(\frac{d_t^2 \overline{\tau}_w}{\rho v^2}\right) \tag{9}$$

Calibrations of different-diameter Preston tubes in fully developed turbulent pipe flows have been carried out by Preston [4], Rechenberg [51], Patel [50], and

Head and Rechenberg [54]. On the basis of these measurements, Head and Ram [55] have developed the "universal calibration" given in Table 9.1.

The application of the Preston tube and this calibration to turbulent boundary layer flows depends on the accuracy of the law of the wall. For turbulent flow in pipes or turbulent boundary layers with small pressure gradients, the law of the wall is usually assumed to hold, provided $2d_t/d$ or d_t/δ is less than 0.1, where δ is the boundary layer thickness. For turbulent boundary layers with large positive or negative pressure gradients, the law of the wall holds over a smaller distance, so that d_t/δ must be an even smaller quantity in order to use the universal calibration. For many situations, an example of which would be a separated region, the law of the wall is not correct, and the Preston tube cannot be used if it extends beyond the region where $\overline{U} = (\overline{\tau}_w/\mu)y$.

Preston [4] demonstrated the applicability of Preston tubes to turbulent boundary layers by comparing measurements with tubes of different diameter. Head and Rechenberg [54] did, essentially, the same type test by comparing measurements with Preston tubes to measurements with a Stanton tube or with a sublayer fence. Tests with Preston tubes in turbulent boundary layers on flat plates in a number of laboratories [54] have yielded results that differ by as much as 11%. This could be the type of accuracy to be expected from the Preston tube, but more than likely, it reflects the accuracy of the methods used to evaluate the wall shear stress in these different tests. This matter has not, as yet, been satisfactorily resolved. Other work with Preston tubes was done by Holmes and Luxton [56], Hsu [57], Nitsche [58] and Relf et al. [59].

Patel [50] has presented guidelines for estimating the influence of pressure gradient on the accuracy of Preston tubes. For this purpose he uses the parameter

$$\Delta = \frac{\nu(dP/dx)}{\rho u^{*3}} \tag{10}$$

He recommends that $d_t^+ \leq 250$ in adverse pressure gradients with $\Delta < 0.015$ and that $d_t^+ \leq 200$ in favorable pressure gradients with $\Delta < -0.007$ for the maximum error to be less than 6%. Operation outside the range $-0.007 < \Delta < 0.015$ is not recommended.

A number of investigators have developed methods for using Preston tubes in compressible flows [34, 60]. The simplest of these appears to be that proposed by Bradshaw and Unsworth [61],

$$\frac{\overline{\Delta P}}{\overline{\tau}_w} = f_i\left(\frac{d_t u^*}{\nu_w}\right) + f_c\left(\frac{d_t u^*}{\nu_w}, \frac{u^*}{a_w}\right) \tag{11}$$

where f_i is the calibration for incompressible flow and f_c is the correction to take account of compressibility effects. Bradshaw found that the following functions gave the best fit to available data for $50 < d_t u^*/\nu < 1000$:

Table 9.1 Calibration of Preston tube developed by Head and Ram [55]

$\dfrac{\Delta P\, d^2}{\rho v^2} \times 10^{-2}$, $\dfrac{\Delta P}{\tau_W}$

$\dfrac{\Delta P\, d^2}{\rho v^2}$	$\dfrac{\Delta P}{\tau_W}$
4.0	9.18
4.2	9.41
4.4	9.63
4.6	9.85
4.8	10.06
5.0	10.27
5.2	10.47
5.4	10.67
5.6	10.87
5.8	11.06
6.0	11.25
6.2	11.44
6.4	11.62
6.6	11.80
6.8	11.98
7.0	12.16
7.2	12.33
7.4	12.50
7.6	12.67
7.8	12.83
8.0	12.99
8.2	13.15
8.4	13.31
8.6	13.47
8.8	13.63
9.0	13.78
9.2	13.93
9.4	14.08
9.6	14.23
9.8	14.38

$\dfrac{\Delta P\, d^2}{\rho v^2} \times 10^{-3}$, $\dfrac{\Delta P}{\tau_W}$

$\dfrac{\Delta P\, d^2}{\rho v^2}$	$\dfrac{\Delta P}{\tau_W}$
1.0	14.53
1.02	14.67
1.06	14.95
1.10	15.23
1.14	15.51
1.18	15.78
1.22	16.04
1.26	16.30
1.30	16.56
1.34	16.81
1.38	17.06
1.42	17.31
1.46	17.55
1.50	17.79
1.54	18.02
1.58	18.25
1.62	18.48
1.66	18.71
1.70	18.94
1.74	19.16
1.78	19.38
1.82	19.59
1.86	19.80
1.90	20.01
1.95	20.27
2.00	20.53
2.05	20.79
2.10	21.05
2.15	21.30
2.20	21.54
2.25	21.78
2.30	22.02
2.40	22.49
2.50	22.96
2.6	23.41
2.7	23.86
2.8	24.30
2.9	24.73
3.0	25.08
3.1	25.43
3.2	25.78
3.3	26.13
3.4	26.48
3.5	26.82
3.6	27.16
3.7	27.50
3.8	27.83
3.9	28.15
4.0	28.46
4.2	29.07
4.4	29.66
4.6	30.23
4.8	30.79
5.0	31.33
5.2	31.84
5.4	32.34
5.6	32.84
5.8	33.31
6.0	33.78
6.2	34.23
6.4	34.68
6.6	35.11
6.8	35.52
7.0	35.94
7.2	36.34
7.4	36.72
7.6	37.11
7.8	37.50
8.0	37.87
8.5	38.74
9.0	39.58
9.5	40.40

$\dfrac{\Delta P\, d^2}{\rho v^2} \times 10^{-4}$, $\dfrac{\Delta P}{\tau_W}$

$\dfrac{\Delta P\, d^2}{\rho v^2}$	$\dfrac{\Delta P}{\tau_W}$
1.00	41.18
1.05	41.93
1.10	42.65
1.15	43.34
1.20	44.00
1.25	44.64
1.30	45.27
1.35	45.87
1.40	46.45
1.45	47.01
1.50	47.56
1.55	48.09
1.60	48.61
1.65	49.12
1.70	49.62
1.8	50.56
1.9	51.46
2.0	52.32
2.1	53.15
2.2	53.93
2.3	54.68
2.4	55.40
2.5	56.62
2.6	56.80
2.7	57.45
2.8	58.07
2.9	58.68
3.0	59.28
3.2	60.40
3.4	61.46
3.6	62.47
3.8	63.43
4.0	64.34
4.2	65.20
4.4	66.01
4.6	66.80
4.8	67.57
5.0	68.32
5.2	69.05
5.5	70.00
6.0	71.55
6.5	73.00
7.0	74.35
7.5	75.60
8.0	76.80
8.5	77.95
9.0	78.95
9.5	79.90

$\dfrac{\Delta P\, d^2}{\rho v^2} \times 10^{-5}$, $\dfrac{\Delta P}{\tau_W}$

$\dfrac{\Delta P\, d^2}{\rho v^2}$	$\dfrac{\Delta P}{\tau_W}$
1.0	80.80
1.05	81.70
1.10	82.55
1.15	83.35
1.20	84.1
1.30	85.5
1.4	86.8
1.5	88.0
1.6	89.1
1.7	90.2
1.8	91.2
1.9	92.1
2.0	93.0
2.2	94.7
2.4	96.2
2.6	97.6
2.8	98.8
3.0	99.9
3.2	100.9
3.5	102.4
4.0	104.5
4.5	106.4
5.0	108.0
5.5	109.3
6.0	110.4
6.5	111.4
7.0	112.2
7.5	113.0
8.0	113.7
9.0	114.9

$\dfrac{\Delta P\, d^2}{\rho v^2} \times 10^{-6}$, $\dfrac{\Delta P}{\tau_W}$

$\dfrac{\Delta P\, d^2}{\rho v^2}$	$\dfrac{\Delta P}{\tau_W}$
1.0	116.0
1.1	117.1
1.2	118.1
1.4	119.8
1.6	121.5
1.8	123.1
2.0	124.6
2.2	126.0
2.5	127.8
3.0	130.7
3.5	133.3
4.0	135.6
4.5	137.7
5.0	139.5
6.0	142.7
7.0	145.4
8.0	147.8
9.0	150.0

$\dfrac{\Delta P\, d^2}{\rho v^2} \times 10^{-7}$, $\dfrac{\Delta P}{\tau_W}$

$\dfrac{\Delta P\, d^2}{\rho v^2}$	$\dfrac{\Delta P}{\tau_W}$
1.0	151.9
1.2	155.3
1.4	158.1
1.6	160.6
1.8	162.8
2.0	164.9
2.2	166.8
2.5	169.3
3.0	172.8
3.5	175.8
4.0	178.5
4.5	180.9
5.0	183.1
6.0	186.8
7.0	190.0
8.0	192.8
9.0	195.3

$\dfrac{\Delta P\, d^2}{\rho v^2} \times 10^{-8}$, $\dfrac{\Delta P}{\tau_W}$

$\dfrac{\Delta P\, d^2}{\rho v^2}$	$\dfrac{\Delta P}{\tau_W}$
1.0	197.5
1.2	201.4
1.4	204.7
1.6	207.5
1.8	210.1
2.0	212.5
2.5	217.4
3.0	221.6
3.5	225.0

$$f_i = 96 + 60 \log_{10} \frac{d_t u^*}{50 \nu} + 23.7 \left(\log_{10} \frac{d_t u^*}{50 \nu} \right)^2 \tag{12}$$

$$f_c = 10^4 \left(\frac{u^*}{a_w} \right)^2 \left[\left(\frac{d_t u^*}{\nu_w} \right)^{0.26} - 2 \right] \tag{13}$$

The Preston tube appears to offer no real advantages (and, in fact, offers disadvantages) over the force balance with respect to accurate measurement of wall shear stress. Because it presents an obstruction to the flow, it can interfere with the flow field. In three-dimensional boundary layers the direction of flow at the wall has to be known in order to orient the Preston tube properly. The accuracy of the Head and Ram calibration is uncertain unless the flow is an incompressible, two-dimensional, turbulent boundary layer with a moderate pressure gradient. The main reason for the choice of a Preston tube over a wall balance is that it is easier to fabricate and to operate.

9.4 STANTON GAGE

The Stanton gage has been an attractive method for measuring wall shear stress because of its small size. The original design by Stanton has proved to be awkward to use, and a number of alternate arrangements have been suggested. Hool [62] proposed a form of surface channel that was formed on one side by the solid surface and on the other by the lower side of the tapered cutting edge of a segment of razor blade. The static pressure developed in this small enclosure was measured by a normal static-pressure hole in the solid surface. With the razor blade removed, this same static-pressure hole could be used to measure the true static pressure, and the difference between these two pressures calibrated against wall shear stress. Bradshaw and Gregory [63] applied this technique by using a 0.002-in. steel shim with one edge chamfered. A 0.2 in. × 0.1 in. piece was held over a 0.04-in.-square hole with cellulose adhesive. The leading edge of the shim was aligned so that it was just over the front side of the static-pressure hole. Brown [64] used a similar design to Bradshaw and Gregory's, but with a 0.005-in.-thick shim over a 0.020-in. hole.

The values of ΔP determined with a Stanton gage can be quite small. Consequently, it is necessary to use a micromanometer with an accuracy of ± 0.01 mm.

Extensive tests were carried out by East [65] with a Stanton gage having the design shown in Fig. 9.3. This has the unique feature that it is held in place by a magnet that surrounds the static hole. This procedure avoided the use of an unknown thickness of glue between the blade and the surface and allowed for the determination of the height of the blade edge above the surface h as simply the half thickness of the blade. East recommended the following dimensions be used:

$$\frac{d_h}{h} = 6 \qquad \frac{b}{h} = 36 \qquad \frac{\ell}{b} = 1 \qquad \frac{\Delta x}{h} = 0$$

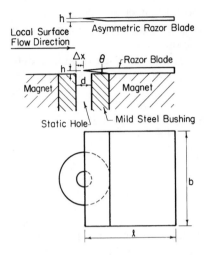

Figure 9.3 Design of a Stanton gage as given by East [65]

where ℓ = length of blade
b = breadth of blade
d_h = diameter of static hole.

East used two commercially available blades with $h = 0.002$ in. and 0.005 in. and a fabricated blade with $h = 0.0153$ in. The blade angle was $11°$–$13°$. He tested the gages against a Preston tube that used the calibration given in Table 9.1. The field consisted of the turbulence flow of air over a flat plate.

Calibrations of Stanton gages have been carried out in laminar flows by Bradshaw and Gregory [63], Hool [62], and Taylor [66]. These results can be correlated by an equation either of the form $\Delta P h^2 \rho / \mu^2$ versus $\tau_w h^2 \rho / \mu^2$ or, if Eq. (6) is used, of the form K_t versus $\tau_w h^2 \rho / \mu^2$ (see [67]). Taylor's experiments were carried out at extremely low Reynolds numbers, where a creeping flow approximation can be made. He showed from dimensional reasoning that in the Stokesian region ($\log_{10} h_s^2 \tau_w \rho / \mu^2 < -0.6$),

$$\Delta P = k_s \tau_w \tag{14}$$

His experiments indicate that $k_s \approx 1.2$. Taylor argued that at large values of the Reynolds number, $h^2 \tau_w \rho / \mu^2$, the results are interpreted in the manner suggested by Eq. (6) and that K_t = const. Experiments by Hool clearly indicate that this is not the case and that K_t varies as $(h^2 \tau_w \rho / \mu^2)^{-1/5}$ at large values of the gage Reynolds number. The effective center of the Stanton gage is found to be less than $d_t/2$ at large $h^2 \tau_w \rho / \mu^2$, as has been observed by Preston [4] for flattened Pitot tubes touching a surface.

These results are quite different from what is found for a Preston tube. They suggest that the interpretation of the reading of the Stanton gage as resulting from a stagnation flow at $y = \frac{1}{2} K_t d_t$ is not appropriate. Rather, it should be recognized that a recirculating flow region exists in front of the gage and that the details of the flow in this region could be exerting an important effect on the readings (see [63, 68]). The pressure measured by the Stanton gage might thus be interpreted as similar to that observed on a rearward facing step on a wall.

The calibration of a Stanton gage in a turbulent flow can use the same dimensionless groups as for laminar flow. East [65] presents the following calibration for the gage shown in Fig. 9.3:

$$y^* = -0.23 + 0.618x^* + 0.0165x^{*2} \qquad 2 < x^* < 6 \tag{15}$$

$$x^* = \log_{10}\left(\frac{\overline{\Delta P}h^2\rho}{\mu^2}\right) \tag{16}$$

$$y^* = \log_{10}\left(\frac{\overline{\tau}_w h^2\rho}{\mu^2}\right) \tag{17}$$

This calibration cannot be expected to be usable for another Stanton gage unless the design and dimensions closely match that used by East, who pointed out that readings can be particularly sensitive to deviations from the recommended zero value of the distance of the leading edge of the blade from the leading edge of the static hole. A discussion of the effect of design variables on the calibration has been presented by Haritonidis [69].

In the flow range where the gage covers distances from the wall where $U = (\tau_w/\mu)y$, the calibration is different for laminar and turbulent flows [66] and cannot be explained by the type of reasoning presented in Eq. (7). This is not too surprising, since the circulating zone in front of the gage for a turbulent flow would be expected to be quite different from that for a steady flow. Calibrations obtained in steady laminar flows therefore should not be used in turbulent flows.

In tests comparing the performance of Stanton gages and heated film gages in laminar boundary layers, Brown [64] has found that Stanton gages give inaccurate readings in large unfavorable pressure gradients. He interpreted these results by suggesting that the region of the flow field "seen" by a Stanton gage is many times its height and that this effect becomes greater as the wall shear stress decreases in regions of unfavorable pressure gradient. The inaccurate readings result because they are being influenced by regions of the field where the velocity is not varying linearly with distance from the wall. The velocity field "seen" by the gage under test conditions was therefore not the same as that under calibration conditions. Brown [64] also points out that variations in the calibrations of different investigators for laminar flows can be caused by the same effect.

These results of Brown would indicate that Stanton gages cannot be regarded as reliable devices in boundary layers with large pressure gradients. It is possible that this conclusion is only applicable to laminar boundary layers. Consequently, there is a need for tests in turbulent boundary layers similar to those carried out by Brown. Experiments by Head and Rechenberg [54] for turbulent flows with moderately favorable pressure gradients that exist in the entry region of a pipe have yielded encouraging results.

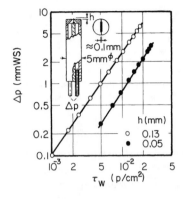

Figure 9.4 Assembly and calibration curve for sublayer fence presented by Rechenberg [51]

Studies with Stanton gages in compressible flows have been carried out by Winter and Gaudet [39] and by Gadd et al. [67]. However, no generally accepted method of correcting for compressibility effects has been developed. Work with Stanton gages by Abarnel et al. [70] and Thom [71] should also be cited.

9.5 SUBLAYER FENCE

Because of the larger pressure readings with a sublayer fence, the height of the fence can be somewhat smaller than that of the Stanton gage. A sketch of the fence used by Rechenberg [51] is shown in Fig. 9.4. The design presented by Konstanti-nov and Dragnysh [6] is more elaborate but has the advantage that the height of the fence can be varied by a screw mechanism.

The flow over a fence consists of a recirculation zone both in front of and in back of the fence. The observations regarding the influence of a two-dimensional obstacle on a wall made by Brown [53], with reference to Stanton gages, and the studies by Good and Joubert [72] indicate the region of disturbed flow is much larger than the height of the fence. Consequently, the height would have to be much smaller than the thickness of the viscous sublayer for it to be influenced, in a turbulent flow, only by the region where $\overline{U} = \tau_w y/\mu$. As has already been pointed out for the Stanton gage, the advantage of the sublayer fence, over a Preston tube, might not be as great as originally anticipated.

Head and Rechenberg [54] have calibrated a sublayer fence and a Preston tube in a turbulent flow with zero pressure gradient, and compared their readings in a flow with unfavorable pressure gradients. Agreement was noted in moderately unfavorable pressure gradients. However, in strongly unfavorable pressure gradients the two instruments indicated different values of the wall shear stress. It is quite possible that both instruments were giving erroneous readings. Until more data are available, the sublayer fence is of unknown accuracy in flows with strong pressure gradients. Achenbach [73, 74] has used the sublayer fence in studies of

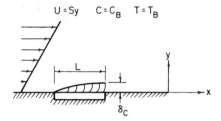

$$U = Sy \quad C = C_B \quad T = T_B$$

Figure 9.5 Description of mass or heat transfer technique for the case of negligible diffusion in the x direction

flow around a cylinder and a sphere. Vagt and Fernholz [75] showed the directional sensitivity of fences. Higuchi and Rubesin [76] used two fences at 20° to each other to study wall stress direction and magnitude in a three-dimensional boundary layer.

9.6 ANALYSIS OF HEAT OR MASS TRANSFER PROBES

9.6.1 Design Equation for a Two-Dimensional Mass Transfer Probe

The difficulties cited above in the use of floating heads, Preston tubes, Stanton gages, and sublayer fences have stimulated considerable research on the application of flush-mounted thermal or mass transfer (electrochemical) probes. These have the advantages that they can be used in a wide variety of flows, they do not interfere with the flow, they offer the possibility of measuring time-varying flows and, in the case of the electrochemical probes, a calibration is not necessary.

The principle of operation is illustrated in Fig. 9.5 for the case of a two-dimensional element aligned with its long side perpendicular to the direction of mean flow and negligible molecular transport in the x direction. The fluid at the surface of the wall element is controlled at a concentration C_w or temperature T_w that is different from that in the bulk field. The rate of mass or heat transfer between the fluid and the wall element is then measured. If the element is small enough in the flow direction, the concentration or thermal boundary layer will be so thin that it lies within a region for which the velocity is given by

$$U = Sy \tag{18}$$

where S is the magnitude of the velocity gradient at the wall. A calibration is established between the measured mass transfer or heat transfer rate and the velocity gradient S. If the viscosity of the fluid is known, then the wall shear stress can be evaluated from the measured S, since $\tau_w = \mu S$.

An analytical expression for the calibration can be derived, provided the following conditions are satisfied:

1. The scalar boundary layer is within the region where $U = Sy$.
2. The flow is homogeneous over the surface of the element.

3. Parameter δ_c is small enough compared to the width of the electrode W that diffusion in the spanwise direction can be neglected.
4. Forced convection is large enough that diffusion in the x direction can be neglected.
5. Natural convection is small in comparison with forced convection.
6. The scalar boundary layer is small enough that the influence of turbulent transport in the y direction can be neglected.

The mass balance equation for a two-dimensional field is given as

see pg 87 for conversion

$$\frac{\partial C}{\partial t} + Sy\frac{\partial C}{\partial x} + V\frac{\partial C}{\partial y} = D\left(\frac{\partial^2 C}{\partial y^2} + \frac{\partial^2 C}{\partial x^2} + \frac{\partial^2 C}{\partial z^2}\right) \tag{19}$$

Because of assumption 1, Sy is substituted for U. Assumptions 2, 5, and 6 allow for $V\,\partial C/\partial y$ to be neglected. Assumptions 3 and 4 allow for $D\,\partial^2 C/\partial z^2$ and $D\,\partial^2 C/\partial x^2$ to be neglected. Equation (19) can therefore be simplified to

$$\frac{\partial C}{\partial t} + Sy\frac{\partial C}{\partial x} = D\frac{\partial^2 C}{\partial y^2} \tag{20}$$

If a pseudo–steady state assumption is made,

$$Sy\frac{\partial C}{\partial x} = D\frac{\partial^2 C}{\partial y^2} \tag{21}$$

This implies that the flow is steady or that it is changing slowly enough with time that the concentration field is described by the steady state solution to Eq. (20). Equation (21) is solved using the boundary conditions

$$C = C_w \qquad y = 0 \tag{22}$$

$$C = C_B \qquad \text{large } y, x = 0 \tag{23}$$

The solution for $C(y, x)$ is given by Mitchell and Hanratty [19]. From it, the average mass transfer can be calculated, since

$$\langle N \rangle = \frac{1}{L}\int_0^L D\left(\frac{\partial C}{\partial y}\right)_{y=0} dx \tag{24}$$

Mitchell and Hanratty [19] give the following expression for the mass transfer coefficient $K = \langle N \rangle/(C_B - C_w)$:

$$\frac{KL}{D} = 0.807Z^{1/3} \tag{25}$$

where

$$Z = \left(\frac{SL^2}{D}\right) = L^{+2}\,Sc \tag{26}$$

Equation (25) is the basic design relation for mass transfer wall gages. It indicates a disadvantage of this type of instrument, in that the mass transfer coefficient varies with only the cube root of S.

The use of boundary condition (22) is not quite correct. For the case of an electrochemical probe the voltage V_0 on the electrode and not the concentration is kept constant. The rate of reaction at the electrode surface is a function of both the electrode voltage and the concentration, $R(V_0, C_w)$. Thus boundary condition (22) should be replaced by

$$y = 0 \qquad -D\left(\frac{\partial C}{\partial y}\right)_w = C_w k_R(V_0) \qquad (27)$$

if the rate equation is first order. The rate constant $k_R(V_0)$ is a strong function of V_0, so that at large voltages, $C_w \to 0$ in order that $-D(\partial C/\partial y)_w$ remain finite. This method maintains C_w at an approximately zero value over the whole surface, except for a region at $x = 0$, where the solution presented by Mitchell indicates an infinite local mass transfer rate. It would therefore be of interest to solve (21) using boundary condition (27) in order to examine, in more detail, if the kinetics of the surface reaction places any limitations on the choice of an electrode length.

9.6.2 Limitations of the Design Equation

The assumptions made in the derivation of design Eq. (25) place limitations on its application to mass transfer and heat transfer probes. These will now be explored.

First, consider the condition that $U = Sy$. If δ_v is the thickness of the region over which this is valid, then it is necessary that $\delta_c/\delta_v < 1$. The thickness of the concentration boundary layer can be estimated using the concept of a Nernst diffusion layer, $\delta_c = D/K$. If K is calculated from Eq. (25), the condition $\delta_c/\delta_v < 1$ requires

$$0.807\frac{S^{1/3}\delta_v}{L^{1/3}D^{1/3}} > 1 \qquad (28)$$

For laminar flows this can usually be satisfied for mass transfer probes, except possibly in regions where there is a large pressure gradient. Spence and Brown [77] have explored the influence of pressure gradient for laminar flows by solving Eq. (21) using

$$U = Sy + \frac{y^2}{2\mu}\frac{dP}{dx} \qquad (29)$$

They obtained the relation

$$\left(\frac{KL}{D}\right)^3 = (0.807)^3 Z + \frac{25}{171}\left(\frac{KL}{D}\right)^{-1}\left(\frac{L^3\,dP/dx}{D\mu}\right) \qquad (30)$$

They found that the second term on the right-hand side of Eq. (30) can be important in boundary layers with large unfavorable pressure gradients and that, for fully developed laminar flow in a channel, the corrections are less than 2% of S, provided $Z^{1/3}(h_c/L) > 28$, where h_c is the channel height.

For turbulent flows the region where $U = Sy$ can be quite thin, and condition (28) is more restrictive. If $\delta_v = 5\nu/u^*$ is substituted into Eq. (28), the following limitation on the length of the mass transfer surface is obtained:

$$L^+ < 64Sc \tag{31}$$

Another limitation on the length is imposed by the requirement that turbulent transport of mass be negligible if design Eq. (25) is used. Son and Hanratty [78, 79] and Shaw and Hanratty [80] have carried out studies of turbulent mass transfer to surfaces with different electrode lengths at large Sc and found that Eq. (25) describes the results provided $L^+ < 700$. For Sc = 1 it is expected that turbulent transport would start to be important when $\delta_c > \delta_v$, i.e., when the thickness of the concentration boundary layer is larger than the region where $U = Sy$. Thus it is estimated that turbulent transport will not be important, provided the length of the electrode is chosen to satisfy the more restrictive of the two conditions $L^+ < 64Sc$ and $L^+ < 700$.

Fluctuations in the wall shear stress should not be measured with mass transfer probes with $L^+ > 700$, since measured fluctuations in the mass transfer rate could be coupled with the turbulent velocity fluctuations associated with the turbulent transport as well as with fluctuations in S. However, if it is only desired to measure the time-averaged wall velocity gradient $\bar{S}$, then the condition that turbulent transport effects be negligible is not a limitation, provided the design Eq. (25) is modified so as to take into account the additional effect of turbulent mass transport described by Son and Hanratty [78, 79] and Shaw and Hanratty [80].

A lower limit on the length of the mass transfer surface is imposed by the requirement that forced convection dominates over molecular diffusion in determining the rate of mass transfer. The errors involved in neglecting diffusion in the streamwise direction have been assessed by solving the equation

$$Sy \frac{\partial C}{\partial x} = D\left(\frac{\partial^2 C}{\partial x^2} + \frac{\partial^2 C}{\partial y^2}\right) \tag{32}$$

using the boundary conditions at $y = 0$ of

$$\begin{aligned} C = C_w \qquad & 0 < x < L \\ \frac{\partial C}{\partial y} = 0 \qquad & x < 0 \qquad x > L \end{aligned} \tag{33}$$

Ling [81] presents the following numerical solution valid for $Z \geq 50$:

$$\frac{KL}{D} = 0.807Z^{1/3} + 0.19Z^{-1/6} \tag{34}$$

On the basis of this equation, it is seen that for $Z > 200$ the error will be less than 5% if the second term is ignored. This requires that

$$L^{+2} \, Sc > 200 \tag{35}$$

Py and Gosse [27] have also carried out numerical solutions for Eq. (32) with boundary conditions (33) and have found, for $Z < 50$, that the mass transfer rate becomes insensitive to variations in S and that it is not practical to use this instrument even if the design equation is modified to take account of the effect of streamwise diffusion.

Limitations are also placed on the width of the electrode in order to minimize the effects of diffusion in the spanwise direction. To estimate these, it is assumed that $W/\delta_c > 10$. Using the estimate of $\delta_c = L^{1/3} D^{1/3}/0.807 S^{1/3}$, already derived, the following condition is obtained:

$$0.807 Z^{1/3} W/L > 10 \tag{36}$$

For $W/L = 2$, this requires a lower limit of $Z = 200$; for $W/L = 10$, Z should be greater than 2.

To summarize, for surfaces with $W/L > 2$, the length of the electrode for laminar flows should be chosen so that $200 < L^{+2} \, Sc < 64 Sc$. For typical electrochemical reactions where $Sc \cong 1000$ this requires $0.5 < L^+ < 64,000$. With turbulent flow it is probably wise to use $W/L > 10$ if the surface is to respond only to flow in the x direction. For the measurement of the time average of the turbulent wall shear stress, the design equation is not applicable for $L^+ > 700$, and it would be necessary to calibrate. However, it is recommended that $L^+ < 700$ if the turbulent fluctuations in the wall shear stress are to be measured.

With electrochemical mass transfer probes it is usually easy to operate within the above specified operating conditions. The chief limitation usually involves chemical considerations. At very high flow rates the mass transfer rate becomes large in comparison with the reaction velocity constant and it is therefore not possible to maintain boundary condition (22).

9.6.3 Nonhomogeneous Two-Dimensional Laminar Flows

Usually the mass transfer or heat transfer element is small enough that the assumption of uniform flow is acceptable for two-dimensional laminar flows. However, it might not hold close to a stagnation point or a separation point, where the wall shear stress vanishes. The application of mass transfer and heat transfer probes under these circumstances has been considered by Dimopoulos and Hanratty [21], Jolls [82], and LeBouche [83, 84].

The velocity gradient is considered to be varying over the mass transfer surface so that, from the continuity equation, the velocity normal to the surface is given by $V = -y^2/2 \; \partial C/\partial x$, and the steady mass balance equation is

$$Sy \frac{\partial C}{\partial y} - \frac{y^2}{2} \frac{\partial S}{\partial x} \frac{\partial C}{\partial y} = D \frac{\partial^2 C}{\partial y^2} \tag{37}$$

The solution of the equation for the case where $S = S_0 + \gamma x$ is given by

$$K = 0.807 \frac{D^{2/3}}{L^{1/2}} \left(\frac{2}{3}\right)^{2/3} \left[\frac{(S_0 + \gamma)^{3/2} - S_0^{3/2}}{\gamma L}\right]^{2/3} \tag{38}$$

At a stagnation or separation point $S_0 = 0$, $\gamma = (\partial S/\partial x)_s$ and

$$K = 0.807 (D^2 \gamma)^{1/3} \left(\frac{2}{3}\right)^{2/3} L^{-1/6} \tag{39}$$

For $\gamma L/S_0 << 1$, Eq. (38) simplifies to

$$K = 0.807 \frac{D^{2/3} S_0^{1/3}}{L^{1/3}} \left(1 + \frac{1}{6} \frac{\gamma L}{S_0} + \dots \right) \tag{40}$$

Now $\gamma L/S_0$ is approximately equal to the ratio of the electrode length to the distance from stagnation or separation. It is therefore concluded that for two-dimensional steady flows the error from nonuniform flow will be less than 10%, provided measurements are made at distances of five gage lengths or greater from stagnation or separation. However, even if this precaution is observed, it is necessary, close to stagnation or separation points, to take into account the influence of the pressure gradient and to ensure that Z is large enough for the design equation to be valid.

9.6.4 Frequency Response

The time response of a mass transfer probe to a fluctuating flow field is primarily associated with the capacitance effect of the concentration boundary layer. For a slowly varying velocity field a pseudo–steady state assumption can be made whereby the instantaneous mass transfer coefficient is related to the instantaneous velocity by the same equation as for a steady flow:

$$\frac{K(t)L}{D} = 0.807 \frac{L^2}{D} [S_x(t)]^{1/3} \tag{41}$$

Mitchell and Hanratty [19] and Sirkar and Hanratty [25] defined

$$K(t) = \overline{K}(t) + k(t) \tag{42}$$

$$S_x(t) = \overline{S}_x(t) + s_x(t) \tag{43}$$

and showed that for sufficiently small s_x/S_x, Eq. (41) can be written in the form

$$\frac{\overline{KL}}{D} = 0.807\left(\frac{L^2}{D}\right)^{1/3} \overline{S}_x^{1/3}\left(1 - \frac{1}{9}\frac{\overline{s_x^2}}{\overline{S}_x^2} + \cdots\right) \tag{44}$$

$$\frac{k}{K} = \frac{1}{3}\frac{s_x}{\overline{S}_x} - \frac{1}{9}\left(\frac{s_x^2}{\overline{S}_x^2} - \frac{\overline{s_x^2}}{\overline{S}_x^2}\right) + \cdots \tag{45}$$

Thus, to zeroth order, this analysis indicates that $\overline{K}$ is related to $\overline{S}_x$ by the same equation derived for steady flow and that the measured mass transfer fluctuations are related to the measured fluctuations in the velocity gradient by

$$\frac{\overline{k}}{\overline{K}} = \frac{1}{3}\frac{s_x}{\overline{S}_x} \tag{46}$$

It would be expected that for large frequency fluctuations in S_x that the concentration field would not change rapidly enough so as to maintain approximately the same conditions as exist at steady state. Consequently, the fluctuations in the mass transfer coefficient predicted by the above equation could be too large. Mitchell and Hanratty [19], Fortuna and Hanratty [20], Mao and Hanratty [85], Ambari et al. [86], Py [28], Nakoryakov et al. [87], Bogolyubov et al. [88], Dumaine [89], and Vorotyntsev [90] have carried out analyses of the time response of electrochemical probes by solving Eq. (20) for the case where $s_x/\overline{S}_x$ is small. Because of this assumption, the following linearized form of Eq. (20) can be used:

$$\frac{\partial c}{\partial t} + \overline{S}_x y \frac{\partial c}{\partial y} + s_x y \frac{\partial \overline{C}}{\partial x} = D \frac{\partial^2 c}{\partial y^2} \tag{47}$$

where

$$\overline{C} = \frac{C_B}{\Gamma\frac{4}{3}} \int_0^\eta e^{-t^3}\, dt \tag{48}$$

$$\eta = y\left(\frac{\overline{S}_x}{9Dx}\right)^{1/3} \tag{49}$$

If the fluctuating velocity gradient is given as a harmonic function,

$$s_x = \hat{s}_R e^{i\omega t} \tag{50}$$

then the solution of Eq. (47) can be given as

$$\hat{c}(y, x) = [\hat{c}_R(y, x) + i\hat{c}_I(y, x)]e^{i\omega t} \tag{51}$$

If Eq. (51) is substituted into Eq. (47), the following equation for $\hat{c}$ is obtained

$$i\omega\hat{c} + \overline{S}_x y \frac{\partial \hat{c}}{\partial x} + \hat{s}_R y \frac{\partial \overline{C}}{\partial x} = D \frac{\partial^2 \hat{c}}{\partial y^2} \tag{52}$$

This was solved by Fortuna and Hanratty [20] using the boundary conditions

$$\hat{c} = 0 \qquad y = 0 \text{ and large } y \tag{53}$$

For $\omega \to 0$ the solution is the pseudo–steady state approximation

$$\hat{c} = \frac{\hat{s}_R}{S_x} \frac{1}{3} y \frac{\partial \overline{C}}{\partial y} \tag{54}$$

$$\frac{k}{K} = \frac{1}{3} \frac{s_x}{\overline{S}_x} \tag{46'}$$

Any general time-varying s_x can be represented as a Fourier series. Because of the linearization assumption, the function describing the time-varying mass transfer coefficient can be constructed from the solution presented by Fortuna and Hanratty [20] as the sum of a number of Fourier components having the same frequencies as the forcing function $s(t)$.

For the case of a turbulent field, the mean squared values of the fluctuating velocity gradient and the fluctuating mass transfer coefficient can be represented in terms of spectral density functions:

$$\overline{k^2} = \frac{1}{2\pi} \int_0^\infty W_k \, d\omega \tag{55}$$

$$\overline{s_x^2} = \frac{1}{2\pi} \int_0^\infty W_{s_x} \, d\omega \tag{56}$$

Mitchell and Hanratty [19] represented the relation between W_{s_x} and W_k in the following way:

$$W_{s_x} = 9 \frac{\overline{S_x^2}}{\overline{K^2}} \frac{W_k}{A_d^2} \tag{57}$$

where $1/A_d^2 \geq 1$. Values of A_d^2 and the phase angles between k and s_x are given in Figs. 9.6a and 9.6b. These were calculated by Mao and Hanratty [85] by using the techniques of Fortuna. As suggested by W. J. McMichael (private communication, 1972), the ordinate is $\omega^* L^{+2/3}$, where $\omega^* = \omega^+ Sc^{1/3}$. At $\omega^* L^{+2/3} = 1$ the amplitude correction $1/A_d^2$ is only about 1.07; however, at this same condition the phase lag correction is close to $16°$. Ambari et al. [86] have expressed a concern that the kinetics of the reaction could be playing a role at large frequencies, but this has not been explored.

Another way of presenting the solution of Eq. (47) is a transfer function $H(i\omega)$ defined as

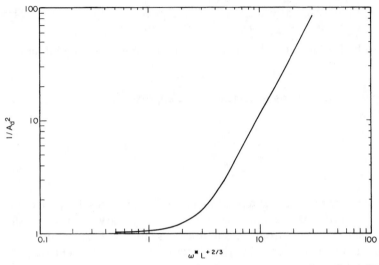

Figure 9.6a Amplitude correction factor for frequency response of an electrochemical wall shear stress probe

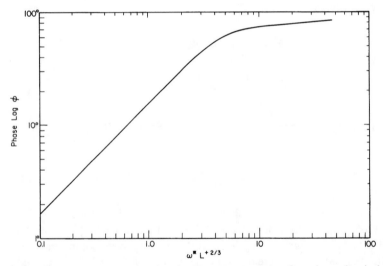

Figure 9.6b Phase angle correction for frequency response of an electrochemical wall shear stress probe

$$\hat{k} = H(i\omega)\hat{s}_x \qquad (58)$$

Expressions for $H(i\omega)$ have been presented by Nakoryakov et al. [87]. The evaluation of the transfer function offers the possibility of developing an electronic circuit or a computer algorithm to represent $H(i\omega)$ and therefore to convert time-varying currents from the electrolysis cell to a signal that directly represents the time-varying s_x. This approach would be needed to obtain amplitude probability distributions of a fluctuating quantity.

The above results indicate that the larger the Schmidt number, the smaller the frequency at which the behavior of the mass transfer probe departs from the pseudo–steady state solution because of damping in the concentration boundary layer. Electrochemical mass transfer probes usually operate at a Schmidt number of about 1000. Since heat transfer probes can be used in air, with Prandtl number about 0.7, they are superior to electrochemical probes with respect to the frequency response of the concentration boundary layer.

Situations involving large-amplitude oscillations cannot be analyzed by the methods developed by Mitchell and Hanratty [19] and Fortuna and Hanratty [20]. For low-frequency large-amplitude oscillations a pseudo–steady state assumption can be made, analogous to Eq. (41). By solving Eq. (41) for S_x, one obtains the following relations, valid for small frequencies:

$$S_x = \beta K^3 \qquad (59)$$

$$\beta = \left(\frac{L}{D}\right)^3 \left(\frac{1}{0.807}\right)^3 \frac{D}{L^2} \qquad (60)$$

$$\overline{S}_x = \beta \overline{K^3} \qquad (61)$$

$$s_x = \beta(K^3 - \overline{K^3}) \qquad (62)$$

It is seen from Eq. (59) that the measured signal must be cubed in order to determine S_x. This is cumbersome if analog methods are used and can be avoided if it is possible to use the linearization assumption, Eqs. (47) and (48).

Some progress has been made in recent years in dealing with large-amplitude oscillations when a pseudo–steady state assumption cannot be made. Pedley [91], Kaiping [92], and Mao and Hanratty [30, 31] have considered large-amplitude sinusoidal oscillations. The use of inverse mass transfer techniques explored by Mao and Hanratty [30–32] appears promising. This approach has been made possible because of the availability of dedicated work stations with relatively large computer capacity.

As shown by Pedley, Kaiping, and Mao and Hanratty, efficient numerical techniques can be developed to solve Eq. (20) for a time-varying mass transfer rate $K(t)$. The inverse problem involves the determination of $S(t)$ from a measured $K(t)$. This can be done with single electrodes, such as considered in this section, providing $S(t)$ does not change sign. The $K(t)$ is digitized. Suppose that $K(t)$ and $S(t)$ are known as time t. A value of $S(t + \Delta t)$ is assumed at Δt later. A finite difference method is used to solve Eq. (20) for $K(t + \Delta t)$. This is compared with the measurement. If agreement is obtained, the assumed $S(t + \Delta t)$ is correct. If not, a new value is assumed.

9.6.5 Turbulence Measurements

Turbulent flows are one of the most important type fluctuating flows for which mass transfer probes are used. Equations (44) and (45) allow an estimate of the errors involved in using linear theory. From Eq. (44) it is seen that an error of only 3–4% would be made in using

$$\frac{\overline{KL}}{D} = 0.807 \left(\frac{L^2}{D}\right)^{1/3} \overline{S}_x^{1/3} \tag{63}$$

to evaluate $\overline{S}_x$. Mitchell and Hanratty [19] have shown that if the pseudo–steady state approximation is applicable and if the probability function describing s_x is Gaussian, the error in calculating $(\overline{s_x^2/S_x^2})^{1/2}$ from $(\overline{k^2/K^2})^{1/2}$ by the linear relations

$$\frac{k}{\overline{K}} = \frac{1}{3}\frac{s_x}{\overline{S}_x}$$
$$\frac{\overline{k^2}}{\overline{K^2}} = \frac{1}{9}\frac{\overline{s_x^2}}{\overline{S}_x^2} \tag{64}$$

is less than 3% if $(\overline{s_x^2/S_x^2})^{1/2} < 0.5$.

These considerations would suggest that it is not necessary to use the more complicated nonlinear relations (61) and (62) for mass transfer probes in turbulent flows to evaluate $\overline{S}_x$ and $(\overline{s_x^2/S_x^2})^{1/2}$. The chief difficulties in applying Eq. (46) are spatial averaging of the fluctuations over the sensor surface and the frequency response of the concentration boundary layer. Both of these effects will lead to an underprediction of $\overline{s_x^2}$.

Reiss and Hanratty [7, 8] discovered that turbulent velocity fluctuations close to a wall are dominated by flow structures that have small spanwise dimensions and that are greatly elongated in the flow direction. These results suggest that spatial averaging in the flow direction is not important but that, if the spanwise dimension of the probe is not small in comparison with the spanwise dimension of the wall eddies, the assumption of a homogeneous flow over the sensor surface will no longer be valid. Under these circumstances, the following equation, from [19], relates the measured mean square value of the mass transfer fluctuations to the true local value $\overline{k^2}$:

$$\overline{k_m^2} = \frac{2\overline{k^2}}{W^2} \int_0^W (W - z)R_k(z)\,dz \tag{65}$$

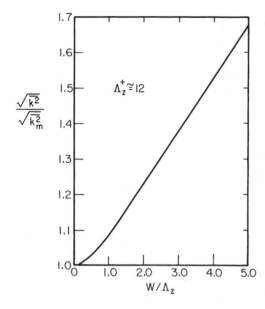

Figure 9.7 Nonuniform flow correction for electrochemical probes of varying widths

where $R_k(z)$ is the circumferential correlation coefficient for the mass transfer fluctuations. Mitchell and Hanratty [19] developed from Eq. (65) the curve shown in Fig. 9.7 for $(\overline{k^2})^{1/2}/(\overline{k_m^2})^{1/2}$, where Λ_z is the circumferential integral scale of the mass transfer fluctuations. Mitchell and Hanratty [19] have suggested that for turbulent pipe flow $\Lambda_z^+ \cong 12$. Therefore it is estimated that W^+ must be less than 8 for the error associated with spatial averaging to be less than 5%. This is quite restrictive for rectangular electrodes, so in many cases, it might be necessary to use electrodes with $W^+ > 8$ and Fig. 9.7 to correct the measurements.

From the analysis already presented, it was shown that damping by the concentration boundary layer would be unimportant for $(L^{+2}\omega^{+3} Sc)^{1/2} \leq 1$. For pipe turbulence the median frequency of s_x is $\omega^+_{median} \cong 2\pi(0.009)$ [93]. This suggests that, for no serious damping to occur at the median frequency, $Z = L^{+2} Sc$ should be less than 5600. For $Sc = 1$ this requires $L^+ < 75$. However, since electrochemical mass transfer probes operate at $Sc = 1000$, the much more restrictive condition of $L^+ < 2.4$ is obtained for them. For such probes it will probably be necessary to correct for frequency response by using Eqs. (55)–(57) and Fig. 9.6. This will require the measurement of the frequency spectrum of k and not just $\overline{k^2}$.

The methods just outlined, for correcting for spatial averaging and for frequency response, have been developed for the situation where either of these effects is separately affecting the determination of $\overline{s_x^2}$. The usual case is that both effects are simultaneously operative. No well-established methods for dealing with this situation have been developed. The approach taken at present is to assume the effects are independent and to apply both corrections to the measurement.

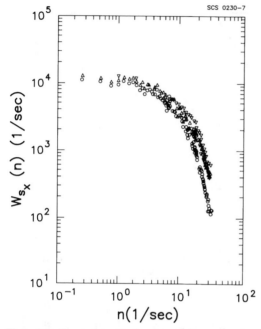

Figure 9.8 Spectral density function of the streamwise component of shear stress fluctuations at the wall for Re = 29,270: (o) quasi-steady theory, (Δ) corrected for frequency response with linear theory, and (∇) calculated with the inverse method

Measurements $(\overline{s_x^2})^{1/2}$ by the methods outlined above have yielded values of 0.35–0.36$\overline{S_x}$ [20, 94]. The measurements of the probability distribution show a large positive skewness, so that large positive excursions in s_x, of the order of $\overline{S_x}$, are observed. This provides some concern about inaccuracies in the determination of the spectral density function and the probability density function, in the evaluation of $\overline{S_x}$. The calibration of the electrode probes can be calculated from the dimensions and the active area of the probe. For very precise results, it is best to calibrate by measuring $\overline{K}$ for a known $\overline{S_x}$ under steady flow conditions. However, more often it is convenient to measure $\overline{K}$ in a turbulent flow for a known $\overline{S_x}$. Equation (63) is then used to evaluate an effective value of $D^{2/3}L^{-1/3}$. This procedure is valid only if $(\overline{s_x^2})^{1/2}/\overline{S_x}$ is small. These difficulties can be avoided by using the inverse mass transfer techniques.

Mao and Hanratty [31] carried out studies in a 5-cm pipe with an aqueous solution 0.1 M in KI and 2.0×10^{-3}MI_2 over a range of Reynolds numbers of 9000 to 30,000. The kinematic viscosity of the solution was 0.0085 cm²/s, and the Schmidt number characterizing the electrochemical reaction was 850. The platinum test electrode was 0.0127 cm, i.e., small enough that the dimensionless diameter d_e^+ was less than 5 over the whole flow range, so spatial averaging was not

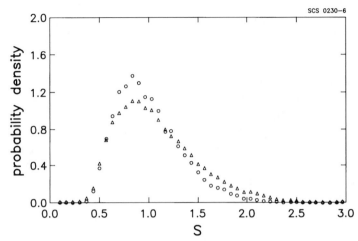

SCS 0230–6

Figure 9.9 The amplitude probability density function of the velocity gradient at the wall of a pipe for Re = 29,270: (o) quasi-steady method and (Δ) inverse method

important. Values of $\overline{S}_x$ calculated from Eq. (63) were found to be in error by about 3.5%. Correct values of $\overline{S}_x$ could be calculated from Eq. (63) if the constant was changed from 0.807 to 0.795. This is almost in exact agreement with the value calculated from Eq. (44):

$$\frac{\overline{KL}}{D} = 0.807 \left(\frac{L^2}{D}\right)^{1/3} \overline{S}_x^{1/3} \left(1 - \frac{1}{9}\frac{\overline{s_x^2}}{\overline{S}_x^2}\right) \tag{66}$$

Amplitude probability density functions of s_x were determined from $K(t)$. These are characterized by average values of $(\overline{s_s^2})^{1/2}/\overline{S}_x = 0.37$, $(\overline{s_x^3})/(\overline{s_x^2})^{3/2} = 0.96$, and $(\overline{s_x^4})/(\overline{s_x^2})^2 = 4.2$ for the three Reynolds numbers that were studied. The spectral density function of s_x is given in Fig. 9.8 for the run at Re = 29,270. The circles were obtained from quasi-steady theory, Eq. (57) with $A_d = 1$. The triangles used the linear theory correction, Eq. (57) with A_d given by Fig. 9.6a. The inverted triangles were calculated by the inverse method. It is noted that the linear theory correction increases the energy in the high-frequency range but not as much as calculated by the inverse method. For this particular experiment, values for $(\overline{s_x^2})^{1/2}/\overline{S}_x$ of 0.33, 0.37, and 0.38 were obtained with the quasi-steady method, the linear theory correction, and the inverse method, respectively. The amplitude probability density function obtained with the inverse method is given by the triangles in Fig. 9.9. The usual method for evaluating this function from $K(t)$ is with Eq. (41). It is noted from Fig. 9.9 that this approach introduces considerable error, in that it underestimates the large positive fluctuations in S_x. Figure 9.10 gives the spectral density function for three Reynolds numbers, normalized with the law of wall scaling.

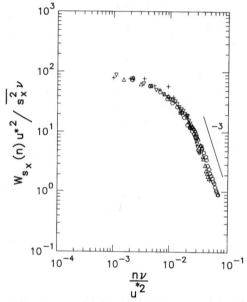

Figure 9.10 Spectral density function of the x component of the velocity gradient normalized with wall parameters ($\circ$) Re = 9900, (Δ) Re = 19,600, (∇) Re = 29,700 LDV measurements at y^+ = 3.4, Re = 11,110 in channel flow

9.7 EFFECT OF CONFIGURATION OF MASS TRANSFER PROBES

9.7.1 Circular Probes

Circular mass transfer probes offer considerable advantage and are much more widely used than rectangular probes, since they are more compact and more easily fabricated. They can be constructed simply by gluing a wire in a hole in the wall and by sanding the end of the wire so that it is flush with the surface.

Reiss and Hanratty [8] analyzed the performance of a circular mass transfer surface as indicated in Fig. 9.11. The rate of mass transfer to the strip shown is assumed to be given by the same equation derived earlier for a two-dimensional flow:

$$\frac{N_i}{C_B - C_w} = 0.807 \frac{D^{2/3} S^{1/3}}{(2R_e \sin \alpha)^{1/3}} (2R_e \sin \alpha) \, dz \tag{67}$$

The total mass transfer rate is then given as

$$K = \frac{0.807 D^{2/3} S^{1/3}}{\pi R_e^2} \int_0^\pi 1/2 (2R_e \sin \alpha)^{5/3} \, d\alpha \tag{68}$$

If an equivalent length for the surface is defined as

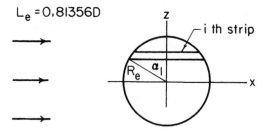

Figure 9.11 Evaluation of the effective length of a circular electrode

$$L_e = \left[\pi R_e^2 \Big/ \int_0^\pi 1/2 (2R_e \sin \alpha)^{5/3} \, d\alpha \right]^3 = 0.81356(2R_e) \tag{69}$$

then the design equation for a circular surface is obtained as

$$K = \frac{0.807 D^{2/3} S^{1/3}}{L_e^{1/3}} \tag{70}$$

A disadvantage of the circular surface over an approximately two-dimensional rectangular surface is that diffusion in the spanwise direction can be more important. Numerical solutions of the mass balance equations that include molecular diffusion in the flow and spanwise directions have been carried out by Py [95]. These suggest that Eq. (70) can be used, provided $Z = L_e^{+2}$ Sc > 1000. For Sc = 1000 an operating range of $1 < L_e^+ < 700$ is therefore recommended for a turbulent flow to ensure that the design equation holds.

Recent work by Phillips [96] and by Stone [97] shows how relations between the Nusselt number (KL/D) and Z can be calculated for general shapes of the transfer surface.

In a turbulent flow the instantaneous S is given by

$$S = S_x \left[1 + \left(\frac{S_z}{S_x} \right)^2 \right]^{1/2} \tag{71}$$

where the x coordinate is in the direction of mean flow. The quantity s_z/S_x may be defined in terms of the angle between the instantaneous direction of the flow at the wall and the x axis,

$$\tan \theta = \frac{S_z}{S_x} \tag{72}$$

Measurements by Sirkar [98] show that $\theta < 17°$ for 99.5% of the time. This indicates that the circular surface is sensitive to S_x and not to s_z; i.e., the equation

$$K = 0.807 \frac{D^{2/3} S_x^{1/3}}{L_e^{1/2}} \tag{73}$$

will be in error by less than 5% for 99.5% of the time.

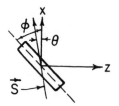

Figure 9.12 Single slanted electrode

9.7.2 Slanted Transfer Surface

Mitchell [99] was the first to consider the use of a wall transfer surface to measure the direction of the velocity gradient vector at a wall. For this purpose, he proposed the use of a slant surface of the type shown in Fig. 9.12 and showed that such a surface would be sensitive to the two components of the wall velocity gradient, S_x and S_z. This type of surface was tested by Sirkar and Hanratty [25]. On the basis of these tests, they developed the chevron-electrode arrangement shown in Fig. 9.13. Karabelas and Hanratty [100] have suggested an interesting modification of the chevron that could be useful in three-dimensional boundary layer flows.

By using the Reiss method, the slant gage shown in Fig. 9.12 can be analyzed to give

$$K = 0.807D^{2/3}\left[\frac{S \sin (\phi - \theta)}{L}\right]^{1/3}\left[1 + \frac{L \cot (\phi - \theta)}{5W}\right] \tag{74}$$

For $L/W \to \infty$ it is seen from the above relation that K depends only on the component of S perpendicular to the leading edge of the surface.

For the chevron arrangement shown in Fig. 9.13, the following results are obtained for $-\phi + \psi \le \theta \le \phi + \psi$, where $\tan \psi = L/W$:

$$K_1 = 0.807D^{2/3}\left[\frac{S \sin (\theta - \phi)}{L}\right]^{1/3}\left[1 + \frac{L}{5W} \cot (\phi - \theta)\right] \tag{75}$$

$$K_2 = 0.807D^{2/3}\left[\frac{S \sin (\phi + \theta)}{L}\right]^{1/3}\left[1 + \frac{L}{5W} \cot (\phi + \theta)\right] \tag{76}$$

From measurements of K_1 and K_2 the quantities S and θ can be calculated from Eqs. (75) and (76) to give the magnitude and the direction of the velocity gradient at the wall. This calculation can be implemented by combining the measurements in the following way, as suggested by Tournier and Py [101]:

$$\frac{K_1 - K_2}{K_1 + K_2} = g_1(\theta) \tag{77}$$

$$K_1 + K_2 = \frac{S^{1/3}D^{2/3}}{L^{1/3}}g_2(\theta) \tag{78}$$

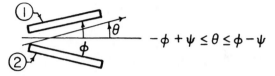

$$-\phi + \psi \le \theta \le \phi - \psi$$

Figure 9.13 Pair of slanted electrodes used for measurement of the direction of the wall shear stress vector, as well as its magnitude

The functions $g_1(\theta)$ and $g_2(\theta)$ are evaluated from Eqs. (77) and (78) as follows:

$$g_1(\theta) = \frac{f_1 - f_2 + \tau_1 (f_1 f_3 - f_2 f_4)}{f_1 + f_2 + \tau_1 (f_1 f_3 + f_2 f_4)} \qquad (79)$$

$$g_2(\theta) = \frac{1.5}{\Gamma(4/3)9^{1/3}} [f_1 + f_2 + \tau_1 (f_1 f_3 + f_2 f_4)]$$

$$f_1 = [\sin (\phi - \theta)]^{1/3} \qquad f_3 = \cot (\phi - \theta) \qquad (80)$$
$$f_2 = [\sin (\phi + \theta)]^{1/3} \qquad f_4 = \cot (\phi + \theta)$$

$$\tau_1 = \frac{L}{5W}$$

From measurements of K_1 and K_2, the angle θ is first calculated from Eq. (77). Then, using this value of θ and the measurement of $K_1 + K_2$, the magnitude of the velocity gradient S can be evaluated from Eq. (78). One of the difficulties with this approach is that the function $g_1(\theta)$ is multivalued and therefore carries the restriction of $\theta < \phi$. In designing the chevron pair, the selection of ϕ is a compromise. Small values give greater sensitivity to s_z and give better spatial resolution in the spanwise direction. However, ϕ cannot be so small that it violates the above restriction.

For a turbulent flow where K_1 and K_2 are functions of time it is necessary to use a computer in order to solve (77) and (78) for instantaneous values of S and θ. An approach taken by Sirkar and Hanratty [26] can avoid this difficulty. Nonlinear terms in the fluctuating quantities are ignored. The mean flow is assumed to be in the x direction. It is assumed that

$$\frac{|s_z| \cot \theta}{|\bar{S}_x + s_x|} < 1$$

Then Eqs. (77) and (78) can be rearranged in the following form after substituting $S_x = \bar{S}_x + s_x$, $S_z = s_z$, $S^2 = [(\bar{S}_x + s_x)^2 + s_z^2]$ and $\theta = \tan^{-1} [s_z / (\bar{S}_x + s_x)]$:

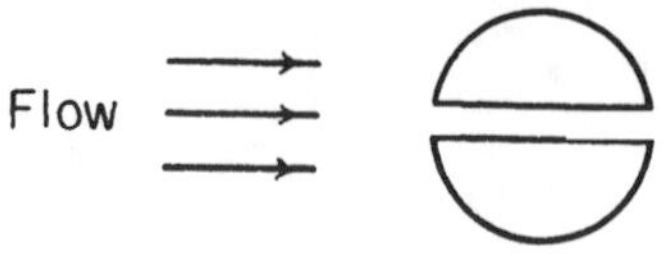

Flow →

Figure 9.14 Twin semicircular electrodes

$$\overline{K} = 0.807D^{2/3} \left(\frac{\overline{S}_x \sin \phi}{L} \right)^{1/3} \left(1 + \frac{L \cot \phi}{5W} \right) \tag{81}$$

$$\frac{K_1 - K_2}{2\,\overline{K}} = \frac{1}{3} \frac{s_z}{\overline{S}_x} \left(\frac{\cot \phi - (2L \cot^2 \phi / 5W)}{1 + (L/5W) \cot \phi} \right) \tag{82}$$

$$\frac{(K_1 + K_2) - 2\overline{K}}{2\overline{K}} = \frac{1}{3} \frac{s_x}{\overline{S}_x} \tag{83}$$

Equations (82) and (83) show that s_z is directly proportional to the difference in the fluctuating signals and that s_x is directly proportional to sum of the fluctuating signals. These operations can be easily implemented on analog circuits.

9.7.3 Other Methods to Measure Direction

Py and his coworkers have developed configurations other than the chevron for measuring the direction of the wall velocity gradient. Py and Gosse [27] suggested the use of two semicircular surfaces separated by insulation, as sketched in Fig. 9.14. Menzel et al. [102] and Wein and Pokryvaylo [103] have described a circular electrode with three segments that allows the measurement of the magnitude and direction of τ_w and of the limiting behavior of the velocity component normal to the wall. The sensor with a pair of electrodes is arranged in such a way that the insulation lies along the direction of mean flow. For the case of negligible insulation thickness,

$$\frac{s_z}{S} = 4.20 \frac{K_1 - K_2}{K_1 + K_2} \qquad\qquad Z > 50 \tag{84}$$

$$K_1 + K_2 = 0.807 \left(\frac{D^2}{L_e} \right)^{1/3} (S)^{1/3} \qquad Z > 10^3 \tag{85}$$

This arrangement is quite attractive because it is more compact than the pair of slant surfaces and because the relations between the measured mass transfer rates and the velocity field are so simple, $g_1(\theta) = $ const and $g_2(\theta) = $ const. Py [95] has also shown that the frequency response to s_z is better than for pairs of slant surfaces. However, this probe has not been used in experimental studies because of perceived difficulties in its fabrication and in controlling the thickness of insulation.

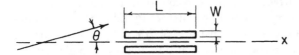

Figure 9.15 Twin rectangular electrodes

The pair of rectangular surfaces [28, 104] shown in Fig. 9.15 is easier to fabricate than a pair of semicircles and can also be made quite compact in the spanwise direction. The equations describing its performance are of the same form as Eqs. (77) and (78). Calculations presented by Py [28, 95] show that rectangular pairs of surfaces with large L/W ratios are not attractive choices because the functions $g_1(\theta)$ and $g_2(\theta)$ indicate too strong a dependence on θ at small values of θ and too weak a dependence at large values of θ. Because of possible difficulties in controlling the thickness of the insulation and the dimensions of the surfaces, it might be necessary to determine the forms of $g_1(\theta)$ and $g_2(\theta)$ empirically. In studies of turbulent spanwise velocity fluctuations and of flow around slanted cylinders, Py and Tournier [23, 28, 95] used pairs of rectangular probes with an L/W ratio of 2.5. These were made from two platinum ribbons, 0.5×0.1 mm, separated by a strip of Mylar, 8 μm thick. Labraga et al. [105] have recently presented a theoretical and experimental study of the frequency response of these electrodes.

9.7.4 Sandwich Elements

The use of two elements separated by a thin layer of insulation with the configuration shown in Fig. 9.16 was suggested by Son and Hanratty [22] to detect the direction of a two-dimensional flow. If the flow is in the forward direction, surface 2 will give a smaller value of K than surface 1, since it lies in the wake of the concentration boundary layer from surface 1. Work done since then by Py, Tournier, LeBouche, and Martin has indicated that such a probe design holds other advantages than just detecting flow direction.

Py [28, 95] has shown that

$$2L(K_1 - K_2) = g_0\left[\frac{S_x(2L)^2}{D}\right]^{1/3} \tag{86}$$

where g_0 is a constant dependent on the insulation thickness. Thus measurements $K_1 - K_2$ provide a means of determining S_x. Py has shown that Eq. (86) is valid

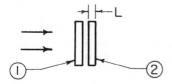

Figure 9.16 Sandwich electrode

for $Z > 10$, so that a pair of surface elements appears to have an advantage over a single element, in that they can be used at smaller Z. Py [28, 95] also demonstrated that the use of a composite signal consisting of 2/3 of $K_1 + K_2$ and 1/3 of $K_1 - K_2$ improves the frequency response over that obtained by a single element by almost an order of magnitude.

These types of designs have been applied extensively to the study of flow around a cylinder [23, 24, 84, 101]. For this purpose, LeBouche [84] derived the following equation for the case of negligible insulation thickness, analogous to Eq. (38) already presented for single elements:

$$K_1 - K_2 = 0.420 \left(\frac{D^2 S_x}{2L}\right)^{1/3} \left(1 - \frac{0.803L \; \partial S_x/\partial x)}{S_x}\right)$$

9.7.5 Time-Varying and Reversing Flows

Usual practice is to calculate the frequency response of circular electrodes as if they are rectangular electrodes with $L_e = 0.81 d_e$. Direct analyses of the transfer function have been presented by Nakoryakov et al. [87] for circular electrodes, for differential electrodes that are two halves of a disc, or for two strip electrodes at an angle to one another. Ambari et al. [86] have also derived the transfer function for circular electrodes and suggested that $L_e = 0.756 d_e$ rather than $0.81 d_e$. Py [28, 95] has used linear theory to calculate the frequency response for electrode pairs, as well as for circular electrodes.

One of the more interesting applications of the sandwich electrode, shown in Fig. 9.16, is the measurement of $S_x(t)$ oscillations that are large enough that the flow is reversing direction at the wall. Mao and Hanratty [32] used two 0.011×0.093 cm electrodes separated by a layer of 0.024-cm-thick insulation. Measurements of $K_1 - K_2$ were related to $S(t)$ with inverse mass transfer techniques. The system studied was turbulent flow through a 5-cm pipe with large imposed flow oscillations. The results of this study are quite encouraging and suggest that electrochemical techniques can be used to measure wall shear stress in haphazard conditions that exist in separated flows. However, there is need to improve the iteration technique close to times at which the flow is reversing direction and to check on the influence of molecular diffusion in the x direction.

9.8 HEAT TRANSFER PROBES

9.8.1 Analysis

A sketch of a wall heat transfer probe is given in Fig. 9.17. It can be a metal film deposited on the wall or a very small cylindrical heating element resting on the wall or embedded in it. Fluid flowing over the probe cools it. The electric current needed to keep the probe at a constant temperature is measured. The analysis will

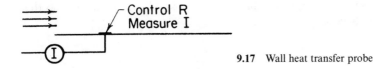

9.17 Wall heat transfer probe

assume a rectangular heating element flush with the wall. The results already derived for mass transfer probes are directly transferable if $\rho C_p T$ is substituted for C, $\alpha_T = k_T/\rho C_p$ for D, and hL/k_T for KL/D. Equation (25) can therefore be written as

$$\frac{q}{\Delta T} = 0.807\frac{C_P^{1/3}k_T^{2/3}}{L^{1/3}\mu^{1/3}}(\rho\tau_w)^{1/3} \tag{87}$$

Since the heat loss from the thermal element per unit area is related to the heating current I and the resistance by the relation $q = I^2R/A_e$, Eq. (87) can be written in the form

$$\frac{I^2R}{\Delta T} = A(\rho\tau_w)^{1/3} \tag{88}$$

where $A = 0.807A_e C_p^{1/3}k_T^{2/3}/L^{1/3}\mu^{1/3}$ is a weak function of temperature.
Experiments [3, 14, 64, 106] give the following result:

$$\frac{I^2R}{\Delta T} = A(\rho\tau_w)^{1/3} + B \tag{89}$$

both for heated films and for heated wires lying on the wall. The term B, which represents the heat loss to the substrate, can often be larger than $A(\rho\tau_w)^{1/3}$, the heat loss to the fluid. The effective length of the thermal element calculated from experimentally determined values of A can be many times greater than the actual length of the heating element. This arises because heat is being transferred to the fluid both from the heating element and from the substrate.

This large heat loss to the substrate greatly complicates the use of wall heat transfer gages. The sensitivity of the instrument to changes in τ_w is diminished with respect to a mass transfer probe, particularly at small τ_w. Since the heat loss to the substrate and the effective length of the thermal element are not predictable, the thermal gage must be calibrated in order to determine A and B.

From the mass transfer analysis, it is found that $SL_{\text{eff}}^2/\alpha_T = L_{\text{eff}}^{+2}$ Pr should be greater than 200 for forced convection to dominate, and that L_{eff} should be less than $(0.807)^3 S\delta_v^2/\alpha_T$ to ensure that the thermal boundary layer is within a region where the velocity is varying linearly with distance from the wall. For a turbulent fluid this later restriction leads to the condition of $L_{\text{eff}}^+ < 64$Pr. Because of losses from the heating element, it is often quite difficult to satisfy the above condition for gas flows. It is then necessary to calibrate the instrument in a turbulent flow using the law of wall assumptions discussed in section 9.3. Under these circumstances, it can no longer be expected that the heat loss will vary with $\bar{\tau}_w^{1/3}$, so that the design equation becomes

$$\frac{I^2 R}{\Delta T} = A_t (\overline{\rho \tau_w})^{1/n} + B_t \tag{90}$$

where A_t, B_t, and n must be determined empirically and $\bar{\tau}_w$, $\bar{I}$ are the time-averaged values of the wall shear stress and current.

In circumstances where $L_{\text{eff}}^+ < 64\text{Pr}$, the exponent $1/n$ equals $1/3$. Even for these cases it is desirable to calibrate the gage in a turbulent flow. For reasons given in the next section the constants A_t and B_t determined in a turbulent flow need not be equal to the constants A and B determined in a steady flow. For experiments in which the calibration has been carried out in a laminar flow, equations developed in the next section should be used when the instrument is to be applied to a turbulent flow for $L_{\text{eff}}^+ < 64\text{Pr}$.

9.8.2 Use in Turbulent Flows

If $L_{\text{eff}}^+ < 64\text{Pr}$, then $\delta_c < \delta_v$, and it is possible to use thermal wall probes to measure fluctuations in τ_w. If the resistance of the heating element is held constant and the current or the voltage drop $E = IR$ is measured, Eq. (89) can be rearranged as follows for an incompressible fluid:

$$\tau_w^{1/3} = A^\dagger E^2 + B^\dagger \tag{91}$$

$$A^\dagger = \frac{1}{A \Delta T R \rho^{1/3}} \tag{92}$$

$$B^\dagger = -\frac{B}{A \rho^{1/3}} \tag{93}$$

The simplifications that result in the case of mass transfer probes, from using linearized equations, may not be realized for heat transfer probes. Sandborn [17, 18], who was the first to point out this difficulty, indicates that errors of the order of 10% in $\bar{\tau}_w$ and of the order of 50% in $(\tau_w')^{1/2}$ may be experienced if linearization techniques are used.

Consequently, $\bar{\tau}_w$ and τ_w' should be evaluated with the following relations:

$$\bar{\tau}_w = \overline{(A^\dagger E^2 + B^\dagger)^3} \tag{94}$$

$$\tau_w' = (A^\dagger E^2 + B^\dagger)^3 - \overline{(A^\dagger E^2 + B^\dagger)^3} \tag{95}$$

Because these equations involve evaluating the sixth power of E, it is not convenient to use analog methods. It is necessary to digitize the measured function $E(t)$ and to use a computer to evaluate $\bar{\tau}_w$ and τ_w' (t) from Eqs. (94) and (95).

Quite often it is necessary to use a turbulent flow for which $\bar{\tau}_w$ is known (such as fully developed flow in a pipe) to calibrate the wall heat transfer probe, rather than a laminar flow. This introduces additional problems if it is desired to use the probe to measure τ'_w. Expansion of Eq. (94) with $E = \overline{E} + e$ gives the following relation:

$$\overline{\tau}_w = A^{\dagger^3}[C1] + 3A^{\dagger^2}B^{\dagger}[C2] + 3A^{\dagger}B^{\dagger^2}[C3] + B^{\dagger^3} \tag{96}$$

where

$$[C1] = \overline{E^6} + 15\overline{E^4 e^2} + 20\overline{E^3 e^3} + 15\overline{E^2 e^4} + 6\overline{E e^5} + \overline{e^6}$$
$$[C2] = \overline{E^4} + 6\overline{E^2 e^2} + 4\overline{E\ e^3} + \overline{e^4} \tag{97}$$
$$[C3] = \overline{E^2} + \overline{e^2}$$

It is convenient to fit an equation of the form

$$\overline{\tau}_w^{1/3} = A_t^{\dagger}\overline{E^2} + B_t^{\dagger} \tag{98}$$

to the calibration measurements of $\overline{\tau}_w$ versus $\overline{E}$. From Eq. (96), it is seen that $\overline{\tau}_w$ is a function of $\overline{e^2}$, $\overline{e^3}$, $\overline{e^4}$, $\overline{e^5}$, $\overline{e^6}$, in addition to $\overline{E}$. Consequently, if the moments of the fluctuating voltage cannot be ignored, then $A_t^{\dagger}$ and $B_t^{\dagger}$ will not be equal to the constants $A^{\dagger}$ and $B^{\dagger}$ that would be determined in a steady flow.

Sandborn [17, 18] has discussed two methods for evaluating $A^{\dagger}$ and $B^{\dagger}$ from calibration measurements in a turbulent flow. In the first of these, $A_t^{\dagger}$ and $B_t^{\dagger}$ are determined from the best fit of Eq. (98) to measurements of $\overline{E}$ versus $\overline{\tau}_w$. It is initially assumed that $A^{\dagger} \cong A_t^{\dagger}$ and $B^{\dagger} \cong B_t^{\dagger}$. The calibration measurements are digitized, and τ_w is calculated for each of the calibration points by using Eq. (91). This gives a distribution function for τ_w. If the value of $\overline{\tau}_w$ calculated from this distribution function is in error, then it is evident that the above assumption is not correct. New values of $A^{\dagger}$ and $B^{\dagger}$ are assumed, and the calculation repeated until the values of $\overline{\tau}_w$ obtained from the distribution function for τ_w agree with the measured values of $\overline{\tau}_w$.

A second approach is to use Eq. (96) directly. Values of $\overline{E}$ and of the moments of e are determined at each of the values of $\overline{\tau}_w$ at which the calibration was conducted. The constants $A^{\dagger}$ and $B^{\dagger}$ are obtained by a least squares fit of Eq. (95) that relates the determined values of [C1], [C2], and [C3] to $\overline{\tau}_w$. Sandborn has suggested that a very close approximation to the correct values of $A^{\dagger}$ and $B^{\dagger}$ can be obtained by ignoring the much more difficult-to-determine $\overline{e^3}$, $\overline{e^4}$, $\overline{e^5}$, and $\overline{e^6}$.

Since the design equation for the heat transfer wall probes cannot be linearized when using them to determine turbulent fluctuations, it does not appear feasible to use the linear methods developed for the mass transfer probes to correct for frequency response. The best practice at present would be to design the experiment so that the pseudo–steady state approximation is acceptable.

A review of a number of carefully executed experiments in which wall heat transfer probes were used to measure the turbulent fluctuations in τ_w has been presented by Alfredsson et al. [107]. These experiments show a strong sensitivity of the determination of $\tau_x'^2$ to the design of the probe. Haritonidis [69] has pointed out that accurate measurements require that the ratio of the heat transfer to the substrate and the fluid be kept small. Thus the use of a cylindrical hot wire at, or very close to, the wall reduces heat transfer to the substrate. The use of liquid flows instead of gas flows also reduces the heat transfer to the substrate (coefficient B is small) relative to the fluid.

9.8.3 Design Considerations for Turbulent Flows

The observation that B in Eq. (89) is constant over a relatively large range of flows indicates that the average heat transfer to the substrate remains unchanged. This suggests that if the temperature of the heat transfer element is kept constant, then the temperature environment around the element also remains relatively constant with changing flow. According to this rough assumption, the chief effect of a substrate is to increase the effective length of probe L_e. If the probe is visualized as rectangular with dimensions of $L_e \times W$, the analysis presented for mass transfer probes can be used.

The length of the probe is limited by the need to keep heat transfer in the flow direction negligible (L^{+2} Pr > 200), to keep the thickness of the thermal boundary layer less than the thickness of the viscous sublayer ($L^+ < 64$Pr), and to limit the influence of normal velocity fluctuations at large Pr ($L^+ < 700$). For Pr $= 0.7$ this provides a range of operation, $16.9 < L_e^+ < 44.8$. For water with Pr $= 7$, the range is $5.3 < L_e^+ < 448$. For an electrochemical system with Sc $= 850$, $0.48 < L_e^+ < 700$. These considerations lead to the conclusion that the use of wall probes to measure wall shear stress fluctuations in gas flows greatly limits the range of L_e that can be used and requires large enough probes that spatial averaging of the turbulence will be important ($W^+ < 8$).

From the linear analysis of the frequency response of mass transfer probes, damping of the fluctuations due to the capacitance of the thermal boundary layer becomes important for ω^+ Pr$^{1/3}$ $L_e^{+2/3} < 1$. This would seem to suggest that frequency response is better for low-Pr fluids. However, because of limitations on the length of the element to avoid streamwise diffusion, the above criterion is actually independent of Pr. Thus, if the minimum probe length is used, one obtains ω^+ Pr$^{1/3}$ $(200$Pr$^{-1})^{1/3} < 1$ or $5.8\omega^+ < 1$. Thermal probes will obtain better measurements of turbulent fluctuations in high-Pr fluids because probes can be made smaller to avoid spatial averaging, heat losses to the substrate become less important, and the range of operation is increased.

9.8.4 Analysis of Unsteady Probe Performance

Bellhouse and Rasmussen [108] carried out a one-dimensional and Beasley et al. [109], a two-dimensional analysis of the behavior of a constant-temperature wall subjected to a time-varying heat transfer rate. Cole and Beck [110, 111] used a two-dimensional model, and Liang and Cole [112] a three-dimensional model for a heated substrate over which a gas is flowing. The wall shear stress was allowed to vary with time, and the heat flux to the substrate was calculated by solving a thermal balance for the fluid that uses by a pseudo–steady state approximation, the thermal analog of Eq. (21). This differed from previous studies, in that there is a spatial variation of the heat transfer to the surface film and the substrate.

The analyses cited above show that there is a range of low frequencies for which the phase lag between the sensor signal (the voltage to keep it at a constant temperature) and a sinusoidally modulated airflow can be quite large. Cole [113] has shown that for quartz-substrate sensors this range is 0.2–1 Hz. At higher frequencies the phase lag becomes zero, and temperature fluctuations in the substrate are highly damped. This suggests that the frequency response at high frequencies is controlled by the capacitance effect of the thermal boundary layer and that the simple model outlined in section 9.8.3 might be useful.

An inverse heat transfer analysis of the type carried out by Mao and Hanratty [30–32] for electrochemical probes would be most useful to understand the relation between the fluctuating wall shear stress to the measured fluctuating heating current needed to keep the probe at constant temperature. The analysis should include both the time response of the thermal boundary layer and the time response of the substrate. Encouraging work in this direction has been carried out by Cole [113]. However, this calculation made a pseudo–steady state assumption for the fluid phase heat transfer. This work needs to be extended to include the capacitance effect of the thermal boundary layer and to include a test of the numerical schemes in actual test facilities.

As mentioned in section 9.8.3, the problem of heat transfer to the substrate is less important in liquid flows. These simplifications are discussed in a paper by Menendez and Ramaprian [114].

9.8.5 Compressible Flows

Because the constant A is a weak function of temperature, Eq. (91) or the following modification may be used to determine the quantity $\rho\tau_w$ in compressible flows:

$$(\rho\tau_w)^{1/3} = \frac{1}{A\Delta TR}E^2 - \frac{B}{A} \tag{99}$$

Experiments by Owen and Bellhouse [115], Diaconis [116], Bellhouse and Schultz [13], and Liepmann and Skinner [12] have demonstrated that probes calibrated under subsonic conditions can be used to carry out measurements in supersonic flows.

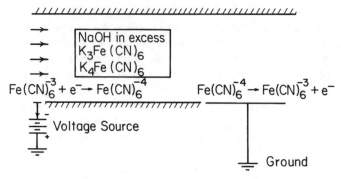

Figure 9.18 Electrochemical cell

9.9 EXPERIMENTAL PROCEDURES FOR MASS TRANSFER PROBES

9.9.1 Electrochemical Cell

The mass transfer probe used in shear stress measurements is part of an electrochemical cell. A voltage applied to the cell drives a reaction at the probe, or test electrode. The reverse reaction occurs at the counterelectrode. The current produced by the reaction can be related to the molar flux $<N>$ at the probe by using Faraday's law,

$$I = n_e F A_e <N> \tag{100}$$

where F = Faraday's constant
A_e = probe area
n_e = numer of electrons involved in the stoichiometric equation

In Fig. 9.18 the test electrode is operated cathodically by applying a negative voltage. The ferrocyanide produced at the cathode is oxidized at the anode, or counterelectrode, to ferrocyanide. The current resulting from this transfer of electrons flows through the circuit made by the solution and the ground in a clockwise direction. The applied voltage is controlled so that the reaction at the test electrode is diffusion controlled, i.e., polarized.

Figure 9.19 [117] is a plot of the cell current I as a function of applied voltage V_0 for two different systems. Position B, the plateau, represents an operating condition under which the test electrode is polarized. Since an increase in applied voltage results in the same current, the reaction rate is independent of the kinetics. The concentration of the reacting specie at the wall is approximately zero because the reaction rate is large. The portion of the solid curve represented by position A indicates operating conditions where the wall concentration is not zero. The kinetics cannot keep pace with the mass transfer, and the reaction is said to be kinet-

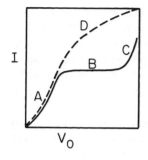

Figure 9.19 Electrode polarization. A, kinetically controlled; B, mass transfer controlled; C, side reactions; and D, above critical K_{limit}

ically limited. Operating conditions in the C portion of the curve are typically due to additional reactions, such as the hydrolysis of water. If the onset of an additional reaction occurs at a low enough applied voltage, the electrode may not polarize. The dashed line, labeled D, represents a system in which the kinetics are so slow or the mass transfer rate so great that a polarization plateau does not exist.

The application of a voltage to the test electrode and the deficit of charge at the electrode, caused by the reaction, set up an electric field in the region of the probe. Ions migrate when acted upon by this field. This effect is characterized by the transference number T_R defined as the fraction of the current carried by an ion in the absence of concentration variations [117]:

$$<N> = \frac{I}{A_e n_e F}(1 - T_R) \tag{101}$$

It is necessary to minimize the effect of migration, so that the transfer of the reacting species to the test electrode is controlled by molecular diffusion. This is accomplished by adding an excess of neutral electrolyte, so as to have a small transference number. In the system shown in Fig. 9.18 the use of a two-molar solution of sodium hydoxide results in a transference number of about 0.001 [118]. The test electrode is commonly operated near the center of the plateau region. Doubling the concentration of the reacting species doubles the driving force for mass transfer and therefore should double the current characterizing the plateau. An increase in the flow rate increases the mass transfer coefficient and therefore the plateau current. However, if the flow rate is too large, the mass transfer rate can be too rapid to polarize the electrode. The limiting flow rate for polarization can be increased by decreasing the concentration of the reacting species in the electrolyte.

With no reacting species present, a small current (tenths of a microampere) may appear at voltages below that at which water hydrolyzes. Ranz [119] attributes this to capacity and double-layer effects (see [117]). When operating with low concentrations or low currents, one should determine this residual current and subtract it from the measured current.

In the cell shown in Fig. 9.18, the voltage is expressed with respect to the ground and not to a reference voltage (as is common in electrochemistry). With a large counterelectrode, it may be assumed that the reference voltage is constant. Because of this choice for a reference electrode, there could be some variation of the polarization voltage with system design if an appreciable IR drop through the solution exists. Therefore it is necessary to determine the polarization curve to define the operating voltage.

If conditions are maintained so that (1) migration effects are negligible, (2) only the test reaction is occurring, and (3) the rate of reaction is large enough that $C_w \cong 0$, then Eqs. (100) and (25) can be combined to give the following relation between the cell current and the velocity gradient:

$$I = 0.807 \frac{Dn_e FA_e C_B}{L} Z^{1/3} \tag{102}$$

9.9.2 Electrolyte

The selection of an electrolyte is based on the following requirements:

1. It must react electrochemically at voltages where other reactions do not occur.
2. It must have a high reaction rate constant.
3. It must contain nonreacting ions that eliminate migration effects.
4. It should be easy to use in a flow loop (nontoxic, nonflammable, easy to store, etc.).
5. It must not produce negative effects at the electrodes or on the system (such as probe poisoning or corrosion).

In principle, there should be numerous electrolytic reactions that could be used [120]. The reduction of oxygen has been used by Lin and coworkers [121], Ranz [119], and Reiss [122]. Lin et al. [121] also reduced quinone in a strongly buffered solution. The reduction of cuprous ion is mentioned by Mizushina [123]. Reiss [122] and Lin et al. [121] have studied the oxidation of ferrocyanide. The most popular reaction seems to be the reduction of ferricyanide [119, 121, 124–127, and others]:

$$Fe(CN)_6^{3-} + e^- \rightarrow Fe(CN)_6^{4-} \tag{103}$$

Concentrations of 0.01–0.1 M are commonly used. The reduction of triiodide, which was first suggested by Py [28], has been used extensively at the University of Illinois [128–130 and others]:

$$I_3^- + 2e^- \rightarrow 3I^- \tag{104}$$

Potassium chloride and sodium hydroxide (0.5–2 M) are commonly used in the ferricyanide system as neutral electrolytes to make migration effects negligible.

They do not interfere with the reaction and are good conductors. Potassium iodide is used in the triiodide system in 0.02–0.5 M concentrations [128].

Although the ferricyanide system has been widely used, there are many problems associated with it. Ferricyanide decomposes slowly with light to form hydrogencyanide [118]:

$$Fe(CN)_6^{4-} + 2H_2O \overset{light}{\rightarrow} Fe(CN)_5H_2O^{3-} + OH^- + HCN \qquad (105)$$

The cyanide ions poison the electrodes [127]. Ferricyanide is also known to decompose with oxygen and light but not as rapidly as ferrocyanide [127]. Alkaline solutions may not only produce cyanide but also ferric hydroxide, which could foul the electrodes [131]:

$$2Fe(CN)_6^{3-} + 6OH^- \underset{\leftarrow}{\overset{\rightarrow}{}} 2Fe(OH)_3 + 12CH^- \qquad (106)$$

Jenkins [128] also mentions the possibility of an Fe^{2+}/O_2 cell being set up when dissolved oxygen is present. Oxygen in the presence of light may also decompose the ferrocyanide ion to iron oxide, resulting in surfaces having an oxide film [127]. Most of these problems can be overcome by frequently making up fresh solutions and operating within an opaque system under an atmosphere of nitrogen [123]. Despite these difficulties, a number of investigators have used this system to study air-liquid flows [124, 132–135].

The fouling of the electrode surface necessitates frequent cleaning. This is done normally with mild soap and a soft rag, followed by an alcohol or carbon tetrachloride rinse [129]. Mizushina [123] suggests buffing with soft paper to remove the oxide coat. Jenkins [127] and Mizushina [123] advocate operating the test electrode as a cathode in a 5% NaOH solution at 10–20 mA/cm^2 (cathodic cleaning).

An additional disadvantage of the ferricyanide–sodium hydroxide system is that glycerols and sucrose cannot be used to increase the range of solution viscosities because they react with the ferricyanide ion. This is not the case with the iodine system (see theses by Chorn [130] and McConaghy [136]). The iodine reaction has also been used successfully in the presence of drag-reducing agents [130, 136, 137]. Shaw [128] has added sodium hydroxide to vary the viscosity of the ferricyanide system, but the use of highly caustic concentrations increases the dangers associated with the experiment.

Problems common to both the ferricyanide and the triiodide systems are the reduction of dissolved oxygen [99], the hydrolysis of water at high applied voltages, high corrosivity, and the need to be enclosed. The enclosure of the triiodide solution is necessary to minimize the loss of iodine by vaporization and to avoid corrosion caused by iodine vapors. Periodic additions of iodine must be made to make up for the vapor losses. The high vapor pressure of iodine also limits its use in gas-liquid flow studies.

Because of corrosiveness of the electrolytes, special care has to be taken in

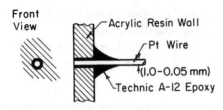

Figure 9.20 Circular electrode

designing the experiment. Polyvinyl chloride, acrylic resin, and stainless steel are usual materials of construction for the flow loop. Hicks and Pagotto [138] report on the effect of plastics on the decomposition of ferricyanide. Epoxy glue and cellulose acetate in the presence of NaOH were found to affect decomposition strongly. Ferricyanide decomposed about 1.5 times as fast when in contact with methyl methacrylate or polyvinyl chloride sheet in the presence of NaOH. Decomposition rates with KCl as the neutral electrolyte were found to be half as great as for NaOH. A potentiometric titration using isoniazid is recommended as an analytical procedure to follow changes for ferricyanide concentration.

9.9.3 Counterelectrode

In both the ferricyanide and triodide systems the counterelectrode is the anode. It has a much larger area than the test electrode to ensure that the anodic reaction is not limiting. It is normally a portion of the flow loop, such as a section of stainless steep pipe, and is usually located downstream of the test electrode to diminish interference effects. Stainless steel and nickel are suitable for the construction of an anode for the ferricyanide system with NaOH; however, nickel is not suitable with KCl. Stainless steel is preferred in the triiodide system.

9.9.4 Test Electrode

Platinum is the material used for the construction of the test electrode for the triiodide system. It has also been used in the ferricyanide system [98, 128, 137], but Jenkins [127] claims that platinum cathodes produce chemical polarization. The use of nickel cathodes is more popular in the ferricyanide system; however, platinum must be used when KCl is the neutral electrolyte.

The geometry of the test electrode is determined by the constraints outlined in section 9.6. Figure 9.20 depicts a circular test electrode. A hole is drilled through a section of acrylic resin pipe perpendicular to the surface at which the stress is to be measured. The hole is filled with epoxy and a platinum wire is inserted through the hole, so it protrudes slightly on the solution side. After the epoxy has hardened, the probe surface is sanded with progressively finer grades of emery cloth until it is flush with the wall. The probe is then cleaned as discussed in section 9.9.2. The manufacture of the slant or chevron electrodes is described by Fortuna [137] and Sirkar [98]; the sandwhich electrodes, by Son [22] and Mao and Hanratty [32];

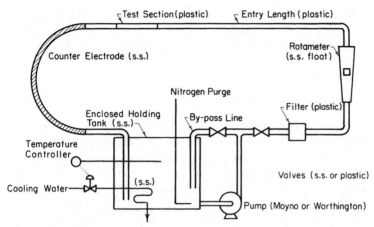

Figure 9.21   Flow loop of 1-in.-diameter pipe

the split-circle electrode, by Chorn [130]; the split-rectangular electrode, by Py [28]; and multiple slant electrodes, by Hogenes [129].

The determination of the area of the test electrode is critical, especially when it is very small. The sanding of small probes tends to smear the area of the electrode exposed to the fluid. Contamination may also alter the effective area. Son [22] and Mao and Hanratty [32] measured electrode dimensions with a calibrated microscope. Mitchell [19] assumed the electrode had the same area as the edge of the sheet used in its manufacture and reported results for shear stress in pipe flow that varied only a few percent from the accepted values. The electrodes used were rectangular, 0.0075–0.052 cm long, and 0.05–0.155 cm wide.

The probe area can be determined by calibration against a large probe of known area. If the flow geometry is one in which the flow field is known (e.g., pipe flow, couette flow), the probe may also be calibrated by using empirical or theoretical correlations for the velocity gradient at the wall.

A very interesting recent development is the use of photolithography techniques to fabricate arrays of microelectrodes by the group working with Desolouis and Tribollet at the University of Paris [139, 140].

9.9.5 Flow Loop

The chemical complications associated with the use of the electrochemical technique require that the measurements be taken under controlled conditions. At present, the use of electrochemical techniques is limited to experimental laboratory studies mainly because of the lack of suitable naturally occurring electrolytes for field studies. The fluid properties (D, v, C_B) must be held constant to ensure precise results. This is accomplished by recirculation and temperature control. The loop shown in Fig. 9.21 is similar to one used by Reiss [141], Chorn [130], and Mitchell

[19]. One should notice that the system is completely enclosed, and clear pipe is painted black for the ferricyanide studies. Most materials are plastic or stainless steel to resist corrosion. A filter is included to remove any particles that might foul the electrodes.

9.9.6 Measurement of Fluid Properties

Precise measurement of fluid properties is essential if Eq. (102) is to be used and a calibration avoided. Equation (102) requires the values of the diffusion coefficient and the bulk concentration of the reacting species. If the diffusion coefficient is to be determined from correlations in the literature, the fluid viscosity, density, and temperature must be known. An alternative procedure would be to calibrate the probe with the electrolyte in a system where the velocity gradient is known.

The literature contains much information on the diffusion coefficients of the ferricyanide and the triiodide ions. Chin [142] determined the diffusivities of both the ferricyanide and triiodide ions by studying mass transfer with a rotating disk. Bazan and Arvia [143] also used a rotating disc for ferricyanide, as did Newson and Riddiford [144] for triiodide. Lin et al. [121] used the Nernst equation for ions at infinite dilution:

$$D = \frac{GT\lambda_i^0}{z_i F^2} \qquad (107)$$

where G = universal gas constant
z_i = ionic charge on species i

along with tables of equivalent conductance λ_i^0 to calculate the diffusion coefficient for ferricyanide. The capillary method of Anderson and Saddington [145] was used by Eisenberg et al. [125, 146] for ferricyanide. Shaw [147] and Fortuna [137] used a Couette flow apparatus to study triiodide and ferricyanide. Bazan and Arvia [143] showed that the diffusion coefficient is only very weakly dependent on the concentration of the diffusing entities. Fortuna [137] showed that the addition of drag-reducing agents does not significantly affect the diffusion coefficient.

In most cases the results of these studies can be represented as

$$D \propto \frac{T}{\mu} \qquad (108)$$

The following correlations are fits of the data and apply only to the range of conditions cited.

$$D_{\text{ferricyanide}} = 2.5 \times 10^{-10} \frac{T}{\mu} \quad \text{cm}^2 \text{ P/s K} \qquad (109)$$

with

$$T = 10°\text{–}40°C \qquad C_{\text{NaOH}} = 0.5\text{–}2.1 \; M$$

$$C_{\text{ferricyanide}} = 0.38\text{–}300 \times 10^{-3} \; M \tag{110}$$

$$D_{\text{triiodide}} = 7.4 \times 10^{-8}\left(\frac{\rho}{\mu}\right)^{1.054}$$

with μ/ρ and D in cm²/s and

$$T = 25°C \qquad C_{\text{KI}} = 0.1 \text{ to } 0.5 \; M$$
$$C_{\text{I}_3^-} = 0.12\text{–}3.7 \times 10^{-3} \; M$$
$$C_{\text{sucrose}} = 0\text{–}1.58 \; M$$

With electrolytes or conditions other than cited above, the diffusion coefficient should be measured using one of the methods described in the references.

The iodometric titration to a clear end point discussed by Kolthoff and Furman [131] is used to obtain both the ferricyanide and the triiodide concentrations. They react with thiosulfate ion to produce tetrathionate:

$$I_3^- + 2S_2O_3^{2-} \rightarrow 3I^- + S_4O_6^{2-} \tag{111}$$

$$2Fe(CN)_6^{3-} + 2S_2O_3^{2-} \rightarrow 2Fe(CN)_6^{4-} + S_4O_6^{2-} \tag{112}$$

in the pH range 4.5–9.5 [148]. Kolthoff and Furman [131] suggested the use of a starch indicator when titrating concentrations of less than $5 \times 10^{-4} \; M$. Titration of triiodide in an Erlenmeyer flask is recommended because of volatilization of the iodide and the air oxidation of iodide:

$$4I^- + O_2 + 4H^+ \rightarrow 2I_2 + 2H_2O \tag{113}$$

which is enhanced by sunlight and acidity. Ferricyanide can also be titrated with isonicotinic acid hydrazide with potentiometric indication [128, 138, 149].

9.9.7 Instrumentation

The circuit shown in Fig. 9.22 is used to apply a constant voltage to the electrochemical cell and to convert the current of reaction to a voltage. The Analog Devices 188A operational amplifier sets the voltage at V_1 to a value that is independent of the current. Adjustment of the helipot (variable resistor) selects the voltage V_2. The output of the amplifier V_A is given as

$$V_A = A_A(V_1 - V_2) \tag{114}$$

where $A_A = 10^4\text{–}10^8$ [150] and

$$V_A = V_1 + IR_f \tag{115}$$

Combining these two equations yields

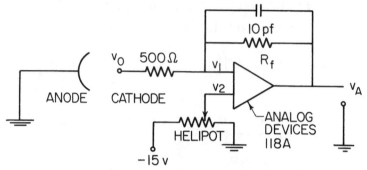

Figure 9.22 Basic instrumentation for electrochemical probe

$$V_A \frac{(1 - A_A)}{A_A} = (-IR_f - V_2) \tag{116}$$

Because A_A is so large,

$$V_A = IR_f + V_2 \tag{117}$$

The output of the amplifier equals the voltage set at V_2 plus the voltage drop through the feedback resistor R_f. By combining Eqs. (115) and (117), it is seen that $V_1 = V_2$ and is independent of the current. The values of V_2 and V_1 can be measured at V when no current is flowing. A subtraction circuit given by Hanratty [151] and Eckelman [152] can be used to remove the portion of the output voltage attributable to the applied voltage. This subtraction may also be done after the output signal has been time averaged.

Generally, an applied voltage of between -0.3 and -0.5 V is needed to polarize an electrode in the ferricyanide or triiodine systems. This voltage is not sensitive to changes in concentration and probe area. The magnitude of the feedback resistor determines the resolution of the current in the output signal. Since cell currents vary from 100 mA to less than 1 μA, resistances from 10 Ω to 2 million Ω are used. The 500-Ω resistor shown in Fig. 9.22 protects the operational amplifier by limiting the current entering it. The 10-pF capacitor corrects for the nonlinear frequency response of the amplifier (A. Saldeer, personal communication, 1980).

9.10 EXPERIMENTAL PROCEDURES FOR HEATED-ELEMENT PROBES

9.10.1 Principles of Operation

The heat transfer analog to the mass transfer probe is operated on principles similar to those for a hot wire (see [153] and Chap. 3 of this book). The resistance of the heated element is temperature sensitive. This enables fixing of the probe

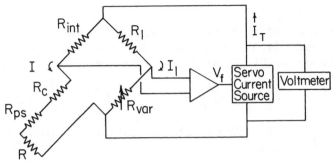

Figure 9.23 Simplified circuitry for heat transfer probe

temperature by fixing the probe resistance (analogous to fixing the wall concentration at zero through polarization). The power needed to keep the probe at that temperature, as the fluid cools it, is related to the shear stress by Eq. (89).

9.10.2 Instrumentation

The sample feedback-controlled Wheatstone bridge shown in Fig. 9.23 can be used to operate the probe. When the bridge is balanced,

$$\frac{R_{var}}{R_1} = \frac{R_c + R_{ps} + R}{R_{int}} \tag{118}$$

and the output of the feedback amplifier V_f is zero. When the probe is cooled, its resistance drops, causing a bridge imbalance. This is sensed by the amplifier, which signals the current source to increase its output to reheat the probe. This return to equilibrium occurs almost instantaneously, so an increase in shear stress is seen as an increase in current or bridge voltage. This type instrumentation is sold commercially as a constant-temperature (constant-resistance) anemometer.

9.10.3 Calibration

Equation (89) has been found to represent calibration data for heated-film probes. The influence of fluid density ρ is usually omitted in noncompressible flows. The constants A and B are weak functions of ΔT over a small range of ΔT values. Bellhouse and Schultz [14] assumed A and B varied linearly with ΔT to represent their measurements with

$$\frac{I^2 R}{\Delta T} = C + D\Delta T + (E + F\Delta T)(\rho\tau)_w^{1/3} \tag{119}$$

where

$$A = C + D\Delta T \qquad B = E + F\Delta T$$

Resistances can be determined by balancing the anemometer bridge with the heating current off. The internal resistances R_1 and R_{int} are given by the manufacturer. The bridge is balanced with R_{var} in the absence of R to determine $R_c + R_{ps}$,

$$R_c + R_{ps} = \frac{R_{int}}{R_1} R_{var} \qquad (120)$$

The cold resistance of the probe R_0 is determined by adding the probe to the bridge at the fluid temperature and balancing:

$$R_0 = \frac{R_{int}}{R_1} R_{var} - (R_c + R_{ps}) \qquad (121)$$

Its value is commonly 5–20 Ω [154].

In the operational mode a heating current is passed through the probe and R_{var} is selected so that the probe resistance needed to balance the bridge is larger than R_0 by a factor of H. The selection of the overheat ratio $H = R/R_0$ sets the temperature difference and determines the sensitivity of the probe. Common overheat ratios for hot films are 1.5 in airflow and 1.1 in water flow, which correspond approximately to temperature differences of 250°C and 50°C, respectively. The development of novel sensors made of materials with high-temperature coefficients of resistance would require a lower overheat ratio to obtain a similar sensitivity.

The temperature difference may be determined if the temperature-resistance relationship of the probe is known. Over a small temperature range the relation

$$\Delta T = \frac{H - 1}{\alpha_R} \qquad (122)$$

is valid, where α_R is the temperature coefficient of resistance. The coefficient of resistance is not only a function of the material of the heated element, but of impurities, defects in the lattice, oxidation of the surface, and stresses on the element [153, 155]. The coefficient should be determined for each probe, or it could be included in the constants A and B:

$$\frac{I^2 R}{H - 1} = \frac{A}{\alpha_R} + \frac{B}{\alpha_R} (\rho\tau)_w^{1/3} \qquad (123)$$

The output of the anemometer is more commonly measured as a voltage, $V_T = I(R + R_{int} + R_c + R_{ps}) = IR_T$. The calibration equation may now be written as

$$\frac{V_T^2 R_0 H}{R_T^2 (H - 1)} = A + B(\rho\tau)_w^{1/3} \qquad (124)$$

This calibration is valid over a small range of temperature differences for a specific probe and can be used knowing only the internal resistances and the overheat ratio.

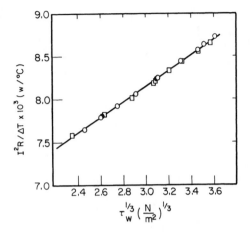

Figure 9.24 Calibration curve of flush-mounted thin film in a turbulent flow. (From [14])

A calibration curve obtained by Bellhouse and Schultz [14] is shown in Fig. 9.24 for a painted on platinum film of small dimensions.

Blackwelder and Eckelmann [156] point out that the deviation from a straight line at low flows is due to natural convection. This indicates that the reading at zero flow is not an appropriate point to use in a calibration to determine the constants A and B.

9.10.4 Insertion of the Probe in the Wall

The positioning of the probe relative to the wall is one of the greatest sources of error in using heated films. Pessoni [157] found that a displacement of a commercial probe of the type shown in Fig. 9.25 from the test section surface of 0.1 mm in a 2.5-cm air tunnel resulted in 30–40% deviations in the calibrations. Pessoni suggests either calibrating the probe in situ or calibrating it in a plug of larger area and transferring the entire plug to the test section. He used a collimated light source directed on the probe at a very shallow angle to position it flush with the surface. Simons et al. [158] also decided to mount their probes in a plug when they discovered that even the slight jarring of the probe upon insertion changed the calibration.

When commercial hot films are mounted in curved surfaces, a discontinuity is unavoidable. This problem may be overcome by the use of flexible glue-on probes [159]. Wylie and Alonso [160] suggest that these contour effects would be negligible in pipes with diameters of 2 in. or more. They cite an agreement within 2% between Foster's [161] results with a 2-in. pipe and a flat wall.

The thermal conductivity of the immediate surroundings will also cause errors if it changes from the calibration site to the test section. This is another advantage of mounting the sensor in a plug.

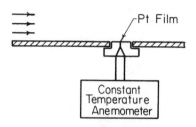

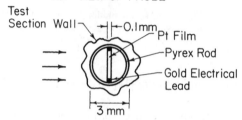

Figure 9.25   Hot-film probe

9.10.5 Basic Probe Design

The element in most commercial flush-mounted probes is a thin layer (film) of platinum that has been deposited by sputtering or vacuum evaporation. Both techniques require very sophisticated equipment. Sputtering involves the bombardment of a source of platinum with positive ions under a moderate vacuum. Platinum atoms are sputtered (from) the source and deposited on the substrate. Vacuum evaporation utilizes strong heat and vacuum to vaporize the platinum. The platinum vapor is then deposited on the substrate. Substrates are made of a good thermal insulator, usually quartz. Two high-conductivity leads run from the film, through the substrate to the anemometer. The element is protected by a quartz film of about 0.5-μm thickness [162]. Thicker films (2 μm) are used for liquid flows to insulate electrically the heated element from the fluid. This has the disadvantage, however, of decreasing the probe frequency response [163].

Hot-film probes are easily damaged by collisions with debris in the flow. Liquid flows, especially, put large stresses on these fragile sensors. Miya [164] filtered his solutions to eliminate particles that could do damage. Like mass transfer probes, the heat transfer probes may also become fouled with grease, dirt, or other debris. Cleaning can be done with successive rinses of 10% acetic acid and distilled water.

In liquid flows the probe should be insulated electrically from the fluid to prevent hydrolysis at the element. Low overheat ratios must be used to avoid boiling or degassing of the field. Thermo Systems, Inc. [165] suggests brushing the probe with a camel hair brush when bubble formation starts.

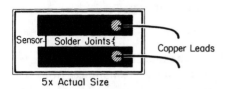

5x Actual Size

Figure 9.26 Glue-on probe as marketed by DISA Electronics

Although recycling of the fluid may not be necessary, as with the mass transfer probe, the fluid properties (T, ρ, μ, C_p, k_p) must still be carefully controlled.

Considerable work has been done to find better substrates for these probes to reduce heat loss to the substrate (reduce L_{eff}); to find cheaper, simpler methods of construction; to develop new designs with more structural integrity; and to manufacture probes whose calibrations are reproducible.

Rubesin et al. [16] have investigated the use of a styrene copolymer substrate with a thermal conductivity one-tenth of that for Pyrex. The melting point, however, was too low for baking on the platinum paint used by Bellhouse and Schultz [14]. Sputtering was also unsuccessful for film sizes of less than 0.2 mm. A probe with a platinum-rhodium wire laid in a groove in the styrene (after [12]) was tried, but the waviness it produced on the surface was too great. In the final design a 25-μm platinum wire was laid on the surface and buried in a thin layer of epoxy, which was hand wiped to expose the wire. This type of probe is easier to fabricate, more reproducible (within 2%), and less likely to be destroyed. In addition, it loses less heat to the substrate than the heated-film probe because of the lower conductivity of its substrate. Higuchi and Peake [166] have used two buried-wire elements to study three-dimensional flows.

Alfredsson et al. [107] describe two interesting designs: one consisted of a cylindrical hot-film sensor with a diameter of 0.05 mm and a length of 1 mm soldered to two prongs reaching approximately 0.1 mm out from the surface. The other was a wire with a length of 0.75 mm and a diameter of 0.002 mm that was welded between two jeweler's broaches that were put through a Plexiglas plug and ground flush with the surface.

Bertelrud [167], Kreplin and Meier [168], and Coney and Simmers [159] have used a "glue-on-probe" (Fig. 9.26) first proposed by McCroskey and Durbin [15]. These are thin polyimide foils onto which a metal film has been sputtered or electrodeposited. These probes are inexpensive and are able to take the shape of curved surfaces. McCroskey and Durbin [15] also claim that printed-circuit sensors are reproducible and suited to the manufacture of multielement devices. Possible disadvantages are the permanence of the probe after it is glued in place and the use of wire leads, which may cause flow disturbances. These were overcome by Coney and Simmers [159] by gluing the probe to a plastic plug and leading the wires down the side of the plug to the outside of the test section. The photolithographic techniques now being explored for the fabrication of mass transfer probes could also find application as heated films.

9.11 APPLICATION OF MASS TRANSFER PROBES

9.11.1 Turbulent Flow in a Pipe

A system with a 1-in.-diameter pipe, similar to the one shown in Fig. 9.21, was used by Reiss [141] to measure wall shear stress at Reynolds numbers from 3000 to 35,000. A 14-gage (0.164cm) nickel wire was the test electrode. The area of the electrode used in the calculations was based on the average wire diameter measured with a vernier scale on a microscope. The effective length ($0.82d_e$) was 0.133 cm. The reacting ion was 0.1 M ferricyanide. The diffusion coefficient was determined from the equation given by Eisenberg [146], which is identical to Eq. (109). The remainder of the solution properties, given in Table 9.2, were determined using standard laboratory procedures. The curves in Fig. 9.27 show that the electrode is polarized by an applied voltage of -0.3 to -0.5 V over a Reynolds number range of about 4000 to 30,000. An operating voltage of -0.4 V was selected by Reiss for all flow rates measured. A feedback resistor of only 100 Ω was used because of the relatively large current resulting from a high flow rate, a high ferricyanide concentration, and a large electrode. The flow conditions and results of a run at a Reynolds number of 34,200 are also given in Table 9.2. The effective length of the electrode (made dimensionless with wall parameters) is 98 at this Reynolds number. The constraints on electrode length discussed in section 9.6 can be represented by

$$14\mathrm{Sc}^{-1/2} < L^+ < 64\mathrm{Sc} \tag{125}$$

In order for turbulent transport to be negligible,

$$L^+ < 700 \tag{126}$$

For spanwise diffusion to be negligible,

$$L^+ > 44 \left(\frac{L}{W}\right)^{3/2} \mathrm{Sc}^{-1/2} \tag{127}$$

The value of the dimensionless length is within these constraints (see Table 9.2). The value of $\bar{S}_x$ calculated from the data using Eq. (73) is within 10% of the value calculated using the Blasius equation.

The calculated value of $\sqrt{\overline{s_x^2}}/\bar{S}_x = 3\sqrt{\overline{k^2}}/\overline{K} = 0.1089$ differs greatly from the accepted value of 0.30–0.37. An averaging of the fluctuations should be expected because the size of the probe is of the order of the scale of the fluctuating flow. The constraints on probe size for the measurement of the shear stress fluctuations are that the flow be uniform over the probe surface,

$$\frac{W^+}{\Lambda_z^+} < 0.67 \quad \text{or} \quad W^+ < 8 \tag{128}$$

(see Fig. 9.7), and that the fluctuations not be damped,

Table 9.2 Summary of the experiment by Reiss [141]

Electrolyte properties		Flow conditions		Results	
$C_{ferricyanide}$	$0.104\ M$	T	$25°C$	$\bar{I}$	1.96×10^{-3} A
$C_{ferrocyanide}$	$0.0116\ M$	d_e	0.1636 cm	$\bar{K}$	9.31×10^{-3} cm/s
C_{NaOH}	$2.168\ M$	L_{eff}	0.133 cm	$\bar{K}L/D$	239
ρ	1.094 g/cm^3	A_e	0.02102 cm^2	Z	2.61×10^7
μ	0.0144 g/(cm·s)	Re	$34{,}200$	$\bar{S}$	7629 s^{-1}
ν	0.01316 cm^2/s	$\bar{U}_B$	176.9 cm/s	u^*	10.02 cm/s
D	5.17×10^{-6} cm^2/s	V_0	-0.4 V	$(\bar{k}^2)^{1/2}/\bar{K}$	0.0363
Sc	2550	R_f	$100\ \Omega$	L^+	98
		$S_{x\,\text{Blasius}}$	6950 s^{-1}	64 Sc	$160{,}000$
				14 Sc$^{-1/2}$	0.28
				$44(L/W)^{3/2}$ Sc$^{-1/2}$	0.87
				W^+	120
				$L^+(\omega^+)^{3/2}$ Sc$^{1/2}$	66

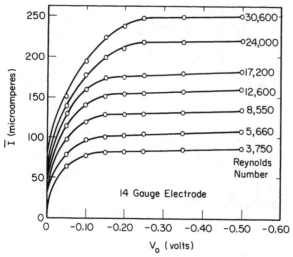

Figure 9.27 Polarization curves for a 14-gage nickel electrode in a 1-in. pipe flow system. (From [7])

$$L^+\omega^{+3/2}\,Sc^{1/2} \leq 1 \tag{129}$$

(see Fig. 9.6). Neither of these conditions were met. To successfully measure the fluctuations without using corrections, a much smaller electrode or a much lower flow is needed. The use of Fig. 9.7 to estimate a correction for the effect of nonuniform flow yields a value of the relative intensity of the wall velocity gradient fluctuations in the x direction of $(\overline{s_x^2})^{1/2}/\overline{S}_x = 0.23$. The frequency damping of the signal could be estimated using Fig. 9.6 if the entire measured frequency spectrum were available.

9.11.2 Flow Around a Cylinder

The water tunnel shown in Fig. 9.28 was used by Son [22] to measure shear stress around a 1-in.-diameter cylinder placed perpendicular to the flow. Circular electrodes of diameter 0.038–0.051 cm were used to measure average shear stresses. A sandwich electrode made of two strips of platinum sheet 2.54 cm by 0.0127 cm separated by a thickness of scotch tape (~0.005 cm) was used to measure the direction of flow and the location of flow separation. The cylinder was rotated to vary the angular location of the test electrode. The cylinder could be removed from the test section, and the proper areas accurately determined with a microscope. Dye was also injected through holes in the cylinder wall to measure qualitatively the flow direction. The measurements were all taken with respect to the front stagnation point, which was located at the position of minimum mass transfer.

Values of $|S|$ calculated from the measurements, using Eq. (25), are presented in Fig. 9.29 for a Reynolds number of 20,000 using a circular probe 0.051 cm in

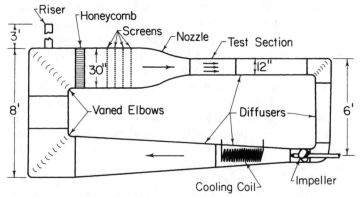

Figure 9.28 Water tunnel used by Son [22] in studies of flow around a cylinder

diameter. In region A the wall shear stress had a periodic oscillation and a maximum at an angle of about 50°. Position S indicates the separation point detected by the sandwich electrodes. Dye studies showed that a reversed flow existed close to the wall in region B. The electrode measurements indicated an unsteady periodic flow characterized by small negative wall shear stresses. Both the dye studies and the electrode measurements showed a highly chaotic flow in region C. The flow at a given location in C changes direction, and the flow fluctuations appear to be greater than the time-averaged flow. The design equation is not applicable in this region, so the calculated values of $|S|$ have no meaning. However, recent work by Mao and Hanratty [32] offer promise of measuring the time variation of S in this haphazard flow.

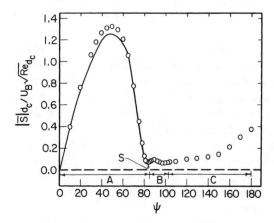

Figure 9.29 Wall velocity-gradient distribution around a 1-in. cylinder at Re = 20,000. The solid curve is the boundary layer solution at Re = 90,000. (From [22])

As discussed in section 9.6.3, the calculations of S by Son and Hanratty might be in error close to the front stagnation point or to the separation point, where S_x is close to zero and the wall shear stress is varying rapidly with changes in angular location. Equation (40) indicates that the use of Eq. (25) to calculate S introduced an error of only about -7.5% at a distance of 5° from these points.

Son and Hanratty found the measurements with the circular electrodes and with a rectangular electrode 0.038 cm long and 2.54 cm wide gave the same results in region A. Further support for the accuracy of the measurements is given in Fig. 9.29, where the solid line represents a boundary layer calculation carried out using the Blasius series [169] and the pressure measurements of Hiemenz [170] at Re = 19,000. The agreement between the measurements and theory is quite good. In fact, Son and Hanratty argued that the slight disagreement near the maximum is more than likely due to errors in determining the pressure gradient rather than in measuring the wall shear stress.

9.12 OTHER METHODS

The focus of this chapter has been on the principal methods for measuring wall shear stress when it cannot be obtained from direct measurements of the velocity gradient at the wall. It seems appropriate to close with discussions of the determination of the wall shear indirectly from velocity measurements and of three new methods that could have promise.

9.12.1 Determination from Velocity Measurements

Usually, it is not possible to measure the fluid velocity close enough to the wall to determine the wall shear stress from the directly measured velocity gradient at the wall. A number of approaches have been used in turbulent flows to obtain wall shear stresses from velocity measurements away from the wall.

One approach is to use measurements of the Reynolds stress, $-\rho \overline{uv}$, and of the viscous stress, $\mu(d\overline{U}/dy)$, to calculate τ at different distances from the wall y. If the functional form of the shear stress variation is known, these total stress measurements can be extrapolated to the wall to obtain τ_w. This technique has been successfully applied in studies of flow in a two-dimensional channel, where the shear stress is varying linearly with distance from the wall [171].

A more common approach, due to Clauser [172], is to use mean velocity measurements at distances away from the wall. The simplest application of this idea is when a well-defined region exists where the mean velocity $\overline{U}$ is varying with the logarithm of the distance from the wall. Law of wall arguments suggest that, in this region,

$$\frac{d\overline{U}}{dy} = \frac{u^*}{\kappa y} \qquad (130)$$

where $u^* = (\tau_w/\rho)^{1/2}$ is the friction velocity and κ is the von Karman constant, which is usually taken as approximately equal to 0.40. Thus the slope of a semilogarithmic plot of $\bar{U}$ versus y gives u^*/κ and therefore τ_w. The difficulties in using this approach are that there is no assurance that $\kappa = 0.40$, that the flow conditions might not be such as to give a well-defined logarithmic region, and that accelerations and decelerations are large enough to contribute to the stress distribution in the fluid. An example of a failure of this technique in a fully developed field is for flow over wavy boundaries, where κ is found to be different from 0.40 [173, 174]. A similar result has been found in some two-phase dispersed flows, where the fluid turbulence is affected by the presence of particles.

Situations for which a well-defined logarithmic layer does not exist are flows with large pressure gradients and flows at low Reynolds numbers. If the wall is flat and an "equilibrium situation," as defined by Clauser [172], exists,

$$\frac{\bar{U}}{u^*} = f\left(\frac{yu^*}{\nu}, \Delta\right) \tag{131}$$

for small enough y. Here $\Delta = |dP/dx| [\nu/u^*\tau_w]$. Comparisons of measured variations of $\bar{U}$ with y for small y with Eq. (131) then gives u^* or τ_w.

This approach has recently been explored by McEligot [175] for a situation where

$$\frac{d\tau}{dy} = \frac{dP}{dx} \tag{132}$$

The application of this idea to evaluate τ_w requires measurements of the velocity and the pressure gradient. However, Eq. (132) can be restrictive, in that it assumes that the velocity measurements are in a region, close to the wall, where acceleration or deceleration of the flow is too small to affect dP/dx.

9.12.2 Oil-Film Gage

Tanner and Blows [176] developed a technique for determining τ_w in airflows that depends on the measurement of the rate of thinning of an oil film on a polished surface. Discussions of its application have been presented by Tanner [177], Monson and Higuchi [178], Monson [179, 180], Westphal et al. [181, 182], and Settles [9]. Problems have arisen in using this technique at very high τ_w, such as exist in supersonic flow, because of the presence of surface waves. A discussion of methods to deal with this problem has been presented by Settles [9] and by Murphy and Westphal [183].

Because of the thinness of the film, special methods are required. All of the above papers used interferometer techniques to measure the rate of change of the height of the film. These methods required a highly polished plate and the absence of particles in the airflow. Bandyopadhyay and Weinstein [184] have developed a method that measures the rate of change of the slope of the water film, thus avoiding the need for a polished surface.

9.12.3 Modified Laser-Doppler Method

The use of standard laser-Doppler techniques to measure velocities close to surfaces is accompanied by a number of difficulties, particularly because of reflection. Naqwi and Reynolds [185] have developed a new method that avoids these problems.

Laser light transmitted through two rectangular slits in the surface forms a "fanlike" fringe pattern in the fluid immediately above the surface. The frequency of scattered light from seed particles is measured with standard laser-Doppler velocimetry (LDV) methods. This frequency is independent of distance from the wall, provided the particles are in a region in which the velocity is varying linearly with distance from the wall; it is directly proportional to the velocity gradient at the wall. An interesting feature of this technique is that it can be used in reversing flows by shifting the frequency of the light incident on one of the slits. Furthermore, it is not hampered by light reflection and a calibration is not required.

9.12.4 Pulsed-Wall Probe

The use of inverse mass transfer methods with electrochemical probes (section 9.7.5) and of special laser-Doppler methods (section 9.12.3) offer the possibility of measuring the time-varying wall shear stress in reversing flow situations, such as exist in separation bubbles. Another method for doing this is the pulsed-wall probe developed by Westphal et al. [182] (see also Eaton et al. [186]).

This is an adaptation of the pulsed-wire anemometer developed by Bradbury and Castro [187]. Three heated wires were mounted on needles that protruded through the wall. The center wire was 12.5-μm heated nickel. The sensor wires at upstream and downstream locations were 5-μm-diameter platinum-plated tungsten. The three wires were located at the same distance from the wall and separated by a distance of 1.8 mm.

The center wire was pulsed with a short-duration (about 5μs) high-amplitude current. This produced a tracer of heated fluid that is detected some time later by the upstream or downstream probe, depending on the direction of flow. The instantaneous velocity is calculated from the distance between the heater and the detector and the time of flight. If the wires are located close enough to the wall that they are in a region where Eq. (1) is applicable, then τ_w can be calculated from the measured velocity. Westphal et al. [181] used this instrument in turbulent flow over a flat plate and in the separated region of a backward facing step. This probe has also been used by Castro and Dianat [188], Castro and Hague [189], Dianat and Castro [190], and Castro et al. [191].

One of the difficulties is that calibrations in a laminar flow lead to measurements of $\bar{\tau}_w$ in a turbulent flow that are too high, similar to what is found for wall films (see section 9.8.2). A possible explanation is the influence of turbulent dispersion in the flow direction. This problem has not yet been resolved. Castro and Hague [189] have carried out calculations in which molecular diffusion in the flow direction was considered. This provided a qualitative explanation but did not account for the numerical difference between experiments in laminar and turbulent flows.

NOMENCLATURE

a	velocity of sound
A	parameter used to characterize heat transfer probe equal to .807 $A_e C_p^{1/3} k_T^{2/3}/L^{1/3}\mu^{1/3}$ (Eq. 88)
$A^\dagger$	parameter defined by Eq. 91 equal to $(A\Delta T R\rho^{1/3})^{-1}$
$A_t^\dagger$	parameter $A^\dagger$ evaluated for a turbulent flow, Eq. 98
A_d	damping factor defined by Eq. 57
A_e	area of electrode
A_t	calibration constant for a thermal probe in turbulent flow defined by Eq. 90
A_A	amplification factor of operational amplifier
b	breadth of the Stanton blade
B	parameter used to characterize heat transfer probe defined by Eq. 89
$B^\dagger$	parameter defined by Eq. 93 equal to $-B/A\rho^{1/3}$
$B_t^\dagger$	parameter $B^\dagger$ evaluated for a turbulent flow, Eq. 98
B_t	calibration constant for a thermal probe in turbulent flow defined by Eq. 90
c	fluctuating part of concentration equal to $C - \overline{C}$
$\hat{c}$	complex amplitude of a simple harmonic fluctuation in c
$\hat{c}_R$	real part of $\hat{c}$
$\hat{c}_I$	imaginary part of $\hat{c}$
C	concentration of the diffusing specie
C_p	heat capacity of the fluid
d	pipe diameter
d_c	diameter of cylinder
d_e	electrode diameter
d_h	hole diameter
d_t	outside diameter of pitot tube or Preston tube
D	diffusion coefficient
e	fluctuating part of E defined by $E = \overline{E} + e$
E	voltage drop through the heat transfer element
f_c	correction to Preston-tube calibration to account for compressibility effects, Eq. 11
f_i	calibration function for a Preston tube for incompressible flow, Eq. 11
F	Faraday's constant, 96487 c/eq

g	gap size
g_1	defined in Eq. 79
g_2	defined in Eq. 80
g_0	constant in Eq. 86
G	universal gas constant, 1.9872 cal/(K·mole)
h	heat-transfer coefficient
h_c	channel height
h_s	height of Stanton blade from wall
H	overheat ratio, R/R_0
I	electric current produced by reaction in electrochemical probe or current used to heat the heat-transfer probe
I_1	electric current in balancing side of anemometer bridge (Fig. 20)
I_T	total current in bridge circuit, $I + I_1$
k	fluctuating part of the mass transfer coefficient defined by $K = \bar{K} + k$
k_R	reaction rate constant
k_s	constant defined in Eq. 14
k_T	thermal conductivity of the fluid or thermal conductivity of the substrate
$\overline{k_m^2}$	measured mean square value of the mass-transfer fluctuations in a turbulent flow for a probe of finite size
K	mass-transfer coefficient defined as $<N>/(C_B - C_w)$
K_t	ratio of effective diameter of Preston tube to outside diameter, d_t
ℓ	length of Stanton blade
L	length of the mass-transfer or heat-transfer element
L_e	equivalent length of a circular electrode surface equal to 0.81356 (d_e) defined in Eq. 69
L_{eff}	effective length of the heated element
Ma	Mach number, U_B/a
n	exponent used in Eq. 90
n_e	number of electrons transferred in stoichiometric reaction
N	local rate of mass transfer per unit area
N_i	local rate of mass transfer per unit area on i^{th} strip in Eq. 67
dP/dx	pressure gradient
ΔP	difference of impact pressure and static pressure for a Preston tube or a Stanton blade, Eq. 6
Pr	Prandtl number, $C_p\mu/k_T$
q	rate of heat transfer per unit area
R	resistance of heat transfer probe at its operating temperature
R_0	resistance of heat transfer probe at the bulk fluid temperature
R_1	internal resistance in anemometer bridge
R_{int}	internal resistance in anemometer bridge
R_{var}	variable resistance in anemometer bridge

R_c	resistance of cable to heat-transfer probe
R_e	electrode radius
R_f	feedback resistance
R_{ps}	resistance of probe support for heat-transfer probe
R_T	defined as $R + R_{int} + R_c + R_{ps}$
R_z	circumferential spatial correlation of the mass-transfer fluctuations
Re	Reynolds number
Re_{d_c}	Reynolds number based on cylinder diameter
s_x	fluctuating part of the wall velocity gradient in the x direction
s_z	fluctuating part of the wall velocity gradient in the z direction
$\hat{s}_R$	amplitude of a simple harmonic fluctuation in s_x assumed to be real
S	magnitude of the velocity gradient at the wall
S_x	velocity gradient at the wall in the x direction
S_0	velocity gradient at the wall evaluated at the center of the wall element
Sc	Schmidt number
t	time
T	temperature or the fluid or temperature of a heated element
T_R	transference number used in Eq. 101
ΔT	temperature difference in Eq. 88, $T_w - T_B$
u^*	friction velocity $(\bar{\tau}_w/\rho)^{1/2}$
U	streamwise velocity of fluid
V	normal component of fluid velocity
V_0	voltage applied to electrode in electrochemical cell
V_1, V_2	voltage inputs to operational amplifier in instrumentation of mass-transfer probe
V_A	output of operational amplifier in instrumentation of mass-transfer probe
V_f	output of amplifier on anemometer bridge
V_T	voltage drop across anemometer bridge equal to $R_T I$
W	width of mass-transfer or heat-transfer element
W_k	spectral-density function of k defined in Eq. 55
W_{s_x}	spectral-density function of s_x defined in Eq. 56
x	distance in the streamwise direction
x^*	defined by Eq. 16
y	distance in the direction normal to wall
y^*	defined by Eq. 15
z	distance in transverse direction
z_i	charge number of ionic species i
Z	defined by Eq. 26
α	angular measurement
α_R	temperature coefficient of resistance
α_T	thermal diffusivity, $k_T/\rho C_p$

β	parameter equal to $L/D^2 0.807^3$ defined in Eq. 60
γ	change of the velocity gradient in the flow direction, $\partial S/dx$, Eq. 37
δ	thickness of the velocity boundary layer
δ_c	thickness of the concentration boundary layer or thermal boundary layer
δ_v	thickness of the region where $U = Sy$
Δ	pressure gradient parameter equal to $(dP/dx)\nu/\rho u^{*3}$
η	dimensionless similarity variable equal to $(y^3 \bar{S}_x/9D_x)^{1/3}$, Eq. 49
θ	angle between the instantaneous direction of the flow at the wall and the x axis
λ_i^0	equivalent conductance of specie i at infinite dilution
Λ_z	circumferential scale of the mass-transfer fluctuations
μ	fluid viscosity
ν	fluid kinematic viscosity, μ/ρ
ρ	fluid density
τ_1	dimensionless parameter for slant surfaces, $L/5W$
τ_w	wall shear stress
τ_w'	fluctuation in the wall shear stress, equal to $\tau_w - \bar{\tau}_w$
ϕ	angle that a slant electrode makes with the x axis
ψ	angular distance around a cylinder measured from the stagnation point or for mass-transfer probes it is defined as $\tan^{-1}(L/W)$
ω	frequency in radians per unit time

Superscripts
$+$	term made dimensionless using wall parameters u^* and ν
$-$	time average

Subscripts
w	fluid or flow property evaluated at the conditions that prevail at the wall
B	fluid or flow property evaluated at bulk conditions
s	at a stagnation or separation point

Other symbols
$< >$	spatial average

REFERENCES

1. T. E. Stanton, D. Marshall, and C. W. Bryant, On the Condition at the Boundary of a Fluid in Turbulent Motion, *Proc. R. Soc. London Ser. A.,* vol. 97, pp. 413–434, 1920.
2. G. Kempf, Neue Ergebnisse der Widerstandforschung, *Werft Reederei Hafen,* vol. 11, pp. 234–249, 1929.
3. H. Ludwieg, Ein Gerat zur messung der Wandschubspannung turbulenter Reibungschichten, *Ing.*

Arch., vol. 17, pp. 207–218, 1940. (Instrument for Measuring the Wall Shearing Stress of Turbulent Boundary Layers, NACA TM 1284, 1950.)

4. J. H. Preston, The Determination of Turbulent Skin Friction by Means of Pitot Tubes, *J. R. Aeronaut. Soc.,* vol. 58, pp. 109–121, 1953.
5. N. I. Konstantinov, Comparative Investigation of the Friction Stress on the Surface of a Body, *Energomashinostroenie,* vol. 176, pp. 201–213, 1955. (DSIR Translation RTS 1500, 1961.)
6. N. I. Konstantinov and G. L. Dragnysh, The Measurement of Friction Stress on a Surface, *Energomashinostroenie,* vol. 176, pp. 191–200, 1955. (DSIR Translation RTS 1499, 1960.)
7. L. P. Reiss and T. J. Hanratty, Measurement of Instantaneous Rates of Mass Transfer to a Small Sink on a Wall, *AIChE J.,* vol. 8, pp. 245–247, 1962.
8. L. P. Reiss and T. J. Hanratty, An Experimental Study of the Unsteady Nature of the Viscous Sublayer, *AIChE J.,* vol. 9, pp. 154–160, 1963.
9. G. S. Settles, Recent Skin-Friction Techniques for Compressible Flows, AINA/ASME 4th Fluid Mechanics, Plasma, and Lasers Conference, Atlanta, Georgia, 1986.
10. A. Fage and V. M. Falkner, On the Relation Between Heat Transfer and Surface Friction for Laminar Flow, Aeronaut. Res. Counc. London, R&M 1408, 1931.
11. H. Ludwieg and W. Tillmann, Investigation of the Wall Shearing Stress in Turbulent Boundary Layers, NACA TM 1285, 1950.
12. H. W. Liepmann and G. T. Skinner, Shearing-Stress Measurements by Use of a Heated Element, NACA TN 3269, 1954.
13. B. J. Bellhouse and D. L. Schultz, The Measurement of Skin Friction in Supersonic Flow by Means of Heated Thin Film Gauges, Aeronaut. Res. Counc. London, R&M, p. 940, 1965.
14. B. J. Bellhouse and D. L. Schultz, Determination of Mean and Dynamic Skin Friction, Separation and Transition in Low-Speed Flow with a Thin-Film Heated Element, *J. Fluid Mech.,* vol. 24, pp. 379–400, 1966.
15. W. J. McCroskey and E. J. Durbin, Flow Angle and Shear Stress Measurements Using Heated Films and Wires, *J. Basic Eng.,* vol. 94, pp. 46–52, 1972.
16. M. W. Rubesin, A. F. Okuno, G. G. Mateer, and A. Brosh, A Hot-Wire Surface Gage for Skin Friction and Separation Detection Measurements, NASA TM X-62, p. 465, 1975.
17. V. A. Sandborn, Surface Shear Stress Fluctuations in Turbulent Boundary Layers, Second Symp. on Turbulent Shear Flows, London, 1979.
18. V. A. Sandborn, Evaluation of the Time Dependent Surface Shear Stress in Turbulent Flows, ASME Preprint 79-WA/FE-17, 1979.
19. J. E. Mitchell and T. J. Hanratty, A Study of Turbulence at a Wall Using an Electrochemical Wall-Stress Meter, *J. Fluid Mech.,* vol. 26, pp. 199–221, 1966.
20. G. Fortuna and T. J. Hanratty, Frequency Response of the Boundary Layer on Wall Transfer Probes, *Int. J. Heat Mass Transfer,* vol. 14, pp. 1499–1507, 1971.
21. H. G. Dimopoulos and T. J. Hanratty, Velocity Gradients at the Wall for Flow Around a Cylinder for Reynolds Numbers Between 60 and 360, *J. Fluid Mech.,* vol. 33, pp. 303–319, 1968.
22. J. S. Son and T. J. Hanratty, Velocity Gradients at the Wall for Flow Around a Cylinder at Reynolds Numbers from 5×10^3 to 10^5, *J. Fluid Mech.,* vol. 35, pp. 353–368, 1969.
23. C. Tournier and B. Py, The Behavior of Naturally Oscillating Three-Dimensional Flow Around a Cylinder, *J. Fluid Mech.,* vol. 85, p. 161, 1978.
24. M. LeBouche and M. Martin, Convection Forcee Autour du Cylindre: Sensibilitie aux Pulsations de l'Ecoulement Externe, *Int. J. Heat Mass Transfer,* vol. 18, pp. 1161–1175, 1975.
25. K. K. Sirkar and T. J. Hanratty, Limiting Behavior of the Transverse Turbulent Velocity Fluctuations Close to a Wall, *Ind. Eng. Chem. Fundam.,* vol. 8, pp. 189–192, 1969.
26. K. K. Sirkar and T. J. Hanratty, The Limiting Behavior of the Turbulent Transverse Velocity Component Close to a Wall, *J. Fluid Mech.,* vol. 44, pp. 605–614, 1970.
27. B. Py and J. Gosse, Sur la Realisation d'une Sonde en Paroi Sensible a la Vitesse et a la Direction de l'Ecoulement, *C. R. Acad. Sci.,* vol. 269A, pp. 401–403, 1969.

28. B. Py, Etude Tridimensionnelle de la Sous-Couche Visqueuse dans une Veine Rectangulaire par des Mesures de Transfert de Matiere en Paroi, *Int. J. Heat Mass Transfer,* vol. 15, pp. 129–144, 1973.

29. B. Py, Sur l'Interet de la Reduction de l'Iode dans l'Etude Polarographique des Ecoulements, *C. R. Acad. Sci.,* vol. 270A, pp. 202–205, 1970.

30. Z. X. Mao and T. J. Hanratty, Analysis of Wall-Shear Stress Probes in Large Amplitude Unsteady Flows, *Int. J. Heat Mass Transfer,* vol. 34, pp. 281–290, 1991a.

31. Z. X. Mao and T. J. Hanratty, Applications of an Inverse Mass-Transfer Method to the Measurement of Turbulent Fluctuations in the Velocity Gradient at the Wall, *Exp. Fluids,* vol. 11, pp. 65–73, 1991b.

32. Z. X. Mao and T. J. Hanratty, Measurement of Wall Shear Rate in Large Amplitude Unsteady Reversing Flows, *Exp. Fluids,* vol. 12, pp. 342–350, 1992.

33. S. Dhawan, Direct Measurements of Skin Friction, NACA TN 2567, 1953.

34. K. G. Winter, An Outline of the Techniques Available for the Measurement of Skin Friction, *Progr. Aerospace Sci.,* vol. 18, pp. 1–57, 1977.

35. M. Acharya, Development of a Floating Element for Measurement of Surface Shear Stress, *AIAA J.,* vol. 24, p. 410, 1985.

36. G. Kempf, Weitere Reibungsergebnisse en ebenen glatten und rauhen Flachen, *Hydromech. Probl. Schiffsantreibs,* vol. 1, pp. 74–82, 1932.

37. F. Schultz-Grunow, Neues Reibungswiderstandsgestz fur glatte Platten, *Luftfahrforschung,* vol. 17, pp. 239–246, 1940. (New Friction Law for Smooth Plates, NASA TM 986, 1941.)

38. D. W. Smith and J. H. Walker, Skin Friction Measurements in Incompressible Flows, NASA TR R 26, 1959.

39. K. G. Winter and L. Gaudet, Turbulent Boundary-Layer Studies at High Reynolds Numbers Between 0.2 and 2.9, Aeronaut. Res. Counc. London, R&M 3712, 1973.

40. D. E. Coles, Measurement of Turbulent Friction on a Smooth Flat Plate in Supersonic Flow, *J. Aeronaut. Sci.,* vol. 21, pp. 433–448, 1954.

41. D. J. Garringer and E. J. Saltzman, Flight Demonstrations of a Skin-Friction Gage to a Local Mach Number of 4.9, NASA TN D-3820, 1967.

42. D. G. Mabey and L. Gaudet, Performance of Small Skin Friction Balances at Supersonic Speeds, *J. Aircraft.,* vol. 12, pp. 819–825, 1975.

43. J. M. Allen, Systematic Study of Error Sources in Supersonic Skin-Friction Balances, NASA TN D-8291, 1976.

44. J. M. Allen, Experimental Study of Error Sources in Skin Friction Balance Measurements, *J. Fluids Eng.,* pp. 197–204, 1977.

45. J. M. Allen, Improved Sensing Elements for Skin-Friction Balance Measurements, *AIAA J.,* vol. 18, pp. 1342–1345, 1980.

46. D. Frei, Direkte Wandschubspannungsmessung in der Turbulenten Grenzschicht mit Positiven Druckgradient, Diss. ETH 6392, 1979.

47. D. Frei and H. Thomann, Direct Measurements of Skin Friction in a Turbulent Boundary Layer with a Strong Adverse Pressure Gradient, *J. Fluid Mech.,* vol. 101, pp. 79–95, 1980.

48. S. W. Petri, Development of a Floating Element Wall Shear Stress Transducer, Report 87144-1, Acoustics and Vibration Laboratory, Massachusetts Inst. of Technology, Cambridge, 1984.

49. M. A. Schmidt, R. T. Howe, S. D. Senturia, and J. H. Hanitonidis, Design and Calibration of a Microfabricated Floating-Element Shear Stress Sensor, *IEE Trans. Electron. Devices,* vol. 35, pp. 750–757, 1988.

50. V. C. Patel, Calibration of the Preston Tube and Limitation on Its Use in Pressure Gradients, *J. Fluid Mech.,* vol. 23, pp. 185–208, 1965.

51. I. Rechenberg, Messung der turbulenten Wandschubspannung, *Z. Flugwiss,* vol. 11, pp. 429–438, 1963.

52. A. D. Young and J. N. Mass, The Behavior of a Pitot Tube in a Transverse Total-Pressure Gradient, Aeronaut. Res. Counc. London, R&M 1770, 1936.

53. K. C. Brown and P. N. Joubert, Measurement of Friction in Turbulent Boundary Layers, *J. Fluid Mech.,* vol. 35, pp. 737–757, 1967.
54. M. R. Head and I. Rechenberg, The Preston Tube as a Means of Measuring Skin Friction, *J. Fluid Mech.,* vol. 14, pp. 1–17, 1962.
55. M. R. Head and V. V. Ram, Simplified Presentation of Preston Tube Calibration, *Aeronaut. Q.,* vol. 22, pp. 295–300, 1971.
56. W. E. Holmes and R. E. Luxton, Measurement of Turbulent Skin Friction by a Preston Tube in the Presence of Heat Transfer, *J. Mech. Eng. Sci.,* vol. 9, pp. 159–166, 1967.
57. E. Y. Hsu, The Measurement of Local Turbulent Skin Friction by Means of Surface Pitot Tubes, Report 957, NS 715-102, Navy Department, David W. Taylor Model Basin, Washington, D. C., 1955.
58. Nitsche, Wändschubspannungsmessung mit Prestonrohren in Grenzschiehtströmungen mit Zusätzlichen Einflussparametern, *Z. Flugwiss. Weltraumforsch,* vol. 4, p. 3, 1980.
59. E. F. Relf, R. C. Pankhurst, and W. S. Walker, The Use of Pitot Tubes to Measure Skin Friction on a Flat Plate, Report F.M. 2121, Fluid Motion Sub-Committee, Aeronautical Research Council, London, 1954.
60. J. M. Allen, Evaluation of Compressible Flow Preston Tube Calibration, NASA TN D-7190, 1973.
61. P. Bradshaw and D. Unsworth, A Note on Preston Tube Calibrations in Compressible Flow, *AIAA J.,* vol. 12, no. 9, pp. 1293–1294, 1974.
62. J. H. Hool, Measurement of Skin Friction Using Surface Tubes, *Aircraft Eng.,* vol. 28, p. 52, 1956.
63. B. A. Bradshaw and M. A. Gregory, The Determination of Local Turbulent Skin Friction from Observation in the Viscous Sub-layer, Aeronaut. Res. Counc. London, R&M 3202, 1961.
64. G. L. Brown, Theory and Application of Heated Films for Skin Friction Measurement, *Proc. of 1967 Heat Transfer and Fluid Mech. Inst.,* pp. 361–381, Stanford University Press, 1967.
65. L. F. East, Measurement of Skin Friction at Low Subsonic Speeds by the Razor-Blade Technique, Aeronaut. Res. Counc. London, R&M 3525, 1966.
66. G. I. Taylor, Measurement with a Half-Pitot Tube, *Proc. R. Soc. London Ser. A,* vol. 166, p. 476, 1938.
67. G. E. Gadd, W. E. Cope, and J. L. Attridge, Heat Transfer and Skin-Friction Measurements at a Mach Number of 2.44 for a Turbulent Boundary Layer on a Flat Surface and in Regions of Separated Flow, Aeronaut. Res. Counc. London, R&M 3148, 1959.
68. G. E. Gadd, A Note on the Theory of the Stanton Tube, Aeronaut. Res. Counc. London, R&M 3147, 1958.
69. J. H. Haritonidis, The Measurement of Wall Shear Stress, in M. Gad-el-Hak, ed., *Advances in Fluid Mechanics Measurements,* Lecture Notes in Engineering, pp. 229–261, Springer-Verlag, New York, 1989.
70. S. S. Abarnel, R. J. Hakkinen, and L. Trilling, Use of a Stanton Tube for Skin-Friction Measurements, NASA Memo. 2-17-59W, Washington, D.C., 1959.
71. A. Thom, The Flow of the Mouth of a Stanton Pitot, Aeronaut. Res. Counc. London, R&M 2984, 1956.
72. M. C. Good and P. N. Joubert, The Form Drag of Two-Dimensional Bluff-Plates Immersed in Turbulent Boundary Layers, *J. Fluid Mech.,* vol. 31, pp. 547–582, 1968.
73. E. Achenbach, Distribution of Local Pressure and Skin Friction Around a Circular Cylinder in Cross Flow Up to Re $= 5 \times 10^6$, *J. Fluid Mech.,* vol. 34, pp. 625–639, 1968.
74. E. Achenbach, Experiments on Flow Past Spheres at Very High Reynolds Numbers, *J. Fluid Mech.,* vol. 54, pp. 565–575, 1972.
75. J. B. Vagt and H. Fernholz, Use of Surface Fences to Measure Wall Shear Stress in Three-Dimensional Boundary Layers, *Aeronaut. Q.,* vol. 2, pp. 87–91, 1973.
76. H. Higuchi and M. W. Rubesin, An Experimental and Computational Investigation of the Transport of Reynolds Stress in an Axisymmetric Swirling Boundary Layer, AIAA 19th Aerospace Science Meeting, St. Louis, January 1981.

77. D. A. Spence and G. L. Brown, Heat Transfer to a Quadratic Shear Profile, *J. Fluid Mech.*, vol. 33, pp. 753–773, 1968.
78. J. S. Son, Limiting Relation for the Eddy Diffusivity Close to a Wall, M.S. thesis, Chemical Engineering Department, University of Illinois, Urbana, 1965.
79. J. S. Son and T. J. Hanratty, Limiting Relation for the Eddy Diffusivity Close to a Wall, *AIChE J.*, vol. 13, pp. 689–696, 1967.
80. D. A. Shaw and T. J. Hanratty, Turbulent Mass Transfer Rates to a Wall for Large Schmidt Numbers, *AIChE J.*, vol. 23, pp. 28–37, 1977.
81. S. C. Ling, Heat Transfer from a Small Isothermal Spanwise Strip on an Insulated Boundary, *J. Heat Transfer*, vol. C85, pp. 230–236, 1963.
82. K. R. Jolls, Flow Patterns in a Packed Bed, Ph.D. thesis, Chemical Engineering Department, University of Illinois, Urbana, 1966.
83. M. LeBouche, Transfert de Matiere en Regime de Couche Limite Bidimensionnelle et a Nombre de Schmidt Grand, *C. R. Acad. Sci.*, vol. 270A, pp. 1757–1760, 1970.
84. M. LeBouche, Sur la Mesure Polarographique de Gradient Parietal de Vitesse Dams les Zones d'Arret Amont ou de Decollement du Cylinder, *C. R. Acad. Sci.*, vol. 276A, pp. 1245–1248, 1973.
85. Z. X. Mao and T. J. Hanratty, The Use of Scalar Transport Probes to Measure Wall Shear Stress in a Flow with Imposed Oscillations, *Exp. Fluids*, vol. 3, pp. 129–135, 1985.
86. C. Ambari, C. Deslouis, and B. Tribollet, Frequency Response of the Mass Transfer Rate in a Modulated Flow at Electrochemical Probes, *Int. J. Heat Mass Transfer*, vol. 29, pp. 35–45, 1986.
87. V. Ye. Nakoryakov, A. P. Burdukov, O. N. Kashinsky, and P. I. Geshev, Electrodiffusion Method of Investigation into the Local Structure of Turbulent Flows, Thermosphysics Inst., Novosibirsk, 1986.
88. Yu. E. Bogolyubov, P. I. Geslev, V. E. Nakoryakov, and I. A. Ogorodnikov, Theory of the Electrodiffusion Method of Measuring the Spectral Characteristics of Turbulent Flows, *Prikl. Mekh. Teckln. Fiz.*, vol. 13, no. 14, 1972.
89. J. Y. Dumaine, Etude Numerique de la Réponse en Fréquence des Sordes Electrochimiques, *Lett. Heat Mass Transfer*, vol. 8, p. 293, 1981.
90. M. A. Vorotyntsev, S. A. Martem'Yanov, and B. M. Grafov, Temporal Correlation of Current Pulsations at One or Several Electrodes: A Technique to Study Hydrodynamic Fluctuation Characteristics of a Turbulent Flow, *J. Electroanal. Chem.*, vol. 179, p. 1, 1984.
91. T. J. Pedley, Hot Film in Reversing Flow, *J. Fluid Mech.*, vol. 78, pp. 513–584, 1976.
92. P. Kaiping, Unsteady Forced Convective Heat Transfer from a Hot Film in Non-reversing and Reversing Flow, *Int. J. Heat Mass Transfer*, vol. 24, pp. 545–557, 1983.
93. T. J. Hanratty, L. G. Chorn, and D. T. Hatziavramidis, Turbulent Fluctuations in the Viscous Wall Region for Newtonian and Drag Reducing Fluids, *Phys. Fluids Suppl.* pp. S112–S119, 1977.
94. Z. X. Mao and T. J. Hanratty, Studies of the Wall Shear Stress in a Turbulent Pulsating Pipe Flow, *J. Fluid Mech.*, vol. 177, pp. 545–564, 1986.
95. B. Py, Les Proprietes Generales des Transducteurs Electrochimiques Scindes, *Euromech. 90 Proc.*, 1977.
96. C. G. Phillips, Heat and Mass Transfer from a Film into Steady Shear Flow, *Q. J. Mech. Appl. Math.*, vol. 43, pp. 135–159, 1990.
97. H. A. Stone, Heat/Mass Transfer from Surface Films to Shear Flows at Arbitrary Péclet Numbers, *Phys. Fluids A*, vol. 1, pp. 1112–1122, 1989.
98. K. K. Sirkar, Turbulence in the Immediate Vicinity of a Wall and Fully Developed Mass Transfer at High Schmidt Numbers, Ph.D. thesis, Chemical Engineering Department, University of Illinois, Urbana, 1969.
99. J. E. Mitchell, Investigation of Wall Turbulence Using a Diffusion-Controlled Electrode, Ph.D. thesis, Chemical Engineering Department, University of Illinois, Urbana, 1965.
100. A. J. Karabelas and T. J. Hanratty, Determination of the Direction of Surface Velocity Gradients in Three-Dimensional Boundary Layers, *J. Fluid Mech.*, vol. 34, pp. 159–162, 1968.

101. C. Tournier and B. Py, Analyse et Reconstitution Spatiotemporelle de la Composante Circonferentille de Vitesse Instantanee a Proximite Immediate de la Paroi d'ur Cylindre, *C. R. Acad. Sci.*, vol. 276A, pp. 403–406, 1973.

102. Th. Menzel, V. Sobolik, O. Wein, and U. Onken, Segmentierte Elektrodiffusionssonden zur Messung des Wandschergeschwindigkeitsvekton, *Chem. Ing. Tech.*, vol. 59, pp. 492–493, 1987.

103. O. Wein and N. A. Pokryvaylo, Method of Manufacture of Segmented Probe with Circular Cross-Section, Czechoslovak Patent AO 262 823, 1989.

104. P. Duhamel and B. Py, Caractere Intermittent dela Sous-Couche Visqueuse, *AAAE 9e Colloque D'Aerodynamique Appliquee*, Paris, 1972.

105. L. Labraga, C. Tournier, and P. Florent, Experimental Study of the Frequency Response of Electrochemical Split Probes to Transerval Velocity Fluctuations, *Exp. Fluids*, vol. 11, pp. 325–332, 1991.

106. M. W. Rubesin, A. F. Okono, G. G. Mateer, and A. Brosh, A Hot-Wire Surface Gauge for Skin Friction and Separation Detection Measurements, NASA TM X-62465, 1975.

107. P. H. Alfredsson, A. V. Johansson, J. H. Haritonidis, and E. Eckelmann, The Fluctuating Wall-Shear Stress and the Velocity Field in Viscous Sublayer, *Phys. Fluids*, vol. 31, pp. 1026–1033, 1988.

108. B. J. Bellhouse and C. G. Rasmussen, Low Frequency Characteristics of Hot-Film Anemometers, *DISA Inf. Bull.*, no. 6, pp. 3–11, 1968.

109. D. E. Beasley and R. S. Fegliola, A Generalized Analysis of a Local Heat Flux Probe, *J. Phys. E.*, vol. 21, pp. 316–322, 1979.

110. K. D. Cole and J. V. Beck, Conjugated Heat Transfer from a Strip Heater with the Unsteady Surface Element Method, *AIAA J. Thermophys. Heat Transfer*, vol. 1, pp. 348–354, 1987.

111. K. D. Cole and J. V. Beck, Conjugated Heat Transfer from a Hot-Film Probe for Transient Air Flow, *J. Heat Transfer*, vol. 110, pp. 290–296, 1988.

112. P. W. Liang and K. D. Cole, Conjugated Heat Transfer from a Rectangular Hot Film with the Unsteady Surface Element Method, *AIAA J. Thermophys. Heat Transfer*, vol. 6, pp. 349–355, 1992.

113. K. D. Cole, Shear Stress from Hot-Film Sensors in Unsteady Gas Flow Numerical Experiments, *Heat Transfer in Unsteady Flows*, HTD, vol. 158, National Heat Transfer Conference, Minneapolis, Minn., July 1991.

114. A. N. Menendez and B. R. Ramaprian, The Use of Flush-Mounted Hot-Film Gages to Measure Skin Friction in Unsteady Boundary Layers, *J. Fluid Mech.*, vol. 161, pp. 139–159, 19 .

115. F. K. Owen and B. J. Bellhouse, Skin Friction Measurements at Supersonic Speeds, *AIAA J.*, vol. 8, pp. 1358–1360, 1970.

116. N. S. Diaconis, The Calculation of Wall Shearing Stresses from Heat-Transfer Measurements in Compressible Flow, *J. Aeronaut. Sci.*, vol. 21, pp. 201–202 (errata, p. 499), 1954.

117. J. S. Newman, *Electrochemical Systems*, pp. 1–25, Prentice-Hall, Englewood Cliffs, N.J., 1973.

118. T. J. Hanratty, Study of Turbulence Close to a Solid Wall, *Phys. Fluids Suppl.*, pp. S126–S133, 1967.

119. W. E. Ranz, Electrolytic Methods for Measuring Water Velocities, *AIChE J.*, vol. 4, pp. 338–342, 1958.

120. I. M. Kolthoff and J. J. Lingane, *Polarography*, Interscience, New York, 1952.

121. C. S. Lin, E. B. Denton, H. S. Gaskill, and G. L. Putnam, Diffusion-Controlled Electrode Reactions, *Ind. Eng. Chem.*, vol. 43, pp. 2136–2143, 1951.

122. L. P. Reiss, Turbulent Mass Transfer to Small Sections of a Pipe Wall, M.S. thesis, Chemical Engineering Department, University of Illinois, Urbana, 1960.

123. T. Mizushina, The Electrochemical Method in Transport Phenomena, *Adv. Heat Transfer*, vol. 1, pp. 87–159, 1971.

124. A. M. Sutey and J. G. Knudsen, Mass Transfer at the Solid-Liquid Interface for Climbing Film Flow in an Annular Duct, *AIChE J.*, vol. 15, p. 719, 1969.

125. M. Eisenberg, C. W. Tobias, and C. R. Wilke, Mass Transfer at Rotating Cylinders, *Chem. Eng. Prog. Symp. Ser.,* vol. 51, no. 16, pp. 1–16, 1955.
126. E. A. Vallis, M. A. Patrick, and A. A. Wragg, Radial Distribution of Wall Fluxes in the Wall Jet Region of a Flat Plate Held Normal to an Axisymmetric, Turbulent, Impinging Jet, *Euromech. 90 Proc.,* Session IIIc, pp. 1–14, 1977.
127. J. D. Jenkins and B. Gay, Experience in the Use of the Ferri-Ferrocyanide Redox Couple for the Determination of Transfer Coefficients in Models in Shell and Tube Heat Exchangers, *Euromech. 90 Proc.,* Session IIIa, pp. 1–20, 1977.
128. D. A. Shaw, Mechanism of Turbulent Mass Transfer to a Pipe Wall at High Schmidt Number, Ph.D. thesis, Chemical Engineering Department, University of Illinois, Urbana, 1976.
129. J. G. A. Hogenes, Identification of the Dominant Flow Structure in the Viscous Wall Region of a Turbulent Flow, Ph.D. thesis, Mechanical Engineering Department, University of Illinois, Urbana, 1979.
130. L. G. Chorn, An Experimental Study of Near-Wall Turbulence Properties in Highly Drag Reduced Pipe Flows of Pseudoplastic Polymer Solutions, Ph.D. thesis, Chemical Engineering Department, University of Illinois, Urbana, 1978.
131. I. M. Kolthoff and N. H. Furman, *Volumetric Analysis,* vol. 1, p. 231, Wiley, New York, 1928.
132. A. M. Sutey and J. C. Knudsen, Effect of Dissolved Oxygen on the Redox Method for the Measurement of Mass-Transfer Coefficients, *Ind. Eng. Fundam.,* vol. 6, pp. 133–139, 1967.
133. V. E. Nakoryakov, O. N. Kashinsky, A. V. Petukhnov, and R. S. Gorelik, Study of Local Hydrodynamic Characteristics of Upward Slug Flow, *Exp. Fluids,* vol. 7, pp. 560–566, 1989.
134. V. E. Nakoryakov, O. N. Kashinsky, A. P. Burdukov, and V. P. Odnoral, Local Characteristics of Upward Gas-Liquid Flows, *Int. J. Multiphase Flow,* vol. 7, pp. 63–81, 1981.
135. V. E. Nakoryakov, O. N. Kashinsky, and B. K. Kozmenko, Experimental Study of Gas-Liquid Slug Flow in a Small-Diameter Vertical Pipe, *Int. J. Multiphase Flow,* vol. 12, pp. 337–355, 1986.
136. G. A. McConaghy, The Effect of Drag Reducing Polymers on Turbulent Mass Transfer, Ph.D. thesis, University of Illinois, Urbana, 1974.
137. G. Fortuna, Effect of Drag-Reducing Polymers on Flow near a Wall, Ph.D. thesis, Chemical Engineering Department, University of Illinois, Urbana, 1971.
138. R. E. Hicks and N. Pagotto, CSIR Rep. CENG M-024, Pretoria, South Africa, 1974.
139. S. Robin, Etude de la Turbulence Pariétale par la Méthode Electrochimique, Thèse de Doctorat de l'Université Paris 6, 1987.
140. O. Gil, Réponse en Fréquence de Microsondes Electrochimiques, Thèse de Doctorat de l'Université Paris 6, 1990.
141. L. P. Reiss, Investigation of Turbulence near a Pipe Wall Using a Diffusion Controlled Electrolytic Reaction on a Circular Electrode, Ph.D. thesis, Chemical Engineering Department, University of Illinois, Urbana, 1962.
142. D. T. Chin, An Experimental Study of Mass Transfer on a Rotating Spherical Electrode, *J. Electrochem. Soc.,* vol. 118, p. 1764, 1971.
143. J. C. Bazan and A. J. Arvia, The Diffusion of Ferro- and Ferricyanide Ions in Aqueous Solutions of Sodium Hydroxide, *Electrochim. Acta,* vol. 10, p. 1025, 1965.
144. J. D. Newson and A. C. Riddiford, Limiting Currents for the Reduction of the Tri-iodide Ion at a Rotating Platinum Disk Cathode, *J. Electrochem. Soc.,* vol. 108, p. 695, 1961.
145. J. S. Anderson and K. Saddington, The Use of Radioactive Isotopes in the Study of the Diffusion of Ions in Solution, *J. Chem. Soc.,* pp. S381–S386, 1949.
146. M. Eisenberg, C. W. Tobias, and C. R. Wilke, Selected Physical Properties of Ternary Electrolytes Employed in Ionic Mass Transfer Studies, *J. Electrochem. Soc.,* vol. 103, pp. 413–416, 1956.
147. D. A. Shaw, Measurement of Frequency Spectra of Turbulent Mass Transfer Fluctuations at a Pipe Wall, M.S. thesis, Chemical Engineering Department, University of Illinois, Urbana, 1973.
148. G. Nickless (ed.), *Inorganic Sulfur Chemistry,* pp. 200–239, Elsevier, New York, 1968.
149. J. Vulterin and J. Zyka, Investigation of Some Hydrazine Derivatives as Reductimetric Titrants, *Talanta,* vol. 10, pp. 891–898, 1963.

150. H. V. Malmstadt, C. G. Enke, and S. F. Crouch, *Electronic Measurements for Scientists,* pp. 94–97, Benjamin, Reading, Mass., 1974.

151. T. J. Hanratty, The Use of Electrochemical Techniques to Study Flow Fields and Mass Transfer Rates, in Z. Zaric (ed.), *Heat and Mass Transfer in Flows with Separated Regions,* pp. 139–159, Pergamon, New York, 1972.

152. L. D. Eckelman, The Structure of Wall Turbulence and Its Relation to Eddy Transport, Ph.D. thesis, Chemical Engineering Department, University of Illinois, Urbana, 1971.

153. V. A. Sandborn, *Resistance Temperature Transducers,* pp. 361–365, Metrology Press, Fort Collins, Colo., 1972.

154. DISA Electronics, *DISA Probe Manual,* Denmark, 1976.

155. M. W. Rubesin, A. F. Okuno, L. L. Levy, J. B. McDevitt, and H. L. Seegmiller, An Experimental and Computational Investigation of the Flow Field About a Transonic Airfoil in Supercritical Flow with Turbulent Boundary Layer Separation, NASA TM X-73, p. 157, 1976.

156. R. F. Blackwelder and H. Eckelmann, Influence of the Convection Velocity on the Calibration of Heated Surface Elements, *Euromech. 90 Proc.,* Session I.a.1, pp. 1–10, 1977.

157. D. H. Pessoni, An Experimental Investigation into the Effects of Wall Heat Flux on the Turbulence Structure of Developing Boundary Layers at Moderately High Reynolds Numbers, Ph.D. thesis, Mechanical and Industrial Engineering Department, University of Illinois, Urbana, 1974.

158. D. S. Simons, R. M. Li, and F. D. Schall, Spatial and Temporal Distribution of Boundary Layer Shear Stress in Open Channel Flows, NSF Final Rep., Civil Engineering Department, Colorado State University, Fort Collins, 1979.

159. J. E. Coney and D. A. Simmers, The Determination of Shear Stress in Fully Developed Laminar Axial Flow and Taylor Vortex Flow, Using a Flush-Mounted Hot Film Probe, *DISA Inf. Bull.,* no. 24, pp. 9–14, 1979.

160. K. F. Wylie and C. V. Alonso, Some Stochastic Properties of Turbulent Tractive Forces in Open Channel Flow, *Proc. of the Fifth Biennial Symp. on Turbulence,* p. 181, Rolla, Missouri, 1979.

161. G. R. Foster, Hydraulics of Flow in a Rill, Ph.D. thesis, Purdue University, 1977.

162. O. Christensen, New Trends in Hot-Film Probe Manufacturing, *DISA Inf. Bull.,* no. 9, pp. 30–36, 1970.

163. K. E. W., Coated Hot Film Probes for Measurement in Liquids, *DISA Inf. Bull.,* no. 3, p. 28, 1966.

164. M. Miya, Properties of Roll Waves, Ph.D. thesis, Chemical Engineering Department, University of Illinois, Urbana, 1970.

165. Thermo Systems, Inc., *TSI Operating Manual for Model 1050 and 1050A Anemometer Modules,* pp. 2–8, St. Paul, Minn., 1973.

166. H. Higuchi and D. J. Peake, Bi-Directional Buried-Wire Skin-Friction Gage, NASA TM 78531, 1978.

167. A. Bertelrud, Measurement of Shear Stress in a Three-Dimensional Boundary Layer by Means of Heated Films, *Euromech. 90 Proc.,* Session I.b.2, pp. 1–14, 1977.

168. H. P. Kreplin and H. U. Meier, Application of Heated Element Techniques for the Measurement of the Wall Shear Stress in Three-Dimensional Boundary Layers, *Euromech. 90 Proc.,* Session I.B.3, pp. 1–21, 1977.

169. H. Schlichting, *Boundary Layer Theory,* 6th ed., pp. 154–162, McGraw-Hill, New York, 1960.

170. K. Hiemenz, Die Grenzschicht en einem in den Gleichformigen Flussiqkeitsstrom Eingetauchten Geraden Kreiszylinder, *Dingl. Polytechn. J.,* vol. 326, pp. 321–324, 1911.

171. M. A. Niederschulte, R. J. Adrian, and T. J. Hanratty, Measurements of Turbulent Flow in a Channel at a Low Reynolds Number, *Exp. Fluids,* vol. 9, pp. 222–230, 1990.

172. F. H. Clauser, Turbulent Boundary Layers in Adverse Pressure Gradients, *J. Aeronaut. Sci.,* vol. 21, pp. 91–108, 1954.

173. L. S. Cohen and T. J. Hanratty, Effect of Waves at a Gas-Liquid Interface on a Turbulent Air Flow, *J. Fluid Mech.,* vol. 31, pp. 467–479, 1968.

174. J. L. Hudson, Effect of Wavy Walls on Turbulence, Ph.D. thesis, University of Illinois, Urbana, 1993.

175. D. M. McEligot, Measurement of Wall Shear Stress in Favorable Pressure Gradients, *Lecture Notes in Physics,* vol. 235, pp. 292–303, 1985.
176. L. H. Tanner and L. G. Blows, A Study of the Motion of Oil Films in Surfaces in Air Flow, With Applications to the Measurement of Skin-Friction, *J. Phys. E Sci. Instrum.,* vol. 9, pp. 194–202, 1976.
177. L. H. Tanner, The Application of Fizeau Interferometry of Oil Films to Study of Surface Flow Phenomena, *Opt. Lasers Eng.,* vol. 2, pp. 105–118, 1981.
178. D. J. Monson and H. Higuchi, Skin-Friction Measurements by a Plural-Laser-Beam Interferometer Technique, *AIAA J.,* vol. 19, pp. 739–744, 1981.
179. D. J. Monson, A Nonintrusive Laser Interferometer Method for Measurement of Skin Friction, *Exp. Fluids,* vol. 1, pp. 15–22, 1983.
180. D. J. Monson, A Laser Interferometer for Measuring Skin Friction in Three-Dimensional Flows, *AIAA J.,* vol. 22, pp. 557–559, 1984.
181. R. V. Westphal, J. Eaton, and K. Johnston, A New Probe for Measurement of Velocity and Wall Shear Stress in Unsteady, Reversing Flow, Report IL-16, Heat Transfer and Turbulence Mechanics Group Thermosciences Division of Mechanical Eng., Stanford Univ., Stanford, Calif., April 1980. (Presented at the ASME Winter Annual Meeting, Chicago, Ill., 1980.)
182. R. V. Westphal, W. D. Bachalo, and M. H. Houser, Improved Skin-Friction Interferometer, NASA TM 88216, 1986.
183. J. D. Murphy and R. V. Westphal, The Laser Interferometer Skin-Friction Meter: A Numerical and Experimental Study, *J. Phys. E Sci. Instrum.,* vol. 19, pp. 744–751, 1986.
184. P. R. Bandyopadhyay and L. M. Weinstein, A Reflection-Type Oil-Film Skin Friction Meter, *Exp. Fluids,* vol. 11, pp. 281–292, 1991.
185. A. A. Naqwi and W. C. Reynolds, Measurement of Turbulent Wall Velocity Gradients Using Cylindrical Waves of Laser Light, *Exp. Fluids,* vol. 10, pp. 257–266, 1991.
186. J. K. Eaton, R. V. Westphal, and J. Johnston, Two New Instruments for Flow Direction and Skin Friction Measurements in Separated Flows, *ISA Trans.,* vol. 21, p. 69, 1981.
187. L. J. S. Bradbury and I. P. Castro, A Pulsed-Wire Technique for Velocity Measurements in Highly Turbulent Flow, *J. Fluid Mech.,* vol. 49, pp. 657–691, 1971.
188. I. P. Castro and M. Dianat, Surface Flow Patterns on Rectangular Bodies in Thick Boundary Layers, *J. Wind Eng. Ind. Aeronaut.,* vol. 11, p. 107, 1983.
189. I. P. Castro and A. Hague, The Structure of a Turbulent Shear Layer Bounding a Separation Region, *J. Fluid Mech.,* vol. 179, pp. 439–468, 1987.
190. M. Dianat and I. P. Castro, Turbulence in a Separated Boundary Layer, *J. Fluid Mech.,* vol. 226, pp. 91–123, 1991.
191. I. P. Castro, M. Dianat, and L. J. S. Bradbury, The Measurement of Fluctuating Skin Friction with a Pulsed Wall Probe, in F. Durst, B. E. Launder, F. W. Schmidt, and J. H. Whitelaw (eds.), *Turbulent Shear Flows,* vol. V, p. 278, Springer, New York, 1987.

ACQUIRING AND PROCESSING TURBULENT FLOW DATA

T. W. Simon

10.1 INTRODUCTION

Obtaining useful results from an experiment involves appropriately processing the instrument signals. From the myriad of information in a hot-wire signal, for instance, we want to extract quantities that characterize important aspects of the flow, e.g., transport rates and evidence of coherent structure. These quantities may be mean values, mean-square values, correlation functions, probability distributions, turbulence scales, spectra, etc.

The foremost prerequisite to fluid mechanics measurements is an understanding of the physics of fluid mechanics. There are too many pitfalls for a person who uses, without this understanding, even such common instruments as the Pitot tube or hot-wire anemometer.

In the following section, the perspective is taken of one who wishes to process a signal from a constant-temperature hot wire in a boundary layer to evaluate several of the familiar turbulence quantities. This setting is to serve as an example and a vehicle for presenting some of the processing techniques documented in the literature. Many of the techniques could be applied equally well to laser-Doppler anemometer (LDA) measurements. With the LDA, however, velocity values are available only when a particle resides within the measurement volume. Thus the signal is not continuous. Obtaining a velocity waveform will therefore require extra care with the LDA. In particular, there must be a high rate of particle seeding, so

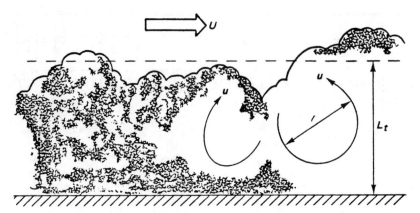

Figure 10.1 Large eddies in a turbulent boundary layer. The flow above the boundary layer (dashed line) has a velocity U; the largest eddy size l is comparable to the boundary layer thickness L_t. The interface between the turbulence and the flow above the boundary layer is quite sharp. (From [1])

that a continuous waveform can be inferred. Once this has been done, the continuous-wave techniques to follow, presented for the hot-wire signal, are applied. Of course, discussions below concerning linearizing the signals will apply to hot-wire signals only (the LDA signal is linear with velocity). The discussion begins by reviewing the nature of a hot-wire signal while noting some assumptions typically used in signal analysis. Then, many of the familiar quantities measured in turbulent flow are presented. Finally, hardware that could be used to effect this processing is discussed.

10.2 NATURE OF THE HOT-WIRE SIGNAL

The signal to be processed is generated by inserting a hot-wire anemometer probe into a turbulent flow, perhaps the turbulent boundary layer flow shown schematically in Fig. 10.1. Power is supplied to maintain the wire at a predetermined temperature. The convective cooling effect is sensitive to the velocity perpendicular to the wire; hence the power demand varies with the flow velocity. If the thermal and electronic response times are short, the instantaneous power demand, or the square of the supply voltage, corresponds with the instantaneous value of the velocity normal to the wire averaged over the active length of the wire. The flow is composed of an array of eddies of varying orientation, with sizes ranging from nearly the scale of the flow (the boundary layer thickness in the case of Fig. 10.1) to less than, typically, 10^{-4} times this size—this ratio depends on the Reynolds number of the flow. To a stationary observer, e.g., a hot wire, this array of eddies convected through the sensing zone appears as a fluctuating velocity field that is quite random and has frequency components ranging, typically, from several hertz to as high as

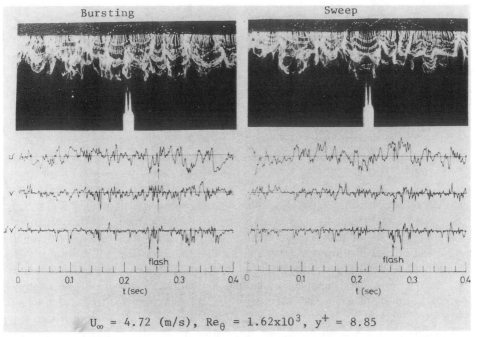

Figure 10.2 Combined smoke-wire/hot-wire technique in a flat-plate turbulent boundary layer. (From [2])

50 kHz—this also depends on Reynolds number. Examples are shown in Figs. 10.2–10.5 for cases of near-wall boundary layer flow (Fig. 10.2); laminar, transitional, and turbulent boundary layers (Fig. 10.3); intermittent turbulence at the edge of a shear layer (Fig. 10.4); and viscous sublayer, buffer, and inertial sublayer regions of turbulent boundary layers (Fig. 10.5). The hot-wire anemometer voltage will respond to these fluctuations except for possible filtering of high-frequency fluctuations by the thermal inertia of the wire, intentional filtering to eliminate electronic circuit noise, or averaging of fluctuations characteristic of eddies smaller than the wire active length. The hot-wire voltage signal then appears similar to the velocity traces (e.g., Fig. 10.6).

10.2.1 Random Versus Deterministic

Any observed data representing a physical phenomenon can be classified as random or deterministic. Deterministic data can be described by an explicit mathematical relationship. This is in contrast to random data, where exact values cannot be predicted into the future and the data must be described in terms of probability statements and statistical averages rather than explicit relationships. Although

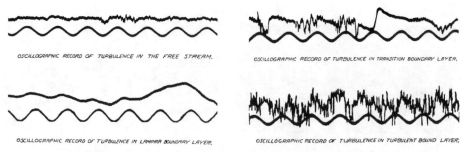

Figure 10.3 Oscillographic record of turbulence in the free stream, laminar boundary layer, transitional boundary layer, and turbulent boundary layer. Lower traces are 60-Hz sine waves. (From [3])

studies of turbulent flows (e.g., [7, 8]) have shown coherency in the flow and it can be argued that the process is completely described by the Navier-Stokes equation, we are presently not able to make the requisite prediction of future values. Even if we could predict the flow, given the specified boundary and initial conditions, we would not be able to specify the initial conditions with the accuracy and resolution required to have unambiguous "future values" computed. Turbulent flow is highly dependent upon the smallest variation in the initial conditions. This is a characteristic of a "chaotic" process [9], such as turbulence. For these reasons, turbulent fluctuations are considered random, and hot-wire signals are processed as random signals.

10.2.2 Stationary Versus Nonstationary Processes

The next classification of the signal is between stationary and nonstationary behavior. To make this evaluation, a time record of the hot-wire signal for a particular event is made, then repeated from the same starting point (beginning of the event—such as the beginning of a cycle) N times, as shown in Fig. 10.7. Each is called a sample record and may be thought of as one physical realization of a random process. From each of these traces the value at time t_1 is taken; then an ensemble average is calculated as

$$<u(t_1)> = \lim_{N\to\infty} \frac{\sum_{i=1}^{N} u_i(t_1)}{N} \tag{1}$$

where the angle brackets denote ensemble average, and an autocorrelation function is calculated as

$$<u(t_1)u(t_1 - \tau)> = \lim_{N\to\infty} \frac{1}{N}\sum_{i=1}^{N} u_i(t_1)u_i(t_1 - \tau) \tag{2}$$

where τ is the delay time, an independent parameter to the autocorrelation function (discussed below). Similarly, ensemble average and autocorrelation values are

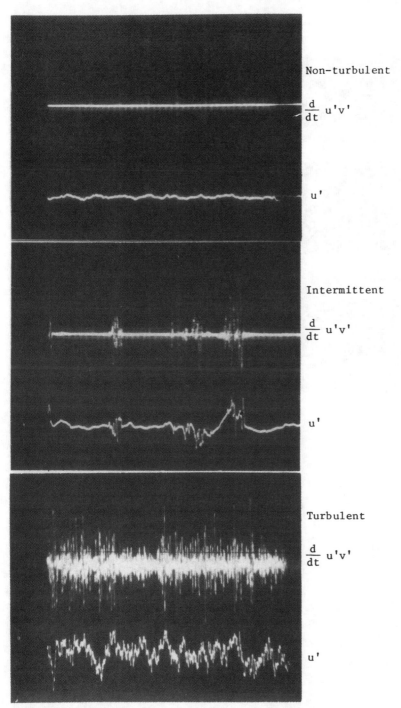

Figure 10.4 Response of criterion function [$d(u'v')/dt$]. Oscillographic record for laminar, intermittent, and turbulent boundary layer flows. (From [4])

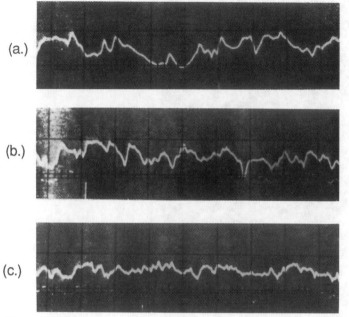

Figure 10.5 Oscillographs of $u(t)$ signal for turbulent boundary layer flow: (*a*) outer viscous sublayer, $y^+ = 5.01$; (*b*) outer buffer layer, $y^+ = 17.9$; and (*c*) log-linear region, $y^+ = 159$. (From [5])

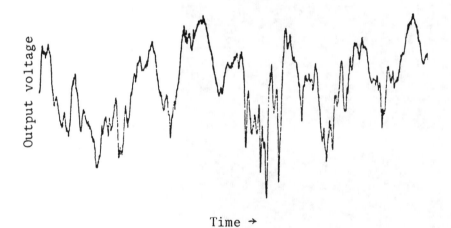

Figure 10.6 Trace of a hot-wire signal from a random turbulent flow. (From [6])

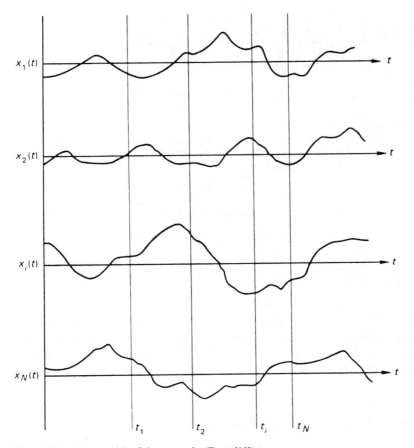

Figure 10.7   An ensemble of time records. (From [10])

computed for $t_2, t_3, \ldots, t_N$. If plots of ensemble averages and autocorrelations show no significant variations with time (Fig. 10.8), the process is stationary. It is usually assumed in processing turbulence data that the process is stationary. There are exceptions to this assumption in practical applications. For instance, suppose turbulence measurements are to be taken within a reciprocating or rotating engine. This would certainly be an unsteady (nonstationary) flow. The ensemble-averaged velocity is given by

$$<u(\theta)> = \lim_{N\to\infty} \frac{1}{N} \sum_{i=1}^{N} u_i(\theta) \tag{3}$$

where θ is a particular position from the beginning of an "event," i is the event number, and N is the number of "events" used to compute the average. Figure 10.9

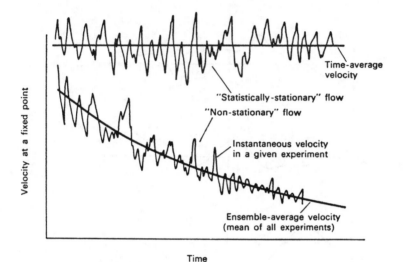

Figure 10.8 Comparison of ordinary and time-dependent turbulent flow (e.g., pipe flow with steady or decreasing supply pressure). (From [11])

shows, for oscillating (zero time-mean velocity) flow in a pipe, the axial velocity versus crank position for an ensemble average over 500 cycles. The event in this case is a revolution of the flywheel, and the beginning of each event in this case was an indication of top dead center given by a photodetector on the flywheel of the facility.

10.2.3 Ergodic Versus Nonergodic Processes

Another form of averaging is found by looking at only one of the sample records i of Fig. 10.7 and averaging over time. The mean is given as

$$u = \bar{u}_i = \lim_{T \to \infty} \frac{1}{T} \int_0^T u_i(t)\, dt \qquad (4)$$

and the autocorrelation is

$$\overline{u(t)u(t - \tau)} = \lim_{T \to \infty} \frac{1}{T} \int_0^T u_i(t)u_i(t - \tau)\, dt \qquad (5)$$

If the random process is stationary and $\bar{u}_i$ and $\overline{u_i(t)u_i(t - \tau)}$ do not differ from record to record, the random process is said to be ergodic. Therefore in an ergodic random process, the time-averaged values are equal to the ensemble-averaged values. Usually, it is assumed that the turbulent flow signal has the ergodic property

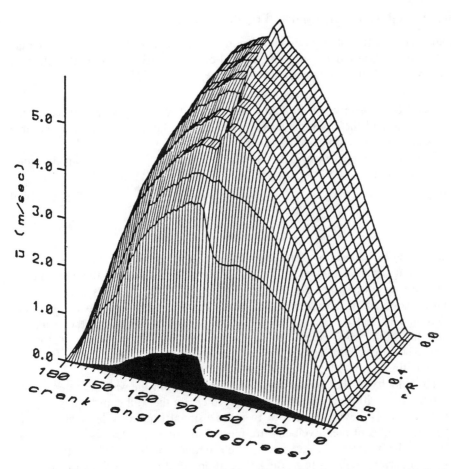

Figure 10.9 Ensemble-averaged velocity traces in an oscillatory flow through a pipe in which the time-mean is zero. Traces are for a half cycle; the second half is a reflection of the first. (From [12])

and time averages are made on one sample record. Exceptions would be the non-stationary engine measurements mentioned above. When measuring in such flows, one of two techniques may be employed: (1) only ensemble averages are used for characterizing the signal—the ergodic assumption is not employed, and (2) if the engine cycle is long relative to the turbulence events (largest turbulence timescales), a quasi-stationary assumption can be imposed and the time of integration T is made sufficiently short that little change of nonturbulence quantities takes place. The first option is usually exercised under these circumstances. Only average quantities are available under the first option, however, and such quantities as spectra (to be discussed) are sacrificed.

10.2.4 Analysis of the Random Ergodic Process

Analysis of the turbulent flow data may take many paths, depending on the purpose of the measurement program. Several of the quantities frequently extracted from the turbulent flow data are discussed below.

10.2.4.1 Mean velocity. In the following section, two averages are discussed, the time average and the ensemble average. For an ergodic random process, the time average and the ensemble average are equal.

The time-averaged signal (in this case, velocity) is processed from the time-varying signal as

$$\bar{u}(x,y,z) = \lim_{T \to \infty} \frac{1}{T} \int_0^T u(t,x,y,z)\, dt \tag{6}$$

In averaging turbulent flow velocities, the averaging time must be long enough to average out several rollover times of the largest features in the flow. For a laboratory-scale turbulent boundary layer flow of velocity 15 m/s and Reynolds number based upon development length of 10^6, averaging times of 30–60 s are common. This can be checked for a stationary flow by increasing the averaging time from sample to sample until no further sensitivity to averaging time is observed. The required integration time will increase with the magnitude of the velocity fluctuations and with the size of the largest features in the flow.

The velocity in a boundary layer and distance from the wall are often nondimensionalized with the wall shear stress τ_w to create the wall coordinates u^+ ($= u/\sqrt{\tau_w/\rho}$) and y^+ [$= (y/\nu)\sqrt{\tau_w/\rho}$]. In two-dimensional boundary layer studies, for example, profiles of mean velocity are plotted in inner coordinates to investigate effects on the near-wall behavior (e.g., the roughness effect of Fig. 10.10), on the viscous sublayer thickness (e.g., the acceleration effect of Fig. 10.11), or the effect on the turbulent core (e.g., the streamwise curvature effect of Fig. 10.12). Smooth-wall profiles plotted in inner coordinates display similar behavior (Fig. 10.13). In the turbulent core the velocity distribution is given as $u^+ = A + B \ln (y^+)$—called the log-linear relationship or the law of the wall. For a smooth wall without a strong pressure gradient effect on the near-wall flow, the values of A and B are known ($A = 2.44$, $B = 5.0$). This can be exploited to deduce values of the skin friction coefficient from mean velocity data taken in the turbulent core region; an example of data deduced for the technique is given in Fig. 10.14, where such data characterize the effect of streamwise convex curvature on the wall shear stress. If the wall shear stress is unknown, but log-linear behavior is expected, one can determine the value of wall shear stress τ_w, which forces the processed data to fall on the log-linear curve. This technique was first proposed by Clauser [18]. The case of streamwise curvature without streamwise acceleration was believed to follow this log-law, and therefore it was a candidate for the Clauser technique. Profiles of mean

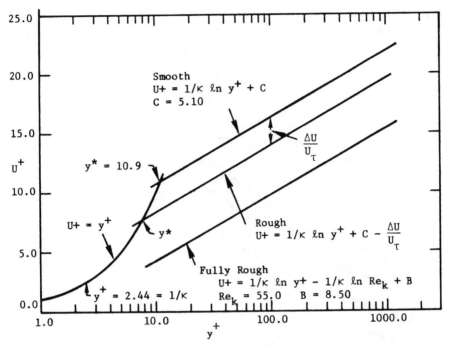

Figure 10.10 Fully rough, transitionally rough and smooth mean velocity profiles. (From [13])

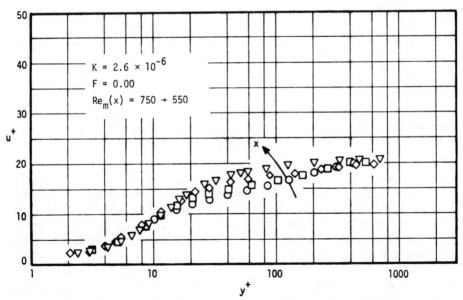

Figure 10.11 Sequential velocity profiles within a strong acceleration ($K = 2.6 \times 10^{-6}$) boundary layer. (From [14])

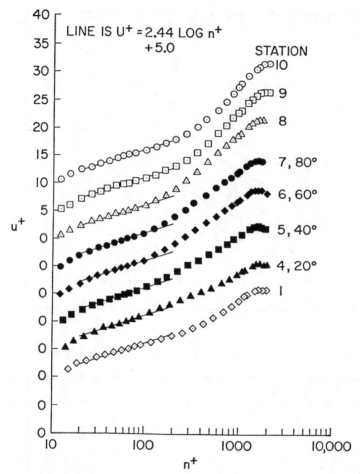

Figure 10.12 Mean velocity profiles plotted in wall coordinates. Station 1 is upstream of convex curvature, stations 4–7 are within the curve, and stations 8–10 are downstream of the curve, on a flat wall. (From [15])

velocity may also be plotted in outer, or defect, coordinates to investigate effects on the wake region (Fig. 10.15). This set of coordinates would be used to characterize the effects of events in the history of the development of the flow that affect the outer region, namely, effects of pressure gradients, transpiration through the wall, upstream bluff bodies, etc. [19].

10.2.4.2 Probability density. The probability density function shows the relative amount of time that $u(t)$ is at particular velocity levels (Fig. 10.16). It may be

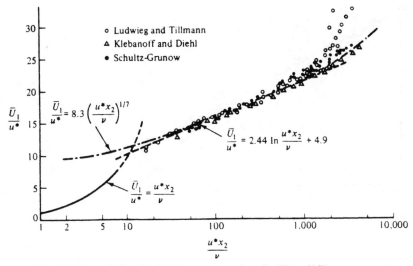

Figure 10.13 Mean velocity distribution near smooth walls. (From [16])

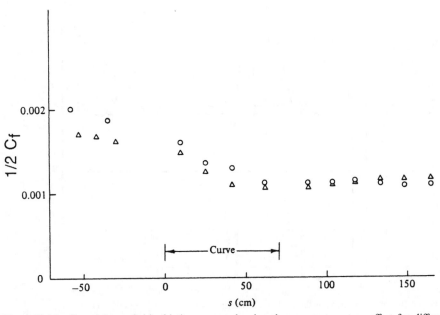

Figure 10.14 Comparison of skin friction curves showing the convex curvature effect for different ratios of boundary layer thickness δ to radius of curvature R: ($\triangle$) $\delta/R = 0.10$; ($\circ$) $\delta/R = 0.05$. (From [17])

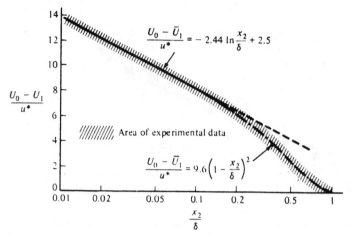

Figure 10.15 Logarithmic plot of velocity distributions in the outer part of a turbulent boundary layer. (From [16])

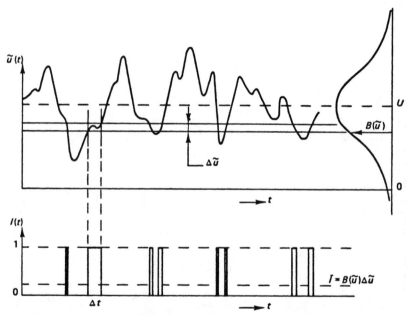

Figure 10.16 Measurement of the probability density of a stationary function. (From [1])

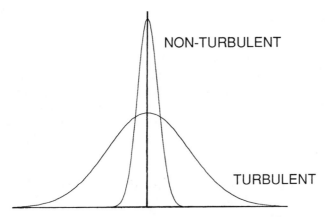

NON-TURBULENT

TURBULENT

Figure 10.17   Probability distribution of a fluctuating quantity. (Redrawn from [4])

computed by dividing the velocity range into increments Δu, then computing the fraction of the time, $\Sigma \Delta t / T$, that u is within each increment, $u - \Delta u/2$, $u + \Delta u/2$:

$$B(u) = \lim_{\Delta u \to 0} \left(\frac{1}{\Delta u} \lim_{T \to \infty} \frac{1}{T} \Sigma \Delta t \right) \tag{7}$$

The shape of this probability density function may be described by its moments of increasing order: variance, skewness, kurtosis [1]. It has been shown that probability density functions of most turbulent flows that are not within intermittent regions (wake of a turbulent boundary layer or within a transitional boundary layer) are nearly Gaussian in shape. The width of the probability density distribution, the variance (which is proportional to the u range for a given B value), is an indication of the intensity of turbulence (Fig. 10.17), as discussed below.

10.2.4.3 Turbulence intensities.   Turbulence intensity is the standard deviation s of the fluctuations of velocity about the mean:

$$s = \sqrt{\overline{u'^2}} = \sqrt{\lim_{T \to \infty} \frac{1}{T} \int_0^T (u')^2 \, dt} \tag{8}$$

Similarly, turbulence intensities are computed from v and w as $\overline{v'^2}$ and $\overline{w'^2}$. Note that the turbulence intensities are square roots of autocorrelation functions for u', v', and w', respectively, when τ, the separation time, is equal to zero. These turbulence intensity components reflect the magnitudes of the velocity fluctuations in the x, y, and z directions and the variance (or second moment) values of each respective probability density function, such as

$$s^2 = \overline{u'^2} = \int_{-\infty}^{\infty} u'^2 \, B(u') \, du' \tag{9}$$

Examples are shown in Figs. 10.18 and 10.19 for smooth-wall turbulent boundary layer flows. They have been nondimensionalized with the free-stream velocity and local mean velocity (Fig. 10.18) and shear velocity $U_\tau \, (= \sqrt{\tau_w/\rho})$ (Fig. 10.19). Note that the dependent variables in Figs. 10.18 and 10.19 are the standard deviation expression and the variance (the square of the standard deviation) expression, respectively. Figure 10.20 shows similar plots for a rough-wall boundary layer flow.

10.2.4.4 Joint probability density. The development of the probability density function (discussed in section 10.2.4.2) will now be extended to two variables, u and v. The joint probability density $B(u, v)$ is given by

$$B(u,v) = \lim_{\Delta u \to 0} \left\{ \lim_{\Delta v \to 0} \frac{1}{\Delta u \, \Delta v} \left[\lim_{T \to \infty} \left(\frac{\Sigma \Delta t}{T} \right) \right] \right\} \tag{10}$$

where $\Sigma \Delta t$ is the sum of all the time increments in which

$$u - \frac{\Delta u}{2} < u(t) < u + \frac{\Delta u}{2} \qquad v - \frac{\Delta v}{2} < v(t) < v + \frac{\Delta v}{2}$$

10.2.4.5 Covariance. The moment of the joint probability density function is called the covariance or the correlation. For the two variables u' and v', the correlation is given by

$$\overline{u'v'} = \int_{-\infty}^{\infty} \int_{-\infty}^{\infty} u'v' B(u',v') \, du' \, dv' \tag{11}$$

This is the Reynolds turbulent shear stress term responsible for turbulent transport of momentum to the wall in boundary layer flows. It appears as an additional transport term (added to molecular transport) in the time-averaged Navier-Stokes equation written with boundary layer assumptions:

$$\frac{\partial u}{\partial t} + u\frac{\partial u}{\partial x} + v\frac{\partial u}{\partial y} = -\frac{1}{\rho}\frac{dp}{dx} + \frac{\partial}{\partial y}\left(-\overline{u'v'} + v\frac{\partial u}{\partial y} \right) \tag{12}$$

Another way of visualizing this correlation comes from multiplying the Navier-Stokes equations by the velocity (mean plus fluctuating components) and time averaging [1] to get the turbulent kinetic energy budget equation. From this, one can see that a significant term that represents the production of the streamwise component of turbulent kinetic energy $\overline{u'^2}$ in a two-dimensional boundary layer flow is $\overline{u'v'}(\partial U/\partial y)$. The correlation $\overline{u'v'}$ then shows the effectiveness of converting mean strain $\partial U/\partial y$ into streamwise turbulent kinetic energy $\overline{u'^2}$. Other off-diagonal elements of the Reynolds stress tensor, $\overline{u'w'}$ and $\overline{v'w'}$, are the moments of joint

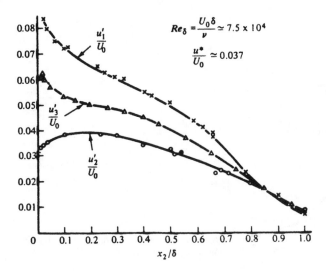

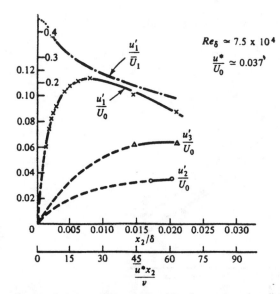

Figure 10.18 Relative turbulence intensities along a smooth wall with zero pressure gradient; u' is the root-mean-square of the velocity fluctuation. (From [16])

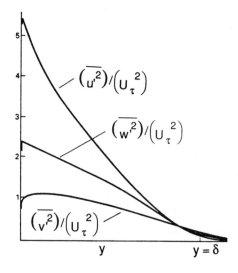

Figure 10.19 Sketch of turbulent intensities in a constant-pressure boundary layer. (From [11])

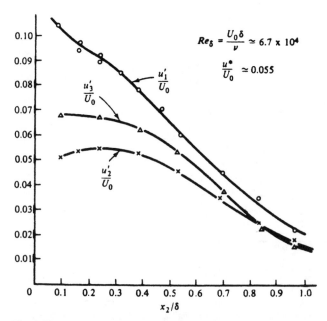

Figure 10.20 Relative turbulence intensities in a boundary layer along a rough wall with a zero pressure gradient; u' is the root-mean-square fluctuation. (From [16])

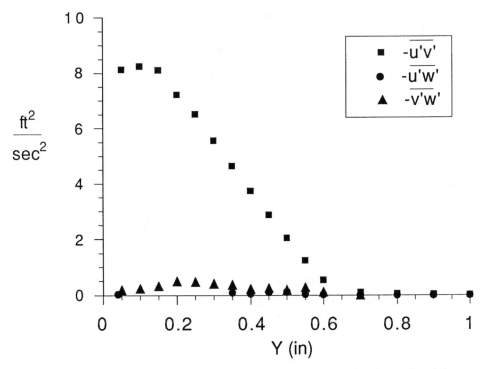

Figure 10.21 Reynolds shear stress components for a smooth-wall boundary layer. Plotted data from [20]

probability density functions for u' and w' and v' and w' traces, respectively (Fig. 10.21).

Another covariance, important in heat transfer studies, is the turbulent cross-stream transport term for thermal energy, $\overline{v'T'}$. It is analogous to the Reynolds turbulent shear stress term and appears in the thermal energy equation—for a turbulent boundary layer with properties that do not vary with temperature:

$$\frac{\partial T}{\partial t} + u\frac{\partial T}{\partial x} + v\frac{\partial T}{\partial y} = \frac{\partial}{\partial y}\left(-\overline{v'T'} + \alpha\frac{\partial T}{\partial y}\right) \tag{13}$$

Other covariance terms associated with heat transfer are $\overline{u'T'}$ (streamwise turbulent transport of thermal energy) and $\overline{w'T'}$ (cross-span turbulent transport of thermal energy).

10.2.4.6 Intermittency. If the hot-wire probe were moved to the outer part of the boundary layer that resides in an otherwise low-turbulence flow, $y = L_t$ in Fig. 10.1, or the edge of any other turbulent shear layer that resides in an otherwise

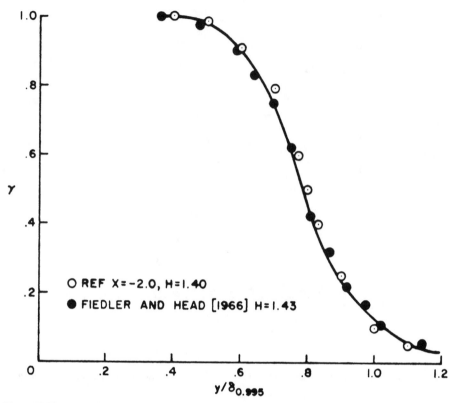

Figure 10.22 Intermittency of a turbulent boundary layer. (From [4])

low-turbulence flow, it would alternately experience highly turbulent and low-turbulence flow, producing a signal similar to the center trace of Fig. 10.4. A descriptor of this situation is the fraction of time the probe is experiencing turbulent flow—the intermittency. A profile of intermittency (Fig. 10.22) shows that irrotational flow resides, intermittently, within about 60% of the turbulent boundary layer thickness. A sketch that shows the edge of the boundary layer and the separation between rotational and irrotational flow is given in Fig. 10.1.

Intermittency is measured in three steps. First, the appropriate signal for intermittency discrimination is generated and reviewed to establish threshold values. The flow is regarded as turbulent (boundary layer flow) if the signal exceeds the threshold value; otherwise it is laminar (free stream flow). Threshold values are generally determined by visual inspection of the signal and are set to be just greater than the background noise level. In the third step, the intermittency function is generated; assigning 1 when the flow is turbulent and zero when the flow is laminar. The intermittency is then found by time averaging the intermittency function over a sufficient time. This is called the retail scheme. In an alternate scheme, called the wholesale scheme, all short-duration turbulent signals are ignored as being due to noise, eccentricities of the algorithm, or glancing encounters with turbulent regions. Comparisons of intermittency distributions processed under these two schemes show that this rather subjective processing should not affect the intermittency value significantly.

The choice of the signal to analyze to determine intermittency depends upon the purpose of the intermittency measurement and the ease of measurement. Streamwise velocity and local temperature (in a heated-wall flow) are the two main signals chosen, although the cross-stream velocity v, the uv correlation, or the time derivatives of u, v, or uv are also used. The quantity uv, or its time derivative, is an attractive quantity upon which to base intermittency, for it represents the fluctuation due to vorticity; u and v are well correlated in turbulent flow and poorly correlated in laminar flow. Thus the uv signal provides maximum separation between laminar and turbulent flow. In the temperature scheme of intermittency measurement, the fluid is heated slightly, while a constant-current anemometer is used to monitor temperature levels above the free stream temperature. A review of velocity and temperature signals reveals that the temperature signal shows the difference between turbulent and nonturbulent flow more distinctly than does the velocity signal.

More on intermittency can be found in [21–23].

10.2.4.7 Power spectral density. In working with the autocorrelation function, one must deal with the waveform in the time domain. Recall that the full waveform was required to process the autocorrelation. Thus the autocorrelation contains information concerning the frequency distribution, or eddy size distribution, of the signal. Referring to Eq. (5) and a representative waveform such as Fig. 10.6, one can see that if the autocorrelation decreases significantly as τ is increased (the flow rapidly loses correlation with its past), high frequencies dominate, but when $u(t)u(t - \tau)$ changes slowly with τ, low frequencies must be dominant. Hence the autocorrelation contains information about the "memory" of the flow. Referring to the waveforms of Fig. 10.5, one would expect that the waveform in Fig. 10.5c, which is rich in fine-scale structure, would have a more rapidly dropping (with time) autocorrelation function than that of the coarser-structure flow of Fig. 10.5b.

An alternate approach to the time domain is to work in the frequency domain and separate the signal into its frequency components. This is done by taking the Fourier transform of the autocorrelation function:

$$s(\omega) = \frac{1}{2\pi} \int_{-\infty}^{\infty} e^{-i\tau\omega} \, \rho(\tau) \, d\tau \qquad (14)$$

The function $s(\omega)$ is called the spectral function, or simply spectrum. The variable $\rho(\tau)$ is the normalized autocorrelation coefficient given by

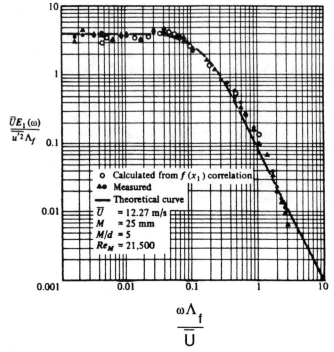

Figure 10.23 Spectral distribution $E_1(n)$ of the longitudinal turbulence velocity component. (Taken from Hinze [16])

$$\rho(\tau) = \frac{\overline{u'(t)u'(t - \tau)}}{\overline{u'^2}} \tag{15}$$

so that $\rho(\tau) \leq 1 = \rho(0)$. Evaluation of a spectrum can be thought of as a filtering process for which the entire frequency domain of the signal is partitioned into bands of frequency width $d\omega$. Imagine passing a signal $u(t)$ through a filter that admits frequencies only within the frequency band $d\omega$, and recording the mean-square amplitude (power) of that filtered signal $(\overline{u'^2})_{\omega, d\omega}$. Then repeat for each increment of $d\omega$ over all frequencies observable in the signal. The power spectral density curve could then be computed as that power of the signal, centered on ω and filtered over $d\omega$, divided by the filter width $d\omega$, $E_1(\omega) = \overline{u'^2}_{\omega, d\omega}/d\omega$ versus ω. Such a one-dimensional power spectrum is shown in Fig. 10.23. It displays the partitioning of power, or velocity fluctuation, according to frequency. The power spectrum that this figure presents is nondimensionalized with the local mean velocity $\overline{U}$, the turbulence intensity $\overline{u'^2}$, and a length scale of the flow Λ_f.

The frequency of a particular turbulent fluctuation will depend on the velocity at which the eddy that is responsible for the fluctuation is convected past the probe, and on the eddy size. For convection at a local velocity U, an eddy size l (same as l in Fig. 10.1) can be computed from the measured frequency as $l = 2\pi u/\omega$, or a wavenumber k_1, which varies as the inverse of the eddy size, can be defined as

$$k_1 = \frac{\omega}{u} \qquad 1 = \frac{2\pi}{k_1}$$

The subscript 1 for k_1 indicates a wavenumber based upon the main flow direction. Relating the eddy size l to the frequency and convection velocity in this fashion employs Taylor's hypothesis, $\tau = l/u$. Often, the spectrum is expressed in terms of wavenumber instead of frequency:

$$s(k) = \frac{E(k)}{\overline{u'^2}l} = \frac{1}{2\pi l}\int_{-\infty}^{\infty} e^{-i\tau k u}\rho(\tau)\, d\tau \qquad (16)$$

The length scale l is some appropriate scale; it could be δ_{995} in a boundary layer flow. Note that Fig. 10.24, another spectral distribution, is plotted in "wavenumber space." The identifier $E_{22}(k_1)$ in the figure indicates that the spectrum is based upon velocity fluctuations $v\ (= u_2)$ and that the wavenumber is in the x direction (k_1). The spectrum then will show the partitioning of the root-mean-square velocity fluctuations according to frequency, wavenumber, or eddy size.

Spectra may be constructed from autocorrelations of the velocities u' (Fig. 10.23), v' (Fig. 10.24), or w', or cross-spectral densities may be constructed from cross correlations of velocities, i.e., $u'v'$,* $u'w'$, and $v'w'$. Spectra based upon the autocorrelation of velocity v' would indicate a partitioning of the energy of fluctuations of velocity normal to the wall in boundary layer flow (Fig. 10.24). Spectra based upon cross correlations of velocity components indicate the partitioning of turbulent transport (e.g., $u'v'$) according to eddy size.

The wavenumber, the separation distance associated with eddy size, could be assigned in the x direction k_1 (Figs. 10.23 and 10.24), the y direction k_2, and the z direction k_3. Note that with three k values (k_1, k_2, k_3), three autocorrelations (u, v, w), and three cross correlations ($u'v'$, $v'w'$, $u'v'$), 18 separate one-dimensional spectra could be computed. Since the wavenumber has components in the x, y, and z directions, it may be considered a vector $\mathbf{k} = k_1 i + k_2 j + k_3 k$. A three-dimensional spectrum, $s(\mathbf{k})$, is the spectral density of eddies that have the same wavenumber magnitude $|\mathbf{k}| = \sqrt{k_1^2 + k_2^2 + k_3^2}$ regardless of direction. An example of a three-dimensional spectrum is shown in Fig. 10.25. The total turbulent kinetic energy $[0.5(\overline{u'^2} + \overline{v'^2} + \overline{w'^2})]$ is partitioned according to wavenumber magnitude $|\mathbf{k}|$. More on power spectral density can be found in [1, 11, 16].

---

*For example, $\rho_{uv}(\tau) = \dfrac{\overline{u'\,(t_1)v'(t_1 - \tau)}}{\overline{u'v'}} = \lim\limits_{T\to\infty} \dfrac{1}{T}\int_0^T u'(t)v'(t - \tau)\, dt.$

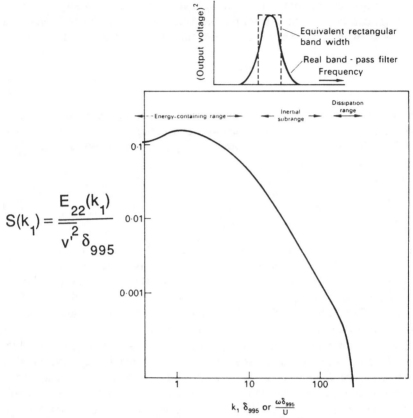

$$S(k_1) = \frac{E_{22}(k_1)}{\overline{v'}^2 \, \delta_{995}}$$

Figure 10.24 Typical $E_{22}(k_1)$ wavenumber (or frequency) spectrum in a boundary layer at about $y/\delta_{995} = 0.5$. (From [11])

10.3 TECHNIQUES OF SIGNAL PROCESSING

Processing of the signal is tailored to the intent of the measurement program and must be done with some knowledge of the nature of the signal, particularly the distribution of energy with frequency. There are three main objectives in signal processing: (1) to exclude unwanted components of the signal, as in filtering to remove low- or high-frequency components that are not related to the turbulent flow signal or, as in conditional sampling, to exclude signals unrelated to particular events; (2) to modify the signal, making it easier to interpret, as in linearization of the anemometer output; and (3) to perform operations on the signal that result in particular quantities, such as time averages, root-mean-square values, correlations, or power spectra. Several signal processing steps are discussed.

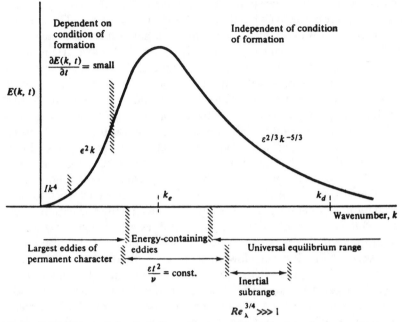

Figure 10.25 Form of the three-dimensional spectrum $E(k, t)$ in the various wavenumber ranges. (From [16])

10.3.1 Filtering

Signals contain noise. In a hot-wire signal this noise may be produced by resistors and electronics in the anemometer or by less than perfect control of the test [11]. A typical noise spectrum is shown in Fig. 10.26. At very low frequencies, noise is due to drift, mostly from temperature effects or transients in the flow delivery

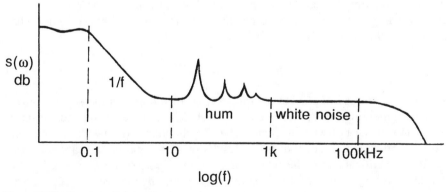

Figure 10.26 Spectrum for signal noise

system (such as fluctuations in the supply pressure due to reservoir cycling). These are minimized with proper instrumentation and test facility design. The spectral range from 0.1 to 10 Hz is dominated by electronic circuitry noise, the magnitude of which is inversely proportional to frequency. The range 10 Hz to 1 kHz is dominated by hum at the line frequency (60 Hz) and harmonics of the line frequency, by electrostatic or magnetic coupling with external fields such as solenoids or electric motors, or by ground loop hum (see [25] for a discussion of ground loop disturbances). The high-frequency end of the spectrum is white noise having a flat spectral density over the full bandwidth of the amplifier. This noise arises from thermal vibration of charged particles in solids and is often referred to as resistor or Johnson noise. Care is taken in the design of the anemometer to minimize these unwanted signals. Noise of frequencies higher than an expected turbulent flow signal frequency is filtered out with a low-pass filter (allows lower frequencies to pass without attenuation). This eliminates contributions of such noise to measurements of time-averaged fluctuation quantities, i.e., u'^2, but does not change spectra data in the range of interest. Typical values of low-pass filter cutoff frequencies range from 5 to 20 kHz. "Noise" at very low frequencies is filtered by the choice of sampling time or by high-pass filtering (perhaps dc coupling). To elaborate on sampling time filtering, if the signal were sampled over, for instance, τ seconds, frequencies in the signal of a longer period would appear as "drift" of the mean rather than cyclic behavior. In so doing, a decision must be made about which frequencies are important and are to be processed and which frequencies represent unimportant and perhaps uncontrolled low-frequency unsteadiness that should be eliminated. This choice must be tied to the frequencies associated with the events under study. Noise from frequencies less than the high-frequency cutoff and more than the low-frequency cutoff cannot be separated from the signal. Care must be taken to keep the signal-to-noise ratio large at these frequencies.

10.3.2 Linearization

The voltage required to maintain a constant temperature of a hot-wire anemometer varies primarily with the velocity perpendicular to the wire. Analysis of small cylinders in cross flow [26] shows the relationship

$$\frac{(V^2/R)}{(T_w - T_a)} = A + BU^n \tag{17}$$

where V is the voltage, applied to the wire of resistance R and temperature T_w, in an ambient temperature T_a and flow velocity U; A, B, and n are considered constant. Calibration of the hot wire in a known and well-controlled flow shows that this relationship is approximately correct. Results of such a calibration are shown in Fig. 10.27. Unfortunately, the relationship is nonlinear; therefore averaging the fluctuation voltage signal and then converting to an effective velocity could lead to a small error, the magnitude of which would be proportional to the degree of nonlinearity of the calibration curve over the velocity fluctuation range and the

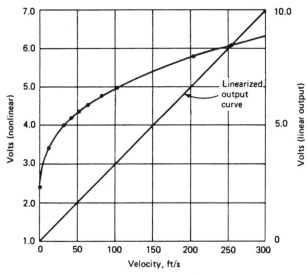

Calibration for a 0.002-in (0.051-mm) diameter hot-film sensor in atmospheric air; 0 — 300 ft/s (0 — 91m/s)

Figure 10.27 Calibration for a 0.002-in. (0.051-mm) diameter hot-film sensor in atmospheric air; 0–300 ft/s (0–91 m/s). Similar to a hot-wire calibration. (From [27])

magnitude of the signal fluctuation. Thus one may need to convert the signal into one in which the voltage is proportional to the velocity (linearize the signal) before averaging. One would also like to linearize because a voltage that has a linear relationship with velocity would be a more convenient quantity to process, particularly if the processing is by analog circuitry. Linearization can be performed by either digital or analog means.

10.3.2.1 Digital linearization. If the anemometer signal is digitized, a straightforward method of linearization can be implemented. In the computer, a look-up table or an equation, of whatever form fits the calibration data best (including different equations over different segments of the operation domain), can be used to calculate the velocity from the anemometer output voltage. It has been found, for standard hot wires, that an equation in the form of $U^{0.45} = A' + B'V^2$ usually fits the entire range of velocities from 1 to 30 m/s. Note that distortion of the turbulence signals can appear when the curve is segmented. This harmonic distortion of the signal can create a high-frequency apparent turbulence signal that could erroneously affect the spectra. This is similar to the many high-wavenumber terms that appear in a Fourier series expansion of a curve that has a knee (discontinuity in slope). Although digital linearization is more convenient, it may be slower than its analog counterpart and can slow the data processing if many data points are to be taken, i.e., when computing spectra. If turbulence intensities are small (e.g., $\sqrt{\overline{u'^2}}/u_\infty < 10\%$), processing time can be shortened by assuming a locally linear

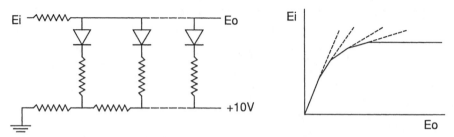

Figure 10.28   Biased-diode linearizer circuit and characteristic curve

relationship between voltage and velocity. If so, linearization could be made on the processed quantities rather than on each reading, e.g., $\sqrt{\overline{u'^2}} = (du/dV)\sqrt{\overline{V'^2}}$, where du/dV is the slope of the calibration curve centered about the mean velocity $\bar{u}$. Care should be taken in doing this, however, to assure that the technique is not employed inadvertently at excessive turbulence levels or when a decreased velocity moves the segment into a severely nonlinear portion of the calibration curve.

10.3.2.2 Analog linearization. An early analog linearizer used a triode operating close to its cutoff, where its nonlinearity matched fairly closely the nonlinear behavior of the hot-wire voltage/velocity relationship. These linearizers gave 1–2% inaccuracy but were susceptible to drift and were difficult to adapt to the different characteristics of different wires. Biased-diode linearizers later became more popular than the triode primarily because of their ease of adjustment. In one form, the input voltage is applied through a series of resistors to a number of reverse biased diodes in parallel. Each diode begins to conduct when the applied voltage rises above its bias voltage, giving an output voltage characterized by a series of segments (Fig. 10.28). With this scheme, any monotonic curve can be matched by a suitable choice of resistor values. The number of segments is usually about 10. Although the output voltage is nearly linear with velocity, there are discontinuities

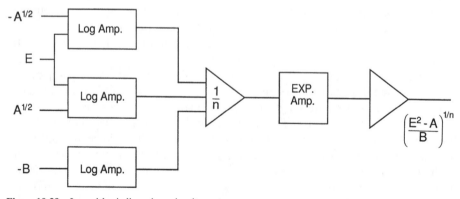

Figure 10.29   Logarithmic linearizer circuit

in slope. If segment fits are used, there may be false high-frequency fluctuations due to discontinuities in slope that would distort portions of the spectra, as discussed above. For this reason, biased-diode linearizers are not often used today.

The biased-diode linearizer has been replaced by two different analog linearization schemes. The first scheme linearizes with a fourth-order polynomial fit. The values of the coefficients are calculated from a least squares fit of the calibration curve. These values are normalized and entered as coefficient settings. The other analog linearization scheme attempts to match the law, $E^2 = A + BU^n$, through the use of logarithmic and exponential amplifiers. Three logarithmic amplifiers produce outputs whose sums are $\log[(E^2 - A)/B]$; these sums are amplified in a linear amplifier of adjustable gain $1/n$. The output passes through an exponential amplifier, giving $[(E^2 - A)/B]^{1/n} = V_{out}$ (Fig. 10.29). An advantage of this system is that it is specially tailored to hot wires.

10.3.3 Corrections

Corrections to hot-wire signals may be for effects of finite length, mismatch in oncoming flow direction, wire orientation, changes in ambient temperature between calibration and test, free convection, or nearness to the wall. It is usually presumed that wire length effects are eliminated in the probe design and/or corrected in calibration, that the probe is properly aligned in the flow, and that the hot wire is operated so that free convection effects are minimal. When the probe is used too near the wall, the eddy sizes become similar to the probe dimensions. Then error due to eddy averaging occurs, and the signal may become unreliable. Little can be done in the form of correction. Usually the near-wall data are discarded, a smaller probe is used, or the scale of the flow is increased. More details on probe size effects are given in the chapters of this text on hot-wire anemometry and LDA. The following discussion on corrections is limited to correcting the hot-wire data for changes in the ambient temperature. The effect of probe alignment during shear stress measurements is also reviewed.

Variations in the ambient temperature of the hot wire can be the source of the most serious errors in experimental measurements with hot wires. Most commercial probes use wires made of tungsten, which oxidizes rapidly at temperatures above 300°C. To prevent oxidation, many probes are coated with platinum and are operated below 250°C. With lower wire temperatures (smaller wire-to-ambient temperature difference), there is an increased sensitivity to ambient temperature. In air, a 1°C temperature change may produce an error of 2% in measured velocity. The wire is not only sensitive to wire-to-ambient temperature difference but also to temperature level by way of variable fluid property effects, particularly thermal conductivity and viscosity.

If the temperature varies slowly with time, as in the gradual heating of the room, there are several correction techniques available. The first and most obvious is to install a heat exchanger in the test loop, so that the ambient temperature of the test is controlled to always equal the calibration temperature. The second option is to use a temperature-compensated probe that has a temperature-sensitive element, which is a compensating arm of a constant-temperature anemometer (Wheatstone) bridge. Commercial probes presently have this feature, but the element has a sufficiently long time constant that it cannot follow high-frequency turbulent temperature fluctuations, as will occur in nonisothermal turbulent flows, but would follow slow drifts in temperature. A third scheme for compensation when the drift is small is to calibrate at one fluid temperature and use theoretical or empirical data to deduce calibrations at other temperatures. If analog linearizers are used, the setting must be updated with the corrected calibration. If digital linearization is used, the correction could be calculated and stored each time a new ambient temperature is measured.

The sensitivity of the hot-wire voltage to changes in the ambient air temperature can be approximated as [28]:

$$\left(\frac{V^2_{\text{actual}}}{V^2_{\text{calibration}}}\right)_{\text{same velocity}} = \left(\frac{(T_{\text{w}} - T_\infty)_{\text{actual}}}{(T_{\text{w}} - T_\infty)_{\text{calibration}}}\right) \times \left(\frac{U^m_\infty - AT^n_{\infty,\text{actual}}}{U^m_\infty - A\,T^n_{\infty,\text{calibration}}}\right) \quad (18)$$

where A and B are found in calibration, i.e.,

$$U^m_\infty = AT^n_\infty + \frac{B}{(T_{\text{w}} - T_\infty)}V^2_{\text{anemometer}}$$

The property variation exponent n was 0.76 (T has the units of K), and m, the exponent on velocity in the hot-wire calibration, was 0.435, in [28].

If the temperature is fluctuating rapidly, as in a heated turbulent boundary layer, more sophisticated correction schemes are needed. One such scheme is to make an independent temperature measurement with a fast response temperature probe, e.g., a resistance thermometer—a low-current, constant-current anemometer wire, which is positioned in the flow very near the hot wire. If the current is low, the sensor resistance is affected by ambient temperature only. Unfortunately, the response of the sensor can be slow. This is partially compensated by (1) making the sensor of very fine (and fragile) wire and (2) electronically compensating the sensor based upon the measured voltage, measured rate of change of voltage, and flow velocity as measured by a nearby sensor [29]. If analog linearization is used, a circuit must be devised to continuously compensate the anemometer signal for the sensed change in ambient temperature. Examples of such circuits are shown in Fig. 10.30. If the anemometer signal and temperature signal are digitized, the correction could be incorporated into the data reduction program at the expense of considerable computation time. More detail is available in [31, 32]. A second scheme for high-frequency compensation is to operate two heated sensors at different temperatures. The sensors are located very near each other and have the same orientation. They thus are presumed to be sensing the same velocity and the same ambient temperature. Since one is operated at a higher temperature (higher resistance) than the other, their voltages will differ according to the magnitude of the ambient temperature. With two parallel sensors at two different temperatures

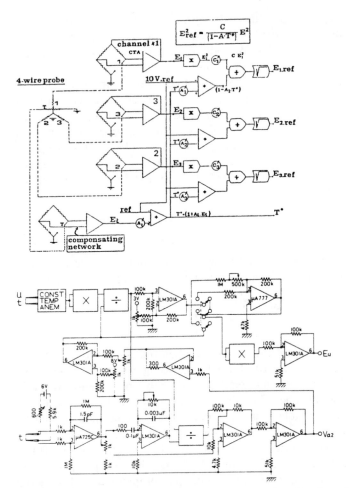

Figure 10.30 Analog circuits for temperature compensation: (*a*) schematic of a system [30] and (*b*) compensating circuit. (From [29])

[28, 33], there is enough information to determine the velocity and temperature separately. A two-wire probe could be used for *u* and *t* measurements; three wires for *u, v,* and *t* measurements; and four wires for *u, v, w,* and *t* measurements. Processing of the signals to deduce velocity and temperature could be by either analog or digital means. The advantage of this multi-hot-wire scheme is that the temperature frequency response would be good without the use of extremely fine wires. The disadvantage is that additional processing is required and the functional relationships seem to be sufficiently stiff that large errors can occur with very small uncontrolled variations [28]. Increased temperature sensitivity would be gained by

increasing the difference in temperature between the two hot wires. This is constrained by the oxidation temperature on the high side and the frequency response of the lower-temperature wire on the low side. A reasonable choice for the two overheat ratios, given these constraints, is 1.3 and 1.9 [33].

An analysis was made to determine the effect of wire-wire nonperpendicularity on data taken with X-wire probes. It was found that 2° nonperpendicularity results in an error of about 2% in measurements of u'^2 and v'^2, and no significant error in $\overline{u'v'}$. Another analysis was made to determine the effect of probe misalignment with the flow on measurements using X-wire probes. A misalignment in the plane of the wires of 2° would result in an error in $\overline{u'v'}$ measurements of ~10% when the local turbulence intensity $\sqrt{\overline{u'^2}}/U$ is about 20%. This error decreases with decreasing local turbulence intensity.

10.3.4 Spectral Analysis

Spectral data may be processed by introducing the $u(t)$ signal to a band-pass filter that, ideally, passes only frequency components within the range

$$\omega_0 - \frac{\Delta\omega}{2} < \omega < \omega_0 + \frac{\Delta\omega}{2}$$

where ω_0 is the nominal setting of the band-pass filter and $\Delta\omega$ is the bandwidth. This is shown schematically in Fig. 10.24. Such processing can be done electronically in a device called a spectral analyzer. Spectral analyzers come in two basic varieties: real time and swept tuned. The simplest to envision, but most clumsy, of the real-time schemes employs a large array of narrow-range band-pass filters that, together, cover the full range of frequencies simultaneously. Another real-time analysis is based on digital Fourier analysis, commonly using the Cooley-Tukey fast Fourier transform (FFT) [34, 35]. With this scheme, the analog signal is first digitized. Then a special-purpose computer program computes the FFT of the signal. Since this method looks at all frequencies simultaneously, it has excellent resolution. It is particularly good for low-frequency signals approaching dc, where analog spectrum analyzers tend to have difficulty. The drawback is that this scheme requires a digitizer, a computer, and adequate computation time to make the transform. Presently, digital spectrum analyzer units that are dedicated to this purpose can be purchased. They are called "spectrum analyzers" or "waveform analyzers" and are marketed primarily for vibration analyses. Because they are designed for this purpose, they can be much faster than the typical laboratory computer.

The second main category of analog spectral analyzers uses sweep tuning. These analyzers consist of a superheterodyne receiver and local oscillator that can be swept using an internally generated ramp waveform (Fig. 10.31). As the oscillator is swept through its range of frequencies, different input frequencies are successively mixed to pass through the amplifier and through a filter that is tuned to a

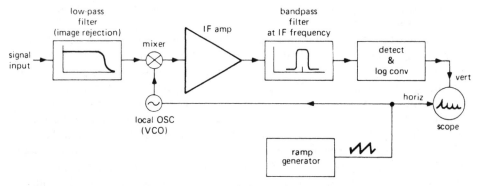

Figure 10.31 Swept-lo spectrum analyzer. (From [25])

particular frequency, ω_0. When the local oscillator is at $\omega_0 + \omega_1$, input signals at frequency ω_1 pass through the band-pass filter. Signals at $2\omega_0 + \omega_1$ would also pass, which is why the input signal has a low-pass filter with a cutoff frequency below ω_0. Input frequencies ω_0 less than the local oscillator frequency are processed. The complete spectrum is found by sweeping in time. The sweep can be obtained almost instantaneously, but often time averages of about 30 s are necessary to achieve spectra that are stationary in time, indicating that all frequencies of the stationary signal have been properly sampled. When scanning with a narrow bandwidth, the sweep rate must be slow. This can be a serious disadvantage if spectral analysis is to be made on transient events. Presently, digital processing, either in the digital spectrum analyzer or in the laboratory computer, is far more common.

10.3.5 Digital Signal Processing

The various quantities discussed in section 10.2 can be processed by either digital or analog means. With digital processing, the voltage signal from the hot-wire anemometer is first converted into digital form compatible with the computer. The analog-to-digital (A/D) converter that makes this conversion is discussed in section 10.4.1. The output of the converter is an integer that corresponds with one of the increments into which the range of the analog signal has been divided. This integer is expressed as a binary number. The number of possible combinations of digits is 2^n, where n is the number of digits (bits of binary switches) available for each reading (word). Many A/D converters have 8-bit words; therefore the voltage range is divided into 2^8 (= 256) increments. Eight-bit converters can sample as fast as 10^7–10^8 samples/s, although at this rate some of the less significant of the 8 bits may not be accurate (as will be discussed in section 10.4.1). Presently 10-, 12-, and 16-bit converters are also common. These allow for partitioning of the analog signal into 1024, 4096, and 65,536 increments, respectively. There is a trade-off between resolution and sample rate, however. The resolution decreases as the sampling rate increases.

One must be aware that at fast sampling rates, the inaccuracy may be degraded from the specified number of bits. One manufacturer offers a "12-bit" A/D converter that is capable of sampling at 10^5 samples/s. Little warning is given that, at the maximum sampling rate, 12 bits of precision may not be achieved; there is not enough time for the less significant bits to settle. The manufacturer states an aperture uncertainty δT of ± 1 ns. From this, one can estimate a worst-case error of $\pi f \delta T$, where f is the sampling rate. In terms of computing, the total error band is 0.0628% of the range when sampling at 10^5 samples/s. Equating the error to half the bin size, $1/(2 \cdot 2^n)$, yields $n = 9.6$ effective bits. This worst-case estimation hints that there is a considerable degradation from the low-speed precision of 12 bits.

Once the anemometer signal has been digitized, the signal processing becomes a series of computer statements in the data reduction program. Linearization is by calculation, as discussed in section 10.3.2, and spectral analysis is by use of the FFT, discussed in section 10.3.4. Note that the rate at which data are taken for FFT spectral analysis must be sufficiently fast that the waveform of the highest frequency signal can be reproduced, consistent with the Nyquist criterion, which will be introduced in section 10.4.2. For a turbulent boundary layer flow with Reynolds number $\mathrm{Re}_x = 10^6$ and a free stream velocity of 15 m/s, these criteria may require 50–60 thousand readings/s for 30–60 s, or a total of about 4 million points. If each reading is a 16-bit word, this would require computer storage of 8 Mbytes per sample. For this reason, the spectrum is usually calculated using FFT for only a portion of the original spectral frequency range. The remainder is filtered; the high frequencies are filtered with a low-pass filter, and the low frequencies are filtered by the choice of sampling period. This partial-range sampling is repeated until the entire range has been computed. When generating the high-frequency portion of the spectrum, long sampling times are not necessary; when generating the low-frequency portion, high sampling rates are unnecessary. The overall spectrum is then constructed by combining these partial spectra.

Processing of several other turbulence quantities from digitized data is next discussed.

10.3.5.1 Mean velocity. The mean velocity is given as

$$U = \frac{1}{N} \sum_{i=1}^{N} u_i(t) \tag{19}$$

where N is sufficiently large that a representative sample is available; 3000–5000 readings would probably suffice. The number of points required for a sufficiently large sample could be evaluated by taking several samples over the same length of time (to separate the time effect from the sample size effect) but with increasing N. When U ceases to be dependent upon N, N is sufficiently large. A smaller number of readings per sample would give a less accurate measure of the influence of the large-deviation fluctuations in the signal. Therefore accurate measurement of higher moments would require large-N samples. As discussed in section 10.2.4.1, the time over which the N samples are taken must be sufficiently long to properly represent the statistics of the low-frequency components in the flow. This can be tested by increasing the length of time over which the N samples are taken (to separate the sample size effect) until the result is no longer dependent upon the sample period.

Note that, for a stationary random function, the time between readings need not be short enough to capture the waveform of the smallest eddies when statistical quantities are evaluated; this is necessary only when an FFT spectral analysis of the data is to be made or when doing any other processing that requires knowing the waveform. However, it is necessary when measuring mean or other statistical quantities of a stationary signal that each sample be taken more quickly than the characteristic time for the highest frequency, significant event in the signal. Otherwise, there will be an averaging, "slewing," of the signal over these high-frequency events, and the maxima of the signal will be missed. This becomes more important when higher-order moments (e.g., skewness and kurtosis) are taken. This is a frequently committed error. One could probably be assured that the sampling was fast enough if the A/D converter had an up-front sample-hold device so that the conversion was done on a "frozen" sample rather than on a changing sample.

10.3.5.2 Ensemble-averaged velocity. The ensemble-averaged velocity is given as

$$<u(\theta)> = \frac{1}{N} \sum_{i=1}^{N} u_i(\theta) \tag{20}$$

where θ is a particular time after the event initiation position within the cycle, such as the flywheel position beyond top dead center of an engine. In the case of a cyclic event, i would represent the cycle number and N would be the number of cycles used for evaluating the average. As with the time average, N should be sufficiently large that the sample is representative of the mean. Tests with increasing N could be conducted to discover the minimum acceptable N, the value at which the average becomes independent of N.

10.3.5.3 Probability density and turbulence intensity. The probability density function is easily calculated from the digitized flow data using Eq. (7). This is a difficult term to obtain by analog techniques. Once the probability density function is found, the various moments of the probability density function, the variance (turbulence intensity), skewness, and kurtosis can be calculated. These moments are usually found separately with specially tailored circuitry when processing is done with analog components (discussed below).

The turbulence intensity is computed directly from the data more frequently than it is computed as the moment of the probability density function. Direct computation would require that Eq. (8) be written in a discretized form:

$$s \cong \sqrt{\frac{1}{N-1}\left[\sum_N (u_i - U)^2\right]} \qquad (21)$$

where u_i is the sample velocity and U is the mean velocity over the N samples. An equivalent and more convenient form of Eq. (21) is

$$s \cong \sqrt{\frac{1}{N-1}\left[\sum_N (u_i)^2 - \frac{1}{N}(\sum_N u_i)^2\right]} \qquad (22)$$

With Eq. (22), updated values of the series $\sum(u_i)^2$ and $\sum u_i$ are found as each sample is read and processed; thus only two numbers remain in storage when sampling. Direct application of Eq. (21) would require storing all u_i so that U could be calculated after all the samples were taken. Then the summation would be computed by retrieving each u_i from storage and processing.

10.3.5.4 Joint probability density function and Reynolds shear stress. The joint probability density function requires simultaneously (or nearly simultaneously— faster than the characteristic time of the most rapid, significant event in the signal) digitizing the $u(t)$ and $v(t)$ signals, then computing $B(u,v)$ using Eq. (10). The first moment is the Reynolds shear stress given by Eq. (11).

The computation is straightforward; the difficulty is that now two signals must be digitized. A digitizer could be used on each signal for this purpose, and simultaneous readings could be taken. Because the computer interfaces with two devices, it must continually switch back and forth. This switching can slow the reading rate. If the A/D converters had buffer memory (Fig. 10.32), each channel could sample and store in unison; then the computer could collect data in "fast handshake" or direct memory access (DMA) mode after the data were taken, one device at a time.

If there are fewer A/D converters than channels to be sampled, a multiplexer must be used (Fig. 10.33). The multiplexer allows the A/D converter to sample one line at a time. This slows the reading rate and cannot be used when simultaneous readings must be taken (unless each line has a front-end, sample-hold circuit upstream of the multiplexer, such that the readings are simultaneous and the A/D conversion is multiplexed). When there is no front-end sample-hold, the maximum time allowed between samples that can be claimed to be "essentially simultaneous" depends on the cross correlation between the two variables:

$$\rho_{u,v}(\tau) = \left[\lim_{T\to\infty} \frac{1}{T}\int_0^T u'(t)v'(t-\tau)\,dt\right](\overline{u'v'})^{-1} \qquad (23)$$

The time between samples must be less than the separation time τ at which the cross correlation begins to drop perceptibly below 1.0.

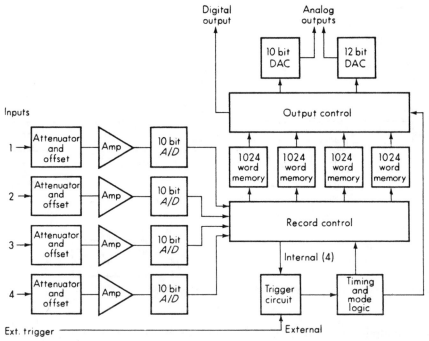

Figure 10.32 Digital-memory waveform recorder. (From [27])

10.3.5.5 Autocorrelation. In computing the autocorrelation by digital means, one must sample an integral number of readings M per time interval τ. The autocorrelation for τ is then

$$<u(t)u(t - \tau)> = \lim_{N\to\infty} \frac{1}{N} \sum_{i=1+M}^{N+M} (u_i u_{i-M}) \qquad (24)$$

One must store only $M + 1$ readings plus the value of the summation.

10.3.6 Analog Signal Processing

Processing the various quantities discussed in section 10.2 by analog means involves using the various components one would find in analog computers. These components are constructed using operational amplifiers with feedback [37]: amplification, summation, multiplication, division, and integration. An example of a circuit for finding the mean-square velocity $\overline{u^2(t)}$ is shown in Fig. 10.34. Here the high-frequency "noise" is filtered, leaving only the turbulent energy-containing frequencies; then a multiplier is used to give $[u(t)]^2$, and an integrator gives $\int_0^T [u(t)]^2\, dt$. A timer is needed to set the predetermined integration time, or alternatively, a filter is used to record only the dc component of the signal. The same circuit, but with-

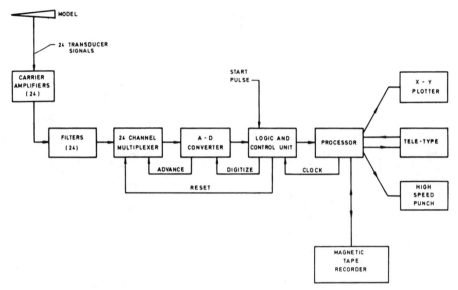

Figure 10.33 Typical unsteady flow measurement chain. (From [36])

out the multiplier, could be used to compute the integral of the signal $\int_0^T [u(t)]\, dt$ and the mean velocity $(1/T)\int_0^T [u(t)]\, dt$. The variance can be calculated from the two integrated values using

$$u(t)^2 = [U + u'(t)]^2 \tag{25}$$

where $u'(t)$ is the fluctuating component of the signal about the mean velocity U. Squaring and time averaging give

$$u(t)^2 = U^2 + 2u'(t)U + u'^2 \tag{26}$$

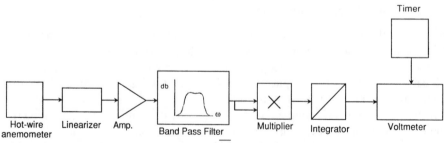

Figure 10.34 Analog processor for evaluation of $\overline{u'^2}$. (Redrawn from [24])

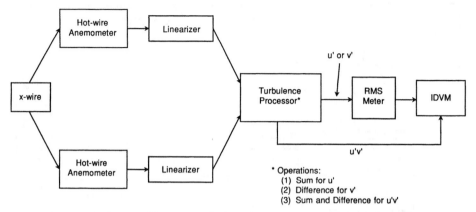

Figure 10.35   Block diagram of instruments for hot-wire measurement. (Redrawn from [4])

Note that the second term on the right goes to zero when time averaged, by the definition of $u'(t)$. Thus

$$u'^2 = s^2 = u(t)^2 - U^2 \qquad (27)$$

Other analog processing schemes are now discussed. In Fig. 10.35 is shown a "turbulent processor," a device that contains analog circuitry for filtering out the dc components, then computing the sum, difference, and sum times difference for two signals. The two signals are produced by two anemometer bridges driving wires that cross at 90°. Each wire is inclined 45° to the flow. One wire is sensitive to the component of the velocity vector $(u + v)/2$, and the other is sensitive to $(u - v)/2$. Adding the signals, which have had the mean flow (dc component) removed by high-pass filtering in the processor, gives u', subtracting gives v', and multiplication of the two gives $u'v'$. The time average of these signals is measured using the integrating digital voltmeter and a timer.

The scheme shown in Figs. 10.36 and 10.37 is for analog processing of signals from three anemometer bridges, each driving a wire on the three-wire probe sketched in Fig. 10.38. The signals are linearized (see Fig. 10.36), then introduced to a matrix transformation circuit, which gives the velocities in the wire coordinates x', y', and z' of Fig. 10.37. Another transformation circuit is then used to convert the velocities into laboratory coordinates u, v, and w, and the three velocities are processed to give the mean velocities U, V, W; the turbulence intensities u'^2, v'^2, w'^2; the cross correlations $u'v'$, $v'w'$, $u'w'$; and the turbulence kinetic energy $q^2 = u'^2 + v'^2 + w'^2$. The advantage of this analog system is that all this information is continuously available. The disadvantage is the tedious preparation, which involves design, construction, calibration, and setup of this analog circuit.

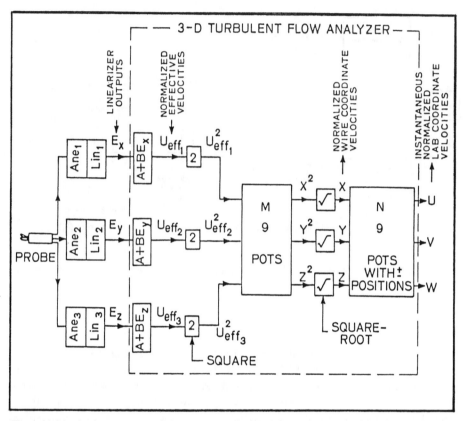

Figure 10.36 Analog processor of three-wire signals. (From [38])

10.3.7 Uncertainty

The measurement program is not complete until best estimates of the errors are determined. No person is in a better position to assume this responsibility than the one performing the measurements. There is a growing awareness of the need for such reporting [39]; many journals, such as the *ASME Journal of Fluids Engineering* or the *ASME Journal of Heat Transfer,* require such documentation. In one method of uncertainty analysis [40], it is presumed that every attempt has been made to eliminate all recognized bias errors in the measurement by way of calibration and calculated corrections. It is then assumed that the only errors that remain are random errors, which are equally likely to be positive or negative; the measurement is "zero centered." The term "uncertainty" is used to quantify an estimate of reasonable bounds on this random variation. Usually, the uncertainty is chosen to be the error band bounding 95% of all independent readings that would hypotheti-

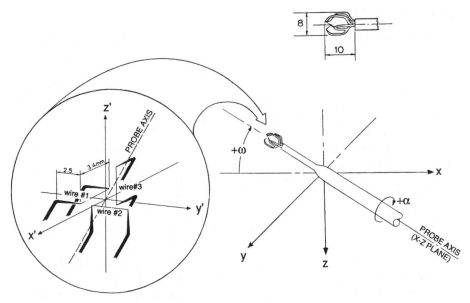

Figure 10.37 Triaxial hot-wire probe. (Redrawn from [30])

cally be taken. It is a statement saying that the evaluator believes that if the true value of the measured quantity could be found, it would be within plus or minus the uncertainty interval of the measured quantity 95% of the time (or with 95% confidence, or 20:1 odds). Methods for computing the uncertainty are discussed in [40–44].

The uncertainty of fluid mechanics measurements varies with the test conditions, quality of instrumentation, and care with which the measurements are taken and the experiment is controlled. Typical reported values for careful experiments in low-speed flows are as follows:

±2–5% for mean velocity with a total pressure probe and wall static port

±5–7% for mean velocity or u'^2 measurements with a single, horizontal hot-wire probe

±8–10% for Reynolds shear stresses, u'^2, w'^2 or v'^2, away from a wall with a cross-wire or triple-wire probe

±10–15% for near-wall v'^2 with a cross-wire or triple-wire probe
±15–20% for third-order correlations with multiwire probes

A discussion of uncertainties of hot-wire measurements can be found in [44].

10.4 SAMPLING

In digital processing the first step is to convert the anemometer output into digital form compatible with the computer and interface bus. Digital processing systems generally have 8-, 12-, 16-, or 32-bit character lengths. Computers generally use 16- and 32-bit words ("word" reflects the natural width of the computer's CPU registers). In computer jargon, a 4-bit array is a "nibble" and an 8-bit array is a "byte." A bit is a binary integer, and a binary word is a sequence of bits representing a number. For instance, 101 is a 3-bit word representing the decimal number 5. The right-hand bit, which is called the least significant bit, is the units column, the next bit is the 2^1 column, and the next bit, the most significant bit in this case, is the 2^2 column. An 8-bit array can represent any decimal integer between 0 and 256; a 12-bit array, 0–4095; a 16-bit array, 0–65535; etc. An 8-bit array allows maximum resolution of 1 part in 256 (1/2%); a 12-bit, 0.025%; a 16-bit, 0.0015%; etc. Eight-bit resolution may require offsetting an unlinearized hot-wire signal before digitizing, whereas 12-bit resolution would allow digitizing the signal directly.

Bits are presented by two voltage levels corresponding to zero and 1. The actual voltage level depends on the type of circuitry, with 0 V and 5 V most common for 0 and 1 states, respectively. The different bits may be represented by these voltages in individual wires, one wire per bit (parallel data), or as a train of voltage pulses on a single wire (serial data).

10.4.1 Analog-to-Digital Conversion

In the first of the two A/D schemes to be discussed, the binary equivalent of the sampled voltage is found by comparing it against a reference voltage of which the digital equivalent is known. The comparison signal is made by summing analog voltages equivalent to each bit of a binary signal (the output of a D/A converter). The binary signal continues to increase in "least increments" until the nearest voltage to the sampled voltage is found (see Fig. 10.38). The bit pattern of the digital signal, corresponding to the analog voltage, is then read into the output buffer of the A/D converter to be read by the computer. This requires some time, during which the signal may change slightly, thus leading to error in setting the less significant bits. Such error can be eliminated by putting a sample/hold device upstream of the converter. Upon receipt of the trigger signal, the output of the sample/hold device is "frozen" at the value that the input of the sample/hold has at the time of trigger. Fig. 10.39 shows the analog signal, the sample/hold output, and the A/D output for a representative signal. The A/D conversion cycle is then repeated at equal trigger-pulse intervals until the desired number of samples has been input to the computer. Equally spaced data samples make the computation of spectra easier; otherwise, individual sample intervals must also be recorded, since this time information would be required for processing the spectra.

In the second A/D scheme, the analog signal is integrated for a fixed period,

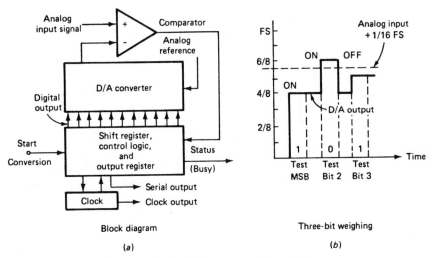

Figure 10.38   Successive-approximation A/D converter. (From [27])

then the signal input to the integrator is removed, and a negative reference voltage input is connected. The integrator output then begins to decrease with a constant slope, since its input voltage is constant (Fig. 10.40). A clock/counter circuit times the downslope portion of the cycle in binary code. When the integrator output signal passes through zero, a comparator trips and the counter is stopped. The binary number in the counter, then, is representative of the magnitude of the input analog signal to the A/D converter.

If more than one signal is required (e.g., cross-wire probes), the A/D converter can be switched to each channel in turn (multiplex the signals) at its fastest possible rate, often 50 kHz, on receipt of the trigger pulse. With multiplexed A/D conversion, there is a small shift in the time between readings. If this shift is not acceptable (correlations of high-frequency signals are to be made), then separate but synchronized sample-and-hold amplifiers [27] should be used, with the A/D multiplexed to them instead of multiplexing the single sample/hold. Alternatively, each channel may have its own A/D converter. Such A/D converters for 8-, 12-, and 16-bit data are available with and without multiplexers from various manufacturers. The cost rises steeply with the number of bits and less steeply with sampling speed (to 100 MHz). Over the last two decades, strong competition has driven the cost of A/D converters down, as many new brands have appeared on the market. Significant recent improvements, in addition to the decreasing cost, have been in the size of buffer memory available in A/D converter systems and the increase in maximum sampling rates. Also, cards have become available as plug-in modules to personal computers that allow fast sampling and storing—essentially allowing the PC to become a digital oscilloscope. Multichannel A/D converters or waveform recorders

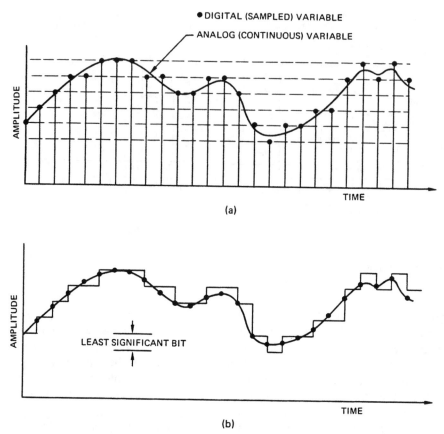

Figure 10.39 Digital variables showing errors: (*a*) digitized variable and (*b*) output of sample-and-hold amplifier. (From [10])

(with a display would be called digital oscilloscopes) are presently available with 1 kbyte memory per channel, 4 kbytes per channel, and more.

10.4.2 Sampling Rate

The required sampling rate depends on the highest frequency that is to be reproduced in the subsequent computer analysis. If mean or averaged values (any of the quantities discussed in section 10.3 except those processed from the waveform—such as spectra) are to be computed and the signal can be considered stationary and random, any sampling rate will do. The only constraints are that a sufficiently large sample size be taken, so that it may be considered statistically representative, that the time over which the samples are taken be long enough that statistical information from the lowest frequencies is captured, and that no averaging of the

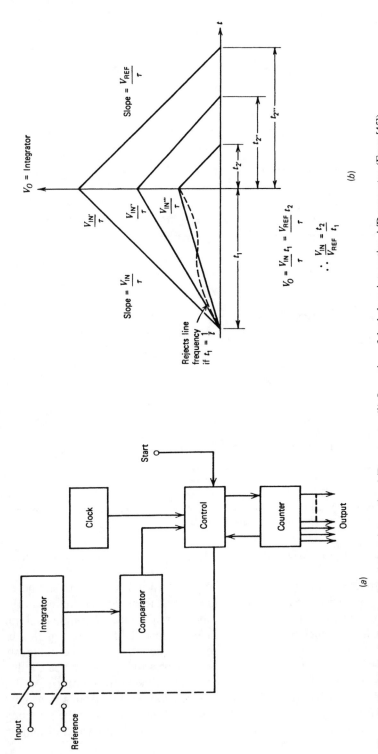

Figure 10.40 (a) Block diagram of dual-slope integrating A/D converter. (b) Operation of dual-slope integrating A/D converter. (From [45])

693

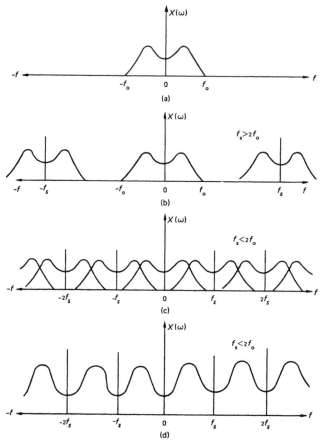

Figure 10.41 Spectra of (a) analog (continuous) data, (b) properly sampled digital data, (c) components of aliased digital data; and (d) aliased digital data. (From [10])

signal is done during the A/D conversion. These points were discussed in section 10.3.5.

If power spectra are to be computed from the waveform, the data should be sampled at at least twice the frequency of the highest frequency of interest in the signal (this is the Nyquist criterion). To be safe, it is advisable to sample at about 3 times the highest frequency to avoid "aliasing," since the hot-wire signal does not have a sharp cutoff at some particular high frequency, but falls continuously in power level with increasing frequency. Also, low-pass filters do not create a distinct and complete dropoff at the chosen "nominal" value of the cutoff frequency. In the sampled signal it is impossible to separate the contribution from a sinusoidal component at frequency f from one at frequency $f \pm nf_s$, where f_s is the sampling frequency. This phenomenon, shown in Fig. 10.41c, is known as frequency folding

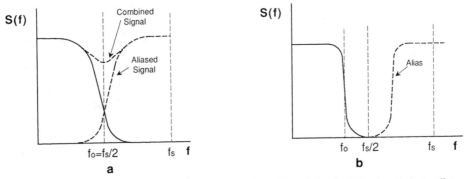

Figure 10.42 (*a*) Spectrum with alias error; (*b*) spectrum with anti-aliasing filter to eliminate alias error

and leads to aliasing error. If the cutoff frequency f_0 (the frequency where, due either to the nature of the signal or to low-pass filtering, the signal strength is essentially zero) is less than half the sampling frequency, $f_s/2$, the apparent spectra of the sampled signal do not interfere and the waveform for $0 < f < f_0$ is not contaminated (Fig. 10.41*b*). However, if the waveform attenuates more slowly and $f_0 > f_s/2$, the extraneous spectra due to discrete sampling contaminate the waveform (Fig. 10.41*c*) and alias error is visible at high frequency. Figures 10.41*c* and 10.41*d* show rather extreme examples. Certainly errors of this magnitude would be perceptible. The more typical alias error is shown in Fig. 10.42, where folding frequency contamination is seen only in the very highest frequency portion of the spectrum. Alias error can be eliminated by passing the signal through a low-pass filter with a cutoff frequency higher than the highest frequency of interest in the original signal but lower than half the sampling frequency. A check for alias error would be to verify that, in the spectrum, the power decreases as the cutoff frequency is approached. If the power were to level off with increasing frequency in the vicinity of the cutoff frequency, there probably is a significant alias error component. Greater detail on sampling techniques can be found in [10, 45–48].

10.5 DIGITAL COMPUTER SYSTEMS

Relatively inexpensive, small, but rather powerful computers are available for control of experiments and collection and processing of data. The computer system configuration that one would choose depends on special needs, budgets, personal preferences, and other factors. Since there are many variables to this selection, it is not possible to present the "optimum" system—there is no such optimum. In the following section, elements of a computer system are listed. The internal architecture and logic of the computer are not discussed herein, for the manufacturer has taken care of these details and has spared the user from doing so. Users need only worry about the details of special-purpose interfacing and user programming (which is usually in a high-level language like BASIC, C, or FORTRAN). Since these details vary from system to system, documentation is best found in the computer manuals.

A laboratory computer system may consist of a central processing unit (CPU), memory unit, mass storage devices, a terminal for user interface, a printer, a plotter, and analog and digital interfaces for communication with the experiment. All these devices communicate via the data bus.

The central processing unit is the heart of the machine, where computation takes place. The data are organized as 8-, 16-, or 32-bit words. All computers have some fast random-access memory (RAM). The size of this memory typically ranges from 250 kbytes to 100 Mbytes in small laboratory computers. Computers also have read-only memory (ROM) for special-purpose memory, typically the "bootstrap" program for getting the system started and operating system or library programs. Computer systems usually have some mass storage like disks or tapes. Typical laboratory computers would have mass storage ranging from a floppy disk at 250 kbytes to hard disk storage of 40–80 Mbytes (or larger). Mass storage is generally slow in relation to computer cycle times; tapes are the slowest and least expensive, with access times of tens to hundreds of milliseconds, and hard disks are fastest, with access times in fractions of milliseconds. Once data have been located, data transfer is rapid, typically 10,000–100,000 bytes/s or faster. Tape storage is rapidly being displaced by disk storage, which continues to decrease in price, precipitously. Peripherals such as an alphanumeric keyboard, line printer, and plotter allow interaction with the experimenter. Access to the experiment is via digital or analog (through a D/A converter) signals sent to switches or actuators and via digital or digitized analog signals received from the sensors. The data bus provides an interface between the CPU and the peripherals and between the CPU and memory. Since the components on the data bus may be from different manufacturers, it is important that the bus (and interfacing) be standardized. Standard buses are IEEE-488, RS-232, S100, Q-bus, and Multibus. Most scientific equipment is designed for communication on the IEEE-488 or RS-232 interface bus. The data bus contains lines for transferring bits of data and bits of logic information needed to coordinate the data transfer.

10.6 CONCLUSIONS

Many of the familiar quantities processed from turbulent flow data were discussed, and techniques used to effect this processing were presented. An important point to echo in conclusion is that an understanding of the physics of fluid mechanics is the foremost prerequisite of these measurements; no high-speed integrated-circuit gadgetry will compensate for this basic knowledge.

From the discussion, it is clear that choices must be made between processing by analog means or digitally. Each has assets and liabilities. The field of signal processing is a rapidly changing one due to the many new instruments entering the market. The rapid reduction in price and increase in speed of digital circuitry have resulted in a clear trend toward digital processing.

Continued activity in the field of fluid mechanics measurements will require a continuing effort to keep informed of the advances in instrumentation.

NOMENCLATURE

B	probability density function, Eq. (7)
C_f	skin friction coefficient, $C_f/2 = \tau_w/\rho U^2$
d	diameter
E	power dissipated by wire; power of the fluctuating signal
I	integral in Fig. 10.25; see ref. 16
k	wave number; roughness element size, turbulence kinetic energy
K	acceleration parameter, $= (\nu/U_\infty^2)dU_\infty/dx$
l	eddy size
n	distance normal to wall
n^+	distance normal to wall in inner coordinates, $= nU^*/\nu$
p	static pressure
q^2	turbulence intensity, $= (\overline{u'^2} + \overline{v'^2} + \overline{w'^2})/3$
R	resistance; radius of curvature
Re_k	Reynolds number based on roughness element height, $= U_\infty k/\nu$
$\mathrm{Re}_m, \mathrm{Re}_\theta, \mathrm{Re}_M$	momentum thickness Reynolds number, $\mathrm{Re}_\theta = U_\infty\theta/\nu$
s	standard deviation; streamwise distance; power spectral density, Eq. (14)
t	time
T	period; temperature
u	streamwise velocity
$U, U_1, \overline{U}$	streamwise mean velocity
U^*, U_τ, U^*	shear velocity, $U^* = U_\tau = U^* \sqrt{\tau_w/\rho}$
$\overline{u'v'}, \overline{u'w'}, \overline{v'w'}$	Reynolds turbulent shear stress components
$\overline{u'^2}, \overline{v'^2}, \overline{w'^2}$	turbulent kinetic energy components
U^+, V^+	mean velocity in inner coordinates, $U^+ = U/U_\tau$
v, V	velocity in y direction
V	voltage
w, W	velocity in z direction
x	streamwise distance
Y, y, X_2	distance normal to the wall
y^+	distance normal to the wall in inner coordinates, $= yU_\tau/\nu$
z	cross-stream direction

δ, δ_{995}	boundary layer thickness
ε	turbulence energy dissipation
θ, M	momentum thickness
κ	von Karman's constant
λ	dissipation scale
Λ_f	length scale for turbulence in the flow
ν	kinematic viscosity
ρ	density
$\rho(\tau)$	autocorrelation coefficient, Eqs. (15) and (16)
τ	time delay
τ_w	wall shear stress
ω	frequency

Subscripts

a	ambient
w	wire; wall
$0, \infty$	far beyond the boundary layer
1, 2, 3	index notation

Superscripts

‾	time average
'	fluctuation about the mean, $u' = u - U$, or rms value, according to context

REFERENCES

1. H. Tennekes and J. L. Lumley, *A First Course in Turbulence,* MIT Press, Cambridge, Mass., 1972.
2. M. Hirata and N. Kasagi, Studies of Large-Eddy Structures in Turbulent Shear Flow with the Aid of Flow Visualization Techniques, J. P. Harnett, T. F. Irvine, Jr., E. Pfender, and E. M. Sparrow, eds., *Studies in Heat Transfer: A Festschrift for E.R.G. Eckert,* Hemisphere, New York, pp. 145–164, 1979.
3. M. Clauser and F. Clauser, The Effect of Curvature on the Transition from Laminar to Turbulent Boundary Layer, NACA TN-613, 1937.
4. J. Kim, S. J. Kline, and J. P. Johnston, Investigation of Separation and Reattachment of a Turbulent Shear Layer: Flow Over a Backward-Facing Step, Mechanical Engineering Department Report MD-37, Stanford University, Stanford, Calif., 1978.
5. H. Ueda and J. O. Hinze, Fine-Structure Turbulence in the Wall Region of a Turbulent Boundary Layer, *J. Fluid Mech.,* vol. 67, no. 1, pp. 125–143, 1975.
6. V. A. Sandborn, *Resistance Temperature Transducers,* Metrology Press, Fort Collins, Colo., 1972.
7. M. Hirata, H. Tanaka, H. Kawamura, and N. Kasagi, Heat Transfer in Turbulent Flows, *Heat Transfer 1982,* vol. I, pp. 31–57, Hemisphere, New York, 1982.
8. B. J. Cantwell, Organized Motion in Turbulent Flow, *Annu. Rev. Fluid Mech.,* vol. 13, pp. 457–515, 1981.
9. J. M. Ottino, Mixing, Chaotic Advection, and Turbulence, *Annu. Rev. Fluid Mech.,* vol. 22, pp. 207–253, 1990.
10. R. E. Uhrig, *Random Noise Technique in Nuclear Reactor Systems,* Ronald, New York, 1970.

11. P. Bradshaw, *An Introduction to Turbulence and Its Measurement*, Pergamon, New York, 1975.
12. G. T. Friedman, Experimental Investigation of Fluid Mechanics in Oscillating Flow, Master's thesis, Mechanical Engineering Department, University of Minnesota at Minneapolis St. Paul, 1991.
13. P. M. Ligrani, W. M. Kays, and R. J. Moffat, The Thermal and Hydrodynamic Behavior of Thick, Rough-Wall, Turbulent Boundary Layers, Mechanical Engineering Department Report HMT-29, Stanford University, Stanford, Calif., 1979.
14. W. M. Kays and R. J. Moffat, The Behavior of Transpired Boundary Layers, Mechanical Engineering Department Report HMT-20, Stanford University, Stanford, Calif., 1975.
15. J. C. Gillis, J. P. Johnston, W. M. Kays, and R. J. Moffat, Turbulent Boundary on Convex, Curved Surface, Mechanical Engineering Department Report HMT-31, Stanford University, 1980.
16. J. O. Hinze, *Turbulence,* McGraw-Hill, New York, 1975.
17. J. C. Gillis and J. P. Johnston, Turbulent Boundary Layer Flow and Structure on a Convex Wall and Its Redevelopment on a Flat Wall, *J. Fluid Mech.,* vol. 135, pp. 123–153, 1983.
18. F. H. Clauser, The Turbulent Boundary Layer, in *Advances in Applied Mechanics,* vol. IV, pp. 1–51, Academic, New York, 1956.
19. T. Cebeci and P. Bradshaw, *Physical and Computational Aspects of Convective Heat Transfer,* Springer-Verlag, New York, 1984.
20. R. M. C. So and G. L. Mellor, An Experimental Investigation of Turbulent Boundary Layers Along Curved Surfaces, NASA CR-1940, 1972.
21. K. C. Muck, Comparison of Various Schemes for the Generation of the Turbulent Intermittency Function, Imperial College Aero Report 80-03, 1980.
22. T. B. Hadley and J. F. Keffer, Turbulent/Non-Turbulent Decisions in an Intermittent Flow, *J. Fluid Mech.,* vol. 64, p. 675, 1974.
23. P. Bradshaw and J. Murlis, On the Measurement of Intermittency in Turbulent Flow, Imperial College Aero. Report 74-04, 1974.
24. J. A. B. Wills, Hot-Wire Anemometry, Mechanical Engineering Department Lecture Notes, Stanford University, Stanford, Calif., 1979.
25. P. Horowitz and W. Hill, *The Art of Electronics,* Cambridge University Press, New York, 1981.
26. L. V. King, On the Convection of Heat from Small Cylinders in a Stream of Fluid: Determination of the Convection Constants of Small Platinum Wires with Application to Hot-Wire Anemometry, *Philos. Trans. R. Soc.,* vol. A214, pp. 373–432, 1914.
27. E. O. Doebelin, *Measurement Systems, Application and Design,* McGraw-Hill, New York, 1983.
28. J. Kim, The Development of a Turbulent Heat Flux Probe and Its Use in a 2-D Boundary Layer over a Convex Surface, M.S. thesis, University of Minnesota at Minneapolis St. Paul, 1986.
29. M. Hishida and Y. Nagano, Simultaneous Measurements of Velocity and Temperature in Nonisothermal Flows, *J. Heat Transfer,* vol. 100, pp. 340–345, 1978.
30. M. N. Frota and R. J. Moffat, Triple Hot-Wire Technique for Measurements of Turbulence in Heated Flows, *Heat Transfer 1982,* vol. 4, pp. 491–496, Hemisphere, New York, 1982.
31. J. Kim, Free-Stream Turbulence and Concave Curvature Effects on Heated, Transitional Boundary Layers, Ph.D. thesis, University of Minnesota at Minneapolis St. Paul, 1990.
32. J. Kim, T. W. Simon, and M. Kestoras, Fluid Mechanics and Heat Transfer Measurements in Transitional Boundary Layers Conditionally Sampled on Intermittency, in *Heat Transfer in Convective Flows,* ASME HTD, vol. 107, pp. 69–81, 1989.
33. M. F. Blair and J. C. Bennett, Hot-Wire Measurements of Velocity and Temperature Fluctuations in a Heated Turbulent Boundary Layer, ASME paper presented at the 29th ASME Int. Gas Turbine Conf., June 1984.
34. G. D. Bergland, A Guided Tour of the Fast Fourier Transform, *IEEE Spectrum,* pp. 41–52, 1969.
35. W. H. Press, B. P. Flannery, S. A. Teukolshy, and W. T. Vetterling, *Numerical Recipes: The Art of Scientific Computing,* Cambridge University Press, New York, 1987.
36. B. E. Richards, *Measurement of Unsteady Fluid Dynamic Phenomena,* Hemisphere, Washington, D.C., 1977.

37. G. A. Korn and T. M. Korn, *Electronic Analog and Hybrid Computers,* McGraw-Hill, New York, 1964.

38. S. Yavuzkurt, M. E. Crawford, and R. J. Moffat, Real-Time Hot-Wire Measurements in Three-Dimensional Flows, *Proc. of the 5th Biennial Symposium on Turbulence,* Rolla, Missouri, 1977.

39. S. J. Kline, B. J. Cantwell, and G. W. Lilly (eds.), *The 1980–81 AFOSR-HTTM-Stanford Conference on Complex Turbulent Flows: Comparison of Computation and Experiment,* vol. 1, Mechanical Engineering Department, Stanford University, Stanford, Calif., 1981.

40. R. J. Moffat, Contributions to the Theory of Single-Sample Uncertainty Analysis, *J. Fluids Eng.,* vol. 104, pp. 250–260, 1982.

41. R. J. Moffat, Contributions to the Theory of Uncertainty Analysis for Single-Sample Experiments, *The 1980–81 AFOSR-HTTM-Stanford Conference on Complex Turbulent Flows: Comparison of Computation and Experiment,* vol. 1, pp. 40–56, Mechanical Engineering Department, Stanford University, Stanford, Calif., 1981.

42. S. J. Kline and F. A. McClintock, Describing Uncertainties in Single-Sample Experiments, *Mech. Eng.,* vol. 75, pp. 3–8, 1953.

43. H. Schenck, *Theories of Engineering Experimentation,* Hemisphere, Washington, D.C., 1979.

44. S. Yavuzkurt, A guide to Uncertainty Analysis of Hot-Wire Data, *J. Fluids Eng.,* vol. 106, no. 2, p. 181, 1984.

45. E. R. Hnatek, *A User's Handbook of D/A and A/D Converters,* Wiley and Sons, New York, 1966.

46. J. S. Bendat and A. G. Piersol, *Measurement and Analysis of Random Data,* Wiley and Sons, New York, 1966.

47. R. B. Blackman and J. W. Tukey, *The Measurement of Power Spectra,* Dover, New York, 1959.

48. R. H. Cerni and L. E. Foster, *Instrumentation for Engineering Measurement,* Wiley and Sons, New York, 1966.

49. H. Fiedler and M. R. Head, Intermittency Measurements in Turbulent Boundary Layer, *J. Fluid Mech.,* vol. 25, part 4, pp. 719–735, 1966.

INDEX

Aberration coma, 463
Acoustic flowmeters, 319–320
Acoustic radiation, 47–48
A/D converter (*see* Analog-to-digital converter)
Advance ratios, 394
Aerodynamic flow visualization, 369–418
Aerodynamic heating, 106–113
Aerosols, 211, 230, 372, 375
Airfoils
    canard design, 415, 417, 423, 427–431
    delta wings, 389, *390, 424–425*
    flow visualization, 135, 369–418
    forward-swept, 423, *427–431*
    Karman-Trefftz design, 411, *412*
    leading edge of, 416, 443
    noise and, 9, 11
    oscillating, 394, *396,* 442–443
    smoke-wire method, 400, 402
    swept-back, *423*
    vortices and, 389–391, *393*
Air–glass interface, 209
Alias error, 695
Almost-periodic data, 30
Ambiguity bandwidth, 224, 226, 227, 228, 233,
    235, 236, 257
Ambiguity broadening, 226–227
Ambiguity noise, 232–233, 236
Ammonium chlorate, 372
Amplitude correlators, 237, 238–240, 252
Amplitude factor, 215–220

Analog methods, 676–677, 680, 685–688
Analog-to-digital (A-D) converter, 126, 237, 243,
    684, 690–692
Anemometers, thermal, 115–167
    (*See also* Hot-film sensors; Hot-wire anemom-
    eters)
Angle sensitivity, 134–136, 156
Anomalous effects, 322, 334
Antiturbulence screens, 369
Arcsine law, in LDV, 239–240
ART solution, 498
Aspect ratio, 130
Aspirating probe, 164–165
Attenuation, 150–152, 166
Autocorrelation, 260, 656, 685
    analog methods, 688
    averaging and, 656–657
    bandwith, 40
    defined, 34–35, 685
    Fourier transform and, 37–39, 225, 669
    function, 652
    power spectrum and, 38, 260
    spectral density and, 37–38
Auxiliary signals, 239–240
Axisymmetry, 43

Back-lighting techniques, 379
Back scatter, 185
Baffles, 322

Ballistic calibrator, 330–332, *331*
Ball prover, 330–332
Beam splitter, 184, 201
Beam waist, 189
Bearings, 109
Bellhouse-Schultz model, 160, 163
Bell-type provers, 329–330
Benard cavity, 498
Bernoulli equation, 4, 15, 16, 27, 65, 70, 94
Biased-diode linearizers, 676
Bias effect, 258
Bidirectional meters, 316
Bingham plastics, 529
Biot number, 160
Bird-Carreau model, 527
Birefringence, 566–568
Blasius equation, 630
Blowdown wind tunnel, 497
Boiling, two-phase flow and, 92–93
Boundary layer, 19, 79
  constant-temperature hot wire, 649
  damping and, 599, 601
  errors and, 491
  flows, 377, 578
  flow separation and, 94
  growth of, 464
  injection suction, 580
  lag and, 150
  natural convection and, 99
  no slip condition, 559
  recovery factor and, 109
  separation and, 94–95, 578
  spinning, 382
  thickening of, 485
  thin, 95
  transition and, 473, *488*
  turbulence and, 93, 650–651, 666
  wall and, 16, 575–581, 665–666
Bragg cell, 182, 208
Brownian motion, 175, 215, 227, 236, 562
Bubbles, 47, 50–55, 85, 155, 432–445, 551–552, 556
Buoyancy, 28
  error of, 324
  natural convection and, 89, 98
  Navier-Stokes equations, 94
  separation bubble and, 102
Burst processors, 228, 231, 238, 248–251, *254,* 444

Calibration, 131
  ballistic, 330–332

calibration curves, 125–126, 318, 347
hot-film sensors, 579–580, 585–586, 610–612, 615, 625–627
hot-wire anemometer and, 125–128, 130, 135, 369–370
hot-wire sensors, 125–128, 130, 135, 369–370
methods of, 123–127, 318–320
practice of, 351
Preston tube and, 587, 590, 610–611
Stanton tube and, 586–589
uncertainty and, 323
wall shear stress and, 576, 589, 627
Cameras (*see* Photography)
Canard, 414–417, 423, 427–431
Capacitance effect, 595, 615
Capillary method, 622
Carreau model, 527
Cauchy-Green strain, 511
Cavitation, 47, 50–55, 155, 159, 435
CEF equation (*see* Criminale-Erickson-Filbey equation)
Central limit theorem, 229
Channel flow, Reynolds numbers, 558–559
Characteristic function, 221–231
Characteristic length, 21
Characteristic time constant, 206
Circular probes, 604–605
Clauser method, 634–635, 658
Clipped correlation, 239
Close-coupled canard, 412, *414, 415*
Coaxial systems, 185
Coherence, 47–48, 185, 196–199
Collection interval, 353
Collis-Williams expression, 128
Complex flows, measurements in, 559–568
Composition instability, 560
Compressible flows, 509–574, 584, 615–617
Computer modeling, 309, 318–319
  (*See also specific models*)
Concentration boundary layer, 601
Conditioners, 321–322, 334
Conductivity, 118–119, 127, 152
  similarity and, 101, 103–104, 110–112
  supports and, 147–149, 158–160
Cone sensor, *162*
Constant-current operation, 163–164
Constant-temperature systems, 143–154, 164–166, 625
Constitutive relations, 523–541
Continuous wave (CW) laser, 184–185
Contraction flows, 556
Contraction ratio, 369

Convection, 19, 127–129, 132–133, 485, 590, 592, 611
Coriolis-acceleration flowmeters, 321–322
Coriolis forces, 67, 68
Corotational derivative, 536
Correlation analysis, 220–228, 239
Corrosion, 618
Couette flow, *561, 567*–568
Counterelectrode, 630–631
Covariance, 664, 667
Creep, 109, 517, *518*
Criminale-Ericksen-Filbey (CEF) equation, 537
Crossed-beam systems, 469
Crossflow, 385, 568, 674
Cross-power spectrum, 40
Cutoff frequency, 156
CW laser (*see* Continuous wave laser)
Cylinder flow, 81–84, 102, 435–437, 632–633, 674
Cylindrical sensors, 154–158

Damping, 539, *539,* 599, 601
Dark currents, 208
Data classification, 29–33
Data processing, 252–262
Deborah number, 540
Deformation, of non-Newtonian fluids, 509–510, 521–522, 525–527, 530
Degassing, 435
DeGrande technique, 140
Delta function, 36–37
Delta wings, 389, *390*
Demodulators, 207
Density, and refraction, 452–456, 463, 471–472, 478–480, 489–491
Density functions, 220–221, 228–234, 256
Detectability, 282
Deterministic data, 29, 651
Differential pressure, 301–309
Diffraction gratings, 182
Diffraction, spot size and, 270
Diffusion coefficient, 620–623, 630–631
Diffusion equation, 20
Digital computer systems, 117, 319, 695–696
Digital signal processing, 377–378, 483, 681–685
Dimensional analysis, 583
Dirac delta function, 36, 223
Direct injection, flow visualization and, 367–450
Direct inversion technique, 496
Directionality, 130
Direct memory analysis (DMA), 684

Discharge coefficients, 357
Dispersion, 509–510, 513
Dissipation, 21, 24, 207, 110
Diverter systems, 323, 331, 353
DMA (*see* Direct memory analysis)
Doppler bursts, 196, 221–228, 240–254
Doppler light flux, 177–179, 187, 214–215, 224
Doppler shift, laser systems and, 177–247
Double-clipped correlators, 239
Double-diffusive plumes, 474
Double refraction (*see* Birefringence)
Drag, 2, 6, 543
  coefficient of, 84, *97*
  of marine vehicles, 50
  reduction of, 619
  rheometers, 550
  wake and, 389
Droplets, oils and, 212
Dropout, 244, 246, 256
D-taps, 304–305
D-type particles, 215, 216
Dual-beam system, 180, 193–196
Dual-scatter system, 180, 183, 200
Ducts (*see* Tubes)
Dye method, 418–431
Dynamic flow effects, 150
Dynamic mechanical analysis, 518
Dynamic stall, 394, *396,* 442, 443
Dynamic traceability, 337
Dynamic-weighing procedures, 326

Eckert number, 108
Eddies, 20–21, 677
Effective velocity, 235
Elasticity, 518, 537
Elbow meter, 309–310, *311*
Electrochemical techniques, 578, 590–604
Electrodes, types of, 579, 592–597, 600, 616–620, 630–631
Electrolyte, wall probe, 618–619
Electronic pickup, 151
Electronic testing, 144, 161
Ellis model, 526–527
End effects, 491, 544
End-loss effects, 163
Energy separation factor, 111
Entrance flow, *556*
Entropy, 105
Ergodic process, 636, 658–672
Errors
  diverter systems and, 353, 354, 356

Errors (*continued*)
flowmetering and, 323–325
hot-film probes, 600, 612, 632
interferometers, 475
in LDV, 188, 238, 246, 248
pressure transducer mounts, 560–561
rheometers and, 541, 542, 544
sensors and, 118, 127, 136, 146, 151, 560–561, 632
sublayer fence and, 610
Euler equation, 70
Eulerian velocity field, 234
Euler number, 50
Evaporation, 324
Expansion coefficient, 89
Expansion-deflection nozzle, 394
Expansion fan, 464
Explosive events, 464
Extensional flow, 520, 537, 551–559

Factorized memory function, 537–539
Faraday's law, 616
Fast-Fourier-transform, (FFT), 243, 250, 680, 682
Ferrocynanide, 616–620
FFT (*see* Fast-Fourier-transform)
Fiber-optic systems, 262–265
Fiber spinning, 520, 551, *557*
Filament stretching method, 558
Film blowing, 551
Film boiling, 92
Film sensors (*see* Hot-film sensors)
Filtered photoemission, 228, 673–674
Fine wires, 126–131
Finger tensor, 511–512, 531, 533
Finite resolution, 150–151
Flame phenomena, 452
Flange taps, 307
Floating head devices, 581
Flow coefficient, 307
Flow conditioners, 321–322, 334, 345–346
Flow fields
optical testing in, 451–452, 464–466, 483–485
shear stress in, 575–579, 584, 594, 630–631
Flow-loop, 621–622
Flows (*see specific types, topics*)
Fluid mechanics, laws of, 65–114
Fluid meter function, 353
Flumes, 317–318
Fluorescent-particle technique, 422

Foaming, 520
Forward-scatter system, 185
Fouling, 562, 619
Fourier, J., equation in, 87
Fourier series, 30, 35
Fourier transforms
autocorrelation and, 669–670
delta function and, 36
FFT (*see* Fast-Fourier-transform)
inversion process, 497
power spectrum and, 221
two dimensional, 277, 483, 496
Fredholm equation, 253
Free convection, 98–102, 119
Free molecular flow, 128
Frequency counters, 247–249
Frequency demodulation, 205, 233
Frequency folding, 694
Frequency-locked loop (FLL), 244, *245*
Frequency response, 595–600
Frequency shifting, 182
Frequency trackers, 243–247
Freymuth theory, 144, 146, 157, 160, 163
Friction factor, 84
Fringe biasing, 253, 262
Fringe model, 194–196
Fringe pattern, 478–483
Fumes, 166, 371
Fungicides, 560

Gas, 301
Gas flow, dynamic procedure, 329–330
Gaussian beams, 188–189
Gaussian distribution, 32, 222, 224
Generalized-function transforms, 225
Generalized Newtonian fluid (GNF) model, 525–526
Gladstone-Dale equation, 452–453
Glass, 568
Glue-on probe, 629
GNF model (*see* Generalized Newtonian fluid model)
Gradient detection algorithms, 444
Grashof number, 89
Gravity effects, 580

HBD (*see* High-burst-density signals)
Head-Ram calibration, 583–584

Heated-film probes, 625
Heat transfer, 87–89, 126–131, 155, 609–610, 624–629
Heat waves, attenuation and, 150
Heaviside function, 254, 516
Heisenberg principle, 271
Helicopters, 394, 474, 498
Helium bubble method, 361–362, 408–418
Helium jet, 464, *465, 474*
Helmholtz resonator, 26
Henky strain, 555
Heterodyne detection method, 179, *180, 209*
High-burst-density signals (HBD), 231–233, 255–257, 276
High-pass filtering, 247–248
High-photon-density signal, 207
High-speed flows, 117, 130, 134, 163
Hole error, 561
Holography, 267, 470, 493–495, *494–495,* 498
Hot-film sensors, 117, *153, 157, 158,* 562, 577
    calibration of, 579–580, 585–586, 610–612, 615, 625–627
    cylindrical, 154–156
    frequency response, 154
    noncylindrical, 159–161
    nonlinear effects, 158
    square-wave testing, 158
Hot-wire anemometer, 116, 124, 143, *153,* 692
    calibration and, 125–128, 130, 135, 369–370
    constant-temperature, 166
    control circuit, 123–125
    electronic testing and, 144
    film sensors and, 152–153
    jitter and, 13
    noise and, 151
    non-linear effects, 158
    optimization and, 144
    turbulent flow and, 118, 650
    Wollaston process, 122
Hot-wire materials, *121*
Hydrafoil, tip vortex and, *52*
Hydrocarbons, 332, 372
Hydrodynamic and cavitation research tunnel (HYKAT), 55, 56, *58*
Hydrodynamic flow visualization, 418–445
Hydrofoil, 50–51, *52*
Hydrogen bubble method, 432–445, 562, *563*
Hydrosols, 211, 230
HYKAT (*see* Hydrodynamic and cavitation research tunnel)

Illuminating beams in velocimetry, 177–203, 228
Image-plane approach, 272, 277
Image processing technique, 377
Impinging helium, *485*
Incoherent signals, 185
Incompressible fluids, 125, 127, 220, 304, 351, 509–511, 521, 523–524
Indexing techniques,557
Inertia problems, 544
Infinite fringe setting, 480–481
Injection methods, 367–450
Integral length scale, 42
Interaction effects, 56
Interface, phase change at, 90
Interference effects, 620
Interferograms, *456,* 483, *486–489,* 490
Interferometric techniques, 474–500
Intermittency, 667–669
Interpolation schemes, 261
Interrogation analysis, 272–273, *276–278, 281,* 282
Intersecting beams, *480,* 481
Inviscid flow, 13–15, 68–72, 102–106
Isentropic relationship, 352
Isoniazid, 620
Isotropic turbulence, 24, 146, 525

Jacob number, 90
Jets, 19
    impinging, *283, 286*
    mixing and, 165
    noise scales, 48, *49*
    Reynolds numbers and, 464
Johnson noise, 151, 208
Juamann derivative, 536

Karman-Trefftz airfoil, 411, *412*
Karman vortex sheet, 82, 395
Kelvin-Helmholtz instability, 391
Kempf technique, 576
Kerosene smoke technique, 372–375
Kerr cells, 182
Ketchup, 529–531
King's law, 128
Kistler gauge, 581
Knife-edge, 459, *460,* 463, 465, 494
Knudsen number, 128
Kool technique, 140
Kramer equation, 129

Kurtosis, 663, 683
Kutta-Joukowski condition, 416

Lag effect, 150, 215
Lagrangian fluid, 192
Lagrangian scales, 219
Laminar boundary layers, 110, *111, 578*
Laminar flow, 20, 74, *75,* 84, 312, *313,* 386, 587, 594–595
  differential pressures, 575, 581, 588
  high burst density, 231
  hydrodynamic visualization, 418
  LDV for, 175, 220–221, 231
  smokeline visualization, 382, 402, 413
  transition in, 81, 381–382, 402, 451
  wall shear stress and, 578, 588–589, 592, 594, 599, 610–611
Laminar flowmeters, 310, 357
Laplace equation, 14
Large cavitation channel (LCC), 55–56
Laser diode systems 262–265
Laser-Doppler anemometer (LDA), 649
  birefringence and, 566
  index of refraction and, 562
Laser-Doppler velocimetry (LDV), 175–177, 180, 636
  aerosols and, 211
  aperture diameter and, 199
  applications of, 211, 287–288
  Doppler shift and, 117–121, 177–179
  dual-beam, 180–181, 184–198, 209
  focal length of, 189
  frequency shifting, 282
  frequency trackers, 247
  fringe model, 194
  geometry of, *186*
  hydrosols and, 211
  illumination-beam geometries and, 200, *201*
  laser flowmeters and, 320
  measurement volume, 190
  optical system types, 180
  reference-beam, 182, 183, 198–200
  two-color system, *202*
  typical signals, 192
Laser flowmeters, 320
Laser speckle, 267, 271
LBD (*see* Low-burst-density data)
LCC (*see* Large cavitation channel)
LDA (*see* Laser-Doppler anamometer)
LDD (*see* Low-data-density signal)

LDV (*see* Laser-Doppler velocimetry)
Leading-edge effects, 391, 402–403, *405,* 416
Leaks, 580
Lekakis probe, *140*
Lighthill theory, 9
Linearization, 240, 242, 597, 612, 674–677
Linear viscoelastic model, 532
Lip shock, 464
Liquid flow, 325–327
Lock loss, 244
Long-time averaging, 237, 252
Lorentz-Lorentz relation, 452
Low-burst-density (LBD) data, 257–261
Low-data-density (LDD) signal, 255
Low-image-density limit, 273–274
Lubrication, 75, 509
Lumley method, 42

Mach number, 6, 48–49, 103
Mach-Zehnder interferometer, 98, *99,* 480, *481,* 493
Magnetic flowmeters, 318–319, *319*
Magnetohydrodynamics (MHD), 3
Magnus force, 386
Manometer, 302
MAP (*see* Measurement Assurance Programs)
Margarine, 529
Marine engineering, 47
Markov chain, 256
Mass flow transducer, *167*
Mass transfer probes, 590–604
  application of, 630
  configuration of, 604–610
  experimental procedures for, 616–624
Material functions, 510–523
Max-min method, 308
Maxwell model, 531
Mayonnaise, 529
Mean gradient broadening, 227
Measurement Assurance Program (MAP), 337–338, 346–347, 350–351
Measurement volume, 190–192
Mechanics, 13
Melt fracture, 560
Memory, material, 524, 531, 537
Meter-factor variation, 343
Metering devices, 301–363
  (*See also specific types, topics*)
MHD (*see* Magnetohydrodynamics)
Microphone techniques, 320, 369

Mie scattering theory, 195–196, *196*
Migration, 617
Misalignment angle, 491–492
Mists, 371
Mixing
  conditioner and, 322
  cross-sectional, 317
  efficiency of, 197
  measurements and, 40
  nozzle for, 42
  parallel streams and, 475
  turbulence and, 11–12
Modulation method, 567
Moire fringes, 492
Molecular diffusivity, 21
Molecular flow regime, 130
Mounting holes, effects of, 560
Multi-hot-wire scheme, 679
Multimode fibers, 263
Multipoint measurements, 267
Multiposition measurements, 140–141
Multivelocity-component systems, 200–203

Nahme number, 108
National Bureau of Standards (NBS), 308, 332,
  336
Natural convection, *87,* 98–102
  boundary layer and, 99, 464
  buoyancy and, 89, 98
  instability waves of, *101*
  open cavity and, 485
Natural gas, 302
Navier-Stokes equations, 18, 72, 85, 98, 267, 664
NBS (*see* National Bureau of Standards)
N-cycle mode, measurement of, 248
Nernst diffusion layer, 592
Nernst equation, 622
Newton's law, 66–67
Nickel, films and, 155
Noise, 151–152, 159
  constant-current and, 163
  constant-temperature systems and, 163
  hot-wire anemometry and, 151
  immunity, 245
  Johnson noise, 151–208
  randomness and, 208, 213
  shot noise, 203–208, 231
  signal-to-noise ratio, 203, 208–210
Nonisothermal flows, 141–143
Nonlinearity effects, 126

Nonlinear models, 535–537
Non-Newtonian fluids, 509–574
No-slip condition, 559
Nozzles, 48, 125, *398,* 551, *558*
Nuclear magnetic resonance, 267
Nusselt number, 87, 127, 130
Nyquist criterion, 682, 694

Ocean engineering, 2
Oil films, 109, 635
Oils, droplets and, 212
Oil smoke, 379
One-component fluids, 81–85
Open cavity convection, 485
Open-channel metering, 318
Opposed nozzles, 557
Optical fibers, 262
Optical heterodyne detection, 179–180
Optical mixing, 179–180
Optical path, 477, 496, 497
Optical systems, 451–508
  (*See also specific types, topics*)
Optimization, 144
Orifice discharge coefficient, 357
Orifice flow constant, 353
Orifice meters, 302, 304–309
Orthogonal decomposition, 42
Oscillating airfoil, 394
Overheating, 132, 163, 680
Oxidation temperature, 680

Paddle-type orifice plate, 306
Paint, 529
Paraffin, 375
Parallel-beam system, 469–472
Parallel-plate rheometer, 546, *547*
Parshall flume, 318
Particle image velocimetry (PIV), 118, 267, 271–
  272, 276, 287–288
Particle tracer methods, 562
Particle tracking velocimetry (PTV), 267,
  274–275
Path lines, 69, 75, 77, 368
PDV (*see* Phase doppler velocimetry)
Peclet number, 87
Pedestal correlation, 241
Pedestal flux, 187
Pedestal signals, 211, *214*
Perfect gas equation, 454

Period-averaged signal, 210
Periodic functions, 34–36
Petroleum industry, 301
Phase doppler velocimetry (PDV), 265–266
Phase-locked loop (PLL), 238, 243
Phase noise, 231
Photodetectors, 203–208, 223
Photoelasticity, 564
Photoelectric effect, 203
Photoemission, 204–205, 228
Photography, 376–377
    dye injection and, 422
    helium bubbles and, 409
    hydro-bubbles and, 435
    schlieren optics, 463–467
    smokelines, 369, 371, 372, 383–385, 395, 399,
        400, 402
Photomultiplier tube (PMT), 203, *205*
Photon correlation, 205, 237, 240–242, *241,* 252
Pinhole cavity system, 26
Piston force, 550
Pitot coefficient, 359
Pitot tube, 65, *105,* 106, 118, *310–312,* 583
PIV (*see* Particle image velocimetry)
Pixelization, 270
Planck's constant, *271*
Plastic, 211, *529–531,* 620
Platinum, 121, 155, 620–621
PLL (*see* Phase-locked loop)
Plug-in tip, 122
Plug nozzle, 394
PLV (*see* Pulsed light velocimetry)
PMT (*see* Photomultiplier tube)
Pockels cells, 182
Poisson distribution, 205, 218
Polarization, 202, 264, 492, 618
Polydisperse suspensions, 213
Polyethylene, 514, 522, 533, *533–535, 541, 566*
Polymers, *523, 526*–528, 533, 559–560, 564–568
Polymide foils, 629
Pool boiling, 91–92
Powders, 211
Power law model, 212, 220, 525–526, 529
Power spectrum, 38, 220–228
Prandtl, L., 71, 78–79, 94
Prandtl number, 66, 87, 90, 108, 361, 599
Prandtl tube, 72
Pressure, fluid
    anemometry and, 115, 125, 165
    basic laws, 68–73
    metering of, 302–308, 309, 353, 357–358

    non-Newtonian fluids and, 509–510, 524, 526,
        560–561, 567
    optical systems and, 452–453
    pressure drops, 82, 312
    similarity analysis, 68–73, 81–82
    transducers, 121, 320, 397, 550, 558, 560
    wall shear stress and, 575, 579–589
Preston method, 577
Preston tube, 575, 582–586
Probability density function, 31–32
Probes, 452
    bubble-wire, 437
    circular, 604
    damage to, 152–153, 561, 626
    design, 618, 628–630
    glue-on, 630
    mass transfer, 604–610, 630
    supports, 122
Propellers, 7–8, 50–51
Pseudoplastics, 529
Pseudo-steady state assumption, 591, 597, 599,
    613, 615
PTV (*see* Particle tracking velocimetry)
Pulsed light velocimeter (PLV), 267–268
Pulsed-wall probe, 636–649

Quadrature spectrum, 40
Quantum efficiency, 204

Rake system, 372–374
Random light flux, 213–237
Randomness, 29, 205, 651–652
    noise and, 208, 213, 483
    particle motion, 175, 215, 227, 236, 562
    phase shift and, 215
    random functions, 34
    signal analysis, 30
    spectral analysis, 30
Rankine vortex, 50
Reattachment, 404
Recirculation, 589, 621
Recompression shock, 397, *398,* 464
Recovery factor, 107–108, 129, 360–361
Reference-beam system, 180
Refraction, 379, 451, 454, 457, *462,* 483, 491,
    496, 562
(*See also* Optical systems)
Reiner-Rivlin fluid, 524–525
Reiss method, 606

Relaxation, 517, 530–533
Repeatability testing, 349
Residence time weighing technique, 258
Resin, 372
Resistance coefficient, 626
Resolution, 25–27, 120
Retail scheme, 669
Reversed flow, 141, 443, 610
Reynolds numbers
  angle of attack and, 6
  channel flow and, 558
  critical value of, 93
  cylinder flow and, 96
  defined, 74–75, 582
  discharge coefficient and, 307
  Grashof number and, 89
  high-speed flows and, 130
  jet geometry and, 49, 464, *465, 483*
  non-Newtonian fluids and, 559
  Pitot tubes, 583
  similarity and, 83–84
  viscosity and, 72, 92, 583
  vortex shedding and, 320
Reynolds, O., 74–76, 81, 85
Reynolds stress, 24, 40, 75, 634, 664, 667, 684
Rheology, 68, 510, 541–559
Rod climbing effect, 537
Ronchi system, 470, 492
Room controls, 166
Rotameters, 314–315
Rotary wing, 417
Round jet, 41–42
Rouse model, 533
Rudd model, 193

Sampling methods, 692–697
Sandwich elements, 609–610
Scaling laws, 308
Scanning laser light sheet, 377
Scattering models, 196, 199–200, 210–213, 469
Schlieren methods, 103, 379–380, 397, 451–508
Schmidt number, 78, 578, 599, 602
Schmidt trigger, 247
Second-order fluid model, 537
Self-cleaning features, 318
Self-illuminated system, 470
Sensing volume, 321
Separation, 595
  boundary and, 7, 95, 485
  bubbling and, 53, 402, 432

buoyancy and, *102*
cavitation and, 53
leading edge and, *405*
reattachment and, 404
turbulence and, 95, 471
Series arrangements, 362
Series expansion methods, 497
Settling, 211
Shadowgraphs, 13, 451–508
Shape factor, 7
Sharkskin, 560
Shear flows, 24, *528, 534, 543,* 552
Shearing interferometer, 492–493
Shear layers, 94, 378, 381–386, 402, 404, 526
Shear rheometers, 542
Shear stress, 86–87, 512–515, *528–529,* 542–550,
  558, 575–637, 664
Sheet stretching, 556
Shock waves, 105–106, 453, 464, 473–474
Shot noise, 46, 203–208, 231
Side losses, 160
Signal analysis, 29–31, 40, 47
Signal processing, 672–689
Signal-to-noise ratio (SNR), 203, 208–210
Silicon carbide particles, 211
Silt rheometer, *550–551*
Similarity analysis, 40, 66, 81–92
Similitude, law of, 6
Sine wave testing, 163
Single-clipped correlators, 239
Single-leg manometers, 308
Single-particle signal, *185*
Skin friction, 581
Slanted transfer surface, 606–608
Slewing signal, 683
Sliding plate rheometer, 551
Slip flow, 128–129, 575
Smoke photographs, 371, *373, 374, 379, 383,*
  384
Smoke-tube methods, 369–397
Smoke-wire method, 391, 397–398, 400–402,
  405–407, 651
SNR (*see* Signal-to-noise ratio)
Soap bubbles, 408–410, 527
Solids testing, 517
Spatial coherence, 196, 198
Spatial resolution, 25–27, 270
Speckle, 267, 271–272, *272*
Spectral broadening, 227
Spectral density, 37–38
Spectral function, 669

Spectrum analysis, 35, 176, 242–243, 252, 680–681
Split-film sensors, 139, 165–166, *166*
Split-rectangular electrode, 620
Splitter plate, schlieren, 469, 474, 456
Spot size, diffraction and, 270–271
Springs, dashpot viscoelasticity and, 535
Sputtering, 628–629
Square-law devices, 179, 242
Square wave testing, 143, 158
Squeezing technique, 555
Stagnation, 130, 359–360, 595, 632
Stalling, 442–444
Standpipe system, 327
Stanton gauge, 575–576, 584, 586–589
Stanton tube, 586–589
Static pressure, 310, 321, 396–397
    basic laws, 93, 102–103
    sensing, 357–359, 375–383
    tap geometry, 357, 375–376, 582
    (*See also* Pitot tubes)
Static traceability, 337
Static-weighing method, 325
Stationarity, 30, 221, 652–656
Steady shear flows, 512–515
Stereoscopic schlieren, 470
Strain, 509–568 *passim*
Streak lines, 77, *77, 80,* 368, 379, 402
    (*See also specific methods*)
Streamlines, *80, 82,* 368, 552
Stress, in non-Newtonian fluids, 509–568
Stress optical coefficient, 565
Stress-relaxation, 516, *517, 538*
Stress, shear and, 512–513
Strobe, bubble illumination and, 437
Strouhal number, 408
Sublayer fence, 575, 577, 589–590
Supersonic flow, 104, *105,* 378–381
Supports, conduction and, 134–136, 147–148, 158–159
Surface acoustic-wave (SAW) analyser, 243
Surface tension, 551–553, 557
Suspensions, 513, 545
Sweep tuning, 680
Swell ratio, 550
Swirling, 322, 474
Switching error, 324
Syphon, 557

Tandem meters, 340
Tapping hole, 358–359

Target meters, 315–316
Taylor's hypothesis, 26, 671
Taylor microscale, 24, 246, 255
Temperature gradient, in optical systems, 471, 472–473, 476–477, 493, 494
Temperature recovery factor, 108
Temperature, refraction and, 455
Temperature sensor, 104, 111
Thermal anemometers, 115–173
    constant-temperature, 151
    laser velocimeter and, 119
    probes for, 122–124
    turbulence and, 118
    (*See also* Anemometers; Hot-film sensors; Hot-wire anemometry)
Thermal diffusivity, 150, 361
Thermal flowmeters, 316
Thermal stress problem, 568
Thermodynamic state, 106, 452
Thick-window effects, 493
Thin lens, 188
Three-dimensional measurements, 495–499
Time-averaging, 175, 205, 237–238, 252–255, 579, 581, 591, 610, 636–637
Time constant, 212
Time-resolving, 252–260
Time-varying flows, 579, 610
Tip vortex, 52, *52*
Titanium tetrachloride, 371, *375,* 391
Tollmien-Schlichting waves, 383, *385,* 385–386
Total-burst-mode measurement, 249
Total-pressure sensing, 357–358
Traceability, 336–337
Tracers, 562
Trailing edge, 411
Transducers, 121, 320, 397, 579
Transference number, 617
Transient effects, 515–518
Transit-time broadening, 225
Transition, 432
Transmission function, 236
Transonic wind tunnel, 378–379
Transportation, 660
Trapezoidal-rule approximation, 261
Triiodide system, 619–622
Tubeless syphon, 557
Tubes (pipes), 303, 304, 309–318, 34, 600–601, 630–632
Tungsten, 121
Turbines, 212. *213,* 312–314, 485
Turbomachinery, 21, *22,* 50, 111

Turbulence, 12, *16,* 20, *41, 75,* 84, 206, 222, 246
    acoustic radiation and, 47–48
    antiturbulence screens and, 369
    boundary layers and (*see* Turbulent boundary
        layers)
    broadening and, 227
    bursts, 470–471
    coherent structures in, 42, 47–48, 382
    covariance and, 667
    data processing, 649–696
    diffusion and, 20
    digitized data and, 680–685
    flow visualization and, 378
    hot-wire probe and, 650
    intensity of, 663, 683–684
    isotropic, 23
    jets and, 10, 41, 44
    measurements of, 600–604
    mixing layer and, 11–12, 12,
    signal analysis and, 40
    similarity law and, 40
    slip and, 575
    smoke-wire technique and, 405–407
    sound and, 8
    Taylor hypothesis and, 26
    Taylor microscale and, 24–26
    thermal anemometers and, 118
    transition from laminar, 382, 405
    viscous flow and, 18–20
Turbulent boundary layer, 582–615 *passim*
    eddies and, 650
    flat-plate, 651
    flows, 653
    intermittency and, *668*
    law of the wall and, 576
    separation and, 95
Two-color design, 201, *202*
Two-component fluids, 85–87
Two-phase flow, 90–91, 92

UCM model (*see* Upper convected Maxwell
        model)
Ultrasonic flowmeter, 318, *319*
Uncertainty, 224, 323, 688–689
Uniaxial extension, 522
Unsteady flow, 394
Upper convected Maxwell (UCM) model, 536
U.S. Army Aeromechanics Laboratory, 434–435,
        443
U-tube manometer, 305, 308

Variance analysis, 362
Vaporization, 50, 371–372
Venturi tubes, 308, *310*
Vertical water tunnel, 419–421
Video cameras, 270, 410–412, 434–440
Viscoelasticity, 531–535, *536*
Viscosity, 19, 361, *523, 527, 541*
    dissipation, 20, 23
    effects of, 68
    elasticity and, 518
    extensional, 522
    heating and, 106–113
    Navier-Stokes equation and, 72
    non-Newtonian fluids, 510, 540, 559
    ratio, 87
    Reynolds number and, 72
    Reynolds stress and, 634
    shear rate and, 526
    stress coefficients and, 513
    turbulence and, 18–20
    viscous fluids, 92–93, 524
Visibility, of signal, 187
Visometric material functions, 513
Volatility, error of, 324
Volume flow measurements, 301–367
Von Karman effects, 96, 389, 635
Von Mises criterion, 531
Vorticity, 267, 669
    breakdowns, 389
    cavitation and, 48–49
    core tagging, 392
    formation, 386
    Reynolds number and, 320
    shedding, 320, 410–411, 440
    smoke tube and, 392
    tensor, 511

Wagner model, 539
Wakes, 140, 387, 389, 399, 432, 660
Walls, *16,* 603–604
    boundary layer and, 666–667
    conduction and, 130–131
    Couette flow, 561
    eddy sizes, 677
    film gages, 577
    law of, 576–577, 603, 634
    no slip condition, 559
    probe in, 627
    rough, 86
    turbulence and, 16, 665
Wall shear stress, 575–637

Wastewater discharge, 301
Water, scattering in, 211
Water tunnels, 47, 55, 79, *79*, 419–421, *420–422, 425–427*, 434
Waveform analysers, 680
Waveguiding, 322
Wave-shearing interferometer, 492
Weber number, 86
Wedge fringes, 481, 482, *489*
Weirs, 317–318
Weissenberg number, 541, 561
White noise, 39
Wholesale scheme, 669
Wind tunnels, 369–397

WLF equation, 528
Wollaston prism, 122, 492

X probe, 136, 137–138

Yaw sensitivity, 135
Yield stress, 529
Youden analysis, 279, 348, 350

Zehnder-Mach interferometer, 98, *99*, 480, *481*, 493

2

3

14